Colony-Stimulating Factors

Molecular and Cellular Biology

Second Edition, Revised and Expanded

edited by

John M. Garland

Institute of Clinical Science
University of Exeter
Exeter, England

Peter J. Quesenberry

University of Massachusetts Medical Center
Worcester, Massachusetts

Douglas J. Hilton

The Walter and Eliza Hall Institute of Medical Research
Melbourne, Victoria, Australia

CRC Press
Taylor & Francis Group
Boca Raton London New York

CRC Press is an imprint of the
Taylor & Francis Group, an **informa** business

CRC Press
Taylor & Francis Group
6000 Broken Sound Parkway NW, Suite 300
Boca Raton, FL 33487-2742

First issued in paperback 2019

© 1997 by Taylor & Francis Group, LLC
CRC Press is an imprint of Taylor & Francis Group, an Informa business

ISBN-13: 978-0-8247-9492-7 (hbk)
ISBN-13: 978-0-367-40095-8 (pbk)

This book contains information obtained from authentic and highly regarded sources. Reasonable efforts have been made to publish reliable data and information, but the author and publisher cannot assume responsibility for the validity of all materials or the consequences of their use. The authors and publishers have attempted to trace the copyright holders of all material reproduced in this publication and apologize to copyright holders if permission to publish in this form has not been obtained. If any copyright material has not been acknowledged please write and let us know so we may rectify in any future reprint.

Except as permitted under U.S. Copyright Law, no part of this book may be reprinted, reproduced, transmitted, or utilized in any form by any electronic, mechanical, or other means, now known or hereafter invented, including photocopying, microfilming, and recording, or in any information storage or retrieval system, without written permission from the publishers.

For permission to photocopy or use material electronically from this work, please access www.copyright.com (http://www.copyright.com/) or contact the Copyright Clearance Center, Inc. (CCC), 222 Rosewood Drive, Danvers, MA 01923, 978-750-8400. CCC is a not-for-profit organization that provides licenses and registration for a variety of users. For organizations that have been granted a photocopy license by the CCC, a separate system of payment has been arranged.

Trademark Notice: Product or corporate names may be trademarks or registered trademarks, and are used only for identification and explanation without intent to infringe.

The Library of Congress Cataloging-in-Publication Data

Colony-stimulating factors: molecular and cellular biology / edited
   by John M. Garland, Peter J. Quesenberry, Douglas J. Hilton. — 2nd ed.
      p. cm.
   Includes index.
   ISBN 0-8247-9492-3 (hc: alk. paper)
      1. Colony-stimulating factors (Physiology)   I. Garland, John M.
   II. Quesenberry, Peter J.   III. Hilton, Douglas J. (Douglas James).
   QP92.C65   1997
   616.07'9—dc21

                                                        97-12099
                                                        CIP

Visit the Taylor & Francis Web site at
http://www.taylorandfrancis.com

and the CRC Press Web site at
http://www.crcpress.com

Preface

Thirty years have passed since the studies of Bradley and Metcalf and Pluznik and Sachs provided the first evidence for the existence of secreted proteins—colony-stimulating factors—that regulate blood cell production. In the first decade after these seminal observations, the major advances were cell-biological in nature, with the relationships between hematopoietic progenitor cells and the mature progeny becoming clarified. In the second 10 years, the first colony-stimulating factors and their cousins—the interleukins (ILs) and cytokines—were purified and their genes were cloned.

In the last 10 years, the number of cytokines that have been cloned and that have begun to be characterized have come to number 20 or more. Many have been tested in clinical settings and a few, such as granulocyte colony-stimulating factor (G-CSF) and erythropoietin, have found routine and widespread clinical use. Our understanding of the physiological role of colony-stimulating factors has also burgeoned during this time.

An amazing number of cell surface receptors for the hematopoietic regulators have also been cloned. In addition to explaining some aspects of biological redundancy, isolation of receptor components has led to a better understanding of the structure–function relationship between ligands and their receptors. The intracellular signal transduction pathways used by many cytokines have also been determined in great detail, with common elements becoming readily apparent.

BIOLOGY AND CLINICAL UTILITY

Perhaps the most exciting aspect of the story of colony-stimulating factors to develop over the last five years has been their routine clinical use in a number of disease settings. The use of G-CSF, for example, has expanded from its initial use, as an agent to overcome neutropenia associated with cancer therapy, to an agent that is commonly used as a treatment for congenital neutropenia and as a means of mobilizing hematopoietic stem cells into the peripheral blood.

The latter clinical use of G-CSF is likely to lead to the use of peripheral blood rather than bone marrow as the prime source of cells with which to reconstitute the hematopoietic system.

Important developments have occurred in our understanding of the physiological role as well as the therapeutic role of colony-stimulating factors. This has occurred through the generation of "knockout mice," whereby targeted mutations can be introduced into the germline of mice to allow a detailed analysis of growth factor function in vivo. Because of the pleiotropic and redundant nature of growth factors, an inability to produce a single growth factor often has a limited phenotypic effect. More complex knockout mouse models are therefore needed, whereby the genes for several key regulators are removed from a pathway to allow determination of the role of each.

RECEPTORS AND SIGNAL TRANSDUCTION

There has also been an explosion in our understanding of colony-stimulating factor receptors and signal transduction. Receptor components for almost every colony-stimulating factor and cytokine have been cloned. This has led to the definition of a new receptor superfamily: the hematopoietin/interferon or cytokine receptor family. Although many cytokines bind to specific receptor subunits, most also share receptor subunits required for generating high-affinity functional receptors. For instance, IL-3, IL-5, and GM-CSF have specific α-subunits that allow low-affinity binding to their specific ligands; however, high-affinity ligand binding and signal transduction is achieved through the recruitment of a common β-subunit. In a more complex situation, IL-6, leukemia inhibitory factor (LIF), oncostatin M, ciliary neurotrophic factor, and IL-11 receptors all share the gp130 molecule for signaling, and some also share the LIF receptor α-chain. For other colony-stimulating factors, such as erythropoietin and G-CSF, receptor homodimerization rather than heterodimerization is important for effective signaling. Likewise, cytokines, such as macrophage colony-stimulating factor (M-CSF) and stem cell factor (SCF), bind to homodimeric receptors. However, unlike the majority of hematopoietic receptors, these are receptor tyrosine kinases.

Dissection of the molecular signaling events that follow receptor oligermerization and ligand binding has also highlighted common signaling pathways. From somatic cell genetic and biochemical studies, it now appears that activation of receptors that have intrinsic tyrosine kinase activity and those that belong to the hematopoietin/interferon receptor family may lead to a very similar sequence of events. The initial consequence of receptor dimerization appears to be activation of the kinase activity of receptor tyrosine kinases, or cytoplasmic kinases of the JAK family that are noncovalently associated with members of the hematopoietin/interferon receptor family. Four members of the JAK kinase family have been described to date: Jak1, Jak2, Jak3, and Tyk2. JAK kinases lead to tyrosine phosphorylation and the eventual activation of STAT molecules (signal transducers and activators of transcription). Six different STAT molecules have been defined. Once activated, STAT molecules dimerize and translocate to the nucleus to activate transcription via specific cytokine-responsive elements. Each member of the JAK family binds to a different subset of cytokine receptor subunits, and similarly each member of the STAT family is activated by distinct ligand/receptor complexes. For instance, STAT1 and STAT2 appear to be functionally specific to the interferons, STAT4 to IL-12 signaling, and STAT6 to IL-4 and IL-13 signaling. Other STATS, such as STAT3 and STAT5, appear to be activated by a wider group of cytokines. Although JAKs and STATs appear to be important mediators of cytokine signal transduction, they are clearly not the entire story. It should be kept in mind that cytokines exert effects at a number of levels, on proliferation, differentiation, cell survival, and functional activation. Determining the signal transduction events that mediate these diverse biological outcomes remains a challenge.

In addition to increasing our understanding of how extracellular signals are translated into intracellular responses, the cloning of the cytokine receptors, the JAKs, and the STATs offers exciting targets for the discovery of the second generation of cytokines—small, orally administrable molecules that mimic or antagonize cytokine action.

John M. Garland
Peter J. Quesenberry
Douglas J. Hilton

Contents

Contributors

Ahmad-Samer Al-Homsi, M.D. Senior Fellow, Cancer Center, University of Massachusetts Medical Center, Worcester, Massachusetts

Jacques Banchereau, Ph.D. Director, Laboratory for Immunological Research, Schering-Plough, Dardilly, France

Stephen H. Bartelmez, Ph.D. Assistant Professor, Department of Pathobiology, University of Washington, Seattle, Washington

Ivan Bertoncello, M.Sc., Ph.D. Head, Stem Cell Biology Laboratory, Sir Donald and Lady Trescowthick Research Laboratories, Peter MacCallum Cancer Institute, East Melbourne, Victoria, Australia

Peter Besmer, Dr.Sc.Nat. Professor, Cornell University Graduate School of Medical Sciences, and Member, Molecular Biology Program, Memorial Sloan-Kettering Cancer Center, New York, New York

Gillian B. Bradford, B.Sc.(Hons) Stem Cell Biology Laboratory, Sir Donald and Lady Trescowthick Research Laboratories, Peter MacCallum Cancer Institute, East Melbourne, Victoria, Australia

Mary K. L. Collins, Ph.D. Team Leader, CRC Centre for Cell and Molecular Biology, Institute of Cancer Research, London, England

Qing Ping Dou, Ph.D. Assistant Professor, Department of Pharmacology, University of Pittsburgh School of Medicine, Pittsburgh, Pennsylvania

Claire M. Dubois, Ph.D. Associate Professor, Department of Pediatrics, Immunology Division, University of Sherbrooke, Sherbrooke, Quebec, Canada

John M. Garland, MB.Ch.B., B.Sc., Ph.D. Division of Cancer Cell and Molecular Biology, FORCE Cancer Research Laboratory, Institute of Clinical Science, University of Exeter, Exeter, England

M. Y. Gordon, Ph.D., D.Sc., F.R.C.Path. Professor and Deputy Director for LRF Centre for Adult Leukaemia, Department of Hematology, Royal Postgraduate Medical School, London, England

Krzysztof J. Grzegorzewski, M.D. Senior Scientist, IRSP, Hematopoiesis Regulation Section, Laboratory of Experimental Immunology, SAIC—Frederick and National Cancer Institute, Frederick Cancer Research and Development Center, Frederick, Maryland

Douglas J. Hilton, B.Sc.(Hons), Ph.D. QEII Fellow, Cancer Research Unit, The Walter and Eliza Hall Institute of Medical Research, and Program Leader, Cytokine Research, AMRAD Operations, Melbourne, Victoria, Australia

Sten Eirik W. Jacobsen, M.D., Ph.D.* Department of Immunology, Institute for Cancer Research, The Norwegian Radium Hospital, Oslo, Norway

Jonathan R. Keller, Ph.D. Senior Scientist, Hematopoiesis and Gene Therapy Section, SAIC—Frederick and National Cancer Institute, Frederick Cancer Research and Development Center, Frederick, Maryland

Khandan Keyomarsi, Ph.D. Research Scientist and Assistant Professor of Molecular Medicine, Laboratory of Diagnostic Oncology, Wadsworth Center, Albany, New York

David S. Latchman, M.A., Ph.D., D.Sc., M.R.C.Path. Professor, Department of Molecular Pathology, University College London Medical School, London, England

Michael W. Long, Ph.D. Professor, Department of Pediatrics and Communicable Diseases, University of Michigan, Ann Arbor, Michigan

Abelardo Lopez-Rivas, Ph.D. Instituto de Parasitologia y Biomedicina CSIC, Granada, Spain

Darryl W. Maher, M.B., B.S., Ph.D. Deputy Head, Clinical Research Programme, Ludwig Institute for Cancer Research, Melbourne Tumour Biology Branch, Parkville, Victoria, Australia

Robert C. Moen, M.D., Ph.D. Vice President, Clinical and Regulatory Affairs, Gene Therapy Unit, Baxter Healthcare Corporation, Round Lake, Illinois

---

*Current affiliation: Associate Professor and Head, Stem Cell Laboratory, Department of Internal Medicine, University Hospital of Lund, Lund, Sweden.

Malcolm A. S. Moore, D.Phil. Enid A. Haupt Professor of Cell Biology and Head, James Ewing Laboratory of Developmental Hematopoiesis, Memorial Sloan-Kettering Cancer Center, New York, New York

Nicos A. Nicola, B.Sc.(Hons), Ph.D. Director, Cooperative Research Centre for Cellular Growth Factors, and Assistant Director, The Walter and Eliza Hall Institute of Medical Research, Melbourne, Victoria, Australia

Joost J. Oppenheim, M.D. Chief, Laboratory of Molecular Immunoregulation, Division of Basic Sciences, National Cancer Institute, Frederick Cancer Research and Development Center, Frederick, Maryland

Arthur B. Pardee, Ph.D. Professor, Division of Cell and Growth Regulation, Dana-Farber Cancer Institute, Boston, Massachusetts

Mark A. Plumb, B.Sc., Ph.D. Doctor, Radiation and Genome Stability Unit, Medical Research Council, Harwell, Didcot, Oxfordshire, England

Peter J. Quesenberry, M.D. Professor of Medicine and Chair, Cancer Center, University of Massachusetts Medical Center, Worcester, Massachusetts

John E. J. Rasko, B.Sc.(Med), M.B.B.S., Ph.D., F.R.A.C.P., F.R.C.P.A.* The Walter and Eliza Hall Institute of Medical Research and The Cooperative Research Centre for Cellular Growth Factors, Melbourne, Victoria, Australia

Andrew W. Roberts, M.B.B.S., F.R.A.C.P., F.R.C.P.A. Cancer Research Unit, The Walter and Eliza Hall Institute of Medical Research, Melbourne, Victoria, Australia

Claudius E. Rudin, M.R.C.P. Registrar, Department of Haematology, Royal Devon and Exeter NHS Trust, Exeter, England

Francis W. Ruscetti, Ph.D. Chief, Laboratory of Leukocyte Biology (Microbiologist), Division of Basic Science, SAIC—Frederick and National Cancer Institute, Frederick Cancer Research and Development Center, Frederick, Maryland

Sem Saeland, Ph.D. Staff Scientist, Laboratory for Immunological Research, Schering-Plough, Dardilly, France

Colin J. Sanderson, Sc.D. Professor, TVWT Institute for Child Health Research, Perth, Western Australia, Australia

Ewa Sitnicka, Ph.D. Senior Fellow, Department of Pathology, University of Washington, Seattle, Washington

---

*_Current affiliation_: Fred Hutchinson Cancer Research Center, Seattle, Washington.

Erlend B. Smeland, M.D., Ph.D.   Head, Department of Immunology, The Norwegian Radium Hospital, Oslo, Norway

Paul M. Waring, M.B.B.S, F.R.C.P.A., Ph.D.*   Cancer Research Unit, The Walter and Eliza Hall Institute of Medical Research, Melbourne, Victoria, Australia

Stephanie S. Watowich, Ph.D.[†]   The Whitehead Institute for Biomedical Research, Cambridge, Massachusetts

Robert H. Wiltrout, Ph.D.   Head, Experimental Therapeutics Section, Laboratory of Experimental Immunology, SAIC—Frederick and National Cancer Institute, Frederick Cancer Research and Development Center, Frederick, Maryland

Eric Wright, Ph.D., M.R.C.Path.   Professor, Experimental Haematology, Radiation and Genome Stability Unit, Medical Research Council, Harwell, Didcot, Oxfordshire, England

Yu-Chung Yang, Ph.D.   Professor, Department of Medicine, Indiana University School of Medicine, Indianapolis, Indiana

James R. Zucali, Ph.D.   Professor, Department of Medicine, University of Florida, Gainesville, Florida

---

Current affiliations:
*Department of Pathology, Melbourne University, Parkville, Victoria, Australia.
†Assistant Professor, Department of Immunology, The University of Texas M.D. Anderson Cancer Center, Houston, Texas.

1

Introduction to the Hematopoietic System

John M. Garland
University of Exeter, Exeter, England

Claudius E. Rudin
Royal Devon and Exeter NHS Trust, Exeter, England

I. HEMATOPOIESIS: AN OVERVIEW

Hematology is the study of blood and its disorders. In this book it will be shown that the individual cell and whole-system renewal is highly complex and involves an astronomical array of biological variables, ranging from how single genes are regulated, to apparently stereotyped patterns of group cell behavior. The challenge is to understand how these are integrated and what mechanisms control each part. This is far from being solved. Hematopoiesis is concerned with cell division, adaptation to particular functions, and regulation of the overall system. Its approach has become more systematic owing to the rapid advances in molecular biology.

A. Cell Renewal and Lineage Commitment

Replacement Versus Renewal

Cell replacement is not necessarily the same as cell renewal. Cell renewal implies the replacement of cells with the same function. Only if cells in a given tissue are replaced by those of the same function and response to control mechanisms will the regeneration be productive. Furthermore, replacement cells must follow the normal tissue architecture. For example, in myelofibrosis, matrix is replaced, but with hematopoietically nonfunctional tissue.

Determination of Cell Function: Differentiation and Definitions of Cell Lineage
and Maturation

Individual cell function exists at two levels: the first is the innate restriction of a cell to express only certain genes (phenotype); the second is the ability to respond to some kind of external stimulation (inducible function). Innate gene restriction is determined during differentiation that irreversibly commits a cell to its phenotype. The basic repertoire is generated during differentiation and occurs as an orderly progression of changes in gene expression, but constitutive and inducible characteristics are coordinated such that a mature cell will not only have the basic repertoire that defines it, but will also have the definitive inducible properties of that lineage.

Maturation is the full realization of function in cells that are already differentiated or committed to a lineage. Thus, end-stage red cells "mature" by completing synthesis of hemoglobin and losing their nuclei. The lines between maturation and differentiation are not always clear, however; for instance, B cells mature by changing the types and amounts of immunoglobin expressed on the surface (which could be considered part of differentiation). The important question is: What determines differentiation and the choice of lineage?

B. Differentiation and Stem Cells

Differentiation and Lineage Commitment

Differentiation is almost always irreversible. However, during differentiation cells become more differentiated as they undergo successive divisions. In the B-cell lineage, for example, pro-B cells first rearrange their immunoglobulin (Ig) genes, then express IgM heavy chains cytoplasmically, followed by surface IgM expression. Later, surface IgM is replaced by IgG. However, expression of specific genes does not necessarily mean that the entire cell changes its expressible complement; hence, there is ambiguity in how cells decide to differentiate and whether the process is gradual, or accomplished within, and linked to, a defined set of cell cycles. Two general models have been proposed:

1. *Deterministic.* In this, a particular signal results in activation of specific differentiation-inducible genes, which successively "program" the cell to reach the end-stage; once initiated, the program runs to completion.
2. *Stochastic.* In this, differentiation is the result of small changes in gene expression, each one amplifying the difference between the new daughter(s) and the original mother cell, resulting in a final stage greatly different from the original. Here, the cell "wanders" into full differentiation. These models are not mutually exclusive.

Stem Cells, Self-Renewal, and Differentiation

The paradox that cells need to divide to differentiate and renew or replace lost cells, but differentiation loses the ability of cells to extensively divide, has been solved by ensuring that within almost every tissue there are a few undifferentiated immature cells that not only can divide and differentiate into end-cells, but also they can replace themselves ("stem" cells). Stem cells are able to (1) renew themselves (self-renewal potential; SRP) and (2) differentiate (differentiation potential; DP). For a stem cell pool to balance, its regeneration must equal that at which its daughters differentiate; total differentiation (0.0 SRP, 1 DP) effectively removes all stem cells. Divisions without any differentiation (1.0 SRP, 0 DP) increases only the stem cell pool. An SRP of 0.5 thus allows equal-rate differentiation and maintenance of the stem cell pool; less than 0.5 progressively diminishes it, more than 0.5 increases it. By variously manipulating the SRP to either side of 0.5, tissues can satisfy the demand for new differentiated cells or for maintaining stem cells; providing both SRP and DP remain balanced, the tissue can be perpetually regenerated. The concept of SRP at the individual cell level, however, is somewhat abstract: Can a stem cell undergo unequal division? If so, how, considering that cell contents are deemed to be randomly distributed?

Cell Loss and Replacement

Regeneration requires a supply of cells that can replace those lost and, therefore, involves cell division (*proliferation*) and acquisition of function (*differentiation–maturation*). Control of cell division is crucial to success in replacing lost cell numbers; differentiation or maturation is equally crucial in determining whether those cells will be of any use. Although, usually, the two arms are interrelated, they can be dissociated, for example, in leukemias during which cells abnormally proliferate, but do not differentiate.

Cell Loss

All blood cells have a finite life span. Granulocytes have a half-life of about 1 day; some lymphocytes (immune "memory" cells) last 10–20 years. Rates of loss are also greatly influenced by infection or illness. What determines a natural life span is unknown. Dead, damaged, or old cells are removed by the reticuloendothelial system in the liver and spleen and by circulating cells of the mononuclear phagocyte series (macrophages). Both reprocess the cell materials. Replacement needs to take into account the rates of cell loss and the differentiated cell lineage to which they belong to maintain an accurate balance. A small deviation from this balance (e.g., overproduction of only 0.001% in granulocytes) will generate a leukemic profile in only a few weeks. Thus, the feedback between progenitor cells and the peripheral demand must be very accurately set.

Cell Replacement

In common with many others, hematopoietic cells undergo a process of programmed cell death in response to a variety of conditions. Apoptosis is a mechanism for removing damaged or unwanted cells and for controlling the balance of cell populations. For example, immature T cells in the thymus are selected by an apoptotic process that removes potentially self-reactive cells, ensuring that only those cells capable of reacting appropriately to non–self-antigens are selected. Mature red cells, which in humans do not have a nucleus, have a finite life span owing to their inability to renew mRNA coding for proteins. In the hematopoietic system, cells die at very different rates; granulocytes have a half-life of some 12 hr, red cells several months. The bone marrow, therefore, must not only respond rapidly to any stress that increases the rate of cell loss (e.g., infections involving granulocytes) but also it must ultimately return the output to normal. Furthermore, its capacity for renewing peripheral blood cells must be practically limitless throughout life.

Mechanisms, therefore, must exist for

1. Producing and accurately regulating cell replacement
2. Taking into account the different rates of cell replacement needed in different tissues
3. Responding rapidly to stress
4. Ensuring maintenance of its renewal capacity

C. Hematopoiesis in Development

Development of the hematopoietic system has recently been reviewed in detail (191). Formation of blood cells first occurs in extraembryonic sites. Cells destined to produce hematopoietic tissue first arise from extraembryonic mesoderm in the yolk sac in association with other cells which form vascular endothelium. These two form embryonic blood islands. The precise derivation of fetal or adult hematopoiesis is still uncertain; either yolk sac progenitors migrate to the fetal liver, or a separate population of progenitor cells arises from the dorsal mesoderm and migrates through the yolk sac to the fetal liver. Subsequently, a small number of these progenitors seed into the developing marrow cavities through the fetal circulation, whereas fetal liver hematopoiesis gradually disappears. These multipotential stem cells regenerate the entire hematopoietic system and maintain adult hematopoiesis. The proposed common origin of the vascular endothelial cells and hematopoietic progenitors has raised questions of whether a common progenitor for blood and vascular cells that can produce either still persists. Despite work that reported this (67), it was later found to be flawed (65), and there remains no evidence that purified hematopoietic progenitors can, in fact, generate both vascular stroma and hematopoietic cells (175).

In adult primates, blood cells are produced exclusively in the marrow cavity. Such is its specificity for hematopoiesis that if the intact marrow is destroyed (e.g., by radiation or drugs), it can regenerate from stem cells injected directly into the bloodstream. This implies that there are "homing" mechanisms by which marrow cells find the right environment. In adult mice, there is a further potential site of hematopoiesis, the spleen. The reason for this is unknown, but it can be usefully used to measure the numbers of engrafted marrow stem cells (CFU-S).

D. Immature and Progenitor Cells

Marrow contains a bewildering array of dividing and "intermediate" or "immature" cell types. These immature types can be recognized by morphology, antibodies to specific surface markers, or stains. In this way, it is possible to arrange the immature cells into lineages and construct a path by which an immature cell becomes fully mature (Fig. 1). However, the more primitive a cell is, the more difficult it is to put it into a lineage. The only definitive way of deciding a cell's lineage is to see what it becomes. This is largely determined by the environment; hence, the major problem in hematology: by the time a progenitor cell's identity is known, it has lost its original characteristics.

Importantly, proliferating and immature cells do not usually appear in peripheral blood; and in the marrow, cells of a particular lineage, but at different stages of maturation, are often clumped together. Therefore, marrow is somehow able to select out fully mature cells for passage into blood. This physical discrimination is important, for in blood disorders immature or proliferating cells may appear in the blood. Whether such cells can continue to proliferate outside the marrow is still an unresolved question.

Lineage Fidelity

Despite the apparent blurring of lineage commitment at the stem cell level, lineage fidelity is extremely accurate. Often, several early markers may be shared among lineages, but once, for example, a cell develops certain erythroid markers, it cannot become a granulocyte. This indicates the tightness of lineage fidelity. However, in leukemia, this fidelity may break down, and cells may express spurious markers. Sometimes this is due to genuine genetic dysregulation, other times it may represent expansion of a clone of progenitor cells so transient or in such low numbers that it has not been recognized before as an immature type.

Committed Stem Cell Progenitors

Early studies in the mouse showed that mature cells of discrete lineages (macrophages or granulocytes) could be grown from apparently undifferentiated single cells. However, they could produce only progeny restricted to one or at most two lineages, and their ability to self-renew was very limited (i.e., their SRP was less than 0.5). The bone marrow thus has two effective tiers of stem cells: A truly multipotential pool, and various pools of committed progenitors. The physiological organization for marrow replenishment thus becomes clearer:

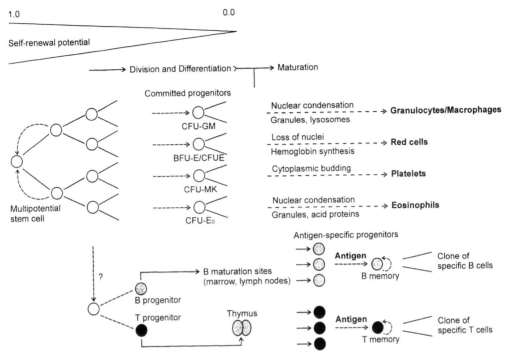

Figure 1 Stem-cell heirarchy in the bone marrow: All the final mature blood cells (extreme right) are derived from a multipotential stem cell (MPSC; extreme left). The first MPSC division may generate two similar MPSCs (both dotted arrows; SRP = 1.0); one MPSC (one dotted arrow), and one different daughter (SRP = 0.5); or two different daughters (SRP = 0.0; and the stem cell effectively "disappears"). Successive divisions of the MPSC generate a series of lineage-limited divisions, and differentiation generates the end cells. End cells complete differentiation through a process of maturation, during which the final complement of genes determining end-function is expressed (e.g., synthesis of hemoglobin or formation of lysosomes). Stem cells for B and T cells appear to diverge early and follow separate modes and sites of differentiation: progeny of B-committed stem cells differentiate partly in marrow, then migrate to lymph nodes; T-stem cells migrate to the thymus where independent proliferation, differentiation, and selection for immune function takes place. Self-renewal potential decreases from left to right coordinately with increasing differentiation; therefore, cells on the left are more "primitive" than those on the right.

1. All lineages are derived initially from a multipotential stem cell with an SRP of at least 0.5.
2. The multipotential stem cell pool generates pools of committed progenitors for different lineages: their SRPs are less than 0.5.
3. End cells are derived from a lineage-committed stem cell. This arrangement is illustrated in Figure 1.

The advantage of this tiered system is that committed pools can be regulated independently of the multipotential stem cells and other lineages, and because committed progenitors are already partially differentiated along a particular lineage, rapid replenishment of mature cells can be delivered on demand; the committed progenitors can then be replaced from the multipotential pool (7).

E. Local Versus Systemic Control of Hematopoiesis: Growth Factors

Despite the ability of marrow to respond very rapidly to changes in the blood (in infection, granulocyte production may increase tenfold within a day), cell proliferation in any given marrow site appears to be basically autonomous. Thus, destruction of large amounts of marrow by limb irradiation has no direct effect on intact marrow in other unirradiated limbs, but paradoxically wholesale destruction of proliferating marrow cells by cytotoxic drugs (e.g., fluorouracil) results in rapid increases in proliferating "early" cells in all marrow sites. Therefore, there must be mechanisms (presently unknown) by which marrow senses both the complete peripheral blood picture and its own status locally, which together regulate correct cell production for each lineage. Systemic regulation is brought about through hematopoietic growth factors and cytokines. A large array of growth factors and cytokines exist, which has led to the concept of "cytokine networks" in which various growth factors control the production of others in complex feedback systems. Growth factors are a vital component of all body systems, and because many may also be cell-bound, play an equally important role in local cell regulation.

Peripheral Maturation

Not all cells fully differentiate or mature in the marrow. In particular, lymphoid cells need to be processed elsewhere to become functional: T cells need to be "educated" in a thymus, which determines their ability to respond to foreign and self-proteins and to cooperate in immune responses; B cells (immunoglobin-producing cells) need lymphoid follicles in which to finally mature, although many mature in the marrow itself. These are two special cases, however, because, first, their ability to function properly requires selection of an ability to recognize foreign antigens while being refractory to self-antigens; and, second, because their sites of functional activity are also peripheral (i.e., exposure to antigen is peripheral, and this determines their clonal expansion). The extent of this can be controlled only by short-range effectors; hence, an elaborate system of antigen-capture (macrophages, specific antigen receptors, and cell–cell cooperation) is conducted within special structures that bring these together (lymph nodes).

Environmental Cells

Damage to the marrow environment results in inability of the hematopoietic cells to proliferate or mature, leading to cytopenias or aplasias. These may affect only one lineage (e.g., primary red cell aplasia) or all of them (e.g., myelofibrosis). How such differential effects can be brought about is still unknown. However, the relative resistance of environmental cells to toxicity (radiation or drugs), compared with the hematopoietic cells themselves, can be used in therapies to delete the hematopoietic cells while still leaving the environment intact. These observations raise questions about the specificity of environmental cells for lineage commitment or development: Can any part of the marrow support any lineage?

F. Cell Migration and Trafficking

Blood is a highly dynamic system, and cells are continually entering and leaving it. Lymphocytes migrate out of blood through specialized endothelium in lymphoid follicles (high-capillary endothelium–high-endothelial venules; HEV) and circulate through lymph back into blood. Granulocytes migrate between endothelial cells, and macrophages and other cells migrate into extravascular tissues by mechanisms not wholly understood. Cell trafficking involves specific proteins on endothelial cells, specific recognition sequences in matrix proteins, and receptors for such extracellular matrix proteins on cells. In an immune response, massive cell trafficking takes place in draining lymph nodes, during which response both macrophages and, particularly, activated T cells, secrete very powerful hematopoietic growth factors. In infection, for example, there is a major burst of neutrophil production. However, the major source of "constitutive" hematopoietic growth factors appears to be the reticuloendothelial system (RES) in general. Because the RES is also involved in the clearance of unwanted or damaged peripheral mature cells, a potential unitary system for overall hematopoiesis control by RES-derived growth factors could be envisaged.

G. Assays for Progenitor Cells

Assays for progenitor cells are described in more detail in Section III. Basically, progenitor cells are recognized by their ability to form colonies derived from single cells. In vitro, these are recognized by plating cell suspensions in semisolid media in the presence of specific growth factors. The ability to form colonies is a function of the replicative potential of all daughter cells and the self-renewal potential of the progenitor. Committed progenitors are recognized by the differentiated progeny; therefore, the identity of the progenitor is retrospective. This is important, for there are few specific markers for such progenitors. In vivo, lethally irradiated mice produce colonies in their spleens after engraftment with bone marrow cells. This is particular to mice: few other animals will do so. Spleen colonies (CFU-S) have been shown to be derived from single cells by chromosome marking, and CFU-S assays are a standard means of enumerating the content of multipotential stem cells in a marrow sample.

II. THE BIOLOGY OF BONE MARROW

A. Anatomy and Microenvironment

Bone marrow is a specialized site for hematopoiesis. During embryogenesis, cartilage models of long bones are excavated by vascular endothelial cells that eventually coalesce to form networks of vascular sinuses. These sinuses are lined with endothelium ("littoral" cells) that, however, lack the basement membrane associated with capillary endothelium (179). On the nonluminal (abluminal) side of the sinuses, a layer of reticular cells develops that projects large extensions into the intersinus space (Fig. 2). Hematopoiesis develops in these spaces (i.e., in association with the reticular cells) and is, therefore, extravascular. To enter the bloodstream, cells must pass through the endothelial layer. The segregation of hematopoietic cells within the intersinus spaces creates a specialized microenvironment essential for stem cell survival and proliferation. This microenvironment is complex, and functions through cell–cell contact, local production of soluble growth factors, and growth factors adsorbed to extracellular matrix proteins. That it is specialized is illustrated by the failure of hematopoiesis to survive without the stromal elements and by the existence within the marrow itself of areas where lineage-specific generation of mature cells occurs.

B. Stromal Cells and Long-Term Marrow Cultures

If marrow is cultured in medium with calf serum at 37°C, a variety of cells adhere and some grow. Short-term adherent layers contain macrophages, granulocytes, T cells, some stem cells, and fibroblast-like stromal cells. At low density, these fibroblast-like stromal cells from colonies (CFU-F) and are derived from the marrow sinuses, which include vascular endothelium of all types, reticular cells, dendritic macrophages, and adipocytes. However, no hematopoiesis occurs. The importance of the stromal compartment and its relation with hematopoiesis was elucidated through ability to grow murine marrow in long-term cultures (Dexter-type cultures; long-term bone marrow cultures; LTBMC; 29). In a fundamental advance, Dexter showed that marrow

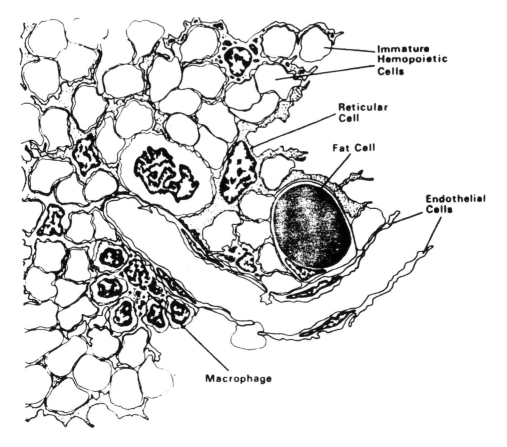

Figure 2 Diagrammatic representation of bone marrow stromal elements in vivo. The venous sinus in the center is lined with endothelial cells and is covered on its adventitial surface by an incomplete sheath of reticular cells. Processes of reticular cells branch into the extravascular space forming a scaffolding on which hematopoietic cells are arranged. An adventitial reticular cell that has accumulated fat is also shown. Macrophages occur through the marrow and are intimately associated with developing hematopoietic cells. (From Ref. 4.)

could be grown in liquid culture, provided that horse serum or hydrocortisone was present and the temperature kept at 33°C. Such cultures maintain hematopoiesis for many months, with continued production of mature granulocytes, erythroid progenitors, and multipotential colony-forming cells, showing that stem cells are present and continue to self-renew. The essential component of LTBMC is the stromal layer, without which all hematopoiesis rapidly stops. The explanations for the success of LTBMC include that stem cells need to interact with an established adherent cell layer to prevent either rapid differentiation or death; that hydrocortisone (present in horse serum) modifies the nature of adherent cells for a hematopoietic capability; and that the lower temperature regulates cell division, allowing stem-cell self-renewal to be balanced with differentiation. Although the first LTBMCs were performed with a two-stage inoculum separated by 1 week (to allow stem cell "niches" to form?), equally successful cultures could be grown using single inocula, provided that extra hydrocortisone was added. During development of LTBMC, many stromal cells acquire the characteristics of adipocytes, and although adipocytes do not seem to be directly involved in hematopoiesis, they are an indication of a successful culture. Removing hydrocortisone and raising the temperature to 37°C results in a rapid decline of myelopoiesis, but cells of the B lineage appear and continue to be produced for several weeks (Whitlock-Witte cultures; 180). These observations show that stromal cells are an integral part of the hematopoietic environment and that this environment is crucial to lineage determination and survival. Although LTBMC generate erythroid precursors, as shown by development of erythroid burst-forming units (BFU-E), they do not generate erythrocytes. However, if anemic mouse serum is added, such cultures will produce mature erythrocytes. Finally, although platelets are not produced, occasional megakaryocytes, eosinophils, and basophils can be seen. These variations on the LTBMC show that the full capacity to generate all hematopoietic lineages is contained in them, and that LTBMCs faithfully replicate marrow hematopoiesis.

C. LTBMC Architecture

An extensive review of LTBMC architecture has been published (4). In murine LTBMC, the stromal layer develops over about 7 days. Macrophages, identified by the antibody F4/80, persist, but the fibroblast-like cells extend and become large, spread cells (Fig. 3). These cells, termed blanket cells, cover islands of granulopoiesis. These granulocytic islands can be recognized as clusters of close-packed polygonal cells, hence the term "cobblestone" areas (Fig. 4g). The blanket cells separate the cultures into areas below and above, and above them similar granulocytic and erythropoietic islands develop (Fig. 5). Adipocytes are situated both above and below blanket cells. Considerable cell movement occurs in LTBMC, with many cells (e.g., macrophages) migrating underneath the blanket cells where they become relatively fixed, and granulocytes migrating out from islands covered by blanket cells (see Fig. 5).

Macrophages appear to play an integral part in the hematopoietic process. In the myeloid series, granulocytes in cobblestone areas interact with macrophages through membrane contact and, subsequently, migrate out when mature. In the erythroid series, late erythroid progenitors associate with the surface of a macrophage on the blanket cell layer, where they undergo synchronous division to produce reticulocytes (Fig. 6). Nuclei ejected during the final stages of erythrocyte maturation are engulfed by the macrophage. Macrophages also serve another important function; namely, the removal of dead cells undergoing apoptosis.

D. Growth Regulation of Hematopoietic Cultures

Although Dexter cultures are highly hematopoietic, concentrations of growth factors in the medium are usually very low. Certain growth factors (e.g., interleukin-3; IL-3) are present in such low amounts that complementary DNA–polymerase chain reaction (cDNA–PCR) is needed to detect them. Addition of growth factors granulocyte–macrophage colony-stimulating factor (GM-CSF) and IL-3 also have minimal effects. For example, GM-CSF added to LTBMC generates only a transient increase in mature granulocytes and CFU-GM. The probable reason for the lack of effects is that cell regulation is chiefly provided by cell–cell contact and, therefore, soluble factors do not have access to cell receptors, or the cell-associated signaling takes preference. Support for cell-associated signaling mediated by growth factors has been provided by observations that soluble growth factors can be adsorbed to extracellular matrices where they are able to stimulate hematopoietic cell growth. However, the K_m of such adsorptions is rather low (10^{-4} to 10^{-6}), compared with 10^{-14} for the respective receptors. Nevertheless, sequestration of soluble factors is a potential mechanism for regulation, and the low K_m may be an advantage in that the distribution of adsorbed factor would be sensitive to concentration.

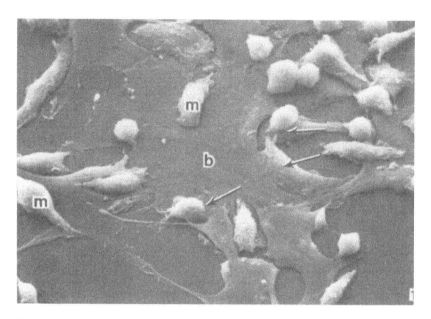

Figure 3　(1) Scanning electron micrograph (SEM) of an area of developing adherent layer cells from mouse bone marrow culture at 7 days after inoculation. Stromal precursor cells that were initially of typical fibroblastoid morphology are now adopting a less bipolar and increasingly spread morphology, typical of blanket cells (b). The majority of the nonspread cells are macrophages (m) some of which (arrowed) are migrating underneath the central blanket cell. (2) SEM detail from adherent layer of "mature" blanket cell in a 4-week culture. The cell is approximately 300 μm in diameter [compare peripheral macrophages (m) and granulocytes (g)]. Because of the highly spread morphology and veillike nature of the blanket cell cytoplasm, many macrophage profiles (three arrowed as examples) are clearly visible beneath the blanket cell, visualized by the penetration of the electron beam in the scanning electron microscope. (From Ref. 4.)

E.　The Role of Stromal Cells in Hematopoiesis

Stromal cells appear to be a prerequisite for maintenance of hematopoiesis in vitro. Without them, the most primitive stem cells are lost through accelerated differentiation (59,61,146,171,172). Matrix is needed for both homing of progenitor cells to inductive sites for hematopoiesis and for establishment of hematopoietic islands (niches) on which balanced production of cells and progenitors occur (32,52,155,170).

Stromal Phenotypes

Few studies have characterized in detail marrow stromal cells. Schweitzer et al. (140) used a battery of endothelial cell-specific antibodies to show that cultured stromal cells derived from fractionated human marrow expressed markers typical of vascular endothelium, particularly EN4, HEC19, BNH9, Endo-1, and UEA-1. Such markers were also found on marrow sinus endothelium in situ. A similar antibody study analyzing stromal cells from human long-term cultures (20) found a variety of types represented, including "stromal fibroblasts," adipocytes, and macrophages, but not vascular endothelial cells. However, the focus has been on the mechanisms by which they support hematopoiesis, rather than to dissect out their precise lineage. The questions asked may be summarized: (1) Are stromal cells lineage-restricted for their support of hematopoiesis? (2) Do they function through producing soluble growth factors, or through intercellular signaling?

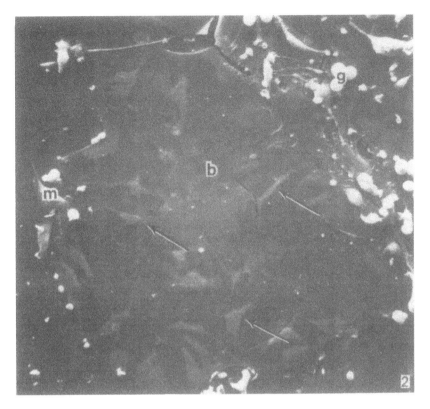

Figure 3 Continued

Stromal Cell Restriction

In LTBMC there are areas of lineage-specific development (e.g., erythropoiesis in association with macrophages and granulopoiesis in cobblestone areas). Although Yanai (184) found that stromal cells from murine spleen supported selective erythropoiesis, there is no definitive evidence that stromal cells are innately lineage-restricted, whereas there is evidence that stroma does restrict or modulate the renewal or amplification of stem cells at different stages of maturation. The reason for the uncertainty is that hematopoietic cultures are highly complex, heterogeneous systems, and removing a part may affect its ability to perform other functions. Stromal lines offer a means to test lineage or maturation selectivity. A series of murine fetal liver stromal lines obtained using SV40 transformation was heterogeneous in its ability to support murine hematopoiesis, with only a low incidence of cells capable of maintaining primitive stem cells (LTC-IC) (181). Bovine fetal liver cell lines also selectively supported murine erythropoiesis (131), which reflected their topographical association in vivo (117). However, other cell lines variously support in vitro growth of several lineage-restricted progenitors. For example, the S17 line from LTBMC maintains both myeloid and granulocytic cultures (23). A series of stromal lines (TBR series), derived from mice transgenic for SV40 temperature-sensitive T antigen, could support both myeloid and lymphoid development (118). However, neither could support long-term maintenance of very primitive progenitors. On the other hand, another mouse stromal line MS-5, derived from mouse LTBMC, could expand early human progenitor cells (HPP-CFC) for 5–10 weeks, whereas equivalent human stromal cells could not (68). Overall, there appears

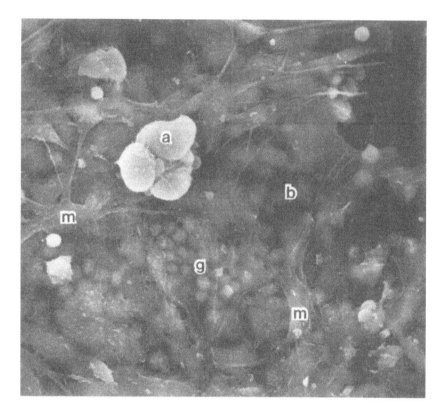

Figure 4 Scanning electron micrograph detail of a similar area to that shown in Fig. 3, where electron beam penetration also shows a group of developing granulocytes (g) beneath the blanket cell cytoplasm (b). Adipocytes (a) and macrophages (m) also occur on the surface of the blanket cell. (From Ref. 4.)

to be great heterogeneity in both ability to maintain or increase primitive progenitors and to support lineage development among the variety of stromal cells investigated.

What is the reason for this heterogeneity? First, the precise identity of "the" primitive multipotential, self-renewing stem cell in marrow remains uncertain; therefore what is being maintained is uncertain. Second, purified populations containing enriched stem cells all require accessory cells for optimal survival, which function may be divorced from self-renewal and replication. Third, cells in tissue culture cannot absolutely duplicate marrow anatomy and, although some cell lines may be able to more faithfully present or express key regulators than others, none alone can do everything. Cell lines inevitably are selected. Part of the answer to the problem lies in how stem cells are regulated by their environment and what this consists of at the molecular level.

Stem Cell Homing and Niches

Marrow stem cells injected into the blood naturally home to preferential sites (e.g., bone marrow, spleen, and liver). In mice, sites of homing can be recognized by the formation of multilineage colonies; hence, the CFU-S assay in spleen. Are there specific mechanisms for this homing, and do stem cells occupy an anatomical or physiological niche outside of which they cannot develop?

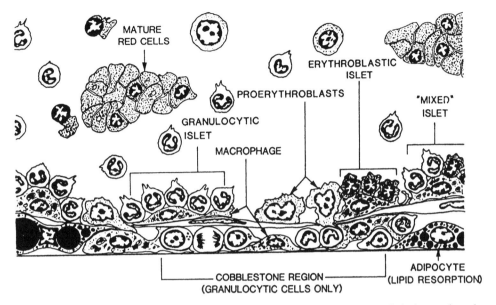

Figure 5 Diagram summarizing the cell types and their stromal/hematopoietic interactions in mouse long-term bone marrow culture. (From Ref. 4.)

Using synthetic glycoproteins as competitive inhibitors, homing of stem cells appears to involve lectin-type receptors on stem cells that recognize sugars, mostly those containing galactose, mannose, and N-acetylglucosamine expressed by stromal cells (156). However, ingress into the marrow requires stem cells to migrate from the vascular compartment to the marrow extravascular one through the vascular endothelium. Here, the situation appears to be reversed, with stem cell membrane glycoproteins interacting with receptors on the vascular endothelium, allowing stem cells to migrate through. Once in the marrow, do stem cells attach to specific sites (niches), or is the formation of hematopoietic island simply a functional aggregation? Many molecules have been implicated in adhesion of stem cells to stroma (186,189). These include P-selectins (189); heparan sulfate (24,47); chondroitin sulfate; fibronectin–VCAM–PECAM, and associated receptors (integrins) (69,113,139,177); growth factor receptors, such as c-kit (85) and CD45 (24); and collagen (84). Of particular interest is CD34. CD34 is a heavily glycosylated sialomucin found on both hematopoietic progenitor cells and vascular endothelium (88). It binds to L-selectin, which may also be present on adult vascular endothelial cells (7) and similarly during early embryonic hematopoiesis (187). However, L-selectin may discriminate between different glycosylated forms of CD34 (107), suggesting that endothelium may selectively adhere different subsets of cells. Similarly, binding of committed granulocyte–macrophage progenitors in cobblestone areas appears to be related to a cell lectin with galactosyl and manosyl specificities (2). Many cells appear to express both adhesion molecules and their cognate receptors. Subsequent stabilization of this interaction could involve other adhesion molecules, such as VLA-4 receptor and VCAM-1 ligand. In support of a role for CD34 in regulating hematopoiesis (but not determining it), transgenic mice lacking CD34 have significantly lower levels of progenitors, although their peripheral blood picture is normal (21).

None of these studies answers whether sites of stem cell lodgment and hematopoiesis are predetermined (anatomical niches) or are induced by mutual cell interactions (physiological niches). The hematopoietic microenvironment is complex and probably a mixture of both. Its

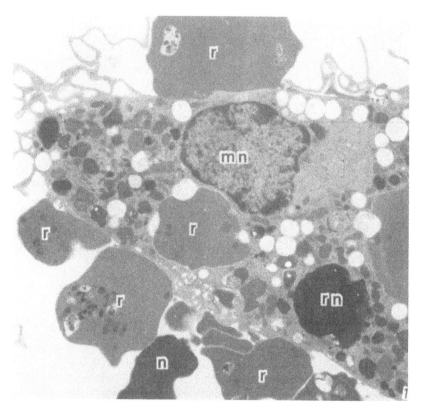

Figure 6 Macrophage/erythroblast interaction in murine long-term bone marrow culture. In (1) an erythroblastic islet has been sectioned parallel to the growing surface, showing six reticulocytes (r) attached to the surface of the central macrophage, and also an adjacent normoblast (n) probably in the process of enucleation. The macrophage nucleus (MN) is indicated, but an expelled reticulocyte nucleus (rn) is also present within a phagocytic vacuole (2). A detail of the interfaced membranes in another erythroblastic islet reveals coated pits and vesicles (v) indicative of the process of receptor-mediated endocytosis at both the surface of the erythrocyte (e) and the macrophage (m). (From Ref. 4.)

ability to sustain stem cells and differentiation of all lineages must rely on various modulators, of which the growth factors play a major role.

Growth Factors and Stromal Hematopoiesis

Stromal cells supporting in vitro hematopoiesis do not release significant amounts of growth factors in in situ LTBMC (30). In murine LTBMC, CSF-1 is constitutively present in the supernate, (3,55), and mRNA for certain cytokines (IL-6, GM-CSF, G-CSF, KL) is detectable (83). An increase in cytokine production can be induced by stimulation with inflammatory proteins, such as IL-1α. By RT-PCR, human stromal cells lines and primary LTBMC stroma express similar cytokine mRNAs: G-CSF, GM-CSF, IL-6, IL-7, IL-11, c-*kit*, KL, LIF, M-CSF, MIP-1α, TGF-β, and TNF-α (157). Secretion of cytokines and their support for stroma-independent expansion of primitive multipotential stem cells is of crucial importance for large-scale amplification of stem cells in clinical transplants. Soluble growth factors can undoubtedly help sustain stem cells in liquid culture (e.g., in stroma-free bioreactor cultures; 86). In these, frequent medium change promotes LTC-IC retention, but stroma is still required to

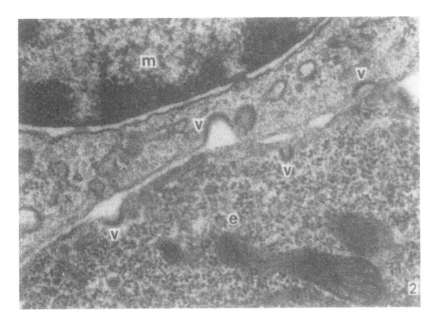

Figure 6 Continued

provide maintenance (87). Therefore, although stromal contact appears to be very efficient in LTBMC in maintaining primitive stem cell numbers, recent work suggests that noncontact cultures, created by separating CD34+ve cells from stromal layers by a permeable membrane, are also quite efficient at retaining multipotential stem cells, although the cultures overgrow with differentiated myeloid cells (169). These cultures were supplemented with growth factors. Rationalization of these apparently conflicting results may be through the release of stromal matrix components that bind growth factors and present them to stem cells (49). In the noncontact cultures, stroma-derived factors are able to maintain LTC-IC in levels of cytokines too low to be supportive by themselves, but are sensitive to agents that digest heparan sulfate (56). Whether the cytokines are physically adsorbed to this matrix has not been determined.

Some ligands normally considered to be soluble may also act as adhesion or niche factors, for example, c-*kit* (85).

III. HEMATOPOIETIC PROGENITOR CELLS

The scale of the hematopoietic system is quite remarkable, considering that daily some 200×10^9 erythrocytes and 50×10^9 granulocytes, in addition to platelets, lymphocytes, and monocytes, enter the circulation (35). Despite the magnitude of this production system, dysregulation is uncommon and external influences can rapidly induce changes in the blood cell count of a specific lineage. For example, hypoxia induces an increase in erythrocyte production, but does not affect the neutrophil count, whereas the opposite is true in an acute bacterial infection. Likewise, and just as importantly, the system is sufficiently sensitive to return blood cell production to normal levels when required (e.g., after blood loss caused by trauma). Clearly, the regulation of hematopoiesis must be exquisitely fine to be able to maintain blood cell counts within a relatively narrow range of normal.

In the 1940s and 1950s, the advent of ionizing radiation for military, peaceful, and clinical purposes made the investigation of the hematopoietic tissue essential, because this appeared to be the process most critically affected by radiation (90,101). Since then, advances in the treatment of malignancies and the prospect of gene therapy have brought issues in the physiology of hematopoiesis related to blood stem cell transplantation to the fore (5,63,80).

At the outset, partial body shielding and marrow grafts were found to protect against radiation injury (70,102). It was then shown that it was the cells in the graft that were responsible for the restoration of hematopoiesis (36). Because the spleen remains an active hematopoietic organ in the adult mouse, this led to the development, in 1961, of the first assay for primitive murine progenitor cells, the spleen colony or CFU-S assay (158), and the search for the hematopoietic stem cell. Following the first description of an in vitro assay for blood progenitor cells in 1966 (13), a great number of other assay systems for progenitor cells, as well as techniques for their isolation, have been described.

Attempts at the purification of the most highly enriched population of clonogenic hematopoietic stem cells possible have been driven by several considerations. First, hematopoiesis offers a relatively easily accessible system to investigate how a single cell can generate diverse cell populations. Second, the phenotypic definition of a cell population, the behavior of which can be reliably predicted, is essential for the investigation of hematopoietic control. Third, the prospect of gene therapy requires the isolation of populations of hematopoietic progenitor cells capable of receiving and expressing the genetic information of interest in the relevant blood cell lineages, including self-renewing stem cells. Fourth, the concepts and technologies used to isolate blood progenitor cells could, in principle, be applied to the study of other developing organ or tissue systems. Finally, it is of great clinical importance to be able to isolate blood stem cells free of tumor cells and separate them from immunocompetent cells for the prevention of graft-versus-host disease in the setting of bone marrow transplantation.

A. Stem Cell Purification Strategies

The many attempts to isolate hematopoietic stem cell populations undertaken so far have been based on a combination of physical, biological, and immunological techniques (174).

Physical Properties

Early work on murine bone marrow used a variety of density gradients and filtration through glass-wool columns to separate subpopulations of cells. Density-gradient separation with dextrans (144) or sedimentation with methylcellulose, gelatin, or starch (27), are methods still routinely used as preenrichment steps in stem cell purification protocols, and counterflow centrifugal elutriation has been used to fractionate mouse bone marrow cells (25,75,164) as well as peripheral blood progenitor cells on a clinical scale (54). Similarly, in flow cytometry, the forward scatter (a measure of cell size) and side scatter (a reflection of the internal complexity of the cell) are commonly used characteristics to serve as a first enrichment step in the isolation of progenitor cells (150). The use of these techniques has led to the notion that pluripotent stem cells are morphologically indistinguishable from lymphocytes, whereas the majority of committed clonogenic progenitor cells resemble small lymphoblasts (150).

Biological Properties

Because most primitive blood progenitor cells are in the G_0 phase of the cell cycle and, therefore, are insensitive to cycle-specific cytotoxic drugs (97,122,133,134,138) agents, such as tritiated thymidine (128), cytarabine (cytosine arabinoside; 130), hydoxyurea (16), and fluorouracil (10), as well as supravital DNA dyes, such as Hoechst 33342 (96,183), have long been employed in the isolation of noncycling progenitor cells. The use of another drug, 4-hydroperoxycyclophosphamide, or its stable analogue mafosfamide, for the isolation of primitive precursor cells (142), is based on the loss early on during the maturation of blood progenitor cells of the enzyme aldehyde dehydrogenase, which catalyzes the conversion of vitamin A to its active metabolite retinoic acid and confers resistance to the cytotoxic action of these alkylating agents (45,77,136). More recently, a fluorescent molecular probe for the presence of aldehyde dehydrogenase itself has been developed that can serve as a marker for a very primitive class of blood progenitor cells (72,74).

Vital Dyes. Rhodamine-123 (Rh-123) is a supravital dye that binds specifically to the mitochondria of living cells. Uptake of this permeant cationic fluorochrome is believed to reflect transmembrane potential, and cellular staining intensity with rhodamine-123 is, therefore, thought to be related to the metabolic state of the cells, as reflected in their mitochondrial mass and activation (26). It is uncertain to what degree the enrichment of Rh-123^{dull} cells for primitive blood progenitor cells (96,115,124,165) is also influenced by the differential expression among hematopoietic progenitor cells of P-glycoprotein, a membrane pump that confers multidrug resistance and for which Rh-123 can act as a substrate (19,82,143,167,178).

A P-glycoprotein-like mechanism also accounts for the specific staining profile of very primitive hematopoietic progenitor cells, which, when analyzed for their Hoechst 33342 fluorescence, displayed simultaneously at two emission wavelengths, segregate into a region of both low green and low red fluorescence (40). This population of cells (referred to as side population) was enriched at least 1000-fold for in vivo reconstitution activity.

Reflecting other phenotypic changes that occur during their ontogeny (91), progenitor cells from murine fetal liver with high totipotentiality and long-term repopulating potential retain more rhodamine-123 (Rh-123^{bright} phenotype) than their counterparts in terms of engraftment potential, from adult murine bone marrow, which show a Rh-123^{dull} phenotype (132).

Substrate Specificity. Hematopoietic cells can also be characterized according to their adherence to various substrates. These differences in adhesive phenotype are thought to be related to the homing of transplanted stem cells across vascular endothelial cells into the marrow cavity (8,39) and have been exploited to separate progenitor cells from their progeny. Cultured marrow-derived stromal layers (33), as well as tissue culture plastic (42,79,81,106), have been used to enrich for more primitive progenitors, whereas a system using nylon wool for large-scale negative selection has recently been described (31).

Lectins are molecules of animal as well as plant origin that bind to specific cell surface carbohydrate residues that are also thought to be related to stem cell homing (76). Soybean (95) and wheat germ agglutinin (WGA) (132,173) have been proposed for use in the initial steps of human and murine blood progenitor cell enrichment protocols, respectively. Within the murine progenitor cell population, the affinity of hematopoietic cells for WGA has been reported to be inversely related to their long-term repopulating ability (LTRA) allowing the concentration of LTRA without significant enrichment of CFU-S day-12 (126).

Immunophenotype

Advances in hybridoma technology, coupled with the possibilities offered by flow cytometry and fluorescent-activated cell sorting, have created the most pronounced advances in the understanding of the physiology of hematopoiesis.

CD34. The identification of CD34 as a stem cell marker in both mice and humans was achieved by following a strategy designed to develop antibodies that specifically recognized small subsets of human marrow cells, but not mature blood or lymphoid cells (22), and has greatly aided the study of human hematopoiesis (28,149). The CD34 molecule (reviewed in Refs. 51,88) is a transmembrane glycophosphoprotein that is expressed on immature lymphohematopoietic cells, some fibroblasts, vascular endothelium, and the high endothelial venules of lymph nodes. Human, murine, and canine CD34 cDNAs are well conserved in evolutionary terms. Homology is highest in the intracellular domains, which carry a region that is a substrate for phosphorylation by protein kinase C, suggesting a possible role in transmembrane signaling (34). In its amino-terminal extracellular portion, CD34 is a sialomucin, with heavily sialylated glycan chains. Its extended bulky conformation may provide a scaffold on which specific recognition interactions may occur. Endothelial CD34 binds to the lectin-like adhesion molecule L-selectin which is involved in lymphocyte trafficking through lymph nodes and neutrophil rolling, and it appears that posttranslational carbohydrate modifications of CD34 are critical in determining its capability to act as a ligand for L-selectin (6). However, CD34 on high endothelial venules appears to be glycosylated differently from CD34 on other endothelia or on hematopoietic cells (6,51), and a ligand for the CD34 molecule expressed by hematopoietic cells has not yet been identified. Nevertheless, there is evidence that suggests that CD34 on blood progenitor cells functions as a regulator of hematopoietic cell adhesion, conferring increased binding to human stromal cells (60). That this effect appears to be species-specific (60) implies that the adhesive function of CD34 is not simply attributable to the negatively charged rod-like conformation of its amino-terminus. An indirect mechanism through which cellular adhesion by hematopoietic progenitor cells is promoted by CD34 is suggested by the enhanced cytoadhesiveness provoked by the engagement of particular epitopes of CD34 and their activation of the β_2-integrin cytoadhesion pathway, similar to that triggered by CD44 between follicular dendritic cells and B cells (104). However, CD34 is not the only cell surface glycoprotein thought to be related to hematopoietic progenitor cell homing and mobilization (163,188). Studies in CD34-deficient mice have shown no abnormalities in hematopoiesis or interactions between CD34-negative progenitor cells and a stromal cell line in vitro (153).

Many epitopes exist on the CD34 molecule, and more than 30 monoclonal antibodies have been classified according to the susceptibility to digestion with different glycoproteases of the epitopes they recognize on human CD34 (38). Class I epitopes are sensitive to cleavage with neuraminidase, chymopapain, and a glycoprotease from *Pasteurella haemolytica*. Class II epitopes are sensitive to degradation with chymopapain and glycoprotease, but are resistant to neuraminidase, whereas the class III epitopes on CD34 are resistant to the action of all three enzymes. Even though there is a linear relation between class II and class III epitope expression in populations of unselected CD34$^+$ cells from peripheral blood progenitor cell collections and umbilical cord blood samples (137), flow cytometric light-scatter analysis suggests that class I and II epitopes may be down-regulated before class III epitopes during normal hematopoietic progenitor cell differentiation (148). Likewise, CD34$^+$ cells that express high levels of the antigen (CD34^{bright}) are more primitive than the CD34^{dull} cells (150). Recent evidence, however, suggests that, in the mouse and rhesus macaques at least, the CD34$^-$ cell population may conceal a very small population of hematopoietic stem cells that appear to be even more primitive than their immunophenotypically otherwise similar CD34$^+$ counterparts when assayed in terms of their competitive long-term repopulating potential (71,74,114,120). Cell-tracking studies, using the lipophilic dye PKH26, have shown that the CD34$^-$ cells in the Hoechst 33342 side population of rhesus macaque bone marrow give rise to CD34$^+$ cells when cultured on irradiated bone marrow stroma (71).

Other Cell Surface Antigens.   Several other membrane-bound molecules have been useful for the separation of early from late human hematopoietic progenitor cells. They include Thy-1 (CD90), a member of the immunoglobulin superfamily expressed on a variety of tissues, that shows an inhibitory function on the proliferation of primitive hematopoietic cells (108,109); the transferrin receptor CD71 (a cell activation marker; 92,162); CD38 (37,66,175); and the CD45 isoform recognized by anti-CD45RO antibodies (93). Expression of the class II histocompatibility antigen HLA-DR appears to be dependent on the hematopoietic tissue that gives rise to the progenitor cells: whereas only those CD34$^+$ cell populations from adult human bone marrow that lack HLA-DR contain the primitive long-term culture-initiating cells (LTC-IC) (15), LTC-IC activity in human cord blood appears to be contained in the CD34$^+$HLA-DR$^+$ cell population (18,161).

Monoclonal antibodies can also be used to deplete blood progenitor cell populations from lineage-committed cells. The expression of the human lineage-associated molecules CD2 and CD7 (T lymphocytes), CD10 and CD19 (B lymphocytes), and CD33 (granulocytes) does not segregate with lineage commitment, whereas CD34$^+$CD13^{bright}, CD34$^+$CD14$^+$, and CD34$^+$CD15$^+$ cells (in order of appearance during ontogeny) generate only myeloid progeny, and CD34$^+$CD20$^+$ cells do not to differentiate along the T-lymphoid and granulocytic pathways (160). A consensus phenotype for the most primitive adult CD34$^+$ human progenitor cells is CD34^{bright}, Lineage$^-$, CD71^{low}, CD45RO$^+$, Thy-1$^+$, CD38$^-$, HLA-DR$^-$, and Rh-123^{dull}. Stem and progenitor cells in the mouse can be categorized immunophenotypically along similar lines (100,145).

B. Primary Progenitor Cell Assays

Cell marker studies in transplant recipients support the concept of a hematopoietic stem cell with the capacity to reconstitute all lymphomyeloid lineages in a recipient and retain the ability to self-renew, but in vivo studies of human hematopoietic stem cells require that recipients be observed for months to years to establish that a complete and stable reconstitution has occurred. To accelerate studies on the biology of blood progenitor cells, short-term in vivo and in vitro assays have been developed. These can be broadly divided into three classes: (1) direct colony-forming (clonogenic) assays, (2) assays in which early precursor cells are detected by their ability to produce more mature colony-forming progeny, and (3) transplantation assays.

Direct In Vitro Clonogenic Assays

In its simplest form, an assay system for mammalian hematopoietic progenitor cells consists of a semisolid medium (most often agarose or methylcellulose) to which hematopoietic growth factors, either in the form of cell-conditioned media or recombinant proteins, are added. Such clonogenic assays are the most direct quantitative means of measuring hematopoietic progenitor cells in vitro (112). The colonies observed are clones of cells produced by a single progenitor. The progeny can be analyzed morphologically to determine the lineage attribution of the colony-forming unit assayed. The self-renewal and differentiation potential of the colony-forming cells can be assessed by replating the cells contained in the colonies. This has led to the establishment of a hierarchical relation between hematopoietic progenitor cells measured directly by in vitro colony formation. Thus, CFU-GEMM, defined as colony-forming cells which give rise to colonies that contain cells from all myeloid lineages (i.e., cells of the granulocytic lineage, erythroid cells, monocytes–macrophages, and megakaryocytes), under certain culture conditions, possess an extensive replating (self-renewal) capacity (17) that is not seen with the more mature lineage-restricted colony-forming progenitor cells.

Spleen Colony-Forming Cells

The earliest assay for hematopoietic progenitor cells followed a report in 1961 on the ability of injected murine bone marrow cells to form macroscopic colonies in the spleen of irradiated mice (158) and the demonstration that each colony was derived from a single pluripotent progenitor cell, defined as the spleen colony-forming unit (CFU-S; 9). However, several lines of evidence supported the view that the CFU-S population was heterogeneous, and the notion of a stem cell hierarchy was deduced. Most of the nodules on day 8 after transplantation (CFU-S day-8) consisted of cells belonging to a single lineage, the erythroblastic one being predominant, whereas CFU-S day-12 contained cells of several myeloid lineages. CFU-S day-12 were not the progeny of CFU-S day-8 (103). Furthermore, when cells contained in nodules of late-appearing, but not of early, CFU-S were transplanted into secondary myeloablated hosts, secondary spleen colonies could be observed in certain of these recipients, demonstrating the self-renewal capacity of that class of CFU-S (159). Finally, the administration of cycle-specific chemotherapeutic agents that spare cells in the G_0/G_1, phase before bone marrow harvesting showed that late, but not early, CFU-S were dormant noncycling cells (62,135). That CFU-S do not belong to the most primitive class of murine hematopoietic progenitor cells was demonstrated by the transplantation of bone marrow cell populations that were physically depleted of CFU-S day-12, yet retained marrow repopulating ability (75). Also, in limiting-dilution experiments, the frequency of stem cells assayed by their ability to cure the anemia of W/W^- mice was significantly less than that of the CFU-S population (12).

High-Proliferative Potential Colony-Forming Cells

Not least because of the need of large numbers of experimental animals to meet statistical requirements in CFU-S assays, several in vitro assays for primitive clonogenic hematopoietic progenitor cells have been developed. High-proliferative potential colony-forming cells (HPP-CFC) give rise to very large in vitro macrophage colonies, with diameters larger than 0.5 mm, that contain more than 50,000 cells per colony after prolonged exposure to cytokine combinations in semisolid media (110). Their very early position in the blood progenitor cell hierarchy is substantiated by their high resistance in vivo to treatment with fluorouracil, their self-renewal potential (147), and in mice, their capacity to give rise to lineage-restricted colony-forming cells of the entire myeloid series and repopulate hematopoiesis after myeloablative irradiation (111). Macroscopic (> 1 mm) multilineage colonies termed CFU type A or CFU-A have also been described in other culture systems (64), but these may represent more mature precursor cells than the most primitive HPP-CFC (130).

Blast Colony-Forming Cells

Hematopoietic progenitor cells remain as small blast cells without signs of cytoplasmic maturation during all stages of hematopoietic development except for the last few cell divisions. However, blast cell colonies that are adequately restricted in terms of their cell number and time of appearance in culture to include only those colonies that show high-replating and multilineage capacity (78,94) have been an important tool for the investigation of mechanisms underlying hematopoiesis at a level that is more primitive than that occupied by CFU-GEMM, for they are amenable to serial observations and micromanipulation (116).

C. Secondary Clonogenic Cell Assays

Primitive progenitor cells that do not themselves form colonies in semisolid colony-forming assays can be detected indirectly by measuring the number of clonogenic progeny they produce. Because individual precursor cells produce an unknown number of cells in the daughter population that can be assayed as colony-forming units or cobblestone areas, precursor cell frequencies cannot be obtained from the results of bulk cultures, but need to be measured by limiting-dilution analysis. This follows the principle that when a constant volume of a dilute cell suspension is used to initiate a large number of cultures, the number of clonogenic cells inoculated per culture will follow a Poisson distribution (i.e., the chance of no growth being observed in a given culture vessel is e^{-x}). If, for example, out of 100 inoculated cultures 37 fail to develop progeny, the mean number of progenitor cells in the cultured volume would be calculated as the solution of $e^{-x} = 0.37$, in this case $x = 1.0$ (129). For the economical design of limiting-dilution experiments a single-dilution protocol is adequate (11). Limiting-dilution analysis will also yield a measurement of the number of progeny formed per individual progenitor cell. Thus, in one assay system, the number of clonogenic cells per long-term culture-initiating cell (LTC-IC) in 5-week-old cultures ranged from 1 to 30, the average being 4.2 ± 0.5 (152). The variability in the proliferative capacity of individual progenitor cells during the culture period is thought to reflect the operation of a probabilistic mechanism in the regulation of hematopoiesis (1,89,159), rather than heterogeneity in the progenitor cell population expressing different degrees of maturity and ability to generate more mature progeny. Because there is experimental evidence that the stochastic process in hematopoiesis can be influenced by exogenous factors (99,105), the average number of colony-forming cells produced per LTC-IC can not be assumed to be the same in assays with different culture conditions.

Long-Term Culture-Initiating Cell Assay

When long-term bone marrow cultures are set up in two stages by first growing a stromal feeder layer to confluence before adding a hematopoietic cell suspension, the culture system can be used to enumerate primitive progenitor cells by measuring the clonogenic cells produced 5–6 weeks after the initiation of the culture. This follows the observation that when untreated control marrow samples are used to initiate long-term cultures, the production of colony-forming cells decreases rapidly during the first 2 weeks, whereas when human marrow samples are layered onto stromal feeder layers after treatment with 4-hydroperoxycyclophosphamide, which from clinical transplant experience is known to leave intact hematopoietic stem cell activity (185). Colony-forming cells appear in the cultures after a delay of 2–3 weeks. Their numbers reach those of the controls after 4 weeks, indicating that, after this time, all of the colony-forming cells are the progeny of the primitive progenitor cell population (182). Supporting the validity of this assay system, a linear relation between the input cell number and the number of clonogenic cells present after 5 weeks has been demonstrated over a 1000-fold range (152,166). Week-5 LTC-IC numbers in normal human bone marrow have been calculated at $7.2 \pm 1.3/10^4$ mononuclear cells (150). Recent modification of the LTC-IC assay include the use of a murine fibroblast cell line that is genetically engineered to produce human IL-3, stem cell factor, and G-CSF, as well as the assessment of the colony-forming content of the nonadherent cells, combined with the trypsinized adherent cell population at the end of the sixth week of incubation (128). Such standardized culture conditions make it possible to rely on the high degree of consistency seen in the average clonogenic cell output per LTC-IC (151) to calculate the LTC-IC content in the analyzed sample. The greater delay in the assessment of the endpoint is related to the perceived structure of the stem cell population. It is assumed that the more primitive progenitor cells produce colony-forming units later on during the culture period. The LTC-IC assay has also been adapted to include the lymphoid differentiation series in the endpoint, by changing the long-term cultures to conditions permissive for B-cell lymphopoiesis 1 week before harvesting. The progeny is then assayed separately in semisolid cultures for its content in pre-B and myeloid colony-forming units (98).

Cobblestone Area-Forming Cell Assay

The physiological role of stromal cells in hematopoiesis has also been exploited in the cobblestone area-forming cell (CAFC) assay, in which the occurrence of hematopoiesis in distinct foci beneath the stromal blanket cells of long-term bone marrow cultures is being used as another endpoint for the measurement of LTC-IC (123). In long-term cultures of human bone marrow, cord blood, and blood stem cells, the weekly CAFC numbers were linearly related to the nonadherent clonogenic cells produced at the same time points, as well as to the number of clonogenic cells in the supernatant and adherent layer combined at week 5. It was concluded, therefore, that the CAFC assay is capable of enumerating LTC-IC using a direct visual endpoint (14,121). Phase-contrast microscopy is used to discriminate between the cobblestone areas that are covered by stroma (large, dark, often polyclonal cells), or other cells situated on top of the stroma (small, bright, spherical) (125). The behavior of CAFC of different degrees of maturation is reminiscent of the appearance of erythroid burst-forming units in semisolid cultures (53). Immature CAFC show highly migratory behavior such that cells produced during the first five to ten divisions lead to the formation of fields of cobblestone areas, whereas the more mature CAFC typically form single, large cobblestone areas (123). With increasing maturity, the cells migrate to the surface of the adherent layer and are shed into the growth medium (106). Similar to CFU-S, cobblestone area formation is time-dependent, the variation in time required for the clonal expansion of CAFC reflecting their doubling time and primitiveness (125). CAFC day-10 frequency in murine marrow is closely correlated to the frequency of CFU-S day-12, whereas marrow-repopulating ability, defined as a cell capable of generating new granulocyte–macrophage CFU (CFU-GM) after 12–13 days in lethally irradiated mice, showed close correlation with CAFC day-28 ($r^2 = 0.79$), but not with CFU-S day-12 ($r^2 = 0.24$) (127). From a large number of samples, the median week-5 CAFC frequencies per 10^4 mononuclear cells were 0.75 in human bone marrow, 0.80 in umbilical cord blood, and 0.97 in leukapheresis material (121).

Delta Assays

Because two-stage long-term bone marrow cultures are technically complex, they are thought to be prone to significant interlaboratory variations. LTC-IC assays also require prolonged incubation periods. Therefore, stroma-free, indirect colony-forming assays, in which hematopoietic cells are incubated for 1 week in suspension cultures in the presence of cytokine combinations, before their colony-forming progenies are assayed, have been proposed for the enumeration of early blood progenitor cells (pre-CFUs) (141,142). These assays are named in analogy to the Greek letter Δ at the apex of which is placed the pre-CFU and at its base the clonogenic progeny. Apart from allowing the pre-CFUs to develop into clonogenic cells, the incubation of the sample containing the pre-CFUs is designed to drive the CFUs initially present to terminal maturation and, thereby, exclude them from being measured together with the colony-forming progeny at the end of the liquid-culture period. The absence of stromal cells in the suspension cultures of the delta assays allows the primitive progenitor cells to mature more rapidly in response to the added cytokines than they would otherwise do on stromal feeder layers, such as in the LTC-IC assay (46,168).

*P*Δ *Assay.*   In the plastic-adherent progenitor (PΔ) cell assay, primitive progenitor cells are first selected for their plastic adherent phenotype by a 2-hour incubation period in flat-bottomed plastic tissue culture vessels, followed by removal of the nonadherent cells by several washing steps. Enrichment of early hematopoietic progenitors in the plastic-adherent population is shown by their high content of Thy-1$^+$, CD38$^-$, CD33$^-$, HLA-DR$^-$ cells (46); their resistance to treatment in vitro with fluorouracil (48); their ability to form cobblestone areas in long-term bone marrow cultures (42); and in mice, their radioprotective (8) as well as competitive repopulation activity (81). In the PΔ assay, these selected cells are incubated during 1 week, either in the presence of a bladder carcinoma cell line-conditioned medium, or a combination of IL-1α and

GM-CSF, after which the colony-forming progeny in the supernatant of the liquid culture is assayed in a conventional manner (41). Finally, a linear relation between the input cell number and the result of the PΔ assay has been demonstrated (42). The PΔ cell frequency in normal adult human bone marrow has been measured at 0.25–1.0/10^4 mononuclear cells (42).

4-HCΔ Assay. Although similar in design to the PΔ assay, the 4-HCΔ assay employs the resistance of primitive progenitor cells to the actions of the alkylating agent 4-hydroperoxy-cyclophosphamide (see foregoing) to select an early stem cell population from the hematopoietic tissue analyzed. This population is then allowed to undergo maturation and expansion during 1 week in liquid culture before their clonogenic progenies are again assayed as the endpoint. Good correlations between 4-HCΔ assay readings and the HPP-CFC assay as well as week-4 LTC-IC assay results have been reported (142).

D. In Vivo Hematopoietic Progenitor Cell Assays

Apart from the CFU-S, which are assayed 7–14 days after transplantation of the hematopoietic cells, there are several other endpoints that can be assessed for the measurement of progenitor cell number and function by transplantation into animals. The repopulation ability of murine bone marrow progenitor cells is generally assayed by performing a colony-forming unit count per mouse femur on day-13 after transplantation. Another short-term in vivo assay has as its endpoint the radioprotective ability of the transplanted cells, assessed as survival for a minimum of 30 days after otherwise lethal irradiation and transplantation. It is assumed that it is the rapid recovery of the neutrophil and platelet count that is the determinant of the radioprotective ability observed.

Supported by the multilineage nature of the progeny produced, these short-term in vivo assays were initially considered to quantitate diverse properties of a homogeneous primitive hematopoietic progenitor cell population. However, this concept was cast into doubt by the demonstration that serial transplantation, while maintaining large numbers of short-term re-populating cells, exhausted cells able to repopulate secondary irradiated recipients, considered to be the hallmark of hematopoietic stem cells (73).

Competitive Long-Term Repopulation Assays. A generally applicable in vivo assay proce-dure, which allows absolute numbers of stem cells with long-term lymphohematopoietic re-population ability to be quantitated, measures hematopoietic progenitor cell activity based on the ability of individual cells to reconstitute the hematopoietic tissues of lethally irradiated histo-compatible recipients when transplanted at limiting-dilution concentrations in competition with a defined population of cotransplanted cells that ensure the survival of all the recipients subjected to that radiation dose. Greater than 5% repopulation by test cells more than 5 weeks after transplantation is counted as a positive outcome (154). Unique retroviral integration sites and chromosome, isoenzyme, or hemoglobin markers can be used to distinguish the progenies of the stem cells of interest from those of the cotransplanted progenitor cells. Transplantation into nonirradiated W/W$^-$ mice in which wild-type hematopoietic cells have a growth advantage over the mutant phenotype, which is heterozygous for c-*kit*, the receptor for stem cell factor, has also been used for the quantitation of competitive repopulation units (CRU) (81). CRUs defined in this way are found in the bone marrow of adult mice at a frequency of approximately 1:10^4 nucleated cells (81,154). For the measurement of human CRUs, transplantation into im-munodeficient NOD/SCID mice is being used. There frequency (95% confidence interval) per 10^6 mononuclear cells was 0.33 (0.19–0.55) in normal bone marrow, 1.2 (0.7–1.9) in umbilical cord blood, and 0.17 (0.08–0.32) in leukapheresis material (176).

IV. CONCLUSION

The understanding of how a single cell can generate diverse cell populations remains a fundamental challenge in developmental biology. Findings showing that cells with long-term hematopoietic activity in vivo and in vitro can be selectively enriched stimulated a search for *the* stem cell. Whereas the clinical definition of a hematopoietic stem cell calls for a cell that is capable of reconstituting myeloablated hosts, providing enough mature progeny of all the lymphoid and myeloid lineages to replenish the hematopoietic organs and reverse the pancytopenia, the biological definitions of a primitive hematopoietic progenitor cell population are aimed at identifying populations for which behavior can be reliably predicted, such as high-proliferative, multipotent cells with self-renewal capacity (119). Failure to purify prospectively homogeneous cell populations, defined by retrospective functional analysis, is to be expected because to display in vivo stem cell behavior, a cell needs to undergo several self-renewal divisions, each of which occurs only with a certain probability, before the generation of committed progenitors. Therefore, as the number of required self-renewal divisions increases, cells showing a particular type of stem cell behavior become increasingly rare (clonal deletion) (43). By limiting the numbers of stem cells in a (noncompetitive) transplant experiment, progenitor cells may be forced to display a higher proliferative and smaller self-renewal potential (58). This may explain the results of some cell separation and transplantation experiments, which have not been confirmed in other settings (44,57,190), that indicate the existence of subpopulations of stem cells responsible for early and long-term reconstitution (73). On the basis of data from the observation of long-term engraftment kinetics in cats, computer simulations of hematopoiesis, taking into account stem cell decisions relative to replication, apoptosis, and initiation of a differentiation and maturation program, have been undertaken. From these studies it is suggested that the frequency of feline hematopoietic stem cells is near $1.7/10^6$ nucleated marrow cells and that in steady-state hematopoiesis these cells do not, on average, replicate more frequently than once every 3 weeks (1).

ACKNOWLEDGMENT

This work was supported by the Exeter Leukaemia Fund and Friends of the Oncology Department at Exeter. C. Rudin is the recipient of an Educational Grant from AMGEN UK.

REFERENCES

1. Abkowitz JL, Catlin SN, Guttorp P. Evidence that hematopoiesis may be a stochastic process in vivo. Nature Med 1996; 2:190–197.
2. Aizawa S, Tavassoli M. Interaction of murine granulocyte–macrophage progenitors and supporting stroma involves a recognition mechanism with galactosyle and mannosyle specificities. J Clin Invest 1987; 80:1698–1705.
3. Alberico T, Ihle JN, Liang C, McGrath HE, Quesenberry PJ. Stromal growth factor production in irradiated lectin-exposed long-term murine bone marrow cultures. Blood 1987; 69:1120–1127.
4. Allen TD, Dexter TM, Simmons PJ. Marrow biology and stem cells. In: Dexter TM, Garland JM, Testa NG, eds. Colony-Stimulating Factors; Molecular and Cell Biology. New York: Marcel Dekker, 1990:1–38.
5. Armitage JO. Bone marrow transplantation. N Engl J Med 1994; 330:827–835.
6. Baumhueter S, Dybdal N, Kyle C, Lasky LA. Global vascular expression of murine CD34, a sialomucin-like endothelial ligand for L-selectin. Blood 1994; 84:2554–2565.

7. Baumhueter S, Singer MS, Henzel W, Hemmerich S, Renz M, Rosen SD, Lasky LA. Binding of L-selectin to the vascular sialomucin CD34. Science 1993; 262:436–438.

8. Bearpark AD, Gordon MY. Adhesive properties distinguish sub-populations of haemopoietic stem cells with different spleen colony-forming and marrow repopulating capacities. Bone Marrow Transplant 1989; 4:625–628.

9. Becker AJ, McCulloch EA, Till JE. Cytological demonstration of the clonal nature of spleen colonies derived from transplanted mouse marrow cells. Nature 1963; 197:452.

10. Berardi AC, Wang A, Levine JD, Lopez P, Scadden DT. Functional isolation and characterization of human hematopoietic stem cells. Science 1995; 267:104–108.

11. Blackett NM, Gordon MY. Optimizing limiting dilution assays: frequency and "ability" measurements of haemopoietic progenitor cells. Br J Haematol 1996; 92:507–513.

12. Boggs DR, Boggs SS, Saxe DF, Gress LA, Canfield DR. Hematopoietic stem cells with high proliferative potential. Assay of their concentration in marrow by frequency and duration of cure of W/Wv mice. J Clin Invest 1982; 70:242–253.

13. Bradley TR, Metcalf D. The growth of mouse bone marrow cells in vitro. Aust J Exp Biol Med Sci 1996; 44:287–300.

14. Breems DA, Blokland EAW, Neben S, Ploemacher RE. Frequency analysis of human primitive haematopoietic stem cell subsets using a cobble stone forming cell assay. Leukemia 1994; 8:1095–1104.

15. Briddell RA, Broudy VC, Bruno E, Brandt JE, Srour EF, Hoffman R. Further phenotypic characterization and isolation of human hematopoietic progenitor cells using a monoclonal antibody to the c-*kit* receptor. Blood 1992; 79:3159–3167.

16. Byron JW. Comparison of the action of ^{3}H-thymidine and hydroxyurea on testosterone-treated hemopoietic stem cells. Blood 1972; 40:198–203.

17. Carrow CE, Hangoc G, Broxmeyer HE. Human multipotent progenitor cells (CFU-GEMM) have extensive replating capacity for secondary CFU-GEMM: an effect enhanced by cord blood plasma. Blood 1993; 81:942–949.

18. Caux C, Favre C, Saeland S, Duvert V, Mannoni P, Durand I, Aubry JP, de Vries JE. Sequential loss of CD34 and class II MHC antigens on purified cord blood hematopoietic progenitors cultured with IL-3: characterisation of CD34$^-$, HLA-DR$^+$ cells. Blood 1989; 73:1287–1294.

19. Chaudhary PM, Roninson IB. Expression and activity of P-glycoprotein, a multidrug efflux pump, in human hematopoietic stem cells. Cell 1991; 66:85–94.

20. Chen ZZ, Bockstaele D, Buyssens N, Hendrics D, DeMeester I, Vanhoof G, Scharpe SL, Peetermans M, Berneman Z. Stromal populations and fibrosis in human long-term marrow cultures. Leukemia 1991; 5:772–781.

21. Cheng J, Baumhueter S, Cacalano G, Carver-Moore K, Thibodeaux H, Thomas R, Broxmeyer HE, Cooper S, Hague N, Moore M, Lasky LA. Haematopoietic defects in mice lacking the sialomucin CD34. Blood 1996; 87:479–490.

22. Civin CI, Strauss LC, Brovall C, Fackler MJ, Schwartz JF, Shaper JH. Antigenic analysis of hematopoiesis. III. A hematopoietic progenitor cell surface antigen defined by a monoclonal antibody raised against KG-1a cells. J Immunol 1984; 133:157–165.

23. Collins L, Dorshkind K. A stromal cell line from long-term bone marrow cultures can support myelopoiesis and B lymphopoiesis. J Immunol 1987; 138:1082–1087.

24. Coombe DR. Mac-1 (CD11b/CD18) and CD45 mediate adhesion of hematopoietic progenitor cells to stromal cells via recognition of stromal heparan sulphate. Blood 1994; 84:739–752.

25. Cooper S, Broxmeyer HE. Purification of murine granulocyte–macrophage progenitor cells (CFU-GM) using counterflow centrifugal elutriation. In: Freshney RI, Pragnell IB, Freshney MG, eds. Culture of Hematopoietic Cells. New York: Wiley-Liss, 1994:223–234.

26. Darzynkiewicz Z, Traganos F, Staiano-Coico L, Kapuscinski J, Melamed MR. Interactions of rhodamine 123 with living cells studied by flow cytometry. Cancer Res 1982; 42:799–806.

27. Denning-Kendall P, Donaldson C, Nicol A, Bradley B, Hows J. Optimal processing of human umbilical cord blood for clinical banking. Exp Hematol 1996; 24:1394–1401.

28. de Wynter EA, Coutinho LH, Pet X, Marsh JCW, Hows J, Luft T, Testa NG. Comparison of purity and enrichment of CD34$^+$ cells from bone marrow, umbilical cord and peripheral blood (primed apheresis) using five separation systems. Stem Cells 1995; 13:524–532.

29. Dexter, TM, Allen TD, Lajtha LG. Conditions controlling the proliferation of haemopoietic stem cells in vitro. J Cell Physiol. 1977; 91:335–344.

30. Dexter, TM, Coutinho LH, Spooncer ES, Heyworth CM, Daniel CP, Schiro R, Chang J, Allen TD. Stromal cells in haematopoiesis. In: Bock G, March J, eds. Molecular Control of Haematopoiesis. Chichester: Wiley & Sons, 1990.

31. di Nicola M, Bregni M, Siena S, Ruffini PA, Milanesi M, Ravagnani F, Gianni AM. Combined negative and positive selection of mobilized CD34$^+$ blood cells. Br J Haematol 1996; 94:716–721.

32. Dorshkind K. Regulation of hemopoiesis by bone marrow stromal cells and their products. Annu Rev Immunol 1990; 8:111–137.

33. Dowding CR, Gordon MY. Physical, phenotypic and cytochemical characterisation of stroma-adherent blast colony-forming cells. Leukemia 1992; 347–351.

34. Fackler MJ, Krause DS, Smith OM, Civin CI, May WS. Full-length but not truncated CD34 inhibits hematopoietic cell differentiation of M1 cells. Blood 1995; 85:3040–3047.

35. Finch CA, Harker LA, Cook JD. Kinetics of the formed elements of human blood. Blood 1977; 50:699–707.

36. Ford CE, Hamerton JL, Barnes DWH, Loutit JF. Cytological identification of radiation chimeras. Nature 1956; 177:452.

37. Funaro A, Roggero S, Horenstein A, Calosso L, Dianzani U, de Monte LB, Zocchi E, Franco L, Guida L, Ausiello CM, Drach J, Metha K, Bargellesi A, Malavasi F. CD38: a transmembrane glycoprotein with pleiotropic ectoenzyme function. In: Schlossman SF, Boumsell L, Gilks W, et al., eds. Leucocyte Typing V. Oxford: Oxford University Press, 1995:380–383.

38. Gaudernack G, Egeland T. Epitope mapping of 33 CD34 mAb, including the Fifth Workshop panel. In: Schlossman SF, Boumsell L, Gilks W, et al., eds. Leucocyte Typing V. Oxford: Oxford University Press, 1995:861–864.

39. Gibbs RV, Lewis JL, Gordon MY. Expression of cell-surface lectins on haemopoietic progenitor cells. Br J Biomed Sci 1995; 52:249–256.

40. Goodell MA, Brose K, Paradis G, Conner AS, Mulligan RC. Isolation and functional properties of murine hematopoietic stem cells that are replicating in vivo. J Exp Med 1996; 183:1797–1806.

41. Gordon MY. Culture of plastic-adherent and stroma-adherent hemopoietic progenitor cells. In: Doyle A, Griffiths JB, Newell DG, eds. Cell and Tissue Culture: Laboratory Procedures. Chichester: John Wiley & Sons, 1994:Module 21B:10.

42. Gordon MY. Plastic-adherent cells in human bone marrow generate long-term hematopoiesis in vitro. Leukemia 1994; 8:865–870.

43. Gordon MY, Blackett NM. Routes to repopulation—a unification of the stochastic model and separation of stem-cell subpopulations. Leukemia 1994; 8:1068–1073.

44. Gordon MY, Blackett NM, Lewis JL, Goldman JM. Evidence for a mechanism that can provide both short-term and long-term haematopoietic repopulation by a seemingly uniform population of primitive human haematopoietic precursor cells. Leukemia 1994; 9:1252–1256.

45. Gordon MY, Goldman JM, Gordon-Smith EC. 4-Hydroxycyclophosphamide inhibits proliferation by human granulocyte–macrophage colony-forming cells (GM-CFC) but spares more primitive progenitor cells. Leuk Res 1985; 9:1017–1021.

46. Gordon MY, Lewis JL, Grand FH, Marley SB, Goldman JM. Phenotype and progeny of primitive adherence hematopoietic progenitor. Leukemia 1996; 10:1347–1353.

47. Gordon M, Riley GP, Clarke D. Heparan sulphate is necessary for adhesive interactions between human early hematopoietic progenitor cells and the extracellular matrix of the bone marrow microenvironment. Leukemia 1988; 2:804–809.

48. Gordon MY, Riley GP, Greaves MF. Plastic-adherent progenitor cells in human bone marrow. Exp Hematol 1987; 15:772–778.

49. Gordon MY, Riley GP, Watt SM, Greaves MF. Compartmentalisation of a haematopoietic growth

factor (GM-CSF) by glycosaminoglycans in the bone marrow microenvironment. Nature 1987; 326:403–405.

50. Grand FH, Marley SB, Lewis JL, Goldman JM, Melo JV, Gordon MY. Phenotype and function of plastic-adherent progenitor cells in human hematopoietic tissue [abstract]. Exp Hematol 1995; 23:817.

51. Greaves MF, Titley I, Colman SM, Bhring H-J, Campos L, Castoldi GL, Garrido F, Gaudernack G, Girard J-P, Ingls-Esteve J, Invernizzi R, Knapp W, Lansdorp PM, Lanza F, Merle-Bral H, Parravicini C, Razak K, Ruiz-Cabello F, Springer TA, van der Schoot CE, Sutherland DR. CD34 cluster workshop report. In: Schlossman SF, Boumsell L, Gilks W, et al., eds. Leucocyte Typing V. Oxford: Oxford University Press, 1995:840–846.

52. Greenberger J. The hematopoietic environment. Crit Rev Oncol Hematol 1991; 11:65–84.

53. Gregory CJ, Eaves AC. Human marrow cells capable of erythropoietic differentiation in vitro: definition of three erythroid colony responses. Blood 1977; 49:855–864.

54. Grimm, J, Zeller W, Zander AR. Separation and characterization of mobilized and unmobilized peripheral blood progenitor cells by counterflow centrifugal elutriation. Exp Hematol 1995; 23:535–544.

55. Gualtieri RJ, Shadduck RK, Baker DG, Quesenberry PJ. Haematopoietic regulatory factors produced in long-term murine marrow cultures and the effect of in vitro irradiation. Blood 1984; 64:516–525.

56. Gupta P, McCarthy JB, Verfaillie CM. Stromal fibroblast heparan sulfate is required for cytokine-mediated ex vivo maintenance of human long-term culture-initiating cells. Blood 1996; 87:3229–3236.

57. Harrison DE, Jordan CT, Zhong RK, Astle CM. Primitive hemopoietic stem cells: direct assay of most productive populations by competitive repopulation with simple binomial, correlation and covariance calculations. Exp Hematol 1993; 21:206–219.

58. Harrison DE, Lerner CP. Most primitive hematopoietic stem cells are stimulated to cycle rapidly after treatment with 5-fluorouracil. Blood 1991; 78:1237–1240.

59. Haydock DN, To LB, Dowse TL, Juttner CA, Simmons PJ. Ex vivo expansion and maturation of peripheral blood CD34$^+$ve cells into the myeloid lineage. Blood 1992; 80:1405–1411.

60. Healy L, May G, Gale K, Grosveld F, Greaves M, Enver T. The stem cell antigen CD34 functions as a regulator of hemopoietic cell adhesion. Proc Natl Acad Sci USA 1995; 92:12240–12244.

61. Heimfeld S, Hudak S, Weissman I, Rennick D. The in vitro response of phenotypically defined mouse stem cells and myeloerythroid progenitors to single or multiple growth factors. Proc Natl Acad Sci USA 1991; 88:9902–9907.

62. Hodgson GS, Bradley TR. Properties of haematopoietic stem cells surviving 5-fluorouracil treatment: evidence for a pre-CFU-S cell? Nature 1979; 281:381–382.

63. Holyoake TL, Franklin IM. Bone marrow transplantation from peripheral blood. Br Med J 1994; 309:4–5.

64. Holyoake TL, Freshney MG, Konwalinka G, Haun M, Petzer A, Fitzsimons E, Lucie NP, Wright EG, Pragnell IB. Mixed colony formation in vitro by the heterogeneous compartment of multipotential progenitors in human bone marrow. Leukemia 1993; 7:207–213.

65. Huang S, Terstappen LW. Correction. Nature 1994; 368:664–668.

66. Huang S, Terstappen LWMM. Lymphoid and myeloid differentiation of single human CD34$^+$, HLA-DR$^+$, CD38$^-$ hematopoietic stem cells. Blood 1994; 83:1515–1526.

67. Huang S, Terstappen LW. Formation of haematopoietic microenvironment and haematopoietic stem cells from single human bone marrow stem cells. Nature 1992; 360:745.

68. Issaid C, Croisille L, Katz A, Vainchenker W, Coulombel L. A murine stromal cell line allows the proliferation of very primitive CD34$^+$/CD38$^-$ progenitor cells in long-term cultures and semi-solid assays. Blood 1993; 81:2916–2924.

69. Jacobsen K, Kravitz J, Kincade P, Osmond DG. Adhesion receptors on bone marrow stromal cells: in vivo expression of vascular cell adhesion molecule-1 by reticular cells and sinusoidal endothelium in normal and γ-irradiated mice. Blood 1996; 87:73–82.

70. Jacobson LO, Marks EK, Gaston EO, Robinson M, Zirkle RE. The role of the spleen in radiation injury. Proc Soc Exp Biol Med 1949; 70:740.

71. Johnson RP, Rosenzweig M, Goodell MA, Marks DF, Demaria M, Mulligan RC. Identification of a candidate hematopoietic stem cell population that lacks CD34 in rhesus macaques. Blood 1996; 88:629a.

72. Jones RJ, Barber JP, Vala MS, Collector MI, Kaufmann SH, Ludeman SM, Colvin OM, Hilton J. Assessment of aldehyde dehydrogenase in viable cells. Blood 1995; 85:2742–2746.

73. Jones RJ, Celano P, Sharkis SJ, Sensenbrenner LL. Two phases of engraftment established by serial bone marrow transplantation in mice. Blood 1989; 73:397–401.

74. Jones RJ, Collector MI, Barber JP, Vala MS, Fackler MJ, May WS, Griffin CA, Hawkins AL, Zehnbauer BA, Hilton J, Colvin OM, Sharkis SJ. Characterization of mouse lymphohematopoietic stem cells lacking spleen colony-forming activity. Blood 1996; 88:487–491.

75. Jones RJ, Wagner JE, Celano P, Zicha MS, Sharkis SJ. Separation of pluripotent haematopoietic stem cells from spleen colony-forming cells. Nature 1990; 347:188–189.

76. Joshi SS, Connolly J, Mann SL, Sharp JG. Altered in vivo and in vitro behavior of butanol-modified bone marrow cells. Exp Hematol 1995; 23:1284–1288.

77. Kastan MB, Schlaffer E, Russo JE, Colvin OM, Civin CI, Hilton J. Direct demonstration of elevated aldehyde dehydrogenase in human hematopoietic progenitor cells. Blood 1990; 75:1947–1950.

78. Katayama N, Ogawa M. Assay for murine blast cell colonies. In: Freshney RI, Pragnell IB, Freshney MG, eds. Culture of Hematopoietic Cells. New York: Wiley-Liss, 1994:41–53.

79. Kerk DK, Henry EA, Eaves AC, Eaves CJ. Two classes of primitive pluripotent hemopoietic progenitor cells: separation by adherence. J Cell Physiol 1985: 125:127–134.

80. Kessinger A. Circulating stem cells—waxing hematopoietic. N Engl J Med 1995; 333:315–316.

81. Kiefer F, Wagner EF, Keller G. Fractionation of mouse bone marrow by adherence separated primitive hematopoietic stem cells from in vitro colony-forming cells and spleen colony-forming cells. Blood 1991; 78:2577–2582.

82. Kim MJ, Spangrude GJ. Correlation of P-glycoprotein function and rhodamine-123 staining in long-term repopulating stem cells. Blood 1996; 88:42a.

83. Kittler EL, McGrath H, Temeles D, Crittenden RB, Kister VK, Quesenberry PJ. Biologic significance of constitutive and subliminal growth factor production by bone marrow stroma. Blood 1992; 79:3168–3178.

84. Klein G, Muller CA, Tillet E, Chu M-L, Timpl R. Collagen type VI in the human bone marrow microenvironment: a strong cytoadhesive component. Blood 1995; 86:1740–1748.

85. Kodama H, Nose M, Niida S, Nishikawa S, Nishikawa S-I. Involvement of the c-kit receptor in the adhesion of hematopoietic stem cells to stromal cells. Exp Hematol 1994; 22:979–984.

86. Koller MR, Bender JG, Papoutsakis ET, Miller WM. Effects of synergistic cytokine combinations, low oxygen and irradiation on the expansion of human cord blood progenitors. Blood 1992; 80:403–411.

87. Koller MR, Palsson MA, Manchel I, Palsson B. Long-term culture-initiating cell expansion is dependent on frequent medium exchange combined with stromal and other accessory cell effects. Blood 1995; 86:1784–1793.

88. Krause DS, Fackler MJ, Civin CI, May WS. CD34: structure, biology, and clinical utility. Blood 1996; 87:1–13.

89. Kurnit DM, Matthysse S, Papayannopoulou T, Stamatoyannopoulos G. Stochastic branching model for hemopoietic progenitor cell differentiation. J Cell Physiol. 1985; 123:55–63.

90. Lajtha LG. The common ancestral cell. In: Wintrobe MM, ed. Blood, Pure and Eloquent. New York: McGraw-Hill, 1980:81.

91. Lansdorp PM. Developmental changes in the function of hematopoietic stem cells. Exp Hematol 1995; 23:187–191.

92. Lansdorp PM, Dragowska W. Maintenance of hematopoiesis in serum-free bone marrow cultures involves sequential recruitment of quiescent progenitors. Exp Hematol 1993; 21:1321–1327.

93. Lansdorp PM, Sutherland HJ, Eaves CJ. Selective expression of CD45 isoforms on functional

subpopulations of CD34$^+$ hematopoietic cells from human bone marrow. J Exp Med 1990; 172:363–366.

94. Leary AG, Ogawa M. Assay for human blast cell colonies. In: Freshney RI, Pragnell IB, Freshney MG, eds. Culture of Hematopoietic Cells. New York: Wiley-Liss, 1994:55–65.

95. Lebkowski JS, Schain LR, Okrongly D, Levinsky R, Harvey MJ, Okarma TB. Rapid isolation of human CD34$^+$ hematopoietic stem cells: purging of human tumor cells. Transplantation 1992; 53:1011–1019.

96. Leemhuis T, Yoder MC, Grigsby S, Agero B, Eder P, Srour EF. Isolation of primitive human bone marrow hematopoietic progenitor cells using Hoechst 33342 and rhodamine 123. Exp Hematol 1996; 24:1215–1224.

97. Leitner A, Strobel H, Frischmeister G, Kurz M, Romanakis K, Haas OA, Printz D, Buchinger P, Bauer S, Gadner H, Fritsch G. Lack of DNA synthesis among CD34$^+$ cells in cord blood and in cytokine-mobilized blood. Br J Haematol 1996; 92:255–262.

98. Lemieux, ME, Rebel VI, Lansdorp PM, Eaves CJ. Characterization and purification of a primitive hematopoietic cell type in adult mouse marrow capable of lymphomyeloid differentiation in long-term marrow "switch" cultures. Blood 1995; 86:1339–1347.

99. Lewis JL, Marley SB, Goldman JM, Gordon MY. Interleukin-3 (IL-3) alters the self renewal and stem cell factor (SCF) alters the amplification of erythropoiesis in burst-forming units [abstract]. Br J Haematol 1996; 93:213.

100. Li CL, Johnson GR. Murine hematopoietic stem and progenitor cells: I. Enrichment and biologic characterization. Blood 1995; 85:1472–147.

101. Lord BI, Dexter TM. Which are the hematopoietic stem cells. Exp Hematol 1995; 23:1237–1241.

102. Lorenz E, Uphoff D, Reid TR, Shelton E. Modification of irradiation injury in mice and guinea pigs by bone marrow injections. J Natl Cancer Inst 1951; 12:197.

103. Magli MC, Iscove NN, Odartchenko N. Transient nature of early haematopoietic spleen colonies. Nature 1982; 295:527–529.

104. Majdic O, Stöckel J, Pickl WF, Bohuslav J, Strobel H, Scheinecker C, Stockiger H, Knapp W. Signaling and induction of enhanced cytoadhesiveness via the hematopoietic progenitor cell surface molecule CD34. Blood 1994; 83:1226–1234.

105. Marley SB, Amos TAS, Gordon MY. Kinetics of colony formation by BFU-E grown under different culture conditions in vitro. Br J Haematol 1996; 92:559–561.

106. Mauch P, Greenberger JS, Botnick L, Hannon E, Hellman S. Evidence for structured variation in self-renewal capacity within long-term bone marrow cultures. Proc Natl Acad Sci USA 1980; 77:2927–2930.

107. May G, Healy L, Gale K, Greaves M. Functional analysis of the stem cells antigen CD34 in transgenic mice. Blood 1994; 84:871.

108. Mayani H, Kay R, Smith C, Lansdorp PM. Thy-1 mediates inhibition of human haemopoietic cell proliferation. In: Schlossman SF, Boumsell L, Gilks W, et al., eds. Leucocyte Typing V. Oxford: Oxford University Press, 1995:968–970.

109. Mayani H, Lansdorp PM. Thy-1 expression is linked to functional properties of primitive hematopoietic progenitor cells from human umbilical cord blood. Blood 1994; 84:2410–2417.

110. McNiece IK, Briddell RA. Primitive hematopoietic colony forming cells with high proliferative potential. In: Freshney RI, Pragnell IB, Freshney MG, eds. Culture of Hematopoietic Cells. New-York: Wiley-Liss, 1994:23–39.

111. McNiece IK, Williams NT, Johnson GR, Kriegler AB, Bradley TR, Hodgson GS. The generation of murine hematopoietic precursor cells from macrophage high proliferative potential colony-forming cells. Exp Hematol 1987; 15:972–977.

112. Metcalf D. Clonal Culture of Hempoietic Cells: Techniques and Applications. Amsterdam: Elsevier, 1984.

113. Minguell JJ, Hardy C, Tavassoli M. Membrane-associated chondroitin sulphate proteoglycan and fibronectin mediate the binding of haematopoietic progenitor cells to stromal cells. Exp Cell Res 1992; 291:200–207.

114. Morel F, Galy A, Chen B, Szilvassy SJ. Characterization of CD34 negative hematopoietic stem cells in murine bone marrow. Blood 1996; 88:629a.

115. Mulder AH, Visser JWM. Separation and functional analysis of bone marrow cells separated by rhodamine-123 fluorescence. Exp Hematol 1987; 15:99–104.

116. Ogawa M. Differentiation and proliferation of hematopoietic stem cells. Blood 1993; 81:2844–2853.

117. Ohneda O, Yanai N, Obinata M. Microenvironment created by stromal cells is essential for rapid expansion of erythroid cells in fetal liver. Development 1990; 110:379–384.

118. Okuyama R, Koguma M, Yanai N, Obinata M. Bone marrow stromal cells induce myeloid and lymphoid development of the sorted hematopoietic stem cells in vitro. Blood 1995; 86:2590–2597.

119. Orlic D, Bodine DM. What defines a pluripotent hematopoietic stem cell (PHSC): will the real PHSC please stand up! Blood 1994; 84:3991–3994.

120. Osawa M, Hanada K, Hamada H, Nakauchi H. Long-term lymphohematopoietic reconstitution by a single CD34-low/negative hematopoietic stem-cell. Science 1996; 273:242–245.

121. Pettengell R, Luft T, Henschler R, Hows JM, Dexter TM, Ryder D, Testa NG. Direct comparison by limiting dilution analysis of long-term culture-initiating cells in human bone marrow, umbilical cord blood, and blood stem cells. Blood 1994; 84:3653–3659.

122. Pietrzyk ME, Priestley GV, Wolf NS. Normal cycling patterns of hematopoietic stem cell sub-populations: an assay using long-term in vivo BrdU infusions. Blood 1985; 66:1460–1462.

123. Ploemacher RE. Cobblestone area forming cell (CAFC) assay. In: Freshney RI, Pragnell IB, Freshney MG, eds. Culture of Hematopoietic Cells. New York: Wiley-Liss, 1994:1–21.

124. Ploemacher RE, Brons NHC. Cells with marrow and spleen repopulating ability and forming spleen colonies on day 16, 12 and 8 are sequentially ordered on the basis of increasing rhodamine 123 retention. J Cell Physiol. 1988; 136:531–536.

125. Ploemacher RE, van den Sluijs JP, Voerman JSA, Brons NHC. An in vitro limiting-dilution assay of long-term repopulating hematopoietic stem cells in the mouse. Blood 1989; 74:2755–2763.

126. Ploemacher RE, van der Loo JCM, van Beurden CAJ, Baert MRM. Wheat-germ-agglutinin affinity of murine hematopoietic stem-cell subpopulations is an inverse function of their long-term repopulating ability in vitro and in vivo. Leukemia 1993; 7:120–130.

127. Ploemacher RE, van der Sluijs JP, van Beurden CAJ, Baert MRM, Chan PL. Use of limiting-dilution type long-term marrow cultures in frequency analysis of marrow-repopulating and spleen colony-forming hematopoietic stem cells in the mouse. Blood 1991; 78:2527–2533.

128. Ponchio L, Conneally E, Eaves C. Quantitation of the quiescent fraction of long-term culture-initiating cells in normal human blood and marrow and the kinetics of their growth factor-stimulated entry into S-phase in vitro. Blood 1995; 86:3314–3321.

129. Porter EH, Berry RJ. The efficient design of transplantable tumour assays. Br J Cancer 1963; 17:583–595.

130. Pragnell IB, Freshney MG, Wright EG. CFU-A assay for measurement of murine and human early progenitors. In: Freshney RI, Pragnell IB, Freshney MG, eds. Culture of Hematopoietic Cells. New York; Wiley-Liss, 1994:67–79.

131. Quinggang L, Congote F. Bovine fetal liver stromal cells support erythroid colony formation: enhancement by insulin-like growth factor II. Exp Hematol 1995; 23:66–73.

132. Rebel VI, Miller CL, Thronbury GR, Dragowska WH, Eaves CJ, Lansdorp PM. A comparison of long-term repopulating hematopoietic stem cells in fetal liver and adult bone marrow from the mouse. Exp Hematol 1996; 24:638–648.

133. Reems JA, Torok-Storb B. Cell cycle and functional differences between $CD34^+/CD38^{hi}$ and $CD34^+/CD38^{lo}$ human marrow cells after in vitro cytokine exposure. Blood 1995; 85:1480–1487.

134. Roberts AW, Metcalf D. Noncycling state of peripheral blood progenitor cells mobilized by granulocyte colony-stimulating factor and other cytokines. Blood 1995; 86:1600–1605.

135. Rosendaal M, Hodgson GS, Bradley TR. Haemopoietic stem cells are organised for use on the basis of their generation age. Nature 1976; 264:68–69.

136. Rowley SD, Colvin OM, Stuart RK. Human multilineage progenitor cell sensitivity to 4-hydroperoxycyclophosphamide. Exp Hematol 1985; 13:295–298.

137. Rudin CE, Garland JM, Joyner MV. Four-parameter cytometric analysis with CD45/CD34 dual staining shows concordant expression of class II and class III CD34 epitopes on haemopoietic progenitor cells in leukapheresis and umbilican cord blood samples. Br J Haematol 1997; in press.

138. Rutela S, Rumi C, Teofili L, Etuk B, La Barbera EO, Leone G. RhG-CSF mobilized peripheral blood haemopoietic progenitors reside in G_0/G_1 phase of cell cycle independently of the expression of myeloid antigens. Br J Haematol 1996; 93:737–741.

139. Ryan DH, Nuccie BL, Abboud CN, Winslow JM. Vascular cell adhesion molecule-1 and the integrin VLA-4 mediate adhesion of human B cell precursors to bone marrow adherent cells. J Clin Invest 1991; 88:995–1004.

140. Schweitzer CM, van der Schoot CE, Drager AM, van der Valk P, Zevenbergen A, Hooibrink B, Westra AH, Langenhuisjsen MM. Isolation and culture of human bone marrow endothelial cells. Exp Hematol 1995; 23:41–48.

141. Scott MA, Apperley JF, Jestice HK, Bloxham DM, Marcus RE, Gordon MY. Plastic-adherent progenitor cells in mobilized peripheral blood progenitor cell collections. Blood 1995; 86:4468–4473.

142. Smith C, Gasparetto C, Collins N, Gillio A, Muench MO, O'Reilly RJ, Moore MAS. Purification and partial characterization of a human hematopoietic precursor population. Blood 1991; 77:2122–2128.

143. Spangrude GJ. Two mechanisms of discrimination between stem cells and progenitors by rhodamine-123: mitochondrial activation and multi-drug resistance. Blood 1995; 86:461a.

144. Spangrude GJ. Enrichment of murine hemopoietic stem cells. In: Doyle A, Griffiths JB, Newell DG, eds. Cell and Tissue Culture: Laboratory Procedures. Chichester: John Wiley & Sons, 1994:Module 21B:9.

145. Spangrude GJ, Smith L, Uchida N, Ikuta K, Heimfeld S, Friedman J, Weissman IL. Mouse hematopoietic stem cells. Blood 1991; 78:1395–1402.

146. Spooncer E, Lord BI, Dexter TM. Defective ability to self-renew in vitro of highly purified primitive haematopoietic stem cells. Nature 1985; 316:62–65.

147. Srour EF, Brandt JE, Briddell RA, Grigsby S, Leemhuis T, Hoffman R. Long-term generation and expansion of human primitive hematopoietic progenitor cells in vitro. Blood 1993; 81:661–669.

148. Steen R, Tjønnfjord GE, Gaudernack G, Brinch L, Egeland T. Difference in the distribution of CD34 epitopes on normal haemopoietic progenitor cells and leukaemic blast cells. Br J Haematol 1996; 94:597–605.

149. Sutherland DR, Keating A, Nayar R, Anania S, Stewart AK. Sensitive detection and enumeration of $CD34^+$ cells in peripheral and cord blood by flow cytometry. Exp Hematol 1994; 22:1003–1010.

150. Sutherland HJ, Eaves CJ, Eaves AC, Dragowska W, Lansdorp PM. Characterization and partial purification of human marrow cells capable of initiating long-term hemopoiesis in vitro. Blood 1989; 74:1563–1570.

151. Sutherland HJ, Eaves CJ, Lansdorp PM, Phillips GL, Hogge DE. Kinetics of committed and primitive blood progenitor mobilization after chemotherapy and growth factor treatment and their use in autotransplants. Blood 1994; 83:3808–3814.

152. Sutherland HJ, Lansdorp PM, Henkelman DH, Eaves AC, Eaves CJ. Functional characterisation of individual human hematopoietic stem cells cultured at limiting dilution on supportive marrow stromal layers. Proc Natl Acad Sci USA 1990; 87:3584–3588.

153. Suzuki A, Andrew DP, Gonzalo J-A, Fukumoto M, Spellberg J, Hashiyama M, Takimoto H, Gerwin N, Webb I, Molineux G, Amakawa R, Tada Y, Waheham A, Brown J, McNiece I, Ley K, Butcher EC, Suda T, Gutierrez-Ramos J-C, Mak TW. CD34-deficient mice have reduced eosinophil accumulation after allergen exposure and show a novel crossreactive 90kD protein. Blood 1996; 87:3550–3562.

154. Szilvassy SJ, Humphries RK, Lansdorp PM, Eaves AC, Eaves CJ. Quantitative assay for totipotent reconstituting hemopoietic stem cells by a competitive repopulation strategy. Proc Natl Acad Sci USA 1990; 87:8736–8740.

155. Tavassoli M. Handbook of the Hematopoietic Microenvironment. Clifton NJ: Humana Press, 1989.

156. Tavassoli M, Hardy CL. Molecular basis of homing of intraveneously transplanted stem cells to the marrow. Blood 1990; 76:1059–1070.

157. Thalmaier K, Meissner P, Reisbach G, Hultner L, Mortensen BT, Brechtel A, Oostendorp RA, Dormer P. Constitutive and modulated cytokine expression in two permanent human marrow stromal cell lines. Exp Hematol 1996; 24:1–10.

158. Till JE, McCulloch EA. A direct measurement of the radiation sensitivity of normal mouse bone marrow cells. Radiat Res 1961:213–222.

159. Till JE, McCulloch EA, Siminovitch L. A stochastic model of stem cell proliferation, based on the growth of spleen colony-forming cells. Proc Natl Acad Sci USA 1964; 51:29–36.

160. Tjúnnfjord GE, Steen R, Veiby OP, Egeland T. Lineage commitment of CD34$^+$ human hematopoietic progenitor cells. Exp Hematol 1996; 24:875–882.

161. Traycoff CM, Abboud MR, Laver J, Brandt JE, Hoffman R, Law P, Ishizawa L, Srour EF. Evaluation of the in vitro behavior of phenotypically defined populations of umbilical cord blood hematopoietic progenitor cells. Exp Hematol 1994; 22:215–222.

162. Trowbridge IS. Overview of CD71. In: Schlossman SF, Boumsell L, Gilks W, et al., eds. Leucocyte Typing V. Oxford: Oxford University Press, 1995:1139–1141.

163. Turner ML, Sweetenham JW. Haemopoietic progenitor homing and mobilisation. Br J Haematol 1996; 94:592–596.

164. Uchida N, Jerabek L, Weissman IL. Searching for hematopoietic stem cells II: the heterogeneity of Thy-1.1^{0}Lin$^-$/loSca-1$^+$ mouse hematopoietic stem cells separated by counterflow centrifugal elutriation. Exp Hematol 1996; 24:649–659.

165. Udomsakdi C, Eaves CJ, Sutherland HJ, Lansdorp PM. Separation of functionally distinct sub-populations of primitive human hematopoietic cells using rhodamine-123. Exp Hematol 1991; 19:338–342.

166. Udomsakdi C, Lansdorp PM, Hogge DE, Reid DS, Eaves AC, Eaves CJ. Characterization of primitive human cells in normal human peripheral blood. Blood 1992; 80:2513–2521.

167. van der Keuer M, Zijlmans JM, Kluin-Nelemans JC, Willemze R, Fibbe WE. Rhodamine efflux capacity of murine hematopoietic progenitor cells [abstract]. Cytometry 1996; 23:80.

168. Verfaillie CM. Regulation of human hematopoiesis by the bone marrow environment [abstract]. Exp Hematol 1994; 22:714.

169. Verfaillie CM. Direct contact between human primitive hematopoietic progenitors and bone marrow stroma is not required for long-term in vitro hematopoiesis. Blood 1992; 79:2821–2826.

170. Verfaillie C, Hurley R, Bhatia R, McCarthy J. Role of bone marrow matrix in normal and abnormal hematopoiesis. Crit Rev Oncol Hematol 1994; 16:201–204.

171. Verfaillie CM, Miller JS. A novel single-cell proliferation assay shows that long-term culture initiating cells (LTC-IC) maintenance over time results from extensive proliferation of a small fraction of LTC-IC. Blood 1995; 86:2137–2143.

172. Verfaillie CM, Miller S. CD34$^+$/CD33$^-$ cells reselected from macrophage inflammatory protein 1 and interleukin-3 supported "stroma-noncontact" cultures are highly enriched for long-term bone marrow culture intitiating cells. Blood 1994; 84:1442–1449.

173. Visser JWM. Isolation and biological characterization of bone marrow stem cells and progenitor cells. In: Laerum OD, Bjerknes R, eds. Flow Cytometry in Hematology. London: Academic Press, 1992:9–30.

174. Visser JWM, van Bekkum DW. Purification of pluripotent hemopoietic stem cells: past and present. Exp Hematol 1990; 18:248–256.

175. Waller EK, Olweus J, Lund-Johansen F, Huang S, Nguyen M, Guo G-R, Terstappen L. The "common stem cell" hypothesis re-evaluated: human fetal bone marrow contains separate populations of hematopoietic and stromal progenitors. Blood 1995; 85:2422–2435.

176. Wang JCY, Doedens M, Dick JE. Primitive human hematopoietic cells are enriched in cord blood compared to normal adult bone marrow or G-CSF mobilized peripheral blood. Blood 1996; 88:628a.

177. Watt SM, Williamson J, Genevier H, Nesbitt SA, Hatzfeld A, Fawcett J, Simmons PDL, Edwards JA, van Dongen J, Coombe DR. The heparin binding PECAM-1 adhesion molecule is expressed by

CD34 positive hematopoietic precursor cells with early myeloid and lymphoid cell phenotypes. Blood 1993; 82:2649–2663.

178. Webb M, Raphael CL, Asbahr H, Erber WN, Meyer BF. The detection of rhodamine 123 efflux at low levels of drug resistance. Br J Haematol 1996; 93:650–655.

179. Weiss L. The haematopoietic microenvironment of the bone marrow: an ultrastructural study of the stroma in rats. Blood Cells 1975; 1:617–638.

180. Whitlock CA, Witte ON. Long-term culture of lymphocytes-B and their precursors from murine bone marrow. Proc Natl Acad Sci USA 1982; 79:3608–3611.

181. Wineman J, Moore K, Lemischka I, Muller-Sieberg C. Functional heterogeneity of the haematopoietic microenvironment: rare stromal elements maintain long-term repopulating stem cells. Blood 1996; 87:4082–4090.

182. Winton EF, Colenda KW. Use of long-term human marrow cultures to demonstrate progenitor cell precursors in marrow treated with 4-hydroperoxycyclophosphamide. Exp Hematol 1987; 15:710–714.

183. Wolf NS, Kon A, Priestly GV, Bartelmez SH. In vivo and in vitro characterization of long-term repopulating primitive hematopoietic cells isolated by sequential Hoechst 33342–rhodamine 123 FACS selection. Exp Hematol 1993; 21:614–622.

184. Yanai N, Matsuya Y, Obinata M. Spleen stromal cell lines selectively support erythroid colony formation. Blood 1989; 74:2391–2397.

185. Yeager AM, Kaizer H, Santos GW, Saral R, Colvin MO, Stuart RK, Braine HG, Burke PJ, Ambinder RF, Burns WH, Fuller DJ, Davies JM, Karp JE, May WS, Rowley SD, Sensenbrenner LL, Vogelsang GB, Wingard JR. Autologous bone marrow transplantation in patients with acute nonlymphocytic leukemia, using ex vivo treatment with 4-hydroperoxycyclophosphamide. N Engl J Med 1986; 315:141–147.

186. Yoder MC, Williams DA. Matrix–molecule interactions with hematopoietic stem cells. Exp Hematol 1995; 23:961–967.

187. Young P, Baumhueter S, Lasky LA. The sialomucin CD34 is expressed on haematopoietic cells and blood vessels during murine development. Blood 1995; 85:96–105.

188. Zannettino ACW, Berndt MC, Butcher C, Butcher EC, Vadas MA, Simmons PJ. Primitive human hematopoietic progenitors adhere to P-selectin (CD62P). Blood 1995; 85:3466–3477.

189. Zern M, Reid L. Extracellular Matrix. New York: Marcel Dekker, 1995.

190. Zijlmans JMJM, Kleiverda K, Heemskerk DPM, Kluin PM, Visser JWM, Willemze R, Fibbe WE. No role for committed progenitor cells in engraftment following transplantation of murine cytokine-mobilized blood cells. Blood 1995; 86:112a.

191. Zon LI. Developmental biology of haematopoiesis. Blood 1995; 86:2876–2891.

2

Surrogate Assays
for Hematopoietic Stem Cell Activity

Ivan Bertoncello and Gillian B. Bradford
Sir Donald and Lady Trescowthick Research Laboratories, Peter MacCallum Cancer Institute, East Melbourne, Victoria, Australia

I. INTRODUCTION

The normal hematopoietic system generates approximately 5×10^3 functional mature blood cells per second in the mouse (27), and 4×10^6 cells per second in humans (51) throughout life. This prodigious output of mature blood cells is ultimately dependent on the precise regulation of primitive hematopoietic stem cells (PHSC) which, in turn, give rise to an ordered series of transit populations of progenitor and precursor cells of progressively more restricted proliferative and differentiative potentiality. In these models of hematopoietic organization, PHSC are operationally defined as transplantable, multipotential cells, with extensive proliferative capacity that, in addition to maintaining normal blood cell production in the steady state, are able to clonally regenerate the entire hematopoietic system of myeloablated mice long-term.

The PHSC and their immediate progeny are morphologically indistinguishable. Despite continued refinement of multiparameter cell separative strategies (5,41,54,57,58,62,73,82,89,92), unique phenotypic markers that are able to unmistakably characterize, resolve, and purify PHSC have not been identified. Consequently, the formulation of models of hematopoietic organization and regulation has relied entirely on the direct measurement of the proliferative and differentiative potential of candidate stem and progenitor cells in transplantation models and in cell culture.

The observation that a subset of bone marrow cells exists, which is capable of reconstituting hematopoiesis in a marrow-ablated recipient (42), has led to the development of a variety of in vivo assays with which to measure PHSC. However, although the measurement of long-term hematopoietic-reconstituting potential following transplantation into myeloablated recipients is the most stringent assay for PHSC, the impracticality of using such assays for routine assessment of the functional properties of manipulated stem cells has led to the development of short-term in vivo and in vitro surrogate assays. The validation of these assay systems is based on the correlation of short-term assay endpoints with long-term multilineage hematopoietic reconstitution in animal transplant models.

II. IN VIVO TRANSPLANTATION ASSAYS

In vivo transplantation is generally considered to be the most appropriate assay for PHSC, because the ability of the cells to reconstitute blood cell production in a myeloablated recipient measures their capacity for both extensive proliferation and multilineage differentiation over extended time periods.

However, analysis of transplantation models has shown that a large spectrum of hematopoietic cells with widely divergent proliferative histories and potentialities are capable of engrafting and restoring normal blood cell production in myeloablated animals. Radiation protection, for example, is not indicative of long-term stem cell engraftment because, in the early stages following transplantation, multilineage cells of limited potentiality have the ability to provide short-term support, preventing early death of the recipient, but are then exhausted and replaced by more primitive endogenous host PHSC that may have survived the radiation treatment (10,74). Engraftment following transplantation is often biphasic, with initial blood cell production resulting from the contribution of relatively mature subsets, followed later by the proliferation of earlier PHSC (23,24,31,88). Retroviral-marking studies have confirmed that clonal instability in the initial period of engraftment following transplantation persists for 4–6 months (32), and Mauch and Hellman (47) have demonstrated that stem cell deficits can take up to a year to manifest themselves following transplantation of unfractionated bone marrow cell suspensions.

In vivo assays for PHSC with long-term hematopoietic reconstituting ability include analysis of competitive repopulation following transplantation using estimates of variance or covariance (25), and limiting-dilution transplant analysis (81).

Competitive repopulation assays measure the relative ability of donor PHSC to compete with standard doses of marrow cells of host phenotype injected into stem cell-deficient, or lethally irradiated hosts. Analysis of variance in this system is based on the assumption that all injected PHSC contribute equally to the outcome; in the event of unequal contribution, only the progeny of dominant clones will be measured, and the frequency of competitive repopulating units (CRU) will be considerably underestimated (23,25). However, although applicable only in models for which there is no overt interference caused by graft-versus-host reactions or histocompatibility differences, these measurements, nevertheless, have proved to be an extremely sensitive PHSC assay.

The PHSC can also be enumerated by analysis of the proportion of donor-negative transplant recipients in in vivo limiting-dilution assays. These assays assume a Poisson distribution of injected cells, and a single-hit model, in which only one cell of only one type is necessary for a positive response. Cell suspensions are serially diluted for transplantation, and the frequency of repopulating PHSC is determined using Poisson statistics. The threshold level of donor cell engraftment that is set for a positive response is critical to these calculations (20,83). This varies between laboratories (66,71,81,82), and the ability of transplanted cells to attain this level is dependent on factors such as preconditioning regimens, the level of hematological support and competition, mouse strain and sex, and posttransplant care.

III. IN VIVO SURROGATE ASSAYS

Although long-term transplantation is the gold-standard against which short-term surrogate assays of stem cell activity are evaluated, short-term in vivo spleen colony-forming assays

(CFU-S), and pre-CFU-S assays have played a pivotal role in defining and ordering PHSC subpopulations.

A. CFU-S

The spleen colony-forming unit (CFU-S) assay of Till and McCulloch was the first quantitative short-term in vivo assay of PHSC (84). This assay measures the generation of hematopoietic colonies in the spleen of lethally irradiated mice at time points from 7 to 16 days following transplantation. Heterogeneity in the capacity of CFU-S for both proliferation and differentiation was noted in early experiments (13,69), but it was the documentation of the transience of spleen colonies by Magli and co-workers (46) that provided the first convincing evidence that CFU-S assayed at different times following transplantation represented distinct subpopulations of differing potentiality and proliferative capacity. CFU-S developing at early time points represent a more mature progenitor population than later-appearing CFU-S. They have a more-limited proliferative potential (46,69), are more sensitive to cytotoxic agents (26,67), and have faster cycling kinetics than CFU-S detected at later time points after transplantation (28,56). These CFU-S populations have been hierarchically ordered on the basis of lectin binding (57), and differential uptake of Hoechst 33342 (1) and rhodamine 123 (5,92). Although CFU-S detected at later time points following transplantation are an index of PHSC with long-term–reconstituting potential, a significant proportion of these cells have reseeded from the bone marrow during regeneration and, therefore, include cells of more restricted potentiality and differentiative capacity (91).

B. Pre-CFU-S Assays

Experiments in mice treated with near-lethal doses of 5-fluorouracil (5-FU) showed that whereas CFU-S cells were ablated, the long-term repopulating ability of the marrow was preserved. This indicated that the CFU-S assay was not a true index of long-term–reconstituting PHSC, and that a pre-CFU-S compartment must exist (26). These pre-CFU-S cells preferentially home to the marrow, rather than to the spleen (5), and have the capacity for long-term regeneration of both the CFU-S marrow compartment and the entire hematopoietic system.

Pre-CFU-S are quantified by analyzing regeneration of CFU-S or committed progenitor cell compartments in the marrow and spleen of myeloablated recipients following transplantation (27,57,58). These marrow- and spleen-repopulating ability assays (MRA and SRA) are an example of an in vivo amplification assay and, therefore, are a powerful tool for revealing the existence of subsets of PHSC with high-proliferative potential; however, they are not clonal. Therefore, they provide only a relative measure of the reconstituting ability of a cell suspension, and do not distinguish between populations containing rare cells with high-repopulating potential and those highly enriched with cells of lower-repopulating ability (72).

Peripheral blood parameters can also be used as a readout in pre-CFU-S assays. Measurement of circulating platelets in the peripheral blood of primary transplant recipients (platelet-regenerating ability; PRA) is closely correlated with MRA cells, whereas reticulocyte levels (erythrocyte-regenerating ability; ERA) are correlated with cells that form CFU-S on day-13 (5,27).

IV. IN VITRO SURROGATE ASSAYS

There are primarily three categories of in vitro surrogate assays of PHSC activity: (1) clonogenic assays in cytokine-supplemented semisolid agar culture, including the high-proliferative potential colony-forming cell (HPP-CFC) assay; (2) long-term bone marrow culture-based assay systems, including long-term culture-initiating cell (LTC-IC) assays and cobblestone area-forming cell (CAFC) assays; and (3) delta assays in which the potentiality and proliferative capacities of defined target cells are evaluated by their relative ability to generate mature progeny in stromal adherent or liquid culture systems.

Since Pluznik and Sachs (63) and Bradley and Metcalf (7) first described the clonogenic growth of murine myeloid progenitor cells in semisolid agar culture, advances in molecular biology and protein chemistry have led to the identification and characterization of at least 40 cytokines and various other factors that affect the proliferative capacity and potentiality of stem and progenitor cells in vivo and in vitro (65). PHSC in vitro are distinguished from more mature multipotential and lineage-restricted progenitor cells by their dependence on stromal support for long-term maintenance in culture; their obligatory requirement for multiple cytokines acting in synergy to express their full potential in clonal assays or liquid culture; and their inability to proliferate in the presence of any single known cytokine.

Growth factor cocktails that stimulate PHSC in culture reflect their requirement for factors that primarily promote survival and induce deeply quiescent cells to cycle [stem cell factor (SCF), Flt3, interleukin-6 (IL-6), IL-11]; factors inducing proliferation of multipotential cells [IL-3, granulocyte–macrophage colony-stimulating factor (GM-CSF)]; and factors directing differentiation and expansion of lineage restricted progenitors [CSF-1, G-CSF, erythropoietin (Epo)] (53,86). Also SCF + IL-3 + IL-6 + Mpl can induce formation of colonies of 10^5 cells in single-cell deposition, indicating that Mpl ligand not only stimulates platelet progenitors but also acts at the stem cell level (35,70).

A. HPP-CFC Assay

High-proliferative potential colony-forming cells (HPP-CFC) are among the most primitive hematopoietic cells yet identified in vitro (8). They are operationally defined by their relative resistance to 5-FU; their obligatory requirement for multiple cytokines acting in synergy; and their ability in agar culture to form macroscopic colonies larger than 0.5 mm in diameter that contain at least 50,000 cells (4).

The HPP-CFC compartment shares many of the characteristics of PHSC with long-term repopulating ability. They are multipotential; their pattern of regeneration following 5-FU treatment is closely correlated with that of pre-CFU-S; and they cofractionate with PHSC populations capable of long-term hematopoietic-reconstituting potential (9,44,92). However, cell separative studies have demonstrated that whereas PHSC with long-term hematopoietic reconstituting ability are HPP-CFC, the reverse is not necessarily true (6).

The HPP-CFC compartment is heterogeneous, comprising at least four subpopulations that can be hierarchically ordered on the basis of their growth factor preferences (34). Broadly, the more primitive the progenitor, the more growth factors are required for its stimulation. HPP-CFC stimulated by combinations of two cytokines are relatively mature, and are precursors of the lineage-restricted progenitor cell compartment. Combinations of three factors are the minimal requirement for the growth of PHSC with long-term–reconstituting potential in the HPP-CFC assay (2,4), and combinations of up to seven cytokines are required for optimal colony growth of these cells (55,92). It should be noted that growth of HPP-CFC in the presence of multiple growth factor combinations does not indicate an obligatory requirement for all of those factors, and that HPP-CFC responsive to all possible permutations and combinations of these growth factor cocktails will be stimulated. Although HPP-CFC stimulated by combinations of two or three cytokines are separable, it is not yet possible to resolve HPP-CFC subpopulations requiring three or more factors by any cell separative technique.

Target cells stimulated in the CFU-A assay (19,64), and the blast colony-forming cell (CFU-Bl) assay (52), share many similarities with HPP-CFC. CFU-A can be considered as a relatively mature HPP-CFC subpopulation (4), whereas CFU-Bl cells, like more primitive HPP-CFC, require multiple cytokines acting in synergy for optimal stimulation (29,30,33,93). In addition, the time course of CFU-Bl colony growth (29,30,78) is similar to that observed for IL-1α + IL-3 + CSF-1–responsive HPP-CFC in the first week of culture. Similar to IL-1α + IL-3 + CSF-1–responsive HPP-CFC, CFU-Bl are also highly quiescent, being relatively resistant to 5-FU (29,30,78).

B. LTC-IC and CAFC Assays

Recognition of the requirement for stroma to sustain long-term–reconstituting cells in liquid culture (15–17) has led to the development of long-term bone marrow culture-based assays. These surrogate assays measure the ability of PHSC to generate either foci of phase-dark cells (cobblestone areas; CAFC assay) (60,61) or granulocyte–macrophage colony-forming cells (LTC-IC assay) (18) following culture in association with preformed irradiated stroma (either primary explanted bone marrow cells or cloned stromal cell lines). When set up at limiting dilution, these assays are suitable for frequency analysis using Poisson statistics.

Examination of CAFC has demonstrated that the PHSC compartment includes cells that form cobblestone areas at different times following seeding, which are both transient and persistent, resulting in a nomenclature similar to that used for CFU-S. A relation between the time to clone formation in the CAFC assay and various in vivo PHSC subsets has been suggested, in which cells that form cobblestone areas after 7 days in culture (CAFC-7) represent relatively mature progenitor subsets, and cells that form cobblestone areas after 35 days (CAFC-35) are an indicator of cells with long-term hematopoietic-reconstituting ability (62,87). However, it is likely that, as with primitive HPP-CFC, even though CAFC-35 may be equivalent to a subset of primitive PHSC, they may not be predictive of the long-term reconstituting ability of marrow samples in all instances.

Similarly, there is a correlation between the length of time required for LTC-IC formation and the proliferative potential and differentiative capacity of hematopoietic stem cells (18), demonstrating that, similar to the situation in vivo, longer-term in vitro assays are more accurate as predictive assays for primitive PHSC. The LTC-IC assay has recently been modified to allow a subset of murine LTC-IC to express both their myeloid (M) and lymphoid (L) differentiation potentials (LTC-ICML), thereby allowing quantitation of multipotent cells in vitro (38). Limiting-dilution analysis showed that LTC-ICML are present in normal murine bone marrow at a frequency (1:10^5 cells) which is in the order of previous estimates of PHSC by in vivo competitive repopulation assays (23). Furthermore, LTC-ICML, similar to LTC-IC and CAFC (62), are 5-FU–resistant and cofractionate with transplantable PHSC, suggesting that these assays are also indicative of primitive PHSC capable of long-term hematopoietic reconstitution.

C. DELTA Assay

Delta assays measure the relative ability of a cell to generate lineage-restricted progenitor cells within a fixed time period in vitro: usually 1 week (48–50). Variations of this assay are performed in stromal-adherent or plastic-adherent culture, or in growth factor-supplemented nonadherent liquid culture systems. Although delta assays have a role in the analysis of stem cell function and have proved particularly useful in the development, refinement, and optimization of ex vivo expansion systems, they should be treated with caution when used to measure stem cell potential. Analysis of highly purified, sorted stem cell populations in single-cell deposition experiments has shown that there is considerable variation in the range and rate of clonal expansion of these cells and in the time taken for PHSC to enter into first division. The use of tracking dyes (85), limiting-dilution analysis (36), and the incubation of cell suspensions in the presence of 5-FU (3) have demonstrated that not all PHSC divide in culture, leading to an underestimation by these assays of progenitor cell output and, consequently, relative stem cell potential. In addition, as in pre-CFU-S assays in vivo, delta assays are amplification assays and, therefore, provide only a relative measure of the proliferative potential of a target cell suspension.

V. DO SURROGATE ASSAYS PREDICT?

It is clear, particularly in the mouse model, that various analytical and cell separative strategies are capable of identifying and resolving candidate stem cell subpopulations with the ability to clonally reconstitute the entire hematopoietic system following myeloablation (41,92). The ultimate test of a gold-standard surrogate hematopoietic stem cell assay is its ability to accurately quantify these most primitive stem cells. It must predict their engraftment potential and reconstituting ability following transplantation, and predict the extent to which the PHSC compartment may be compromised by myeloablative treatment, perturbation, or disease.

In reality, each of the surrogate assays described in this chapter represents a window in passing traffic, monitoring hematopoietic cell maturation and differentiation at different levels along the stem and progenitor cell hierarchy. The further removed from the stem cell compartment, the less reliable the surrogate assay becomes in its measurement of stem cell activity. In the normal, unperturbed, steady-state bone marrow, in which the ratio of primitive and committed progenitor cells to stem cells with long-term–repopulating ability is more or less constant, short-term surrogate assays are capable of predicting engraftment and transplant potential. However, following perturbation, myeloablation, or disease, during which the relation between the stem cell compartment and later compartments is disrupted, the ability of surrogate assays to measure stem cell activity is compromised.

This is illustrated schematically in Figure 1, and in data obtained comparing surrogate assays in a serial transplantation model (Fig. 2). Lethally irradiated mice were transplanted with 0.1 femur equivalents of normal bone marrow. One month following transplantation, the femoral bone marrow of these transplanted mice was harvested, and various surrogate assays of hematopoietic stem cell activity were performed. A 0.1 femur equivalent of marrow from these primary recipients was then transplanted into lethally irradiated secondary recipients, and surrogate assays were again performed 1 month later. This procedure was again repeated in tertiary and quaternary recipients. Although the ability of the transplant recipients to generate MRA cells (MRA-generating ability; GA-MRA) progressively declined with each transfer, on each occasion, peripheral white blood cell counts returned to normal. A significant deficiency in bone marrow cellularity or committed progenitor cell content was not detected until the fourth transplant generation, and the CFU-S assay also failed to predict a deficit of earlier stem cell subpopulations until the third transplant generation.

However, even assays of long-term–repopulating ability represent surrogate assays of the most primitive PHSC. There is evidence that these assays may be approaching the limits of detection and resolution, testing the ingenuity of investigators to refine their measurements to further subset PHSC. Our own studies, like those of others (40,41,75,96), have shown that even fractions of highly enriched PHSC demonstrate functional heterogeneity when tested in long-term competitive repopulation assays in vivo (Bradford GB, et al., submitted). It is unclear at this stage whether these differences in transplantation potential are due to inherent differences

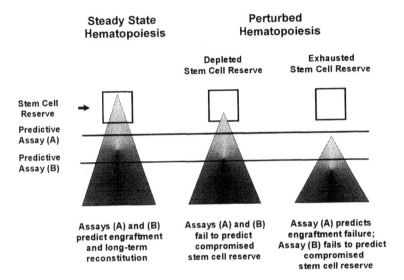

Figure 1   A schematic diagram, showing the ability of surrogate assays to predict perturbation in the stem cell reserve of the marrow. The farther removed from the stem cell compartment, the less reliable the surrogate assay becomes in its measurement of stem cell activity.

in the functional potential of the cells; due to the different behavior of identical cells in different states of activation; or whether these differences are intrinsic to the assay system.

Therefore, although these assays are the best available, by their nature, they are imperfect tools. All of these surrogate assays have similar shortcomings:

1.   They are indirect functional assays. Consequently, the target cell must be induced

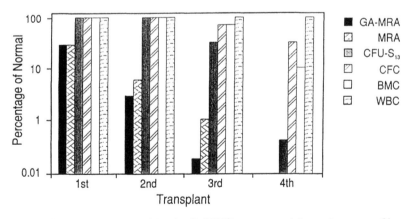

Figure 2   Peripheral white blood cell (WBC) counts; and femoral content of bone marrow cells (BMC), myeloid lineage-restricted low-proliferative potential colony-forming cells (CFC), 13-day spleen colony-forming units (CFU-S$_{13}$), marrow-repopulating ability (MRA), and MRA-generating cells (GA-MRA) following serial retransplantation of 0.1 femur equivalent of bone marrow cells at 4-week intervals, expressed as a percentage of normal values. (Data provided by GS Hodgson.)

to proliferate and differentiate, irreversibly altering the properties of the cell in the process.

2. They measure functional characteristics that are likely to be shared by closely related cells of differing potentiality and proliferative capacity. Therefore, these assays measure overlapping cell populations, imposing limitations on their ability to discriminate, resolve, and quantify specific PHSC subsets.

3. The assay conditions do not recapitulate the conditions under which stem cells normally grow and function in situ. The artificial proliferative and differentiative pressures exerted on the cells in these assays are likely to modify their behavior and influence their fate.

4. Being subject to the laws of deterministic chaos (12), which predict that the fate of individual target cells are determind by the "initial conditions" to which they are exposed, the proliferative potential and differentiative capacity of individual target cells will be determined by the probability of interaction with regulatory factors.

VI. SUMMARY

The lack of short-term functional in vivo or in vitro assays capable of reliably quantitating PHSC and accurately measuring their potentiality and proliferative capacity is a major impediment to the study of stem cell biology. Functional stem cell assays of varying integrity, complexity, sensitivity, and specificity are available, particularly in the mouse model. Short-term in vitro assays can be used to quantify primitive progenitors of high-proliferative capacity and potentiality and to define their growth factor requirements. However, there is as yet no adequate method of verifying the purity, or of predicting the long-term hematopoietic-reconstituting ability of either murine or human PHSC, other than by demonstrating long-term regeneration of the hematopoietic system by limited numbers of transplanted donor cells.

This review has focused on surrogate hematopoietic assays in the mouse model; similar assays exist for human stem cells and have been reviewed (22,90). Broadly, the same caveats apply in the human system, but there is a higher degree of difficulty, owing to the lack of a genuine gold-standard assay for human PHSC with long-term–reconstituting ability. Human LTC-IC, CAFC, and delta assays are the most commonly used surrogate assays of human PHSC activity (11,18,21,50,68,80). Human HPP-CFC assays are available, but it is unclear how these are related to the true stem cell pool (45). However, the requirement of these cells for multiple cytokines for optimal growth in culture are similar to those of murine PHSC (37). Even in the human system, transplantation remains the most rigorous assay of stem cell potential. In vivo transplantation assays have been developed using SCID mouse models (14,43), and transplantation into the immune-compromised environment of fetal sheep (76,77,94). PHSC capable of generating human blood cells in the NOD-SCID mouse model are a minor subset of LTC-IC (14), and highly enriched human PHSC are capable of engrafting and generating further PHSC capable of successfully engrafting secondary fetal sheep recipients (95).

ACKNOWLEDGMENT

We are indebted to George Hodgson for permission to use Figure 2, and acknowledge George Hodgson and Ray Bradley for many invaluable discussions about the nature of hematopoietic stem cells and stem cell assay.

REFERENCES

1. Baines P, Visser JWM. Analysis and separation of murine bone marrow stem cells by H33342 fluorescence-activated cell sorting. Exp Hematol 1983; 11:701–708.

2. Bartelmez SH, Bradley TR, Bertoncello I, Mochizuki DY, Tushinski RJ, Stanley ER, Hapel AJ, Young IG, Kriegler AB, Hodgson GS. Interleukin 1 plus interleukin 3 plus colony-stimulating factor 1 are essential for clonal proliferation of primitive myeloid bone marrow cells. Exp Hematol 1989; 17:240–245.

3. Berardi AC, Wang A, Levine JD, Lopez P, Scadden DT. Functional isolation and characterization of human hematopoietic stem cells. Science 1995; 267:104–108.

4. Bertoncello I. Status of high proliferative potential colony-forming cells in the hematopoietic stem cell hierarchy. Curr Top Microbiol Immunol 1992; 177:83–94.

5. Bertoncello I, Hodgson GS, Bradley TR. Multiparameter analysis of transplantable hemopoietic stem cells. I. The separation and enrichment of stem cells homing to marrow and spleen on the basis of rhodamine 123 fluorescence. Exp Hematol 1985; 13:999–1006.

6. Bertoncello I, Hodgson GS, Bradley TR. Multiparameter analysis of transplantable hemopoietic stem cells. II. Stem cells of long-term bone marrow reconstituted recipients. Exp Hematol 1988; 16:245–249.

7. Bradley TR, Metcalf D. The growth of mouse bone marrow cells in-vitro. Aust J Exp Biol Med Sci 1966; 44:287–299.

8. Bradley TR, Hodgson GS. Detection of primitive macrophage progenitor cells in mouse bone marrow. Blood 1979; 54:1446–1450.

9. Bradley TR, Hodgson GS, Bertoncello I. Characteristics of primitive macrophage progenitor cells with high proliferative potential: their relationship to cells with marrow repopulating ability in 5-fluoroura-cil treated bone marrow. In: Baum SG, Ledney G, van Bekkum DW, eds. Experimental Hematology Today, 1979. New York: Karger, 1980:285–297.

10. Brecher G, Neben S, Yee M, Bullis J, Cronkite EP. Pluripotential stem cells with normal or reduced self renewal survive lethal irradiation. Exp Hematol 1988; 16:627–630.

11. Breems, DA, Blokland EAW, Neben S, Ploemacher RE. Frequency analysis of human primitive haematopoietic stem cell subsets using a cobblestone area forming cell assay. Leukemia 1994; 8:1095–1104.

12. Cross SS, Cotton DWK. Chaos and antichaos in pathology. Hum Pathol 1994; 25:630–637.

13. Curry JL, Trentin JJ. Haemopoietic spleen colony studies. I. Growth and differentiation. Dev Biol 1967; 15:395–400.

14. Dick JE. Normal and leukemic human stem cells assayed in SCID mice. Semin Immunol 1996; 8:197–206.

15. Dexter TM. Cell interactions in-vitro. Clin Haematol 1979; 8:453–468.

16. Dexter TM, Lajtha LG. Proliferation of haemopoietic stem cells in-vitro. Br J Haematol 1974; 28:525–530.

17. Dexter TM, Allen TD, Lajtha LG. Conditions controlling the proliferation of hemopoietic stem cells in-vitro. J Cell Physiol 1977; 91:335–344.

18. Eaves CJ, Sutherland HJ, Udomsakdi C, Lansdorp PM, Szilvassy SJ, Fraser CC, Humphries RK, Barnett MJ, Phillips GL, Eaves AC. The human hematopoietic stem cell in vitro and in vivo. Blood cells 1992; 18:301–307.

19. Eckmann L, Freshney M, Wright EG, Sproul A, Wilkie N, Pragnell IB. A novel in-vitro assay for murine haematopoietic stem cells. Br J Cancer 1988; (suppl 9):36–40.

20. Fazekas de St Groth S. The evaluation of limiting dilution assays. J Immunol Methods 1982; 49:R11–23.

21. Goldstein NI, Moore MA, Allen C, Tackney C. A human fetal spleen cell line, immortalised with SV40 T-antigen will support the growth of $CD34^+$ long-term culture-initiating cells. Mol Cell Diff 1993; 1:301–321.

22. Gordon MY. Human haemopoietic stem cell assays. Blood Rev. 1993; 7:190–197.

23. Harrison DE, Astle CM, Lerner C. Number and continuous proliferative pattern of transplanted immunohematopoietic stem cells. Proc Natl Acad Sci USA 1988; 85:822–826.

24. Harrison DE, Zhong RK. The same exhaustible multilineage precursor produces both myeloid and lymphoid cells as early as 3–4 weeks after marrow transplantation. Proc Natl Acad Sci USA 1992; 89:10134–10138.

25. Harrison DE, Jordan CT, Zhong RK, Astle CM. Primitive hemopoietic stem cells: direct assay of most productive populations by competitive repopulation with simple binomial, correlation and covariance calculations. Exp Hematol 1993; 21:206–219.

26. Hodgson GS, Bradley TR. Properties of haemopoietic stem cells surviving 5-fluorouracil treatment: evidence for a pre-CFU-S cell? Nature 1979; 281:381–382.

27. Hodgson GS, Bradley TR, Radley JM. The organisation of hemopoietic tissue as inferred from the effects of 5-fluorouracil. Exp Hematol 1982; 10:26–35.

28. Hodgson GS, Bradley TR. In vivo kinetic status of hematopoietic stem and progenitor cells as inferred from labeling with bromodeoxyuridine. Exp Hematol 1984; 12:683–687.

29. Ikebuchi K, Wong GG, Clark SC, Ihle JN, Hirai Y, Ogawa M. Interleukin 6 enhancement of interleukin 3-dependent proliferation of multipotential hemopoietic progenitors. Proc Natl Acad Sci USA 1987; 84:9035–9039.

30. Ikebuchi K, Ihle JN, Hirai Y, Wong GG, Clark SC, Ogawa M. Synergistic factors for stem cell proliferation: further studies of the target stem cells and the mechanism of stimulation by interleukin-1, interleukin-6, and granulocyte colony-stimulating factor. Blood 1988; 72:2007–2014.

31. Jones RJ, Celano P, Sharkis SJ, Sensenbrenner LL. Two phases of engraftment established by serial bone marrow transplantation in mice. Blood 1989; 73:397–401.

32. Jordan CT, Lemischka IR. Clonal and systemic analysis of long-term hematopoiesis in the mouse. Genes Dev 1990; 4:220–232.

33. Koike K, Nakahata T, Takagi M, Kobayashi T, Ishiguro A, Tsuji K, Naganuma K, Okano A, Akiyama Y, Akabane T. Synergism of BSF-2/interleukin-6 and interleukin-3 on development of multipotential hemopoietic progenitors in serum-free culture. J Exp Med 1988; 168:879–890.

34. Kriegler AB, Verschoor SM, Bernardo D, Bertoncello I. The relationship between different high proliferative potential colony-forming cells in mouse bone marrow. Exp Hematol 1994; 22:432–440.

35. Ku H, Yonemura Y, Kaushansky K, Ogawa M. Thrombopoietin, the ligand for the Mpl receptor, synergizes with steel factor and other early acting cytokines in supporting proliferation of primitive hematopoietic progenitors of mice. Blood 1996; 87:4544–4551.

36. Lansdorp PM, Dragowska W. Maintenance of hematopoiesis in serum-free bone marrow cultures involves sequential recruitment of quiescent progenitors. Exp Hematol 1993; 21:1321–1327.

37. Leary AG, Hirai Y, Kishimoto T, Clark SC, Ogawa M. Survival of hemopoietic progenitors in the G_0 period of the cell cycle does not require early hemopoietic regulators. Proc Natl Acad Sci USA 1989; 86:4535–4538.

38. Lemieux ME, Rebel VI, Lansdorp PM, Eaves CJ. Characterization and purification of a primitive hematopoietic cell type in adult mouse marrow capable of lymphomyeloid differentiation in long-term marrow "switch" cultures. Blood 1995; 86:1339–1347.

39. Li CL, Johnson GR. Rhodamine 123 reveals heterogeneity within murine Lin$^-$, Sca-1$^+$ hemopoietic stem cells. J Exp Med 1992; 175:1443–1447.

40. Li CL, Johnson GR. Long-term hemopoietic repopulation by Thy-1^{lo}, Lin$^-$, Ly6A/E$^+$ cells. Exp Hematol 1992; 20:1309–1315.

41. Li CL, Johnson GR. Murine hematopoietic stem and progenitor cells: I. Enrichment and biologic characterisation. Blood 1995; 85:1472–1479.

42. Lorenz E, Uphoff D, Reid TR, Shelton E. Modification of irradiation injury in mice and guinea pigs by bone marrow injections. J Natl Cancer Inst 1951; 12:197–201.

43. McCune JM. Development and application of the SCID-hu mouse model. Semin Immunol 1996; 8:187–196.

44. McNiece IK, Williams NT, Johnson GR, Kriegler AB, Bradley TR, Hodgson GS. Generation of

murine hematopoietic precursor cells from macrophage high-proliferative-potential colony-forming cells. Exp Hematol 1987; 15:972–977.

45. McNiece IK, Bertoncello I, Kriegler AB, Quesenberry PJ. Colony-forming cells with high proliferative potential (HPP-CFC). Int J Cell Cloning 1990; 8:146–160.

46. Magli MC, Iscove NN, Odartchenko N. Transient nature of early haematopoietic spleen colonies. Nature 1982; 295:527–529.

47. Mauch P, Hellman S. Loss of hematopoietic stem cell self-renewal after bone marrow transplantation. Blood 1989; 74:872–875.

48. Meunch MO, Moore MAS. Accelerated recovery of peripheral blood cell counts in mice transplanted with in-vitro cytokine-expanded hematopoietic progenitors. Exp Hematol 1992; 20:611–618.

49. Meunch MO, Firpo MT, Moore MAS. Bone marrow transplantation with interleukin-1 plus *kit*-ligand ex-vivo expanded bone marrow accelerates hematopoietic reconstitution in mice without the loss of stem cell lineage and proliferative potential. Blood 1993; 81:3463–3473.

50. Moore MAS. Clinical implications of positive and negative hematopoietic stem cell regulators. Blood 1991; 78:1–19.

51. Moore MAS. Overview of hemopoiesis and hemopoietic reconstruction. In: Cell Therapy: Stem Cell Transplantation, Gene Therapy and Cellular Immunotherapy. Morstyn G, Sheridan S, eds. New York: Cambridge University Press, 1996:3–17.

52. Nakahata T, Ogawa M. Identification in culture of a class of hemopoietic colony-forming units with extensive capability to self-renew and generate multipotential hemopoietic colonies. Proc Natl Acad Sci USA 1982; 79:3843–3847.

53. Ogawa M. Differentiation and proliferation of hematopoietic stem cells. Blood 1993; 11:2844–2853.

54. Okada S, Nakauchi H, Nagayoshi K, Nishikawa S, Miura Y, Suda T. In vivo and in vitro stem cell function of c-*kit*- and Sca-1 positive murine hematopoietic cells. Blood 1992; 80:3044–3050.

55. Peters SO, Kittler ELW, Ramshaw HS, Quesenberry PJ. Murine marrow cells expanded in culture with IL-3, IL-6, IL-11, and SCF acquire an engraftment defect in normal hosts. Exp Hematol 1995; 23:461–469.

56. Pietrzyk ME, Priestley GV, Wolf NS. Normal cycling patterns of hematopoietic stem cell subpopulations: an assay using long term in vivo BrdU infusion. Blood 1985; 66:1460–1462.

57. Ploemacher RE, Brons NHC. Isolation of hemopoietic stem cell subsets from murine bone marrow: I. Radioprotective ability of purified cell suspensions differing in the proportion of day-7 and day-12 CFU-S. Exp Hematol 1988; 16:21–26.

58. Ploemacher RE, Brons NHC. Isolation of hemopoietic stem cell subsets from murine bone marrow: II. Evidence for an early precursor of day-12 CFU-S and cells associated with radioprotective ability. Exp Hematol 1988; 16:27–32.

59. Ploemacher RE, Brons NH. Cells with marrow and spleen repopulating ability and forming spleen colonies on day 16, 12, and 8 are sequentially ordered on the basis of increasing rhodamine 123 retention. J Cell Physiol 1988; 136:531–536.

60. Ploemacher RE, van der Sluijs JP, Voerman JSA, Brons NHC. An in vitro limiting-dilution assay of long-term repopulating hematopoietic stem cells in the mouse. Blood 1989; 74:2755–2763.

61. Ploemacher RE, van der Sluijs JP, van Beurden CAJ, Baert M, Chan PL. Use of limiting-dilution type long-term marrow cultures in frequency analysis of marrow-repopulating and spleen colony-forming hematopoietic stem cells in the mouse. Blood 1991; 78:2527–2533.

62. Ploemacher RE, van der Loo JCM, van Beurden CAJ, Baert MRM. Wheat germ agglutinin affinity of murine hemopoietic stem cell subpopulations is an inverse function of their long-term repopulating ability in vitro and in vivo. Leukemia 1993; 7:120–130.

63. Pluznik DH, Sachs L. The cloning of normal "mast" cells in tissue culture. J Cell Comp Physiol 1965; 66:319–324.

64. Pragnell IB, Wright EG, Lorimore SA, Adam I, Rosendaal M, Delamarter JF, Freshney M, Eckmann L, Sproul A, Wilkie N. The effect of stem cell regulators demonstrated with an in-vitro assay. Blood 1988; 72:196–201.

65. Quesenberry PJ. Too much of a good thing. "Reductionism ran amok" [editorial]. Exp Hematol 1993; 21:193–194.

66. Rebel VI, Dragowska W, Eaves CJ, Humphries RK, Lansdorp PM. Amplification of Sca-1$^+$ Lin$^-$ WGA$^+$ cells in serum-free cultures containing steel factor, interleukin-6 and erythropoietin with maintenance of cells with long-term in vivo reconstituting potential. Blood 1994; 83:128–136.

67. Rosendaal M, Hodgson GS, Bradley TR. Haemopoietic stem cells are organised for use on the basis of their generation-age. Nature 1976; 264:68–69.

68. Shapiro F, Yao T-J, Raptis G, Reich L, Norton L, Moore MA. Optimization of conditions for ex vivo expansion of CD34$^+$ cells from patients with stage IV breast cancer. Blood 1994; 84:3567–3574.

69. Siminovitch L, McCulloch EA, Till JE. The distribution of colony-forming cells among spleen colonies. J Cell Comp Physiol 1963; 62:327–336.

70. Sitnicka E, Lin N, Priestley GV, Fox N, Broudy VC, Wolf NS, Kaushansky K. The effect of thrombopoietin on the proliferation and differentiation of murine hematopoietic stem cells. Blood 1996; 87:4998–5005.

71. Smith LG, Weissman IL, Heimfeld S. Clonal analysis of hematopoietic stem-cell differentiation in vivo. Proc Natl Acad Sci USA 1991; 88:2788–2792.

72. Spangrude GJ. The pre-spleen colony-forming unit assay: measurement of spleen colony-forming unit regeneration. Curr Top Microbiol Immunol 1992; 177:31–39.

73. Spangrude GJ, Heimfeld S, Weissman IL. Purification and characterization of mouse hematopoietic stem cells. Science 1988; 241:58–62.

74. Spangrude GJ, Scollay R. A simplified method for enrichment of mouse hematopoietic stem cells. Exp Hematol 1990; 18:920–926.

75. Spangrude GJ, Brooks DM, Tumas DB. Long-term repopulation of irradiated mice with limiting numbers of purified hematopoietic stem cells: in vivo expansion of stem cell phenotype but not function. Blood 1995; 85:1006–1016.

76. Srour EF, Zanjani ED, Brandt JE, Leemhuis T, Briddell RA, Heerema NA, Hoffman R. Sustained human hematopoiesis in sheep transplanted in utero during early gestation with fractionated adult human bone marrow cells. Blood 1992; 79:1404–1412.

77. Srour EF, Zanjani ED, Cornetta K, Traycoff CM, Flake AW, Hedrick M, Brandt JE, Leemhuis T, Hoffman R. Persistence of human multilineage, self-renewing lymphohaematopoietic stem cells in chimeric sheep. Blood 1993; 82:3333–3342.

78. Suda T, Suda J, Ogawa M. Proliferative kinetics and differentiation of murine blast cell colonies in culture: evidence for variable G_0 periods and constant doubling rates of early pluripotent hemopoietic progenitors. J Cell Physiol 1983; 117:308–318.

79. Sutherland HJ, Eaves CJ, Eaves AC, Dragowska W, Lansdorp PM. Characterization and partial purification of human marrow cells capable of initiating long-term hematopoiesis in vitro. Blood 1989; 74:1563–1570.

80. Sutherland HG, Lansdorp PM, Henkelman DH, Eaves AC, Eaves CJ. Functional characterisation of individual human hematopoietic stem cells cultured at limiting dilution on supportive marrow stromal layers. Proc Natl Acad Sci USA 1990; 87:3584–3588.

81. Szilvassy SJ, Lansdorp PM, Humphries RK, Eaves AC, Eaves CJ. Quantitative assay for totipotent reconstituting hematopoietic stem cells by a competitive repopulation strategy. Proc Natl Acad Sci USA 1990; 87:8736–8740.

82. Szilvassy SJ, Cory S. Phenotypic and functional characterization of competitive long-term repopulating hematopoietic stem cells enriched from 5-fluorouracil-treated murine marrow. Blood 1993; 81:2310–2320.

83. Taswell C. Limiting dilution assays for the determination of immunocompetent cell frequencies. J Immunol 1981; 126:1614–1619.

84. Till JE, McCulloch EA. A direct measurement of the radiation sensitivity of normal mouse bone marrow cells. Radiat Res 1961; 14:216–222.

85. Traycoff CM, Kosak ST, Grigsby S, Srour EF. Evaluation of ex vivo expansion potential of cord blood

and bone marrow hematopoietic progenitor cells using cell tracking and limiting dilution analysis. Blood 1995; 85:2059–2068.

86. Tsuji K, Lyman SD, Sudo T, Clark SC, Ogawa M. Enhancement of murine hematopoiesis by synergistic interactions between steel factor (ligand for c-*kit*), interleukin-11, and other early acting factors in culture. Blood 1992; 79:2855–2860.

87. van der Loo JCM, van den Bos C, Baert MRM, Wagemaker G, Ploemacher RE. Stable multilineage hematopoietic chimerism in α-thalassemic mice induced by a bone marrow subpopulation that excludes the majority of day-12 spleen colony forming units. Blood 1994; 83:1769–1777.

88. Van Zant G, Thompson BP, Chen JJ. Differentiation of chimeric bone marrow in vivo reveals genotype-restricted contributions to hematopoiesis. Exp Hematol 1991; 19:941–949.

89. Visser JWM, Bol SJL, van den Engh G. Characterisation and enrichment of murine hemopoietic stem cells by fluorescence activated cell sorting. Exp Hematol 1981; 9:644–655.

90. Watt SM, Visser JWM. Recent advances in the growth and isolation of primitive haemopoietic progenitor cells. Cell Prolif 1992; 25:263–267.

91. Wolf NS, Priestley GV. Kinetics of early and late spleen colony development. Exp Hematol 1986; 14:676–682.

92. Wolf NS, Koné A, Priestley GV, Bartelmez SH. In vivo and in vitro characterization of long-term repopulating primitive hematopoietic cells isolated by sequential Hoechst 33342–rhodamine 123 FACS selection. Exp Hematol 1993; 21:614–622.

93. Wong GG, Witek-Gianotti JS, Temple PA, Kriz R, Ferenz C, Hewick RM, Clark SC, Ikebuchi K, Ogawa M. Stimulation of murine hemopoietic colony formation by human IL-6. J Immunol 1988; 140:3040–3044.

94. Zanjani ED, Pallavincini MG, Ascensao JL, Flake AW, Langlois RG, Reitsma M, MacKintosh FR, Stutes D, Harrison MR, Tavassoli M. Engraftment and long-term expression of human fetal hematopoietic stem cells in sheep following transplantation in utero. J Clin Invest 1992; 89:1178–1188.

95. Zanjani ED, Srour EF, Hoffman R. Retention of long-term repopulating ability of xenogeneic transplanted purified adult human bone marrow haematopoietic stem cells in sheep. J Lab Clin Med 1995; 126:24–28.

96. Zhong R, Astle CM, Harrison DE. Distinct developmental patterns of short-term and long-term functioning lymphoid and myeloid precursors defined by competitive limiting dilution analysis in vivo. J Immunol 1996; 157:138–145.

3

Receptors for Hematopoietic Regulators

Douglas J. Hilton
The Walter and Eliza Hall Institute of Medical Research, Melbourne, Victoria, Australia

I. INTRODUCTION

A bewildering number of secreted proteins, termed cytokines, regulate the production and function of lymphoid and myeloid cells (1). Each exerts its biological effects by binding to receptors expressed on the surface of responsive cells. The receptors for cytokines that regulate hematopoiesis fall into two major families, the hematopoietin–interferon receptors and the receptor tyrosine kinases, based on primary amino acid sequence similarity and predicted tertiary structure. This review will summarize the structural features of both receptor families and highlight some surprising similarities in their signal transduction pathways.

II. RECEPTOR STRUCTURE

A. Hematopoietin–Interferon Receptors

Most hematopoietic cytokines bind to type I transmembrane glycoproteins of the hematopoietin–interferon receptor family (Fig. 1). The extracellular domains of hematopoietin–interferon receptors are modular, containing various numbers of immunoglobulin, fibronectin, and hematopoietin–interferon receptor domains. The latter define members of this family. Conserved residues of hematopoietin–interferon receptor domains include a pair of cysteines and a series of aromatic residues (2,3). Other regions of sequence similarity, notably a second pair of cysteines and the five amino acid motif Trp-Ser-Xaa-Trp-Ser, are found in the hematopoietin receptor domains, but not in the interferon receptor domains.

Hematopoietin–interferon receptor domains contain approximately 200 amino acids (D200) and are composed of two 100-amino acid subdomains (SD100), which are themselves structurally related (2,3). Each SD100 was predicted to form seven β-strands arranged as a β-sandwich similar in structure to those found in fibronectin, papD, and CD4 (2,3). These predictions were confirmed with solution of the structure of the growth hormone–growth hormone receptor complex (4).

The structure of growth hormone–growth hormone receptor has become the paradigm for

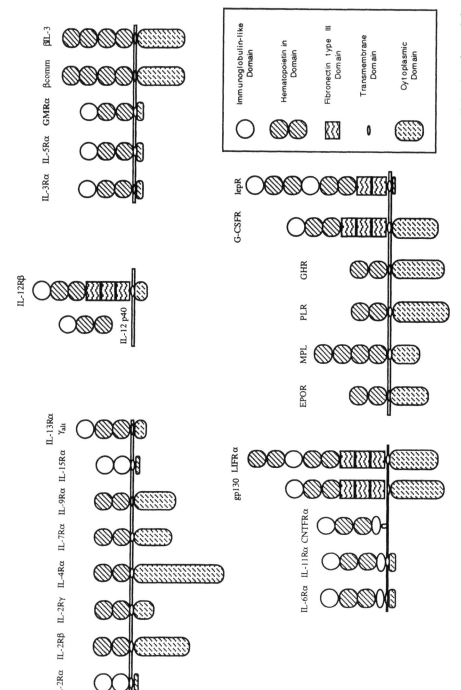

Figure 1   Schematic illustration of the hematopoietin receptor family highlighting the modular nature of the extracellular domains and the utilization of shared components in the receptors for different cytokines.

the hematopoietin receptor family not only in terms of its secondary and tertiary structure, but also because it highlights the importance of ligand-induced receptor dimerization. Fascinatingly, with similar residues, two receptors bound to different faces of a single growth hormone molecule (4). As discussed later, receptor multimerization and the sequential nature of cytokine binding appear to be important characteristics of all hematopoietin receptors (Table 1; see Fig. 1).

Compelling evidence exists that the receptors for granulocyte colony-stimulating factor (G-CSF; 5) and erythropoietin (EPO) also form homodimers. For the EPO receptor, mutants have been derived that contain cysteine substitutions in a region of the receptor predicted, by analogy with the growth hormone receptor, to lie in the dimer interface (6–8). These point-mutants form disulfide-linked homodimers and deliver a proliferative signal in the absence of EPO, suggesting constitutive dimerization recapitulates a normally EPO-driven event. The EPO receptor mutants with truncations of the cytoplasmic tail also support the notion that receptor dimerization is important, because these biologically inactive receptors function in a dominant negative manner when coexpressed with wild-type EPO receptors (8).

Other hematopoietin receptors are heterodimers or heterotrimers. Such receptors appear to comprise a specific α-subunit and one or more shared subunits. Three classes of receptors can be defined on the basis of shared components.

IL-3, IL-5, and GM-CSF Receptors

Interleukin (IL)-3, IL-5, and granulocyte–monocyte colony-stimulating factor (GM-CSF) bind to specific receptor α-subunits with a relatively low affinity (9–13). The generation of a

Table 1 Cytokines, Receptor Components, and Signal Transducers

Cytokine	Specific receptor component	Common receptor components	Jak	STAT
IL-3	IL-3Rα/β-IL-3	β-common	Jak1, Jak2	? STAT5
IL-5	IL-5Rα	β-common	Jak1, Jak2	? STAT5
GM-CSF	GM-CSFRα	β-common	Jak1, Jak2	? STAT5
IL-6	IL-6Rα	gp130	Jak1, Jak2, Tyk2	STAT3, STAT1
IL-11	IL-11Rα	gp130	Jak2	STAT3, STAT1
CNTF	CNTFRα	gp130, LIFR	Jak1, Jak2, Tyk2	STAT3, STAT1
OSM	?	gp130, LIFR	Jak1, Jak2, Tyk2	STAT3, STAT1
LIF	?	gp130, LIFR	Jak1, Jak2, Tyk2	STAT3, STAT1
IL-2	IL-2Rα (tac)	IL-2Rγ, IL-2R	Jak1, Jak3	? STAT5
IL-4		IL-4Rα, IL-2Rγ, IL-13Rα	Jak1, Jak3	STAT6
IL-7	IL-7Rα	IL-2Rγ	Jak1, Jak3	?
IL-9	IL-9Rα	IL-2Rγ?	Jak1, Jak3	?
IL-13	?	IL-4Rα, IL-13Rα	?	STAT6
IL-15	IL-15Rα	IL-2Rγ, IL-21Rβ	Jak1, Jak3	?
IL-12	IL-12p40, IL-12Rβ	?	Jak2, Tyk2	STAT4
EPO	epoR	?	Jak2	STAT5
G-CSF	G-CSFR	?	Jak1, Jak2	
Thrombopoietin	mpl	?	Jak2	?
Growth hormone	GHR	PLR	Jak2	?
Prolactin		PLR	Jak2	STAT5

References to the observations summarized in Table 1 are to be found in the body of the text.

high-affinity receptor–ligand complex capable of signal transduction requires interaction with a receptor β-subunit (11–14). The same β-subunit, termed β-common or AIC2B, is responsible for the generation of high-affinity IL-3, IL-5, and GM-CSF receptors (10,13,15–18).

In the mouse, but apparently not the human, an additional IL-3-specific β-subunit (AIC2A) also exists (19–21). The role of the IL-3-specific β-subunit in the mouse is unclear. Although the common β-subunit is unable to bind IL-3, IL-5, or GM-CSF when expressed alone, the IL-3-specific β-subunit binds IL-3 with low affinity (21). In vitro, however, both β-subunits appear to produce analogous signals (19). The targeted disruption of the genes encoding these receptor components may more clearly illuminate their specific roles.

As described in the following section, mutants of GM-CSF, IL-3, and IL-5, in addition to β-common have been generated that are incapable of forming a high-affinity receptor complex, but do retain the capacity to transduce a proliferative signal (22,23). The separation of high-affinity binding from signal transduction was also shown with the IL-2-dependent cell line CTLL, which expresses neither the α- nor β-subunits of the GM-CSF receptor. Expression of the α-subunit of the human GM-CSF receptor in CTLL cells resulted in the capacity to bind human GM-CSF with a low affinity, but not the ability to proliferate in response to GM-CSF. As expected, the coexpression the of human β-common subunit with the human GM-CSF receptor α-subunit generates a high-affinity receptor for human GM-CSF and allows the cells to proliferate in response to human GM-CSF as well as IL-2. The surprise was found when the human GM-CSF receptor α-subunit was expressed with murine common β-subunit. In this case, no high-affinity receptors could be detected, but at high concentrations of human GM-CSF, CTLL cells could proliferate (24).

IL-2, IL-4, IL-7, IL-9, IL-13, and IL-15 Receptors

The α-subunit of the IL-2 receptor was described initially as the *tac*-antigen, and it is unusual in that it is not a member of the hematopoietin receptor family; rather, the extracellular domain belongs to the immunoglobulin superfamily (25,26). The α-subunit of the IL-2 receptor binds IL-2 with low affinity and does not transduce a biological signal. The IL-2 receptor β-chain, similar to the α-chain, was cloned using monoclonal antibodies directed against it. This β-chain is unable to bind IL-2 when expressed in COS cells, but can bind IL-2 with an intermediate affinity when expressed in other cell types, suggesting the existence of a third element of the receptor (27). The γ-subunit of the IL-2 receptor was recently cloned on the basis of its ability to bind to a complex of IL-2 and the IL-2 receptor β-subunit (28).

An additional complexity to IL-2 receptor function was suggested by the realization that human X-linked agammaglobulinemia was caused by a mutation in the γ-subunit of the IL-2 receptor (29; reviewed in Ref. 30). If the γ-subunit of the IL-2 receptor functioned only in the IL-2 receptor complex, it would be predicted that null mutants of IL-2 would be similar in phenotype to null mutants in an essential receptor component. This, however, was not true, because mice producing no IL-2 have a relatively mild phenotype in comparison with humans with defective IL-2 receptor γ-subunits. This difference suggested that the IL-2 receptor γ-subunit may be a component of other cytokine receptors; recent results have confirmed this notion.

Ligand-binding components of the IL-4, IL-7, and IL-9 receptors have each been cloned by direct expression (31–33). When expressed in cells, such as COS fibroblasts, these receptors are specific for their cognate cytokine. Similarly, the complex pattern of cross-competition between IL-13 and IL-14 on cells that are responsive to both cytokines has suggested that the IL-4 and IL-13 receptors share a common component (34). Indeed, the IL-4 receptor α-chain and a recently cloned IL-13 receptor α-chain appear to be components of both high-affinity IL-4 and IL-13 receptors (217). This result was extended to show that IL-2, IL-4, IL-7, IL-9, and IL-15 each use the γ-subunit of the IL-2 receptor for both signal transduction as well as the generation of a high-affinity receptor (35–40).

IL-6, IL-11, LIF, OSM, CNTF, and IL-12

Although the GM-CSF and IL-2 receptor groups discussed in the foregoing share a single component, the group of receptors for IL-6, IL-11, leukemia inhibitory factor (LIF), oncostatin M (OSM), and ciliary neurotropic factor (CNTF) utilize two common components: gp130 and the LIF receptor polypeptide.

IL-6, IL-11, and CNTF bind to specific α-subunits with low affinity. Similar to the α-subunits of the GM-CSF, IL-5, and IL-3 receptors, those for IL-6 and IL-11 contain a relatively short cytoplasmic domain, and they are incapable of delivering a proliferative signal when expressed alone (41–43). The situation for CNTF and IL-12 is even more extreme. The CNTF receptor α-subunit is not tethered to the membrane by a conventional transmembrane domain, but instead, contains a glycophosphatidylinositol (GPI) anchor (42). More remarkably, the IL-12 receptor α-subunit is secreted and forms part of the IL-12 cytokine itself. Hence, this molecule is composed of a p40 subunit (the receptor α-subunit) and a p35 subunit, which shares a low level of sequence similarity to IL-6 and IL-11 (44,45).

The cytoplasmic domain of the IL-6 receptor α-subunit has been shown to be dispensable, for the addition of IL-6 together with the soluble IL-6 receptor α-subunit stimulates certain cells to respond (46). The critical factor in determining whether a response occurred was the presence of an additional transmembrane glycoprotein gp130. gp130 was purified from various human sources, and monoclonal antibodies were raised to this molecule. These, in turn, were used to clone gp130 cDNAs by expression in λgt11. Analyses of the cDNA revealed that gp130 was a member of the hematopoietin receptor family and was most similar to the G-CSF receptor (47). Expression of gp130 in the factor-dependent cell line Ba/F3 confirmed that gp130 was essential both for the generation of high-affinity IL-6 receptors and for IL-6–induced stimulation of Ba/F3 proliferation. The stoichiometry of subunits in the IL-6 receptor appears to be one α-subunit to two gp130 molecules (48). The latter become disulfide-linked after association with the complex between the ligand and the receptor α-subunit (48).

In addition to G-CSF, the receptor for LIF shows significant sequence similarity to gp130, both in its extracellular and cytoplasmic domains. The LIF receptor was cloned by direct expression in COS cells and binds LIF with a low affinity (49). When expressed alone, the LIF receptor appears unable to transduce a proliferative signal (50). As with the IL-6 receptor system, expression of gp130 with the LIF receptor generated high-affinity receptors for LIF capable of signal transduction (51). Unlike the α-subunits of the IL-6, IL-11, and CNTF receptors, the low-affinity LIF receptor is not specific for LIF, and its cytoplasmic domain appears necessary for signal transduction.

gp130 and the LIF receptor polypeptide are also components of the OSM and CNTF receptors. Similar to LIF, a classic receptor α-subunit has yet to be described for OSM. Oncostatin M binds to gp130 with low affinity and requires association with the LIF receptor for the generation of a high-affinity receptor capable of signal transduction (51,52). CNTF, as has been described earlier, binds with low affinity to the CNTF receptor α-subunit (42). Formation of this complex triggers association of gp130 and then the LIF receptor (53,54). Recent results suggest that CNTF is also capable of binding with low affinity to the LIF receptor and gp130, in the absence of the α-subunit (50). The physiological relevance of this observation is unclear.

The IL-12 p35 and p40 subunits appear to be analogous to IL-6 and the soluble IL-6 receptor α-subunit (44,45). In the same way the latter binds to and triggers its biological effect through gp130, IL-12 binds to a recently described cell surface receptor that is similar in sequence to gp130, the LIF receptor and the G-CSF receptor (55). The existence of an additional IL-12 receptor component was suggested by the observation that the affinity of IL-12 for the cloned IL-12 receptor was lower than the affinity for the receptor on IL-12 responsive cells (55).

The complexity of the LIF receptor–gp130 receptor system is likely to increase with further study. By analogy with IL-6, IL-11, and CNTF, it might be predicted that receptor α-subunits exist for LIF and OSM. Indeed, the observation that certain cells (e.g., lines derived from Kaposi's sarcoma cells) bind OSM with high affinity, and that they respond to OSM, but not LIF, argues for the existence of an OSM receptor that does not simply contain gp130 and the LIF receptor polypeptide (56–58). Similarly, given the homology of the G-CSF receptor to gp130 and the LIF receptor, the existence of an α-subunit for this cytokine would not be surprising.

As with the observation that mutation of the IL-2 receptor γ-subunit led to defects that could not be ascribed to IL-2 alone, experimental disruption of the gp130 gene in mice resulted in phenotypes that were not expected on the basis of the known actions of IL-6, IL-11, CNTF, or LIF. gp130-deficient mice die in utero at approximately day 14 of gestation and, strikingly, there is abnormal development of the heart. This suggests the presence of a novel cytokine that plays a role in cardiogenesis and that uses gp130 as part of its cell surface receptor (Kishimoto and colleagues, meeting presentations).

B. *fms*, *kit*, and flk3: Tyrosine Kinase Receptors for M-SCF, Stem Cell Factor, and flk3-Ligand

The M-CSF receptor, stem cell factor (SCF) receptor, and flk3 are class III receptor tyrosine kinases. Although most hematopoietin receptors were cloned on the basis of their ability to bind their corresponding cytokine. The M-CSF, SCF, and flk3 receptor tyrosine kinases were cloned in quite different ways, with no prior knowledge of their cognate ligands.

The elucidation of the biology and molecular details of the stem cell factor and its receptor has occurred through insights gained from two separate streams of research. One, into anemic mutant mice, and the second, into oncogenes carried by acutely transforming retroviruses.

Two groups of mice carrying mutations in the *white spotting (W)* and *steel (SL)* loci have similar phenotypes (60–62). The most striking characteristics of these mice include areas of nonpigmented skin, sterility, and anemia, which are due to defects in melanocytes, primordial germ cells, and hematopoietic stem cells (reviewed in Refs. 63–65). At the cellular level, however, the nature of the defects in white spotting and steel mice appear reciprocal; that is the hematopoietic defect of white spotting mice could be cured by transplantation of bone marrow from wild-type or steel mice, whereas the defect in steel mice was independent of the source of hematopoietic cells (reviewed in Ref. 64). This observation led to the suggestion that the defect in white spotting mice was at the level of a cellular receptor, whereas that of the steel mice was at the level of the corresponding ligand (64)

The gene affected in white spotting mice was discovered in a completely separate set of studies. In 1986, the acutely transforming oncogene of the feline sarcoma virus (HZ4-FeSV) was cloned and named v-*kit* (66). The cellular homologue of v-*kit* was isolated and shown to be putative receptor tyrosine kinase with homology to the M-CSF and platelet-derived growth factor (PDGF) receptors (67,68). The cellular homologue of this oncogene, c-*kit*, was isogenic with the *white spotting* locus (69–71).

As predicted, the gene involved in the steel defect was cloned and shown to encode a cytokine that was expressed both as a cell surface or secreted regulator capable of binding to and activating the c-*kit* receptor tyrosine kinase (72–79). This cytokine was termed stem cell factor (SCF), steel factor (SF or SLF), *kit* ligand (KL), or mast cell growth factor (MGF).

Analyses of SCF transcripts and protein revealed that the ligand is produced in both a cell surface and a secreted form (80,81).

Similar to c-*kit*, the M-CSF receptor was initially identified as the oncogene v-*fms* of an acutely transforming retroviruses—the McDonough strain of the feline sarcoma viruses. Subsequently, v-*fms* and its cellular homologue c-*fms* were shown to contain a tyrosine kinase domain (82) with particular similarity to the PDGF receptor: c-*fms*, was then shown to be the receptor for M-CSF (83). As with SCF, in which the secreted and membrane-bound forms are generated from alternatively spliced transcripts, multiple M-CSF transcripts are also found (84,85). Interestingly, the proteins produced from each transcript contain a transmembrane region and are thus expressed at the cell surface. Proteolysis and liberation of secreted M-CSF occurs at the membrane in the extracellular environment (86–88, reviewed in Ref. 89).

The receptor for the third ligand in this family, flt3/flk2, was cloned by an entirely different route: a polymerase chain reaction (PCR)-based screen that used primers from conserved elements of the kinase domain (90,91). The ligand for flt3/flk2 was then cloned using a fusion protein containing the immunoglobulin constant region and the flt3/flk2 receptor as a probe for cell surface ligands (92,93).

As with members of the hematopoietin receptor family, dimerization of receptor kinases appears to be of central importance. Numerous *white spotting* alleles and, in particular those with a defective kinase, such as *W42*, are strongly dominant negative (64,94). That is, nonfunctional receptors are generated when the mutant allele is coexpressed with a wild-type allele. This suggested that inactive heterodimers were formed between mutant and wild-type receptors. Similar mutations, when engineered into the M-CSF receptor, also act in a dominant negative manner when expressed with wild-type M-CSF receptor (95). Direct biochemical evidence for dimerization has also been obtained for both the SCF and M-CSF receptor (96,97). The biochemical consequences of dimerization of receptor kinases and hematopoietin–interferon receptors are discussed in the following section.

III. CYTOKINE–RECEPTOR INTERACTIONS

The structure of many hematopoietic cytokines have been solved by X-ray crystallography or nuclear magnetic resonance (NMR) spectroscopy (98–105). Despite a lack of obvious sequence similarity, each hematopoietic regulator forms a four–α-helical bundle. The arrangement of helices in IL-5, which exists as a noncovalent dimer, is particularly interesting; the dimer is clearly composed of two, four–α-helical bundles; however, in each bundle, three helices are contributed to by one IL-5 molecule and the fourth by the other IL-5 molecule. M-CSF binds to a receptor tyrosine kinase, rather than a member of the hematopoietin receptor family, yet it also forms a four–α-helical bundle similar to growth hormone (101). Similar to IL-5, M-CSF is a dimer. In contrast with IL-5, however, the four helics in each bundle of M-CSF are formed by a single M-CSF polypeptide (101).

In addition to becoming a paradigm for cytokine and receptor structure, the growth hormone–growth hormone receptor complex has provided a model on which to base studies on the interaction of cytokines with members of the hematopoietin–interferon receptor families (103).

For the growth hormone, formation of a ternary complex, containing a single hormone bound to two identical receptors, proceeds in a strict sequence (106). The hormone first interacts with a receptor using residues in its A helix, the D helix, and the loop between helices A and B. Only then does the second receptor interact with epitopes on the A and C helices of the hormone (103,106). Similar sequential binding of receptor components occurs in other cytokine systems.

A great deal of work has been carried out to define the residues of GM-CSF, IL-3, and IL-5 that are important in binding to their respective α-subunits and common β-subunit. In summary, residues in the D helix of these cytokines are important for binding to receptor α-subunits (107,108), whereas binding to the β-subunit utilizes residues in the A and C helices (109). With human GM-CSF, Glu-21 in helix A appear to be particularly important in the generation of high-affinity complex, because mutants with a substitution of this residue to arginine are unable to bind to receptors with high affinity, yet retain the capacity to bind with low affinity (22,23). Surprisingly, such GM-CSF mutants are still biologically active. Corresponding mutations in the A helix of IL-3 (Glu-23) and IL-5 (Glu-11) yield cytokines with analogous properties (110). A similar effect is observed when mutating the reciprocal reside on the receptor β-subunit, His-367, which interacts with Glu-12 (111). Again this receptor is unable to contribute to the generation of a high-affinity GM-CSF, IL-3, or IL-5 complex, but is capable of delivering a proliferative signal in response to high concentrations of ligand (111).

The interaction of cytokines with receptors containing the IL-2 receptor γ-subunit and gp130 is more complex because these receptors are composed of at least three separate proteins. Residues important for IL-2 binding are perhaps the best studied. Interaction with the IL-2 receptor α-subunit (*tac*) appears to involve resides in the A–B and C–D loops, whereas interaction with the β-subunit centers on residues in the A-helix (112,113). The Asp-34, which is in a similar position to Glu-12 in GM-CSF, is of particular importance here. Finally, Gln-128, in the D-helix of IL-2 has been identified as a site of interaction with the γ-chain of the IL-receptor.

IV. SIGNAL TRANSDUCTION

A. SH2s, GRBs, SOS, STATs, and JAKs

Ligand-driven association of receptor subunits is a central step in signaling through both hematopoietin receptors (35) and receptor tyrosine kinases (114). Receptor multimerization is important because it allows information to pass from the extracellular environment into the cytoplasm without the necessity of a conformational change passing through the transmembrane domain. The initial events in signal transduction by receptor tyrosine kinases and hematopoietin receptors have been clearly elucidated in the past few years. Surprisingly, there exist a number of steps in common.

The binding of cytokines to receptor tyrosine kinases results in dimerization and rapid receptor tyrosine phosphorylation. Phosphorylation of each receptor appears to be catalyzed by the other receptor in the dimer. Phosphorylated receptor tyrosine residues are the site of interaction for a large number of proteins containing *src* homology type 2 (SH2) domains. Residues surrounding the phosphotyrosine determine the spectrum of SH2–domain-containing protein that can bind to a particular site. These sites have been mapped, in some detail, for both the M-CSF and SCF receptors. In the M-CSF receptor, for example, Grb2 interacts with Tyr-697 and the residues surrounding it, whereas the p85 subunit of phosphatidylinositol-3 (PI3) kinase interacts with Tyr-721, and Tyr-809 is the site of interaction for members of the *src* kinase family (115–117). The SCF receptor *c-kit* likewise associates with the p85 subunit of PI3 kinase, phospholipase-Cγ and the hematopoietic cell phosphates, but not GAP (118–122). Similarly, it is likely than flt3/flk2 also associates with the p85 subunit of PI3 kinase and Grb2 (123).

The consequences of the interaction of Grb2 with phosphorylated tyrosine residues in the cytoplasmic domain of receptor kinases have recently been eluciated in the *Drosophila* species sevenless and mammalian EGF receptor systems (reviewed in 124). Grb2 not only contains an SH2 domain, but also two SH3 domains, through which it appears to be constitutively associated with the mammalian homologue of son of sevenless (sos) 125–133; reviewed in Ref 134) The sos is a guanine nucleotide exchange protein, that activates Ras by promoting the exchange of GTP for GDP (133,135,136). The interaction of the Grb2–sos complex with the cytoplasmic domains of receptor kinases does not appear to result in activation of sos by covalent modification, but rather, increases the local concentration of sos, and thereby increases the proportion of *ras* molecules that are in the activated state. Activation of *ras*, among other things, leads to a serine–threonine kinase cascade, involving Raf, MEK, and MAP kinase. Activated MAP kinase moves from the cytoplasm to the nucleus where it phosphorylates and activates transcription factors, including *fos, jun*, and the serum response factor. These factors, in turn, increase the rate of transcription of genes, such as *cdc-2*, that are involved in initiating the cell cycle (137,138).

Although receptors of the hematopoietin–interferon family do not contain an intrinsic tyrosine kinase activity, they induce rapid tyrosine phosphorylation of several intracellular proteins (see, for example, 139–146). The search for cytoplasmic tyrosine kinases that are coupled to hematopoietin–interferon receptors has focused on the *src*-related kinases and, more recently, on the Jak family of kinases (35). The Jak kinases were originally cloned in PCR-based screens for novel tyrosine kinases. Four members have been described; Jak1, Jak2, Jak3, and Tyk2 (147–150). Each have a molecular weight of 120,000 to 130,000 and contain a consensus tyrosine kinase domain as well as a "kinase-like" domain. The evidence for a role of Jak kinases in signal transduction emerged from a series of genetic studies into interferon signaling. By generating mutant interferon-unresponsive cell lines (151,152) and complementing their defects, Pellegrini and colleagues demonstrated that Jak2 and Jak1 were necessary for signaling through the interferon α/β receptor and that Tyk2 and Jak1 were required for interferon-γ signaling (153–155). The arrangement of each kinase within the signaling pathway, however, remained unclear, because in response to interferon-γ both Tyk2 and Jak1 are phosphorylated, whereas in cells with mutant Tyk2, neither Tyk2 nor Jak1 were phosphorylated. Likewise, in cells with a mutant Jak1, neither Tyk2 nor Jak1 are phosphorylated. Some light was shed on this problem with the recognition that the importance of Jak kinases was not limited to interferon signaling (reviewed in Refs. 156,157).

Receptor dimerization occurs rapidly after binding of cytokines and results in the juxtaposition of Jak kinases. The next event, cross-phosphorylation and activation of Jak kinases appears to be analogous to the cross-phosphorylation observed in dimers of receptor tyrosine kinases, such as the EGF receptor, SCF receptor, and M-CSF receptor (156–158). Because receptor dimerization is also important for interferon signal transduction, the cross-phosphorylation model may explain why mutation of, for example, Jak1 abolishes phosphorylation of Tyk2 in response to interferon-γ (159).

The different members of the Jak kinase family appear important for signaling through different groups of receptors (see Table 1). The EPO receptor and prolactin receptor bind and activate Jak2 (160–162), G-CSF activates Jak1 (163), whereas IL-2, 4, 7, 9, 13, and 15 are likely to act through Jak3 (150,164). LIF, OSM, CNTF, IL-6, and IL-11 are different because Jak1, Jak2, and Tyk2 bind to gp130 and become activated in response to cytokine (165,166).

One possible class of substrates of the Jak kinases have been identified by Darnell's group, using biochemical techniques, and also by Pelligrini and colleagues using their genetic approach: these are the signal transducer and activators of transcription or STATs (reviewed in 156). The cloning of cDNAs for the STAT proteins, p91/84 (STAT1a/STAT1b) and p113 (STAT2), revealed that they share significant sequence similarity. Strikingly, each protein contained SH2 and SH3 domains. Evidence exists that the SH2 domains present in the STATs allow them to bind to phosphorylated tyrosine residues present in the cytoplasmic domains of receptors. Stimulation of cells with interferon-α and INF-β then results in phosphorylation of p91 (STAT1a) and p84 (STAT1b), which are generated by alternative transcription of the same gene, and p113 (STAT2).

The three proteins then translocate to the nucleus and associate with p48. This complex (insulin-stimulated gene factor-3; ISGF-3) then binds to the interferon-stimulatable response element (ISRE) in the promoters of genes such as 6–16 (167–178). The situation for INF-γ is simpler, with p91 (also termed gamma-activated factor; GAF) alone being phosphorylated and migrating to the nucleus to activate the gamma-activated sequence (GAS) of IFN-γ responsive genes, such as those encoding the class I and II major histocompatibility complex (MHC) (179–181).

Growth factors that bind to receptor tyrosine kinases, such as EGF, PDGF, and M-CSF are also capable of stimulating the phosphorylation of p91/STAT1a (182–185). The consequences of phosphorylation of p91 in response to EGF appears to be similar to the situation for INF-γ. The p91 migrates to the nucleus, where it is capable of interacting with promoters (e.g. the *c-fos* promoter) containing a c-*sis*-inducible element (SIE) (182).

Recently, more members of the STAT family have been isolated and appear to be involved in signaling in response to other cytokines (see Table 1). STAT 3 or acue-phase response factor (APRF) is phosphorylated when hepatocytes and other cells are stimulated with IL-6. Similar to STAT1 and STAT2, when STAT3 is phosphorylated, it migrates from the cytoplasm to the nucleus, where it binds to a sequence (type 2 IL-6 responsive element) in the promoters of IL-6 responsive genes such as α1-acid glycoprotein (165, 186–189). The activation of STAT3 is not limited to IL-6, because other cytokines including LIF, OSM, and CNTF, that utilize gp130 as a receptor component are also capable of inducing STAT3 phosphorylation (189). STAT molecules have also been described that are activated in response to IL-4 [IL-4 STAT (190)] and prolactin (mammary gland-specific nuclear factor; MGF, or STAT5 [191]).

B. Truncations of Receptor Cytoplasmic Domains Reveal Multiple Routes of Signal Transduction

The redundant nature of cytokine biology is partly explained by the use of shared receptor components and by the presence of conserved receptor cytoplasmic domain sequences necessary for the activation of Jak kinases. These elements, termed box 1 and box 2, are found close to the transmembrane domain, and their deletion diminishes the ability of receptors to phosphorylate members of the Jak kinase family and stimulate a proliferative response (192–203).

In addition to exhibiting redundancy, many cytokines are biologically pleiotropic. This suggests that signal transduction pathways, in addition to those involving Jaks may be important. Support for this notion has come from the generation of receptors with truncations of their cytoplasmic tails. These studies have shown that, similar to receptor tyrosine kinases, in which residues surrounding different phosphotyrosines bind to distinct SH2-domain–containing proteins, discrete regions of cytokine receptor cytoplasmic tails are necessary for triggering different biological and biochemical responses (e.g. see Refs.193, 199, 204–213). Particularly interesting results have been found with the receptors for G-CSF and EPO, for which mutations that result in truncations of the receptor may play a role in the disease pathogenesis.

Different regions of the G-CSF receptor appear to be required for stimulating proliferation and in inducing neutrophilic granulocyte differentiation (210–213). Notably, although the first 76 amino acids of the G-CSF receptor cytoplasmic domain (including box1 and box2) are capable of mediating G-CSF-stimulated proliferation of factor-dependent hematopoietic cell lines, both the first 99 amino acids of the cytoplasmic tail and the COOH-terminal 73 amino acids were necessary for full induction of the myeloperoxidase gene in FDCP-1 cells by G-CSF (21). A nonsense mutation, resulting in the production of a short G-CSF receptor, has been implicated, in a single case, in the pathogenesis of some cases of Kostmann's syndrome or severe congenital neutropenia (212, see also 214 for additional reports of normal G-CSF receptors in patients with Kostmann's syndrome). The truncated receptor, lacking the COOH-terminal 98 amino acids was capable of transducing a proliferative signal in response to G-CSF, but not a differentiation signal. In the affected patient, the truncated receptor was suggested to function as a dominant negative mutant. Dominant negative mutants containing truncated cytoplasmic domains have been described for the EPO, SCF, and M-CSF receptors (8,94,95). Box 1 and box 2 of the EPO receptor are also essential for EPO-stimulated proliferation (215); however, deletion of the COOH-terminal 42 amino acids renders cells hypersensitive to EPO (6,215). A similar mutation in humans with familial polycythemia has also been described (216).

V. CONCLUSIONS

In the past decade, cDNAs encoding cell surface receptors for almost every known hematopoietic regulator have also been cloned. This work has led to the recognition that there are two major families of receptors: the hematopoietin–interferon receptors and the receptor tyrosine kinases. Given the intense interest in hematopoietic regulators, it is certain that other cytokines and their corresponding receptor components will soon be discovered.

In this review I have attempted to stress the importance of receptor dimerization in the initiation of signal transduction from both receptor kinase and hematopoietin receptors. Common elements in the signal transduction pathways used by these two receptor families have also become evident quite recently. These include phosphorylation and activation of the STAT family of transcription factors and the activation of the ras–MAP kinase pathway. Exciting questions concerning signal transduction still remain to be answered. For example, hematopoietic regulators clearly effect a wide range of cellular functions, in addition to stimulating proliferation. Whether the same signaling molecules are also responsible for stimulating self-renewal, promoting cell survival, and inducing differentiation down the various hematopoietic lineages remains to be determined.

ACKNOWLEDGMENTS

The author is supported by a Queen Elizabeth II Postdoctoral Fellowship from the Australian Research Council.

REFERENCES

1. Metcalf D. Haemopoietic regulators. Trends Biochem Sci 1992; 17:286–289.
2. Thoreau E, Petridou B, Kelly PA, Djiane J, Mornon JP, Structural symmetry of the extracellular domain of the cytokine/growth hormone/prolactin receptor family and interferon receptors revealed by hydrophobic cluster analysis. FEBS Lett 1991; 282:26–31.
3. Bazan JF. Structural design and molecular evolution of a cytokine receptor superfamily. Proc Natl Acad Sci USA 1990; 87:6934–6938.
4. de-Vos AM, Ultsch M, Kossiakoff AA. Human growth hormone and extracellular domain of its receptor: crystal structure of the complex. Science 1992; 255:306–312.
5. Fukunga R, Ishizaka-Ikeda E, Nagata S. Purification and characterization of the receptor for murine granulocyte colony-stimulating factor. J Biol Chem 1990; 265:14008–15.
6. Yoshimura A, Longmore G, Lodish HF. Point mutation in the exoplasmic domain of the erythropoietin receptor resulting in hormone-independent activation and tumorigenicity. Nature 1990; 348:647–9.
7. Watowich SS, Yoshimura A, Longmore GD, Hilton DJ, Yoshimura Y, Lodish HF. Homodimerization and consecutive activation of the erythropoietin receptor. Proc Natl Acad Sci USA 1992; 89:2140–4.

8. Watowich SS, Hilton DJ, Lodish HF. Activation and inhibition of erythropoietin receptor function: role of receptor dimerization. Mol Cell Biol 1994; 14:3535–3549.

9. Gearing DP, King JA, Gough NM, Nicola NA. Expression cloning of a receptor for human granulocyte–macrophage colony-stimulating factor. EMBO J 1989; 8:3667–76.

10. Kitamura T, Sato N, Arai K, Miyajima A. Expression cloning of the human IL-3 receptor cDNA reveals a shared beta subunit for the human IL-3 and GM-CSF receptors. Cell 1991; 66:1165–74.

11. Devos R, Plaetinck J, Van der Heyden J, Cornelis S, Vandekerckhove J, Fiers W, Taverneir J. Molecular basis of a high affinity murine interleukin-5 receptor. EMBO J 1991; 10:2133–7.

12. Taverneir J, Devos R, Cornelis S, Tuypens T, Van der Heyden J, Fiers W, Plaetinck G. A human high affinity interleukin-5 receptor (IL5R) is composed of an IL5-specific alpha chain and a beta chain shared with the receptor for GM-CSF. Cell 1991; 66:1175–84.

13. Takaki S, Mita S, Kitamura T, Yonehara S, Yamaguchi N, Tominaga A, Miyajima A, Takatsu K. Identification of the second subunit of the murine interleukin-5 receptor: interleukin-3 receptor-like protein, AIC2B is a component of the high affinity interleukin-5 receptor. EMBO J, 1991; 10:2833–8.

14. Miyajima A, Mui A, Ogorochi T, Sakamaki K. Receptors for granulocyte–macrophage colony-stimulating factor interleukin-3, and interleukin-5. Blood 1993; 82:1960–1974.

15. Hayashida K, Kitamura T, Gormant DM, Arai K, Yokota T, Miyajima A. Molecular cloning of a second subunit of the receptor for human granulocyte–macrophage colony-stimulating factor (GM-CSF): reconstitution of a high-affinity GM-CSF receptor. Proc Natl Acad Sci USA 1990; 87:9655–9.

16. Gorman DM, Itoh N, Kitamura T, Schreurs J, Yonehara S, Yahara I, Arai K, Miyajima A. Cloning and expression of a gene encoding an interleukin 3 receptor-like protein: identification of another member of the cytokine receptor gene family. Proc Natl Acad Sci USA 1990; 87:5459–63.

17. Kitamura T, Miyajima A. Functional reconstituion of the human interleukin-3 receptor. Blood 1992; 80:84–90.

18. Takaki S, Murata Y, Kitamura T, Miyajima A, Tominaga A, Takatsu K. Reconstitution of the functional receptors for murine and human interleukin 5. J Exp Med 1993; 177:1523–9.

19. Hara T, Miyajima A. Two distinct functional high affinity receptors for mouse interleukin-3 (IL-3). EMBO J 1992; 11:1875–84.

20. Schreurs J, Hung P, May WS, Arai K, Miyajima A. AIC2A is a component of the purified high affinity mouse IL-3 receptor: temperature-dependent modulation of AIC2A structure. Int Immunol 1991; 3:1231–42.

21. Itoh N, Yonehara S, Schreurs J, Gorman DM, Maruyama K, Ishii A, Yahara I, Arai K, Miyajima A. Cloning of an interleukin-3 receptor gene: a member of a distinct receptor gene family. Science 1990; 247:324–7.

22. Shanafelt AB, Kastelein RA. High affinity ligand binding is not essential for granulocyte–macrophage colony-stimulating factor receptor activation. J Biol Chem 1992; 267:25466–72.

23. Lopez AF, Shannon MF, Hercus T, Nicola NA, Cambareri B, Dottore M, Layton MJ, Eglinton L, Vadas MA. Residue 21 of human granulocyte–macrophage colony-stimulating factor is critical for biological activity and for high but not low affinity binding. EMBO J 1992; 11:909–16.

24. Kitamura T, Hayashida K, Sakamaki K, Yokota T, Arai K, Miyajima A. Reconstitution of functional receptors for human granulocyte/macrophage colony-stimulating factor (GM-CSF): evidence that the protein encoded by the AIC2B cDNA is a subunit of the murine GM-CSF receptor. Proc Natl Acad Sci USA 1991; 88:5082–6.

25. Nikaido T, Shimizu A, Ishida N, Sabe H, Teshigawara K, Maeda M, Uchiyama T, Yodoi J, Honjo T. Molecular cloning of cDNA encoding human interleukin-2 receptor. Nature 1984; 311:631–5.

26. Leonard WJ, Depper JM, Crabtree GR, Rudikoff S, Pumphrey J, Robb RJ, Kronke M, Svetlik PB, Peffer NJ, Waldmann TA, Greene WC. Molecular cloning and expression of cDNAs for the human interleukin-2 receptor. Nature 1984; 311:626–31.

27. Hatakeyama M, Tsudo M, Minamoto S, Kono T, Doi T, Miyata T, Miyasaka M, Taniguchi T. Interleukin-2 receptor beta chain gene: generation of three receptor forms by cloned human alpha and beta chain cDNA's. Science 1989; 244:551–6.

28. Takeshita T, Asao H, Ohtani K, Ishii N, Kumaki S, Tanaka N, Munakata H, Nakamura M, Sugamura K. Cloning of the gamma chain of the human IL-2 receptor. Science 1992; 257:379–82.

29. Noguchi M, Yi H, Rosenblatt HM, Filipovich AH, Adelstein S, Modi WS, McBride OW, Leonard WJ. Interleukin-2 receptor gamma chain mutation results in X-linked severe combined immunodeficiency in humans. Cell 1993; 73:147–57.

30. Voss SD, Hong R, Sondel PM. Severe combined immunodeficiency, interleukin-2 (IL-2), and the IL-2 receptor: experiments of nature continue to point the way. Blood 1994; 83:626–35.

31. Mosley B, Beckmann MP, March CJ, et al. The murine interleukin-4 receptor: molecular cloning and characterization of secreted and membrane bound forms. Cell 1989; 59:335–48.

32. Goodwin RG, Friend D, Ziegler SF, et al. Cloning of the human and murine interleukin-7 receptors: demonstration of a soluble form and homology to a new receptor superfamily. Cell 1990; 60:941–51.

33. Renauld JC, Druez C, Kermouni A, Houssiau F, Uyttenhove C, Van Roost E, Van Snick J. Expression cloning of the murine and human interleukin 9 receptor cDNAs. Proc Natl Acad Sci USA 1992; 89:5690–4.

34. Aversa G, Punnonnen J, Cocks BG, de Waal Malefyt R, Vega F, Zurawski Jr SM, Zurawski G. de Vries JE. An interleukin 4 (IL-4) mutant protein inhibits both IL-4 or IL-13-induced human immunoglobulin G4 (IgG4) and IgE synthesis and B cell proliferation: support for a common component shared by IL-4 and IL-13 receptors. J Exp Med 1993; 178:2213–8.

35. Kishimoto T, Taga T, Akira S. Cytokine signal transduction. Cell 1994; 76:253–262.

36. Kondo K, Takeshita T, Ishii N, Nakamura M, Watanabe S, Arai KI, Sugamura K. Sharing of the interleukin-2 (IL-2) receptor γ chain between receptors for IL-2 and IL-4. Science 1993; 262:1874–1877.

37. Kondo M, Takeshita T, Higuchi M, Nakamura M, Sudo T, Nishikawa SI, Sugamura K. Functional participation of the IL-2 receptor γ chain in IL-7 receptor compelxes. Science 1994; 263:1453–1454.

38. Noguchi M, Nakamura Y, Russel SM, Ziegler SF, Tsang M, Cao X, Leonard WJ. Interleukin-2 receptor chain: a functional component of the interleukin-7 receptor. Science 1993; 262:1877–1880.

39. Giri JG, Ahdieh M, Eisenman J, Shanebeck K, Grabstein K, Kumaki S, Namen A, Park LS, Cosman D, Anderson D. Utilization of the beta and gamma chains of the IL-2 receptor by the novel cytokine IL-15. EMBO J 1994; 13:2822–30.

40. Russell SM, Keegan AD, Harada N, et al. Interleukin-2 receptor gamma chain: a functional component of the interleukin-4 receptor. Science 1993; 262:1880–3.

41. Yamasaki K, Taga T, Hirato H, Kawanishi Y, Seed B, Taniguchi T, Hirano T, Kishimoto T. Cloning and expression of the human interleukin-6 (BSF-2/IFNB2) receptor. Science 1988; 241:825–828.

42. Davis S, Aldrich TH, Valenzuela DM, Wong VV, Furth ME, Squinto SP, Yancopoulos GD. The receptor for ciliary neurotrophic factor. Science 1991; 253:59–63.

43. Hilton DJ, Hilton AA, Raicevic A, Rakar S, Harrison-Smith M, Gough NM, Begley CG, Metcalf D, Nicola NA, Willson TA. Cloning of a murine interleukin-11 receptor alpha-chain; requirement for gp130 for high affinity binding and signal transduction. EMBO J 1994; 13:4765–4775.

44. Wolf SF, Temple PA, Kobayashi M, et al. Cloning of cDNA for natural killer cell stimulatory factor, a heterodimeric cytokine with multiple biologic effects on T and natural killer cells. J Immunol 1991; 146:3074–81.

45. Gearing DP, Cosman D. Homology of the p40 subunit fo natural killer cell stimulatory factor (NKSF) with the extracellular domain of the interleukin-6 receptor. Cell 1991; 66:9–10.

46. Taga T, Hibi M, Hirata Y, Yamasaki K, Yasukawa K, Matsuda T, Hirano T, Kishimoto T. Interleukin-6 triggers the association of its receptor with a possible signal transducer, gp130. Cell 1989; 58:573–81.

47. Hibi M, Murakami M, Saito M, Hirano T, Taga T, Kishimoto T. Molecular cloning and expression of an IL-6 signal transducer, gp130. Cell 1990; 63:1149–57.

48. Murakami M, Hibi M, Nakagawa N, Nakagawa T, Yasukawa K, Taga T, Kishimoto T. IL-6 induced homodimerization of gp130 and associated activation of a tyrosine kinase. Science 1993; 260:1808–1810.

49. Gearing DP, Thut CJ, VandeBos T, Gimpel SD, Delaney PB, King J, Price V, Cosman D, Beckmann

MP. Leukemia inhibitory factor receptor is structurally related to the IL-6 signal transducer, gp130. EMBO J 1991; 10:2839–48.

50. Gearing DP, Ziegler SF, Comeau MR, Friend D, Thoma B, Cosman D, Park L, Mosley B. Proliferative responses and binding properties of hematopoietic cells transfected with low-affinity receptors for leukemia inhibitory factor, oncostatin M, and ciliary neurotrophic factor. Proc Natl Acad Sci USA 1994; 91:1119–1123.

51. Gearing DP, Comeau MR, Friend DJ, Gimpel SD, Thut CJ, McGourty J, Brasher KK, King JA, Gillis S, Mosley B, Ziegler SF, Cosman D. The IL-6 signal transducer, gp130: an oncostatin M receptor and affinity converter for the LIF receptor. Science 1992; 255:1434–7.

52. Gearing DP, Bruce AG. Oncostatin M binds the high-affinity leukemia inhibitory factor receptor. New Biol 1992; 4:61–5.

53. Davis D, Aldrich TH, Stahl N, Pan L, Taga T, Kishimoto T, Ip NY, Yancopoulos GD. LIFRβ and gp130 as heterodimerizing signal transducers of the tripartite CNTF receptor. Science 1993; 260:1805–1808.

54. Stahl N, Yancopoulos GD. The alphas, betas and kinases of cytokine receptor complexes. Cell 1993; 74:587–590.

55. Chua AO, Chizzonite R, Desai BB, et al. Expression cloning of human IL-12 receptor component. A new member of the cytokine receptor superfamily with strong homology to gp130. J. Immunol 1994; 153:128–36.

56. Miles SA, Martinez-Maza O, Rezai A, Magpantay L, Kishimoto T, Nakamura S, Radka SF, Linsley PS. Oncostatin M as a portent mitogen for AIDS–Kaposi's sarcoma-derived cells. Science 1992; 55:1432–4.

57. Nair BC, DeVico AL, Nakamura S, Copeland TD, Chen Y, Patel A, O'Neil T, Oroszlan S, Gallo RC, Sarngadharan MG. Identification of a major growth factor for AIDS–Kaposi's sarcoma cells as oncostatin M. Science 1992; 255:1430–1432.

58. Palca J. Kaposi's sarcoma gives on key fronts. Science 1992; 255:1352–4.

59. Yarden Y, Escobedo JA, Kuang W-J, et al. Structure of the receptor for platelet-derived growth factor helps define a family of closely related growth factor receptors. Nature 1986; 323:226–32.

60. de Aberle SB. A study of the hereditary anaemia of mice. Am J Anat 1927; 40:219–247.

61. Sarvella PA, Russell ES. *Steel*, a new dominant gene in the house mouse. J Hered 1956; 47:123–128.

62. Bennett D. Developmental analysis of a mutation with pleiotropic effects in the mouse. J Morphol 1956; 98:199–234.

63. Silvers WK. The Coat Colours of Mice: A Model for Gene Action and Interaction. New York: Springer-Verlag 1979:206–241.

64. Russell ES, Hereditary anemias of the mouse: a review for geneticists. Adv Genet 1979; 20:357–459.

65. Galli SJ, Zsebo KM, Geissler EN. The *kit* ligand, stem cell factor. Adv Immunol 1994; 55:1–96.

66. Besmer P, Murphy JE, George PC, Qiu FH, Bergold PJ, Lederman L, Snyder H Jr, Brodeur D, Zucherman EE, Hardy WD. A new acute transforming feline retrovirus and relationship of its oncogene v-*kit* with the protein kinase gene family. Nature 1986; 320:415–21.

67. Yarden Y, Kuang WJ, Yang-Feng T, Coussens L, Munemitsu S, Dull TJ, Chen E, Schlessinger J, Francke U, Ullrich A. Human proto-oncogene c-*kit*: a new cell surface receptor tyrosine kinase for an unidentified ligand. EMBO J 1987; 6:3341–51.

68. Qiu FH, Ray P, Brown K, Barker PE, Jhanwar S, Ruddle FH, Besmer P. Primary structure of c-*kit*: relationship with the CSF-1/PDGF receptor kinase family—oncogenic activation of v-*kit* involves deletion of extracellular domain and C terminus. EMBO J 1988; 7:1003–11.

69. Chabot B, Stephenson DA, Chapman VM, Besmer P, Bernstein A. The protooncogene c-*kit* encoding a transmembrane tyrosine kinase receptor maps to the mouse *W* locus. Nature 1988; 335:88–9.

70. Geissler EN, Cheng SV, Gusella JF, Housman DE. Genetic analysis of the dominant white-spotting (*W*) region on mouse chromosome 5: identification of cloned DNA markers near *W*. Proc Natl Acad Sci USA 1988; 85:9635–9.

71. Geissler EN, Ryan MA, Housman DE. The dominant-white spotting (*W*) locus of the mouse encodes the c-*kit* proto-oncogene. Cell 1988; 55:185–92.

72. Anderson DM, Lyman SD, Baird A, et al. Molecular cloning of mast cell growth factor, a hematopoietin that is active in both membrane bound and soluble forms. Cell 1990; 63:235–43.

73. Flanagan JG, Leder P. The *kit* ligand: a cell surface molecule altered in steel mutant fibroblasts. Cell 1990; 63:185–94.

74. Williams DE, Eisenman J, Baird A, et al. Identification of a ligand for the c-*kit* protooncogene. Cell 1990; 63:167–74.

75. Zsebo KM, Williams DA, Geissler EN, et al. Stem cell factor is encoded at the *Sl* locus of the mouse and is the ligand for the c-*kit* tyrosine kinase receptor. Cell 1990; 63:213–24.

76. Nocka K, Buck J, Levi E, Besmer P. Candidate ligand for the c-*kit* transmembrane kinase receptor: KL, a fibroblast derived growth factor stimulates mast cells and erythroid progenitors. EMBO J 1990; 9:3287–94.

77. Martin FH, Suggs SV, Langley KE et al. Primary structure and functional expression of rat and human stem cell factor DNAs. Cell 1990; 63:203–11.

78. Zsebo KM, Wypych J, McNiece IK, et al. Identification, purification, and biological characterization of hematopoietic stem cell factor from buffalo rat liver—conditioned medium. Cell 1990; 63:195–201.

79. Copeland NG, Gilbert DJ, Cho BC, Donovan PJ, Jenkins NA, Cosman D, Anderson D, Lyman SD, Williams DE. Mast cell growth factor maps near the steel locus on mouse chromosome 10 and is deleted in a number of steel alleles. Cell 1990; 63:175–83.

80. Flanagan JG, Chan DC, Leder P. Transmembrane form of the *kit* ligand growth factor is determined by alternative splicing and is missing in the Sld mutant. Cell 1991; 64:1025–35.

81. Williams DE, de Vries P, Namen AE, Widmer MB, Lyman SD. The steel factor. Dev Biol 1992; 151:368–76.

82. Hampe A, Gobet M, Sherr CJ, Galibert F. Nucleotide sequence of the feline retroviral oncogene v-*fms* shows unexpected homology with oncogenes encoding tyrosine-specific protein kinases. Proc Natl Acad Sci USA 1984; 81:85–9.

83. Sherr CJ, Rettenmier R, Sacca R, Roussel MF, Look AT, Stanley ER. The c-*fms* proto-oncogene product is related to the receptor for the mononuclear phagocyte growth factor, CSF-1. Cell 1985; 41:665–76.

84. Cerretti DP, Wignall J, Anderson D, Tushinski RJ, Gallis BM, Stya M, Gillis S, Urdal DL, Cosman D. Human macrophage-colony stimulating factor: alternative RNA and protein processing from a single gene. Mol Immunol 1988; 25:761–70.

85. Ladner MB, Martin GA, Noble JA, Nikoloff DM, Tal R, Kawasaki ES, White TJ. Human CSF-1: gene structure and alternative splicing of mRNA precursors. EMBO J 1987; 6:2693–8.

86. Rettenmier CW, Roussel MF. Differential processing of colony-stimulating factor 1 precursors encoded by two human cDNAs. Mol Cell Biol 1988; 8:5026–34.

87. Stein J, Rettenmier CW. Proteolytic processing of a plasma membrane-bound precursor to human macrophage colony-stimulating factor (CSF-1) is accelerated by phorbol ester. Oncogene 1991; 6601–5.

88. Price LK, Choi HU, Rosenberg L, Stanley ER. The predominant form of secreted colony stimulating factor-1 is a proteoglycan. J Biol Chem 1992; 267:2190–9.

89. Stanley ER. Colony-stimulating factor-1 (CSF-1). In: Nicola NA, ed. Guidbook to Cytokines and Their Receptors, Oxford: Oxford University Press, 1994:164–167.

90. Matthews W, Jordan CT, Wiegand GW, Pardoll D, Lemischka IR. A receptor tyrosine kinase specific to hematopoietic stem and progenitor cell-enriched populations. Cell 1991; 65:1143–52.

91. Rosnet O, Marchetto S, deLapeyriere O, Birnbaum D. Murine *Flt3*, a gene encoding a novel tyrosine kinase receptor of the PDGFR/CSF1R family. Oncogene 1991; 6:1641–50.

92. Lyman SD, James L, Vanden Bos T, de Vries P, Brasel K, Gliniak B, Hollingsworth LT, Picha KS, McKenna HJ, Splett RR, Fletcher FA, Maraskovsky E, Farrah T, Foxworthe D, Williams DE, Beckmann MP. Molecular cloning of a ligand fro the flt3/flk-2 tyrosine kinase receptor: a proliferative factor for primitive hemopoietic cells. Cell 1993; 75:1157–1167.

93. Hannum C, Culpepper J, Campbell D, McClanahan T, Zurawski S, Bazan JF, Kastelein R, Hudak S,

Wagner J, Mattson J, Luh J, Duda G, Martina N, Peterson D, Menon S, Shanafelt A, Muench M, Kelner G, Namikawa R, Rennick D, Roncarolo M-G, Zlotnick A, Rosnet O, Dubreuil P, Birnbaum D, Lee F. Ligand for FLT3/FLK2 receptor tyrosine kinase regulates growth of haematopoietic stem cells and is encoded by variant RNAs. Nature 1994; 368:643–8.

94. Nocka K, Tan JC, Chiu E, Chu TY, Ray P, Traktman P, Besmer P. Molecular bases of dominant negative and loss of function mutations at the murine c-*kit*/white spotting locu:*W37*, *Wv*, *W41* and *W*. EMBO J 1990; 9:1805–13.

95. Reith AD, Ellis C, Maroc N, Pawson T, Bernstein A, Dubreuil P. 'W' mutant forms of the *Fms* receptor tyrosine kinase act in a dominant manner to suppres CSF-1 dependent cellular transformation. Oncogene 1993; 8:45–53.

96. Li W, Stanley ER. Role of dimerization and modification of the CSF-1 receptor in its activation and internalization during the CSF-1 response. EMBO J 1991; 10:277–88.

97. Lev S, Yarden Y, Givol D. A recombinant ectodomain of the receptor for the stem cell factor (SCF) retains ligand-induced receptor dimerization and antagonizes SCF-stimulated cellular responses. J Biol Chem 1992; 267:10866–73.

98. Mott HR, Driscoll PC, Boyd J, Cooke RM, Weir MP, Campbell ID. Secondary structure of human interleukin 2 from 3D heteronuclear NMR experiments. Biochemistry 1992; 31:7741–4.

99. Senda T, Shimazu T, Matsuda S, Kawano G, Shimizu H, Nakamura KT, Mitsui Y. Three-dimensional crystal structure of recombinant murine interferon-beta. EMBO J 1992; 11:3193–201.

100. Ealick SW, Cook WJ, Vijay-Kumar S, Carson M, Nagabhushan TL, Trotta PP, Bugg CE. Three-dimensional structure of recombinant human interferon-gamma. Science 1991; 252:698–702.

101. Pandit J, Bohm A, Jancarik J, Halenbeck R, Koths K, Kim SH. Three-dimensional structure of dimeric human recombinant macrophage colony-stimulating factor. Science 1992; 258:1358–62.

102. Hill CP, Osslund TD, Eisenberg D. The structure of granulocyte-colony-stimulating factor and its relationship to other growth factors. Proc Natl Acad Sci USA 1993; 90:5167–71.

103. de Vos AM, Ultsch M, Kossiakoff AA. Human growth hormone and extracellular domain of its receptor: crystal structure of the complex. Science 1992; 255:306–12.

104. Powers R, Garrett DS, March CJ, Frieden EA, Gronenborn AM, Clore GM. Three-dimensional solution structure of human interleukin-4 by multidimensional heteronuclear magnetic resonance spectroscopy. Science 1992; 256:1673–7.

105. Walter MR, Cook WJ, Ealick SE, Nagabhushan TL, Trotta PP, Bugg CE. Three-dimensional structure of recombinant human granulocyte-macrophage colony-stimulating factor. J Mol Biol. 1992; 224:1075–85.

106. Cunningham BC, Ultsch M, De VA, Mulkerrin MG, Clauser KR, Wells JA. Dimerization of the extracellular domain of the human growth hormone receptor by a single hormone molecule. Science 1991; 254:821–5.

107. McKenzie AN, Barry SC, Strath M, Sanderson CJ. Structure–function analysis of interleukin-5 utilizing mouse/human chimeric molecules. EMBO J 1991; 10:1193–9.

108. Lokker NA, Zenke G, Strittmatter U, Fagg B, Movva NR. Structure–activity relationship study of human interleukin-3: role of the C-terminal region for biological activity. EMBO J 1991; 10:2125–31.

109. Kastelein RA, Shanafelt AB. GM-CSF receptor: interactions and activation. Oncogene 1993; 8:231–6.

110. Shanafelt AB, Miyajima A, Kitamura T, Kastelein RA. The amino-terminal helix of GM-CSF and IL-5 governs high affinity binding to their receptors. EMBO J 1991; 10:4105–112.

111. Lock P, Metcalf D, Nicola NA. Histidine 367 of the human common beta chain of the receptor is critical for high affinity binding of human granulocyte macrophage colony stimulating factor. Proc Natl Acad Sci USA 1994; 91:252–256.

112. Zurawski SM, Zurawski G. Mouse interleukin-2 structure–function studies: substitutions in the first alpha-helix can specifically inactivate p70 receptor binding and mutations in the fifth alpha-helix can specifically inactivate p55 receptor binding. EMBO J 1989; 8:2583–90.

113. Imler J-L, Miyajima A, Zurawski G. Identification of three adjacent amino acids of interleukin-2 receptor β chain which control the affinity and the specificity of the interaction with interleukin-2. EMBO J 1992; 11:2047–53.

114. Ullrich A, Schlessinger J. Signal transduction by receptors with tyrosine kinase activity. Cell 1990; 61:203–12.

115. Reedijk M, Liu XQ, Pawson T. Interactions of phosphatidylinositol kinase, GTPase-activating protein (GAP), and GAP-associated proteins with the colony-stimulating factor 1 receptor. Mol Cell Biol 1990; 10:5601–8.

116. Courtneidge SA, Dhand R, Pilat D, Twamley GM, Waterfield MD, Roussel MF. Activation of Src family kinases by colony stimulating factor-1, and their association with its receptor. EMBO J 1993; 12:943–50.

117. van der Geer P, Hunter T. Mutation of Tyr697, a GRB2-binding site, and Tyr721, a PI 3-kinase binding site, abrogates signal transduction by the murine CSF-1 receptor expressed in Rat-2 fibroblasts. EMBO J 1993; 12:5161–72.

118. Reith AD, Ellis C, Lyman SD, Anderson DM, Williams DE, Bernstein A, Pawson T. Signal transduction by normal isoforms and W mutant variants of the Kit receptor tyrosine kinase. EMBO J 1991; 10:2451–9.

119. Serve H, Hsu YC, Besmer P. Tyrosine residue 719 of the c-*kit* receptor is essential for binding of the P85 subunit of phosphatidylinositol (PI) 3-kinase and for c-*kit*-associated PI 3-kinase activity in COS-1 cells. J Biol Chem 1994; 269:6026–30.

120. Yi T, Ihle JN. Association of hematopoietic cell phosphatase with c-*kit* after stimulation with c-*kit* ligand. Mol Cell Biol 1993; 13:3350–8.

121. McGlade CJ, Ellis C, Reedijk M, et al. SH2 domains of the p85 alpha subunit of phosphatidylinositol 3-kinase regulate binding to growth factor receptors. Mol Cell Biol 1992; 12:991–7.

122. Lev S, Givol D, Yarden Y. Interkinase domain of *kit* contains the binding site for phosphatidylinositol 3′ kinase. Proc Natl Acad Sci USA 1992; 89:678–82.

123. Rottapel R, Turck CW, Casteran N, Liu X, Birnbaum D, Pawson T, Dubreuil P. Substrate specificities and identification of a putative binding site for PI3K in the carboxy tail of the murine Flt3 receptor tyrosine kinase. Oncogene 1994; 9:1755–65.

124. Schlessinger J. SH2/SH3 signaling proteins. Curr Opin Genet Dev 1994; 4:25–30.

125. Lowenstein EJ, Daly RJ, Batzer AG, Li W, Margolis B, Lammers R, Ullrich A, Skolnik EY, Bar-Sagi D, Schlessinger J. The SH2 and SH3 domain-containing protein GRB2 links receptor tyrosine kinases to *ras* signaling. Cell 1992; 70:431–42.

126. Gale NW, Kaplan S, Lowenstein EJ, Schlessinger J, Bar-Sagi D. Grb2 mediates the EGF-dependent activation of guanine nucleotide exchange on Ras. Nature 1993; 363:88–92.

127. Egan SE, Giddings BW, Brooks MW, Buday L, Sizeland AM, Weinberg RA. Association of Sos Ras exchange protein with Grb2 is implicated in tyrosine kinase signal transduction and transformation. Nature 1993; 363:45–51.

128. Buday L, Egan SE, Rodriguez Viciana P, Cantrell DA, Downward J. A complex of Grb2 adaptor protein, Sos exchange factor, and a 36-kDa membrane-bound tyrosine phosphoprotein is implicated in *ras* activation in T cells. J Biol Chem 1994; 269:9019–23.

129. Olivier JP, Raabe T, Henkemeyer M, Dickson B, Mbamalu G, Margolis B, Schlessinger J, Hafen E, Pawson T. A *Drosophila* SH2-SH3 adaptor protein implicated in coupling the sevenless tyrosine kinase to an activator of Ras guanine nucleotide exchange, Sos. Cell 1993; 73:179–91.

130. Rozakis-Adcock M, McGlade J, Mbamalu G, et al. Association of the Shc and Grb2/Sem5 SH2-containing proteins is implicated in activation of the Ras pathway by tyrosine kinases. Nature 1992; 360:689–92.

131. Li N, Batzer A, Daly R, Yajnik V, Skolnik E, Chardin P, Bar-Sagi D, Margolis B, Schlessinger J. Guanine-nucleotide-releasing factor hSos1 binds to Grb2 and links receptor tyrosine kinases to Ras signalling. Nature 1993; 363:85–8.

132. Gale NW, Kaplan S, Lowenstein EJ, Schlessinger J, Bar-Sagi D. Grb2 mediates the EGF-dependent activation of guanine nucleotide exchange on Ras. Nature 1993; 363:88–92.

133. Chardin P, Camonis JH, Gale NW, van-Aelst L, Schlessinger J, Wigler MH, Bar-Sagi D. Human Sos1: a guanine nucleotide exchange factor for Ras that binds to GRB2. Science 1993; 260:1338–43.

134. McCormick F. Signal transduction. How receptors turn Ras on. Nature 1993; 363:15–6.

135. Buday L, Downward J. Epidermal growth factor regulates p21ras through the formation of a complex of receptor, Grb2 adapter protein, and Sos nucleotide exchange factor. Cell 1993; 73:611–20.

136. Egan SE, Giddings BW, Brooks MW, Buday L, Sizeland AM, Weinberg RA. Association of Sos Ras exchange protein with Grb2 is implicated in tyrosine kinase signal transduction and transformation. Nature 1993; 363:45–51.

137. Pelech SL, Sanghera JS. MAP kinases: charting the regulatory pathways. Science 1992; 257:1355–6.

138. Blenis J. Signal transduction via the MAP kinases: proceed at your own RSK. Proc Natl Acad Sci USA 1993; 90:5889–92.

139. Stanley IJ, Nicola NA, Burgess AW. Growth factor-induced phosphorylation of c-*ras* p21 in normal hemopoietic progenitor cells. Growth Factors 1989; 2:53–9.

140. Ferris DK, Willet-Brown J, Martensen T, Farrar WL. Interleukin 3 stimulation of tyrosine kinase activity in FDC-P1 cells. Biochem Biophys Res Commun 1988; 154:991–6.

141. Reed JC, Prystowsky MB, Nowell PC. Regulation of gene expression in lectin-stimulated or lymphokine-stimulated T lymphocytes. Effects of cyclosporine. Transplantation 1988; 46:85S–89S.

142. Saltzman EM, Thom RR, Casnellie JE. Activation of a tyrosine protein kinase is an early event in the stimulation of T lymphocytes by interleukin-2. J Biol Chem 1988; 263:6956–9.

143. Morla AO, Schreurs J, Miyajima A, Wang JY. Hematopoietic growth factors activate the tyrosine phosphorylation of distinct sets of proteins in interleukin-3-dependent murine cell lines. Mol Cell Biol 1988; 8:2214–8.

144. Isfort R, Huhn RD, Frackelton A Jr, Ihle JN. Stimulation of factor-dependent myeloid cell lines with interleukin 3 induces tyrosine phosphorylation of several cellular substrates. J Biol Chem 1988; 263:19203–9.

145. Sorensen PH, Mui AL, Murthy SC, Krystal G. Interleukin-3, GM-CSF, and TPA induce distinct phosphorylation events in an interleukin 3-dependent multipotential cell line. Blood 1989; 73:406–18.

146. Sorensen P, Mui AL, Krystal G. Interleukin-3 stimulates the tyrosine phosphorylation of the 140-kilodalton interleukin-3 receptor. J Biol Chem 1989; 264:19253–8.

147. Firmbach-Kraft I, Byers M, Shows T, Dalla-Favera R, Krolewski JJ. *tyk2*, prototype of a novel class of non-receptor tyrosine kinase genes. Oncogene 1990; 5:1329–36.

148. Wilks AF. Two putative protein-tyrosine kinases identified by application of the polymerase chain reaction. Proc Natl Acad Sci USA 1989; 86:1603–7.

149. Harpur AG, Andres AC, Ziemiecki A, Aston RR, Wilks AF. JAK2, a third member of the JAK family of protein tyrosine kinases. Oncogene 1992; 7:1347–53.

150. Johnston JA, Kawamura M, Kirken RA, Chen YQ, Blake TB, Shibuya K, Ortaldo JR, McVicar DW, O'Shea JJ. Phosphorylation and activation of the Jak-3 Janus kinase in reponse to interleukin-2. Nature 1994; 370:151–153.

151. McKendry R, John J, Flavell D, Muller M, Kerr IM, Stark GR. High-frequency mutagenesis of human cells and characterization of a mutant unresponsive to both alpha and gamma interferons. Proc Natl Acad Sci USA 1991; 88:11455–9.

152. Pellegrini S, John J, Shearer M, Kerr IM, Stark GR. Use of a selectable marker regulated by alpha interferon to obtain mutations in the signaling pathway. Mol Cell Biol 1989; 9:4605–12.

153. Watling D, Guschin D, Muller M, Silvennoinen O, Witthuhn BA, Quelle FW, Rogers NC, Schindler C, Stark GR, Ihle JN, Kerr IM. Complementation by the protein tyrosine kinase JAK2 of a mutant cell line defective in the interferon-γ signal transduction pathway. Nature 1993; 366:166–170.

154. Muller M, Briscoe J, Laxton C, Guschin D, Ziemiecki A, Silvennoinen O, Harpur AG, Barbieri G, Witthuhn BA, Schindler C, Pellegrini S, Wilks AF, Ihle JN, Stark GR, Kerr IM. The protein tyrosine

kinase JAK1 complements defects in interferon-α/β and -γ signal transduction. Nature 1993; 366:129–135.

155. Velazquez L, Fellous M, Stark GR, Pellegrini S. A protein tyrosine kinase in the interferon alpha/beta signaling pathway. Cell 1992; 70:313–22.

156. Darnell J Jr, Kerr IM, Stark GR. Jak-STAT pathways and transcriptional activation in response to IFNs and other extracellular signaling proteins. Science 1994; 264:1415–21.

157. Ihle JN, Witthuhn BA, Quelle FW, Yamamoto K, Thierfelder WE, Kreider B, Silvennoinen O. Signaling by the cytokine receptor superfamily: JAKs and STATs. Trends Biochem Sci 1994; 19:222–227.

158. Ullrich A, Schlessinger J. Signal transduction by receptors with tyrosine kinase activity. Cell 1990; 61:203–12.

159. Novick D, Cohen B, Rubinstein M. The human interferon alpha/beta receptor: characterization and molecular cloning. Cell 1994; 77:391–400.

160. Witthuhn BA, Quelle FW, Silvennoinen O, Yi T, Tang B, Miura O, Ihle JN. JAK2 associates with the erythropoietin receptor and is tyrosine phosphorylated and activated following stimulation with erythropoietin. Cell 1993; 74:227–36.

161. Rui H, Kirken RA, Farrar WL. Activation of receptor-associated tyrosine kinase JAK2 by prolactin. J Biol Chem 1994; 269:5364–8.

162. Dusanter-Fourt I, Muller O, Ziemiecki A, Mayeux P, Drucker B, Djiane J, Wilks A, Harpur AG, Fischer S, Gisselbrecht S. Identification of JAK protein tyrosine kinases as signaling molecules for prolactin. Functional analysis of prolactin receptor and prolactin-erythropoietin receptor chimera expressed in lymphoid cells. EMBO J 1994; 13:2583–91.

163. Nicholson SE, Oates AC, Harpur AG, Ziemiecki A, Wilks AF, Layton JE. Tyrosine kinase jak1 is associated with the granulocyte colony stimulating factor receptor and both become tyrosine phosphorylated after receptor activation. Proc Natl Acad Sci USA 1994; 91:2985–2988.

164. Witthuhn BA, Silvennoinen O, Miura O, Lai KS, Cwik C, Liu ET, Ihle JN. Involvement of the Jak-3 Janus kinase in signalling by interleukins 2 and 4 in lymphoid and myeloid cells. Nature 1994; 370:153–157.

165. Lutticken C, Wegenka UM, Yuan J, et al. Association of transcription factor APRF and protein kinase Jak1 with the interleukin-6 signal transducer gp130. Science 1994; 263:89–92.

166. Stahl N, Boulton TG, Farruggella T, Ip NY, Davis S, Witthuhn BA, Quelle FW, Silennionen O, Barbieri G, Pelligrini S, Ihle JN, Yancopoulos GD. Association and activation of Jak/Tyk kinases by CNTF-LIF-OSM-IL-6 β receptor components. Science 1994; 263:92–94.

167. Kessler DS, Veals SA, Fu X-Y, Levy DE. IFN-alpha regulates nuclear transcription and DNA binding affinity of ISGF3, a mulimeric transcriptional activator. Gene Dev 1990; 4:1753–1765.

168. Kessler DS, Levy DE, Darnell J Jr. Two interferon-induced nuclear factors bind a single promoter element in interferon-stimulated genes. Proc Natl Acad Sci USA 1988; 85:8521–5.

169. Kelly JM, Porter AC, Chernajovsky Y, Gilbert CS, Stark GR, Kerr IM. Characterization of a human gene inducible by alpha- and beta-interferons and its expression in mouse cells. EMBO J 1986; 5:1601–6.

170. Porter AC, Chernajovsky Y, Dale TC, Gilbert CS, Stark GR, Kerr IM. Interferon response element of the human gene 6-16. EMBO J 1988; 7:85–92.

171. Levy D, Reich N, Kessler D, Pine R, Darnell J Jr. Transcriptional regulation of interferon-stimulated genes: a DNA response element and induced proteins that recognize it. Cold Spring Harbor Symp Quan Biol 1988; 2:799–802.

172. Levy DE, Kessler DS, Pine R, Reich N, Darnell J Jr. Interferon-induced nuclear factors that bind a shared promoter element correlate with positive and negative transcriptional control. Genes Dev 188; 2:383–93.

173. Levy DE, Kessler DS, Pine R, Darnell J Jr. Cytoplasmic activation of ISGF3, the positive regulator of interferon-alpha-stimulated transcription, reconstituted in vitro. Genes Dev 1989; 3:1362–71.

174. Schindler C, Fu XY, Improta T, Aebersold R, Jarnell J Jr. Proteins of transcription factor ISGF-3:

one gene encodes the 91-and 84-kDa ISGF-3 proteins that are activated by interferon alpha. Proc Natl Acad Sci USA 1992; 89:7836–9.

175. Schindler C, Shuai K, Prezioso VR, Darnell J Jr. Interferon-dependent tyrosine phosphorylation of a latent cytoplasmic transcription factor. Science 1992; 257:809–13.

176. Fu X-Y, Kessler DS, Veals SA, Levy DE, Darnell JE. ISGF3, the transcriptional activator induced by interferon-α, consists of multiple interacting polypeptides. Proc Natl Acad Sci USA 1990; 87:8555–8559.

177. Fu XY, Schindler C, Improta T, Aebersold R, Darnell J Jr. The proteins of ISGF-3, the interferon alpha-induced transcriptional activator, define a gene family involved in signal transduction. Proc Natl Acad Sci USA 1992; 89:7840–3.

178. Fu XY. A transcription factor with SH2 and SH3 domains is directly activated by an interferon alpha-induced cytoplasmic protein tyrosine kinase(s). Cell 1992; 70:323–35.

179. Decker T, Lew DJ, Mirkovitch J, Darnell J Jr. Cytoplasmic activation of GAF, an IFN-gamma-regulated DNA-binding factor. EMBO J 1991; 10:927–32.

180. Decker T, Lew DJ, Darnell J Jr. Two distinct alpha-interferon-dependent signal transduction pathways may contribute to activation of transcription of the guanylate-binding protein gene. Mol Cell Biol 1991; 11:5147–53.

181. Shuai K, Schindler C, Prezioso VR, Darnell J Jr. Activation of transcription by IFN-gamma: tyrosine phosphorylation of a 91-kD DNA binding protein. Science 1992; 258:1808–12.

182. Fu XY, Zhang JJ. Transcription factor p91 interacts with the epidermal growth factor receptor and mediates activation of the c-*fos* gene promoter. Cell 1993; 74:1135–45.

183. Silvennoinen O, Schindler C, Schlessinger J, Levy DE. Ras-independent growth factor signaling by transcription factor tyrosine phosphorylation. Science 1993; 261:1736–9.

184. Sadowski HB, Shuai K, Darnell J Jr, Gilman MZ. A common nuclear signal transduction pathway activated by growth factor and cytokine receptors. Science 1993; 261:1739–44.

185. Sadowski HB, Gilman MZ. Cell-free activation of a DNA-binding protein by epidermal growth factor. Nature 1993; 362:79–83.

186. Zhong Z, Wen Z, Darnell JE. Stat3: a STAT family member activated by tyrosine phosphorylation in response to epidermal growth factor and interleukin-6. Science 1994; 264:95–98.

187. Zhong Z, Wen Z, Darnell J Jr. Stat3 and Stat4: members of the family of signal transducers and activators of transcription. Proc Natl Acad Sci USA 1994; 91:4806–10.

188. Wegenka UM, Lutticken C, Buschmann J, Yuan J, Lottspeich F, Muller-Esterl W, Schindler C, Roeb E, Heinrich PC, Horn F. The interleukin-6-activated acute-phase response factor is antigenically and functionally related to members of the signal transducer and activator of transcription (STAT) family. Mol Cell Biol 1994; 14:3186–96.

189. Akira S, Nishio Y, Inoue M, Wang XJ, Wei S, Matsusaka T, Yoshida K, Sudo T, Naruto M, Kishimoto T. Molecular cloning of APRF, a novel IFN-stimulated gene factor 3 p91-related transcription factor involved in the gp130-mediated signaling pathway. Cell 1994; 77:63–71.

190. Hou J, Schindler U, Henzel WJ, Ho TC, Brasser M, McKnight SL. An interleukin-4-induced transcription factor: IL-4 Stat. Science, 1994; 265:1701–1706.

191. Wakao H, Gouilleux F, Groner B. Mammary gland factor (MGF) is a novel member of the cytokine regulated transcription factor gene family and confers the prolactin response. EMBO J 1994; 13:2182–91.

192. Billestrup N, Allevato G, Norstedt G, Moldrup A, Nielsen JN. Identification of intracellular domains in the growth hormone receptor involved in signal transduction. Proc Soc Exp Biol Med 1994; 206:205–9.

193. Mui AL, Miyajima A. Interleukin-3 and granulocyte–macrophage colony-stimulating factor receptor signal transduction. Proc Soc Exp Biol Med 1994; 206:284–8.

194. Narazaki M, Witthuhn BA, Yoshida K, Silvennoinen O, Yasukawa K, Ihle JN, Kishimoto T, Taga T. Activation of JAK2 kinase mediated by the interleukin 6 signal transducer gp130. Proc Natl Acad Sci USA 1994; 91:2285–9.

195. Seldin DC, Leder P. Mutational analysis of a critical signaling domain of the human interleukin 4 receptor. Proc Natl Acad Sci USA 1994; 91:2140–4.

196. Goujon L, Allevato G, Simonin G, et al. Cytoplasmic sequences of the growth hormone receptor necessary for signal transduction. Proc Natl Acad Sci USA 1994; 91:957–61.

197. Sato N, Sakamaki K, Terada N, Arai K, Miyajima A. Signal transduction by the high-affinity GM-CSF receptor: two distinct cytoplasmic regions of the common beta subunit responsible for different signaling. EMBO J 1993; 12:4181–9.

198. Harada N, Yang G, Miyajima A, Howard M. Identification of an essential region for growth signal transduction in the cytoplasmic domain of the human interleukin-4 receptor. J Biol Chem 1992; 267:22752–8.

199. Baumann H, Symes AJ, Comeau MR, Morella KK, Wang Y, Friend D, Ziegler SF, Fink JS, Gearing DP. Multiple regions within the cytoplasmic domains of the leukemia inhibitory factor receptor and gp130 cooperate in signal transduction in hepatic and neuronal cells. Mol Cell Biol 1994; 14:138–46.

200. Goldsmith MA, Xu W, Amaral MC, Kuczek ES, Greene WC. The cytoplasmic domain of the interleukin-2 receptor beta chain contains both unique and functionally redundant signal transduction elements. J Biol Chem 1994; 269:14698–704.

201. Baumann H, Gearing D, Ziegler SF. Signaling by the cytoplasmic domain of hematopoietin receptors involves two distinguishable mechanisms in hepatic cells. J Biol Chem 1994; 269:16297–304.

202. He TC, Jiang N, Zhuang H, Quelle DE, Wojchowski DM. The extended box 2 subdomain of erythropoietin receptor is nonessential for Jak2 activation yet critical for efficient mitogenesis in FDC-ER cells. J Biol Chem 1994; 269:18291–4.

203. DaSilva L, Howard OM, Rui H, Kirken RA, Farrar WL. Growth signaling and JAK2 association mediated by membrane-proximal cytoplasmic regions of prolactin receptors. J Biol Chem 1994; 269:18267–70.

204. Sakamaki K, Miyajima I, Kitamura T, Miyajima A. Critical cytoplasmic domains of the common beta subunit of the human GM-CSF, IL-3 and IL-5 receptors for growth signal transduction and tyrosine phosphorylation. EMBO J 1992; 11:3541–9.

205. Hanazono Y, Chiba S, Sasaki K, Mano H, Miyajima A, Arai K, Yazaki Y, Hirai H. c-*fps/fes* protein-tyrosine kinase is implicated in a signaling pathway triggered by granulocyte–macrophage colony-stimulating factor and interleukin-3. EMBO J 1993; 12:1641–6.

206. Sato N, Sakamaki K, Terada N, Arai K, Miyajima A. Signal transduction by the high-affinity GM-CSF receptor: two distinct cytoplasmic regions of the common beta subunit responsible for different signaling. EMBO J 1993; 12:4181–9.

207. Watanabe S, Muto A, Yokota T, Miyajima A, Arai K. Differential regulation of early response genes and cell proliferation through the human granulocyte macrophage colony-stimulating factor receptor: selective activation of the c-*fos* promoter by genistein. Mol Biol Cell 1993; 4:983–92.

208. Baumann H, Gearing D, Ziegler SF. Signaling by the cytoplasmic domain of hematopoietin receptors involves two distinguishable mechanisms in hepatic cells. J Biol Chem 1994; 269:16297–304.

209. Touw I, Vanpaassen M, Schelen A, Hoefsloot L, Lowenberg B, Dong F. Distinct cytoplasmic domains of the human G-CSF receptor determine proliferation and maturation signaling—implications for the development of neutropenia and acute myeloid leukemia. Blood 1993; 82 (suppl 1):185a.

210. Fukunaga R, Ishizaka-Ikeda E, Nagata S. Growth and differentiation signals mediated by different regions in the cytoplasmic domain of granulocyte colony-stimulating factor receptor. Cell 1993; 74:1079–87.

211. Bashey A, Healy L, Marshall CJ. Proliferative but not nonproliferative responses to granulocyte colony-stimulating factor are associated with rapid activation of the *p21ras*/MAP kinase signalling pathway. Blood 1994; 83:949–57.

212. Dong F, Hoefsloot LH, Schelen AM, Broeders CA, Meijer Y, Veerman AJ, Touw IP, Lowenberg B. Identification of a nonsense mutation in the granulocyte-colony-stimulating factor receptor in severe congenital neutropenia. Proc Natl Acad Sci USA 1994; 91:4480–4.

213. Ishizaka-Ikeda E, Fukunaga R, Wood WI, Goeddel DV, Nagata S. Signal transduction mediated by

growth hormone receptor and its chimeric molecules with the granulocyte colony-stimulating factor receptor. Proc Natl Acad Sci USA 1993; 90:123–7.

214. Sandoval C, Adamsgraves P, Parganas E, Wang W, Ihle JN. The cytoplasmic portion of the g *csf* receptor is normal in patients with Kostmann syndrome. Blood 1993; 82 (suppl 1):185a.

215. D'Andrea AD, Yoshimura A, Youssoufian H, Zon LI, Koo JW, Lodish HF. The cytoplasmic region of the erythropoietin receptor contains nonoverlapping positive and negative growth-regulatory domains. Mol Cell Biol 1991; 11:1980–7.

216. de-la-Chapelle A, Traskelin AL, Juvonen E. Truncated erythropoietin receptor causes dominantly inherited benign human erythrocytosis. Proc Natl Acad Sci USA 1993; 90:4495–9.

217. Hilton DJ, Zhang JG, Metcalf D, Alexander WS, Nicola NA, Willson TA. Cloning and characterization of a binding subunit of the interleukin 13 receptor that is also a component of the interleukin 4 receptor. Proc Nat Acad Sci USA 1996; 93:497–501.

4

Regulation of the Cell Cycle by Kinases and Cyclins

Arthur B. Pardee
Dana-Farber Cancer Institute, Boston, Massachusetts

Khandan Keyomarsi
Laboratory of Diagnostic Oncology, Wadsworth Center, Albany, New York

Qing Ping Dou
University of Pittsburgh School of Medicine, Pittsburgh, Pennsylvania

I. INTRODUCTION

The purpose of this chapter is to provide current information on molecular events that follow stimulation of mammalian cells by extracellular agents, such as colony-stimulating factors, growth factors, and cytokines. It is in three parts. First is an overview of the biology of cell proliferation, for which the cell cycle provides a framework. Second is a discussion of biochemistry and molecular biology of these processes, in which the roles of cyclins, cdk kinases, and associated molecules in transcriptional controls are emphasized. Finally, important modifications of the basic growth control mechanism are briefly discussed.

The general pattern of eukaryotic cell proliferation is highly conserved through evolution, as shown by its striking similarities at both the physiological and molecular levels in species as widely dissimilar as yeast and humans. Concepts concerning the mitotic cell cycle of mammalian cells are principally based on experiments with fibroblasts and hematopoietic cells, although a great deal of current molecular and genetic information has been obtained from studies with yeasts and the activated eggs of invertebrates and vertebrates, principally the frog *Xenopus laevis.*

The subject of cell cycle control is developing at a great rate; a comprehensive listing of recent publications is far beyond the scope of this brief chapter, which provides only a summary of a great deal of important information. The references provided are thus not comprehensive. In addition to a few germinal papers, they permit an entry into the field through reviews and recent articles (for background see Refs. 1–4).

II. OVERVIEW OF THE BIOLOGY OF CELL PROLIFERATION

A. Cell Proliferation and Quiescence In Vivo

Cell proliferation in vivo is normally tightly controlled; it depends on the stage of growth of an organism, physiological needs to replace or increase quantities of specific kinds of cells and so forth. Complex molecular processes are responsible for these flexible regulations. The cell cycle, consisting of successive major phases, provides a framework for organizing the many events of cell proliferation. Growing cells were initially distinguished with the microscope as periodically briefly undergoing mitosis (M phase). During most of the time they were said to be in interphase. The discovery in 1955 that DNA duplication takes place only during the middle part (S) of each interphase, subdivided interphase into the now familiar G_1 (pre-S), S, and G_2 (pre-M) phases in the cell cycle. Thus, G_1 cells contain unduplicated DNA, in contrast with the duplicated DNA in G_2 and M cells. These phases are followed by M, and then cell division (cytokinesis), to produce two G_1 daughter cells. Transcriptions and translations are required throughout the entire interphase, as shown by early experiments with inhibitors of RNA and protein synthesis (1).

Cells in adult mammals, such as ourselves, are generated after many rounds of proliferation, starting with a fertilized egg and progressing during embryogenesis and early stages of growth of the organism. Orderly embryonic cellular proliferation and differentiation is independent of extracellular stimuli, such as growth factors, and depends on intracellular control mechanisms (1). The cell cycle is simple in embryos during the first dozen rounds after fertilization. These cycles can be very rapid, as short as 30 min in xenopus (4). DNA synthesis, mitosis, and cytokinesis follow one another without long intervening "gaps" (G_1 and G_2). Eggs contain preformed RNAs, and during early stages are independent of transcriptional production of new mRNAs. Transcription is required in subsequent cycles in which G_1 and G_2 phases appear, during which necessary transcriptions and translations take place.

In embryos and in early-life stages, cells rapidly increase in number. Abilities to proliferate diminish during differentiation. In adults, cells are maintained in an approximately steady state, a balance of cell production and loss. Some totally cease to proliferate, for example, nerve and muscle cells. Mature erythrocytes and keratinocytes lose their nuclei after terminal differentiation, and so become unable to proliferate. Other kinds of cells, such as the stem cells of lymphocytes, intestine, and such, are quiescent most of the time, but are periodically stimulated by cytokines to proliferate, thereby replacing lost mature cells. Two daughter cells are produced, one stem and the other differentiating. The cytokine (LIF) leukemia inhibitory factor (inter leukin-6; IL-6) retains the stem cell by preventing its differentiation (5). Eythrocyte stem cells are stimulated by erythropoietin (EPO) when the oxygen supply is inadequate, and B and T lymphocytes are stimulated by antigens during immune responses. Other types of cells are usually quiescent and are reversibly stimulated by extracellular signals. For example, fibroblasts respond to wound healing, which activates release of a stimulatory platelet-derived growth factor (PDGF); after healing the cells return to quiescence. By contrast, defective growth responses cause tumor cells to proliferate at times and places that decrease survival of the organism.

In vivo, most adult cells are quiescent and are not passing through the cell cycle. The quiescent cells are said to be out of cycle in a G_0 phase. When cells become quiescent, their chromatin becomes more condensed, synthesis of their proteins and RNA decrease, and degradations increase. Some new proteins appear, for example, the Gas gene products (6), the functions of which are unknown. Similar to G_1 cells, G_0 cells contain unduplicated DNA, but they differ from G_1 cells in many biochemical properties (1). When a cell terminates replication, it does not stop randomly in the cycle; rather, it specifically leaves the cycle in G_1 phase and becomes quiescent. This key control event, the molecular "switch" that determines the selection between continuing growth versus entering quiescence, must depend on a special biochemical process that takes place at a specific time in G_1: the restriction point in mammalian cells or START in yeast.

B. Universality of the Cell Cycle

The complete four-phase cell cycle is functional in mature cells of all eukaryotes, from yeasts to humans. Many morphological and biochemical findings about the cycle were made initially with animal cells, but much molecular biology was worked out with other organisms. The same general principles apply to all of them.

Saccharomyces cerevisiae, baker's yeast, is a single-cell organism. Yeasts have a haploid form, which provides great advantages for obtaining mutants (7). Before haploid cells of the two mating types conjugate, they are reciprocally arrested in G_1 phase, with unduplicated DNA, by mating factors produced by the other type (8). Mating factors are the only extracellular peptides that regulate their growth, which is also arrested in a quiescent state by nutrient deprivation. Yeasts, similarly to bacteria, have evolved to grow as rapidly as possible under a variety of extracellular conditions. Their basic cell cycle is similar to that of mammalian cells, but is much faster: a 90-min cycle. The nuclear membrane does not break down at M as it does in mammalian cells. And the final separation of the two cells is different from cytokinesis of mammalian cells. In *S. cerevisiae* a bud forms on the oval cell early during the cycle, grows, and is finally pinched off. This morphological progression simplifies determination of a cell's position in the cycle.

Temperature-sensitive cell division cycle (cdc) yeast mutants become reversibly arrested in different stages of the cycle when brought to the nonpermissive temperature of 36°C(9). Studies with the cdc28 mutant, which arrests in G_1, revealed that passing a key genetically controlled event in G_1, named START, is required for subsequent DNA duplication, doubling the spindle pole body of the mitotic apparatus, and release of the budding daughter cell (10). After START, a cell is committed to complete its cycle. The budded daughter is smaller than the mother cell; it requires a longer time to achieve critical mass before it reaches START. This difference provides an example of the frequent and, more or less, stringent requirement of growth to a minimal mass before DNA synthesis can proceed.

Schizosaccharomyces pombe, a fission yeast, is genetically very distantly related to *S. cerevisiae*. Rod-shaped *S. pombe* grows at one of its ends, and its cycle terminates with a constriction that separates the two cells. Work with *S. pombe* led to discovery of several genes and their products that are central to the cycle (11). Entry into M depends on a gene named *CDC2* in *S. pombe*, which is the homolog of the START gene *cdc28* of *S. cerevisiae* (1).

Sea urchin and clam eggs undergo highly synchronous cell cycles after they are fertilized. The latter has only a 30-min cycle time. They can be obtained in very large quantities and, therefore, provide excellent material for identification and isolation of substances during cycle progression, such as the cyclin proteins, which appear and disappear periodically during the cycle (4).

Eggs of the toad *Xenopus laevis* have been used to conduct highly revealing studies, owing to their ordered progression through mitosis and the ease of injecting test substances into them (4). They have no G_1 and a very short G_2 phase in the 12 earliest cycles. Cell-free preparations made from them have been useful in determining mitotic molecular events.

C. Cell Proliferation in Culture

Characteristics of Cultured Cells

Regulated and also defective growth observed in vivo can be approximated by studies in culture, which has numerous advantages over growth in vivo for investigating the molecular basis of cell behavior. Much has been learned from studies with cells in culture, and much remains to be learned with this methodology. Conditions of growth can be experimentally varied—for example, by adding or removing a desired growth factor—and the consequences determined. Normal, cancer, and mutant cells can be cultured separately in sufficient quantities to permit comparative biochemical and molecular investigations of growth and differentiation. Investigations of cell cycle events and comparisons of cells in different phases are facilitated by several physical and pharmacological procedures for synchronizing cells in quiescence and at different cycle positions. However, because behavior of cells in culture is generally similar to, but not identical with the properties of cells in vivo, the differences must be considered in interpreting the results of in vitro experiments.

Cell growth and its control in culture was initially possible mainly with fibroblasts, but now this can be achieved with a variety of cells, such as those from human mammary epithelia (12). However, growth in culture has not yet been achieved with other kinds of cells: for example, normal colon cells. Cells can be stimulated to proliferate by plating them sub-confluently on an appropriate surface and in a culture medium that supplies the essential nutrients and growth factors, which often are supplied by addition of serum. The growing mammalian cell requires approximately 1 day to complete its cycle; typical phase durations are G_1-6 hr, S = 8 hr, G_2 = 2 hr, and M = 0.5 hr, as for 3T3 mouse cells. Cycle durations between cells in the same culture can vary up to twofold. Most of this variability is in G_1 phase, and it only partly depends on cell size.

Cells obtained from in vivo samples and plated to produce primary cultures gradually change their properties as they undergo further rounds of proliferation. Rodent cells pass through a crisis that makes some of the population "immortal," able to proliferate indefinitely. Eventually, some cells achieve a transformed phenotype; these take over the culture; they can form tumors when injected into an immunologically suitable animal. Human cells do not become immortal in culture, but instead, they senesce. Their growth becomes progressively slower and ceases after about 50 doublings (1).

Cells stop growing after they have exhausted or are deprived of growth factors or essential nutrients; they become quiescent (in G_0). Contacts with adjacent cells cause an important negative regulator of normal cell proliferation: "density-dependent" arrest of cell growth in G_0. But contact does not stop tumor cells, which grow to form foci upon a layer of normal cells. Similarly, normal cells cannot grow unless they are attached to a suitable surface, but tumor cells grow in suspension. These cell–cell and substratum contacts involve interactions of extracellular matrix laminin, fibronectin, and other substances with integrin protein receptors in the cell membrane (11), many of which are not produced in tumor cells. Cells emerge from G_0 into G_1 phase of the cycle when the limiting conditions are reversed.

Growth Factors and Cytokines

Whether a mammalian cells grows or is quiescent depends on extracellular factors that impinge on it. In general, cells become quiescent when they are deprived of serum, which provides growth factor proteins, polypeptides, and cytokines. There are many of these that stimulate different kinds of cells. Others, such as tumor necrosis factor-β (TGF-β) provide negative signals for growth (13,14). Growth factors bind to specific receptors on the cell membrane and, thereby, activate associated protein kinases. As an example, the receptor of epidermal growth factor (EGF) is a large cell membrane-spanning protein that has an EGF-binding domain external to the membrane and a nascent kinase activity domain in the cytoplasm. When it binds EGF, it dimerizes, its kinase is activated and phosphorylates tyrosines on its internal end, and also other proteins. This initiates a signal transduction cascade of phosphorylations that activate cell proliferation (15). Other receptors of growth factors act similarly, but some are composed of several nonconvalently associated proteins including kinases. Later in the cycle cells become independent of extracellular factors, and internally controlled signal transduction sequences become operative.

The subject of signal transduction is discussed in detail elsewhere. External stimuli from receptors activate "second-messenger"–signaling pathways within the cell. These signal transduction pathways involve protein kinases, *ras*, Ca^{2+} and others in a regulated cascade of phosphorylation reactions active from G_0 to late G_1.

G_0 to G_1

Cells started from G_0 by addition of growth factors have an interval before reaching S that is longer than is the M to S period (G_1) of cycling cells (12 versus 6 hr), because biochemical syntheses are required to regenerate labile components that were depleted while the cells were quiescent. The longer cells are retained in G_0 the longer the time required to emerge into cycle after they are stimulated. Activation of the *myc* protooncogene that appears in early G_1 is delayed by long quiescence (16). Emergence from quiescence induced by serum starvation is at least a two-step process. Fibroblasts cannot proliferate if they are exposed to PDGF or to plasma from which PDGF has been removed. They are stimulated if these components are sequentially provided, but only if PDGF is added first (17). The first stimulation makes the cells "competent," and the second lets them "progress." Similarly, hematopoietic cells first require stimulation by an antigen or mitogen, followed by interleukin-2 (IL-2) (18).

A set of immediate-early–response genes in fibroblasts is transiently activated by competence factors, such as PDGF (19). These include *fos, jun,* and *Egr*-1, the mRNAs of which appear within an hour and do not require new protein synthesis. They have short half-lives and are soon degraded. These gene products appear to activate transcription of delayed response genes involved in G_1 progression, such as *myc*, ornithine decarboxylase, cyclin D, cyclin E, and E2F.

Progression Through G_1

After cells have recovered from quiescence, they enter G_1, which is divided into early and late portions. Cycling cells are highly dependent on growth factors during the first 3–4 hr postmitosis; this period is quite constant in duration (20,21). EGF and insulin-like growth factor-1 (1GF-1) are required for progression of fibroblasts from competence to S phase. Quiescent cells provided with serum, but without essential amino acids, arrest in mid-G_1, at the V point (17); after which they need only IGF-1. Progression though early G_1 is blocked by the drug lovastatin (22) that inhibits production of isoprenoids, which are post-translationally attached to numerous proteins, including *ras* (23). Other inhibitors block progression at different stages of G_1 (24). Ki-67, an antigen of unknown function, is absent in human G_0 cells and appears during early G_1; it serves as a marker of proliferation (25). Polyamines are required in G_1; synthesis of the related enzyme ornithine decarboxylase in mid-G_1 is *myc*-dependent (16). Calcium and calmodulin have important G_1 roles (26).

The Restriction Point

A critical biochemical control switch, the restriction point, determines whether or not a mammalian cell can complete its cycle in response to external conditions, such as availability of growth factors (3,20). Normal cells located in the cycle before the restriction point cannot initiate DNA synthesis after removal of growth factors; they exit from G_1 into quiescence (G_0). Those beyond this point complete their cycle and undergo cytokinesis. The restriction point is similar to the critical control point START to *S. cerevisiae* (9). The restriction point is located 2–3 hr before onset of S phase (20,21). Partial inhibition of protein synthesis with cycloheximide also is effective in blocking cycling of nontumorigenic cells located before the restriction point. Cycloheximide is much less effective in blocking growth of tumorigenic cells, which are defective in the restriction point mechanism. Conversely, increasing the rate of general protein synthesis by introducing the gene for translation factor transforms normal cells into controlled proliferation and also modifies differentiation (27).

These results are the basis for the proposal that accumulation of a critical amount of a short-lived (R) protein is essential for passing the restriction point (28). This protein appears to be overproduced or is stabilized in tumor cells. Kinase inhibitors, such as staurosporin, block G_1 cells from entering S, showing that kinases are involved, and there is evidence that the restriction point protein is a cyclin (see under cyclin and cdk in Sec. II). At the restriction point, cells begin to transcribe various genes that code for enzymes involved in DNA synthesis (1,29–31). Studies of these transcriptions provide biochemical and molecular biological, as contrasted with cell biological, information that is important for elucidation of control of the cell cycle.

From the Restriction Point to S

The restriction point to S interval of a few hours is not block by transcriptional inhibitors, and it may be a time required for assembly of the DNA synthetic apparatus in the nucleus. Enzymes related to DNA replication are synthesized at the end of G_1 and are moved into the nucleus by an unknown mechanism; they exit in G_2 (26). Mimosine is an amino acid analogue that reversibly blocks cells at the G_1–S boundary (32).

S Phase

Extracellular factors do not determine progression of cells through S, nor through G_2 and M phases. Rather, intracellular regulatory reactions control the smooth progression of the numerous events involved in these phases of the cycle. Genes are duplicated in a definite temporal order (33). The S phase is not a uniform process; numerous stages have been proposed. Although origins of DNA replication (ori) have been identified in yeast (34), progress in finding them in mammalian cell is recent (35,36). The limiting factors required for initiating and continuing DNA synthesis have not yet been determined.

The hallmark of S phase is DNA synthesis. Major enzymes for DNA synthesis (37) are primase, five polymerases, proliferating cell nuclear antigen (PCNA) (38,39), ligase (40,41), and helicases (42). Helicases appear to have a critical role, as found with in vitro systems that utilize SV40 DNA as the template. RF-A, an unwinding protein, is activated in S, probably by cyclin A–cdk2 kinase (4). Replication protein A is a single-stranded binding protein involved in DNA replication and repair (43,44).

Replication complexes of these enzymes are formed at the start of S phase (45). Replication of DNA by this complex takes place on the nuclear matrix (46). There is evidence for a larger complex, named *replitase*, which includes enzymes of both replication and precursor synthesis (26).

Several major syntheses other than DNA take place during S. Most striking is production of histone proteins, which is closely correlated with DNA synthesis. Histone mRNA is rapidly degraded in the presence of DNA synthesis inhibitors. The mechanisms of histone transcription have been worked out in detail (47). When cells are prevented from entering and progressing through S by hydroxyurea, aphidicolin, or other chemicals that inhibit enzymes required for DNA synthesis, related proteins and histones are also not synthesized.

G_2 Phase

During this interval following completion of DNA synthesis and before initiation of mitotic prophase, most cells do not require growth factors. Studies with inhibitors have demonstrated that protein and RNA syntheses are necessary, so G_2 represents the time required for synthesis of macromolecules and assembly of the mitotic apparatus. Little is known at the molecular level about early G_2 events (1). Kinase inhibitors, such as staurosporin, and protein synthesis inhibitors prevent passage through G_2 (48).

Mitosis

The stages of mitosis—prophase, metaphase, anaphase, and telophase—are defined by changes in visible structures of the chromosomes. RNA and protein metabolism is sharply decreased. This final phase of the cycle is not dependent on extracellular factors, but is initiated by intracellular maturation-promoting factor (MPF) (49). This factor is produced periodically in fertilized eggs, independently of periodic DNA synthesis. Its properties and mechanism, worked out in detail, particularly by studies with yeasts and xenopus, have been reviewed extensively (4). MPF initiates the numerous biochemical events of mitosis. Three lamin proteins located on the inner nuclear membrane are phosphorylated, and the nuclear membrane breaks down. Other phosphorylations include histone H_1, involved in chromosomal condensation, and microtubules, involved in chromosomal movements. Kinases and phosphatases coded by *cdc* genes are essential elements of the mitotic machinery (see later discussion). Colchicine, nocodazole, paclitaxel (taxol), and related compounds affect microtubule structures and block cells at prophase. They are useful to collect and count cells that reach mitosis.

At the end of M phase, initiation of cell division (cytokinesis) depends on inactivation of MPF by actions of phosphatases and ubiquitin-activated proteases. Actin and myosin filaments form a constricting cleavage furrow.

D. Coordination of Cell Cycle Phases

An important question concerns mechanisms by which the overall cycle is coordinated (50). Mechanisms, now named "checkpoints," have become clearer through biochemical and molecular biological studies. Early experiments demonstrated that factors controlling cycle phases are located in the cell cytoplasm. When cells in different stages of the cycle were fused to one another, one phase demonstrated dominant behavior over the other (see Ref. 51). An S phase cell caused the nucleus of a G_1 cell fused with it to synthesize DNA, indicating a dominant S phase cytoplasmic factor sufficient to allow DNA synthesis. Cytochalasin B specifically blocks cytokinesis, but not nuclear events, in mammalian cells; the two nuclei remaining in one cell initiate S phase at the same time, showing that a cytoplasmic factor controls S phase entry (52).

However, S phase cells do not induce DNA synthesis in nuclei of G_2 cells fused to them, showing that G_2 cells cannot be made to initiate another round of DNA synthesis under these conditions. To initiate S phase, not only metabolic events in G_1, but also completion of the prior M phase is required (4,50). This control is bypassed in a few special cases, as when polytenic chromosomes are formed in drosophila. Not only is the initial synthesis of DNA tightly regulated, but one must account for the precise order of successive gene replication initiations that follow during the lengthy S period (33). To explain why cells cannot perform a second round of chromosome replication during one S phase, a "licensing factor" (LF) is proposed to come from the cytoplasm and attach to chromatin when the nuclear membrane breaks down at M. LF cannot enter the nucleus after the membrane is re-formed at the end of M. Thus, if new LF is needed to start DNA replication, only one round would be possible after each M phase (53).

Conversely, mitosis is not initiated until after completion of the prior S phase. This checkpoint blocks all mitotic events—chromosome condensation, lamin phosphorylation, nuclear membrane breakdown, and chromosome partitioning. Thus, inhibitors of DNA synthesis, such as hydroxyurea, applied during S phase prevent the subsequent mitosis. Caffeine permits several mitoses of the inhibited rodent cells (54), but human cells undergo apoptotic cell death (55,56). The mechanism is unclear but may involve cdc2 phosphorylation and cyclin A accumulation. The *rum-1* gene product of *S. pombe*, which is made in G_1 phase and has a role in G_1 progression, also prevents mitosis until S phase is completed (57,58). Furthermore, cells with maloriented chromosomes cannot complete mitosis (59).

Mitosis is dominant over G_2, because fusion of cells in these phases caused premature chromosomal condensation in the G_2 nuclei. Finally, fusion of M and S phase cells condenses and fragments the S chromosomes (see Ref. 51).

III. MOLECULAR MECHANISMS OF CYCLE REGULATION

By the mid-1980s, the biology and considerable descriptive biochemistry of the cell cycle and its regulation had been explored. Many of the main events of cell proliferation were defined. The evident question then was to identify the key molecules that control the phase transitions. The central discovery was the finding of *cyclins*, a family of proteins that are produced transiently in sequence at defined times during the cycle (60). Cyclins are the regulatory subunits of complexes that they form with and activate called cycle-dependent protein kinases (cdks). The activated cdks phosphorylate various proteins (61), thereby regulating critical steps in the cell cycle. This subject is currently being extremely actively investigated (4).

A. Cyclins

Cyclins were initially identified in fertilized marine invertebrate eggs and oocytes of the sea urchin *Arbacia* and the clam *Spisula*, in which early embryonic cleavages occur with a high degree of synchrony. Two proteins showed dramatic fluctuations in abundance during cell cycle progression (61). Their levels peaked at the onset of mitosis, after which they were rapidly degraded, and again began to accumulate during the subsequent interphase. They were classified into cyclins of A and B types. These investigations led to the proposal that cyclins are rate-limiting elements involved in mitotic induction (61,62).

Not long after cyclin sequences from clam and urchin were published, three cyclin homologues (names CLN1, 2, and 3) were discovered in *S. cerevisiae* (63–65). Their integrity is essential for cell cycle progression because the knockout of all three is lethal, but not of any one. These are known as "start" or "G_1" cyclins because they are required to START in budding yeast, for G_1 and S progression. In S phase two more C1b cyclins are produced, and two more in G_2/M. The G_1 cyclins act in the cell cycle before DNA replication (S phase; 66), whereas the G_2 or mitotic cyclins act after S phase (60).

From these yeast genes, mammalian homologues were identified. More than 30 sequences of cyclin polypeptides have been deduced from cDNA and genomic clones. The main cyclins in mammalian somatic cell lines presently are A, B1, B2, C, D1, D2, D3, and E (67). All cyclins share a 150-amino acid region of structural homology called the cyclin box (60). It is this part of the protein that is responsible for binding to the cyclin-dependent kinase subunit. Clearly, the existence of many classes of cyclins in the same cell and the great diversity among different classes is suggestive of their multiple functions at different points in the cell cycle.

Cyclins in G_1

Cyclins appear in proliferating cells (69) and are induced by growth factors (68). They are not generally present in G_0 cells. G_1 cyclins were first identified in *S. cerevisiae*. Expression of only one CLN protein can be sufficient to drive the cell cycle. This property was used successfully to identify novel cyclins from other species by heterologous complementation;

CLN3, cloned under the control of a galactose-inducible promoter, was turned off on glucose media and the G_1 arrest was rescued by overexpression of cloned genes from a complementary DNA library. By using this strategy, the human cyclin D1 gene has been identified (70). D1 was also simultaneously isolated in a search to identify a candidate oncogene involved in a benign tumor of the parathyroid gland and was found at a chromosomal breakpoint in parathyroid adenomas (71) where it was previously studied as an anonymous transcription unit.

Studies with murine macrophages, regulated by colony-stimulating factor 1 (CSF-1), led to the identification of three novel cyclin-like genes called *CYL1*, *CYL2*, and *CYL3* that appear during G_1 phase. *CYL1* is the murine homologue of human cyclin D (72). Subsequently, cyclins D2 and D3 were cloned, based on sequence homology and apparent times of appearance in the cell cycle (73,74). The three cyclin D loci, now officially designated *CCND1*, *CCND2*, and *CCND3*, were mapped to human chromosomes 11q13, 12p13, and 6p21, respectively (75,76).

In human cells, changes of cyclin D1 mRNA suggest that cyclin D1 plays a role in G_1-phase progression or in the G_1–S transition (71,74). Overexpression of cyclin D1 in mammalian fibroblasts shortens G_1, reduces cell size, and decreases the cells' serum dependency (77). Cells microinjected during the G_1 interval with cyclin D1 antisense vectors, or with specific antibodies to the D1 protein, are inhibited from entering S phase, but injections performed at or after the G_1–S transition are without effect (77,78). The oncogene *myc* can repress cyclin D1 (79). Cyclin D1 is a candidate for the restriction point protein because of its frequent modification in cancers (80).

Cyclin D1 is able to associate with cdk4 during G_1 in CSF-1-stimulated murine macrophage lysates (81), and with cdk2, cdk4, cdk5, cdk6, and with proliferating cell nuclear antigen (PCNA) in WI38 human fibroblast lysates (81–84). Although cdk4 appears to be the major partner of cyclin D1 in macrophages and fibroblasts (81,82), complexes with cdk6 predominate in peripheral blood T cells (83). When cyclin–cdk complexes are reconstituted in insect cells, each of the D-type cyclins can activate cdk4 and cdk6, whereas D2 and D3, but not D1, can also functionally interact with cdk2 (83,85,86). Histone H1 is a relatively poor substrate for reconstituted cyclin D–cdk4 and cyclin D–cdk6 complexes, which preferentially phosphorylate the retinoblastoma protein, pRb. Phosphorylation of pRb induced by cyclin D–cdk4 was seen in vitro (81) and, more recently, in intact mammalian cells (84).

Cyclin E, another G_1 cyclin present in animal cells, was originally identified by its ability to rescue G_1 cyclin-defective yeast (87,88). It forms complexes almost exclusively with cdk2 and activates serine–threonine kinase at the restriction point shortly before entry into S phase (89–91). The observation, suggesting that the cyclin E–cdk2 complexes might be directly involved in regulating G_1 to S transition, is supported by the finding that microinjection of cdk2-specific antibodies results in a G_1 block (92). In addition, cells overproducing cyclin E have a decreased cell size, a decreased requirement for growth factors, and a shorter G_1 phase (93–95). Furthermore, cyclin E releases a pRb-induced G_1 block (30). The level of cyclin E appears to be limiting in G_1, and it is high in many tumor cell lines and tissue samples (94,95), suggesting it as a candidate for the restriction point protein (91).

Cyclin C is the least characterized of all G_1 cyclins. Human cyclin C cDNA was cloned by its ability to rescue yeast mutants defective in G_1 cyclins (87). Its level oscillate only minimally throughout the cell cycle, with very modest increases observed in early G_1 (87,95). No physiological partner cdk for cyclin C has yet been identified.

Cyclins in S

The human cyclin A gene was first identified as a protein associated with adenovirus *E1A* protein (98,99), and as an integration site of hepatitis B virus in a hepatocellular carcinoma (42).

p60 is the human homologue of cyclin A and is complexed with cdc2 (99). The addition of cyclin A to a G_1 extract devoid of DNA replicating activity is sufficient to start replication (101). Cyclin A appears at the beginning of S; its transcriptional control is under investigation (103). The blocking of cyclin A synthesis inhibits DNA synthesis (105). A cdc2 kinase activity is needed to activate DNA replication in xenopus eggs (106). This kinase activity might release negative control of the DNA polymerase complex activity (107,108). The cdc2 family kinases phosphorylate a DNA replication factor (RPA) in human cells and activate DNA replication (105).

Cyclin A interacts with p107, a pRB-related protein (109), which simultaneously complexes with the transcription factor E2F that binds to specific DNA sequences. The resultant cdk-2–cyclin A–p107–E2F complex appears at about the G_1–S transition and disappears after S phase, suggesting a role in S-phase progression (110,111).

Cyclins in G_2/M

The series of mitotic events are regulated by cyclins A and B (112). DNA synthesis and its entry into mitosis in human cells are both inhibited by microinjected anit-cyclin A antibodies. At distinct times cyclin A binds both cdk2 and cdc2, giving two different cyclin A kinase activities; the former appears in S phase and the latter in G_2 (113). Both A- and B- type cyclin immunocomplexes display in vitro protein kinase activity and have very similar in vitro substrate specificities (114).

Whereas A-type cyclins appear to be involved in both S phase and mitosis, the B-types are restricted to mitosis; hence, they are referred to as classic mitotic cyclins. Although both cyclins A and B interact with p34cdc2, the synthesis and destruction of cyclin A oscillates in advance of cyclin B during the cell cycle; cyclin A levels increase throughout S and G_2 phases and decline as cells begin metaphase, by way of a ubiquitin-associated pathway (115). In mammalian cells, cytological studies have shown that cyclin A is located in the nucleus in interphase and disappears in metaphase, whereas cyclin B is located mainly in the cytoplasm during interphase and is translocated to the nucleus in mitosis (116).

The B-type cyclins have been found in all eukaryotes from yeast to humans. Mitotic cyclins are believed to act through association with the highly conserved protein–serine–threonine kinase p34cdc2 (117). The two classes of B-type cyclins (B1 and B2) both bind to p34cdc2. Both are active in the G_2 to M transition in xenopus, and injection of M–phase-promoting factor (cdc2–cyclin B) causes xenopus oocytes to mature in the absence of protein synthesis (118).

As soon as cyclin B is synthesized in late S phase p34cdc2 forms a complex, but it is kept inactive by phosphorylations of p34cdc2 on tyrosine-15 and threonine-14 in higher eukaryotes. At the end of G_2 phase p34cdc2 is activated by dephosphorylation of Tyr-15 by a phosphatase encoded by *Cdc25*. Structures of cdk2 obtained by X-ray diffraction show this tyrosine is located on an inhibitory T-loop that is moved by phosphorylation (119). The wee1 protein of *S. pombe* catalyzes Tyr-15 phosphorylation. Mutation of wee1 permits mitosis at half the normal yeast cell size, but the total cycle time is unchanged because the normally very brief G_1 phase is lengthened to compensate for the smaller mass. This adjustment of cycle phases thereby connects cell mass with growth rate (120). Many mechanisms regulate M; for example, still other phosphorylations of Cdc25 are involved. Drosophila eggs keep MPF active during early nuclear divisions, but later when the string gene mRNA is degraded, Cdc25 drops. The MPF is in inactive state during most of the cycle (124).

Exit from mitosis depends on the abrupt ubiquitin-mediated degradation of cyclin B during anaphase, resulting in the release of p34cdc2 as an inactive monomer (118). This proteolytic destruction of MPF (cyclin B) prevents another mitosis. Benomyl, a microtubule depolymerizing drug, blocks proteolysis of MPF in yeast and progression through M (B-60, 91; see Ref. 4 for details). In anaphase, the addition of multiple ubiquitins (a 76-amino acid protein that bonds to lysines) to a "destruction box" of 9 amino acids near cyclin B's NH_2-terminus marks it for destruction by proteolytic protein complexes named proteosomes (124). At the end of mitosis, phosphatases remove phosphates from lamin, and the nuclear membrane re-forms (123). Prenylation and myristylation of lamin B are also necessary for completing mitosis. In turn, degradation of MPF permits initiation of a subsequent DNA synthesis and, also, formation of microtubule-organizing centers for the next mitosis.

B. The cdk Family

The discovery of cyclins coincided with studies of maturation-promoting factor (MPF) and meiosis in xenopus oocytes in which the reappearance of MPF at only the second (but not first) meiosis requires protein synthesis (124). By purifying the active components from a cell-free system, MPF was isolated from xenopus eggs; it consists of two protein subunits, p34 and p45. This protein dimer has kinase activity. The sequence of p34 suggested that it is the catalytic subunit that phosporylates serine and threonine residues in target proteins. The other regulatory subunit of MPF is cyclin B. Homologous subunits can be found in all eukaryotic cells. In *S. pombe*, p34 is encoded by the cell cycle gene *cdc2*; in *S. cerevisiae*, p34 is encoded by the homologous gene *CDC28* (125,126).

Either *cdc2* or *CDC28* can stimulate both the G_1–S transition and the G_2/M transition in yeast when associated with cell cycle-specific cyclins (127). In contrast, a large family of *cdc2*-related genes in higher eukaryotic cells have been identified, several of which are involved in cell cycle regulation (128).

The cyclin family's recent expansion is somewhat matched by that of the cyclin-dependent kinases (132). In all, six different protein kinases have been demonstrated to bind to a cyclin molecule in human cells, and have been disignated cdk1 through 6(128). Distinct cdks from this family are believed to regulate discrete cell cycle transitions in vertebrate cells. Three of these, cdk1 (i.e., cdc2), cdk2, and cdk3, all can substitute for a defective cdc2 protein when introduced into yeast, suggesting that these kinases perform a role analogous to that of yeast cdk (129). The cdc2 in higher eukaryotes acts at the G_2–M transition and is required for entry into mitosis, whereas cdk2 is essential for G_1–S transition. The cdk3 is also necessary for cell cycle progression, for expression of a dominant negative cdk3 mutant in human cells can arrest cells in G_1 (130). The cdk4 (previously known as PSK-J3) and cdk5 kinases, as well as cdc2 and cdk2, associate with D-type cyclins (82). The cdk5 kinase is active only in terminally differentiated neurons and is not active in cycling cells, suggesting that cdk5 is not involved in promoting cell cycle progression (131). Recently, cdk6 was found to associate with and activate D-type cyclins during mid-G_1, and it can phosphorylate pRb before the activation of cdk2 kinase, suggesting that cdk6 and cdk4 may be the earliest cdks activated in response to growth factor stimulation (129). Several other structurally related proteins are known, and some of these may be reclassified as cdks if they are demonstrated to have cyclin-dependent kinase activity.

C. Inhibitors of cdk Kinase Activity

Activation of a cdk requires association of a cyclin. The activated cdk can be inactivated in several ways: (1) cyclin levels can be reduced by turning down their transcription; (2) by degradation of the protein itself; (3) the catalytic subunit can be dephosphorylated at the activating tyrosine residue, or conversely, by phosphorylation of the tyrosine residue; (4) the active cyclin–cdk complex can be rendered inactive by the binding of a kinase inhibitor (134,135).

Recent studies have identified several regulatory subunits for cdks; proteins that bind to the cyclin–cdk complexes and inhibit their activity (133). These inhibitory proteins, together with positively acting kinases, provide a balanced control mechanism such as is very frequently seen in biological regulation. They may provide a "ceiling" that must be exceeded by increased cyclin-dependent kinase production during G_1 to abruptly accomplish the restriction point event.

The first of these inhibitory proteins, p21, was discovered almost simultaneously using four very different approaches and, accordingly, was given several names. SDI-1 (senescent cell-derived inhibitor) increased in senescent cells that are growth arrested at the restriction point (137). Cip-1 (cyclin-dependent kinase-interacting protein) binds to cdk–cyclin complexes (138), and the mouse CAP20 homologue (139) is an inhibitor of cyclin kinases and an integral component of cell cycle control. WAF-1 (wild-type p53-activated factor-1) gene was identified following induction of wild-type p53 protein expression in a human glioblastoma multiforme cell line (140). This protein was also named mda-6 when it was isolated during the process of terminal cell differentiation in human melanoma cells (141). p21 is a nuclear-localized protein that is induced by DNA damage and during apoptosis in specific cell types as a function of wild-type p53 activation. These studies suggest that p21 is an important downstream mediator of wild-type p53-induced growth control in mammalian cells. Recently, mda-6/WAF-1/CIP-1/SDI1 expression was also induced by mechanistically diversely acting agents causing differentiation of macrophage to monocyte (TPA and vitamin D_3) or granulocyte (retinoic acid or dimethyl sulfoxide), of human HL60 promyelocytic leukemia cells that lack endogenous p53 genes (141).

p21 blocks DNA elongation in a cell-free synthesis system, even in the absence of cyclin A and cdk2. Thus, it appears to inhibit cycle progression both before and during S phase. Its inhibitory action is overcome by PCNA, a subunit of DNA polymerase-δ that activates polymerase activity (39). PCNA is found in complexes with cdk–cyclins and the negatively regulatory p21 in untransformed cells, but not in transformed cells (142).

A related protein, p27 which inhibits cdk–cyclins, is activated by transforming growth factor (TGF)-β and stops cell growth at the restriction point (13). Another inhibitory protein is p15, which is induced by the negative "growth" factor TGF-β (142). Yet another protein, p16, identified by its association with human cdk4 in a yeast two–hybrid protein interaction screen, specifically inhibits cdk4–cyclin D kinase activity in vitro (69,143). The importance of p16 is suggested by the recent discovery that a tumor suppressor gene (MTS1) codes for p16 (144). Although it is currently disputed whether loss of p16 is a frequent event in primary tumors, defects in the MTS1 gene are found in a wide variety of tumor cell lines, and germline mutations in the p16 gene have now been found in familial melanoma (80). Similarly in yeast, FAR1 (which is phosphorylated by Fus3) is an inhibitor of G_1 cyclin–CDC28 kinases, and is induced by mating factor, which stops yeast at START. The SIC1 protein of S. cerevisiae also inhibits cdk activity (145); (see also ref. 146).

D. Transcriptional Promoters and Their Transactivating Proteins

How are cyclins related to progression through the cell cycle at the G_1–S restriction point? Cyclins and cdks form complexes (147) with several other very important proteins. These complexes bind specifically to DNA sequences in promoters of genes that encode enzymes involved in DNA synthesis. These enzymes are not produced until they are needed to activate initiation of S phase, when a cell approaches the G_1/S boundary. These DNA-binding complexes form at specific times during G_1 after their corresponding cyclins are synthesized.

This problem of molecular analysis of cycle progression has been studied in detail with the gene for thymidine kinase (TK) (148). This enzyme, involved in DNA synthesis, is transcribed at a time very close to the restriction (R) point, and provides a convenient endpoint to assay. For the investigation of *TK* transcription, the promoter of the murine *TK* gene was searched for DNA motifs that are needed for the S phase-specific induction of a reporter gene. A DNA sequence located between −174-base pairs (bp) upstream and +159 bp downstream from the mouse TK translation initiation site was sufficient to permit transcription in an S phase-specific manner (149). Further analysis of this promoter sequence using DNase 1 footprinting, a DNA–protein interaction assay, revealed three separate protein-binding motifs (named MT1, MT2, and MT3) each of which contains a 6- to 10-bp consensus DNA sequence (150).

Several protein complexes that bind to these three DNA motifs were discovered by using another DNA–protein interaction assay, gel retardation (150–155). Each sequence binds different protein complexes, and each binds several complexes in succession during the G_0 to S transition. MT1 binds a complex that contains murine transcription factor Sp1, which is present throughout G_1 phase (150) and that is critical to promoter activity because mutations of MT1 abolish *TK* transcription (149). MT1, and also MT2 and MT3, bind another cyclin-containing complex when the cells approach the G_1–S boundary (150).

A series of complexes appear early in G_1. A complex, containing the EGR-1 protein, binds to the MT3 motif very early in G_1 phase (153). Then in early G_1 phase MT3 binds a complex Yi1 which contains cyclin D1–cdk2 kinase, a murine retinoblastoma (pRb)-related protein and a DNA-binding protein (151,154). The target molecule of cyclin D1–cdk2 in the Yi1 complex should be the pRb protein because cyclin D1-and D3-dependent kinases bind to and phosphorylate pRb in vitro (81), and kinases containing cyclin D2, but not D1 and D3, phosphorylate pRb in vivo and override its transcriptional suppression mediation (85). The finding of cyclin D1–cdk2 kinase in Yi1 (154) links the oncogenesis-related D1 cyclin (71) to a cycle-specific DNA-binding activity that is involved in regulating *TK* transcription and probably DNA synthesis.

Later in G_1 phase the MT2 motif also successively bound several complexes (155). Various proteins including E2F have been identified as components of these complexes. DNA-binding proteins of the E2F family, E2F-1, E2F-2, DP1, and others have been found in multiprotein transcriptional complexes with pRb, cyclins, and cdk2 (4,156,157). The MT2-binding complexes include E2F, cyclin E- and cyclin A-dependent cdk2 kinases, and p107 (155), a protein related to pRb (158). Mutation of MT2 causes constitutive expression of a reporter gene (159).

A current model is that E2F is the cycle-controlling factor. In G_1 phase (4), p107 (as well as pRb) complexes with and inactivates E2F protein. This E2F–p107-containing complex binds to MT2 DNA and blocks gene transcription. As evidence of this mechanism, Rb-minus cells do not arrest in G_1; transfected pRb stops them, and cotransfection with cyclin E or A reverses this block (96). Furthermore, a negative-feedback control loop by which pRb production is regulated by cyclins E and D has been proposed (102,103). At the restriction point, p107 in these E2F-containing complexes (160) is phosphorylated, similar to pRb (161), by cyclin E–cdk2 and cyclin A–cdk2 kinases (162). This reaction releases "free" E2F, which is proposed to be responsible for transcription of several genes, the products of which are required for DNA synthesis, including TK and dihydrofolate reductase (163).

Other aspects of this regulation are being discovered. E2F is synthesized in mid-to-late G_1. DP-1, which suppresses these promoters (157), is constitutively produced (163). The E2F–DP1 complex binds strongly to pRb, and inhibits E2F function. Phosphorylation of pRb in late G_1 releases this inhibition of the dependent transcriptions. In S and G_2, cyclin A-kinase binds to E2F–DP-1 heterodimers and specifically phosphorylates DP-1.

In interactions of these complexes, binding of Yi1 possibly has a down-regulatory effect on MT3 function, which works in cooperation with inhibition by pRb in the E2F complex on the

MT2 site. This double suppression could counteract activation by the permanently occupied Sp1 site during much of the G_1 phase. Consistent with this model, transformed cells express low levels of Yi1, but high constitutive levels of Yi2 (152).

IV. VARIATIONS OF THE BASIC CYCLE SCHEME

The cell cycle provides the basic pattern for events involved in cell proliferation and their regulation. It is altered in important ways during a number of physiological and disease processes. Some of these are briefly summarized in this section.

A. Differentiation

When cells undergo terminal differentiation, they withdraw from the cycle during G_1 phase and proliferation ceases (164,165). Extracellular factors, such as interferons and retinoids, induce differentiation, as can small molecules, including inhibitors of DNA methyltransferases. For example, removal of growth factors not only causes myogenic cells to stop growing, but also they fuse and differentiate into myotubes with production of special proteins, such as muscle creatine kinase. Intracellular myogenic factor proteins, including myoD, myogenin, myf5, and MRF4 are required (166). The Id protein sequesters myoD and, thereby, blocks differentiation of muscle; a balance of plus and minus regulatory factors again operates (167). A basis for the inverse relation between growth and differentiation of muscle cells is proposed to involve hypophosphorylation of pRb or p107, thereby maintaining inhibition of the E2F function needed for cell proliferation and also pRb's interaction with myoD and MEF2 binding to DNA (168). Growth factors hyperphosphorylate pRb and, thereby, overcome this exit from cycle and entry into differentiation. Cyclin D2- and D3-dependent kinases are involved; their overexpression prevents differentiation of granulocytes (169).

B. Senescence

Human cells progressively grow more slowly as they are passaged in culture; they stop growing after about 50 cycles, arresting late in the G_1 phase (170). The p53-inducible 21-kD a protein (SDI-1) that blocks cyclin–cdk activities rises strongly in senescent human cells (137). The senescent cells contain high levels of hypophosphorylated pRb and of inactive cyclin E–cdk2 and cyclin D–cdk2 complexes (171). The SV40 virus' T protein, and HPV E6 and E7 from human papilloma virus inactivate p53 and pRb; cells infected by these viruses become "immortal" (170). Telomerase, an enzyme that extends DNA at chromosomal ends, becomes inactive in senescent cells; it is high in tumor cells (172,173).

C. Cancer

Cancer is a vast subject. In relation to this chapter, it is characterized by defective growth control and differentiation. Cycle controls are defective in cancer cells.

Mutations are the main events that initiate human cancers, which require several accumulated genetic alterations. These mutations have many effects; some directly modify growth controls and others alter growth by activating angiogenesis, invasion, adhesion, apoptosis, or block cell senescence. The proliferative mutations fall into two general groups: mutated oncogenes stimulate proliferation, and tumor suppressor genes negatively regulate proliferation, and their regulation is lost with mutation (174). Genes that regulate mRNA and protein production of other genes often mutate, rather than the mutations being directly in the structural genes measured (175).

Viruses can cause cancer by introducing their genes into cells, thereby modulating growth controls (1,4). RNA viruses introduce oncogenes, such as v-*src*, v-*ras*, v-*myc*, on others that are mutated forms of normal genes that they supplant to activate G_1 events, in part replacing needs for stimulation by growth factors. More than one oncogene is required to transform the phenotype of a normal cell, somewhat in parallel with the requirement of several growth factors for proliferation through the G_1 phase.

DNA viruses, in contrast with RNA viruses, introduce inactivators of the tumor suppressor genes *p53* and *pRb*. The large T protein of SV40, and E6 and E7 of human papilloma virus, release the inhibition by *pRb* of DNA synthesis. Adenovirus *E1A* protein binds to multiprotein complexes that contain *pRb* and *E2F*, among the first to be described (176). Middle T protein of polyoma virus forms complexes with several proteins involved in growth control, including 14-3-3 (177).

Very many metabolic changes have been observed in cancer cells (1). Most of these are secondary to the basic alterations in control mechanisms. "Reversions" to more embryonic cycle control patterns are observed. As expected, requirements for growth factors and hormones are diminished or lost. These changes can be caused by autocrine production by the tumor cell, of a growth factor, such as transforming growth factor-α, which replaces epidermal growth factor or PDGF from the V-*sis* oncogene, by alterations in growth factor receptors, such as *erbB2*, or in signal transduction pathways, as by v-*src*.

Requirements for both the G_0 to G_1 transit and the G_1 to S restriction point are relaxed in tumors (3,174). The cells become resistant to partially inhibited protein synthesis, indicating that the proposed unstable restriction point protein can no longer be rate-limiting (3,28). Kinetic data are consistent with cyclin E or A being the altered restriction point protein (91). Cyclin E protein is both overexpressed and altered in size in nearly all of the many human tumor biopsies examined (94). Cyclin D is amplified in some tumors, and is overexpressed in others because of translocations (178–181). The molecular genetics of defective control in cancer has been reviewed (182).

Tumor suppressor genes block development of cancers. An example is maspin which inhibits cell invasion and the metastatic potential of breast tumors (183). Many tumors have mutated *p53* or *pRb* suppressor genes, with numerous consequences (184). For example, wild-type *p53* up-regulates thrombospondin, which blocks the angiogenesis necessary for extended tumor growth and is involved in the G_1 checkpoint and in angiogenesis (185).

D. Effects of DNA Damage on the Cycle

Checkpoints are important in coordinating the cell cycle and also after a cell's DNA is damaged. The G_1 and G_2 checkpoints maintain genetic stability because the damaged cells stop cycling, either before DNA is duplicated or before they enter mitosis. Thereby checkpoints prevent fixation of the damage in mutations caused by errors in DNA replication or partitioning, respectively. Progression through the cycle resumes after repair of the lesions (186).

Inducible protein, such as GADD45, which suppresses growth at the R point (188), appear after exposure to X-radiation (187) or alkylating agents. Normal, but not mutant *p53*, causes G_1 arrest after treatment of cells in G_1 phase with a DNA-damaging drug (189). *p53* rises after DNA damage (190) because of greater stability of the protein. It induces p21, a negative regulator or cyclin–cdk2 (188), and blocks activation of DNA polymerase by PCNA (39). Transcription of *mdm2* an inactivator of p53 is induced by p53, creating a negative-feedback loop (108).

The G_2 checkpoint delays entry into mitosis of damaged post-G_1 cells (191,192). DNA damage in *S. cerevisiae* blocks the cdc25 phosphatase, which prevents activation of cyclin

B–cdc2 kinase activity. In *S. pombe*, two genes *rad24* and *rad25* required for the G_2 checkpoint encode the 14-3-3 protein, which is also essential for cell proliferation and limits normal entry into M phase (193,194). Human papilloma virus genes *E6* and *E7* did not affect the G_2 checkpoint, indicating p53 and pRb do not have roles in this mechanism (195). Damage also blocks progression through S, by a p53-independent mechanism (33).

Caffeine eliminates the G_2 checkpoint; it lets damaged cells proceed through mitosis. Because insufficient time is available for repair before mitosis, chromosomes are shattered and the cells die (196). The hamster cell RCC1 mutant is defective in G_2 checkpoint control at nonpermissive temperature (197). RCC1 protein, together with Ran, is central to repair and many other nuclear processes (198). The rad mutants of *S. cerevisiae* were discovered owing to their greater ultraviolet radiation sensitivity (199). Rad 9 is defective in connecting DNA damage to G_2 arrest; the cells die after passing through M (200).

The DNA damage checkpoint mechanisms are defective in many cancers. *p53* is frequently mutated in cancers, a defect that results in rapid further mutational changes, owing to loss of G_1 checkpoint control (201). Cancer-prone mutations cause checkpoint defects. Li-Fraumeni syndrome is caused by a *p53* mutation; ataxia telangiectasia causes defects in checkpoints that make these cells radiation-sensitive and prone to develop into cancers (202). In contrast, xeroderma mutants have actual defects in DNA repair enzymology, as do novel repair mutants found in yeast and in colon cancer, which act at repetitive sequences (203). The relative importance of G_1 versus G2 checkpoints has been discussed (204).

E. Apoptosis

The process of apoptosis, also called programmed cell death, occurs naturally in cells that are deleted during development. It is also induced when two conflicting growth signals are activated. An example is seen on expression of the positive *myc* oncogene in serum starvation-blocked cells (205–207). Cyclin A is involved (56,208) as is tyrosine phosphorylation (209). Activation of *E2F* causes G_0 arrested cells to enter S phase and then undergo apoptosis (210). Unrepaired DNA damage induces p53-dependent apoptosis at the G_1 checkpoint (211). This is countered by the *Bcl-2* system, the overexpression of which prevents apoptosis caused by diverse cytotoxic insults (212). Acting together with *myc*, *Bcl-2* creates poorly controlled growth and produces lymphomas. Apoptosis thus has an anticancer role (213,214), and the loss of *p53* permits tumor progression by preventing apoptosis.

REFERENCES

1. Baserga R. The Biology of Cell Reproduction. Cambridge: Harvard University Press, 1985.
2. Cross F, Roberts J, Weintraub H. Simple and complex cell cycles. Annu Rev Cell Biol 1989; 5:341–395.
3. Pardee AB. G_1 events and regulation of cell proliferation. Science 1989; 246:603–608.
4. Murray A, Hunt T. The Cell Cycle an Introduction. New York: WH Freeman, 1993.
5. Ernst M, Gearing DP, Dunn AR. Functional and biochemical association of Hck with the LIF/IL-6 receptor signal transducing subunit of gp130 in embryonic stem cells. EMBO J 1994; 13:1574–1584.
6. Schneider C, King RM, Philipson L. Genes specifically expressed at growth arrest of mammalian cells. Cell 1988; 54:787–793.
7. Forsburg SL, Nurse P. Cell cycle regulation in the yeasts *Saccharomyces cerevisiae* and *Schizosaccharomyces pombe*. Annu Rev Cell Biol 1991; 7:227–256.
8. Oehlen B, Cross FR. Signal transduction in the budding yeast *Saccharomyces cerevisiae*. Curr Opin Cell Biol 1994; 6:836–841.
9. Hartwell LH. Twenty-five years of cell cycle genetics. Genetics 1991; 129:975–980.

10. Reed SI. The role of p24 kinases in the G_1 to S-phase transition. Annu Rev Cell Biol 1992; 8:529–561.

11. Norbury C, Nurse P. Animal cell cycles and their control. Annu Rev Biochem 1992; 61:441–470.

12. Ban V, Sager R. Distinctive traits of normal and tumor-derived human mammary epithelial cells expressed in a medium that supports long-term growth of both cell types. Proc Natl Acad Sci USA 1989; 86:1249–1953.

13. Polyak K, Kato J-Y, Solomon MI, Sherr CJ, Massague J, Roberts JM, Koff A. p27[KIP1], a cyclin–cdk inhibitor, links transforming growth factor-B and contact inhibition to cell cycle arrest. Genes Dev 1994; 8:9–22.

14. Hengst L, Dulic V, Slingerland JM, Lees E, Reed SI. A cell cycle-regulated inhibitor of cyclin-dependent kinases. Proc Natl Acad Sci USA 1994; 91:5291–5295.

15. Ullrich A, Schlessinger J. Signal transduction by receptors with tyrosine kinase activity. Cell 1990; 61:203–213.

16. Soprano KJ. W1-38 cell long-term quiescence model system: a valuable tool to study molecular events that regulate growth. J Cell Biochem 1994; 54:405–414.

17. Stiles CD, Isberg RR, Pledger WJ, Antoniades HN, Sher CD. Control of the BALB/c-3T3 cell cycle by nutrients and serum factors: analysis using platelet-derived growth factor and platelet-poor plasma. J Cell Physiol 1979; 99:395–405.

18. Cantrell DA, Smith KA. The interleukin-2 T-cell system: a new cell growth model. Science 1984; 224:1312–1316.

19. Bravo R. Growth factor inducible genes in fibroblasts. In: Habenicht A, ed. Growth Factors, Differentiation Factors and Cytokines. Heidelberg: Springer Verlag, 1989:324–343.

20. Campisi J, Medrano EE, Morreo G, Pardee AB. Restriction point control of cell growth by a labile protein: evidence for increased stability in transformed cells. Proc Natl Acad Sci USA 1982; 79:436–440.

21. Zetterberg A, Larsson, O. Coordination between cell growth and cell cycle transit in animal cells. Cold Spring Harbor Symp Quant Biol 1991; 56:137–147.

22. Keyomarsi K, Sandoval L, Band V, Pardee AB. Synchronization of tumor and normal cells from G_1 to multiple cell cycles by lovastatin. Cancer Res 1991; 51:3602–3609.

23. Sinensky M, Lutz RJ. The prenlyation of proteins. BioEssays 1992; 14:25–31.

24. Gong JP, Traganos F, Darzynkiewicz A. Use of the cyclin E restriction point to map cell arrest in G(1)—induced by N-butyrate, cycloheximide, staurosporin, lovastatin, mimosine and quercitin. Int J Oncol 1994; 4:803–808.

25. Gerdes J, Lemke H, Baisch H, Wacker HH, Schwab U, Stein H. Cell cycle analysis of a cell proliferation-associated human nuclear antigen defined by the monoclonal antibody Ki-67. J Immunol 1984; 133:1710–1715.

26. Reddy GPV. Cell cycle: regulatory events in G_1–S transition of mammalian cells. J Cell Biochem 1994; 54:379–386.

27. Fredrickson RM, Mushynski WE, Sonnenberg N. Phosphorylation of translation initiation factor eIF-4E is induced in a ras-dependent manner during nerve growth factor-mediated PC12 differentiation. Mol Cell Biol 1992; 12:1239–1247.

28. Rossow PW, Riddle VGH, Pardee AB. Synthesis of labile, serum dependent protein in early G_1 controls animal cell growth. Proc Natl Acad Sci USA 1979; 76:4446–4450.

29. Guda JM, Fridovich-Keil JL, Pardee AB. Posttranslational control of thymidine kinase messenger RNA accumulation in cells released from G_0–G_1 phase block. Cell Growth Regln 1993; 4:421–430.

30. Johnson LF. Posttranscriptional regulation of thymidylate synthase gene expression. J Cell Biochem 1994; 54:387–392.

31. Slansky JE, Li Y, Kaelin WG, Farnham PJ. A protein synthesis-dependent increase in E2F1 mRNA correlates with growth regulation of the dihydrofolate reductase promoter. Mol Cell Biol 1993; 13:1610–1618.

32. Hanauskeabel HM, Park MH, Hanauske AM, Lalande M, Folk JE. Inhibition of the G_1–S transition

of the cell cycle by inhibitors of deoxyhypusine hydroxylation. Biochim Biophys Acta 1994; 1221:115–124.

33. Larner JM, Lee H, Hamlin JL. Radiation effects on DNA synthesis in a defined chromosomal replicon source. Mol Cell Ciol 1994; 14:1901–1908.

34. Diffley JFX, Cocker JH, Dowell SJ, Rowley A. Two steps in the assembly of complexes at yeast replication origins in vivo. Cell 1994; 78:303–316.

35. Falaschi A, Giacca M, Zentilin L, Norio P, Diviacco S, Dimitrova D, Kumar S, Tuteja R, Biamonti G, Perini G, Weighardt F, Brito J, Riva S. Searching for replication origins in mammalian DNA. Gene 1994; 135:125–135.

36. Adachi Y, Laemmli UK. Study of the cell cycle-dependent assembly of the DNA pre-replicative centres in xenopus egg extracts. EMBO J 1994; 13:4153–4164.

37. Kornberg A, Baker TA. DNA Replication, 2nd ed. New York: WH Freeman, 1992.

38. Bell SP, Stillman B. ATP-dependent recognition of eukaryotic origins of DNA replication by a multiprotein complex. Nature 1992; 357:128–134.

39. Waga S, Hannon GJ, Beach D, Stillman B. The p21 inhibitor of cyclin-dependent kinases controls DNA replication by interaction with PCNA. Nature 1993; 369:574–578.

40. Hubscher U, Spadari S. DNA replication and chemotherarpy. Physiol Rev 1994; 74:259–304.

41. Prives C, Manfredi JJ. The p53 tumor suppressor protein; meeting review. Genes Dev 1993; 7:529–534.

42. Matson SW, Bean DW, George JW. DNA helicases: enzymes with essential roles in all aspects of DNA metabolism. BioEssays 1994; 16:13–22.

43. Jessberger R, Podust VN, Hubscher U, Berg P. A mammalian protein complex that repairs double strand breaks and deletions by recombination. J Biol Chem 1993; 268:15070–15079.

44. Hendrickson LA, Umbricht CB, Wold MS. Recombinant replication protein A: expression, complex formation, and functional characterization. J Biol Chem 1994; 269:11121–11132.

45. Li C, Cao L-G, Wang Y-L, Baril EF. Further purification and characterization of a multienzyme complex for DNA synthesis in human cells. J Cell Biochem 1993: 53:405–419.

46. Pardoll DM, Vogelstein B, Coffey D. A fixed site of DNA replication in eukaryotic cells. Cell 1980; 19:527–536.

47. Stein GS, Stein JL, van Wijnen A, Lian JB. Histone gene transcription: a model for responsiveness to an integrated series of regulatory signals mediating cell cycle control and proliferation differentiation interrelationships. J Cell Biochem 54; 1994:393–404.

48. Tobey RA. Different drugs arrest cells at a number of distinct stages in G_2. Nature 1975; 254:245–247.

49. Masui Y, Markert CL. Cytoplasmic control of nuclear behavior during meiotic maturation of frog oocytes. Exp Zool 1971; 177:129–145.

50. Heichman KA, Roberts JM. Rules to replicate by. Cell 1994; 79:557–562.

51. Prescott DM. Reproduction of Eukaryotic Cells. New York: Academic Press, 1976.

52. Fournier RE, Pardee AB. Cell cycle studies of mononucleate and cytochalasin B-induced binucleate fibroblasts. Proc Natl Acad Sci USA 1975; 72:869–873.

53. Coverley D, Laskey R. Regulation of eukaryotic DNA replication. Annu Rev Biochem 1994; 63:745–766.

54. Schlegel R, Pardee AB. Periodic mitotic events induced in the absence of DNA replication. Proc Natl Acad Sci USA 1987 84:9025–9029.

55. Schimke RT, Kung AL, Rush DF, Sherwood SW. Differences in mitotic control among mammalian cells. Cold Spring Harbor Symp Quant Biol 1991; 56:417–425.

56. Meikrantz W, Gisselbrecht S, Tam SW, Schlegel R. DNA replication and chemotherapy. Activation of cyclin A-dependent protein kinases during apoptosis. Proc Natl Acad Sci USA 1994; 91:3754–3758.

57. O'Connell MJ, Nurse P. How cells know they are in G_1 or G_2. Curr Opin Cell Biol 1994; 6:867–871.

58. Moreno S, Nurse P. Regulation of progression through the G_1 phase of the cell cycle by the *rum1+* gene. Nature 1994; 369:236–242.

59. Sluder G, Miller FJ, Thompson EA, Wolf DE. Feedback control of the metaphase–anaphase transition in sea urchin zygotes: role of maloriented chromosomes. J Cell Biol 1994; 126:189–198.

60. Hunt T. Cyclins and their partners: from a simple idea to complicated reality. Semin Cell Biol 1991; 2:213–222.

61. Evans T, Rosenthal E, Youngblom Y, Kistel D, Hunt T. Cyclin: a protein specified by maternal mRNA in sea urchin eggs that is destroyed at each cleavage division. Cell 1983; 33:389–396.

62. Swenson KI, Farrell KM, Ruderman JV. The clam embryo protein cyclin A induces entry into M phase and the resumption of meiosis in xenopus oocytes. Cell 1986; 47:861–870.

63. Richardson HE, Wittenberg C, Cross F, Reed SI. An essential G_1 function for cyclin-like proteins in yeast. Cell 1989; 59:1127–1133.

64. Nash R, Tokiwa G, Anand S, Erickson K, Futcher AB. The *WHI1+* gene of *Saccharomyces cerevisiae* tethers cell division to cell size and is a cyclin homolog. EMBO J 1988; 7:4335–4346.

65. Hadwiger JA, Wittenberg C, Richardson HE, de Barros Lopes M, Reed SI. A novel family of cyclin homologs that control G_1 in yeast. Proc Natl Acad Sci USA 1989; 86:6255–6259.

66. Sherr CJ, Mammalian G_1 cyclins. Cell 1993; 73:1059–1065.

67. Xiong Y, Beach D. Population explosion in the cyclin family. Curr Biol 1991; 1:362–364.

68. Pardee AB, Keyomarsi K. Modification of cell proliferation with inhibitors. Curr Opin Cell Biol 1992; 4:186–191.

69. Draetta JF. Mammalian G_1 cyclins. Curr Opin Cell Biol 1994; 6:842–846.

70. Xiong Y, Connolly T, Futcher B, Beach D. Human D-type cyclin. Cell 1991; 65:691–699.

71. Motokura T, Bloom T, Kim HG, Juppner H, Ruderman JV, Kronenberg HM, Arnold A. A novel cyclin encoded by a *bcl-1* linked candidate oncogene. Nature 1991; 350:512–515.

72. Matsushime H, Roussel MF, Ashman RA, Sherr CJ. Colony-stimulating factor 1 regulates novel cyclins during the G_1 phase of the cell cycle. Cell 1991; 65:701–713.

73. Matsushime H, Roussel MF, Sherr CJ. Novel mammalian cyclins (CYL genes) expressed during G_1. Symp Quant Biol 1991; 56:69–74.

74. Motokura T, Keyomarsi K, Kronenberg HM, Arnold A. Cloning and characterization of human cyclin D3, a cDNA closely related in sequence to the PRAD1/cyclin D1 proto-oncogene. J Biol Chem 1992; 28:20412–20415.

75. Xiong Y, Menninger J, Beach D, Ward DC. Molecular cloning and chromosomal mapping of CCND genes encoding human D-type cyclins. Genomics 1992; 13:575–584.

76. Inaba T, Matsushime H, Valentine M, Roussel MF, Sherr CJ, Look RT. Genomic organization, chromosomal localization, and independent expression of human *CYL* (cyclin D) genes. Genomics 1992; 13:565–574.

77. Quelle DE, Ashmun RA, Shurleff SA, Kato J-Y, Bar-Sagi D, Roussel MF, Sherr CJ. Overexpression of mouse D-type cyclins accelerates G_1 phase in rodent fibroblasts. Genes Dev 1993; 7:1559–1571.

78. Baldin V, Lukas J, Marcote MJ, Pagano M, Barteck J, Draetta G. Cyclin D1 is a nuclear protein required for cell cycle progression in G_1. Genes Dev 1993; 7:812–821.

79. Philipp A, Schneider A, Vasrik I, Finke K, Xiong Y, Beach D, Alitalio K, Eilers M. Repression of cyclin D1: a novel function of *myc*. Mol Cell Biol 1994; 14:4032–4043.

80. Noburi T, Miura K, Wu DJ, Lois A, Takabayashi K, Carson DA. Deletions of the cyclin-dependent Kinase-4 inhibitor gene in multiple human cancers. Nature 1994; 368:753–756.

81. Matsushime H, Ewen ME, Strom DK, Kato J-Y, Hanks SK, Rousssel M, Sherr CJ. Identification and properties of an atypical catalytic subunit (p34PSKJ3/cdk4) for mammalian D type G_1 cyclins. Cell 1992; 71:323–334.

82. Xiong Y, Zhang H, Beach D. D type cyclins associate with multiple protein kinases and the DNA replication and repair factor PCNA. Cell 1992; 71:505–514.

83. Meyerson M, Harlow E. Identification of G_1 kinase activity for cdk6, a novel cyclin D partner. Mol Cell Biol 1994; 14:2077–2086.

84. Matsushime H, Quelle DE, Shurtleff SA, Shibuya M, Sherr CJ, Kato J-Y. D-type cyclin-dependent kinase activity in mammalian cells. Mol Cell Biol 1994; 14:2066–2076.

85. Ewen ME, Sluss HK, Sherr CL, Matsushime H, Kato J-Y, Livingston DM. Functional interactions of the retinoblastoma protein with mammalian D-type cyclins. Cell 1993: 73:487–498.

86. Kato J-Y, Matsushime H, Hiebert SW, Ewen ME, Sherr CJ. Direct binding of cyclin D to the retinoblastoma gene product (pRb) and pRb phosphorylation by the cyclin D-dependent kinase, CDK4. Genes Dev 1993; 7:331–342.

87. Lew DJ, Dulic V, Reed SI. Isolation of three novel human cyclins by rescue of G_1 cyclin (cln) function in yeast. Cell 1991; 66:1197–1206.

88. Koff A, Cross F, Fisher A, Schumacher J, Leguellec K, Philippe M, Roberts JM. Human cyclin E, a new cyclin that interacts with two members of the CDC2 gene family. Cell 1991; 66:1217–1228.

89. Koff A, Giordano A, Desia D, Yamashita K, Harper JW, Elledge SJ, Nishomoto T, Morgan DO, Franza R, Roberts JM. Formation and activation of a cyclin E–cdk2 complex during the G_1 phase of the human cell cycle. Science 1992; 257:1689–1694.

90. Dulic V, Lees E, Reed SI. Association of human cyclin E with a periodic G_1–S phase protein kinase. Science 1992; 257:1958–1961.

91. Dou Q-P, Levin AH, Zhao S, Pardee AB. Cyclin E and cyclin A as candidates for the restriction point protein. Cancer Res 1993; 53:1493–1497.

92. Tsai L-H, Lees E, Faha B, Harlow E, Riabowol K. The cdk2 kinase is required for the G1-to-S transition in mammalian cells. Oncogene 1993; 8:1593–1602.

93. Ohtsubo M, Roberts JM. Cyclin-dependent regulation of G_1 in mammalian fibroblasts. Science 1993; 259:1908–1912.

94. Keyomarsi K, O'Leary N, Molnar G, Lees E, Fingert HJ, Pardee AB. Cyclin E, a potential prognostic marker for breast cancer. Cancer Res 1994; 54:380–385.

95. Keyomarsi K, Pardee AB. Redundant cyclin overexpression and gene amplification in breast cancer cells. Proc Natl Acad Sci USA 1993; 90:1112–1116.

96. Hinds PW, Mittnacht S, Dulic V, Arnold A, Reed SI, Weinberg RA. Regulation of retinoblastoma protein functions by ectopic expression of human cyclins. Cell 1992; 70:993–1006.

97. Gong J, Aradelt B, Traganos F, Darzynkiewicz Z. Unscheduled expression of cyclin B and cyclin E in several leukemic and solid tumor lines. Cancer Res 1994; 54:4285–4288.

98. Harlow E, Whyte P, Franza BR Jr, Schley C. Association of adenovirus early-region 1A proteins with cellular polypeptides. Mol Cell Biol 1986; 6:1579–1589.

99. Giordano A, Whyte P, Harlow E, Franza BR Jr, Beach D, Draetta G. A 60 kd cdc2-associated polypeptide complexes with the EIA protein in adenovirus-infected cells. Cell 1989 58:981–990.

100. Wang J, Chenivesse X, Henglein B, Bréchot C. Hepatitis B virus integration in a cyclin A gene in a hepatocellular carcinoma. Nature 1990; 343:555–557.

101. D'Urso G, Marraccino R, Marshak D, Roberts J. Cell cycle control of DNA replication by a homologue from human cells of the p34cdc2 protein kinase. Science 1990; 250:786–791.

102. Muller H, Lukas J, Schneider A, Warthoe P, Bartek J, Eliers M, Strauss M. Cyclin D1 expression is regulated by the retinoblastoma protein. Proc Natl Acad Sci USA 1994; 91:2945–2949.

103. Henglein B, Chenivesse X, Wang J, Eick D, Brechot C. Structure and cell cycle-regulated transcription of the human cyclin A gene. Proc Natl Acad Sci USA 1994; 91:5490–5494.

104. Guadagno TM, Ohtsubo M, Roberts JM, Assoian RK. A link between cyclin A expression and adhesion-dependent cell cycle progression. Science 1993; 262:1572–1575.

105. Dutta A, Stillman B. cdc2 family kinases phophorylate a human cell DNA replication factor, RPA, and activate DNA replication. EMBO J 1992; 11:2189–2199.

106. Blow JJ, Nurse P. A cdc2-like protein is involved in the initiation of DNA replication in xenopus egg extracts. Cell 1990; 62:855–862.

107. Donehower LA, Bradley A. The tumor suppressor p53. Biochim Biophys Acta 1993; 1155:181–205.

108. Levine A. The tumor suppressor genes. Annu Rev Biochem 1993; 62:623–651.

109. Faha B, Ewen ME, Tsai LH, Livingston DM, Harlow E. Interaction between human cyclin A and adenovirus E1A-associated p107 protein. Science 1992; 255:87–90.

110. Mudryj M, Devoto SH, Heibert SW, Hunter T, Pines J, Nevins JR. Cell cycle regulation of the *E2F* transcription factor involves an interaction with cyclin A. Cell 1991; 65:1243–1253.

111. Shirodkar S, Ewen M, DeCaprio JA, Morgan J, Livingston DM, Chittenden T. The transcription factor *E2F* interacts with the retinoblastoma product and a p107–cyclin A complex in a cell cycle-regulated manner. Cell 1992; 68:157–166.

112. King RW, Jackson PK, Kirschner MW. Mitosis in transition. Cell 1994; 8:563–571.

113. Pagano M, Pepperkok R, Verde F, Ansorge W, Draetta G. Cyclin A is required at two points in the human cell cycle. EMBO J 1992; 11:961–971.

114. Minshull J, Golsteyn R, Hill CS, Hunt T. The A- and B-type cyclin associated cdc2 kinases in xenopus turn on and off at different times in the cell cycle. EMBO J 1990; 9:2865–2875.

115. Glotzer M, Murray AW, and Kirschner MW. Cyclin is degraded by the ubiquitin pathway. Nature 1991; 349:132–138.

116. Pines J, Hunter T. Human cyclins A and B1 are differentially located in the cell and undergo cell cycle-dependent nuclear transport. J Cell Biol 1991; 115:1–17.

117. Pines J, Hunter T. Isolation of a human cyclin cDNA: evidence for cyclin mRNA and protein regulation in the cell cycle and for interaction with p34^{cdc2}. Cell 1989; 58:833–846.

118. Pines J. Cyclins:wheels within wheels. Cell Growth Differ 1991; 2:305–310.

119. DeBondt HL, Rosenblatt J, Jancarik J, Jones HD, Morgan DO, Kim S-H. Crystal structure of cyclin-dependent kinase 2. Nature 1993; 363:595–602.

120. Nurse P. Ordering S phase and M phase in the cell cycle. Cell 1994; 79:547–550.

121. Edgar BA, Sprenger F, Duronio RJ, Leopold P, O'Farrell PH. Distinct molecular mechanisms regulate cell cycle timing at successive stages of drosophila embryogenesis. Genes Dev 1994; 8:440–452.

122. Kuang J, Ashorn CL, Gonzalez-Kuyvenhoven M. cdc25 is one of the MPM-2 antigens involved in the activation of maturation promoting factor. Mol Biol Cell 1994; 5:135–145.

123. Nigg EA. The targets of cyclin-dependent protein kinases. Curr Opin Cell Biol 1993; 5:187–193.

124. Gerhart J, Wu M, Kirschner M. Cell cycle dynamics of an M-phase-specific cytoplasmic factor in *Xenopus laevis* oocytes and eggs. J Cell Biol 1984; 98:1247–1255.

125. Gould KL, Nurse P. Tyrosine phosphorylation of the fission yeast cdc2$_2$ protein kinase regulates entry into mitosis. Nature 1989; 342:39–45.

126. Murray AW, Kirschner MW. Cyclin synthesis drives the early embryonic cell cycle. Nature 1989; 339:275–280.

127. Nurse P. Universal control mechanism regulating onset of M-phase. Nature 1990; 344:503–508.

128. Pines J. The cell cycle kinases. Semin Cancer Biol 1994; 5:305–313.

129. Meyerson M, Harlow E. Identification of G$_1$ kinase activity for cdk6, a novel cyclin D partner. Mol Cell Bio 1994; 14:2077–2086.

130. van den Heuvel S, Harlow E. Distinct roles for cyclin-dependent kinases in cell cycle control. Science 1993; 262:2050–2054.

131. Tsai L-H, Takahashi T, Caviness VS, Harlow E. Activity and expression pattern of cyclin-dependent kinase 5 in the embryonic mouse nervous system. Development 1993; 199:1029–1040.

132. Coleman TR, Dunphy WG. cdc2 regulatory factors. Curr Opin Cell Biol 1994; 6:877–883.

133. Hunter T. Breaking the cycle. Cell 1993; 75:839–841.

134. Peter M, Herskowitz I. Joining the complex: cyclin-dependent kinase inhibitory proteins and the cell cycle. Cell 1994; 79:181–184.

135. Elledge SJ, Harper JW. cdk inhibitors: on the threshold of checkpoints and development. Curr Opin Cell Ciol 1994; 6:847–852.

136. Hunter T, Pines J. Cyclins and cancer II: cyclin D and CDK inhibitors come of age. Cell 1994; 79:573–582.

137. Noda AF, Ning Y, Venable S, Pereira-Smith OM, Smith JR. Cloning of senescent cell-derived inhibitors of DNA sythesis using an expression screen. Exp Cell Res 1994; 211:90–98.

138. Harper JW, Adami GR, Wei N, Keyomarsi K, Elledge SJ. The p21 cdk-interacting protein Cip1 is a potent inhibitor of G_1 cyclin-dependent kinases. Cell 1993; 75:805–816.

139. Gu Y, Turck SW, Morgan DO. Inhibition of cdk2 activity in vivo by an associated 20K regulatory subunit. Nature 1993; 366:707–710.

140. El-Deiry WS, Tokino T, Velculescu VE, Levy DB, Parsons R, Trent JM, Lin D, Mercer WE, Kinzler KW, Vogelstein B. *WAF-1*, a potential mediator of *p53* tumor suppression. Cell 1993; 75:817–825.

141. Jiang H, Lin J, Su A-Z, Collart FR, Huberman E, Fisher PB. Induction of differentiation in human promyelocytic HL-60 leukemia cells activates *p21*, *WAF1/CIP1* expression in the absence of *p53*. Oncogene 1994; 9:3397–3406.

142. Hannon GJ, Beach D. p15^{INK4B} is a potential effector of TGF-β-induced cell cycle arrest. Nature 1994; 371:257–261.

143. Serrano M, Hannon GJ, Beach, D. A new regulatory motif in cell-cycle control causing specific inhibition of cyclin D/CDK4. Nature 1993; 366:704–707.

144. Kamb A, Gruis NA, Weaver-Feldhaus J, Liu Q, Harshman K, Tautigian SV, Stockert E, Day FS, Johnson BE, Skolnick MH. A cell cycle regulator potentially involved in genesis of many tumor types. Science 1994; 264:436–440.

145. Nugroho TT, Mendenhall MD. An inhibitor of yeast cyclin-dependent protein kinase plays an important role in ensuring the integrity of daughter cells. Mol Cell Biol 1994; 14:3320–3328.

146. Schwob E, Bohm T, Mendenhall MD, Nasmith K. The B-type cyclin kinase inhibitor p40^{SIC1} controls the G_1 to S transition in *S. cerevisiae*. Cell 1994; 79:233–244.

147. Tjian R, Maniatis T. Transcriptional activation: a complex puzzle with few easy pieces. Cell 1994; 77:5–8.

148. Coppock DL, Pardee AB. Regulation of thymidine kinase activity in the cell cycle by a labile protein. J Cell Physiol 1985; 124:268–274.

149. Fridovich-Keil JL, Gudas, JM, Dou Q-P, Bouvard I, Pardee AB. Growth–responsive expression from the murine thymidine kinase promoter: genetic analysis of DNA sequences. Cell Growth Differ 1991; 2:67–76.

150. Dou Q-P, Fridovich-Keil JL, Pardee AB. Inducible protein binding to the murine thymidine kinase promoter in late G_1/S phase. Proc Natl Acad Sci USA 1991; 88:1157–1161.

151. Dou Q-P, Markell PJ, Pardee AB. Thymidine kinase transcription is regulated at G_1/S phase by a complex that contains retinoblastoma-like protein and a cdc2 kinase. Proc Natl Acad Sci USA 1992; 89:3256–3260.

152. Bradley DW, Fridovich-Keil, JL, Gudas JM, Pardee AB. Serum-responsive expression from the murine thymidine kinase promoter is specifically disrupted in a transformed cell line. Cell Growth Differ 1994; 5:1137–1143.

153. Molnar G, Crozat A, Pardee AB. The immediate-early gene *Egr-1* regulates the activity of the thymidine kinase promoter at the G_0 to G_1 transition of the cell cycle. Mol Cell Biol 1994; 14:5242–5248.

154. Dou Q-P, Molnar G, Pardee AB. Cyclin D1/cdk2 kinase is present in a G_1 phase-specific protein complex Yi1 that binds to the mouse thymidine kinase gene promoter. Biochem Biophys Res Commun 1994; 205:1859–1868.

155. Dou Q-P, Zhao S, Levin AH, Wang J, Helin K, Pardee AB. G_1/S-regulated E2F-containing protein complexes bind to the mouse thymidine kinase gene promoter. J Biol Chem 1994; 269:1306–1313.

156. Nevins JR. *E2F*: a ling between the *Rb* tumor suppressor protein and viral oncoproteins. Science 1992; 258:424–429.

157. Lam E W-F, La Thangue NB. DP and E2F proteins: corrdinating transcription with cell cycle progression. Curr Opin Cell Biol 1994; 6:859–866.

158. Ewen ME, Faha B, Harlow E, Livington DM. Interaction of p107 with cyclin A independent of complex formation with viral oncoproteins. Science 1992; 255:85–87.

159. Fridovich-Keil JL, Markell PF, Gudas JM, Pardee AB. DNA sequences required for serum-responsive

regulation of expression from the mouse thymidine kinase promoter. Cell Growth Differ 1993; 4:679–687.

160. Cobrink D, Dowdy SF, Hinds PW, Mittnacht S, Weinberg RA. The retinoblastoma protein and the regulation of cell cycling. Trends Biochem Sci 1992; 17:312–315.

161. Hollingsworth RE Jr, Chen P-L, Lee W-H. Integration of cell cycle control with transcriptional regulation by the retinoblastoma protein. Curr Opin Cell Biol 1993; 5:194–200.

162. Krek W, Ewen ME, Shirodkar S, Arany Z, Kaelin WG Jr, Livingston DM. Negative regulation of the growth-promoting transcription factor *E2F-1* by a stably bound cyclin A-dependent protein kinase. Cell 1994; 78:161–172.

163. Li Y, Slansky J, Myers DJ, Drinkwater N, Kaelin WG, Farnham PJ. Cloning, chromosomal location and characterization of mouse *E2F-1*. Mol Cell Biol 1994; 14:1861–1869.

164. Francis GE, Cunningham JM. Growth and differentiation control. Cancer Chemother Biol Response Modif Annu 1990; 11:142–176.

165. Spiegelman B, Watt FM. Cell differentiation. Curr Opin Cel Biol 1994; 6:781–831.

166. Olson EN, Klein WH. bHLH factors in muscle development: dead lines and commitments, what to leave in and what to leave out. Genes Dev 1994; 8:1–8.

167. Benezra R, Davis RL, Lockshon T, Turner DL, Weintraub H. The protein Id: a negative regulator of helix–loop–helix DNA binding proteins. Cell 1990; 61:49–59.

168. Gu W, Schneider JW, Condorelli G, Kaushl S, Mahadevi V, Nadel-Ginard B. Interaction of myogenic factors and the retinoblastoma protein mediates muscle cell commitment and differentiation. Cell 1993; 72:309–324.

169. Kato J-Y, Sherr CJ. Inhibition of granulocyte differentiation by G_1 cyclins D2 and D3 but not D1. Proc Natl Acad Sci USA 1993; 90:11513–11517.

170. Stein GH, Yanishevsky RM, Gordon L, Beeson M. Carcinogen-transformed human cells are inhibited from entry into S phase by fusion to senescent cells but cells transformed by DNA tumor viruses overcome the inhibition. Proc Natl Acad Sci USA 1982; 79:5287–5291.

171. Stein GH, Beeson M, Gordon L. Failure to phosphorylate the retinoblastoma gene product in senescent human fibroblasts. Science 1990; 249:666–668.

172. Kim NW, Piatyszek MA, Prowse KR, Harley CB, West MD, Ho PLC, Coviello GM, Wright WE, Weinrich SL, Shay JW. Specific association of human telomerase activity with immortal cells and cancer. Science 1994; 266:2011–2015.

173. Counter CM, Hirtle HW, Bacchetti S, Harley CB. Telomerase activity in human ovarian cancer. Proc Natl Acad Sci USA 1994; 91:2900–2904.

174. Kohn KW, Jackman J, O'Connor PM. Cell cycle control and cancer chemotherapy. J Cell Biochem 1994; 54:440–452.

175. Sager R. RNA genetics in cancer; shifting the focus from DNA to RNA. 1996; in press.

176. Nevins J. Cell cycle targets of the DNA tumor viruses. Curr Opin Genet Dev 1994; 4:130–134.

177. Pallas DC, Fu H, Haehnel LC, et al. Association of polyomavirus middle tumor antigen with 14-3-3 proteins. Science 1994; 265:535–537.

178. Jiang W, Zhang Y-J, Kahn SM, Hollstein MC, Santella RM, LU S-H, Harris CC, Montesano R, Weinstein IB. Altered expression of the cyclin D1 and retinoblastoma genes in human esophageal cancer. Proc Natl Acad Sci USA 1993; 90:9026–9030.

179. Williams ME, Swerdlow SH, Rosenberg CL, Arnold A. Chromosome 11 translocation break points at the PRAD1/cyclin D gene locus in centrocytic lymphoma. Leukemia 1993; 7:241–245.

180. Buckley MF, Sweeney KJ, Hamilton JA, Sini RL, Manning DL, Nicholson RI, deFazio A, Watts CK, Musgrove EA, Sutherland RL. Expression and amplification of cyclin genes in human breast cancer. Oncogene 1993; 8:2127–2133.

181. Leach FS, Elledge SJ, Sherr CJ, Wilson JK, Markowitz, Kinzler KW, Vogelstein B. Amplification of cyclin genes in colorectal cancer. Cancer Res 1993; 53:1986–1989.

182. Hartwell LH, Kastan MB. Cell cycle control and cancer. Science 1994; 266:1821–1828.

183. Zou Z, Anisowicz A, Hendrix MJC, Thor A, Neveu M. Sheng S, Rafidi K, Seftor E, Sagar R.

Identification of a novel serpin with tumor suppressing activity in human mammary epithelial cells. Science 1994; 263:526–529.

184. Picksley SM, Lane DP. *p53* and *Rb*: their cellular roles. Curr Opin Cell Biol 1994; 6:853–858.

185. Folkman J. Angiogenesis in cancer, vascular, rheumatoid and other disease. Mol Med 1995; 1:27–31.

186. Shivji MKK, Kenny MK, Wood RD. Proliferating cell nuclear antigen is required for DNA excision repair. Cell 1992; 69:367–374.

187. Boothman DA, Bouvard I, Hughes EN. Identification and characterization of X-ray-induced proteins in human cells. Cancer Res 1989; 49:2871–2878.

188. Kastan MB, Zahn Q, El-Deiry W, Carrier F, Jacks RT, Walsh WV, Plunkett BS, Vogelstein B, Fornace AJ Jr. A mammalian cell cycle checkpoint pathway utilizing p53 and GADD45 is defective in ataxia-telangiectasia. Cell 1992; 71:587–597.

189. Lu X, Lane DP. Differential induction of transcription by active p53 following UV or ionizing radiation: defects in chromosomal instability syndromes. Cell 1993; 75:765–778.

190. Dulic V, Kaufmann WK, Wilson SJ, Tlsty TD, Lees E, Harper JW, Elledge ST, Reed SI. p53-dependent inhibition of cyclin-dependent kinase activities in human fibroblasts during radiation-induced G_1 arrest. Cell 1994; 76:1013–1023.

191. O'Conner PM, Ferris DK, Pagano M, Draetta G, Pines J, Hunter T, Longo DL, Kohn KW. G_2 delay induced by nitrogen mustard in human cells affect cyclin A/cdk2 and cyclin B/cdc2-kinase complexes differently. J Biol Chem 1993; 268:8298–8303.

192. Sheldrick KS, Carr AM. Feedback controls and G_2 checkpoints. BioEssays 1993; 15:775–782.

193. Ford JC, Al-Khodairy F, Fotou E, Sheledrick KS, Griffiths DJF, Carr AM. 14-3-3 protein homologs required for the DNA damage checkpoint in fission yeast. Science 1994; 265:533–535.

194. Walworth N, Davey S, Beach D. Fission yeast chk1 protein kinase links the *rad* checkpoint pathway to cdc2. Nature 1993; 363:368–371.

195. Kuerbitz SJ, Plunkett BS, Walsh VW, Kastan MB. Wild-type *p53* is a cell cycle checkpoint determinant following irradiation. Proc Natl Acad Sci USA 1992; 89:7491–7495.

196. Lau CC, Pardee AB. Mechanism by which caffeine potentiates lethality of nitrogen mustard. Proc Natl Acad Sci USA 1982; 79:2942–2946.

197. Dasso M. RCC1 in the cell cycle: the regulation of chromosome condensation takes on new roles. Trends Biochem Sci 1993; 18:96–101.

198. Moore MS, Blobel G. A G protein involved in nucleocytoplasmic transport: the role of Ran. Trends Biochem Sci 1994; 19:211–216.

199. Guzdar SN, Sung P, Bailly V, Prakash L, Prakash S. RAD25 is a DNA helicase required for DNA repair and RNA polymerase II transcription. Nature 1994; 369:578–581.

200. Weinert TA, Hartwell LH. Cell cycle arrest of cdc mutants and specificity of the RAD9 checkpoint. Genetics 1993; 134:63–80.

201. Livinstone LR, White A, Sprouse J, Livanos E, Jacks T, Tlsty TD. Altered cell cycle arrest and gene amplification preferentially accompany loss of wild type p53. Cell 1992; 73:923–935.

202. Hanawalt P, Sarasin A. Cancer prone diseases with DNA processing abnormalities. Trends Genet 1986; 2:124–129.

203. Fishel R, Lescoe MK, Rao MRS, Copeland NG, Jenkins NA, Garber J, Kane M, Kolodner R. The human mutator gene homolog MSH2 and its association with hereditary nonpolyposis colon cancer. Cell 1993; 75:1027–1038.

204. Murnane JP, Schwartz JL. Cell checkpoint and radiosensitivity. Nature 1993; 365:22.

205. Evan G, Wyllie AH, Gilbert CS, Littlewood TD, Land H, Brooks M, Waters CM, Penn LZ, Hancock DC. Induction of apoptosis in fibroblasts by c-*myc* protein. Cell 1992; 69:119–128.

206. Raff MC. Social controls on cell survival and death. Nature 1992; 356:397–400.

207. Harrington EA, Fanidi A, Evan G. Oncogenes and cell death. Curr Opin Genet Dev 1994; 4:120–129.

208. Hoang AT, Cohen KJ, Barrett JF, Bergstrom DA, Dang CV. Participation of cyclin A in *myc*-induced apoptosis. Proc Natl Acad Sci USA 1994; 91:6875–6879.

209. Yousefi S, Green DR, Blaser K, Simon H-U. Protein–tyrosine phosphorylation regulates apoptosis in human eosinophils and neutrophils. Proc Natl Acad Sci USA 1994; 91:10868–10872.

210. Wu X, Levine AJ. *p53* and *E2F-1* cooperate to mediate apoptosis. Proc Natl Acad Sci USA 1994; 91:3602–3606.

211. Kastan MB, Onykwere O, Sidransky D, Vogelstein B, Craig RW. Participation of the p53 protein in the response to cellular damage. Cancer Res 1991; 51:6304–6311.

212. Linette GP, Korsmeyer SJ. Differentiation and cell death: lessons from the immune system. Curr Opin Cell Biol 1994; 6:809–815.

213. Green DR, Bissonnette RP, Cotter TG. Apoptosis and cancer. In: DeVita VT, Hellman S, Rosenberg SA, eds. Important Advances in Oncology 1994. Philadelphia: JB Lippincott 1994.

214. Fisher DE. Apoptosis in cancer therapy: crossing the threshold. Cell 1994; 78:539–542.

5

Gene Organization and Regulation

David S. Latchman

University College London Medical School, London, England

I. INTRODUCTION

Each of the very many different cell types in a mammal or other higher eukaryote synthesizes different proteins; for example, immunoglobulin in antibody-producing B cells, insulin in the pancreas, interleukin-2 (IL-2) and granulocyte–macrophage colony-stimulating factor (GM-CSF) in activated T cells, and so on. In contrast, however, with very few exceptions, the genes encoding these different proteins are identical in all different cell types and in the fertilized egg from which all the different cell types arise in embryonic development. Similarly, addition of factors such as IL-2 or GM-CSF to cells bearing the appropriate receptor results in the synthesis of novel proteins, encoded by specific genes that were previously inactive. Clearly, therefore, eukaryotic gene expression must be regulated so that different genes within the DNA are active in producing proteins in each different cell type or in response to specific stimuli.

II. LEVELS OF GENE REGULATION

Several stages exist between the DNA itself and the production of a particular protein (Fig. 1). Thus, the DNA must first be transcribed into a primary RNA transcript, which is subsequently modified at both ends by the addition of a 5′ cap and a 3′ tail of adenosine residues. Moreover, within this primary transcript, the RNA sequences that actually encode the protein are not present as one continuous block. Rather, they are broken up into segments (exons), which are separated by intervening sequences (introns) that do not contain any protein-coding information. Because these introns interrupt the protein-coding region and would prevent the production of an intact protein, they must be removed by the process of RNA splicing (2) before the mature messenger RNA (mRNA) can be transported from the nucleus to the cytoplasm and translated into protein.

In principle, any of these stages could be regulated so that specific proteins are produced only in certain tissues or in response to specific stimuli. Thus, regulation might take place by deciding which genes are transcribed into an RNA product or by regulating which of these RNA products are correctly spliced by removal of intervening sequences, allowing them to produce a

97

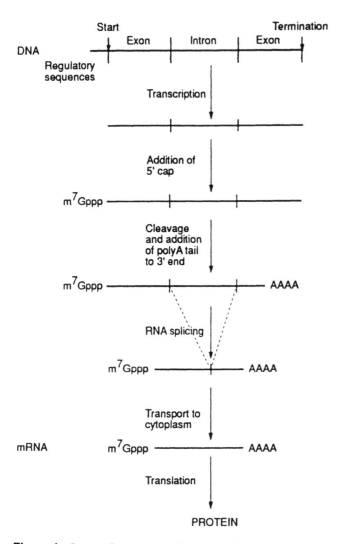

Figure 1 Stages of gene expression that could be regulated.

functional mRNA capable of encoding a protein. Similarly, gene regulation could occur by deciding which of these fully spliced mRNAs are transported to the cytoplasm or by regulating which mRNAs are translated into protein by the cytoplasmic ribosomes.

Indeed, there is evidence that, in specific cases, gene regulation is regulated at each of these stages (3). Thus, the production of many new proteins in the egg following fertilization is dependent on the translation into protein of fully spliced, RNAs that preexisted in the cytoplasm of the unfertilized egg, the translation of which was blocked before fertilization. This form of gene regulation is known as translational control. Similarly, by splicing the protein-coding regions (exons) of a single primary transcript in different combinations, it is possible to produce two or more different mRNAs, encoding different proteins in different tissues. This process of alternative splicing (4) is well illustrated in the single gene that encodes both the calcium-modulating hormone, calcitonin and the potent vasodilator calcitonin gene-related peptide (CGRP). Thus, this gene is transcribed into a primary RNA transcript both in the thyroid gland

and in the brain, but a different combination of exons are spliced together in each cell type to produce the calcitonin mRNA in the thyroid and the CGRP mRNA in the brain (5; Fig. 2).

Although there are several examples of gene regulation after the first stage of transcription, a wide variety of evidence indicates that usually gene regulation is achieved at the initial stage of transcription, by deciding which genes should be transcribed into the primary RNA transcript (2). In these instances, once transcription has occurred, all the other stages in gene expression shown in Figure 1 follow and the corresponding protein is produced. Thus the myosin gene is transcribed only in muscle cells, resulting in myosin being produced only in this cell type, whereas the immunoglobulin gene is transcribed only in B lymphocytes, which produce immunoglobulin.

Similarly, the genes encoding cytokines, such as IL-2 and GM-CSF, are not transcribed in most cell types or in resting T cells, but strongly transcribed following T-cell activation, resulting in the observed production of these cytokines only after cellular activation (6). In turn, the ability of GM-CSF itself to induce increased production of IL-1β by human blood mononuclear phagocytes is dependent, at least partly, on enhanced transcription of the IL-1β gene (7).

Indeed, even in the calcitonin–CGRP example, alternative splicing is acting as a supplement to transcriptional control because the calcitonin–CGRP gene is only transcribed in the thyroid gland and the brain and not in other tissues. Thus, the regulation of gene transcription plays a critical role in the regulation of gene expression and, therefore, the means by which this is achieved is discussed in the remainder of this chapter.

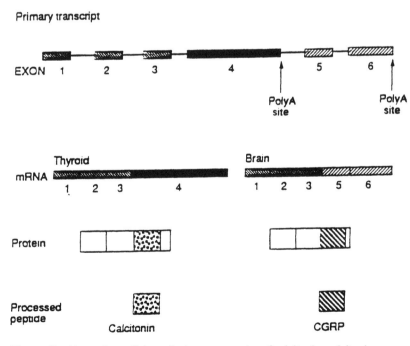

Figure 2 Alternative splicing of primary transcript of calcitonin–calcitonin gene-related peptide (CGRP) gene in brain and thyroid cells. Splicing followed by proteolytic cleavage of precursor protein produced in each tissue yields calcitonin in thyroid and calcitonin gene-related peptide in brain.

III. TRANSCRIPTIONAL CONTROL

A. DNA Sequences

For a gene to be transcribed, it is necessary for specific protein transcription factors (8,9) to bind to particular DNA-binding sites in the regulatory regions of the gene and induce its transcription by the enzyme RNA polymerase.

In most eukaryotic genes, these binding sites are organized into four distinct elements (Fig. 3). Immediately adjacent to the start site of transcription is the promoter, which in many genes contains the TATA box that acts as a binding site for the constitutively expressed transcription factor TFIID. The binding of this factor to the promoter plays a critical role in the assembly of a basal, stable transcriptional complex containing a variety of other transcription factors (TFIIA, B, E, and F) as well as RNA polymerase II itself (10).

The level of transcription directed by the basal transcriptional complex binding at the promoter is greatly enhanced by the binding of other constitutively expressed factors to so-called upstream promoter elements (UPEs), which are usually located immediately upstream from the promoter itself (11). Although the nature of the UPEs varies between different genes, very many genes contain either or both the CCAAT box that binds a variety of different transcription factors and a GC-rich sequence that binds the constitutively expressed transcription factor Sp1.

The DNA sequences discussed so far all bind constitutively expressed factors and, therefore, direct a similar level of transcription in all cell types. In addition, however, very many genes contain other regulatory DNA sequences, which are interdigitated with the UPEs and which bind transcription factors that become active only in specific cell types or in response to a specific signal. Thus, the presence of these sequences can confer a specific expression pattern on a particular gene. For example, the presence of the specific-binding site recognized by the heat-shock transcription factor will render a gene inducible by elevated temperature, whereas the presence of the unrelated site that binds the glucocorticoid receptor will render the gene inducible by treatment with glucocorticoid hormones. Similarly, a fragment of the IL-2 promoter extending from −52 to 326 contains binding sites for the inducible transcription factors NFκB and NF-AT, and this region can activate transcription following T-cell activation (12). Hence, the transcriptional activity of the genes encoding specific cytokines is dependent on the presence of binding sites for specific transcription factors. In turn, the ability of the cytokines themselves to induce the activation of other genes is dependent on these genes possessing binding sites for specific

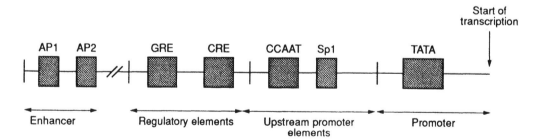

Figure 3 Structure of a typical eukaryotic gene with a TATA box, containing promoter, upstream promoter elements (UPEs) such as the CCAAT and Sp1 boxes, regulatory elements inducing expression in specific cell types or in response to specific treatments such as AMP (CRE) or glucocorticoid hormones (GRE), and other elements within more distant enhancers. Note that, as discussed in the text, UPEs are often interdigitated with regulatory elements, whereas the same regulatory elements are found both upstream from the promoter and in enhancers.

transcription factors that are synthesized or activated following signaling by the cytokine. Thus, for example, treatment of hepatoma cells with IL-6 results in the transcriptional activation of genes, such as those encoding haptoglobulin and hemopexin, that contain binding sites for the transcription factor NF-IL6 which is activated following 1L-6 treatment (13).

All these DNA sequences are located close to the start site of transcription, with the promoter itself being immediately adjacent to the start site, and the UPEs and the regulatory DNA sequences being interdigitated with one another upstream from the promoter. However, the transcription of eukaryotic genes can also be regulated by more distant elements, known as enhancers (14,15). These enhancers contain binding sites for the same constitutively expressed and regulatory factors that are found immediately upstream from the promoter but often contain multiple-binding sites for these factors, or sites for the binding of many different factors. This combination of binding sites allows enhancers to function when located at great distances from the promoter, when placed upstream or downstream from it and when oriented in either direction relative to the promoter. Thus, although the enhancer cannot drive transcription itself, it can enhance the activity of the promoter by several orders of magnitude. Such enhancement may occur in all cell types if the enhancer contains binding sites for constitutively expressed transcription factors, or it may occur only in a specific tissue or in response to a specific signal if the enhancer contains binding sites for factors that are involved in gene regulation.

B. Activation of Transcription

Restricted Expression of Response Element-Binding Proteins

In many genes the binding of constitutively expressed transcription factors, such as Sp1 and the CCAAT box binding factors to UPEs, will result in sufficient activation of the basal transcriptional complex for high-level transcription to occur. These genes, therefore, will be constitutively expressed in all tissues. In other genes, however, such as those encoding tissue-specific cytokines, it will be necessary for other transcription factors to bind to specific regulatory DNA sequences upstream from the promoter or in the enhancer for high-level transcription to occur. Because the factors that bind to these regulatory DNA sequences are active only in specific cell types or following a specific signal, these genes will be expressed only in a limited range of cell types or following exposure to a particular stimulus. The combination of particular binding sites in a particular gene determines the transcription factors that bind to it and, in turn, the presence or absence of these factors determines in which cell type(s) the gene is transcribed.

It is clear, therefore, that a critical role in the control of eukaryotic gene expression is played by the processes that regulate the activity of specific transcription factors such that they stimulate the expression of particular genes only at the appropriate time or place. In some cases, such specific stimulation results from the transcription factor itself being synthesized in only one cell type and being absent in all other cell types (Fig. 4a). Thus, for example, the immunoglobulin genes contain a binding site for the octamer-binding transcription factor Oct-2 upstream from the TATA box. The Oct-2 factor is synthesized only in B lymphocytes and, hence, binds and activates the immunoglobulin genes only in antibody-producing B cells (16).

Another example is seen in genes such as that encoding creatine kinase. These genes are expressed only in muscle cells and are regulated by the MyoD transcription factor, which is also present only in muscle cells. This is an even more dramatic example than immunoglobulin, however, because the artificial expression of MyoD in nonmuscle cells, such as fibroblasts, is sufficient to convert them into muscle cells, indicating that MyoD activates transcription of all the genes whose protein products are necessary to produce a differentiated muscle cell (17).

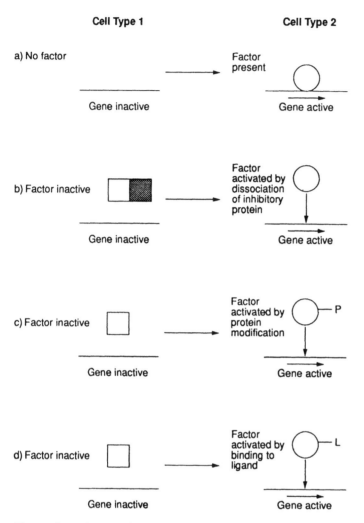

Figure 4 Activation of transcription factors by (a) new synthesis; (b) dissociation of an inhibitory protein (shaded box); (c) protein modification [e.g., by phosphorylation (p)]; or (d) ligand binding.

Unlike the case of MyoD, the expression of Oct-2 alone is not sufficient to produce differentiated B cells. This is because other transcription factors that are specifically active in B cells are also involved in producing the expression of genes specific to B cells, such as those encoding the immunoglobulins. One such factor is NFκB, which binds to a DNA sequence in the regulatory region of the immunoglobulin κ-light–chain gene. Interestingly, unlike Oct-2, the NFκB protein is present in all cell types. In most cells, however, it is present in an inactive form in which it is complexed with an inhibitory protein IκB, resulting in it being restricted to the cell cytoplasm. In mature B cells, however, NFκB is released from the inhibitory protein and moves to the nucleus, where it can bind to its DNA target sequence and activate the transcription of the immunoglobulin κ-light–chain gene (18; see Figure 4b).

Interestingly, this activation of NFκB also occurs when resting T lymphocytes are activated by antigenic stimulation. This activation of NFκB is partially responsible for the transcriptional activation of genes, such as those encoding IL-2 (19) and GM-CSF (20) following T-cell activation when NFκB dissociates from IκB and binds to its specific DNA-binding site in the regulatory region of these genes. Similarly, treatment of human peripheral blood monocytes with IL-2 itself results in activation of NFκB, which binds to the promoter of the gene encoding macrophage colony-stimulating factor thereby allowing IL-2 to activate the transcription of this gene (21).

Hence, the action of transcription factors on gene expression can be controlled not only by regulating their synthesis (see Fig. 4a), but also by the regulation of their activity (see Fig. 4b–d). The combination of these two processes, allows transcription factors to regulate the expression of numerous different genes in different cell types (8,9).

Posttranslational Modification of Response Element-Binding Proteins

As shown in Figure 4 several different mechanisms exist by which transcription factors can be converted from an inactive to an active form in response to a specific stimulus. Thus, in addition to the dissociation of an inhibitory protein, as in NFκB (see Fig. 4b), transcription factors can also be activated by posttranslational modifications, such as phosphorylation (see Fig. 4c).

One example of this is provided by the cyclic-AMP response element-binding protein (CREB) transcription factor. This factor mediates the activation of several cellular genes following AMP treatment. Cyclic-AMP stimulates the protein kinase A enzyme. In turn, this enzyme phosphorylates CREB, which enhances the ability of the CREB protein to activate transcription. Thus, stimulation of gene expression by AMP is mediated by its stimulation of protein kinase A and the consequent phosphorylation of CREB (22). Phosphorylation is also involved in the activation of NKκB in T-cell activation. In this latter event, however, it is the inhibitory protein IκB associated with NKκB (see foregoing) that is phosphorylated, causing it to dissociate from the NFκB protein and thus allowing NFκB to move to the nucleus, bind to DNA, and activate transcription (18).

In addition to dissociation of protein–protein interactions and posttranslational changes, transcription factors can also be activated by conformational changes induced by binding of a specific ligand (see Fig. 4d). This is seen in the ACE1 factor in yeast, which undergoes a conformational change following binding of copper. This conformational change converts ACE1 to an active form and results in the activation of specific ACE1-dependent genes in response to copper (23).

A similar dependence on an activating ligand is also observed in mammalian cells in the glucocorticoid receptor. This receptor must bind the appropriate steroid hormone to activate transcription of target genes. A region at the C-terminus of the receptor, which binds the appropriate hormone, has been identified. It was thought that binding of the hormone to the receptor activates its ability to bind to DNA and, thereby, switch on transcription. It is now clear that, although in vivo the receptor binds to DNA only in the presence of hormone, in vitro it can do so even in its absence. This suggests that in vivo such DNA binding is prevented by some other anchorage protein, with the hormone acting to release the receptor from this factor, thereby allowing it to bind DNA. In support of this hypothesis several steroid receptors, including the glucocorticoid receptor, have been shown to be associated with the 90-kDa heat-inducible protein hsp90 before hormone treatment. The receptor dissociates following the addition of hormone (24).

Therefore, a variety of mechanisms involving both increased synthesis and protein activation by ligand binding, protein modification, or disruption of protein–protein interaction (see Fig. 4) coexist allowing specific factors to become active in response to a particular signal or in a particular cell type. Occasionally, combinations of these mechanisms are used, such that the NFκB factor is activated by a combination of phosphorylation and dissociation of an inhibitory protein, whereas the glucocorticoid receptor is activated by a similar dissociation of an inhibitory protein which, in this instance, is induced by ligand-binding to the receptor. The combination of these different mechanisms allows the regulation of gene expression in a cell type-specific and inducible manner.

C. Repression

Although most constitutively expressed or regulated transcription factors act by activating the transcription of specific genes, it is also possible for transcription to be specifically inhibited by the action of a transcription factor (24,25; Fig. 5). As with activating factors, the synthesis or activation of such an inhibitory factor in particular tissues or in response to a signal can specifically regulate the expression of particular genes.

Interestingly, the same factor can frequently activate certain genes while repressing others. Thus, for example, as well as activating specific steroid-responsive genes by binding to a particular DNA-binding site, the glucocorticoid receptor can also repress other genes by binding to a distinct, but related, DNA-binding site. Following binding to this negative site, the receptor cannot activate transcription itself and acts by preventing another positively acting factor binding to this site, thereby inhibiting transcription (26; see Fig. 5a). In a similar, but distinct, mechanism

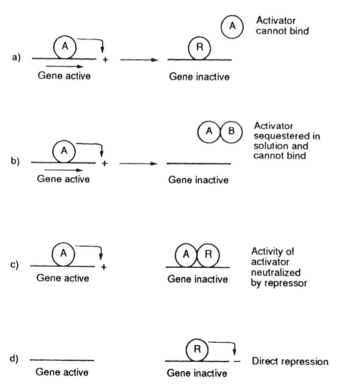

Figure 5 Potential mechanisms by which a transcription factor can inhibit transcription include (a) binding to DNA and preventing a positively acting factor from binding; (b) interacting with the activator in solution and preventing it from binding; (c) binding to DNA with the activator and preventing it from activating transcription; or (d) direct repression.

the glucocorticoid receptor also inhibits genes to which it cannot bind by complexing with the activating transcription factor AP1 in solution. This prevents the AP1 from binding to the DNA and, thereby, inhibits transcription (27; see Fig. 5b).

These examples illustrate two mechanisms by which a factor can inhibit gene expression by preventing the binding to DNA of another activating factor (see Fig. 5a,b). Another way that an inhibitory transcription factor can act is by interfering with the activation of transcription mediated by a bound factor in a phenomenon known as quenching (see Fig. 5c). This is seen in the yeast mating type α_2-protein which binds to a site adjacent to that bound by the MCM1 transcriptional activator protein. This masks the region of the MCM1 protein that is necessary for the activation of transcription and prevents it from activating the α-specific genes (28).

In all these examples, the negative factor exerts its inhibiting effect by neutralizing the action of a positively acting factor, preventing either its DNA binding or its activation of transcription. It is likely, however, that some factors may have an inherently negative effect and may directly inhibit transcription (see Fig. 5d). This direct repression is seen in the thyroid hormone receptor that can activate transcription in the presence of thyroid hormone by specific DNA sequences, known as thyroid response elements (TRE). In the absence of thyroid hormone, however, the receptor can still bind to the TRE, but now it has an inhibitory effect on transcription that appears to represent a direct repression, rather than the neutralization of a positively acting factor (29).

IV. CONCLUSION

The cell type-specific or inducible expression of specific proteins that is commonly observed in higher eukaryotes is dependent on the cell type-specific or inducible transcription of the corresponding genes encoding these proteins. In turn, such transcriptional regulation is dependent on specific transcription factors that are synthesized or activated only in one or a few cell types, or following exposure of cells to a specific stimulus. These factors, therefore, bind to their specific DNA recognition sites in particular genes and activate or repress transcription only in these cells, resulting in a cell type-specific or inducible pattern of gene transcription. Although these basic principles have been worked out using a wide variety of different genes, they are of equal importance in understanding the processes that regulate cytokine gene expression in specific cell types, or in response to specific stimuli, as well as the mechanisms that allow such cytokines themselves to activate the transcription of other genes.

REFERENCES

1. Darnell JE. Variety in the level of gene control in eukaryotic cells. Nature 1982; 297:365–371.
2. Sharp PA. Splicing of messenger RNA precursors. Science 1987; 235:766–771.
3. Latchman DS. Gene Regulation: A Eukaryotic Perspective. London: Chapman and Hall, 2nd ed., 1995.
4. Latchman DS. Cell-type specific splicing factors and the regulation of alternative RNA splicing. New Biol 1990; 2:297–303.
5. Amara SG, Jonas V, Rosenfeld MG, Ong ES, Evans RM. Alternative RNA processing in calcitonin gene expression generates mRNAs encoding different polypeptide products. Nature 1982; 298:240–244.
6. Fujita T, Shibaya H, Ohashi T, Yamanishi K, Taniguchi T. Regulation of human interleukin-2 region functional DNA sequences in the 5′ flanking region for the gene expression in activated T lymphocytes. Cell 1986; 46:401–407.
7. Oster W, Brach MA, Gruss HJ, Mertelsmann R, Hermann F. Interleukin-1 beta (IL-1 beta) expression

in human blood mononuclear phagocytes is differentially regulated by granulocyte–macrophage colony-stimulating factor (GM-CSF), M-CSF and IL-3. Blood 1992; 79:1260–1265.

8. Latchman DS. Eukaryotic Transcription Factors. San Diego: Academic Press, 2nd ed., 1995.

9. Johnson PF, McKnight SL. Eukaryotic transcriptional regulatory proteins. Annu Rev Biochem 1989; 58:799–839.

10. Saltzman AG, Weinmann R. Promoter specificity and modulation of RNA polymerase II transcription. FASEB J 1989; 3:1723–1733.

11. Goodwin GH, Partington GA, and Perkins ND. Sequence specific DNA binding proteins involved in gene transcription. In Adolph KW, ed. Chromosomes: Eukaryotic, Prokaryotic and Viral, Vol. 1. Boca Raton FL: CRC Press, 1990:31–85.

12. Durand D, Bush M, Morgan J, Weiss A, Crabtree G. A 275 base pair fragment at the 5′ end of the interleukin 2 gene enhances expression from a heterologous promoter in response to signals from the T cell antigen receptor. 1987; 165:395–407.

13. Poli V, Mancini FP, Cortese R. IL-6DBP, a nuclear protein involved in interleukin-6 signal transduction, defines a new family of leucine zipper proteins related to C/EBP. Cell 1990; 63:643–653.

14. Hatzopoulous AK, Schlokat U, Gruss P. Enhancers and other cis-acting sequences. In: Hames BD, Glovre DS, eds. Transcription and Splicing. London: IRL Press, 1988:43–96.

15. Muller MM, Gerster T, Schaffner W. Enhancer sequences and the regulation of gene transcription. Eur J Biochem 1988; 176:485–495.

16. Scheidereit C, Heguy A, Roeder RG. Identification and purification of a human lymphoid-specific octamer-binding protein (OTF-2) that activates transcription from an immunoglobulin promoter in vitro. Cell 1987; 51:783–793.

17. Edmondson DG, Olson EN. Helix–loop–helix proteins as regulators of muscle-specific transcription. J Biol Chem 1993; 268:755–758.

18. Grimm S, and Baeurele PA. The inducible transcription factor NFκB. Biochem J 1993; 290:309–312.

19. Granelli-Piperno A, Nolan P. Nuclear transcription factors that bind to elements of the IL-2 promoter. J Immunol 1991; 14:2734–2739.

20. Schrek R, Baeurele PA. NF-kappa B as inducible transcriptional activator of the granulocyte–macrophage colony-stimulating factor gene. Mol Cell Biol 1990; 10:1281–1286.

21. Brach MA, Arnold C, Kiehntopf M, Gruss HJ, Herrmann F. Transcriptional activation of the macrophage colony-stimulating factor gene by IL-2 is associated with secretion of bioactive macrophage colony-stimulating factor by monocytes and involves activation of the transcription factor NF kappa B. J Immunol 1993; 150:5535–5543.

22. Hunter T, Karin M. The regulation of transcription by phosphorylation. Cell 1992; 70:375–387.

23. Furst P, Hu S, Hackett R, Hamer D. Copper activates metallothionein gene transcription by altering the conformation of a specific DNA binding protein. Cell 1988; 55:705–717.

24. Goodbourn S. Negative regulation of transcriptional initiation in eukaryotes. Biochim Biophys Acta 1990; 1032:53–77.

25. Clark AR, Docherty K. Negative regulation of transcription in eukaryotes. Biochem J 1993; 296:521–541.

26. Sakai DD, Helms S, Carlstedt-Duke J, Gustafsson J-A, Rottman FM, Yamamoto KR. Hormone-mediated repression: a negative glucocorticoid response element from the bovine prolactin gene. Genes Dev 1988; 2:1144–1154.

27. Lucibella FC, Slater EP, Joos KU, Beato M, Muller R. Mutual transrepression of Fos and the glucocorticoid receptor: involvement of a functional domain in Fos which is absent in FosB. EMBO J 1990; 9:2827–2834.

28. Keleher CA, Goutte C, Johnson AD. The yeast cell-type specific repressor alpha 2 acts co-operatively with a non cell-type-specific protein. Cell 1988; 53:927–936.

29. Baniahmad A, Kohne AC, Renkawitz R. A transferable silencing domain is present in the thyroid hormone receptor, in the v-*erbA* oncogene product and the retinoic acid receptors. EMBO J 1992; 11:1015–1023.

6

Cytokines and Apoptosis
The Mechanism of Regulation of Cell Death

Mary K. L. Collins
Institute of Cancer Research, London, England

Abelardo Lopez-Rivas
Instituto de Parasitologia y Biomedicina CSIC, Granada, Spain

I. CYTOKINES PROMOTE SURVIVAL OF MANY HEMATOPOIETIC CELLS

Cytokines were first characterized as polypeptide factors that stimulated proliferation, differentiation, and effector functions of cells of the hematopoietic system. Recent work has demonstrated that cytokines also promote survival of many cell types, by rescuing them from a death program known as apoptosis. This was first described in the prescient study of Duke and Cohen (1), which demonstrated that T cells cultured in interleukin-2 (IL)-2 became dependent on this factor and entered apoptosis after its removal. Similar effects of cytokines on a variety of cell lines, primary cells, and tumors have now been reported (summarized in Table 1).

Whether a cytokine promotes only cell survival, or also stimulates other functions, such as cell proliferation, may be controlled by several parameters. For example, the cell state of differentiation can influence its response; progenitor cells will proliferate in response to granulocyte–macrophage colony-stimulating factor (GMCSF; 2), whereas eosinophils will survive but cannot proliferate (15). There is also evidence that there may be specific signals that direct survival, rather than proliferation, which comes from the analysis of different erythropoietin (EPO) receptor isoforms (21). Finally, it is possible that something as simple as the level of receptor occupancy may determine the choice between survival and proliferation (22). The death of many cells in the absence of cytokine provides an attractive mechanism to control cell homeostasis. It allows activated progenitor cells to be eliminated following waves of hematopoiesis, it aids selection of the mature T-4 and B-cell repertoires, and it controls the turnover of aging differentiated cells. Loss of the apoptotic pathway leads to tumorigenesis (see later discussion), although some tumor cells retain a requirement for cytokines for survival (see Table 1).

Table 1   Inhibition of Apoptosis by Cytokines

Cell lines	Cytokine	Ref.
Established/primary T cell	IL-2	1
Bone marrow progenitor	IL-3, GM-CSF	2
	IL-4	3
	IGF-1	4
Primary cells		
Erythroid progenitor	EPO	5
	SCF, IGF-1	6
Thymocyte	IL-1	7
	IL-2	8
	IL-4	9
T-helper cell	IL-2, IL-4	10
Germinal center B cell	IL-1, CD23	11
Monocyte	TNF, IL-1	12
	GM-CSF, IFN-γ	13
Granulocyte	IL-1β, TNF, GM-CSF	14
	GCSF, IFN-γ	
Eosinophil	IL-5, GM-CSF, IL-3	15
Mast cell	IL-3	4
	SCF	16
Tumor cells		
Myeloid leukemia	IL-6, IL-3, M-CSF	17
	G-CSF, IL-1	
CLL	IL-4	18
	IL-2	19
ATL	IL-2	20

II.   INDUCTION OF APOPTOSIS BY CYTOTOXIC CYTOKINES

In contrast to promoting cell survival, cytokines may also induce apoptosis. Tumor necrosis factor (TNF) kills many cell types that express TNF receptors, often by apoptosis (reviewed in Ref. 23). It can also stimulate thymocyte proliferation (24,25). A protein, first identified antigenically as the human APO1 and the murine Fas cell surface antigen, shares significant homology with the TNF receptor (26,27). Antibodies against Fas are cytotoxic; a dramatic demonstration of this is the rapid death of mice following anti-Fas antibody injection, which has been ascribed to apoptosis in the liver (28). However, similar to TNF they can induce proliferation under some circumstances (29). Mutant mice defective in Fas expression develop autoimmune disease (30). A ligand for Fas has recently been identified as a membrane-bound member of the TNF family (31). TNF receptors and Fas may trigger common signals for apoptosis because of their homology. Indeed, a domain in the cytoplasmic tail of Fas, which is conserved in the TNF receptor, has been demonstrated to be essential for Fas signaling (32). Ceramide generation stimulated by TNF has been proposed to be causal in its stimulation of apoptosis (33). Further identification of such signals will greatly advance understanding of the control of apoptosis.

Transforming growth factor-beta (TGF-β) has been described as inducing apoptosis in nonhematopoietic cells and as having antihematopoietic activity. There are now reports that

TGF-β can stimulate apoptosis in germinal center B cells (34), AML cells (35), and eosinophils (36). Finally, several of the cytokines that can promote cell survival in diverse cell types (see Table 1) have also been reported to induce apoptosis under certain circumstances. IL-2 can induce apoptosis if mature T cells are exposed to it before antigen triggering (37); IL-4 induces apoptosis in activated monocytes (38); interferon gamma (IFN)-γ causes apoptosis in pre-B cells (39); IL-6 can induce apoptosis in monocytic cell lines and neutrophils (40,41). In these cases cell status, rather than signal specificity, probably directs the result of cytokine action.

III. HOW DO CYTOKINES INHIBIT APOPTOSIS?

A. Clues from Oncogenes

A potent mechanism of induction of apoptosis was demonstrated by Evan and colleagues, who showed that enforced expression of the c-*myc* gene product in fibroblasts deprived of growth factors induces apoptosis (42). IL-3-dependent cells overexpressing c-*myc* undergo apoptosis more rapidly after factor removal (43) and are unable to grow in suboptimal factor concentrations (44,45). It has been suggested that c-*myc* drives apoptosis in IL-3-dependent cells by inhibiting cell cycle arrest following factor withdrawal (43). However, c-*myc* can induce apoptosis in these cells without altering their cell cycle distribution (45), and parental (46) and c-*myc* overexpressing cells (45) die in all phases of the cell cycle. This suggests that the induction of cell death by c-*myc* is not a secondary consequence of its driving the cell cycle in the absence of IL-3. In the fibroblast system, the domains of the c-*myc* protein required for cell transformation (42) and for transcriptional transactivation following dimerization with Max (47) are required to induce apoptosis. Therefore, c-*myc* appears to induce transcription of genes involved in cell death, the action of which is inhibited in the presence of optimal cytokines. Whether c-*myc* is driving cell death in some of the wide variety of cells that apoptose in the absence of cytokine (see Table 1) remains to be determined. However, it is clear that *myc* expression is not required for the apoptosis of FDCP1 cells (48).

Other oncogenes with the apoptosis-inducing phenotype of c-*myc* remain to be discovered. One such, which has been described, is the *E2A–PBX1* fusion protein, expressed in childhood leukemias (49). To transform hematopoietic cells, which are susceptible to apoptosis, such oncogenes need to cooperate with either cytokines, or further intracellular oncogenes that can suppress the apoptotic pathway. One such cooperating oncogene is the *bcl-2* gene product, which can cooperate with c-*myc* to immortalize primary hematopoietic cells (50). The mechanism of action of *bcl-2* and its possible role in cytokine inhibition of apoptosis are discussed later. Other oncogene products that can suppress apoptosis are v-*raf* (51), v-*abl* (52), and the *PML–RAR-α* fusion protein (53). The former observation is particularly interesting as many cytokines stimulate the signaling pathway that activates *ras*, *raf*, and mitogen-activated protein kinases. Therefore, cytokine stimulation of the *ras* pathway may be necessary and sufficient to inhibit apoptosis.

The *bcl-2* gene product was first shown to inhibit apoptosis when overexpressed in IL-3-dependent cells deprived of factor (54); this result has since been extended to a variety of hematopoietic and other cell types. Some cytokines have been demonstrated to up-regulate expression of *bcl-2* (55,56), which may partly explain their inhibition of apoptosis. The overexpression of *bcl-2* can only partially inhibit the death pathway triggered by TNF and Fas (57). Mice lacking the *bcl-2* protein develop a normal immune system, but this deteriorates owing to apoptosis after birth (58,59). The 32-kDa *bcl-2* protein resides in a number of intracellular membranes (60), this localization appears to be important for its activity (61).

Oncogene *bcl-2* heterodimerizes with a protein that is homologous to *bcl-2*, called Bax; as Bax can stimulate apoptosis the mechanism of action of *bcl-2* may be to complex Bax in an inactive form (62). Also, *bcl-2* complexes with a member of the *ras* gene family, R-*ras*, in vitro (63). The significance of this interaction in *bcl-2* inhibition of apoptosis is unclear; however such interaction of *bcl-2* with signaling molecules may explain the antiproliferative effect of *bcl-2* overexpression (64). Several other *bcl-2* homologues have been isolated, which suggests a complex pattern of regulation of cell death by a family of proteins (65–67), any or all of which may be regulated by cytokines.

The recessive oncogene *p53* is necessary for radiation-induced apoptosis in thymocytes (68) and a variety of other hematopoietic cells (69). Because *p53* is involved in the response of a variety of cells to radiation and other DNA-damaging insults, it is possible that it plays some role in the regulation, by cytokines, of the response of hematopoeitic cells to DNA damage (70,71). An interaction of *p53* with cytokine action in the absence of DNA damage was first shown by introduction of wild-type *p53* into myeloid leukemia cells, which then enter apoptosis when they are deprived of IL-6 (72). Furthermore, cells from *p53*-deficient mice are more resistant to apoptosis induced by deprivation of a variety of cytokines (73), and introduction of a dominant-negative *p53* molecule delays apoptosis induced by IL-3 removal (74). Similar to the action of c-*myc*, the induction of apoptosis by *p53* does not seem to be a secondary result of its modulation of the cell cycle (75). Thus, identification of dominant and recessive oncogenes that can influence apoptosis has identified molecules that are involved in the control of apoptosis by cytokines.

B. How May Cytokine Removal Trigger Apoptosis?

The c-*myc*, *p53*, and *bcl-2* gene products, as well as their relations, are clearly effectors and inhibitors on the apoptotic pathway controlled by cytokines. However, in some cases of apoptosis induced by cytokine removal, de novo gene expression is not required for the induction of the apoptotic program (46). Inhibitory molecules, therefore, may decrease, but in these cases, all the necessary effector molecules must be in place. Many of the rapid intracellular signals induced by cytokines have been implicated in the mechanism of inhibition of apoptosis, including cAMP elevation (76), protein kinase C stimulation (77), tyrosine kinase stimulation (77), and regulation of intracellular pH (77) and calcium (78). However, the onset of apoptosis can be reversed by delayed addition of cytokine at times up to that at which onset of DNA digestion is first observed (10–12h; 70). This suggests that the rapid decay of signals, which follows dissociation of cytokine from its receptor, is not sufficient to trigger apoptosis. The irreversible event that pushes cells into the death pathway must occur many hours after cytokine removal. One candidate for such an event is loss of cellular antioxidant potential, resulting in free radical buildup and cell damage (79). However, this is probably only one in a large number of changes in cellular metabolism that follow cytokine removal.

Several lines of evidence have recently implicated protease activation as a regulatory mechanism in apoptosis. First, the *ced-3* gene product that induces apoptosis during development of the nematode. *Caenorhabditis elegans* is homologous with a mammalian cysteine protease IL-1β–converting enzyme (ICE; 80). Overexpression of a viral inhibitor of ICE, *crmA*, can inhibit neuronal cell death (81). Therefore, a cysteine protease can control apoptosis in mammalian cells, whether or not this is ICE or a relation is unclear, although, in macrophages, increased processing of IL-1 is observed during apoptosis (82). Cytotoxic T cells (CTLs) induce apoptosis in their targets, at least partly, by the introduction of proteases contained in cytotoxic granules (83). Because *bcl-2* cannot inhibit apoptosis induced by CTLs (84), the protease-regulated step may lie downstream from the point of *bcl-2* inhibition (Fig. 1). Finally, there are reports that protease inhibitors can inhibit apoptosis induced by cytotoxic drugs (85), or T-cell triggering (86). It is likely that protease regulation is also important in cytokine regulation of apoptosis.

C. The Final Death Throes

The final stages of apoptosis in all cells involves several characteristic changes in morphology, including cell shrinkage, chromatin condensation, and nuclear fragmentation. This is associated with genome fragmentation, into to 300- or 50-kb fragments (87) and often to oligonucleosomal fragments (88). Chromatin condensation is not dependent on DNA fragmentation (89). Because of some similarities between these changes and nuclear disassembly during mitosis, it has been

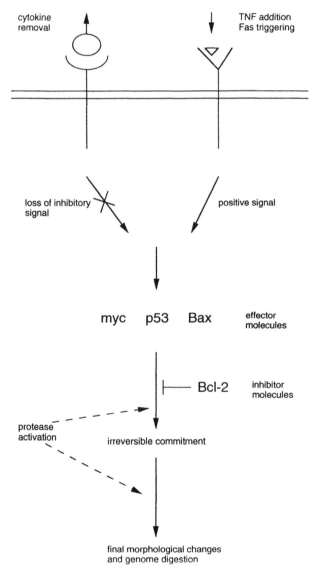

Figure 1   A summary of the mechanisms involved in the regulation of apoptosis by cytokines.

proposed that apoptosis shares some of the mitotic control machinery. Indeed, inhibition of activation of the mitosis-associated kinase p34cdc2 can inhibit apoptosis (90). However, many postmitotic cells enter apoptosis (for those of the hematopoietic system see Table 1) and many cycling cells can enter apoptosis from phases of the cell cycle other than mitosis (e.g., see Ref. 46). It remains unclear what role mitotic regulators play in these circumstances. Several endonucleases have been proposed as candidates for the enzyme responsible for genome digestion during apoptosis, including the digestive enzyme DNase I (91), the lysosomal enzyme DNase II (92), Nuc18 (93), and a Ca^{2+}/Mg^{2+}-dependent enzyme (94). The latter three enzymes are characterized only by their properties, their amino acid sequences have not yet been determined, and DNase I has only been identified antigenically in apoptotic cells. Therefore, the exact nature of the apoptotic endonucleases in various cell types and their mechanism of regulation remains poorly understood. Finally, the apoptotic cell in vivo is rapidly recognized and cleared by neighboring phagocytic cells (95). These characteristic events of the final stages of apoptosis ensure that toxic cell debris is first autodigested as far as possible and then digested within the phagocyte. This is vital in many situations when apoptosis occurs in the hematopoietic system. For example, the clearance of inflammatory cells that have served their purpose prevents secondary tissue damage. Furthermore, potentially immunogenic debris is rapidly cleared from sites at which immune selection is occurring, such as the thymus or lymph node.

IV. SUMMARY

A tentative overall summary of the mechanisms that are involved in regulation of apoptosis by cytokines is presented in Figure 1. Key regulation steps that remain to be elucidated are as follows:

1. Which signals triggered by factor–receptor interaction are involved in the inhibition or stimulation of apoptosis?
2. What is the subsequent intracellular switch that irreversibly commits cells to apoptosis?
3. What are the enzymes that catalyze the final characteristic, morphological changes associated with apoptosis?

REFERENCES

1. Duke RC, Cohen JJ. IL-2 addiction: withdrawal of growth factor activates a suicide program in dependent T cells. Lymphokine Res 1986; 5:289–295.
2. Williams G, Smith C, Spooncer E, Dexter T, Taylor D. Haemopoietic colony stimulating factors promote cell survival by suppressing apoptosis. Nature 1990; 343:76–79.
3. Rodriguez-Tarduchy G, Malde P, Lopez-Rivas A, Collins MKL. Inhibition of apoptosis by calcium ionophores in IL3-dependent bone marrow cells is dependent upon production of IL4. J Immunol 1992; 148:1416–1422.
4. Rodriguez-Tarduchy G, Collins MKL, Garcia I, Lopez-Rivas A. Insulin-like growth factor-I inhibits apoptosis in mouse haemopoietic cells. J Immunol 1992; 149:535–540.
5. Koury MJ, Bondurant MC. Erythropoietin retards DNA breakdown and prevents apoptosis in erythroid progenitor cells. Science 1990; 248:378–381.
6. Muta K, Krantz SB. Apoptosis of human erythroid colony-forming cells is decreased by stem cell factor and insulin-like growth factor I as well as erythropoietin. J Cell Physiol 1993; 156:264–271.
7. McConkey DJ, Hartzell P, Chow SC, Orrenius S, Jondal M. Interleukin-1 inhibits T cell receptor-mediated apoptosis in immature thymocytes. J Biol Chem 1990; 265:3009–3011.

8. Nieto MA, Gonzalez A, López-Rivas A, Diaz-Espada F, Gambon FJ. IL-2 protects against anti-CD3-induced cell death in human medullary thymocytes. J Immunol 1990; 145:1364–1368.

9. Migliorati G, Nicoletti I, Pagliacci MC, Adamio LD, Riccardi C. IL-4 protects double negative and CD4 single positive thymocytes from dexamethasone-induced apoptosis. Blood 1993; 81:1352–1358.

10. Zubiaga AM, Munoz E, Huber BT. IL-4 and IL-2 selectively rescue Th cell subsets from glucocorticoid induced apoptosis. J Immunol 1992; 149:107–112.

11. Liu YJ, Cairs JA, Holder MJ, Abbot SD, Jansen KU, Bonnefoy JY, Gordon J, MacLennan IC. Recombinant 25-kDa CD23 and interleukin 1 alpha promote the survival of germinal center B cells: evidence for bifurcation in the development of centrocytes rescued from apoptosis. Eur J Immunol 1991; 21:1107–14.

12. Mangan DF, Welch GR, Wahl SM. LPS, TNF-α and IL-1β prevent apoptosis in human peripheral blood monocytes. J Immunol 1991; 146:1541–1546.

13. Mangan DF, Wahl SM. Differential regulation of human monocyte apoptosis by chemotactic factors and pro-inflammatory cytokines. J Immunol 1991; 147:3408–3412.

14. Colotta F, Re F, Polentarutti N, Sozzani S, Mantovani A. Modulation of granulocyte survival and programmed cell death by cytokines and bacterial products. Blood 1992; 80:2012–2020.

15. Tai PC, Sun L, Spry CJ. Effects of IL5, GMCSF and IL3 on the survival of human blood eosinophils in vitro. Clin Exp Immunol 1991; 85:312–316.

16. Mekori YA, Oh CK, Metcalfe DD. IL3-dependent murine mast cells undergo apoptosis on removal of IL-3. Prevention of apoptosis by c-*kit* ligand. J Immunol 1993; 151:3775–3784.

17. Lotem J, Sachs L. Regulation of leukaemic cells by IL6 and LIF. Ciba Found Symp 1992; 167:88–99.

18. Dancescu M, Rubio TM, Biron G, Bron D, Delespesse G, Sarfati M. Interleukin 4 protects chronic lymphocytic leukemic B cells from death by apoptosis and upregulates *Bcl-2* expression. J Exp Med 1992; 176:1319–26.

19. Huang RW, Tsuda H, Takatsuki K. IL2 prevents programmed cell death in chronic lymphocytic leukaemia cells. Int J Hematol 1993; 58:83–92.

20. Tsuda H, Huang RW, Takatsuki K. IL2 prevents programmed cell death in adult T-cell leukaemia cells. Jpn J Cancer Res 1993; 84:431–437.

21. Nakamura Y, Komatsu N, Nakauchi H. A truncated erythropoietin receptor that fails to prevent programmed cell death of erythroid cells. Science 1992; 257:1138–41.

22. Collins MKL, Perkins GR, Rodriguez-Tarduchy G, Nieto MA, Lopez-Rivas A. Growth factors as survival factors: regulation of apoptosis. BioEssays 1994; 16:133–138.

23. Golstein P, Ojcius DM, Young D-E. Cell death mechanisms and the immune system. Immunol Rev 1991; 121:29–65.

24. Hurme M. Both IL1 and TNF enhance thymocyte proliferation. Eur J Immunol 1988; 18:1303–1306.

25. Hernandez-Caselles T, Stutman O. Immune functions of TNF. J Immunol 1993; 151:3999–4012.

26. Oehm A, Behrmann I, Falk W, Pawlita M, Maier G, Klas C, Li-Weber M, Richards S, Dhein J, Trauth BC. Purification and cloning of the APO-1 cell surface antigen, a member of the TNF/NGF receptor superfamily. J Biol Chem 1992; 267:10709–10715.

27. Itoh N, Yonehara S, Ishii A, Yonehara M, Mizushima S, Sameshima M, Hase A, Seto Y, Nagata S. The polypeptide encoded by the cDNA for human cell surface antigen Fas can mediate apoptosis. Cell 1991; 66:233–243.

28. Ogasawara J, Watanabe-Fukunaga R, Adachi M, Matsuzawa A, Kasugai T, Kitamura Y, Itoh N, Suda T, Nagata S. Lethal effect of anti-Fas antibody in mice. Nature 1993; 364:806–809.

29. Alderson MR, Armitage RJ, Maraskovsky E, Tough TW, Roux E, Schooley K, Ramsdell F, Lynch DH. Fas transduces activation signals in normal human T lymphocytes. J Exp Med 1993; 178:2231–2235.

30. Watanabe-Fukunaga R, Brannan CI, Copeland NG, Jenkins NA, Nagata S. Lymphoproliferation disorder in mice explained by defects in Fas antigen that mediates apoptosis. Nature 1992; 356:314–317.

31. Suda T, Takahashi T, Golstein P, Nagata S. Molecular cloning and expression of the Fas ligand, a novel member of the TNF family. Cell 1993; 75:1169–1178.

32. Itoh N, Nagata S. A novel protein domain required for apoptosis. J Biol Chem 1993; 268:10932–10937.

33. Obeid LM, Linardic CM, Karolak LA, Hannun YA. Programmed cell death induced by ceramide. Science 1993; 259:1769–1771.

34. Holder MJ, Knox K, Gordon J. Factors modifying survival pathways of germinal center B cells. Eur J Immunol 1992; 22:2725–2728.

35. Taetle R, Payne C, Dos-Santos B, Russell M, Segarini P. Effects of TGF beta 1 on growth and apoptosis of human acute myelogenous leukaemia cells. Cancer Res 1993; 53:3386–3393.

36. Alam R, Forsythe P, Stafford S, Fukuda Y. TGF β abrogates the effects of haematopoietins on eosinophils and induces their apoptosis. J Exp Med 1994; 179:1041–1045.

37. Lenardo MJ. IL2 programmes mouse alpha beta T lymphocytes for apoptosis. Nature 1991; 353:858–861.

38. Mangan DF, Robertson B, Wahl SM. IL4 enhances apoptosis in stimulated human monocytes. J Immunol 1992; 148:1812–1816.

39. Grawunder U, Melchers F, Rolink A. IFN gamma arrests proliferation and causes apoptosis in stromal cell/IL7-dependent normal murine pre-B cell lines and clones in vitro. Eur J Immunol 1993; 23:544–551.

40. Oritani K, Kaisho T, Nakajima K, Hirano T. Retinoic acid inhibits IL-6 induced macrophage differentiation and apoptosis in a murine hematopoietic cell line, Y6. Blood 1992; 80:2298–2305.

41. Afford SC, Pongracz J, Stockley RA, Crocker J, Burnett D. The induction by human IL-6 of apoptosis in the promonocytic cell line U937 and human neutrophils. J Biol Chem 1992; 267:21612–21616.

42. Evan GI, Wyllie AH, Gilbert CS, Littlewood TD, Land H, Brooks M, Waters CM, Penn LZ, Hancock DC. Induction of apoptosis in fibroblasts by c-*myc* protein. Cell 1992; 69:119–128.

43. Askew DS, Ashmun RA, Simmons BC, Cleveland JL. Constitutive c-*myc* expression in an IL-3-dependent cell line suppresses cell cycle arrest and accelerates apoptosis. Oncogene 1991; 6:1915–1922.

44. Askew DS, Ihle JN, Cleveland JL. Activation of apoptosis associated with enforced Myc expression in myeloid progenitor cells is dominant to the suppression of apoptosis by interleukin-3 or erythro-poietin. Blood 1993; 82:2079–2087.

45. Malde P, Collins MKL. Disregulation of *myc* expression in murine bone marrow cells results in an inability to proliferate in sub-optimal growth factor and an increased sensitivity to DNA damage. Int Immunol 1994; 6:1169–1176.

46. Rodriguez-Tarduchy G, Collins M, Lopez-Rivas A. Regulation of apoptosis in interleukin-3-dependent hemopoietic cells by interleukin-3 and calcium ionophores. EMBO J 1990; 9:2997–3002.

47. Amati B, LIttlewood T, Evan G, Land H. The c-*myc* protein induces cell cycle progression and apoptosis through dimerisation with Max. EMBO J 1993; 12:5083–5087.

48. Vaux DL, Weissman IL. Neither macromolecular synthesis nor *myc* is required for cell death via the mechanism that can be controlled by *Bcl*-2. Mol Cell Biol 1993; 13:7000–7005.

49. Dedera DA, Waller EK, LeBrun DP, Sen-Majumdar A, Stevens ME, Barsh GS, Cleary ML. Chimeric homeobox gene E2A-PBX1 induces proliferation, apoptosis and malignant lymphomas in transgenic mice. Cell 1993; 74:833–843.

50. Vaux DL, Cory S, Adams JM. *Bcl*-2 gene promotes haemopoietic cell survival and cooperates with c-*myc* to immortalise pre-B cells. Nature 1988; 335:440–442.

51. Troppmair J, Cleveland JL, Askew DS, Rapp UR. v-*raf*/v-*myc* synergism in abrogation of IL3 dependence: v-*raf* suppresses apoptosis. Curr Top Microbiol Immunol 1992; 182:453–460.

52. Evans CA, Owen-Lynch PJ, Whetton AD, Dive C. Activation of Abelson tyrosine kinase activity is associated with suppression of apoptosis in haematopoietic cells. Cancer Res 1993; 53:1735–1738.

53. Grignani F, Ferrucci PF, Testa U, Talamo G, Fagioli M, Alcalay M, Mencarelli A, Grignani F, Peschle C, Nicoletti I. The acute promyelocytic leukaemia-specific PML–RAR alpha fusion protein inhibits differentiation and promotes survival of myeloid precursor cells. Cell 1993; 74:423–431.

54. Hockenbery D, Nunez G, Miliman C, Schreiber RD, Korsmeyer SJ. *Bcl*-2 as an inner mitochondrial membrane protein that blocks programmed cell death. Nature 1990; 348:334–336.

55. Nunez G, London K, Hockenberry D, Alexander M, McKearn J, Korsmeyer S. De-regulated *bcl2* gene

expression selectively prolongs survival of growth factor-deprived hemopoietic cell lines. J Immunol 1990; 144:3602–3610.

56. Liu YJ, Mason DY, Johnson GD, Abbot S, Gregory CD, Hardie DL, Gordon J, MacLennan IC. Germinal center cells express *bcl-2* protein after activation by signals which prevent their entry into apoptosis. Eur J Immunol 1991; 21:1901–10.

57. Itoh N, Tsujimoto Y, Nagata S. Effect of *bcl-2* on Fas antigen-mediated cell death. J. Immunol 1993; 151:621–627.

58. Veis DJ, Sorenson CM, Shutter JR, Korsmeyer SJ. *Bcl-2* deficient mice demonstrate fulminant lymphoid apoptosis, polycystic kidneys and hypopigmented hair. Cell 1993; 75:229–240.

59. Nakayama K, Nakayama K, Negishi I, Kuida K, Shinkai Y, Louie MC, Fields LE, Lucas PJ, Stewart V, Alt FW. Disappearance of the lymphoid system in *bcl-2* homozygous mutant chimeric mice. Science 1993; 261:1584–1588.

60. Krajewski S, Tanaka S, Takayama S, Schibler M, Reed JC. Investigation of the subcellular distribution of the *bcl-2* oncoprotein. Cancer Res 1993; 53:4701–4714.

61. Tanaka S, Saito K, Reed JC. Structure–function analysis of the *bcl-2* oncoprotein. J Biol Chem 1993; 268:10920–10926.

62. Oltvai Z, Milliman C, Korsmeyer S. *Bcl-2* heterodimerises in vivo with a conserved homolog Bax that accelerates programmed cell death. Cell 1993; 74:609–619.

63. Fernandez-Sarabia MJ, Bischoff JR. *Bcl-2* associates with the *ras*-related protein R-*ras* p23. Nature 1993; 366:274–275.

64. Marvel J, Perkins GR, Lopez-Rivas A, Collins MKL. *Bcl-2* over-expression in murine bone marrow cells induces refractoriness to interleukin-3 stimulation of proliferation. Oncogene 1993; 9:1117–1122.

65. Biose L, Gonzalez-Garcia M, Postema C, Ding L, Lindsten T, Turka L, Mao X, Nunez G, Thompson C. *bcl-x*, a *bcl-2*-related gene that functions as a dominant regulator of apoptotic cell death. Cell 1993; 74:597–608.

66. Kozopas KM, Yang T, Buchan HL, Zhou P, Craig RW. *MCL1* a gene expressed in programmed myeloid cell differentiation has sequence similarity to *bcl-2*. Proc Natl Acad Sci USA 1993; 90:3516–3520.

67. Lin E, Orlofsky A, Berger M, Prystowsky M. Characterisation of *A1*, a novel haemopoietic-specific early response gene with sequence similarity to *bcl-2*. J Immunol 1993; 151:1979–1988.

68. Lowe SW, Schmitt EM, Smith SW, Osborne BA, Jacks T. p53 is required for radiation induced apoptosis in mouse thymocytes. Nature 1993; 362:847–849.

69. Lee JM, Bernstein A. p53 mutations increase resistance to ionizing radiation. Proc Natl Acad Sci USA 1993; 90:5742–5746.

70. Collins MKL, Marvel J, Malde P, Lopez-Rivas A. Interleukin 3 protects murine bone marrow cells from apoptosis induced by DNA damaging agents. J Exp Med 1992; 176:1043–1051.

71. Ascaso R, Marvel J, Collins MKL, Lopez-Rivas A. IL-3 and *bcl-2* co-operatively inhibit etoposide-induced apoptosis in murine bone marrow cells. Eur J Immunol 1994; 24:537–541.

72. Yonish-Rouach E, Resnitzky D, Lotem J, Sachs L, Kimchi A, Oren M. Wild-type *p53* induces apoptosis of myeloid leukaemic cells that is inhibited by interleukin-6. Nature 1991; 352:345–347.

73. Lotem J, Sachs L. Hematopoietic cells from mice deficient in wild-type *p53* are more resistant to induction of apoptosis by some agents. Blood 1993; 82:1092–1096.

74. Gottlieb E, Haffner R, von-Ruden T, Wagner EF, Oren M. Down-regulation of wild-type *p53* activity interferes with apoptosis of IL3-dependent hematopoietic cells following IL3 withdrawal. EMBO J 1994; 13:1368–1374.

75. Yonish-Rouach E, Grunwald D, Wilder S, Kimchi A, May E, Lawrence JJ, May P, Oren M. *p53*-mediated cell death: relationship to cell cycle control. Mol Cell Biol 1993; 13:1415–1423.

76. Berridge MV, Tan AS, Hilton CJ. cAMP promotes cell survival and retards apoptosis in a factor-dependent bone marrow-derived cell line. Exp Hematol 1993; 21:269–276.

77. Rajotte D, Haddad P, Haman A, Cragoe EJ, Hoang T. Role of protein kinase C and the Na^+/H^+ antiporter in suppression of apoptosis by GMCSF and IL3. J Biol Chem 1992; 267:9980–9987.

78. Baffy G, Miyashita T, Williamson JR, Reed JC. Apoptosis induced by withdrawal of IL3 from an IL3-dependent cell line is associated with repartitioning of intracellular calcium and is blocked by enforced *Bcl-2* oncoprotein expression. J Biol Chem 1993; 268:6511–6519.

79. Hockenbery DM, Oltvai ZN, Yin XM, Milliman CL, Korsmeyer SJ. *Bcl-2* functions in an antioxidant pathway to prevent apoptosis. Cell 1993; 75:241–251.

80. Yuan J, Shaham S, Ledoux S, Ellis HM, Horvitz HR. The *C. elegans* cell death gene *ced-3* encodes a protein similar to mammalian interleukin-1β-converting enzyme. Cell 1993; 75:641–652.

81. Gagliardini V, Fernandez P-A, Lee RKK, Drexler HCA, Rotello RJ, Fishman MC, Yuan J. Prevention of neuronal death by the *crmA* gene. Science 1994; 263:826–828.

82. Hoqquist KA, Nett MA, Unanue ER, Chaplin DD. IL1 is processed and released during apoptosis. Proc Natl Acad Sci USA 1991; 88:8485–8489.

83. Shi L, Kam CM, Powers JC, Aebersold R, Greenberg AH. Purification of three cytotoxic lymphocyte granule serine proteases that induce apoptosis through distinct substrate and target cell interactions. J Exp Med 1992; 176:1521–1529.

84. Vaux D, Aguila H, Weissman I. *Bcl-2* prevents death of factor-deprived cells but fails to prevent apoptosis in targets of cell mediated killing. Int Immunol 1992; 4:821–824.

85. Bruno S, Del-Bino G, Lassota P, Giaretti W, Darzynkiewicz Z. Inhibitors of proteases prevent endonucleolysis accompanying apoptotic death of HL-60 leukemic cells and normal thymocytes. Leukemia 1992; 6:1113–1120.

86. Sarin A, Adams DH, Henkart PA. Protease inhibitors selectively block T cell receptor-triggered programmed cell death in a murine T cell hybridoma and activated peripheral T cells. J Exp Med 1993; 178:1693–1700.

87. Oberhammer F, Winter JW, Dive C, Morris ID, Hickman JA, Wakeling AE, Walker PR, Sikorska M. Apoptotic cell death in epithelial cells: cleavage of DNA to 300 and/or 50Kb fragments prior to or in the absence of internucleosomal fragmentation. EMBO J 1993; 12:3679–3684.

88. Wyllie AH. Glucocorticoid-induced thymocyte apoptosis is associated with endogenous endonuclease activation. Nature 1980; 284:555–556.

89. Oberhammer F, Fritsch G, Schmied M, Pavelka M, Printz D, Purchio T, Lassmann H, Sculte-Hermann R. Condensation of chromatin at the membrane of an apoptotic nucleus is not associated with activation of an endonuclease. J Cell Sci 1993; 104:317–326.

90. Shi L, Nishioka WK, Th'ng J, Bradbury EM, Litchfield DW, Greenberg AH. Premature p34cdc2 activation required for apoptosis. Science 1994; 263:1143–1145.

91. Peitsch MC, Polzar B, Stephan H, Crompton T, MacDonald HR, Mannherz HG, Tschopp J. Characterisation of the endogenous deoxyribonuclease involved in nuclear DNA degradation during apoptosis. EMBO J 1993; 12:371–377.

92. Barry MA, Eastman A. Identification of deoxyribonuclease II as an endonuclease involved in apoptosis. Arch Biochem Biophys 1993; 300:440–450.

93. Gaido ML, Cidlowski JA. Identification, purification and characterisation of a calcium-dependent endonuclease (NUC18) from apoptotic rat thymocytes. J Biol Chem 1991; 266:18580–18585.

94. Ribiero JM, Carson DA. $Ca^{2+}/Mg^{(2+)}$-dependent endonuclease from human spleen: purification, properties, and role in apoptosis. Biochemistry 1993; 32:9129–9136.

95. Savill J, Fadok V, Henson P, Haslett C. Phagocyte recognition of cells undergoing apoptosis. Immunol Today 1993; 14:131–136.

7

Hematopoietic Microenvironments

Michael W. Long
University of Michigan, Ann Arbor, Michigan

I. INTRODUCTION

Hematopoietic development is regulated by a complex set of events in which cells interact with each other, with general and specific growth factors, and with the surrounding extracellular matrix (ECM; 1). These interactions are important for a variety of reasons, such as localizing cells within the microenvironment, directing cellular migration, and initiating growth factor-mediated developmental programs. However, simple interactions, such as those between cells and growth factors, are not the only interactions by which developing cells are regulated. Further complexity occurs through interactions of cells and growth factors with the extracellular matrix, or by combined interactions that generate a specific developmental responses.

Developing tissue cells interact with a wide variety of regulators during their ontogeny. Each of these interactions is mediated by defined, specific receptor–ligand interactions necessary to both stimulate the cell's proliferation and its motility. For example, the localization of hematopoietic cells to their "niche" within the microenvironment requires that chemical or extracellular matrix gradients exist that signal the cell to move along "tracks" of molecules into a defined tissue area. High concentrations of the attractant or other signals next serve to "localize" the cell, thereby stopping its nonrandom walk. The signals that stop or regionalize cells in appropriate microenvironments seem to be complexes of cytokines, and extracellular matrix molecules serve to localize progenitor cells (2). Thus, the regulation of hematopoietic cell development is a complex process in which several microenvironmental components work to bring about coordinated blood cell production: stromal and parenchymal cells, growth factors, and extracellular matrix. Each of these is a key component of a localized and highly organized microenvironmental regulatory system.

Blood cell interactions within the microenvironment can be divided into three classes: cell–cell, cell–extracellular matrix, and cell–growth factor. There are several instances in which blood cells interact with each other or with cells in other tissues. Cell–cell interactions occur within the immune system when lymphocytes interact with antigen-presenting cells, whereas neutrophil or lymphocyte egress from the vasculature during inflammation exemplifies blood

117

cell–endothelial cell recognition. Interactions between cells and the extracellular matrix (the complex proteinaceous substance surrounding cells) also play an important role. Matrix components are important in the growth and development of hematopoietic cells, because they serve either as cytoadhesion molecules, or compartmentalize growth factors within specific microenvironmental locales. The most well-known regulatory event in hematopoiesis is cytokine control of developing blood cells. Many experimental studies of the hematopoietic microenvironment have examined simple interactions (e.g., cell–cell, cell–matrix). However, the situation in vivo is undoubtedly much more complex. For example, growth factors are often bound to matrix molecules which, in turn, are expressed on the surface of underlying stromal cells. Thus, very complex interactions occur (e.g., accessory cell–stromal cell–growth factor–progenitor cell–matrix; Fig. 1), and these can be further complicated by a developmental requirement for multiple growth factors.

Given the multiplicity of blood–cell interactions, a highly specialized cell surface structure is required (i.e., receptors) to both mediate cell adhesion and to transmit intracellular signals from other cells, growth factors, or the ECM. Most cell surface receptors are proteins that consist of an extracellular ligand-binding domain, a hydrophobic membrane-spanning region, and a cytoplasmic region that usually functions in signal transduction. The amino acid sequence of these receptors often defines various families of receptors (e.g., immunoglobulin and integrin gene superfamilies). However, some receptors are not linked to the cell surface by protein; certain receptors contain phosphatidylinositol-based membrane linkages. This type of receptor is usually associated with signal transduction events mediated by phospholipase C (PLC) activation (3). Other cell surface molecules important in receptor–ligand interactions are surface proteins that function as coreceptors. Coreceptors function with a well-defined receptor, usually to amplify stimulus–response coupling. The goal of this review is to examine some of the features of the hematopoietic microenvironment (cellular elements, soluble growth factors, and extracellular matrix).

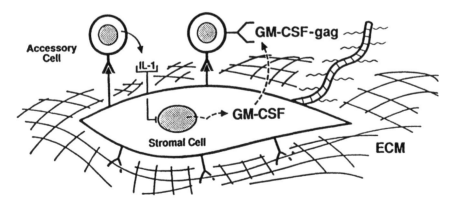

Figure 1 Hematopoietic cellular interactions: the varying complexities of putative hematopoietic cell interactions. A conjectural complex is shown in which accessory cell–stromal cell, hematopoietic cell–stromal cell, and stromal cell–PG–growth factor complexes localize developmental signals. ECM, extracellular matrix; gag, glycosaminoglycan side chain bound to proteoglycan (PG) core protein (indicated by cross-hatched curved molecule); IL-1, interleukin-1; GM-CSF, granulocyte–macrophage colony-stimulating factor. (Modified from Ref. 1.)

II. CELLULAR ELEMENTS

Blood cells develop in a distinct hierarchical fashion. During ontogeny, hematopoietic cells migrate to a variety of organs: the liver, spleen, and eventually, the bone marrow. There they seek out an appropriate region within the nascent tissue and undergo a phase of rapid proliferation and differentiation. In tissues that retain their proliferative capacity (bone marrow, liver, skin, the gastrointestinal lining, muscle, and bone), the complex hierarchy of proliferation cells is retained throughout life. This is best illustrated in the hematopoietic system. Blood cells are constantly produced, such that approximately 70 times an adult human's body weight of blood cells is produced throughout the human life span. This implies the existence of a very primitive cell type that retains the capacity of self-renewal. This cell is called a stem cell, and it is the cell responsible for the engraphment of hematopoiesis in recipients of bone marrow transplantation. Besides a high proliferative potential, the stem cell is also characterized by its multipotentiality, in that it can generate progeny (referred to as progenitor cells) that are committed to each of the eight blood cell lineages. As hematopoietic cells proliferate, they progressively lose their proliferative capacity and become increasingly restricted in lineage potential. As a result, the more primitive progenitor cells in each lineage produce higher colony numbers, and the earliest cells detectable in vitro produce progeny of two to three lineages (there is no in vitro assay for transplantable stem cells).

Cell–cell interactions mediate both cellular development and stimulus–response coupling. When coupled with other interactions (e.g., cell–ECM), these interactions represent a powerful mechanism for directing or localizing blood cell development and can potentially yield lineage-specific or organ-specific information. Much of our understanding of cell–cell interactions comes from the immune system and from the study of developing blood cells and their interactions with adjacent stromal cells (3–5).

The isolation and cloning of immune cell ligands and receptors resulted in the classification of gene families that mediate cell–cell interactions within the immune and hematopoietic system. Three families of molecules are important in the mediation of cell–cell interactions (Table 1). The *integrin family* is a large group of highly versatile proteins that is involved in cell–cell and cell–matrix attachment. The *selectin family* comprises molecules that are involved in lymphocyte, platelet, and leukocyte interactions with endothelial cells. Interestingly, this class of cell surface molecules uses specific glycoconjugates (encoded by glycosyltransferase genes) as their ligands on endothelial cell surfaces. Finally, the *immunoglobulin superfamily* is expressed predominantly on cells mediating immune and inflammatory responses.

A. Integrins

Members of the integrin family mediate interactions between cells and extracellular matrix proteins (6). Cell attachment to these molecules occurs rapidly (within minutes), and is a result of increased avidity, rather than increased expression (7, and references therein). The binding sequence within the ligand for most, but not all, integrins is the tripeptide sequence arganine–glycine–asparagine (RGD; 8). Structurally, integrins consist of two membrane-spanning α- and β-chains. The α-subunits contain three to four tandem repeats of a divalent–ion-binding motif and require magnesium or calcium to function. The α-chains are (usually) distinct and bind with common or related β-subunits to yield functional receptors (6). The β-subunits of integrins have functional significance, and integrins can be subclassified based on the presence of a given β-chain. Thus, integrins containing β1- and β3-chains are involved predominantly in cell–extracellular matrix interactions, whereas molecules containing the β2-subunits function in leukocyte–leukocyte adhesion (Tables 1 and 2). The cytoplasmic domain of many

Table 1 Cell Adhesion Molecule Superfamilies

I. Immunoglobulin superfamily of adhesion receptors	
LFA-2 (CD2)	T-cell receptor (CD3)
LIFA-3 (CD58)	CD4 (TCR coreceptor)
ICAM-1 (CD54)	CD8 (TCR coreceptor)
ICAM-2	MHC class I
VCAM-1	MHC class II

II. Integrins
    β_1-*Integrins* (VLA proteins)
        P150,95 (CD11/CD18)
        VLA 1-3, 6
        VLA 4 (LPAM 1, CD49d/CD20)
        Fibronectin receptor, (VLA 5, CD-/CD20)
        LPAM 2
    β_2-*Integrins*
        LFA 1 (CD11$_a$/CD18)
        Mac 1 or Mol (CD11$_b$/CD18)
    β_3-*Integrins*
        Vitronectin receptor
        Platelet gp-IIb/IIIa

III. Selectin/LEC_CAMS
    Mel 14 (LE-CAM-1, LHR, LAM-1, Leu 8, Ly 22, gp90 MEL)
    ELAM-1 (LE-CAM-2)
    GMP 140 (LE-CAM-3, PADGEM, CD62)

Source: Ref. 1.

integrin receptors interacts with the cytoskeleton. For example, several integrins are known to localize near focal cell contacts at which actin bundles terminate (3,6). As a result, changes in receptor binding offer an important mechanism for linking cell adhesion to cytoskeletal organization.

B. Selectins

The selectin family of cell adhesion receptors contains a single NH$_2$-terminus, calcium-dependent, lectin-binding domain, an EGF receptor (EFGR) domain, and a region of cysteine-rich tandem repeats (from two to seven) that are homologous with complement-binding proteins (3,9,10). Selectins (e.g., ELAM-1, MEL14, gp90MEL, and GMP140/PADGEM) (see Table 2) are expressed on neutrophils and lymphocytes. They recognize specific glycoconjugate ligands on endothelial and other cell surfaces. Early studies of lymphocyte trafficking demonstrated that fucose or mannose inhibited lymphocyte attachment to lymph node endothelial cells (11). The observation that selectins contain a lectin-binding domain (9) led to the identification of the ligands for two members of this family (for review see Ref. 11). Lowe and coworkers first demonstrated that $\alpha(1,3 \rightarrow 1,4)$ fucosyltransferase cDNA converted nonmyeloid COS (or CHO) cells to sialyl-Lewis X-positive cells that bound both HL60 cells and neutrophils in an endo-thelial–leukocyte adhesion molecule (ELAM)-1-dependent manner (12). In an independent study, Goelz and coworkers used a monoclonal antibody, which inhibited ELAM-mediated

Table 2 Cell Surface Molecules Mediating Cell–Cell Interactions

Cell receptor	Receptor–cell expression	Ligand, co-, or counterreceptor	Ligand co-, or counter-receptor cell expression	Ref.
Ig Superfamily				
MHC I	Macroph, T cell	CD8, TCR	T cells	3
MHC II	Macroph, T cell	CD4, TCR	T cells	3
ICAM-1	Endo, Neut HPC, B cells, T cells, macroph	LFA-1	Mono, T and B cells	3
ICAM-2	Endo cells	LFA-1	Mono, T and B cells	3
LFA-2	T cells	LFA-3	T cells, Erythr	3
Integrins				
MAC1	Macroph, Neut	Fibrinogen, C3bi	Endothelial cells, Plts	3
LFA-1	Macroph, Neut	(see above)	(see above)	
VCAM	Endo cells	VLA4	Lymphocytes Monocyte	11
			B-progenitors	79,80
gp150,95	Macroph, Neut	(see above)	(see above)	
FN-R	Erythroid lineage	Fibronectin	N.A.	(see Table 3)
IIb/IIIa	Plts, Mk	Fibrinogen, TSP, VN, vWF	Endothelial cells	3
Selectins				
LEC-CAM-1 (Mel 14)	Endo cells	Addressins, neg.-charged oligo-saccharides	Lymphocytes	11
ELAM-1 (LE-CAM-2)	Endo cells	Sialyl-Lewis X[a]	Endo cells[b] Neut, tumor cells	11,12
LEC-CAM-3 (GMP-140)	Plts, gran, Weible-Palade bodies, endo	Lewis X (CD15)	Endo cells Neut	11

[a]Constitutively expressed by few cells, up-regulated by TNF-β and IL-1.
[b]Sialylated, fucosylated lactosaminoglycans.
Macroph, macrophage; Mono, monocyte; Neut, neutrophil; Endo, endothelial cell; Eryth, erythroid cells; Plts, platelets.
Source: Modified from Ref. 1.

attachment, to screen an expression library that yielded a novel α(1,3)-fucosyltransferase the expression of which conferred ELAM-binding activity on target cells (13).

Information on cell–cell interactions among hematopoietic progenitor cells is less well developed than that concerning mature neutrophils and lymphocytes (Table 3). Most data on the adhesive capacities of early hematopoietic cells deal with the interaction of these cells with underlying, preestablished stromal cell layers (5). Gordon and colleagues documented that primitive hematopoietic human CML blast–colony-forming cells (B1-CFC) adhere to pre-formed stromal cell layers (14,15). They additionally showed that the stromal cell ligand is not one of the known cell adhesion molecules (16). Also, CD34-selected hematopoietic marrow cell populations attach to stromal cell layers, and the attached cells are enriched for granulocyte–macrophage progenitor cells (17). Murine spleen colony-forming cells (CFC-S), also adhere to

Table 3 Hematopoietic Cell–Stromal Cell Interactions
(Unknown Receptor–Ligand)

Cell phenotype	Stromal cell	Ref.
B cells	Fibroblasts	81,82
Pre-B cell	Heterologous stroma	83
BFC-E	Fibroblasts	84,85
CFC-S	Fibroblasts (NIH 3T3)	19
CFC-S, B1-CFC	Heterologous stroma[a]	14,15,86
CFC-GM	Heterologous stroma	43,85
CFC-Mk	Heterologous stroma	43,85

[a]Methylprednisolone-stimulated stromal cells, unstimulated
fail to bind (see 14).
BFC-E, burst-forming cell–erythroid; CFC, colony-
forming cell; S, spleen; B1, blast; GM, granulocyte–macrophage; Mk, megakaryocyte.
Source: Ref. 1.

stromal cells (18). Interestingly, bone marrow stromal cells can be substituted for by NIH 3T3 cells (19), suggesting that these adherent cells supply the necessary attachment ligand for CFU-S attachment (19,20).

C. Immunoglobulins

The immunoglobulin molecules function in both antigen recognition and cell–cell communication. Briefly, the immunoglobulin superfamily (see Table 2) is defined by an immunoglobulin-like domain found within a dimer of two antiparallel β-strands (21,22; for a review see Ref. 3). Two members of this family, the T-cell receptor and immunoglobulin, function in antigen recognition. The T-cell receptor (TCA) recognizes antigenic peptides in the context of two other molecules on the surface of antigen-presenting cells: major histocompatibility complex (MHC) class I and class II molecules (3,21,23). The binding of the T-cell receptor to MHC–antigenic peptide complex is sufficient for cell–cell adhesion. However, cellular activation also requires the binding of either of two coreceptors, CD4 or CD8. Neither coreceptor directly binds the MHC complex but, rather, seems to interact with the T-cell receptors to synergistically amplify intercellular signaling (3,24,25).

III. HEMATOPOIETIC GROWTH FACTORS

The presence of hematopoietic growth factors is an obligate requirement for the in vitro proliferation and differentiation of developing cells. These growth factors differ in their actions from the more typical endocrine hormones, such as anabolic steroids or growth hormone. Endocrine hormones affect general cell function and are required or are important to tissue formation in a variety of organs; their predominant role is one of homeostasis. On the other hand, growth *factors* specifically drive the developmental programs of differentiating cells, although it is unclear whether this is an inductive or permissive phenomena. Whether hematopoietic cytokines function in a permissive or inductive capacity has been the subject of considerable past controversy, particularly relative to blood cell development. However, the large amount of data demonstrating linkages between receptor–ligand interaction and gene activation argue persuasively for a direct action on gene expression and, hence, cell proliferation and differentiation.

Hematopoietic cell proliferation and differentiation is regulated by numerous growth factors (for reviews see Refs. 26,27). Within the last decade, approximately 20 stimulatory cytokines (13 interleukins, M-CSF, erythropoietin, G-CFS, GM-CSF, c-*kit* ligand, interferon gamma, and thrombopoietin) have been molecularly cloned and examined for their function in hematopoiesis. Clearly, this literature is beyond the scope of this review. However, the recent genetic cloning of eight receptors for these cytokines has led to the observation that several of these receptors have amino acid homologies (27), showing that they are members of one or more gene families (Table 4). Hematopoietic growth factor receptors structurally contain a large extracellular domain, a transmembrane region, and a sequence-specific cytoplasmic domain (27). The extracellular domains of interleukin-1 (IL-1), IL-6, and interferon gamma (IFN-γ) are homologous with the immunoglobulin gene superfamily, and weak but significant amino acid homologies exists among the IL-2 (β-chain), IL-6, IL-3, IL-4, erythropoietin, and granulocyte–macrophage colony-stimulating factor (GM-CSF) receptors (27).

IV. EXTRACELLULAR MATRIX

The extracellular matrix (ECM) consists of various molecules, such as fibronectin, thrombospondin, laminin, collagens, proteoglycans, other glycoproteins and varies in its tissue composition throughout the body (28). Early studies in nonhematopoietic systems demonstrated that ECM components stimulated corneal endothelial cell proliferation in vitro (29,30), and that laminin is involved in inductive interactions that give rise to retinal-pigmented epithelium (31). Likewise, mammary epithelial cell differentiation is directly regulated by ECM components, and mammary cell growth in vivo and in vitro requires type IV collagen (28).

Several investigations elucidated a role for ECM and its components in hematopoietic cell function. These studies have identified the function of both previously known and newly identified ECM components in hematopoietic cell cytoadhesion (Table 5). Soluble factors, stromal cells, and extracellular matrix (the natural substrate surrounding cells in vivo), are critical elements of the hematopoietic microenvironment. Wolf and Trenton provided the first evidence that (still-undefined) components of the microenvironment are responsible for the predominance of granulopoiesis in the bone marrow and erythropoiesis in the spleen (32). In

Table 4 Hematopoietic Growth Factor Receptor Families

Receptors with homology to the immunoglobulin gene family
    Interleukin-1 receptor
    Interleukin-6 receptor
    Interferon gamma receptor
Hematopoietic growth factor receptor family
    Interleukin-2 receptor (β-chain)
    Interleukin-3 receptor
    Interleukin-4 receptor
    Interleukin-6 receptor
    Erythropoietic receptor
    GM-CSF receptor

Source: Modified from Ref. 1.

Table 5 Proteins and Glycoproteins Mediating Hematopoietic Cell–Extracellular Matrix Interactions

Matrix component	Cell surface receptor	Cellular expression	Ref.
Fibronectin	FnR	Erythroid	48–51,85
		BFC-E	84,85,87
		B cells	81
		Lymphoid cell lines	88
		HL60 cells	89
	IIb/IIIa	Platelets and mega-karyocytes	90
Thrombospondin	TSP-R	Monocytes and platelets	91,92
		Human CFC	42
		CFC-GEMM	42
Hyaluronic acid	CD44	T and B cells	93–96
		Neutrophils	
		Tumor cells	
Hemonectin	Unknown	CFC-GM, BFC-E	
		Immature neutrophil	97
		BFU-E	98
Proteoglycans: heparan sulfate	Unknown	B1-CFC	60,99
Unfract. ECM	Unknown	B1-CFC,	60
		Bone marrow stroma	43

R, receptor; BFC-E and erythroid progenitor cell, the burst-forming cell–erythrocyte; HL60, a promyelocytic leukemia cell line; CFC, colony-forming cell; GEMM, gran-ulocyte–erythrocyte–macrophage–megakarocyte; GM, granulocyte–macrophage; B1, blast cell.
Source: Modified from Ref. 1.

vitro, the development of adherent cell populations is essential for the sustained proliferation and differentiation of blood cells in long-term bone marrow cell cultures (LTBMC; 33,34). In LTBMC, stromal cells elaborate laminin, fibronectin, various collagens, and proteoglycans, and the presence of these ECM proteins coincides with the onset of hematopoietic cell proliferation (35). The actual roles for extracellular matrix versus stromal cells in supporting cell development remains somewhat obscure, for it is often difficult to disassociate stromal cell effects from those of the ECM—stromal cells are universally observed to be enmeshed in the surrounding extracellular matrix.

The foregoing observations on the general and specific effects of ECM on blood cell development (as well as other systems) have identified certain matrix components that seem to appear as a recurrent theme in tissue development. These are thrombospondin, fibronectin, proteoglycans, and to some degree, the collagens.

A. Thrombospondin

Thrombospondin (TSP), first identified as an α-granule protein of platelets, is a large, trimeric disulfide-linked glycoprotein (molecular weight 450,000, subunit molecular weight 180,000) having a domain-like structure (36). Its protease-resistant domains are involved in mediating various TSP functions, such as cell binding and binding of other extracellular matrix proteins (36). Thrombospondin is synthesized and secreted into extracellular matrix by most cells (for review see Refs. 36,37). Matrix-bound TSP is necessary for cell adhesion (38), cell growth (39), carcinoma invasiveness (40), and is differentially expressed during murine embryogenesis (41). Thrombospondin functions within the hematopoietic microenvironment as a cytoadhesion protein for a subpopulation of human hematopoietic progenitor cells (2,42). Interestingly, immunocytochemical- and metabolic-labeling studies show that hematopoietic cells (both normal marrow cells and leukemic cell lines) synthesize TSP, deposit it within the ECM, and are attached to it (42). Other studies from our laboratories show that bone marrow ECM also plays a major role in hematopoiesis, in that complex ECM extracts greatly augment LTBMC cell proliferation (43) and that marrow-derived ECM contains specific cytoadhesion molecules (2,42,44–46).

B. Fibronectin

Fibronectin is a ubiquitous extracellular matrix molecule that is involved in the attachment of parenchymal cells to stromal cells (35,47). As with TSP, hematopoietic cells synthesize, deposit, and bind to fibronectin (35). Extensive work by Patel and coworkers shows that erythroid progenitor cells attach to fibronectin in a developmentally regulated manor (48–51). In addition to cells of the erythroid lineage, fibronectin is capable of binding lymphoid precursor cells and other cell phenotypes (52,53). Structuraly cells adhere to two distinct regions of the fibronectin molecule, one of which contains the RGD sequence; the other is within the COOH-terminal region and contains a high-affinity binding site for heparan (52).

C. Proteoglycans and Glycosaminoglycans

The role of proteoglycans in blood cell development is poorly understood. Both hematopoietic cells (54) (albeit as demonstrated by cell lines), and marrow stromal cells (55–58) produce various proteoglycans. Proteoglycans (PG) are polyanionic macromolecules located both on the cell surface and within the extracellular matrix. They consist of a core protein containing several covalently linked glycosaminoglycan (GAG) side chains, as well as one or more O- or N-linked oligosaccharides. These GAG side chains consist of nonbranching chains of repeating N-acetylglucosamine or N-acetylgalactosamine disaccharide units. With the exception of hyaluronic acid, all glycosaminoglycans are sulfated. Interestingly, many ECM molecules (fibronectin, laminin, and collagen) contain glycosaminoglycan-binding sites that mediate the complex interactions that occur within the matrix itself.

Proteoglycans play a role in both cell proliferation and differentiation. Murine stromal cells produce hyaluronic acid, as well as heparan sulfate and chondroitin sulfate-containing PGs (55). There is a differential distribution of these between stromal cell surface and the media PGs in vitro, with heparan sulfate PG being the primary cell surface molecule, and chondroitin sulfate PG the major molecular species in the aqueous phase (56). Unlike murine LTBMC cultures, human hematopoietic stromal cell cultures contain small amounts of heparan sulfate, and large amounts of dermatin and chondroitin sulfate PGs, each of which is equally distributed between the aqueous phase and extracellular matrix (57). The stimulation of proteoglycan–GAG synthesis is associated with an increased hematopoietic cell proliferation, as demonstrated by an increase in the percentage of cells in S phase (56). Given the general diversity of proteoglycans, it is reasonable to expect that they may encode both lineage-specific and organ-specific information. For example, marrow-derived ECM proteoglycans directly stimulate differentiation of human progranulocytic cells (HL60), whereas PGs derived from skin fibroblasts lack this inductive capacity (59). Moreover, organ-specific effects are seen in studies of human blood precursor cell adhesion to marrow-derived heparan sulfate, but not to heparan sulfates isolated from bovine kidney (60).

More importantly, stromal cell surface-associated proteoglycans are involved in the compartmentalization or localization of growth factors within the microenvironment. The proliferation of hematopoietic cells in the presence of hematopoietic stroma is associated with a glycosaminoglycan-bound growth factor (GM-CSF; 61), and determination of the precise GAG molecules involved in this process (i.e., heparan sulfate) showed that heparan sulfate side chains bind two hematopoietic growth factors: GM-CSF and IL-3 (62). These data imply that ECM components and growth factors combine to yield lineage-specific information, and also indicates that, when PG- or GAG-bound, growth factor is presented to the progenitor cells in a biologically active form.

D. Collagen

The role of various collagens in blood cell development remains uncertain. In vitro marrow cells produce types I, III, IV, and V collagen (35,63–65), suggesting a role for these extracellular matrix components in the maintenance of hematopoiesis. Consistent with this, inhibition of collagen synthesis with 6-hydroxyproline blocks or reduces hematopoiesis in vitro (66), but little else in known concerning collagen subtypes.

V. EX VIVO EXPANSION OF HEMATOPOIETIC CELLS

The microenvironmental complexities discussed in the foregoing suggest that the ex vivo generation of human (replacement) tissue (e.g., marrow) will be a difficult process. The long-term in vitro growth of hematopoietic cells is routinely performed, albeit at varying degrees of success. When unfractionated human bone marrow is established in long-term culture, the stromal and hematopoietic cells attempt to recapitulate in vivo hematopoiesis. Both growth factors, and ECM proteins are produced (35,42,67), and relatively long-term hematopoiesis occurs. However, when examined closely, it becomes clear that LTBMC do not faithfully reproduce the in vivo microenvironment (18,67,68). Over a period of 8–12 weeks (for human cultures) or 3–4 months (for murine cultures), cell proliferation ceases, and the stromal and hematopoietic cells die. Moreover, the pluripotentiality of these cultures is rapidly lost. In vivo, bone marrow produces a wide variety of cells (erythroid, megakaryocyte–platelet, four types of myeloid cells, and B lymphocytes). Human long-term marrow cultures predominantly produce granulocytes and megakaryocytes for 1–3 weeks, and erythropoiesis is seen only if exogenous growth factors are added. Thereafter, the cultures produce granulocytes and macrophages. These data indicate that current culture conditions are inadequate, and further suggest that other factors, such as the rate of fluid exchange (perfusion) or the three-dimensional structure of these cultures, are limiting.

Recent work by Emerson colleagues demonstrated the effectiveness of altered medium exchange rates in the expansion of blood cells in vitro (69–71). Their studies demonstrated that a 50% medium exchange per day affected stromal cell metabolism and stimulated a transient increase in growth factor production (69). Also, these cultures underwent a tenfold expansion of cell numbers (70). Although impressive, these cultures nonetheless, decay after 10–12 weeks. Recently, this group has used continuous-perfusion bioreactors to achieve a longer ex vivo expansion, and achieved a 10- to 20-fold expansion of specific progenitor cell types (72). These studies demonstrate that bioreactor technology allows significant expansion of cells, presumably by better mimicry of in vivo conditions.

Another aspect of ex vivo blood formation is the physical structure of the developing organ. Tissues, including bone marrow, exist as three-dimensional structures. Thus, the usual growth in tissue culture flasks is far removed from the in vivo setting. Essentially, hematopoietic cells grown in vitro proliferate at a liquid–substratum interface. As a result, LTBMC stromal cell layers usually grow until they reach confluence, and then cease proliferating, a process known as contact-inhibition. This, in turn, severely limits the degree of total cellularity of the system. For example, long-term marrow cultures (which do not undergo as precise a contact inhibition as cells from solid tissues), reach a cell density of $1–4 \times 10^6$ per mL. This is three orders of magnitude less than the average bone marrow density in vivo. Recently, another type of bioreactor has been used to increase the ex vivo expansion of cells. Rotating-wall vessels are designed to result in constant, low-shear suspension of tissue cells during their development. These bioreactors thus simulate a microgravity environment. The studies of Goodwin and associates document a remarkable augmentation of mesenchymal cell proliferation in low-shear bioreactors. Their data show that mesenchymal cell types have an average three- to sixfold increase in cell density in these bioreactors, reaching a cell concentration of approximately 10^7 cells per mL (73,74). Further work by Goodwin et al. shows that the growth of mesenchymal cells (kidney and chondrocyte) under low-shear conditions leads to the formation of tissue-like cell aggregates that is enhanced by growing these cells on collagen-coated microcarriers (74,75). The application of this system to hematopoietic cell growth would be of considerable interest.

VI. SUMMARY

As elegantly demonstrated by the foregoing composite data, the molecular basis and the function of the various components of tissue microenvironments are becoming well understood. However, much remains to be learned about the role of these and other, perhaps as yet unidentified, molecules in tissue development. One of the intriguing questions to be asked is how each molecule helps define the molecular milieu of a given microenvironmental region; for example, the distinct differences in hematopoiesis as it exists in the marrow versus the spleen (32). Another important subject is the dissection of the molecular basis of cell trafficking into tissues (i.e., homing). Again, the immune system offers a model for the assessment of this process. The interaction of the lymphocyte cell surface receptors with specific endothelial cell glyco-conjugates (75–78) implies that similar recognition systems are involved in other tissues, particularly the bone marrow. Finally, the elucidation of the various requirements for optimal progenitor cell growth (cell interactions, specific growth factors and matrix components, or accessory cells that supply them) should allow the improvement of ex vivo culture systems to yield an environment in which the human hematopoietic microenvironment reconstitution is possible. Such a system would have obvious significance in bone marrow transplantation.

REFERENCES

1. Long MW. Blood cell cytoadhesion molecules. Exp Hematol 1992; 20:288–301.
2. Long MW, Briddell R, Walter AW, Bruno E, Hoffman R. Human hematopoietic stem cell adherence to cytokines and matrix molecules. J Clin Invest 1992; 90:251–255.
3. Springer TA. Adhesion receptors of the immune system. Nature 1990; 346:425–434.

4. Gallatin M, St. John TP, Siegleman M, Reichert R, Butcher E, Weissman IL. Lymphocyte homing receptors. Cell 1986; 44:673–680.

5. Dexter TM. Stromal cell associated haemopoiesis. J Cell Physiol 1982; 1:87–94.

6. Giancotti FG, Ruoslahti E. Elevated levels of the alpha 5 beta 1 fibronectin receptor suppress the transformed phenotype of Chinese hamster ovary cells. Cell 1990; 60:849–859.

7. Lawrence MB, Springer TA. Leukocytes roll on a selectin at physiologic shear flow rates: distinction from and prerequisite for adhesion through integrins. Cell 1991; 65:859–873.

8. Ruoslahti E, Pierschbacher MD. New perspectives in cell adhesion: RGD and integrins. Science 1987; 238:491–497.

9. Bevilacqua MP, Stengelin S, Gimbrone MA, Seed B. Endothelial leukocyte adhesion molecule 1: an inducible receptor for neutrophils related to complement regulatory proteins and lectins. Science 1989; 243:1160–1165.

10. Stoolman LM. Adhesion molecules controlling lymphocyte migration. Cell 1989; 56:907–910.

11. Brandley BK, Swiedler SJ, Robbins PW. Carbohydrate ligands of the LEC cell adhesion molecules. Cell 1990; 63:861–863.

12. Lowe JB, Stoolman LM, Nair RP, Larsen RD, Berhend TL, Marks RM. ELAM-1-dependent cell adhesion to vascular endothelium determined by a transfected human fucosyl transferase cDNA. Cell 1990; 63:475–484.

13. Goelz SE, Hession C, Goff D, Griffiths B, Tizard R, Newman B, et al. *ELFT*: a gene that directs the expression of an ELAM-1 ligand. Cell 1990; 63:1349–1356.

14. Gordon MY, Hibbin JA, Dowding, C, Gordon SEC, Goldman JM. Separation of human blast progenitors from granulocytic, erythroid, megakaryocytic, and mixed colony-forming cells by "panning" on cultured marrow-derived stromal layers. Exp Hematol 1985; 13:937–940.

15. Gordon MY, Clarke D, Atkinson J, Greaves MF. Hemopoietic progenitor cell binding to the stromal microenvironment in vitro. Exp Hematol 1990; 18:837–842.

16. Gordon MY, Dowding CR, Riley GP, Greaves MF. Characterisation of stroma-dependent blast colony forming cells in human marrow. J Cell Physiol 1987; 130:150–156.

17. Liesveld JL, Abboud CN, Duerst RE, Ryan DH, Brennan JK, Lichtman MA. Characterization of human marrow stromal cells: role in progenitor cell binding and granulopoiesis. Blood 1989; 73:1794–1800.

18. Spooncer E, Lord BI, Dexter TM. Defective ability to self-renew in vitro of highly purified primitive haematopoietic cells. Nature 1985; 316:62–64.

19. Roberts RA, Spooncer E, Parkinson EK, Lord BI, Allen TD, Dexter TM. Metabolically inactive 3T3 cells can substitute for marrow stromal cells to promote the proliferation and development of multipotent haemopoietic stem cells. J Cell PHysiol 1987; 132:203–214.

20. Yamazaki K, Roberts RA, Spooncer E, Dexter TM, Allen TD. Cellular interactions between 3T3 cells and interleukin-3-dependent multipotent haemopoietic cells: a model system for stromal-cell-mediated haemopoiesis. J Cell Physiol 1989; 139:301–312.

21. Sheetz MP, Turney S, Qian H, Elson EL. Nanometre-level analysis demonstrates that lipid flow does not drive membrane glycoprotein movements. Nature 1989; 340:284–288.

22. Williams AF, Barclay AN. The immunoglobulin superfamily—domains for cell surface recognition. Annu Rev Immunol 1988; 6:381–405.

23. Bierer BE, Sleckman BP, Ratnofsky SE, Burakoff SJ. The biologic roles of CD2, CD4, and CD8 in T-cell activation. Annu Rev Immunol 1989; 7:579–599.

24. Spits H, van Schooten W, Keizer H, et al. Alloantigen recognition is preceded by nonspecific adhesion of cytotoxic T cells and target cells. Science 1986; 232:403–405.

25. Shaw S, Luce GE, Quinones R, Gress RE, Springer TA, Sanders ME. Two antigen-independent adhesion pathways used by human cytotoxic T-cell clones. Nature 1986; 323:262–264.

26. Metcalf D. The molecular control of cell division, differentiation commitment and maturation in haematopoietic cells. Nature 1989; 339:27–30.

27. Arai K, Lee F, Miyajima A, Miyatake S, Yokata T. Cytokines: coordinators of immune and inflammatory responses. Annu Rev Biochem 1990; 59:783–836.
28. Wicha MS, Lowrie G, Kohn E, Bagavandoss P, Mahn T. Extracellular matrix promotes mammary epithelial growth and differentiation in vitro. Proc Natl Acad Sci USA 1982; 79:3213–3217.
29. Gospodarowicz D, Ill C. Extracellular matrix and control of proliferation of vascular endothelial cells. J Clin Invest 1980; 65:1351–1364.
30. Gospodarowicz D, Delagado D, Vlodavsky I. Permissive effect of the extracellular matrix on cell proliferation in vitro. Proc Natl Acad Sci USA 1980; 77:4094–4098.
31. Reh TA, Gretton H. Retinal pigmented epithelial cells induced to transdifferentiate to neurons by laminin. Nature 1987; 330:68–71.
32. Wolf NS, Trentin JJ. Hemopoietic colony studies. V. Effect of hemopoietic organ stroma on differentiation of pluripotent stem cells. J Exp Med 1968; 127:205–214.
33. Dexter TM, Allen TD, Lajtha LG. Conditions controlling the proliferation of haemopoietic stem cells in vitro. J Cell Physiol 1976; 91:335–344.
34. Dexter TM, Lajtha LG. Proliferation of haemopoietic stem cells in vitro. Br J Haematol 1974; 28:525–530.
35. Zuckerman KS, Wicha MS. Extracellular matrix production by the adherent cells of long-term murine bone marrow cultures. Blood 1983; 61:540–547.
36. Frazier WA. Thrombospondin: a modular adhesive glycoprotein of platelets and nucleated cells. J Cell Biol 1987; 105:625–632.
37. Lawler J. The structural and functional properties of thrombospondin. Blood 1986; 67:1197–1209.
38. Varani J, Dixit VM, Fligiel SEG, McKeever PE, Carey TC. Thrombospondin-induced attachment and spreading of human squamous carcinoma cells. Exp Cell Res 1986; 156:1–15.
39. Majack RA, Cook SC, Bornstein P. Control of smooth muscle cell growth by components of the extracellular matrix: autocrine role for thrombospondin. Proc Natl Acad Sci USA 1986; 83:9050–9054.
40. Riser BL, Varani J, Carey TE, Fligiel SEG, Dixit VM. Thrombospondin binding and thrombospondin synthesis by human squamous carcinoma and melanoma cells: relationship to biological activity. Exp Cell Res 1988; 174:319–329.
41. O'Shea KS, Dixit VM. Unique distribution of the extracellular matrix component thrombospondin in the developing mouse embryo. J Cell Biol 1988; 107:2737–2748.
42. Long MW, Dixit VM. Thrombospondin functions as a cytoadhesion molecule for human hematopoietic progenitor cells. Blood 1990; 75:2311–2318.
43. Campbell A, Wicha MS, Long MW. Extracellular matrix promotes the growth and differentiation of murine hematopoietic cells in vitro. J Clin Invest 1985; 75:2085–2090.
44. Campbell AD, Long MW, Wicha MS. Haemonectin, a bone marrow adhesion protein specific for cells of granulocyte lineage. Nature 1987; 329:744–746.
45. Campbell A, Sullenberger B, Bahou W, Ginsberg D, Long M, Wicha M. Hemonectin: a novel hematopoietic adhesion molecule. Prog Clin Biol Res 1990; 352:97–105.
46. Long MW, Williams JL, Mann KG. Expression of bone-related proteins in the human hematopoietic microenvironment. J Clin Invest 1990; 86:1387–1395.
47. Bentley SA, Tralka TS. Fibronectin-mediated attachment of hematopoietic cells to stromal elements in continuous bone marrow cultures. Exp Hematol 1983; 11:129–138.
48. Patel VP, Lodish HF. Loss of adhesion of murine erythroleukemia cells to fibronectin during erythroid differentiation. Science 1984; 224:996–998.
49. Patel VP, Ciechanover A, Platt O, Lodish HF. Mammalian reticulocytes lose adhesion to fibronectin during maturation to erythrocytes. Proc Natl Acad Sci USA 1985; 82:440–444.
50. Patel VP, Lodish HF. The fibronectin receptor on mammalian erythroid precursor cells: characterization and developmental regulation. J Cell Biol 1986; 102:449–456.
51. Patel VP, Lodish HF. A fibronectin matrix is required for differentiation of murine erythroleukemia cells into reticulocytes. J Cell Biol 1987; 105:3105–3118.

52. Bernardi P, Patel VP, Lodish HF. Lymphoid precursor cells adhere to two different sites on fibronectin. J Cell Biol 1987; 105:489–498.
53. Giancotti FG, Comoglio PM, Tarone G. Fibronectin–plasma membrane interaction in the adhesion of hemopoietic cells. J Cell Biol 1986; 103:429–437.
54. Minguell JJ, Tavassoli M. Proteoglycan synthesis by hematopoietic progenitor cells. Blood 1989; 73:1821–1827.
55. Gallagher JT, Spooncer E, Dexter TM. Role of the cellular matrix in haemopoiesis. I Synthesis of glycosaminoglycans by mouse bone marrow cell cultures. J Cell Sci 1983; 63:155–171.
56. Spooncer E, Gallagher JT, Krizsa F, Dexter TM. Regulation of haemopoiesis in long-term bone marrow cultures. IV. Glycosaminoglycan synthesis and the stimulation of haemopoiesis by beta-D-xylosides. J Cell Biol 1983; 96:510–514.
57. Wight TN, Kinsella MG, Keating A, Singer JW. Proteoglycans in human long-term bone marrow cultures: biochemical and ultrastructural analyses. Blood 1986; 67:1333–1343.
58. Kirby SL, Bentley SA. Proteoglycan synthesis in two murine bone marrow stromal cell lines. Blood 1987; 70:1777–1783.
59. Luikart SD, Sackrison JL, Maniglia CA. Bone marrow matrix modulation of HL-60 phenotype. Blood 1987; 70:1119–1123.
60. Gordon MY, Riley GP, Clarke D. Heparan sulfate is necessary for adhesive interactions between human early hemopoietic progenitor cells and the extracellular matrix of the marrow microenvironment. Leukemia 1988; 2:804–809.
61. Gordon MY, Riley GP, Watt SM, Greaves MF. Compartmentalization of a haematopoietic growth factor (GM-CSF) by glycosaminoglycans in the bone marrow microenvironment. Nature 1987; 326:403–405.
62. Roberts R, Gallagher J, Spooncer E, Allen TD, Bloomfield F, Dexter TM. Heparan sulphate bound growth factors: a mechanism for stromal cell mediated haemopoiesis. Nature 1988; 332:376–378.
63. Castro-Malaspina H, Gay RE, Resnick G, Kapoor N, Meyers P, Chairieri D, et al. Characterization of human bone marrow fibroblast colony-forming cells and their progeny. Blood 1980; 56:289–301.
64. Bentley SA. Collagen synthesis by bone marrow stromal cells: a quantitative study. Br J Haematol 1982; 50:491–497.
65. Bentley SA, Foidart JM. Some properities of marrow derived adherent cells in tissue culture. Blood 1981; 56:1006–1012.
66. Zuckerman KS, Rhodes RK, Goodrum DD, Patel VR, Sparks B, Wells J, et al. Inhibition of collagen deposition in the extracellular matrix prevents the establishment of a stroma supportive of hematopoiesis in long-term murine bone marrow cultures. J Clin Invest 1985; 75:970–975.
67. Dexter TM, Spooncer E. Growth and differentiation in the hemopoietic system. Annu Rev Cell Biol 1987; 3:423–441.
68. Schofield R, Dexter TM. Studies on the self-renewal ability of CFU-S which have been serially transferred in long-term culture or in vivo. Leuk Res 1985; 9:305–313.
69. Caldwell J, Palsson BO, Locey B, Emerson SG. Culture perfusion schedules influence the metabolic activity and granulocyte–macrophage colony-stimulating factor production rates of human bone marrow stromal cells. J Cell Physiol 1991; 147:344–353.
70. Schwartz RM, Palsson BO, Emerson SG. Rapid medium perfusion rate significantly increases the productivity and longevity of human bone marrow cultures. Proc Natl Acad Sci USA 1991; 88:6760–6764.
71. Schwartz RM, Emerson SG, Clarke MF, Palsson BO. In vitro myelopoiesis stimulated by rapid medium exchange and supplementation with hematopoietic growth factors. Blood 1991; 78:3155–3161.
72. Koller MR, Emerson SG, Palsson BO. Large-scale expansion of human stem and progenitor cells from bone marrow mononuclear cells in continuous perfusion cultures. Blood 1993; 82:378–384.
73. Goodwin TJ, Schroeder WF, Wolf DA, Moyer MP. Rotating vessel coculture of small intestine as a prelude to tissue modeling: aspects of simulated microgravity. Proc Soc Exp Biol Med 1993; 202:181–192.

74. Goodwin TJ, Prewett TI, Wolf DA, Spaulding GF. Reduced shear stress: a major component in the ability of mammalian tissues to form three-dimensional assemblies in simulated microgravity. J Cell Biochem 1993; 51:301–311.

75. Duke PJ, Daane EL, Montufar-Solis D. Studies of chondrogenesis in rotating systems. J Cell Biochem 1993; 51:274–282.

76. Idzerda RL, Carter WG, Nottenburg C, Wayner EA, Gallatin WM, St John T. Isolation and DNA sequence of a cDNA clone encoding a lymphocyte adhesion receptor for high endothelium. Proc Natl Acad Sci USA 1989; 86:4659–4663.

77. Goldstein LA, Zhou DF, Picker LJ, et al. A human lymphocyte homing receptor, the hermes antigen, is related to cartilage proteoglycan core and link proteins. Cell 1989; 56:1063–1072.

78. Lasky LA, Singer MS, Yednock TA, et al. Cloning of a lymphocyte homing receptor reveals a lectin domain. Cell 1989; 56:1045–1055.

79. Miyamake K, Medina K, Ishihara K, Kimoto M, Auerbach R, Kincade PW. A VCAM-like adhesion molecule on murine bone marrow stromal cells mediates binding of lymphocyte precursors in culture. J Cell Biol 1991; 114:557–565.

80. Miyake K, Weissman IL, Greenberger JS, Kincade PW. Evidence for a role of the integrin VLA-4 in lympho-hematopoiesis. J Exp Med 1991; 173:599–607.

81. Ryan DH, Nuccie BL, Abboud CN, Liesveld JL. Maturation-dependent adhesion of human B cell precursors to the bone marrow microenvironment. J Immunol 1990; 145:477–484.

82. Witte PL, Robinson M, Henley A, Low MG, Stiers DL, Perkins S, et al. Relationships between B-lineage lymphocytes and stromal cells in long-term bone marrow cultures. Eur J Immunol 1987; 17:1473–1484.

83. Palacios R, Stuber S, Rolink A. The epigenetic influences of bone marrow and fetal liver stroma cells on the developmental potential of pre-B lymphocyte clones. Eur J Immunol 1989; 19:347–356.

84. Tsai S, Sieff CA, Nathan DG. Stromal cell-associated erythropoiesis. Blood 1986; 67:1418–1426.

85. Tsai S, Patel V, Beaumont E, Lodish HF, Nathan DG, Sieff CA. Differential binding of erythroid and myeloid progenitors to fibroblasts and fibronectin. Blood 1987; 69:1587–1594.

86. Gordon MY, Bearpark AD, Clarke D, Dowding CR. Haemopoietic stem cell subpopulations in mouse and man: discrimination by differential adherence and marrow repopulation ability. Bone Marrow Transplant 1990; 5:6–8.

87. Coulombel L, Vuillet MH, Tchernia G. Lineage- and stage-specific adhesion of human hematopoietic progenitor cells to extracellular matrices from marrow fibroblasts. Blood 1988; 71:329–334.

88. Borke JL, Eriksen EF, Minami J, Keeting P, Mann KG, Penniston JT, et al. Epitopes to the human erythrocyte CA^{++}–Mg^{++} ATPase pump in human osteoblast-like cell plasma membranes.

89. Van de Water L, Aronson D, Braman V. Alteration of fibronectin receptors (integrins) in phorbol ester-treated human promonocytic leukemia cells. Cancer Res 1988; 48:5730–5737.

90. Giancotti FG, Languino LR, Zanetti A, Peri G, Tarone G, Dejana E. Platelets express a membrane protein complex immunologically related to the fibroblast fibronectin receptor and distinct from GPIIb/IIIa. Blood 1987; 69:1535–1538.

91. Silverstein RL, Nachman RL. Thrombospondin binds to monocytes–marophages and mediates platelet–monocyte adhesion. J. Clin Invest 1987; 79:867–874.

92. Leung LLK. Role of thrombospondin in platelet aggregation. J Clin Invest 1984; 74:1764–1772.

93. Aruffo A, Staminkovic I, Melnik M, Underhill CB, Seed B. CD44 is the principal cell surface receptor for hyaluronate. Cell 1990; 61:1303–1313.

94. Dorshkind K. Hemopoietic stem cells and B-lymphocyte differentiation. Immunol Today 1989; 10:399–401.

95. Miyake K, Medina KL, Hayashi S, Ono S, Hamaoka T, Kincade PW. Monoclonal antibodies to Pgp-1/CD44 block lympho-hemopoiesis in long-term bone marrow cultures. J Exp Med 1990; 171:477–488.

96. Horst E, Meijer CJML, Radaszkiewicz T, Ossekoppele GP, VanKrieken JHJM, Pals ST. Adhesion

molecules in the prognosis of diffuse large-cell lymphoma: expression of a lymphocyte homing receptor (CD44), LFA-1 (DC11a/18), and ICAM-1 (CD54). Leukemia 1990; 4:595–599.

97. Campbell AD, Long MW, Wicha MS. Developmental regulation of granulocytic cell binding to hemonectin. Blood 1990; 76:1758–1764.

98. Ploemacher RE, Brons NHC. Isolation of hemopoietic stem cell subsets from murine bone marrow: II. Evidence for an early precursor of day-12 CFU-S and cells associated with radioprotective ability. Exp Hematol 1988; 16:27–32.

99. Gordon MY. The origin of stromal cells in patients treated by bone marrow transplantation. Bone Marrow Transplant 1988; 3:247–251.

8

Extracellular Matrix- and Membrane-Bound Cytokines

M. Y. Gordon
LRF Centre for Adult Leukaemia, Royal Postgraduate Medical School, London
England

I. INTRODUCTION

Hematopoiesis is regulated by numerous cytokines of which about 20 have been molecularly cloned and characterized, whereas others may await discovery (Fig. 1). This multiplicity of regulators and the fact that many of them share overlapping activities suggest not only that some of them may be redundant, but also that there is a potential for dysregulation (1). The situation is further complicated because many hematopoietic target cells coexpress receptors for several cytokines. Clearly, it is difficult to envisage how hematopoiesis can be precisely regulated if target cells have uncontrolled access to soluble cytokines. Another indication that unrestricted access of target progenitor cells to regulatory cytokines is undesirable is provided by experimental results showing that transgenic mice expressing granulocyte–macrophage colony stimulating factor (GM-CSF; 2) and mice transplanted with cells expressing high levels of GM-CSF (3) experience a lethal myeloproliferative syndrome involving macrophage accumulations in the eye, muscles, liver, lung, and heart. In similar experiments, mice expressing high levels of interleukin-4 (IL-4) develop severe osteoporosis of cortical and trabecular bone (4). That cytokines do not distribute uniformly in vitro is indicated by the finding that intravenous injection of granulocyte colony-stimulating factor (G-CSF) elevates the peripheral blood granulocyte count, whereas GM-CSF increases the peritoneal white cell count, suggesting that the organ distributions of the two regulators are different (1).

The idea that the distribution of cytokines is controlled has found support in studies of their binding properties and modes of expression. Many cytokines involved in the regulation of hematopoiesis can be expressed as membrane-bound proteins, or can bind to components of the extracellular matrix (Table 1). These two properties provide mechanisms for locating cytokines in the vicinity of their sites of production or their target cells and, therefore, can limit their sphere of activity. In addition, soluble extracellular matrix molecules shed by cells can act as carriers of

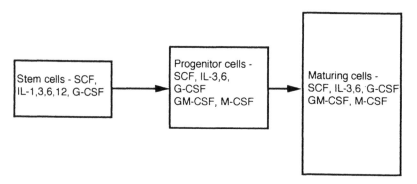

Figure 1 Regulation of different stages of neutrophil development by cytokines that have demonstrable proliferative effects.

cytokines in the bloodstream and, thereby, transport them from their production site to a distant site of action. During transport, the cytokines may be inactivated as a result of complex formation or, alternatively, their activity can be enhanced through association with extracellular matrix molecules. Thus, the location and activity of cytokines can be modified in a variety of ways that are likely to be physiologically important by binding to extracellular matrix molecules or by expression at the cell surface.

The roles of soluble and immobilized cytokines may be distinct. For example, soluble cytokines can act between cells that are not adjacent to one another, signals from spatially distant sources can interact, and a range of biological activities can be achieved by distribution of the cytokine along a concentration gradient. The cellular response to soluble cytokines is dictated by the expression of cell surface receptors, rather than by the position of the cell. In contrast, immobilized cytokines provide spatial control and precise delineation of the signal and, for the potent regulators, their effects are confined. Clearly, localization of the cytokines responsible for regulating hematopoiesis implies that their corresponding target cells must also be positioned correctly. It is thought that target cell localization is accomplished by the expression of cell adhesion molecules (CAMs) and their counterreceptors. The mechanisms of cell immobilization are addressed in another chapter.

Table 1 Cell Surface Expressed and Matrix-Binding Cytokines

Cell surface	Matrix-binding
Macrophage colony-stimulating factor (M-CSF)	Interleukins-1α, 1β, 2,3,6,7
c-kit ligand–stem cell factor (SCF)	Granulocyte–macrophage colony-stimulating factor (GM-CSF)
Interleukin-1 (IL-1)	Stem cell inhibitor/MIP-1α
Transforming growth factor-α (TGF-α)	Differentiation inhibitory factor/
Tumor necrosis factor (TNF)	leukemia inhibitory factor (DIF/LIF)
	Interferon gamma (IFN-γ)

II. MEMBRANE-BOUND GROWTH FACTORS

Several hematopoietic growth factors (e.g., M-CSF, stem cell factor [SCF]/*kit* ligand, IL-1, transforming growth factor [TGF]-α, and tumor necrosis factor [TNF]) can exist as membrane-bound molecules as well as being secreted as soluble molecules. These alternative forms are associated with the different molecular features required for secretion and transmembrane expression. Thus, M-CSF, which is translated from multiple mRNA species resulting from differential gene splicing, exists as a variety of precursor proteins. Two forms (256 and 554 amino acids, respectively) have a hydrophobic transmembrane domain. The 554-amino acid form has in addition a recognition site for an intracellular protease that releases it from the membrane, whereas the 256-amino acid form does not and is stably expressed as a membrane-bound protein (see Ref. 5; Fig. 2). Secretion, on the other hand, is generally associated with a conventional signal sequence, although some cytokines, such as IL-1, lack signal sequences or cleavage sites, and it is unclear how they are transported out of the cell (6).

In the membrane-bound form, it can be assumed that the cytokines can act as cell adhesion molecules for their target cells as well as acting as signaling molecules. For example, the membrane-bound form of M-CSF is biologically active and stimulates the growth of mononuclear phagocytes and macrophages when it is expressed at the surface of NIH-3T3 fibroblasts. Similarly, the biological activities of membrane-bound SCF, IL-1, TNF, and TGF-α have been demonstrated, indicating that several cytokines have the dual potential to act as cell adhesion and as signaling molecules. Because target cells express high-affinity cytokine receptors, it can be assumed that adhesion by membrane-bound growth factors is also likely to be of high affinity.

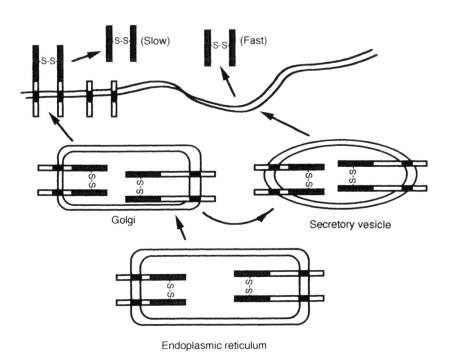

Figure 2 Processing of M-CSF precursor molecules showing soluble and membrane-bound forms. (From Ref. 5.)

One question that has not been fully resolved concerns the fate of membrane-bound cytokines once they have engaged the receptor on their target cells. However, several possible outcomes can be envisaged. First, the cytokine may be cleaved from the membrane of its producer cell, thereby freeing it for action on its target cell and releasing the target cell from its adhesive contact. It is known, for example, that one form of M-CSF expressed as an integral membrane molecule is slowly released by proteolytic cleavage, although the 256-amino acid form lacks a cleavage site. Also, secretion of TGF-α is accomplished by proteolytic cleavage by an elastase-like enzyme of an integral membrane pro-TGF-α glycoprotein (7). Second, interaction with the target cell receptor could alter the conformation of the cytokine or of the receptor, or both, to reduce the affinity of the interaction and release the target cell. Presumably it would be necessary for these changes to occur following transduction of the signal into the cell's interior. Third, mitogenic stimulation may occur in association with cell–cell contact. Anklesaria et al. (8) expressed pro-TGF-α, a cell surface glycoprotein that can interact with epidermal growth factor (EGF) receptors, in a bone marrow stromal cell line. The engineered stromal cell line bound EGF/IL-3-dependent hematopoietic progenitors that proliferated while attached to the stromal layer and formed adherent progenitor cell islands.

Membrane-bound and soluble cytokines may have different effects on hematopoiesis. Williams (9) expressed membrane-bound and soluble SCF in stromal cell lines that do not themselves produce SCF and tested their capacity to support hematopoiesis. The stroma that expressed the soluble form of SCF supported a short burst of hematopoietic activity that subsequently failed, whereas the stroma expressing the membrane-bound form of SCF supported long-term hematopoiesis in vitro.

III. MATRIX-BOUND CYTOKINES

A. Biology

The Importance of Glycosaminoglycans

Various components of the extracellular matrix secreted by a variety of cell types bind hematopoietically active cytokines. These observations have led to concepts concerning the presentation of cytokines by extracellular matrix molecules to their receptors on target cells (10,11). The glycosaminoglycan (GAG) heparan sulfate and its structural analogue heparin have received much attention, although other GAGs (12) and extracellular matrix molecules (13) can operate as binders of cytokines.

Glycosaminoglycans are polysaccharide chains that associate with a core protein to form proteoglycan (PG) molecules (Fig. 3). The structure of heparan sulfate GAG is particularly heterogeneous providing the potential for specificity in its interactions with cytokines and other molecules, and it has been suggested that heparan sulfate GAGs can function as code polymers (14). Heparan sulfate is composed of four different dissacharide units that are present in all heparan sulfates that have been isolated. Significantly, however, the relative proportions of these four disaccharides are tissue-, rather than species-specific suggesting that heparan sulfate has been highly conserved through evolution (15).

The physiological relevance of growth factor binding to GAG can be expected to depend on the affinity of the interaction. The dissociation constants (K_d values) for the binding of several hematopoietic cytokines have been determined using an affinity coelectrophoresis method (16,17). In general, the affinities of cytokines for GAGs are lower than their affinities for their corresponding receptors on their target cells (5). This means that engagement with the target cell

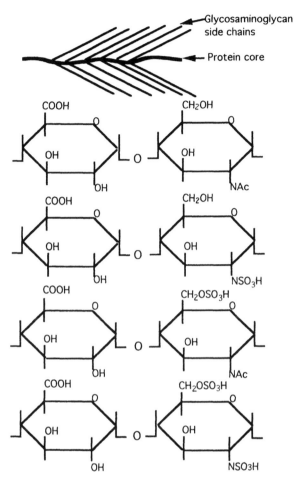

COOH   CH₂OH

OH   OH

OH   NAc

COOH   CH₂OH

OH   OH

OH   NSO₃H

COOH   CH₂OSO₃H

OH   OH

OH   NAc

COOH   CH₂OSO₃H

OH   OH

OH   NSO3H

Figure 3 Generalized structure of heparan sulfate proteoglycans showing the composition of the four glycosaminoglycan disaccharide units.

receptor could competitively dissociate the growth factor from the GAG, which may have implications for any subsequent receptor internalization that is required for signal transduction.

Localization of Cytokines by Glycosaminoglycans

Heparan sulphate is a constituent of the extracellular matrix produced by cultured bone marrow-derived stromal cells. Studies using GM-CSF showed that it could be adsorbed by the extracellular matrix of stromal cells (18). Subsequently, Roberts et al. (19) demonstrated that IL-3 as well as GM-CSF could bind to the matrix and identified the binding molecule as heparan sulfate. These studies led to the concept of cytokine localization in the hematopoietic microenvironment and presentation of the cytokine to its receptor on appropriate target cells (10; Fig. 4).

Proteoglycans can also be expressed by cytokine target cells and bind cytokine in a fashion that facilitates presentation to the cytokine signaling receptor. Associations of this type have been specifically demonstrated for fibroblast growth factor (FGF) and for TGF-β. Yayon et al (20) used Chinese hamster ovary (CHO) cell mutants deficient in glycosaminoglycan biosynthesis.

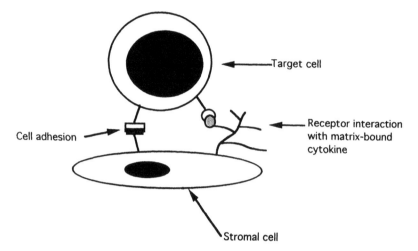

Figure 4   A model for the presentation of matrix-bound growth factors to immobilized target cells.

Wild-type CHO cells were able to bind FGF with low affinity, but mutant cells that did not express cell surface GAGs were not. Comparison of different mutants revealed that the low-affinity binding site was highly sulfated heparan sulfate proteoglycan. Wild-type cells transfected with the murine FGF receptor bound eight to ten times more FGF than nontransfected cells. However, mutant cells transfected with the receptor did not demonstrate significant high-affinity binding. This finding suggested that low-affinity cell surface heparan sulfate proteoglycans are necessary for FGF to bind to its high-affinity receptor and, indeed, binding could be fully restored by adding heparin to cultures of mutant transfected cells.

Lopez-Cassillas et al. (21) showed that the proteoglycan molecule betaglycan, containing both heparan sulfate and chondroitin sulfate side chains (22), which previously was known as the TGF-β type III receptor, presents TGF-β directly to the kinase subunit of the TGF-β receptor, forming a high-affinity complex. The protein kinase receptor for TGF-β has limited binding capacity for TGF-β, but this limitation can be overcome by the presence of betaglycan. Thus, betaglycan is a direct regulator of TGF-β to the signaling receptor and increases TGF-β binding, enhances cell responsiveness to TGF-β, and eliminates biological differences between TGF-β isoforms (21).

Transport and Distribution of Cytokines by Glycosaminoglycans

There is growing evidence that cytokines act locally and that they can be transported from their sites of production to their sites of action in an inactive form. It is now emerging that an important mechanism for the transportation of many cytokines involves their ability to bind to GAGs, particularly heparin and heparan sulfate. Heparin is a normal constituent of plasma (23), and heparan sulfate can be secreted as a soluble molecule. In the GAG-bound form, some cytokines are protected from proteolytic cleavage, thus increasing the likelihood that they will reach their target cells (24). Occasionally, complex formation inactivates the cytokine (24), again increasing the likelihood that the cytokine will reach the target cells. In other cases, cytokine activity is enhanced by complexing with GAG (25), but this pairing is probably inappropriate for a transport mechanism and serves a different physiological function. For IL-8, it has been proposed that heparan sulfate promotes IL-8-dependent transmigration of neutrophils while, at the same time, protecting the tissue microenvironment from damage by lytic enzymes released by the transmigrating cells (25).

Cytokines delivered to their site of action may be used to stimulate target cells immediately, or may be stored in the ECM. For example, GM-CSF stored in the ECM of cultured bone marrow-derived stromal cells can be removed by salt extraction and used to stimulate colony formation by granulocyte–macrophage precursors in semisolid cultures (18). The release of the bound cytokine from storage or carrier may be achieved by competitive dissociation, or by the secretion of heparinase-like enzymes. Such heparinase-like enzymes are secreted by normal neutrophils and lymphocytes (26,27).

A model for the various roles of GAGs in controlling the distribution of hematopoietic growth factor and cytokine activity is shown in Figure 5. Thus, growth factors secreted by producer cells can be sequestered by ECM, which sets up localized concentration gradients. These growth factors may be stored or presented immediately to target cells, which dissociate the growth factor from the GAG by competition or enzymatic degradation. Alternatively, complex formation may inactivate the growth factor and protect it from proteolysis. The GAG is then able to act as a carrier to transport the growth factor to its site of action. On arrival, the complex may be integrated into the ECM, or the growth factor may be released, again by competition or degradation, on encountering the appropriate target cell. Release from the complex could then reactivate the growth factor, enabling it to act on its target cells. Moreover, some cytokines have proadhesive effects on target cells. They include IL-8, which can trigger granulocyte adhesion to endothelial ligands, and macrophage Inflammatory protein (MIP)-1β, which can trigger T-cell adhesion to vascular cells (28).

B. Biochemistry: Structure–Function

Heparin and heparan sulfate are sulfated polysaccharides with highly variable sequences differing in uronic acid epimerization, *N*-deacetylation, and sulfation patterns. Thus, it is possible that sequence variation at the microstructural level of these GAGs provides different binding sites for different cytokines. For basic FGF, specific oligosaccharide sequences are necessary for high affinity binding (29,30), and the oligosaccharide sequences that bind antithrombin III have been identified (31).

Concensus sequences incorporating the basic residues arginine and lysine constitute the heparin-binding sites on four heparin-binding proteins (32). Other reports have implicated histidine-like residues as mediators of protein–heparin interactions (33). Examination of the primary sequences of several hematopoietic cytokines has identified several candidate GAG-binding motifs (Fig. 6). Furthermore, the conformational arrangement of these motifs (e.g., topographic

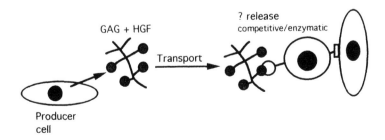

Figure 5 A model for the transport of cytokine–GAG complexes to the site of cytokine action.

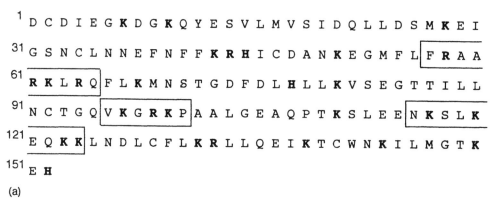


```
1  D C D I E G K D G K Q Y E S V L M V S I D Q L L D S M K E I
31 G S N C L N N E F N F F K R H I C D A N K E G M F L|F R A A
61 R K L R Q|F L K M N S T G D F D L H L L K V S E G T T I L L
91 N C T G Q|V K G R K P|A A L G E A Q P T K S L E E|N K S L K
121 E Q K K|L N D L C F L K R L L Q E I K T C W N K I L M G T K
151 E H
```

(a)

Figure 6 (a) The primary sequence of IL-7 (39) indicating potential glycosaminoglycan-binding sites in the boxed regions. (b) The primary sequence of IL-2 (40) indicating potential glycosaminoglycan-binding sites in the boxed regions. (c) Cα atom backbone trace of IL-2 (41): The basic residues K, R, and H have been highlighted and the potential glycosaminoglycan-binding sites are indicated by the boxed regions. (Courtesy of R.V. Gibbs.)

```
 1  A P T S S S T K K T Q L Q L E H L L L D L Q M I L N G I N N
31  Y K N P K L T R M L T F K F Y M P K K A T E L K H L Q C L E
61  E E L K P L E E V L N L A Q S K N F H L R P R D L I S N I N
91  V I V L E L K G S E T T F M C E Y A D E T A T I V E F L N R
121 W I T F C Q S I I S T L T
```

(b)

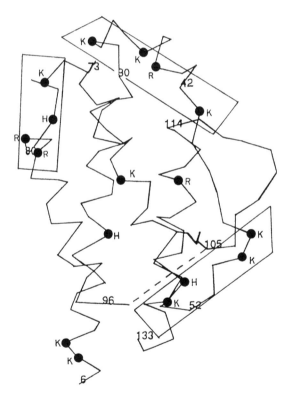

(c)

shape, solvent accessibility) and, hence, their potential to form GAG binding sites, can be assessed by examination of crystallographic structures.

IV. APPLICATIONS

Recombinant hematopoietic cytokines are becoming widely used in the treatment of hematological disease but are limited, particularly IL-1 and IL-2, by the occurrence of severe toxic side effects. The evidence that cytokines may be transported physiologically by complex formation with molecules such as GAGs suggests that systemic administration of unmodified cytokines may be inappropriate. It may be that maximum therapeutic efficacy will be achieved only if the cytokines can be directed specifically to their intended target cells. Moreover, many inflammatory and proliferative disorders are associated with inappropriate or increased levels of cytokines. In these contexts the growth factor-binding GAGs are potentially of clinical value.

Current information raises the possibility that interactions between hematopoietic cytokines and GAGs may be exploited with therapeutic benefit. This view is supported by an existing therapeutic model in the treatment of duodenal ulcers, which respond to recombinant basic FGF. However, a cheaper and more effective treatment uses sucrose octasulfate, which stabilizes endogenous basic FGF in the gastric lining (34). Another indication of the usefulness of this approach is the finding that, in mice, heparan sulfate augments early and suppresses late mixed lymphocyte responses in vitro (35), suggesting a possible role in the moderation of graft-versus-host disease after bone marrow transplantation. A variety of mechanisms could contribute to these effects, including an increased affinity for cytokine in the presence of heparin. Also, the GAG-bound cytokines may be not only less susceptible to proteolysis but also less antigenic or "tolerogenic."

The exploitation of GAG binding in the therapeutic modulation of cytokine activity is clearly in its infancy, and many developments remain to be made. For example, it would be beneficial to identify nonanticoagulant heparin fragments that retain their cytokine-binding properties. However, in the future, it is likely that the clinical use of GAGs could include blocking of excess cytokine levels in such conditions as graft-versus host disease following bone marrow transplantation and in inflammatory disorders.

A variety of malignant cells release matrix-degrading enzymes that can be neutralized in experimental models by the administration of nonanticoagulant heparin species (36). The likelihood that heparinase-like enzymes released by tumor cells can interfere with the transport and distribution of cytokines in vivo remains to be investigated, but potentially provides a further therapeutic target. Thus, administration of competing carbohydrate substrates has the potential for limiting normal tissue damage and restraining the dissemination of malignant disease. Further possibilities are suggested by studies showing that the administration of carbohydrates alters lymphocyte homing, possibly by selectively perturbing the interaction of endothelial proteoglycans with proadhesive cytokines (28,37,38).

ACKNOWLEDGMENTS

The work was supported by the Leukaemia Research Fund of Great Britain.

REFERENCES

1. Metcalf D. Hematopoietic regulators: redundancy or subtlety? Blood 1993; 82:3515–3523.
2. Lange RA, Metcalf D, Cuthbertson RA, Lyons I, Stanley E, Kelso A, Kannourakis G, Williamson DJ,

Klintworth GK, Gonda TJ, Dunn AR. Transgenic mice expressing a hemopoietic growth factor gene develop accumulations of macrophages, blindness and a fatal syndrome of tissue damage. Cell 1987; 51:675–686.

3. Johnson GR, Gonda TJ, Metcalf D, Hariharan IK, Cory S. A lethal myeloproliferative syndrome in mice transplanted with bone marrow cells infected with a retrovirus expressing granulocyte–macrophage colony-stimulating factor. EMBO J 1989; 8:441–448.

4. Lewis DB, Liggitt HD, Effmann EL, Motley ST, Teitelbaum SC, Jepsen KJ, Goldstein SA, Bonadio J, Carpenter J, Perlmutter RM. Osteoporosis induced in mice by overproduction of interleukin-4. Proc Natl Acad Sci USA 1993; 90:11618–11622.

5. Barrett AJ, Gordon MY. Bone Marrow disorders: The Biological Basis of Treatment. Oxford: Blackwell Scientific Publications, 1993.

6. Schindler R, Dinarello CA. Interleukin 1. In: Habenicht A, ed. Growth Factors, Differentiation Factors and Cytokines. Heidelberg: Springer Verlag, 1990:85–102.

7. Wong ST, Winchell EF, McCune BK, Earp HS, Teixido J, Massague J, Herman B, Lee DC. The TGFα receptor expressed on the cell surface binds to the EGF receptor on adjacent cells leading to signal transduction. Cell 1989; 56:495–506.

8. Anklesaria P, Teixido J, Laiho M, Pierce J, Greenberger J, Massague J. Cell–cell adhesion mediated by binding of membrane-anchored transforming growth factor α to epidermal growth factor receptors promotes cell proliferation. Proc Natl Acad Sci USA 1990; 87:3289–3293.

9. Williams DA. Microenvironment regulation of hematopoiesis: the role of stromal derived cytokines in blood cell formation. Exp Hematol 1991; 19:457a.

10. Gordon MY, Greaves MF. Physiological mechanisms of stem cell regulation in bone marrow transplantation and haemopoiesis. Bone Marrow Transplant 1989; 4:335–338.

11. Gordon MY. Hemopoietic growth factors and receptors: bound and free. Cancer Cells 1991; 3:127–133.

12. Ramsden L, Rider CC. Selective and differential binding of interleukin-1α, interleukin-1β, interleukin-2 and interleukin-6 to glycosaminoglycans. Eur J Immunol 1992; 22:3027–3031.

13. Fava RA, McClure DB. Fibronectin-associated transforming growth factor. J Cell Physiol 1987; 131:187–189.

14. Keating A, Gordon MY. Hypothesis: hierarchical organisation of hematopoietic microenvironments: role of proteoglycans. Leukemia 1988; 2:766–769.

15. Dietrich CP. A model for cell–cell recognition and control of cell growth mediated by sulfated glycosaminoglycans. Braz J Med Biol Res 1984; 17:5–15.

16. Lee MK, Lander AD. Analysis of affinity and structural selectivity in the binding of proteins to glycosaminoglycans: development of a sensitive electrophoretic approach, Proc Natl Acad Sci USA 1991; 88:2768–2772.

17. Katoh O, Clarke D, Gibbs RV, Gordon MY. Binding affinities of growth factors for glycosaminoglycans. Exp Hematol 21; 1077a.

18. Gordon MY, Riley GP, Watt SM, Greaves MF. Compartmentalization of a haemopoietic growth factor (GM-CSF) by glycosaminoglycans in the bone marrow microenvironment. Nature 1987; 326:403–405.

19. Roberts R, Gallagher J, Spooncer E, Allen TD, Bloomfield F, Dexter TM. Heparan sulphate bound growth factors: a mechanism for stromal cell mediated haemopoiesis. Nature 1988; 332:376–378.

20. Yayon A, Klagsbrun M, Esko JD, Leder P, Ornitz DM. Cell surface heparin-like molecules are required for binding of basic fibroblast growth factor to its high affinity receptor. Cell 1991; 64:841–848.

21. Lopez-Cassilas F, Wrana JL, Massaue J. Betaglycan presents ligand to the TGFβ signaling receptor. Cell 1993; 73:1435–1444.

22. Lopez-Cassilas F, Cheifetz S, Doody J, Andres JL, Lane WS, Massague J. Structure and expression of the membrane proteoglycan betaglycan, a component of the TGFβ receptor system. Cell 1991; 67:785–795.

23. Israels LG, Foerster J, Zipursky A. A naturally occurring inhibitor of the first stage of blood coagulation. Br J Haematol 1960; 6:275–279.

24. Clarke D, Katoh O, Gibbs RV, Griffiths SD, Gordon MY. Interaction of interleukin-7 (IL-7) with glycosaminoglycans and its biological relevance. Cytokine 1995; 7:325–330.

25. Webb LMC, Ehrengruber MU, Clarke-Lewis I, Baggiolini A, Rot A. Binding to heparan sulfate or heparin enhances neutrophil responses to interleukin 8. Proc Natl Acad Sci USA 1993; 90:7158–7162.

26. Fridman R, Lider O, Naparstek Y, Fuks Z, Vlodavsky I, Cohen IR. Soluble antigen induces T lymphocytes to secrete an endoglycosidase that degrades the heparan sulfate moiety of subendothelial extracellular matrix. J Cell Physiol 1987; 130:85–92.

27. Matzner Y, Bar-Ner M, Yahalom J, Ishai-Michaeli R, Fuks Z, Vlodavsky I. Degradation of heparan sulfate in the subendothelial extracellular matrix by a readily reversible heparanase from human neutrophils. J Clin Invest 1985; 76:1306–1313.

28. Tanaka Y, Adams DH, Shaw S. Proteoglycans on endothelial cells present adhesion-inducing cytokines to leukocytes. Immunol Today 1993; 14:111–115.

29. Turnbull JE, Fernig DG, Ke Y, Wilkinson MC, Gallagher JT. Identification of the basic fibroblast growth factor binding sequence in fibroblast heparan sulphate. J Biol Chem 1992; 267:10337–10341.

30. Tyrrell DJ, Ishihara M, Rao N, Horne A, Kiefer MC, Stauber GB, Lam LH, Stack RJ. Structure and biological activities of a heparin-derived hexasaccharide with high affinity for basic fibroblast growth factor. J Biol Chem 1993; 268:4684–4689.

31. Lindahl U, Thunberg L, Backstrom G, Reisenfeld J, Nordling K, Bjork I. Extension and structural variability of the antithrombin-binding sequence in heparin. J Biol Chem 1984; 259:12368–12376.

32. Cardin AD, Weintraub HJR. Molecular modelling of protein–glycosaminoglycan interactions. Arteriosclerosis 1989; 9:21–32.

33. Burch MK, Blackburn MN, Morgan WT. Further characterisation of the interaction of histidine-rich glycoprotein with heparin: evidence for the binding of two molecules of histidine rich glycoprotein by high molecular weight heparin and for the involvement of histidine residues in heparin binding. Biochemistry 1987; 26:7477–7482.

34. Folkman J, Shing Y. Angiogenesis. J Biol Chem 1992; 267:10931–10934.

35. Wrenshall LE, Cerra FB, Carlson A, Bach FH, Platt JL. Regulation of murine splenocyte responses by heparan sulphate. J Immunol 1991; 147:455–459.

36. Bar-Ner M, Eldor A, Wasserman L, Matziner Y, Cohen IK, Fuks Z, Vlodavsky Y. Inhibition of heparanase-mediated degradation of extracellular heparan sulfate by non-anticoagulant heparin species. Blood 1987; 70:551–557.

37. Brenan M, Parish CR. Modification of lymphocyte migration by sulphated polysaccharides. Eur J Immunol 1986; 16:423–427.

38. Weston SA, Parish CR. Modification of lymphocyte migration by mannans and phosphomannas. J Immunol 1991; 146:4180–4186.

39. Henney CS. Interleukin 7: effects on early events in lymphopoiesis. Immunol Today 1989; 10:170–173.

40. Taniguchi T, Matsui H, Fujita T, Takaoka C, Kashima N, Yoshimoto R, Hamuro J. Structure and expression of a cloned cDNA for human interleukin-2. Nature 1983; 302:305–310.

41. McKay DB. Unravelling the structure of IL-2 response. Science 1992; 257:412–413.

9
Transfection of Stem Cells and Gene Therapy

Robert C. Moen
Baxter Healthcare Corporation, Round Lake, Illinois

I. INTRODUCTION

Gene therapy can be defined as the introduction of genetic material into the cells of an individual with resulting benefits to the individual or others. This transfer of genetic material can be done to increase scientific understanding of basic processes, or it can be done with a therapeutic intent. Gene therapy is a rapidly developing field in which many applications are relevant to the hematopoietic system. Most of the initial uses of gene transfer techniques in both animals and humans have involved bone marrow-derived cells. It is the combination of the advances in molecular biology, cellular biology (most notably in hematopoiesis), and virology that have enabled the development of safe and efficient gene transfer techniques. By integrating the knowledge acquired by each of these fields, it has been possible to develop innumerable gene transfer techniques and to develop a myriad of uses for gene transfer technology. Enterprising scientists are able to develop clinical applications of gene transfer based on these related advances and the increased understanding of the molecular causes of human disease. This chapter will discuss the underlying concepts in gene transfer techniques and, with an emphasis on hematopoietic cells or diseases, how some of these concepts are being brought to the clinic. This technology has already proven useful for expanding our understanding of autologous bone marrow transplantation and hematopoiesis. This understanding will permit the development of improved treatment strategies for a wide variety of human genetic and acquired disorders.

The first authorized use of this new technology in a clinical situation was in May 1989. This first trial used the new technology to *mark* tumor infiltrating lymphocytes expanded ex vivo to enable them to be distinguished from other host cells (1). The first clinical trial of this gene transfer technology designed to be potentially therapeutic *(gene therapy)* began in 1990 (2). Currently there are many dozens of approved gene transfer and therapy trials around the world (see *Human Gene Therapy*, there is a list at the end of each issue; 63 in list as of January 1994). A variety of gene transfer techniques have been employed to conduct these studies. A discussion of some of the methods used to insert new genetic material into eukaryotic cells is warranted to provide the reader with a background of the technologies being used. The rationale for studies using gene transfer will then be discussed along with some of the preliminary findings of the ongoing clinical trials, with an emphasis on trials involving bone marrow cells and bone marrow transplantation. Because the field is new and is rapidly expanding, it will not be possible to discuss all protocols. The goal of this chapter, therefore, is to provide the reader with an overview of the technology used in gene transfer and therapy and an understanding of the vast research and therapeutic potential of this technology.

II.   GENE TRANSFER TECHNIQUES

A large number of methods are available for transferring genetic material either stably or transiently into eukaryotic cells. These methods can be arbitrarily divided into those using physical methods and those using biological methods. The distinction is made in that physical methods do not require the use of biologically produced material(s) to facilitate gene transfer. Most patients involved in gene transfer on therapy experiments to date have undergone biological methods for gene transfer.

A.   Physical Methods of Gene Transfer

Physical methods of gene transfer all have in common that the delivery system results in uptake of added nucleic acid material. The material introduced into the cell is not efficiently integrated into the host genome, nor is it reproduced in the target cell unless it is integrated. Only with cell selection techniques is it possible to fish out the infrequent cell stably expressing the introduced gene. In the vast majority of cells taking up the genetic material, the introduced material is degraded and lost. If the cell divides, the expression of introduced material decreases on a per cell basis as the genetic material is diluted. Thus, physical methods of gene transfer are reasonable to consider if a large number of cells with transient expression is desired, or alternatively, if stable expression is required in a cell that can be selected and expanded. Scientists are developing methods to increase the integration rate, and one clinical trial has been reviewed that relies on Epstein-Barr virus elements in the introduced plasmid to permit intracellular replication of the introduced sequence. Some of the methods of gene transfer using physical methods will be briefly reviewed.

Calcium Phosphate Transfection

Laboratory researchers initially used physical methods of gene transfer into eukaryotic cells. The efficiency of stable gene transfer may have been low, but high-efficiency selection systems in clonogenic cells permitted the scientist to isolate and expand cells with integrated and expressed genes. The most prevalent technique used is calcium phosphate precipitation. Improvements have been made in these methods so that higher-efficiency gene transfer is possible (3,4). It is now possible to obtain moderate efficiency and transient expression of new genetic information in a variety of cells. The stable transfer of genes into the target is still low. A single oriented copy of the transferred gene is seldom seen in the target cell, a tandem array of integrated DNA is normaly seen. Because the gene copy number and physical orientation in a cell are not controlled, appropriate regulation and expression may be difficult to predict and could have significant clinical consequences. In many situations these concerns are not limiting, and several trials have been approved throughout the world that use transfected cell lines.

Microinjection

The other physical method of gene transfer that has been used throughout the world for many years employs the microinjection of genetic material into individual target cells (5,6). This method has been used to produce many transgenic animals. The efficiency of gene transfer into an individual cell is fairly high, but the technique is very labor-intensive. Obviously, a gene therapy trial in which millions or even billions of freshly isolated cells are needed would not use microinjection techniques. If a stem cell were obtained that could be microinjected and maintained during expansion in vitro or in vivo, then the method is viable. In most clinical uses developed to date it has not been possible to identify, purify, and maintain a stem cell that would be of clinical use. Additional problems arise when determining how best to introduce these cells back to the patient. The cell biology field has been rapidly advancing, and it can be envisioned that some of these advances may permit consideration of microinjection techniques for clinical gene therapy uses in the future. For the foreseeable future, this technique remains a valuable research tool, without significant clinical utility.

Electroporation

Other physical methods have been developed for gene transfer that share many of the advantages and disadvantages of calcium phosphate transfection. Electroporation relies on a pulsed electrical field to disrupt the cellular membrane and thus permit uptake of added nucleic acid material into the target cell (7,8). High-efficiency transient expression is possible, but multiple tandem arrays with variable expression are possible, as is seen with calcium phosphate gene transfer. Continued improvements have permitted larger cell numbers to be treated, but no significant clinical usefulness is yet seen for this method that could not be met using transfection methods.

Lipid–DNA Complexes

Lipid-encapsulated–associated nucleic acid material has been used clinically and may be expected to be increasingly used in clinical gene transfer and therapy protocols. This liposomal method of gene transfer (9,10) results in gene transfer difficulties and advantages similar to those seen with the aforementioned physical methods of gene transfer. The method does have advantages in that the efficiency of gene transfer can be very high, and the method appears to be gentle to the target cells. Perhaps its main advantage is that it can be readily used for in vivo gene therapy, indeed, this method was used for the first in vivo gene therapy clinical trial. The use of liposomal gene transfer is increasing in frequency and seems especially suited for transient gene expression of target cells in a physically isolated site.

Naked DNA

Initially surprising to many investigators was the ability of naked nucleic acid to be taken up by eukaryotic cells in vivo. By using naked DNA injected with a syringe and needle, several tissues have demonstrated at least transient DNA uptake in vivo. Certain cell targets, such as muscle cells (11,12), seem especially amenable to gene therapy strategies using naked DNA. Muscle cells have been explored in animal models as targets for gene therapy for muscular genetic disorders, for gene therapy for hemophilia, and as a novel method for eliciting an immune response to infectious agents. How the use of naked DNA will compare with the other physical methods of gene transfer available remains to be resolved.

DNA–Ligand

Besides the use of naked DNA or DNA in liposomal complexes, DNA can be associated with other ligands for use in gene therapy. An example of a DNA–ligand gene transfer is the use of DNA–asialoglycoprotein complexes (13,14). The ligand binds to the cell surface according to the characteristics of the ligand used. The receptor–ligand can then be taken up by the cell and the DNA complexed to the ligand accompanies the ligand into the cell. This system is appealing in that it permits the ligand–receptor association to target the cells available for gene transfer. In the asialoglycoprotein complex discussed, this targeting facet was exploited for in vivo gene transfer specifically into hepatocytes. If the early successes of this procedure can be reliably reproduced, it may become a clinically useful method of gene transfer when in vivo targeted transient gene expression is desired.

Particle Bombardment

Particle bombardment is perhaps the most physical of the methods for gene transfer. Nucleic acid-coated pellets are "shot" into the target cells. These pellets penetrate the host cell membrane and carry the nucleic acid with them. This system was initially developed for use with plant cells, but has proved useful also for eukaryotic cells. Once inside the cell, regardless of the physical method used, the DNA can be transiently expressed, and a small fraction of cells may stably integrate tandem repeats of the introduced genetic material. This method can be used to treat a large number of cells and can be used in vivo as well. As the technology improves it is not yet clear what role this method will have in the clinical use of gene therapy.

B. Biological Methods of Gene Transfer

Many viral vectors have been used to transfer genetic material into target cells. Viruses have evolved the mechanisms to insert their genetic material into cells so that the host cell machinery can be usurped to permit the viruses to replicate themselves. Molecular biology has been used to dissect the processes the viruses use to infect and replicate in eukaryotic cells. The unifying principle behind the use of viruses for gene transfer is to remove the components from the virus vector genome that permit an infectious cycle and replace these components with genetic sequences determined by the molecular biologist. By doing so, the molecular biologist has produced an infectious viral vector that is able to infect only once, thereby enabling the viral vector to transfer the desired genetic sequences into the target cell, without requiring the use of physical methods. The viral vector takes advantage of the scientist's understanding of the life cycle of the virus to produce a vector suitable for gene transfer. One of the major concerns with using viral-based vectors is the possibility that recombinant events may occur, resulting in a fully infectious particle. These concerns will be discussed in a separate section of this chapter. The retroviral vector will be more extensively discussed than the other viral vector systems to demonstrate the application of the principles mentioned in this paragraph.

Retroviral Vectors

Retroviral vectors are RNA viruses containing two copies of a single-stranded RNA genome and represent the best-studied method of transferring genetic material into cells. There are a large variety of retroviruses that could be used for gene transfer, but the Moloney leukemia virus (MoLV) has been the most extensively used. MoLV contains a simple genome containing three genomic elements, *gag*, *pol*, and *env* (Fig. 1). The *gag* portion encodes the structural proteins of the retroviral particle. These structural proteins encapsulate the genome and form the core of the retroviral particle. The *pol* portion of the genome encodes the polypeptide sequence for three functions: reverse transcriptase, DNase H, and integrase. These functions are needed to allow the virus to convert its RNA genome into dsDNA that is integrated into the host cell genome. The third genomic element *env*, encodes the envelope gp70 and p15 proteins. They are translated as one contiguous polypeptide that is cleaved to produce the mature proteins making up the viral envelope. The three genomic elements all share the common feature that because they encode proteins, their functions can be contributed in *trans* to produce an infectious viral particle.

The other portions of the MoLV genome (see Fig. 1) do not encode proteins. These

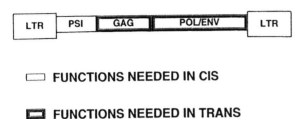

☐ FUNCTIONS NEEDED IN CIS

▭ FUNCTIONS NEEDED IN TRANS

Figure 1 Moloney leukemia virus genome.

sequences are required in *cis* to produce an infectious virus. These sequences include the long terminal repeats (LTR) that serve several functions, including serving important roles in reverse transcriptase priming, integrating the dsDNA into the host genome, providing the enhancer and promoter functions enabling the genome to be reproduced and mRNA for the viral proteins to be transcribed, and the Poly(A) signal for the RNA transcripts. The *Psi* segment encodes the encapsulation signal required for packaging the RNA genome into the retroviral particle. The LTRs and *Psi* segment are required in *cis* to have a replicative infection.

To produce a retroviral vector the *cis* functions are supplied by the LTRs and *Psi* sequence, but the *gag*, *pol*, and *env* sequences, all are eliminated from the vector "genome." Instead, the molecular biologist is able to insert a gene, a promoter, or even several genes and promoters into the region removed. A packaging cell is engineered that contains the *gag*, *pol*, and *env* sequences separated from the *Psi* and LTR sequences. To substitute for the LTR promoter and Poly(A) signal, another of a variety of promoters and Poly(A) signals may be used. These packaging cells are able to produce all the proteins required for an infectious virus, but cannot produce a genome that can be packaged by the retroviral proteins. The insertion of the vector genome into the packaging cell provides a nucleic sequence that can be packaged into retroviral particles (Fig. 2). Thus, the vector genome can be transferred into the target cell instead of the *gag*, *pol*, and *env* sequences that would be transferred by a wild-type retrovirus. Because no *gag*, *pol*, or *env* nucleic acid sequence is present, no further replication can occur. Since these proteins are present in the vector particle, the vector genome can be made into dsDNA and integrated into the host genome. The transferred sequences become part of the host genome. The sequences can be transcribed into mRNA from which proteins can be produced. If the cell replicates, the vector is present in both daughter cells. If the cell dies, the vector genome is destroyed along with the host genome. A stably, genetically altered cell is produced with the use of retroviral vectors. Retroviral gene transfer has been the most frequently applied technique used in clinical studies. The major limitation with retroviral vectors is that efficient gene transfer occurs only into target

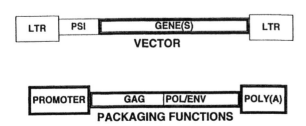

Figure 2 Retroviral vector and packaging functions.

cells that are actively replicating (15,16). Under certain circumstances, described later, this limitation may be exploited in the design of a clinical trial.

Because retroviral vectors (17,18) are the most extensively used, it follows that the most data about the safety of gene transfer have been garnered using these vectors (19–22). From the preclinical and clinical data it appears retroviral vectors are safe. Replication competent retroviruses (RCR) may not exhibit the same safety profile (23). Retroviral vectors need to be designed to reduce the small risk that may be associated with a replication competent retrovirus arising from a preparation of retroviral vector particles. Safeguards in packaging cell and vector design, along with careful testing of materials to be used clinically, are important in reducing the risk associated with RCR (24). The prudent avoidance of replication competent virus contaminating or arising from a vector preparation applies to all viral-based vector systems.

Adenovirus

Adenoviral vectors (25) have also been made and are appealing because they do not require dividing cells. They are especially appealing as vectors to use to target lung epithelial cells. Adenoviral vectors hold promise for gene therapy for lung disorders, such as cystic fibrosis. They have not proved useful for transferring genes into hematopoietic cells.

Adenoassociated Virus

Adenoassociated virus (AAV) has been used to produce vectors capable of transferring genes into many cells (26,27). AAV holds the promise of permitting site-directed integration as the wild-type AAV shows a strong propensity for a region of chromosome 19. AAV may also be able to transfer genes into nonreplicating cells, something retroviruses do very poorly. Once some technical aspects of producing AAV vectors are overcome, the AAV vector system may supplant retroviral vectors in many situations. Many researchers are exploring the use of AAV vectors to transfer genes into hematopoietic cells.

C. Chimeric Methods

The chimeric category includes methods of gene transfer in which components of physical methods of gene transfer are combined with biological methods. By combining methods it can be hoped to include the advantages of one method and combine it with the advantages of another method. As the field grows and more basic understanding of mechanisms of gene transfer is developed, this increased knowledge can be used to develop sophisticated chimeric techniques. Several chimeric methods have been explored, but only one will be discussed in this chapter.

Adenoviral vectors have the advantage of being able to bind ubiquitously to many cell targets. The adenoviral particle binds to receptors on the surface of the cell and is then endocytosed into the cell. The adenoviral particle contains a mechanism to break down the lysosome and efficiently release the contents of the lysosome into the cell. A chimeric method has been developed that combines inactivated wild-type adenovirus with DNA bound up in polylysine (28). This adenoviral–DNA complex is taken up by target cells through the interaction of the adenovirus with cell surface receptors. Once in the cell, the release of the DNA from the lysosome is mediated by adenoviral proteins. The DNA, therefore, is physically transported into the target cell using the biological properties of adenovirus. This chimeric system is hoped to give more efficient transfer of DNA than other physical methods. Early research results have been encouraging, and further developments and refinements in these and similar systems can be expected to be clinically applicable.

D. Summary of Gene Transfer Methods

An ever-expanding variety of gene transfer methods are being developed. Each has unique properties. These properties have distinct advantages and disadvantages. The expectation is that, for any particular contemplated use for gene transfer, a variety of gene transfer methods will be suitable. Each method may ultimately prove to be useful, depending on the needs of the clinician. Factors such as whether a stable integration is necessary, the length of time gene expression is required, ease of use, toxicity concerns, and so on will have to be considered for any given use of gene transfer. Currently, the most widely used method of gene transfer is retroviral-mediated gene transfer. Many other gene transfer methods have begun to be used clinically, including other viral vectors and physical methods of gene transfer. Safety issues have not yet appeared to be a serious drawback to any method, but more patients and follow-up are needed to determine the true risks associated with gene transfer.

Besides the efforts expended to improve existing gene transfer methods and develop new ones, an area of research not discussed extensively in this chapter is the study of the target cells themselves. Physical, biological, and chemical manipulations of the target cells, target organs, or target organism have already been shown to affect gene transfer efficiencies. Through a combination of optimizing gene transfer methods and of altering the receptiveness of target cells, further improvements in gene transfer can be expected.

III. GENE MARKING

Gene marking refers to the use of transferred genes to identify a cell as being different from otherwise indistinguishable cells. Congenic animals have been used in many research experiments to follow cells introduced into an experimental animal. The congenic strains are compatible, but contain subtle differences that allow the scientists to determine which cells are of host origin and which are not. In autologous or syngeneic situations no such distinction is possible. Gene marking enables the analysis of cells that were introduced in an autologous or syngeneic situation. The gene can be inserted by any of the methods previously described, but preference is given to methods, such as retroviral-mediated gene transfer, in which stable insertion occurs.

Other methods of cell marking have limitations that can be overcome with gene marking. Dye labeling of cells or labeling with radioactive materials, such as indium, has long been used to track cells. Unfortunately, indium has a short half-life that prevents long-term follow-up. Cells that are indium- or dye-labeled may die but the "tag" does not disappear with the death of the cell. The signal can be transferred to other cells. If the cell divides the signal is diluted. None of these imitations are seen with retroviral-mediated gene marking. The transferred gene becomes part of the genome of the labeled cell. If the cell dies, the label is destroyed. The gene is stable for the life of the cell, so there are no half-life issues. If the cell divides, the label is not diluted, each daughter cell will contain the same label as the original cell. In fact, genetic cell marking allows the identification of all daughter cells (lineage analysis), each cell originally labeled will have a unique insertion site for the marker gene. Only the daughter cells would contain the same insertion site. Thus, if a stem cell was gene marked, all daughter cells would be tagged with the same unique lineage identifier. Analysis of the insertion site would enable the researcher to determine the variety of cell types that may have descended from the same precursor cell. Gene marking has many advantages over other tagging techniques. These advantages have been used in clinical studies. Indeed the first use of gene transfer and therapy was for such a tagging study.

A.   Tumor-Infiltrating Lymphocytes

Tumor-infiltrating lymphocytes (TIL) are the lymphocytes that are found infiltrating a malignant mass. These lymphocytes are felt to have infiltrated the tumor as part of an immune response to tumor antigens. A hypothesis is that the number of TIL is inadequate to eliminate the tumor and more TIL may be able to overcome the tumor. Clinical studies have implied that isolation of TIL from tumor masses, expansion of the TIL ex vivo, and reinfusion of these cells into patients in conjunction with IL-2 has resulted in encouraging clinical benefits to a subfraction of melanoma and other tumor-bearing patients. Because these cells are autologous, there is no way to determine which are the TIL returned to the patient and which lymphocytes never left the patient. Ex vivo gene marking with a retroviral vector would enable the clinician to identify a lymphocyte that had been expanded with IL-2 ex vivo. The growth of the TIL in the laboratory enabled the insertion of a retroviral marker gene. The fraction of cells labeled with the retroviral vector would serve as a "tracer" to follow the TIL. If the TIL homed to tumor sites, increased marker gene signals would be seen in DNA isolated from tissues. If the TIL died, the signal would disappear so the marker gene could be used to follow the half-life of TIL. If TIL divided, the progeny would all be marked by the presence of the marker gene. Perhaps a difference in those patients showing a response from those not responding to TIL therapy could be discerned. Many interesting and clinically important questions could be explored by using gene-marked TIL.

The use of gene marking is regulated in the United States by two separate government agencies. The Food and Drug Administration (FDA) normally regulates clinical drug development. This review is confidential and open to public scrutiny. Because of concerns over the misapplication of gene therapy, the National Institutes of Health (NIH) had formed an advisory committee to the Director of the NIH. The Recombinant DNA Advisory Committee (RAC) also reviews all clinical protocols involving gene transfer and therapy. The RAC consists of representatives from all facets of society. Scientists and doctors are accompanied by lawyers, ethicists, and patient advocates on the committee. A "Guidelines for the Submission of Human Gene Transfer/Therapy Protocols for Review by the RAC" has been produced to guide the investigator in providing the material needed for RAC review. A similar document entitled "Points to Consider in Human Somatic Cell Therapy and Gene Therapy" has been drafted by the FDA (29). Only after much discussion within the FDA and in public meetings of the RAC and its Human Gene Therapy Subcommittee was the first trial permitted to proceed by the Director of the NIH (30). The first patient was enrolled in this marking study in May 22, 1989. This first trial had to face all the ethical and regulatory hurdles that paved the way for all future gene transfer and therapy trials. Results of the initial five patients were published in the *New England Journal of Medicine* (1). This study served to show that this new technology was feasible and safe.

After this study was begun in 1989, a similar technology was used for a gene therapy trial begun September 13, 1990, in which peripheral blood lymphocytes from a patient with adenosine deaminase deficiency (ADA) were transduced with a retroviral vector containing ADA cDNA encoding a functional gene product (31). This trial demonstrated the feasibility of transferring genes into lymphocytes with a therapeutic intent. By using the experience and knowledge gained from these initial lymphocyte-based experiments, gene transfer and marking have been used with bone marrow cells.

B.   Bone Marrow Transplant Experiment

Allogeneic bone marrow transplantation has been of therapeutic benefit in a vast array of human conditions. The problem of graft-versus-host disease (GVHD) remains a significant source of morbidity and mortality, even when an HLA-matched sibling is available. Unfortunately, most persons will not have an HLA-matched sibling, thus necessitating the use of matched unrelated donors or a less well-matched relative as donor. The GVHD and other associated risks, therefore, are much higher in this less-fortunate patient population. To avoid the difficulties associated with allogeneic marrow transplantation, autologous marrow transplantation was developed for patients with malignant disease.

Autologous bone marrow transplantation (ABMT) has gained wide popularity for the treatment of a variety of malignancies. The results with this technique have been encouraging enough to continue the development of ABMT and try to understand and remedy the problems associated with it. Because ABMT uses marrow harvested from the cancer patient, it is possible that the marrow returned to the patient will contain malignant cells that could cause or contribute to a relapse in the patient. This concern has led many investigators to explore purging techniques to try to remove any residual malignant cells contaminating the harvested marrow. These purging techniques often compromise the patient by decreasing the rapidity of hematological recovery, thereby exposing the patient to increased risks from infection and bleeding. Whether purging procedures are justified has been a point of controversy, as clinical practice could not tell if the autologous marrow was contributing to relapses and thus purging was reasonable to consider. Because the marrow (and cancer cells) removed from the patient for later reinfusion were not distinguishable from the cells remaining in the patient, the source of relapse could not be determined. Gene marking of the harvested marrow would permit the investigator to determine if the marrow reinfused contained malignant cells. This gene marking has been used in ABMT trials for several malignancies.

Acute Myelogenous Leukemia

The initial trial of gene marking with ABMT in acute myelogenous leukemia (AML) patients began in 1991 at St. Jude Children's Research Hospital. In this trial (32), a portion of the marrow harvested from children in remission (no detectable malignant cells using the most sensitive assay available) was exposed to retroviral vectors containing a gene conferring resistance to the neomycin analogue G418. After this 6-hr exposure, the marked marrow sample was cryopreserved along with the remainder of the harvested marrow. After an extensive conditioning regimen, the marked and unmanipulated marrow was infused into the patient. None of the patients exhibited any changes in the rapidity of marrow engraftment. Of the 12 patients so treated, 2 patients relapsed within the first year (33). Both of these patients contained malignant cells that contained the retroviral marker gene. These results unequivocally demonstrated that remission marrow harvested from AML patients can be contaminated with residual malignant cells that contribute to relapse. These results could not determine, nor were they ever expected to be able to determine, if the infused autologous marrow was responsible for all of the malignant cells contributing to relapse or whether residual endogenous disease also contributes. Nonetheless, this finding is very significant and justified the pursuit of purging technologies in at least a fraction of patients undergoing ABMT for AML.

Neuroblastoma

Similar to the trial involving the hematological malignancy AML, a trial (34) involving children in remission from neuroblastoma was initiated at St. Jude. Three of the nine patients enrolled have relapsed within the first year after ABMT, all three patients were found to have malignant cells containing the marker gene (35). One of the patients relapsed in a nonmedullary site, even the solid tumor mass contained marker tumor cells. This finding demonstrates that the infused marrow contained malignant cells that contributed to relapse found at an extramedullary site. These findings are provocative and reaffirm the usefulness of gene marking in understanding tumor biology and ABMT.

Normal Cells

An additional potential use of gene marking with marrow transplantation was also demonstrated in the studies of gene marking to determine whether the rescue marrow contributed at least partially to relapse after ABMT. When the portion of harvested marrow was exposed to the retroviral vector containing supernatant, not only were any residual tumor cells available to be transduced but normal hematopoietic progenitors were also present. If these cells were transduced then they and all their progeny would be marked. If a long-lived precursor or stem cell was so-marked than its progeny would be expected to contain the marker gene. Even differentiated progeny found in circulating blood or in the marrow itself would be expected to be marked months and years later. The results obtained from the initial ABMT studies (32,34) showed that the normal marrow components were also marked with the retroviral vector (36). These results demonstrated that very early precursor or stem cells present in the transfused marrow contributed to long-term hematopoietic reconstitution in these patients. This finding also leads to future use of gene marking to study hematopoiesis.

C.   Dual-Label Marking

Use with BM and PBSC

Several other trials have been initiated in which gene marking is being used in ABMT. These trials have been rescuing the patients with mobilized peripheral blood stem cells (PBSC) or bone marrow cells (BMC). In the studies where both PBSC and BMC are used together to rescue the patient, two distinguishable retroviral vectors have been used. These vectors can be distinguished using polymerase chain reaction (PCR) techniques (37) and allow the investigator not only to determine if relapse can be attributed to the ABMT, but to determine which source of rescuing cells contributed to the relapse. If a clinical trial for a particular malignancy would routinely find labeled tumor cells from one source and not the other, the clinician would have gained valuable information about which source of rescuing cells was "cleaner." On the basis of these findings the protocol could be modified to use only cells obtained from the source that appears free of malignant cells able to contribute to relapse. The ability to use a patient as their own internal control magnifies the importance of the data obtained with even a few patients.

Additional information can also be expected to result from these trials. The normal hematopoietic cells can also be expected to be marked. By following the relative strength over time of the signal from PBMC versus BMC, the clinician will be able to follow kinetically the contribution of each source of transduced cells to the patient's hematopoietic environment. If PBMC marked with vector A were able to contribute to long-term engraftment, as determined by the continued presence of vector A-marked cells in the peripheral blood and marrow of the patient, this would be proved by the use of retroviral marking. Currently, this information is being accumulated, but the follow-up time has not yet been sufficient to determine if one source of cells or the other provides a better source of stem cells. The current trend implies that both sources provide no significant qualitative differences in either short-term or long-term reconstitution. Given the strong interest of clinicians to use mobilized peripheral blood stem cells, gene marking is providing valuable information about the durability and qualitative aspects of PBSC as a source of reconstituting cells in a direct comparison to marrow cells.

Use For Comparing Ex Vivo Techniques

The same use of dual (or more) marker vectors can be used for other comparisons. The advantage of using the patient as an internal control greatly magnifies the power of information obtained. A few examples of other uses of dual gene marking follow.

Data have already been accumulated that ABMT for AML or neuroblastoma can be complicated by the presence of contaminating tumor cells in remission marrow. Various purging technologies have been discussed for removing these residual malignant cells from the rescue marrow. A statistical analysis by the clinicians at St. Jude Children's Research Hospital has determined that gene marking can be used to demonstrate a benefit of purging neuroblastoma marrow versus using unpurged marrow in fewer than 24 patients. This is because dual labeling can be used with the patient serving as their own control. To do a similar trial by randomizing each patient to a single treatment arm with the same statistical power would have required them to enroll a much larger number of patients; therefore, the trial would have taken over 19 years to be completed. Dual marking obviously has a strong scientific basis for use in clinical trials, sometimes enabling studies that could not be done any other way. Such purging studies using more than one vector have begun.

Because it would take a very large trial to show a benefit of purging versus no purging in the treatment of these disorders, the number of patients that would need to be enrolled in a clinical trial comparing the outcome from two different purging procedures would be exorbitant. With a smaller difference between the outcomes of the two groups it would require much larger number of patients to reach statistical significance. By using a much smaller number of patients, with a portion of each patient's marrow treated by one technique and another portion treated by a different technique, a direct, controlled comparison of the two techniques would be possible. From the results obtained by analyzing any patients who relapse with marked tumor cells, the relative effectiveness of the purging techniques could be assessed. One could envision trials using more than two marker gene vectors so that even more than two methods could be tested in each patient. The ability to rapidly test new clinical procedures in comparison with alternative procedures is a strength available from gene-marking experiments that cannot be reproduced by any other method.

Additional uses of gene marking can and will be used to study other clinical interventions in ABMT. If marrow could be exposed ex vivo to various cytokines or growth factors, the kinetics of hematological recovery in both the short-term and long-term could be explored. By directly comparing one treatment with other treatments of portions of marrow from the same patient a direct comparison of the relative contribution of each fraction at various times after ABMT could be determined. This analysis could even determine the relative contribution to each cell lineage (red cell, neutrophil, lymphocyte, or other) after ABMT from each marrow portion. Other experiments could also be introduced to analyze other effects on donor marrow that could not be tested in any other manner. The ability to better understand the biology of hematopoiesis and its possible uses in clinical practice would be greatly enhanced by the use of gene-marking experiments.

Gene marking has also been developed with a goal of understanding other cellular therapies that can be used in conjunction with marrow transplantation. A protocol developed in Seattle (38) looks at the use of cytotoxic T-cell clones that have been developed against cytomegalovirus (CMV). These cells are greatly expanded and returned to the patient to help reduce the risk of CMV-infections in this susceptible population. Once these cells are given, there is no way to follow them, so they have been gene marked with a retroviral vector. All the cells descended from the original cloned cell will contain the retroviral vector used. The location and survival of these cells, therefore, can be directly measured by following the retroviral vector. Unique to this marker protocol is a second function incorporated into the retroviral vector. The vector contains not only a marker gene conferring resistance to hygromycin, but this hygromycin-resistant cDNA has been fused to the cDNA from herpes simplex virus thymidine kinase (39). The inclusion of the thymidine kinase function provides an additional safety factor in the protocol. The researchers were concerned that some unexpected side effects could potentially arise from the infused cytotoxic T cells. If such toxicity were seen, infusion of ganciclovir would specifically kill only the cells containing the herpes simplex thymidine kinase. Thus, the vector serves not only to mark the cells, but also as a potential way to kill the cells if something untoward were to arise. A similar trial to the one just described proposes to use cytotoxic T cells directed against the Epstein-Barr virus (EBV). This trial at St. Jude (40) proposes to mark these cells with a retroviral vector. In this trial, no thymidine kinase function is included. Through these trials with viral-specific cytotoxic T cells, it is hoped to learn if a therapeutic effect is related to survival of the infused cells.

IV. USE OF CYTOKINES TO INCREASE THE EFFICIENCY OF GENE TRANSFER

To improve gene transfer efficiency, many efforts have been directed on methods to improve stem cell cycling. As discussed previously (15,16), retroviral-mediated gene transfer is more efficient in rapidly dividing cells. Efforts to improve gene transfer rates have studied the effects of a myriad array of cytokines to improve these rates. The use of IL-1, IL-3, IL-6, G-CSF, stem cell factor, LIF, GM-CSF, PIXY, and a variety of other growth factors have been examined with varying degrees of success. As new growth factors are discovered, they will also be tested for their effects on gene transfer rates. All these efforts have attempted to duplicate the supportive microenvironment of the marrow cells. To even more closely duplicate the in vivo stem cell environment, bone marrow stroma has also been tested for its effects on gene transfer rates (41). The use of stroma seems to improve gene transfer efficiency. By further exploring cytokines and the marrow microenvironment, gene therapists expect to develop improvements in the overall efficiency of gene transfer into primitive hematopoietic cells. Increasing the rate of gene transfer into these target cells is especially important in the development of gene therapy for therapeutic purposes.

Therapeutic Uses

Besides being a useful tool for understanding malignant relapse and general principles of hematopoiesis, retroviral-mediated gene transfer can also be used for therapeutic goals. Instead of inserting marking genes into target cells, genes with a therapeutic goal could be used. Such experiments have begun and some of the rationales will be quickly discussed.

Multidrug Resistance

The multidrug resistance P-glycoprotein (MDR-1) is a membrane-associated protein that is able to remove toxic materials from a cell. The MDR protein has demonstrated it can result in the removal of chemotherapeutic agents, such as vinblastine or paclitaxel (Taxol), from the cell expressing this gene product. This gene product was recovered from tumor cells exhibiting an enhanced resistance to various agents. When this gene product is transferred into a cell, the cell expressing increased levels of MDR-1 exhibits decreased sensitivity to agents the MDR-1 product "pumps" out of the cell. This increased resistance can be used in vitro to positively select for cells containing the transferred MDR-1 gene.

If a similar resistance to many common therapeutic agents could be transferred to a cancer patient's bone marrow cells, the morbidity and mortality associated with the chemotherapy destroying normal marrow cells could be avoided. This concept of increased in vivo resistance of bone marrow cells has been demonstrated using both transgenic mice (42) as well as gene therapy in mice (43). Several investigators have proposed using retroviral vectors to transfer the MDR-1 cDNA into primitive hematopoietic cells of cancer patients receiving chemotherapy. The initial trials will explore if the frequency of cells containing the MDR minigene increases after exposure to chemotherapeutic agents, such as paclitaxel. Also to be examined is the nadir that ensues after each cycle of chemotherapy to determine if less marrow side effects are seen. The results of these trials will help determine if MDR-1 gene therapy is feasible and practical. The major practical obstacle to be overcome appears to be increasing the frequency of initial transduction of primitive marrow elements. Various cytokines and growth conditions are being explored in attempts to improve the efficiency of gene transfer. Other scientists have been looking at the possible use of other gene transfer systems such as AAV to see if higher efficiencies can be obtained.

Genetic Disorders

Already one genetic disorder involving marrow-related cells is being treated in a gene therapy trial. Adenosine deaminase (ADA) deficiency is one genetic cause of severe combined immunodeficiency (SCID). Patients with SCID can be cured with bone marrow from normal individuals. Unfortunately, only a fraction of ADA-deficient individuals have an HLA-matched sibling to serve as a donor. Alternative donors can be used, but the frequency of morbidity and mortality is increased. If only a small fraction of lymphoid precursor cells could be transduced to contain a normal copy of ADA cDNA, the patients would be expected to be cured. Clinical trials led by Dr. Donald Kohn have begun in which CD34-enriched cells from cord blood or from mobilized peripheral blood are transduced with retroviral vectors containing the ADA cDNA (44). Initial results have been encouraging. This rare disorder may already demonstrate the curative potential of gene therapy for a genetic disorder.

Other gene therapy protocols using bone marrow are being explored. Three groups have proposed treating Gaucher disease patients with retroviral-mediated gene therapy. Research continues on expressing hemoglobin genes in marrow cells. If adequate controlled expression was possible, treatments for such common disorders as sickle cell anemia and thalassemia could be developed. Multiple delivery systems are being explored while a better understanding of hemoglobin gene regulation is being pursued. With continued improvements in both areas, it is not irrational to assume that potential clinical application for hemoglobinopathies may result.

Human Immunodeficiency Virus

Besides the usefulness of gene therapy for malignant and genetic disorders, gene therapy is also being explored for use in combating infectious diseases. Many researchers are exploring the use of gene therapy to treat patients infected with human immunodeficiency virus (HIV). Although many have looked at T-cell–based therapy and clinical trials have been designed (45,46), others have examined the possibility of transducing a stem cell population with a gene vector that would confer resistance to HIV infection. These stem cells would produce progeny that would be expected to be similarly resistant to the effects of HIV. These progeny, especially CD4-positive lymphocytes, would provide the host with hematopoietic elements resistant to HIV. The resistant CD4 cells would be expected to permit the patient to have a functional immune system. Because the morbidity and mortality associated with AIDS appears to be relative to the relative immune deficiency, having a portion of the immune system resistant to HIV may eliminate most if not all of the difficulties these patients have.

Miscellaneous Uses of Gene Therapy

Because the marrow is easily removed from the body, manipulated, and reintroduced into the body, it is especially appealing as a target for gene therapy not only for marrow-related disorders but also for other disorders. If marrow cells could be engineered to release the proteins responsible for the various forms of hemophilia, hemophilia A and B could be treated by marrow gene therapy. Imagination envisions many other such scenarios. If marrow cells could be engineered to physiologically release insulin, a diabetic's lack of pancreatic islet cells would be overcome. An understanding of the basic biology of the disease state and knowledge of gene transfer systems permits the scientist to devise many novel therapies, vaccines, on others, using gene transfer technology.

Nonmarrow Uses of Gene Therapy

Other uses of gene therapy have been seen in which marrow cells are not the target cell population. These will be briefly described as their use will impinge on the overall understanding of gene therapy and may affect hematologists directly.

A clinical trial using gene therapy for hemophilia B has begun in China (47). In this trial autologous fibroblasts have been transduced with a retroviral vector containing the cDNA for factor IX. These cells are greatly expanded ex vivo and then reinjected into the patient. Once injected, these autologous fibroblasts are hoped to secrete enough factor IX that the patients would be converted into a mild form of hemophilia. These trials that are ongoing in China demonstrate that many target cells can be considered for gene therapy, even if the target cell is not normally the one expected to produce the clinical benefit.

Perhaps the largest interest in gene therapy is for "vaccines." Trials have begun in which gene transfer techniques ex vivo or in vivo are used to stimulate an immune response to the HIV virus. Other trials are attempting to make tumor "vaccines" by transferring new genes into or next to tumor cells. Genes introduced have included IL-2, IL-4, TNF, interferon gamma, HLA-B7, and GM-CSF. The target cells have included autologous tumor cells, allogeneic tumor cells, and autologous fibroblasts. Both ex vivo and in vivo approaches have been used in these trials as well as a wide variety of gene transfer methods. All approaches are designed to stimulate the immune system of the patient so that a stronger and, one hopes, therapeutic immune response will be seen against the patient's residual tumor. Some early successes have been reported, but more data need to accumulate to determine if this rationale will be clinically useful.

Another method of stimulating the antitumor immune response uses lymphocytes as the target cell for gene therapy. A cytokine, or other gene, is inserted into TIL with an expectation of improved tumor killing. Another outcome may be that the TIL home to tumor deposits and release cytokines locally giving a vaccine effect, as discussed earlier. Only TNF has been used clinically so far, but other cytokines are being examined. Other methods of manipulating effector lymphocytes have been explored. These include putting chimeric antigen-binding chains fused with T-cell receptor proteins to induce signaling of the effector cell after binding to the tumor antigens on the tumor cell surface. This example demonstrates the ability to manipulate natural molecules with genetic engineering to effect a new purpose. Many similar manipulations of effector cell activity and targeting are being examined, and some can be expected to be explored in clinical trials in the future. Gene therapy for malignant conditions remains one of the most ardently explored uses of gene therapy.

Other uses of gene therapy to modify the immune responses of the host are also being devised. As the molecular immunologists develop an increased understanding of the many molecules involved in regulating the immune system, this knowledge can be used to either increase or decrease the level of immune response. Many autoimmune or organ rejection processes are treated with immunosuppressive agents; the use of gene therapy to release such agents locally may be advantageous in these cases. If an immunosuppressive compound could be made to be produced locally by a transplanted organ, the need for systemic immunosuppression, with its associated toxicities, could be reduced. Similarly, in autoimmune diseases, especially localized processes, such as large joint arthritis, local immunosuppression induced by gene therapy could be advantageous. These and similar uses of gene therapy help express the wide scope of disease processes that might be affected by gene therapy.

Gene therapy has been applied to the treatment of cystic fibrosis. Many investigators are involved using either adenoviral vectors or liposomal systems. Because pulmonary cells cannot be readily manipulated and transplanted, these protocols are all in vivo approaches. The adenoviral approach had some initial concern over the toxicity associated with using an adenoviral vector. It appears that it is indeed possible to transfer and express the cDNA for cystic fibrosis transmembrane regulator protein (25). Newer generations of vectors are being designed, constructed, and tested already. As more experience accumulates, the efficacy of such approaches for the treatment of cystic fibrosis will be known.

Another use of the herpes simplex thymidine kinase gene as a "suicide gene" is seen in a clinical trial treating brain tumors. This trial is an in vivo gene therapy trial in which murine producer cells releasing retroviral vectors are used (48). The vector contains the sequence for herpes simplex thymidine kinase. The rationale is based on the uptake of the retroviral vector into tumor cells so that they can be killed by giving the patient ganciclovir. The gene-altered tumor cells are now able to phosphorylate ganciclovir into the monophosphate form, thus permitting ganciclovir to be toxic to the cell. Within the brain very few cells are dividing. Malignant cells in the brain divide much more rapidly than normal brain cells, and this biological characteristic can be used to "target" the gene therapy. Previous studies (15,16) have shown that retroviruses selectively transduce dividing cells. By injecting the retroviral vector producer cells into the tumor mass, the released vector particles will selectively transduce the more rapidly dividing tumor cells. Subsequent ganciclovir treatment kills the cells containing the herpes thymidine kinase gene product. In addition, there is a "bystander" effect (49) that is responsible for nearby tumor cells, even though they are not transduced, also dying as a result of ganciclovir administration. The mechanism of the bystander effect is not clear, but it may be explained by the transfer of toxic metabolites of ganciclovir from one cell to its neighbors, perhaps through gap junctions. The bystander effect, therefore, allows an amplification of the effects of gene transfer. The normal brain cells are left unaffected, presumably because they are not dividing. The design of the trial is to determine if there is any toxicity associated with the in vivo administration of retroviral vector producer cells and subsequent ganciclovir administration. Effects on the patient's tumor mass are also being examined to provide information about the possible efficacy of this therapy.

Many gene therapy strategies and delivery systems are being developed continuously. This chapter endeavored to concentrate only on those uses and protocols currently approved for clinical trials. The number of approved trials is increasing rapidly as experience and confidence accumulates. Gene therapy appears to be both feasible and safe. The information about clinical protocols presented in this chapter is designed only to give the reader a familiarity with what has already been accomplished while providing a background for understanding future developments in this fast-moving field.

V. SUMMARY

Gene transfer and therapy represent a new advancement in clinical medicine made possible by the progress in molecular biology and in the understanding of virus vectors and other gene delivery systems. Useful clinical knowledge has already been demonstrated with the information gained from the various marker studies. The use of retroviral vectors to study clinical questions and better understand hematopoiesis have and will continue to result in accelerated understanding of complex issues. The development of therapeutic uses of gene therapy will build on this understanding, while providing further knowledge about hematopoiesis and malignancy. As the field of gene transfer and therapy progresses its ultimate usefulness will become clearer. Currently, the field appears to present no special safety issues. After the efficacies of the various uses of gene therapy are tested, clinical medicine will be able to determine the potential usefulness and practicality of gene therapy techniques. Whether these advances will result in the everyday use of gene therapy depends on the findings of these and other studies. Given the demonstrated utility of gene transfer and therapy, it behooves the hematologist to be aware of the progress being made in this exciting new field.

REFERENCES

1. Rosenberg SA, Aebersold P, Cornetta K, et al. Gene transfer into humans—immunotherapy of patients with advanced melanoma using tumor-infiltrating lymphocytes modified by retroviral gene transduction. N Engl J Med 1990; 323:570–578.
2. Blaese RM, Anderson WF, Culver K, et al. Treatment of severe combined immune deficiency (SCID) due to adenosine diaminase (ADA) deficiency with autologous lymphocytes transduced with a human ADA gene. Hum Gene Ther 1990; 1:327–362.
3. Perucho M, Hanahan D, Wigler M. Genetic and physical linkage of exogenous sequences in transformed cells. Cell 1980; 22:390–317.
4. Chen CA, Okayama H. Calcium phosphate-mediated gene transfer: a highly efficient transfection system for stably transforming cells with plasmid DNA. Biotechniques 1988; 6:632–638.
5. Boggs SS. Targeted gene modification for gene therapy of stem cells. Int J Cell Cloning 1990; 8:80–96.
6. Gordon JW. Micromanipulation of embryos and germ cells: an approach to gene therapy? Am J Med Genet 1990; 35:206–214.
7. McNally MA, Lebkowski JS, Okarma TB, Lerch LB. Optimizing electroporation parameters for a variety of human hematopoietic cell lines. Biotechniques 1988; 6:882–886.
8. Kubiniec RT, Liang H, Hui SW. Effects of pulse length and pulse strength on transfection by electroporation. Biotechniques 1990; 8:16–20.
9. Hug P, Sleight RG. Liposomes for the transformation of eukaryotic cells. Biochim Biophys Acta 1991; 1097:1–17.
10. Nabel G, et al. Immunotherapy of malignancy by in vivo gene transfer into tumors. Hum Gene Ther 3:399–410.
11. Wolff JA, Malone RW, Williams P, Chong W, Acsadi G, Jani A, Felgner PL. Direct gene transfer into mouse muscle in vivo. Science 1990; 247:1465–1468.
12. Vitadello M, Schiaffino MV, Picard A, Scarpa M, Schiaffino S. Gene transfer in regenerating muscle. Hum Gene Ther 1994; 5:11–18.
13. Wu GY, Wilson JM, Shalaby F, et al. Receptor-mediated gene delivery in vivo. Partial correction of genetic analbuminemia in Nagase rats. J Biol Chem 1991; 266:14338–14342.
14. Wu GY, Wu CH. Delivery systems for gene therapy. Biotherapy 1991; 3:87–95.
15. Springett GM, Moen RC, Anderson S, et al. Infection efficiency of T lymphocytes with amphotropic retroviral vectors is cell cycle dependent. J Virol 1989; 63:3865–3869.
16. Miller DG, Adam MA, Miller AD. Gene transfer by retrovirus vectors occurs only in cells that are

actively replicating at the time of infection [published erratum appears in Mol Cell Biol 1992; 12(1):433] Mol Cell Biol 10:4239–4242.

17. McLachlin JR, Cornetta K, Eglitis MA, Anderson WF. Retroviral-mediated gene transfer. Prog Nucleic Acid Res Mol Biol 1990; 38:91–135.

18. Miller AD. Retroviral packaging cells. Hum Gene Ther 1990; 1:5–14.

19. Cornetta K, Moen RC, Culver K, et al. Amphotropic murine leukemia virus is not an acute pathogen for primates. Hum Gene Ther 1990; 1:15–30.

20. Cornetta K, Morgan RA, Anderson WF. Safety issues related to retroviral-mediated gene transfer in humans. Hum Gene Ther 1991; 2:5–14.

21. Cornetta K, Morgan RA, Gillio A, et al. No retroviremia or pathology in long-term follow-up of monkeys exposed to a murine amphotropic retrovirus. Hum Gene Ther 1991; 2:215–220.

22. Temin HM. Safety considerations in somatic gene therapy of human disease with retroviral vectors. Hum Gene Ther 1990; 1:111–123.

23. Donahue RE, Kessler S, Bodine D, McDonagh K, Dunbar C, Goodman S, Moen RC, Agricola B, Byrne E, Backer J, Zsebo K, Nienhuis AW. Helper virus induced T-cell lymphoma in non-human primates following retroviral mediated gene transfer. J Exp Med 1992; 176:1125–1135.

24. Anderson WF, McGarrity GJ, Moen RC. Report to the NIH Recombinant DNA Advisory Committee on murine replication-competent retrovirus (RCR) assays. Hum Gene Ther 1993; 4:311–321.

25. Rosenfeld MA, Yoshimura K, Trapnell BC, et al. In vivo transfer of the human cystic fibrosis transmembrane conductance regulator gene to the airway epithelium. Cell 1992; 68:143–155.

26. Lebkowski JS, McNally MM, Okarma TB, Lerch LB. Adeno-associated virus: a vector system for efficient introduction and integration of DNA into a variety of mammalian cell types. Mol Cell Biol 1988; 8:3988–3996.

27. Dixit M, Webb MS, Smart WC, Ohi S. Construction and expression of a recombinant adeno-associated virus that harbors a human beta-globin-encoding cDNA. Gene 1991; 104:253–257.

28. Curiel DT, Wayne E, Cotten M, Birtiel ML, Agarwal S, Li C-M, Loechel S, Hu P-C. High-efficiency gene transfer mediated by adenovirus coupled to DNA-polylinsine complexes. Hum Gene Ther 1992; 3:147–154.

29. Points to Consider in Human Somatic Cell Therapy and Gene Therapy. Hum Gene Ther 1991; 2:215–256.

30. Rosenberg SA, Blaese RM, Anderson WF. The N2-TIL human gene transfer clinical protocol. Hum Gene Ther 1990; 1:73–92.

31. Blaese RM, et al. Treatment of Severe Combined Immune Deficiency (SCID) due to Adenosine Deaminase (ADA) with Autologous Lymphocytes Transduced with a Human ADA Gene. Hum Gene Ther 1991; 1:521–527.

32. Brenner MK, et al. Autologous bone marrow transplant for children with acute myelogeneous leukemia in first complete remission: use of marker genes to investigate the biology of marrow reconstitution and the mechanism of relapse. Hum Gene Ther 1991; 2:137–159.

33. Brenner MK, Rill DR, Moen RC, Krance RA, Mirro J Jr, Anderson WF, Ihle JN. Gene-marking to trace origin of relapse after autologous bone-marrow transplantation. Lancet 1993; 341:85–86.

34. Brenner MK, et al. A phase I trial of high dose carboplatin and etoposide with autologous marrow support for treatment of stage D neuroblastoma in first remission: use of marker genes to investigate the biology of marrow reconstitution and the mechanism of relapse. Hum Gene Ther 1991; 2:257–272.

35. Rill DR, Santana VM, Roberts WM, Nilson T, Bowman LC, Krance RA, Heslop HE, Moen RC, Ihle JN, Brenner MK. Direct demonstration that autologous bone marrow transplantation for solid tumors can return a multiplicity of tumorigenic cells. Blood 1994; 84:380–383.

36. Brenner MK, Rill DR, Holladay MS, Heslop HE, Moen RC, Buschle M, Krance RA, Anderson WF, Ihle JN. Gene marking to determine whether autologous marrow infusion restores long-term haemopoiesis in cancer patients. Lancet 1993; 342:1134–1137.

37. Miller AR, Skotzko MJ, Rhoades K, Belldegrun AS, Moen RC, Economou JS. Simultaneous use of two retroviral vectors in human gene marking trials: feasibility and potential applications. Hum Gene Ther 1992; 3:619–624.

38. Riddell Sr, et al. Phase I study of cellular adoptive immunotherapy using genetically modified CD8+ HIV-specific T cells for HIV-seropositive patients undergoing allogeneic bone marrow transplant. Hum Gene Ther 1992; 3:319–338.

39. Lupton SD, Brunton LL, Kalberg VA, Overell RW. Dominant positive and negative selection using a hygromycin phosphotransferase–thymidine kinase fusion gene. Mol Cell Biol 1991; 11:3374–3378.

40. Heslop HE, et al. Administration of neomycin resistance gene marked EBV specific cytotoxic T lymphocytes to recipients of mis-matched-related or phenotypically similar unrelated donor marrow grafts. Hum Gene Ther 1993; 4:683–685.

41. Moore KA, Deisseroth AB, Reading CL, Williams DE, Belmont JW. Stromal support enhances cell-free retroviral vector transduction of human bone marrow long-term culture initiating cells. Blood 1992; 79:1393.

42. Mickisch G, Licht T, Merlino GT, Gottesman MM, Pasten I. Chemotherapy and chemosensitization of transgenic mice which express the human multidrug resistance gene in bone marrow: efficacy, potency, and toxicity. Cancer Res 1991; 51:5417.

43. Sorrentino BP, Brandt SJ, Bodine D, Gottesman M, Pasten I, Cline A, Nienhuis AW. Selection of drug-resistant bone marrow cells in vivo after retroviral transfer of human MDR1. Science 1992; 257:99–193.

44. Kohn DK. Hum Gene Ther 1993; 4:521–527.

45. Wong-Staal F, et al. A phase I clinical trial to evaluate the safety and effects in HIV-1 infected humans of autologous lymphocytes transduced with a ribozyme that cleaves HIV-1 RNA. Hum Gene Ther 1994; 5:94.

46. Nabel GJ. A molecular genetic intervention for AIDS—effects of a transdominant negative form of Rev. Hum Gene Ther 1994; 5:79–92.

47. Hsueh JL. Treatment of hemophilia B with autologous skin fibroblasts transduced with a human clotting factor IX cDNA. Hum Gene Ther 1992; 3:543–552.

48. Oldfield EH, Ram Z, Culver KW, Blaese RM, DeVroom HL. Gene therapy for the treatment of brain tumors using intra-tumoral transduction with the thymidine kinase gene and intravenous ganciclovir. Hum Gene Ther 1993; 4:36–69.

49. Culver KW, Ram Z, Wallbridge S, Ishii H, Oldfield EH, Blaese RM. In vivo gene transfer with retroviral vector-producer cells for treatment of experimental brain tumors. Science 1992; 256:1550–1552.

10

Granulocyte–Macrophage Colony-Stimulating Factor and Its Receptor

John E. J. Rasko*
The Walter and Eliza Hall Institute of Medical Research and The Cooperative Research Centre for Cellular Growth Factors, Melbourne, Victoria, Australia

I. INTRODUCTION

Granulocyte–macrophage colony-stimulating factor (GM-CSF) holds a preeminent place in the history of cytokines, having been one of the first to be described, purified, cloned, and finally established as a clinically useful therapeutic agent. Initially defined on the basis of its ability to induce myeloid colony formation from hematopoietic progenitor cells plated in semisolid media, the actions of GM-CSF are now known to be strikingly diverse and widespread.

In this review, the biochemistry, molecular biology, genetics, and distribution of GM-CSF and its receptor will be outlined. Recent advances in the signal transduction pathways activated following receptor–ligand interaction will be highlighted. These molecular aspects are followed by an overview of the in vitro and in vivo activities of GM-CSF, with special emphasis on target cell function and differentiation. The evidence supporting a contribution of GM-CSF to the pathogenesis of leukemia and other disease states will be discussed in the light of both transgenic and gene inactivation studies. Finally, the emerging role of GM-CSF in clinical practice will be addressed.

II. THE STRUCTURE OF GM-CSF AND ITS RECEPTOR

A. GM-CSF

The crystal structure of human GM-CSF shares, with several other hematopoietic growth factors, a pair of antiparallel α-helices stabilized by disulfide bonds between the first and third, and second and fourth cysteine residues (1–4). Whereas the first disulfide bridge is essential for

---

Current affiliation: Fred Hutchinson Cancer Research Center, Seattle, Washington.

biological activity, the second is not (5). The amino acid sequences of human (6–8), murine (9–12), gibbon (13), canine (14), ovine and bovine (15) GM-CSF have been obtained either by direct partial sequencing or by inference from the cDNA sequence (Fig. 1). The mature GM-CSF molecule is always preceded by a 25-residue, hydrophobic leader sequence.

The mature murine GM-CSF sequence comprises 124 amino acids with a molecular mass of 14 kDa. The five other known sequences are longer, owing to an insertion near the first α-helix. This region is an important functional domain (16) and may contribute to the lack of cross-species biological activity or receptor binding between murine and human molecules (17). In fact, comparison of murine and human amino acid sequences reveals only 56 and 35% identity between GM-CSF and GM-CSF receptor α-chains, respectively.

The four main regions within GM-CSF that are important for binding to its receptor lie between residues 18 and 22, 34 and 41, 52 and 61, and 94 and 115 (5,17–23). Residues 18–22 within the first α-helix are responsible for GM-CSF's interaction with the β-chain of its receptor to form a high-affinity complex. In particular, glutamate residue 21, which is conserved between murine (m) GM-CSF, murine interleukin-3 (mIL-3), and mIL-5, has been implicated as playing a significant role in the interaction with the shared β-chain. Substitution of GM-CSF Glu-21 with arginine or alanine residues gave rise to mutants in which low-affinity binding to the α-chain was unimpaired, but the ability to bind the β-chain to form a high-affinity complex was abrogated (18,24–26).

Although both glycosylated and nonglycosylated GM-CSF retain biological activity, several differences exist, depending on which expression system is used. GM-CSF isolated from most murine and human sources is glycosylated to approximately 23 kDa (27), because they contain two potential, but differently located, N-linked glycosylation sites. Similar to native hGM-CSF, both yeast-derived (28) and baculovirus-derived (29) recombinant hGM-CSF are glycosylated,

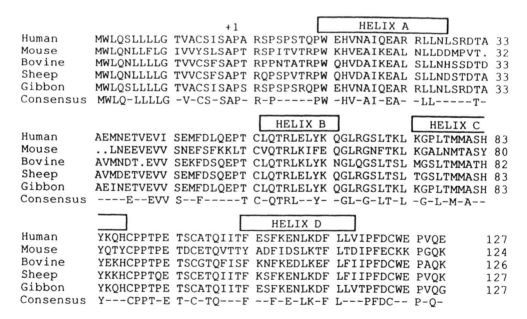

Figure 1 Amino acid sequence comparison of pre-GM-CSFs: Human, mouse, bovine, sheep, and gibbon pre-GM-CSF amino acid sequences were obtained from Genbank and Swissprot databases and Ref. 13. The aligned consensus sequence was derived by applying the algorithm of Needleman and Wunsch. The first residue of native GM-CSF is identified by +1 and the four predicted α-helices are indicated above the sequences.

albeit differently and, in the former, proline at position 23 is replaced by a valine. In contrast, *Escherichia coli*-derived recombinant hGM-CSF is not glycosylated, contains six fewer residues, and an extra initial methionine. The nonglycosylated *E. coli*-derived recombinant hGM-CSF has a significantly faster elimination phase in vivo compared with mammalian Chinese hamster ovary (CHO) cell-derived recombinant hGM-CSF (30), yet it retains high biological activity (4,31,32). However, the nonglycosylated molecule exhibits increased affinity for its receptor (33,34) and, in a review of 32 clinical trials using *E. coli*- or yeast-derived hGM-CSF, it produces more adverse events (35).

B. GM-CSF Receptor

The GM-CSF receptor (GMR) is composed of at least two well-characterized chains, both belonging to the burgeoning cytokine or hematopoietin receptor superfamily (36–39). One unexpected feature of cytokine receptor superfamily members is that, although they are required to mediate critical cell–cell communications, their copy number on the cell surface is typically low, between 100 and 1000 per cell. The 85-kDa α-chain binds GM-CSF with low affinity (K_d = 1–10 nM), whereas the 130-kDa β-chain, despite having no intrinsic binding affinity for GM-CSF when alone, sequentially or simultaneously converts the GM-CSF–α-chain complex to a high-affinity state (K_d = 30–100 pM; 40–44). Consistent with this model, the α-chain exhibits rapid dissociation kinetics (36,45), whereas, in contrast, the α/βGMR displays slow dissociation kinetics (43,46).

Another hallmark of the cytokine receptor superfamily seen in both the α- and β-chains of the GMR, as well as the receptors for growth hormone, prolactin, G-CSF, erythropoietin, IL-2 (β- and γ-chains), IL-3 (α-chain), IL-4, IL-5 (α-chain), IL-6 (α-chain), (LIF; leukemia inhibitory factor α-chain), gp130, ciliary neurotrophic factor (CNTF), IL-7, IL-9, and IL-11 (α-chain), is a conserved approximately 200-amino acid hematopoietin domain (47). This extracellular domain is characterized by a number of conserved amino acid residues, including: four cysteine residues, which form two internal disulfide loops; a Pro–Pro pair, which divides the domain into two subdomains, each of which appears to adopt a seven-stranded β-barrel structure reminiscent of immunoglobulin constant chains; and the largely conserved membrane-proximal motif WSXWS.

In a remarkable example of apparent functional degeneracy, the β-chain of the GMR is shared by the IL-3 and IL-5 receptor α-chains (48–54). That is, GM-CSF, IL-3, and IL-5 bind their particular α-chain with absolute specificity, followed by signaling in association with a common *competable* β-chain. Distinct from receptor trans–down-modulation (55), this phenomenon of cross-competition between the three ligands may partially explain some of the homologous biological activities of these cytokines seen, for example, in eosinophil effector functions (56–58). A curious, as yet unexplained, priority of binding for the three ligands (on eosinophils GM-CSF > IL-3 > IL-5; 52) promises further subtleties in the cross-competition model (55).

Because fibroblasts installed with a functional GMR α-chain show no response to GM-CSF (evidenced by a lack of cell growth and absent phosphorylation on tyrosine; (59), it has generally been held that the α-chain acts solely to confer receptor specificity. However, in *Xenopus* oocytes, installed with only the GMR α-chain, the isolated α-chain is capable of transducing a signal for the activation of glucose transport, without activation of the protein kinase cascade (60).

Several genes that code for cytokine receptors appear to give rise to a variety of alternately spliced transcripts, and the *GMR*α gene is no exception. Two groups have identified potentially soluble GMRα mRNA isoforms that lack a transmembrane region sequence (61,62). In addition, an "α_2-chain," in which the 25 COOH-terminal amino acids of the usual α-chain are replaced by a 35-amino acid serine–proline-rich terminus, exhibits identical low-affinity binding, and effects normal cell proliferation when installed in murine FDC-P1 cells (42,63).

C. Signal Transduction

Cytokine receptors, including GMR, which do not contain an intrinsic, intracytoplasmic protein kinase sequence, require a means of triggering the tyrosine phosphorylation cascade in target cells. The "kinase cascade" is important in mediating GM-CSF–induced signals for proliferation and differentiation (64–66). For example, in differentiated, but not undifferentiated, WEHI-3B cells, GM-CSF treatment leads to a rapid, transient, and concentration-dependent phosphorylation on tyrosine of a number of endogenous proteins (67).

Experiments published during 1993 have left little doubt that the JAK family of protein tyrosine kinases plays a pivotal role in mediating cytokine receptor signaling. Acting in some ways as a "soluble third chain" of the receptor, JAK2 functions as the long sought-after inducible protein tyrosine kinase of GMR, in the same way that it functions for prolactin, growth hormone, erythropoietin, and IL-6 receptors (68–70). The cytoplasmic domain of the GMR β-chain has two distinct regions: an approximately 60-residue membrane proximal region, indispensable for proliferative signaling by c-*myc* and *pim*-1; and an approximately 140-residue distal region necessary for activation of Ras, Raf-1, and further downstream molecules (71,72). The membrane-proximal region is required for JAK2 activation (69). A potentially important negative regulator of the JAK-mediated GM-CSF signal transduction cascade is hematopoietic cell phosphatase (73).

Known intermediates that are phosphorylated and activated consequent to GM-CSF include SHC, GRB2, and *Ras* p21 (72,74), pp100 (75), pim-1 kinase (76), raf-1 kinase (67), p92c-fps/fes kinase (77), p53/p56 lyn kinase, and p62 yes kinase (78), and the p42 mitogen-activated (or microtubule-associated) protein serine–threonine kinase family (MAPK; 79–81). In fact, MAPK appears to transduce a common signaling event elicited by GM-CSF, IL-3, IL-5, and stem cell factor (SCF) (82), but not IL-4 (83). The signaling pathway subsequently involves transcriptional activation of nuclear factors known to be important in the cell cycle—including c-*fos*, c-*jun*, c-*myc* (84), perhaps *p53* (85), and other DNA-binding proteins that recognize the interferon gamma (IFN-γ) response region (86,87). Additionally, a novel early-response gene, *A1*, which exhibits similarity to *bcl-2*, is rapidly and transiently expressed in hematopoietic tissues following induction by GM-CSF (88).

Other events implicated in the signal transduction pathway of GM-CSF include "priming" of neutrophils, possibly by G protein (89) or protein kinase Cε (90) activation of phospholipase D; inositol lipid metabolism (91); and intracellular alkalinization (92); with involvement of the Na$^+$/H$^+$ antiport, glycolysis, and Ca^{2+} (93–95).

In the flourishing field of signal transduction, there exists a growing problem as to how receptor–ligand complexes maintain specificity with such apparent degeneracy and convergence in the machinery. For instance, GM-CSF and IL-3 share similar target cells, have overlapping biological functions, and induce a nearly identical pattern of tyrosine phosphorylation in certain cell lines (77). The identification of features allowing subtle cytokine modulations of cell function, despite a manifestly degenerate signal transduction pathway, is one of the next major hurdles in biology (96).

III. CHROMOSOMAL LOCALIZATION OF GM-CSF AND ITS RECEPTOR

A. GM-CSF

The GM-CSF gene is located on murine chromosome 11, at band A5 to B1 (10,97–100) and on the long arm of human chromosome 5, between bands 23 and 31 (101–105). Gene localization of murine GM-CSF was initially obtained by mouse–Chinese hamster somatic cell hybrids. More specific localization was obtained by interspecies backcross and recombinant inbred strain analysis that demonstrated cosegregation with the independently localized *SPARC* gene. Human GM-CSF gene localization was obtained by in situ hybridization. Both murine and human GM-CSF are encoded by a single-copy gene comprising four exons contained within approximately 2.5 kilobase pairs (kbp) of genomic DNA (106,107).

One of the more intriguing aspects of the genetic localization of GM-CSF is its proximity to the gene for IL-3 (multi-CSF) in both mice and humans. The IL-3 gene is located in the same orientation, approximately 10 and 15 kbp 5′ of the GM-CSF gene in the human (101,108) and murine genomes (100), respectively. The selective constraints that have maintained this close and conserved linkage remain unclear. However, between the GM-CSF and IL-3 genes in both mice and humans, lies an inducible, cyclosporin A-sensitive enhancer that may be part of the signaling mechanism following T-cell receptor (TCR) activation (109,392). Further analysis of a common enhancer could shed light on the linkage of GM-CSF and IL-3 as well as their coexpression in activated T lymphocytes (110).

If one assumes no inversion has occurred on the 5q⁻ chromosome in the HL60 human myeloid leukemia cell line, the order of the GM-CSF–IL-3 gene pair on human chromosome 5 is centromere → 5′ IL-3 → 5′ GM-CSF → qter (105,111). In fact the long arm of chromosome 5 contains a remarkable clustering of hematopoietic regulator genes, including the protooncogene c-*fms*, which codes for the M-CSF receptor; platelet-derived growth factor (PDGF) receptor; IL-3; IL-4; IL-5; M-CSF; and GM-CSF. In the "5q⁻ syndrome" an interstitial deletion of the long arm of chromosome 5 from 5q11–21 to 5q22–34 is associated with therapy-related leukemias and primary and familial myelodysplasias (112).

As 5q contains such a dense group of important genes, there has been much speculation concerning the pathogenesis of the 5q⁻ syndrome (103–105,113,114). However, it is unlikely that deletion of one allele of a CSF or CSF-receptor gene could contribute to neoplasia (see later discussion of the murine GM-CSF gene inactivation model). Furthermore, although lying close to the minimal critical region common to all 5q⁻ chromosomes, the hematopoietic regulator gene cluster may not actually lie within it. A more likely candidate for the pathogenesis of 5q⁻ is the interferon regulatory factor-1 gene (*IRF-1*) mapped to 5q31.1; a gene for which the product is a transcriptional activator with putative antioncogenic properties (115). Although both alleles of *IRF-1* have been inactivated in at least one case of acute leukemia (116), there is evidence that *IRF-1* is not solely responsible for the 5q⁻ syndrome (117).

B. GM-CSF Receptor

The common β-subunit of the human GM-CSF, IL-3, and IL-5 receptors is encoded by a unique gene, located at 22q12–13 (118). This gene is in the vicinity of t(1;22)(p13;q13), which occurs frequently in childhood acute megakaryocytic leukemia (119), although its relation to this disease remains uncertain. By contrast, the murine genome contains two closely related genes with homology to the β-subunit, *Aic2A* and *Aic2B* (71). The *Aic2A* β-subunit is exclusive to the IL-3 receptor and binds mIL-3 with low affinity, whereas *Aic2B* is shared among GM-CSF, IL-3, and IL-5 receptors (53,54). The two genes lie within 250 kbp of one another in a region of murine chromosome 15 around the *Sis* gene, known to be syntenic with region 12–13 on the long arm of human chromosome 22 (120).

The human GM-CSF receptor α-chain gene (*GMR-α*, also called *Csf2ra*) is encoded by at least 45 kbp of DNA within the pseudoautosomal region (PAR) (121,122). The PAR is an approximately 2600-kbp area of homology between the ends of the human sex chromosomes that undergoes obligatory exchange between the X and Y chromosomes during meiosis, thereby behaving "pseudo-autosomaly." As genes in the PAR are present in two copies in both males and females, they escape X-inactivation (123). The GMR-α gene has been localized to the middle of the PAR, approximately 1200 kbp from the telomere (124). Rich in hypervariable repeat sequences, GMR-α is highly polymorphic with a recombination frequency of approximately 20%. Other genes localized to the PAR include: MIC2, encoding a cell surface antigen possibly involved in T-cell adhesion (125); ANT3, encoding the ADP–ATP translocase (126); and the IL-3 receptor α-chain gene (IL3RA; 127). The most likely gene order of these loci is pter → GMR-α → IL3R-α → ANT3 → cent.

In the murine genome the GMR-α gene (Csf2ra) is located at telomeric band D2 of chromosome 19, thus revealing incomplete conservation between human and mouse X chromosomes (128). In contrast, the murine Il3ra gene is located at the proximal end of chromosome 14 (Rakar SJ and Gough NM, personal communication). In summary, whereas GM-CSF and IL-3 are in close physical proximity in both the human and murine genomes, as are the human receptors for these ligands (≤ 190 kbp), in mice the receptors are not linked. The chromosomal localization of the murine and human genes for GM-CSF and its receptor chains are outlined in Table 1.

IV. EXPRESSION AND DISTRIBUTION

A. GM-CSF

The primary GM-CSF transcript expressed from four exons in both mice and humans is processed to a mature mRNA of approximately 780 nucleotides (9,107). In mice the 5'-untranslated region, coding region, and 3'-untranslated region comprise 32–35, 423, and 319–326 nucleotides, respectively; whereas in humans the regions comprise 39–42, 432, and 316 nucleotides. The uncertainty in transcript region sizes may reflect true microheterogeneity or technical artifact (11,106,129).

In various inducible cells, discussed in the following, the accumulation of GM-CSF mRNA is due to both an increase in transcription rate and, more importantly, posttranscriptional stabilization of the mRNA (129–136). Several sequence motifs recognized as transcription-control elements occur within the GM-CSF gene (137). One such Ca^{2+}-responsive motif is the 51-nucleotide adenosine–uridine-rich sequence in the 3'-untranslated region, which is strongly implicated in reducing the stability of mRNA (138,139). Intriguingly, this sequence or closely linked sequences in the GM-CSF mRNA, also seem likely to mediate the transient stabilization of the GM-CSF mRNA after cellular stimulation (140,141).

Two closely related sequence motifs, CK-1 and CK-2, are located between 100 and 300 bp upstream from the transcriptional initiation sites of various cytokine genes, including GM-CSF (142,143). These and other sequence elements close to the GM-CSF promoter have been implicated in GM-CSF transcriptional control, as their removal or mutation ablates GM-CSF

Table 1 Chromosomal Localization of GM-CSF and Its Receptor

	Mouse	Human
GM-CSF	11A5-B1	5q23-31
GMR-α (CSF2R)	19D2	Pseudoautosomal region
GMR-β (AIC2B)	15	22q12.3-13.1

promoter activity (129,144–147). Moreover, specific nuclear proteins, including NFa and NFb, have been identified, which bind to, or in the vicinity of, these motifs (148–153). Another direct repeat-containing element, ICK-1, has been identified in the negative regulatory region of both GM-CSF and IL-3 gene promoters, and its potentially important binding protein, ICK-1A, has been described (154). Whereas some candidate transcription factors are expressed widely, others display a narrow cellular specificity and, in certain instances, are present only after stimulation of the producer cell.

A large variety of cell types and cell lines are capable of producing GM-CSF following stimulation by other cytokines, antigens, or pathological states, as detailed in Table 2. The wide-spread distribution of sources for GM-CSF lends weight to the concept that it plays a significant role in both local inflammatory reactions as well as systemic responses, leading to enhanced numbers of functional inflammatory effector cells. For example, the original purification of GM-CSF was performed on medium conditioned by lungs from endotoxin-primed mice (12,155). In fact, most endotoxin-primed mouse tissues and organs cultured in vitro release GM-CSF into the medium, although the lung remains a bountiful source (27). Production from organ cultures is continuous over several days and appears to reflect de novo synthesis, rather than release of preformed protein, as production can be inhibited by protein synthesis inhibitors.

An exhaustive discussion of each cell known to produce GM-CSF is beyond the scope of this review; however, several examples are particularly noteworthy. T-cell subsets are capable of producing a large number of cytokines, including GM-CSF, IL-2, IL-3, IL-4, IL-5, IL-6, IFN-γ, and tumor necrosis factor (TNF)-β, attesting to the central role they play in immune effector mechanisms (142,187). The major pathway for the activation of GM-CSF synthesis in T cells is induced by ligands of the T-cell antigen receptor, such as lectins, monoclonal antibodies, and antigen (for reviews see Ref. 188). Other cytokines can also induce GM-CSF production in T cells, including IL-1 and the T-cell growth factor IL-2, although the levels of GM-CSF produced are lower (142,189).

Macrophage populations, which are critical for innate immunity, can be induced to produce GM-CSF, in conjunction with G-CSF, TNF-α, and IL-1 (131,158,159). Bacterial cell-wall lipopolysaccharides (LPS), for example, provide a strong stimulus for cytokine production by

Table 2 Cellular Sources of GM-CSF

Cell type	Stimulus	Ref.
T lymphocytes	Antigen, lectins, IL-2	142
B lymphocytes	Antigen, LPS, IL-1	156,157,132
Macrophages	LPS, phagocytosis, adherence	131,158,159
Mesothelial cells		160
Keratinocytes		161
Osteoblasts	Parathyroid hormone, LPS	162,163
Uterine epithelial cells		164
Synoviocytes	Rheumatoid arthritis	165
Mast cells	IgE	166
Fibroblasts	IL-1, TNF-α, NaF, retroviruses	167–174
Various solid tumors	Calcium ionophore	175,176,139
Stromal cells	Mitogens, TNF-α, IL-1	177
Endothelial cells	IL-1, TNF-α, LPS, oxided lipoproteins	178–186

macrophages (190,191). By using short-term cultures of peritoneal exudate cells, substantially enriched for macrophages, several different stimuli have been demonstrated to induce GM-CSF secretion, including LPS, phagocytosis of latex beads, and adherence (131). Similarly, with longer-term cultures of bone marrow-derived macrophages (devoid of potentially stimulatory T lymphocytes, in contrast with the former populations), GM-CSF production by macrophages has also been documented (159). Unlike the enhanced GM-CSF expression seen in T cells (142), stimulation of peritoneal macrophages leads to increased GM-CSF mRNA stability and consequent accumulation of GM-CSF (131).

In the bone marrow microenvironment, stromal fibroblasts and endothelial cells have been increasingly recognized as important in controlling the proliferation, differentiation, and transit of hematopoietic cells (179,182). In cultured endothelial cells, IL-1, TNF-α, LPS, and oxidized lipoproteins, can increase the expression not only of GM-CSF, but also of G-CSF and M-CSF. The IL-1 induced expression of GM-CSF in endothelial cells is regulated post-transcriptionally, and IFN-γ specifically inhibits the increased half-life by destabilization of the GM-CSF message (186). Similarly IL-1, TNF-α, inflammatory agents, tumor promoters, and retroviruses can induce GM-CSF expression in fibroblasts (169,170,172,184). Also, the c-*myb* protooncogene regulates GM-CSF expression in stromal cells capable of generating fibroblast-like colonies (CFU-F) by transactivation of the GM-CSF promoter (174). The production of cytokines, including GM-CSF, following stimulation of such cells by inflammatory mediators, may constitute a pivotal step in the initiation and amplification of the innate immune response.

B. GM-CSF Receptor

The distribution of receptors for GM-CSF in normal cells includes myeloid progenitor cells, neutrophils, dendritic cells, monocytes, and eosinophils, but not lymphocytes or erythroid cells (192,193). Some lymphoid cell lines and a variety of myeloid cell lines, such as HL60 and KG1, also specifically bind iodinated GM-CSF (194,195). Although varying between particular cell types, GMR density is estimated between 50 and 500 per cell. A decline in the expression of GMR occurs as cells mature, but even fully differentiated cells display up to 100 receptors per cell.

In various murine cells examined at 37°C, GM-CSF-induced receptor internalization was followed by lysosomal degradation of ligand (196). Ligand degradation was balanced by new GMR synthesis and recycling leading to reexpression of GMR at the cell surface. The observation that IL-3, G-CSF, and high doses of M-CSF can reduce binding of GM-CSF to normal murine bone marrow cells in a rapid, dose-dependent manner has been called "receptor transmodulation" (192). This phenomenon is distinct from cross-competition owing to common receptor subunits or receptor internalization, and remains incompletely understood.

V. BIOLOGICAL EFFECTS OF GM-CSF

The initial identification of GM-CSF was based on its ability to stimulate the in vitro proliferation and differentiation of hematopoietic progenitors to granulocytes and macrophages (197–199). At least five other glycoproteins (M-CSF, G-CSF, IL-3, IL-6, and SCF) are now known to exhibit related activities on committed progenitors. The widespread and pleiotropic effects of GM-CSF have been even more apparent with analysis of new biological systems, as discussed in the following sections. Figure 2 outlines the four main groups into which the effects of GM-CSF may be categorized.

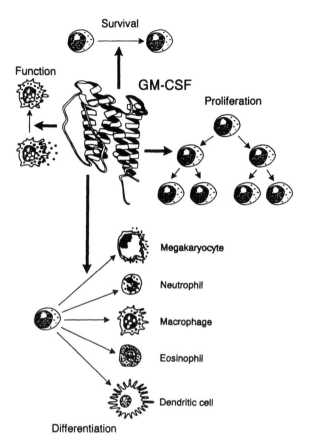

Figure 2 Pleiotropic effects of GM-CSF.

A. Survival

The increasingly recognized importance of apoptotic (programmed cell death) mechanisms in many cells has led to the characterization of a number of proteins including bcl-2, bax, and bcl-x (200–202). Because of the exquisite balances necessary to maintain the normal production of an estimated 3.5×10^7 neutrophils per kilogram per hour (203), any alteration in cell survival effected by such proteins could have catastrophic consequences. Alternatively, prolongation of the survival of granulocytes (204), eosinophils (205), and basophils (206) may substantially facilitate the host inflammatory response. Withdrawal of GM-CSF in vitro from factor-dependent cells or normal marrow progenitor cells led to their loss of viability, with a half-life from 9 to 24 hr (207,208). GM-CSF promotes cell survival by suppression of the process of apoptosis (209). Constitutive overexpression of bcl-2 prolongs the short-term survival of such cells following GM-CSF deprivation (210). However, attesting to the priority of survival, the action of GM-CSF to increase viability and longevity of mature and progenitor cells in vitro is achieved at approximately 1/100 the concentration required to stimulate cell proliferation (211).

The mechanism by which GM-CSF protects against cell death is far from clear, although several systems are available for examination. For example, GM-CSF protected against both transforming growth factor (TGF)-β_1 and chemotherapy-induced apoptosis in 7-M12 leukemic cells, whereas the M1 myeloid cell line did not survive (212). Combinations of several cytokines, including GM-CSF, act to support murine progenitor cell survival in liquid suspension culture (213). Sequential activation of protein kinase C and the Na^+/H^+ antiport have been implicated in the signal transduction pathway protecting against factor withdrawal-induced apoptotic cell death (214). It appears likely that cell survival and proliferation may be separate, dissociable states, with potentially independent cytokine-initiated signaling pathways.

B. Proliferation

Originally thought to be restricted to the stimulation of granulocyte–macrophage colony-forming cells in semisolid culture, GM-CSF is now known to induce proliferation and differentiation of eosinophil, basophil, megakaryocyte, erythroid, dendritic, and some multipotential stem cells (215,216). Although these activities are dose-dependent, clonal heterogeneity in responsiveness and proliferation kinetics, as well as the known up-regulation of GMR induced by low, but not high, concentrations of GM-CSF, have complicated the relation (197,217). Clonogenicity, total progeny number, entry into cell cycle, and its traverse rate, all are dependent on GM-CSF concentrations of between 1 and 100 pM (218,219). Whereas native and recombinant GM-CSF are both effective stimuli for murine granulocyte (CFU-G), macrophage (CFU-M), and granulocyte–macrophage (CFU-GM) colony formation at concentrations between 2 and 80 pg/ml (220), eosinophil colony formation (CFU-Eo) occurs only above this range. Up to tenfold greater concentrations are required to stimulate megakaryocyte (CFU-Meg), erythroid (BFU-E), and mixed progenitor colony formation (CFU-GEMM) (221). In contrast, dose dependency is less apparent in human colony formation (222–224).

The ready availability of both human and murine cell lines able to respond to GM-CSF has facilitated many experiments undertaken in the field of hematopoietic cell biology. Factor-dependent cell lines exhibiting myeloid features include FDC-P1, derived from long-term murine bone marrow culture (225); BAC1, derived from SV40-transformed macrophages (226); DA1 and NFS-60, derived from leukemic mice (227); and DGM36, derived following mutagenesis in vivo by Moloney murine leukemia virus of mice transgenic for GM-CSF (Rasko JEJ, et al, submitted). Other cell lines exhibiting T-lymphocytic features which nevertheless proliferate in the presence of GM-CSF include TALL101, derived from a human leukemia (228), and the murine IL-2-dependent HT2 line (229). Considerable variation in the proliferative response of primary human acute myeloid and erythroblastic leukemia cells to GM-CSF has been demonstrated, with approximately one-quarter of cases responding in vitro (230–234).

The in vitro proliferative activity of GM-CSF when used alone may be quite different from effects resulting from the potential combinatorial interactions of other factors affecting cell growth in vivo. Examples of additive or synergistic effects include GM-CSF- and IL-3-enhanced GM-, E-, Eo-, and Meg-colony formation from murine bone marrow (223,235,236) and human progenitor cell cultures (237); dramatically elevated macrophage colony production following stimulation by M-CSF, combined with very small amounts of GM-CSF (238); and markedly enhanced production of erythroid and myeloid colonies from marrow progenitor cells by the direct action of GM-CSF when combined with stem cell factor (SCF; 239).

Factors known to inhibit the effects of combined GM-CSF and SCF on human myeloid progenitors include MIP-1α, MIP-2α, PF-4, IL-8, and MCAF (240). Even the combination of M-CSF and GM-CSF can produce "synergistic suppression" of macrophage colonies from a cell line, owing to loss of clonogenicity and increased monocyte–macrophage differentiation (236,241). This latter effect is possibly mediated by GM-CSF-induced "transregulation" of the M-CSFR through induction of a ribonuclease (242).

C. Differentiation

A balance must exist between proliferation and commitment to differentiation for the maintenance of steady-state hematopoiesis. If one of two daughter cells resulting from the division of
a stem cell fails to differentiate ("clonally extinguish"), leading to preferential self-renewal, then
cell numbers will rise exponentially (243). Indeed Rudolf Virchow's earliest pathological
description of leukemia ("white blood") as containing a " . . . pus-like mass" attests to the
potential consequences of homeostasis gone awry. It should also be noted that the process
whereby a precursor cell is selected to undergo differentiation commitment is distinct from
the mechanisms by which maturation is achieved. The role of growth factors in a "selective" model (wherein the factor initiates differentiation selection) versus a "stochastic" model
(wherein the factor merely promotes survival), remains contentious, despite considerable experimental effort (244).

One means of examining the effects of cytokines on differentiation induction has been the
use of paired daughter cells resulting from a single cell division (218). Because higher GM-CSF
concentrations led to greater proliferation, it might be inferred that GM-CSF inhibits differentiation commitment and maturation, whereas other evidence points toward the opposite conclusion
(228). In similar paired daughter studies, in which the initial stimulus of either GM-CSF or
M-CSF was then swapped to the other stimulus, GM-CSF irreversibly commited progenitor cells
to granulocytic differentiation (245).

Several examples of model systems currently under study should serve to reveal the breadth
of differentiative effects on primary cells induced by GM-CSF. First, a well-established system
used to examine differentiation commitment in vitro is the GM-CSF-induced appearance
of BFU-Es, which require erythropoietin for complete terminal differentiation into CFU-E
(224,246). Second, GM-CSF promotes the outgrowth of dendritic cells from the nonadherent
class Ia-negative fraction of murine peripheral blood (247,248), human CD34$^+$ progenitor cells
(249), and lymphocyte-depleted human tonsils (250). The GM-CSF-induced dendritic cell
outgrowth is enhanced up to 20-fold by TNF-α (251). In murine brain cell cultures, GM-CSF
drives microglial development toward end cells capable of efficient antigen presentation (252).
Third, CD34$^+$ monkey bone marrow cells, when placed in semisolid culture, differentiate toward
osteoclasts in the presence of GM-CSF and other cytokine combinations (253).

A common concern surrounding experiments performed on primary cells is the potential for
constitutive production of cytokines by heterogeneous subpopulations. Such objections are less
problematic in cultures of cloned cell lines which, therefore, provide the most compelling
evidence for growth factor-induced lineage commitment. Both M1 and WEHI 3B D+ murine
leukemic cell lines, for example, exhibit GM-CSF-induced suppression of proliferation and
macrophage commitment in vitro (254,255). Similar observations may be made following
GM-CSF stimulation of the human promyelocytic cell line HL60, which results in proliferation
and subsequent differentiation (256,257).

D. Function

Several of the pleiotropic functions of GM-CSF are summarized in Table 3. In general the in
vitro functional effects of GM-CSF lead to enhanced innate as well as specific immune responses. At local sites of infection, effector cells, such as neutrophils, macrophages, eosinophils,
and antigen-presenting cells would be recruited, retained, and activated, to facilitate host
defenses. Also, GM-CSF exhibits functional synergy with other cytokines, for example, in enhancing TNF-α-mediated, complement-dependent phagocytosis of the fungal pathogen, *Cryptococcus neoformans* (258). Although the functions listed are bone fide effects, it does not follow

Table 3 Functional Consequences of GM-CSF In Vitro

Function	Cell type	Ref.
Enhanced antibody-dependent cell killing	Neutrophils and macrophages	259–261
Enhanced phagocytosis	Neutrophils	262,263
	Macrophages	258
Enhanced chemotaxis and adhesion/migration inhibition	Neutrophils, mesenchymal, and tumor cells	264–269
Enhanced Fc γ receptor RII and class II MHC	Monocytes and neutrophils	270,271
Enhanced antigen presentation	Dendritic cells	272,273
	Microglia	252
Stimulation of cytokine and leukotriene release	Macrophages/monocytes/eosinophils	274–280
Enhanced viral shedding	Monocytes	281
Suppression of complement production	Monocytes	282,283
Enhanced oxidative burst	Neutrophils/eosinophils/monocytes	284,285
Enhanced migration	Endothelial cells	286,287
Enhanced $\alpha v \beta 3$ integrin expression	Macrophages	

that GM-CSF is necessarily required for efficient host defenses. However, gene "knock-out" experiments, outlined in the following section, strongly support a need for GM-CSF in the prevention of opportunistic bacterial and fungal lung infections.

VI. EFFECTS OF GM-CSF IN VIVO

The question of whether GM-CSF is necessary for constitutive hematopoiesis has been problematic since its discovery. During 1994, reports concerning mice nullizygous for functional GM-CSF have proved that a complete absence of GM-CSF does not preclude normal steady-state hematopoiesis (288,289). Although it remains feasible that other cytokines have substituted for GM-CSF by virtue of "functional redundancy," a more popular interpretation is that GM-CSF is not an important regulator of *normal* hematopoiesis. The latter interpretation does not preclude possible modulating effects of GM-CSF yet to be demonstrated in analyses of GM-CSF knock-out litters. In the future, multiple cytokine and cytokine receptor gene disruption experiments will shed light on these issues (290).

Still, an unexpected, almost universal abnormality present in the GM-CSF nullizygous mice is the accumulation of pulmonary surfactant protein and concomitant opportunistic infection. Elsewhere, GM-CSF has inhibited the steady-state immunosuppressive activity of resident pulmonary macrophages, thereby facilitating local T-cell responsiveness during acute inflammation (291,292). The pulmonary pathology seen in nullizygous GM-CSF mice exhibits many similarities to some forms of human alveolar proteinosis (293).

Another means of addressing the question of GM-CSF's functional importance in vitro is through antisense techniques (294). Incubation with either GM-CSF antisense or GMR antisense led to a reduction in nonerythroid colonies in bone marrow cultures. This effect was seen even in the presence of exogenous GM-CSF, leading the authors to argue for "internal autocrine regulation" of hematopoietic progenitors.

If GM-CSF does have a normal physiological role in modulating the response to infection, then it might be expected that levels would be elevated in patients with bacteremia and febrile neutropenia (295). Surprisingly, this was not so in a well-designed prospective examination that used an immunoassay capable of detecting 20 pg/ml GM-CSF in the serum of such patients (296). It remains feasible that very low levels of GM-CSF could synergize with other cytokines in the circulation during infection, but a paracrine role in the microenvironment seems more likely.

Excess Levels of GM-CSF

Elevation of circulating native or recombinant GM-CSF has been achieved by several means, including treatment of healthy or immunocompromised mice and primates by several routes of delivery; reconstitution of lethally irradiated mice by hematopoietic cells infected with a retrovirus designed to overexpress GM-CSF; and the creation of transgenic mice that constitutively overexpress GM-CSF. A perennial problem for any study of administered cytokines is the question of whether direct causation can be attributed to the agent. Doubts may be raised because in vivo target cells of the administered cytokine may activate a subsidiary network of factors that then generate secondary effects. As will be cited in each of the aforementioned models, in general, excess GM-CSF leads to increased, functionally activated inflammatory effector cells.

Following 6 days of thrice-daily intraperitoneal injections of bacterially synthesized GM-CSF, a dose-related elevation of up to 15-fold in macrophages, and up to 100-fold in neutrophils and eosinophils was demonstrated in the peritoneum of several normal adult mouse strains (297). These peritoneal macrophages exhibited augmentation of phagocytic and mitotic activity (298). The systemic effects of this regimen were less impressive, in that circulating neutrophil counts rose only twofold. However, in distant tissues, such as liver, spleen, and lung, a significant increase in infiltrating neutrophils and macrophages was detected. In bone marrow, levels of neutrophils and monocytes remained constant, whereas a dose-related fall in total cellularity of 40% and in nonerythroid precursors of up to 66% occurred (297).

Continuous intravenous or intermittent subcutaneous infusion of glycosylated or non-glycosylated human GM-CSF in nonhuman primates resulted in leukocytosis—comprising elevations in circulating neutrophils, eosinophils, monocytes, and unexpectedly, lymphocytes (299). As predicted by in vitro experiments, the neutrophils also manifested enhanced phagocytic and bacterial-killing capacity. Platelet and erythrocyte levels did not alter. The effect could be sustained throughout a 1-month infusion, and counts normalized on cessation, without evidence of toxicity (300).

Although the foregoing experiments were designed to examine the effects of pharmacological doses of GM-CSF in animals, alternative studies addressed the question of whether sustained levels of GM-CSF in vast excess would lead to a pathological process or even overt leukemia (243). Enhanced constitutive expression of GM-CSF was achieved by transplanting lethally irradiated mice with hematopoietic cells expressing a GM-CSF gene introduced by retroviral infection in vitro (301). Within a month of transplantation, the mice died, with neutrophil and macrophage infiltration of spleen, lung, liver, peritoneal cavity, skeletal muscle, and heart; the latter two also containing increased eosinophils. GM-CFCs were not elevated in bone marrow or spleen, suggesting that the target population comprised later progenitor cells. Furthermore, the gold standard often used to define leukemia (that is, the ability to yield transplantable tumors in syngeneic recipients) could not be demonstrated in this model.

The transgenic approach to the question of constitutive GM-CSF overexpression has provided a complementary, longer-term model than reconstitution studies (302). In two separate founder lines, GM-CSF was overexpressed exclusively in macrophages, leading to 100-fold elevations in bioactivity detectable in the serum, eye, peritoneal and pleural cavities. Despite marked accumulations of peritoneal, retinal, striated muscle, and pleural macrophages owing to increased local proliferation, peripheral blood and bone marrow progenitor cell counts were normal (303). Peritoneal macrophages in transgenic animals exhibited higher rates of phagocytosis and bacteriolysis compared with macrophages from their normal littermates (304). GM-CSF transgenic animals died prematurely, with weight loss associated with wasting of striated muscle, blindness, and spontaneous intraperitoneal bleeding associated with increased local urokinase-type plasminogen activator activity (305).

A caveat raised earlier in this section relates to whether excess GM-CSF is sufficient to cause the effects described. For example, because GM-CSF is overexpressed in transgenic macrophages that also express its receptor, a potential autocrine loop may have been established (192). GM-CSF induces the in vitro production of TNF-α and IL-1 in monocytes (274,306), and neutrophils (279). Also, analysis by hybridization histochemistry of affected GM-CSF transgenic tissues demonstrated increased expression of TNF-α, IL-1α and β, and basic FGF, compared with littermate controls (307). Thus, the pathological effects of excessive endogenous GM-CSF result from myeloid hyperplasia in certain tissues and functional activation of mature inflammatory effector cells owing to both direct and probable indirect actions. Very high GM-CSF levels do not produce abnormal hematopoietic progenitor cell function or "certifiable" leukemia.

VII. GM-CSF AND ITS RECEPTOR IN THE PATHOGENESIS OF DISEASE

Granulocyte–macrophage colony-stimulating factor is likely involved in the physiological response to infection in the microenvironment; however, to what extent is it implicated in actually causing or exacerbating diseases? In the previous section, several systems of GM-CSF overexpression were discussed—all of which point to the conclusion that GM-CSF is insufficient to produce leukemia. Results presented here from experiments performed on primary leukemic cells and models of leukemic transformation raise the possibility that GM-CSF may, under certain circumstances, act as a protooncogene or, at least, contribute to the pathogenesis of leukemia. Of considerable importance is an understanding of the "autocrine hypothesis" of transformation which, in its simplest form, requires a neoplastic clone to secrete growth factor(s) to which it may respond and, thereby, lead to a self-sustained proliferative loop (243,308).

One prediction of a simple autocrine hypothesis is that spontaneous myeloid leukemias should be growth-factor-independent. However, in vitro analyses of most human myeloid leukemias at diagnosis reveals that they remain factor-dependent and responsive to various cytokines at typical concentrations (230,232,233,309). Genuine cases of autocrine GM-CSF production have been documented in B-lineage acute lymphoblastic leukemia and acute myeloid leukemia (310,311), some expressing large transcripts from rearranged GM-CSF loci (312). Consequently, one group has argued for the existence of intracellular autocrine GM-CSF loops in primary acute myeloid blast cells (313). High rates of myeloid leukemic cell proliferation in serum- and cytokine-free cultures have been correlated with more aggressive disease (314). Because most leukemias appear to be factor-dependent, GM-CSF and other cytokines are likely essential cofactors, either contributing to the in vivo expansion of the transformed myeloid clone itself or amplifying the number of partially transformed cells as targets for further mutations. The latter hypothesis was supported by experiments performed in this laboratory, which depend on the use of the nontumorigenic, factor-dependent FDC-P1 cell line.

After intravenous injection of FDC-P1 cells into irradiated syngeneic (but otherwise normal) mice, transplantable leukemias of donor karyotype developed in all recipients (315,316). In the leukemogenic variants of approximately 80% of the mice, factor-independent cell lines were often established in vitro. One third of the factor-independent FDC-P1 variants were producing either GM-CSF or IL-3 owing to rearrangements associated with transposition of intracisternal A-particles (317). Similar patterns of factor independence were seen with spontaneous and retrovirally induced in vitro mutants of the factor-dependent myeloid precursor cell line D35 (318). In a related series of experiments, enforced expression of GM-CSF by FDC-P1 cells gave rise to autocrine, leukemogenic variants (319). Thus, autocrine production of GM-CSF may make an important contribution to transformation, while being neither necessary nor sufficient for the complete process.

Further support for a role of GM-CSF in facilitating the proliferation of potential target cells (which may then undergo transformation) comes from experiments in which FDC-P1 cells were injected into GM-CSF transgenic mice (320). All GM-CSF transgenic recipients, but no normal littermates, developed transplantable leukemias of FDC-P1 origin, many with rearranged GM-CSF or IL-3 loci, possibly occurring as secondary events. Interestingly, serum GM-CSF was undetectable in GM-CSF transgenic animals that received 10^6 FDC-P1 cells, compared with levels between 500 and 2000 U/ml in transgenic animals that did not receive cells. This suggests that the injected cells may have acted as a GM-CSF "sump," implying that the absence of detectable cytokine in the milieu of autocrine cells may, on occasion, be artifactual.

Constitutive activation of several cytokine receptors may be a means of transformation (321–323). However, the genes and transcripts of GMR-α and GMR-β were intact in a large number of primary human leukemias and cell lines, although the possibility of point mutations was not examined (122). Notably, 65% of the M2 subtype of AML with t(8:21) also display a loss of one sex chromosome (and, therefore, one copy of GMR-α), raising the possibility that GMR-α could act as a recessive oncogene (121,324). Nonetheless, a study of GMR-α nucleotide sequence in several leukemic samples revealed no significant abnormalities (393).

Although no state of GM-CSF hyper- or hyporesponsiveness has as yet been clearly delineated, several clinical disorders implicate GM-CSF in their pathogenesis (325). In polycythemia vera, hyperresponsiveness to GM-CSF and IL-3 (not caused by enhanced receptor binding) was demonstrated in hematopoietic progenitor cells (326). An example of apparent hyporesponsiveness occurs in patients with cyclic neutropenia, in whom peripheral blood and bone marrow progenitor cells oscillate between normal and life-threateningly low, with a periodicity of 3 weeks. Such patients also exhibit cycles in their serum and urine levels of CSFs (309), although the defect may lie in G-CSF or GM-CSF signal transduction. Evidence for a possible signal transduction defect was suggested by the reduced responsiveness of granulocyte–macrophage progenitor cells in these patients, despite normal G-CSF–binding affinity, as well as the cure achieved by bone marrow transplantation (327). Patients with aplastic anemia may also exhibit a reduced responsiveness to GM-CSF (328). In the myelodysplasias, approximately one-third of patients may have elevated GM-CSF serum levels when compared with normal controls, although no abnormality was detected in the levels of serum IL-3 or expression of receptors for GM-CSF and IL-3 (329).

The expression of GM-CSF has been examined in several other nonhematological disorders. Plasma obtained from patients with rheumatoid arthritis and systemic lupus erythematosus contained elevated concentrations of GM-CSF compared with control specimens from patients with noninflammatory arthritis (165,330). In patients with AIDS, modulation of GM-CSF expression was suggested as playing a role in impaired immunity (331) and interstitial lung disease (332). Elevated GM-CSF and other cytokine levels in bronchial epithelial cells and bronchoalveolar lavage fluid may contribute to the persistence of inflammation by eosinophil "priming" in patients with asthma and atopy (333–335).

VIII. CLINICAL APPLICATIONS OF GM-CSF

A. Murine Models

As is often true, murine models for human diseases may provide unexpected insight into therapeutic possibilities. The example of GM-CSF as a potential therapy for human alveolar proteinosis was mentioned earlier in relation to the phenotype seen in GM-CSF nullizygous mice. Mice also have been used to investigate the beneficial radio- and chemoprotective effects of GM-CSF. Although not protective in a single dose following irradiation (336), continuous infusion or multiple injection of GM-CSF before sublethal doses of cytotoxic drugs or irradiation led to accelerated restoration of hematopoiesis (337). Also, GM-CSF was able to synergize with rhIL-1α and rhTNF-α to provide optimal radioprotection, perhaps indicating that cytokine interactions may be important in the hematopoietic recovery following radiation damage.

In murine models of allogeneic bone marrow transplantation, incubation of T-cell–depleted donor graft marrow with GM-CSF ex vivo led to promotion of engraftment (338). One group was unable to demonstrate benefits in posttransplantation mortality, marrow stem cell capacity, or the incidence of graft-versus-host disease (339), whereas later results have been more encouraging (340). In sublethally irradiated rhesus monkeys treated with a continuous infusion of rhGM-CSF following autologous bone marrow transplantation, a sustained and accelerated recovery of neutrophils and platelets was observed (341,342). In monkeys treated with GM-CSF, neutrophils exhibited enhanced superoxide generation and capacity to kill bacteria (299,300).

B. Human Disease

Following phase I and II studies, the initial clinical trials of GM-CSF performed in the mid-1980s concentrated on attempts to hasten myeloid recovery after bone marrow transplantation and chemotherapy. Physicians' increasing experience with GM-CSF, coupled with its pleiotropic activities, have led to examination of GM-CSF as a therapeutic agent in a very broad range of diseases (Table 4). However, by the second half of 1994, licensed indications for GM-CSF in Australia included only chemotherapy-induced neutropenia and bone marrow transplantation. In this section, recent developments concerning potential therapies related to GM-CSF will be introduced, and the reader is referred to excellent reviews summarizing early GM-CSF dose-finding and therapeutic trials (343,356,357).

A major clinical use of GM-CSF is in the context of support following myeloablative therapy before autologous and allogeneic bone marrow transplantation (358). The efficacy is achieved through an acceleration of neutrophil (348) and, possibly, platelet recovery (359), and by reductions in hospital stay (360) and the number of bacterial infections (361,362). Although GM-CSF appears to be cost-effective in the context of bone marrow transplantation, it has not yet been shown to improve survival (363–365).

GM-CSF support during chemotherapy promises the possibilities of higher doses, given more frequently, for longer duration, with lowered morbidity and mortality. The hypothesis of protocols employing this concept of dose intensification is that disease eradication and cure will be facilitated. Such protocols, to date, either with or without peripheral blood progenitor cell support, in solid tumors and hematological malignancies, have been partially successful, and the results of further randomized trials are awaited (see Table 4; 343,344,366).

Novel applications, such as GM-CSF combined with tumor-specific monoclonal antibodies

Table 4 Some Therapeutic Applications of GM-CSF

Application	Ref.
Amelioration of chemotherapy-induced myelotoxicity	343
Dose intensification	344
Radioprotection	345
Allogeneic bone marrow transplantation	346,347
Autologous bone marrow transplantation	348
Peripheral blood stem cell mobilization	349,350
Acquired immunodeficiency syndrome	351,352
Myelodysplasias and aplasias	343
Augmentation of antitumor effects	353
Augmentation of cell-mediated immunity	354
Wound recovery	355

to enhance cell-mediated immunity in the context of bone marrow transplantation and solid tumors, have shown efficacy in early studies (353,367). Of ten different molecules tested in a model of melanoma, GM-CSF was the most potent stimulator of sustained, specific antitumor immunity (368). In a further development, GM-CSF fused to a tumor-derived idiotype from B-cell lymphoma is a strong immunogen—raising the exciting possibility of tailor-made vaccines (369).

C. Cautions

With typical doses of between 5 and 10 µg/kg, administered as a single subcutaneous injection (370–372), patients may experience flushing, fever, bone and muscle pain, tiredness, and skin eruptions (373,374). In general, these side effects are well tolerated and may be treated symptomatically (375). A "first-dose" effect featuring hypoxia and hypotension seen in a small percentage of subjects within several hours of GM-CSF administration may result from activation of secondary cytokines (376–378).

Exacerbations of inflammatory or autoimmune effects have led to the relative contraindication of GM-CSF in diseases such as rheumatoid arthritis (379), thyroiditis (380), hemolytic anemia (381,382), and idiopathic thrombocytopenic purpura (373). In a 2-year-old patient with Hurler's syndrome, who received GM-CSF following failure of allogeneic bone marrow transplantation, and a 6-year-old patient with myelodysplasia, who received combined G-CSF and GM-CSF following high-dose chemotherapy, marked abnormal histiocytic infiltration of bone marrow was noted (383,384). It has also been suggested that GM-CSF used for peripheral blood stem cell mobilization may lead to thrombotic occlusion of apheresis catheters (385).

In addition to the foregoing clinical warnings, theoretical objections to the use of GM-CSF should also be kept in mind. For example, an early concern for recipients of GM-CSF was the potentially catastrophic possibility of complete diversion of their hematopoietic stem cells toward myelopoiesis—which, fortunately, has not been reported in any series. In myelodysplastic syndromes many authors have advocated caution in using GM-CSF for fear of facilitating a proliferative advantage of the malignant clone (386–389). Although this fear has not been sustained by unambiguous clinical examples, increasingly abnormal karyotypes have been noted in some patients with myelodysplasia who received GM-CSF (390,391) and experimental models using GM-CSF transgenic mice also support a prudent approach (320).

NOTE ADDED IN PROOF

Since this review was written, several important advances have been made in the GM-CSF field—notably in signal transduction and vaccine adjuvancy. The reader is referred to recent reviews (394,395).

ACKNOWLEDGMENT

The author thanks Dr. N. A. Nicola for critical reading of the manuscript. Supported by the National Health and Medical Research Council, Canberra, Australia.

REFERENCES

1. La Londe JM, Hanna LS, Rattoballi R, Berman HM, Voet D. Crystallization and preliminary X-ray studies of recombinant human granulocyte–macrophage colony stimulating factor. J Mol Biol 1989; 205:783–785.
2. Diederichs K, Boone T, Karplus PA. Novel fold and putative receptor binding site of granulocyte–macrophage colony-stimulating factor. Science 1991; 254:1779–1782.
3. Minasian E, Nicola NA. A review of cytokine structures. Protein Seq Data Anal 1992; 5:57–64.
4. Schrimsher JL, Rose K, Simona MG, Wingfield P. Characterization of human and mouse granulocyte–macrophage-colony-stimulating factors derived from *Escherichia coli*. Biochem J 1987; 247:195–199.
5. Shanafelt AB, Kastelein RA. Identification of critical regions in mouse granulocyte–macrophage colony-stimulating factor by scanning–deletion analysis. Proc Natl Acad Sci USA 1989; 86:4872–4876.
6. Wong GG, Witek JS, Temple PA, et al. Human GM-CSF: molecular cloning of the complementary DNA and purification of the natural and recombinant proteins. Science 1985; 228:810–815.
7. Cantrell MA, Anderson D, Cerretti DP, et al. Cloning, sequence, and expression of a human granulocyte/macrophage colony-stimulating factor. Proc Natl Acad Sci USA 1985; 82:6250–6254.
8. Lee F, Yokota T, Otsuka T, Gemmell L, Larson N, Luh J, Arai K, Rennick D. Isolation of cDNA for a human granulocyte–macrophage colony-stimulating factor by functional expression in mammalian cells. Proc Natl Acad Sci USA 1985; 82:4360–4364.
9. Gough NM, Metcalf D, Gough J, Grail D, Dunn AR. Structure and expression of the mRNA for murine granulocyte–macrophage colony stimulating factor. EMBO J 1985; 4:645–653.
10. Gough NM, Gough J, Metcalf D, Kelso A, Grail D, Nicola NA, Burgess AW, Dunn AR. Molecular cloning of cDNA encoding a murine haematopoietic growth regulator, granulocyte–macrophage colony stimulating factor. Nature 1984; 309:763–767.
11. DeLamarter JR, Mermod JJ, Liang CM, Eliason JF, Thatcher DR. Recombinant murine GM-CSF from *E. coli* has biological activity and is neutralized by a specific antiserum. EMBO J 1985; 4:2575–2581.
12. Sparrow LG, Metcalf D, Hunkapiller MW, Hood LE, Burgess AW. Purification and partial amino acid sequence of asialo murine granulocyte–macrophage colony stimulating factor. Proc Natl Acad Sci USA 1985; 82:292–296.
13. Wong GG, Witek JS, Temple PA, Wilkens KM, Leary AG, Luxenberg DP, Jones SS, Brown EL, Kay RM, Orr EC. Molecular cloning of human and gibbon T-cell-derived GM-CSF cDNAs and purification of the natural and recombinant proteins. In: Feramisco J, Ozanne B, Stiles C, eds. Cancer Cells. Cold Spring Harbor, NY: Cold Spring Harbor Press, 1985: 235–242.
14. Nash RA, Schuening F, Applebaum F, Hammond WP, Boone T, Morris CF, Slichter SJ, Storb R. Molecular cloning and in vivo evaluation of canine granulocyte–macrophage colony-stimulating factor. Blood 1991; 78:930–937.
15. Maliszewski CR, Schoenborn MA, Cerretti DP, Wignall JM, Picha KS, Cosman D, Tushinski RJ,

Gillis S, Baker PE. Bovine GM-CSF: molecular cloning and biological activity of the recombinant protein. Mol Immunol 1988; 25:843–850.

16. Greenfield RS, Braslawsky GR, Kadow KF, Spitalny GL, Chace D, Bull CO, Bursuker I. Identification of functional domains in murine granulocyte–macrophage colony-stimulating factor using monoclonal antibodies to synthetic peptides. J Immunol 1993; 150:5241–5251.

17. Kaushansky K, Shoemaker SG, Alfaro S, Brown C. Hematopoietic activity of granulocyte/macrophage colony-stimulating factor is dependent upon two distinct regions of the molecule: functional analysis based upon the activities of interspecies hybrid growth factors. Proc Natl Acad Sci USA 1989; 86:1213–1217.

18. Lopez AF, Shannon MF, Hercus T, Nicola NA, Cambareri B, Dottore M, Layton MJ, Eglinton L, Vadas MA. Residue 21 of human granulocyte–macrophage colony-stimulating factor is critical for biological activity and for high but not low affinity binding. EMBO J 1992; 11:909–916.

19. Gough NM, Grail D, Gearing DP, Metcalf D. Mutagenesis of murine granulocyte/macrophage-colony-stimulating factor reveals critical residues near the N terminus. Eur J Biochem 1987; 169:353–358.

20. Shanafelt AB, Johnson KE, Kastelein RA. Identification of critical amino acid residues in human and mouse granulocyte–macrophage colony-stimulating factor and their involvement in species specificity. J Biol Chem 1991; 266:13804–13810.

21. Clark LI, Lopez AF, To LB, Vadas MA, Schrader JW, Hood LE, Kent SB. Structure–function studies of human granulocyte–macrophage colony-stimulating factor. Identification of residues required for activity. J Immunol 1988; 141:881–889.

22. LaBranche CC, Clark SC, Johnson GD, Ornstein D, Sabath DE, Tushinski R, Paetkau V, Prystowsky MB. Deletion of carboxy-terminal residues of murine granulocyte–macrophage colony-stimulating factor results in a loss of biologic activity and altered glycosylation. Arch Biochem Biophys 1990; 276:153–159.

23. Shanafelt AB, Miyajima A, Kitamura T, Kastelein RA. The amino-terminal helix of GM-CSF and IL-5 governs high affinity binding to their receptors. EMBO J 1991; 10:4105–4112.

24. Shanafelt AB, Kastelein RA. High affinity ligand binding is not essential for granulocyte–macrophage colony-stimulating factor receptor activation. J Biol Chem 1992; 267:25466–25472.

25. Meropol NJ, Altmann SW, Shanafelt AB, Kastelein RA, Johnson GD, Prystowsky MB. Requirement of hydrophilic amino-terminal residues for granulocyte–macrophage colony-stimulating factor bioactivity and receptor binding. J Biol Chem 1992; 267:14266–14269.

26. Goodall GJ, Bagley CJ, Vadas MA, Lopez AF. A model for the interaction of the GM-CSF, IL-3 and IL-5 receptors with their ligands. Growth Factors 1993; 8:87–97.

27. Nicola NA, Burgess AW, Metcalf D. Similar molecular properties of granulocyte–macrophage colony-stimulating factors produced by different mouse organs in vitro and in vivo. J Biol Chem 1979; 254:5290–5299.

28. Price V, Mochizuki D, March CJ, Cosman D, Deeley MC, Klinke R, Clevenger W, Gillis S, Baker P, Urdal D. Expression, purification and characterization of recombinant murine granulocyte–macrophage colony-stimulating factor and bovine interleukin-2 from yeast. Gene 1987; 55:287–293.

29. Chiou CJ, Wu MC. Expression of human granulocyte–macrophage colony-stimulating factor gene in insect cells by a baculovirus vector. FEBS Lett 1990; 259:249–253.

30. Hovgaard D, Mortensen BT, Schifter S, Nissen NI. Comparative pharmacokinetics of single-dose administration of mammalian and bacterially-derived recombinant human granulocyte–macrophage colony-stimulating factor. Eur J Haematol 1993. 50:32–36.

31. Wingfield P, Graber P, Moonen P, Craig S, Pain RH. The conformation and stability of recombinant-derived granulocyte–macrophage colony stimulating factors. Eur J Biochem 1988; 173:65–72.

32. Libby RT, Braedt G, Kronheim SR, March CJ, Urdal DL, Chiaverotti TA, Tushinski RJ, Mochizuki DY, Hopp TP, Cosman D. Expression and purification of native human granulocyte–macrophage colony-stimulating factor from an *Escherichia coli* secretion vector. DNA 1987; 6:221–229.

33. Moonen P, Mermod JJ, Ernst JF, Hirschi M, DeLamarter JF. Increased biological activity of

deglycosylated recombinant human granulocyte/macrophage colony-stimulating factor produced by yeast or animal cells. Proc Natl Acad Sci USA 1987; 84:4428–4431.

34. Kaushansky K, O'Hara PJ, Hart CE, Forstrom JW, Hagen FS. Role of carbohydrate in the function of human granulocyte–macrophage colony-stimulating factor. Biochemistry 1987; 26:4861–4867.

35. Dorr RT. Clinical properties of yeast-derived versus *Escherichia coli*-derived granulocyte–macrophage colony-stimulating factor. Clin Ther 1993; 15:19–29.

36. Gearing DP, King JA, Gough NM, Nicola NA. Expression cloning of a receptor for human granulocyte–macrophage colony-stimulating factor. EMBO J 1989; 8:3667–3676.

37. Cosman D, Lyman SD, Idzerda RL, Beckmann MP, Park LS, Goodwin RG, March CJ. A new cytokine receptor superfamily. Trends Biochem Sci 1990; 15:265–270.

38. Bazan JF. Haemopoietic receptors and helical cytokines. Immunol Today 1990; 11:350–354.

39. Bazan JF. Structural design and molecular evolution of a cytokine receptor superfamily. Proc Natl Acad Sci USA 1990; 87:6934–6938.

40. Kitamura T, Sato N, Arai K, Miyajima A. Expression cloning of the human IL-3 receptor cDNA reveals a shared beta subunit for the human IL-3 and GM-CSF receptors. Cell 1991; 66:1165–1174.

41. Tavernier J, Devos R, Cornelis S, Tuypens T, Van DHJ, Fiers W, Plaetinck G. A human high affinity interleukin-5 receptor (IL5R) is composed of an IL5-specific alpha chain and a beta chain shared with the receptor for GM-CSF. Cell 1991; 66:1175–1184.

42. Metcalf D, Nicola NA, Gearing DP, Gough NM. Low-affinity placenta-derived receptors for human granulocyte–macrophage colony-stimulating factor can deliver a proliferative signal to murine hemopoietic cells. Proc Natl Acad Sci USA 1990; 87:4670–4674.

43. Hayashida K, Kitamura T, Gorman DM, Arai K, Yokota T, Miyajima A. Molecular cloning of a second subunit of the receptor for human granulocyte–macrophage colony-stimulating factor (GM-CSF): reconstitution of a high-affinity GM-CSF receptor. Proc Natl Acad Sci USA 1990; 87:9655–9659.

44. Kastelein RA, Shanafelt AB. GM-CSF receptor: interactions and activation. Oncogene 1993; 8:231–236.

45. Park LS, Martin U, Sorensen R, Luhr S, Morrissey PJ, Cosman D, Larsen A. Cloning of the low-affinity murine granulocyte–macrophage colony-stimulating factor receptor and reconstitution of a high-affinity receptor complex. Proc Natl Acad Sci USA 1992; 89:4295–4299.

46. Chiba S, Shibuya K, Piao YF, Tojo A, Sasaki N, Matsuki S, Miyagawa K, Miyazono K, Takaku F. Identification and cellular distribution of distinct proteins forming human GM-CSF receptor. Cell Regul 1990; 1:327–335.

47. Hilton DS, Hilton AA, Raicevic A, Rakar S, Harrison-Smith M, Gough NM, Begley CG, Metcalf D, Nicola NA, Willson TA. Cloning of an interleukin-11 α-subunit: requirement of gp130 for high affinity binding and signal transduction. EMBO J 1994; 13:4765–4775.

48. Lopez AF, Eglinton JM, Gillis D, Park LS, Clark S, Vadas MA. Reciprocal inhibition of binding between interleukin 3 and granulocyte–macrophage colony-stimulating factor to human eosinophils. Proc Natl Acad Sci USA 1989; 86:7022–7026.

49. Lopez AF, Eglinton JM, Lyons AB, Tapley PM, To LB, Park LS, Clark SC, Vadas MA. Human interleukin-3 inhibits the binding of granulocyte–macrophage colony-stimulating factor and interleukin-5 to basophils and strongly enhances their functional activity. J Cell Physiol 1990; 145:69–77.

50. Elliott MJ, Vadas MA, Eglinton JM, Park LS, To LB, Cleland LG, Clark SC, Lopez AF. Recombinant human interleukin-3 and granulocyte–macrophage colony-stimulating factor show common biological effects and binding characteristics on human monocytes. Blood 1989; 74:2349–2359.

51. Park LS, Friend D, Price V, Anderson D, Singer J, Prickett KS, Urdal DL. Heterogeneity in human interleukin-3 receptors. A subclass that binds human granulocyte/macrophage colony stimulating factor. J Biol Chem 1989; 264:5420–5427.

52. Gesner TG, Mufson RA, Norton CR, Turner KJ, Yang YC, Clark SC. Specific binding, internalization, and degradation of human recombinant interleukin-3 by cells of the acute myelogenous, leukemia line, KG-1. J Cell Physiol 1988; 136:493–499.

53. Hara T, Miyajima A. Two distinct functional high affinity receptors for mouse interleukin-3 (IL-3). EMBO J 1992; 11:1875–1884.

54. Takaki S, Mita S, Kitamura T, Yonehara S, Yamaguchi N, Tominaga A, Miyajima A, Takatsu K. Identification of the second subunit of the murine interleukin-5 receptor: interleukin-3 receptor-like protein, AIC2B is a component of the high affinity interleukin-5 receptor. EMBO J 1991; 10:2833–2838.

55. Lopez AF, Elliott MJ, Woodcock J, Vadas MA. GM-CSF, IL-3 and IL-5: cross-competition on human haemopoietic cells. Immunol Today 1992; 13:495–500.

56. Lopez AF, Begley CG, Williamson DJ, Warren DJ, Vadas MA, Sanderson CJ. Murine eosinophil differentiation factor. An eosinophil-specific colony-stimulating factor with activity for human cells. J Exp Med 1986; 163:1085–1099.

57. Lopez AF, To LB, Yang YC, Gamble JR, Shannon MF, Burns GF, Dyson PG, Juttner CA, Clark S, Vadas MA. Stimulation of proliferation, differentiation, and function of human cells by primate interleukin 3. Proc Natl Acad Sci USA 1987; 84:2761–2765.

58. Rothenberg ME, Pomerantz JL, Owen WJ, Avraham S, Soberman RJ, Austen KF, Stevens RL. Characterization of a human eosinophil proteoglycan, and augmentation of its biosynthesis and size by interleukin 3, interleukin 5, and granulocyte/macrophage colony stimulating factor. J Biol Chem 1988; 263:13901–13908.

59. Eder M, Griffin JD, Ernst TJ. The human granulocyte–macrophage colony-stimulating factor receptor is capable of initiating signal transduction in NIH3T3 cells. EMBO J 1993; 12:1647–1656.

60. Ding DX- H, Rivas CI, Heaney ML, Raines MA, Vera JC, Golde DW. The α subunit of the human granulocyte–macrophage colony-stimulating factor receptor signals for glucose transport via a phosphorylation-independent pathway. Proc Natl Acad Sci USA 1994; 91:2537–2541.

61. Ashworth A, Kraft A. Cloning of a potentially soluble receptor for human GM-CSF. Nucleic Acids Res 1990; 18:7178.

62. Raines MA, Liu L, Quan SG, Joe V, DiPersio JF, Golde DW. Identification and molecular cloning of a soluble human granulocyte–macrophage colony-stimulating factor receptor. Proc Natl Acad Sci USA 1991; 88:8203–8207.

63. Crosier KE, Wong GG, Mathey PB, Nathan DG, Sieff CA. A functional isoform of the human granulocyte/macrophage colony-stimulating factor receptor has an unusual cytoplasmic domain. Proc Natl Acad Sci USA 1991; 88:7744–7748.

64. Makishima M, Honma Y, Hozumi M, Nagata N, Motoyoshi K. Differentiation of human monoblastic leukemia U937 cells induced by inhibitors of myosin light chain kinase and prevention of differentiation by granulocyte–macrophage colony-stimulating factor. Biochem Biophys Acta 1993; 1176:245–249.

65. Punt CJ. Regulation of hematopoietic cell function by protein tyrosine kinase-encoding oncogenes, a review. Leuk Res 1992; 16:551–559.

66. Roberts TM. Cell biology. A signal chain of events. Nature 1992; 360:534–535.

67. Hallek M, Druker B, Lepisto EM, Wood KW, Ernst TJ, Griffin JD. Granulocyte–macrophage colony-stimulating factor and steel factor induce phosphorylation of both unique and overlapping signal transduction intermediates in a human factor-dependent hematopoietic cell line. J Cell Physiol 1992; 153:176–186.

68. Argetsinger LS, Campbell GS, Yang X, Witthuhn BA, Silvennoinen O, Ihle JN, Carter SC. Identification of JAK2 as a growth hormone receptor-associated tyrosine kinase. Cell 1993; 74:237–244.

69. Witthuhn BA, Quelle FW, Silvennoinen O, Yi T, Tang B, Miura O, Ihle JN. JAK2 associates with the erythropoietin receptor and is tyrosine phosphorylated and activated following stimulation with erythropoietin. Cell 1993; 74:227–236.

70. Ihle JN, Witthuhn B, Tang B, Yi T, Quelle FW. Cytokine receptors and signal transduction. In: Brenner MK, ed. Cytokines and Growth Factors. London: Bailliere Tindall; 1994:17–48.

71. Miyajima A, Mui A, Ogorochi T, Sakamaki K. The receptors for granulocyte–macrophage colony stimulating factor, interleukin 3, and interleukin 5. Blood 1993; 82:1960–1974.

72. Sato N, Sakamaki K, Terada N, Arai K, Miyajima A. Signal transduction by the high-affinity GM-CSF receptor: two distinct cytoplasmic regions of the common beta subunit responsible for different signalling. EMBO J 1993; 12:4181–4189.

73. Yi T, Mui AL, Krystal G, Ihle JN. Hemopoietic cell phosphatase associates with the interleukin-3 (IL-3) receptor beta chain and down regulates IL-3-induced tyrosine phosphorylation and mitogenesis. Mol Cell Biol 1993; 13:7577–7586.

74. Satoh T, Nakafuku M, Miyajima A, Kaziro Y. Involvement of *ras* p21 protein in signal-transduction pathways from interleukin 2, interleukin 3, and granulocyte/macrophage colony-stimulating factor, but not from interleukin 4. Proc Natl Acad Sci USA 1991; 88:3314–3318.

75. Quelle FW, Quelle DE, Wojchowski DM. Interleukin 3, granulocyte–macrophage colony-stimulating factor, and transfected erythropoietin receptors mediate tyrosine phosphorylation of a common cytosolic protein (pp100) in FDC-ER cells. J Biol Chem 1992; 267:17055–17060.

76. Lilly M, Le T, Holland P, Hendrickson SL. Sustained expression of the pim-1 kinase is specifically induced in myeloid cells by cytokines whose receptors are structurally related. Oncogene 1992; 7:727–732.

77. Hanazono Y, Chiba S, Sasaki K, Mano H, Miyajima A, Arai K, Yazaki Y, Hirai H. c-*fps/fes* protein–tyrosine kinase is implicated in a signaling pathway triggered by granulocyte–macrophage colony-stimulating factor and interleukin-3. EMBO J 1993; 12:1641–1646.

78. Corey S, Eguinoa A, Puyana TK, Bolen JB, Cantley L, Mollinedo F, Jackson TR, Hawkins PT, Stephens LR. Granulocyte macrophage-colony stimulating factor stimulates both association and activation of phosphoinositide 3OH-kinase and *src*-related tyrosine kinase(s) in human myeloid derived cells. EMBO J 1993; 12:2681–2690.

79. Gomez-Cambronero J, Colasanto JM, Huang CK, Sha'afi RI. Direct stimulation by tyrosine phosphorylation of microtubule-associated protein (MAP) kinase activity by granulocyte–macrophage colony-stimulating factor in human neutrophils. Biochem J 1993; 291:211–217.

80. Gomez-Cambronero J, Huang CK, Gomez-Cambronero TM, Waterman WH, Becker EL, Sha'afi RI. Granulocyte–macrophage colony-stimulating factor-induced protein tyrosine phosphorylation of microtubule-associated protein kinase in human neutrophils. Proc Natl Acad Sci USA 1992; 89:7551–7555.

81. Raines MA, Golde DW, Daeipour M, Nel AE. Granulocyte–macrophage colony-stimulating factor activates microtubule-associated protein 2 kinase in neutrophils via a tyrosine kinase-dependent pathway. Blood 1992; 79:3350–3354.

82. Okuda K, Sanghera JS, Pelech SL, Kanakura Y, Hallek M, Griffin JD, Druker BJ. Granulocyte–macrophage colony-stimulating factor, interleukin-3, and steel factor induce rapid tyrosine phosphorylation of p42 and p44 MAP kinase. Blood 1992; 79:2880–2887.

83. Welham MJ, Duronio V, Sanghera JS, Pelech SL, Schrader JW. Multiple hemopoietic growth factors stimulate activation of mitogen-activated protein kinase family members. J Immunol 1992; 149:1683–1693.

84. Yokota T, Watanabe S, Mui AL, Muto A, Miyajima A, Arai K. Reconstitution of functional human GM-CSF receptor in mouse NIH3T3 fibroblasts and BA/F3 proB cells. Leukemia 1993; 7 suppl2, S102–107.

85. Wu J, Wang ML, Sheng Y, Zhu DX. The role of p53 gene in the swith of U937 leukemic cells for growth into differentiation. Shih Yen Sheng Wu Hsueh Pao 1992; 25:311.

86. Larner AC, David M, Feldman GM, Igarashi K, Hackett RH, Webb DS, Sweitzer SM, Petricoin E, Finbloom DS. Tyrosine phosphorylation of DNA binding proteins by multiple cytokines. Science 1993; 261:1730–1733.

87. Silvennoinen O, Schindler C, Schlessinger J, Levy DE. Ras-independent growth factor signalling by transcription factor tyrosine phosphorylation. Science 1993; 261:1736–1739.

88. Lin EY, Orlofsky A, Berger MS, Prystowsky MB. Characterization of *A1*, a novel hemopoietic-specific early-response gene with sequence similarity to *bcl-2*. J Immunol 1993; 151:1979–1988.

89. Bourgoin S, Poubelle PE, Liao NW, Umezawa K, Borgeat P, Naccache PH. Granulocyte–macrophage

colony-stimulating factor primes phospholipase D activity in human neutrophils in vitro: role of calcium, G-proteins and tyrosine kinases. Cell Signal 1992; 4:487–500.

90. Li F, Grant S, Pettit GR, McCrady CW. Bryostatin 1 modulates the proliferation and lineage commitment of human myeloid progenitor cells exposed to recombinant interleukin-3 and recombinant granulocyte–macrophage colony-stimulating factor. Blood 1992; 80:2495–2502.

91. Nishimura M, Kaku K, Azuno Y, Okafuji K, Inoue Y, Kaneko T. Stimulation of phosphainositol turnover and protein kinase C activation by granulocyte–macrophage colony-stimulating factor in HL-60 cells. Blood 1992; 80:1045–1051.

92. Yuo A, Kitagawa S, Azuma E, Natori Y, Togawa A, Saito M, Takaku F. Tyrosine phosphorylation and intracellular alkalinization are early events in human neutrophils stimulated by tumor necrosis factor, granulocyte–macrophage colony-stimulating factor and granulocyte colony-stimulating factor. Biochim Biophys Acta 1993; 1156:197–203.

93. Baxter GT, Miller DL, Kuo RC, Wada HG, Owicki JC. PKC epsilon is involved in granulocyte–macrophage colony-stimulating factor signal transduction: evidence from microphysiometry and antisense oligonucleotide experiments. Biochemistry 1992; 31:10950–10954.

94. Sullivan R. Biochemical effects of human granulocyte–macrophage colony-stimulating factor (GM-CSF) on the human neutrophil. Immunol Ser 1992; 57:485–498.

95. Wada HG, Indelicato SR, Meyer L, Kitamura T, Miyajima A, Kirk G, Muir VC, Parce JW. GM-CSF triggers a rapid, glucose dependent extracellular acidification by TF-1 cells: evidence for sodium/proton antiporter and PKC mediated activation of acid production. J Cell Physiol 1993; 154:129–138.

96. Darnell JE, Kerr IM, Stark GR. Jak–STAT pathways and transcriptional activation in response to IFNs and other extracellular signaling proteins. Science 1994; 264:1415–1421.

97. Barlow DP, Bucan M, Lehrach H, Hogan BL, Gough NM. Close genetic and physical linkage between the murine haemopoietic growth factor genes GM-CSF and multi-CSF (IL3). EMBO J 1987; 6:617–623.

98. Buchberg AM, Bedigian HG, Taylor BA, Brownell E, Ihle JN, Nagata S, Jenkins NA, Copeland NG. Localization of *Evi-2* to chromosome 11: linkage to other proto-oncogene and growth factor loci using interspecific backcross mice. Oncogene Res 1988; 2:149–165.

99. Wilson SD, Billings PR, D'Eustachio P, Fournier RE, Geissler E, Lalley PA, Burd PR, Housman DE, Taylor BA, Dorf ME. Clustering of cytokine genes on mouse chromosome 11. J Exp Med 1990; 171:1301–1314.

100. Lee JS, Young IG. Fine-structure mapping of the murine IL-3 and GM-CSF genes by pulsed-field gel electrophoresis and molecular cloning. Genomics 1989; 5:359–362.

101. Frolova EI, Dolganov GM, Mazo IA, Smirnov DV, Copeland P, Stewart C, O'Brien SJ, Dean M. Linkage mapping of the human CSF2 and IL3 genes. Proc Natl Acad Sci USA 1991; 88:4821–4824.

102. van Leeuwen BH, Martinson ME, Webb GC, Young IG. Molecular organization of the cytokine gene cluster, involving the human IL-3, IL-4, IL-5, and GM-CSF genes, on human chromosome 5. Blood 1989; 73:1142–1148.

103. Huebner K, Isobe M, Croce CM, Golde DW, Kaufman SE, Gasson JC. The human gene encoding GM-CSF is at 5q21-q32, the chromosome region deleted in the 5q⁻ anomaly. Science 1985; 230:1282–1285.

104. Le Beau MM, Westbrook CA, Diaz MO, Larson RA, Rowley JD, Gasson JC, Golde DW, Sherr CJ. Evidence for the involvement of GM-CSF and FMS in the deletion (5q) in myeloid disorders. Science 1986; 231:984–987.

105. Huebner K, Nagarajan L, Besa E, et al. Order of genes on human chromosome 5q with respect to 5q interstitial deletions. Am J Hum Genet 1990; 46:26–36.

106. Stanley E, Metcalf D, Sobieszczuk P, Gough NM, Dunn AR. The structure and expression of the murine gene encoding granulocyte–macrophage colony stimulating factor: evidence for utilisation of alternative promoters. EMBO J 1985; 4:2569–2573.

107. Miyatake S, Otsuka T, Yokota T, Lee F, Arai K. Structure of the chromosomal gene for granulocyte–

macrophage colony stimulating factor: comparison of the mouse and human genes. EMBO J 1985; 4:2561–2568.

108. Yang YC, Kovacic S, Kriz R, Wolf S, Clark SC, Wellems TE, Nienhuis A, Epstein N. The human genes for GM-CSF and IL 3 are closely linked in tandem on chromosome 5. Blood 1988; 71:958–961.

109. Cockerill PN, Shannon MF, Bert AG, Ryan GR, Vadas MA. The *GM-CSF/IL3* locus is regulated by an inducible cyclosporin A sensitive enhancer. Proc Natl Acad Sci USA 1993; 90:2466–2470.

110. Gough NM, Kelso A. GM-CSF expression is preferential to multi-CSF (IL-3) expression in murine T lymphocyte clones. Growth Factors 1989; 1:287–298.

111. Nagarajan L, Lange B, Cannizzaro L, Finan J, Nowell PC, Huebner K. Molecular anatomy of a 5q interstitial deletion. Blood 1990; 75:82–87.

112. Grimwade DJ, Stephenson J, De SC, Dalton RG, Mufti GJ. Familial MDS with 5q⁻ abnormality. Br J Haematol 1993; 84:536–538.

113. Le Beau MM, Pettenati MJ, Lemons RS, Diaz MO, Westbrook CA, Larson RA, Sherr CJ, Rowley JD. Assignment of the GM-CSF, CSF-1, and FMS genes to human chromosome 5 provides evidence for linkage of a family of genes regulating hematopoiesis and for their involvement in the deletion (5q) in myeloid disorders. Cold Spring Harb Symp Quant Biol 1986; 2:899–909.

114. Van den Berghe H, Vermaelen K, Mecucci C, Barbieri D, Tricot G. The 5q⁻ anomaly. Cancer Genet Cytogenet 1985; 17:189–255.

115. Harada H, Kitagawa M, Tanaka N, Yamamoto H, Harada K, Ishihara M, Taniguchi T. Anti-oncogenic and oncogenic potentials of interferon regulatory factors-1 and -2. Science 1993; 259:971–974.

116. Willman CL, Sever CE, Pallavicini MG, et al. Deletion of *IRF-1*, mapping to chromosome 5q31.1, in human leukemia and preleukemic myelodysplasia. Science 1993; 259:968–971.

117. Boultwood J, Fidler C, Lewis S, MacCarthy A, Sheridan H, Kelly S, Oscier D, Buckle VJ, Wainscoat JS. Allelic loss of *IRF1* in myelodysplasia and acute myeloid leukemia: retention of *IRF1* on the 5q⁻ chromosome in some patients with the 5q⁻ syndrome. Blood 1993; 82:2611–2616.

118. Shen Y, Baker E, Callen DF, Sutherland GR, Willson TA, Rakar S, Gough NM. Localization of the human GM-CSF receptor beta chain gene (*CSF2RB*) to chromosome 22q12.2 → q13.1. Cytogenet Cell Genet 1992; 61:175–177.

119. Lion T, Haas OA, Harbott J, Bannier E, Ritterbach J, Jankovic M, Fink FM, Stojimirovic A, Herrmann J, Riehm HJ, Lampert F, Ritter J, Koch H, Gadner H. The translocation t(1;22)(p13;q13) is a nonrandom marker specifically associated with acute megakaryocytic leukemia in young children. Blood 1992; 79:3325–3330.

120. Gorman DM, Itoh N, Jenkins NA, Gilbert DJ, Copeland NG, Miyajima A. Chromosomal localization and organization of the murine genes encoding the beta subunits (*AIC2A* and *AIC2B*) of the interleukin 3, granulocyte/macrophage colony-stimulating factor, and interleukin 5 receptors. J Biol Chem 1992; 267:15842–15848.

121. Gough NM, Gearing DP, Nicola NA, Baker E, Pritchard M, Callen DF, Sutherland GR. Localization of the human GM-CSF receptor gene to the X-Y pseudoautosomal region. Nature 1990; 345:734–736.

122. Brown MA, Gough NM, Willson TA, Rockman S, Begley CG. Structure and expression of the GM-CSF receptor α and β chain genes in human leukemia. Leukemia 1993; 7:63–74.

123. Ellison J, Passage M, Yu LC, Yen P, Mohondas TK, Shapiro L. Direct isolation of human genes that escape X inactivation. Somatic Cell Mol Genet 1992; 18:259–268.

124. Rappold G, Willson TA, Henke A, Gough NM. Arrangement and localization of the human GM-CSF receptor α chain gene within the X-Y pseudoautosomal region. Genomics 1992; 14:455–461.

125. Gelin C, Aubrit F, Phalipon A, Raynal B, Cole S, Kaczorek M, Bernard A. The E2 antigen, a 32 kd glycoprotein involved in T-cell adhesion processes, is the *MIC2* gene product. EMBO J 1989; 8:3253–3259.

126. Slim R, Levilliers J, Ludecke H-J, Claussen U, Nguyen V-C, Gough NM, Horsthemke B, Petit C. A human pseudoautosomal gene encodes the ANT3 ADP/ATP translocase and escapes X-inactivation. Genomics 1993; 16:26–33.

127. Kremer E, Baker E, D'Andrea RJ, Slim R, Phillips H, Moretti PAB, Lopez A, Petit C, Vadas MA, Sutherland GR, Goodall GJ. A cytokine receptor gene cluster in the X-Y pseudoautosomal region? Blood 1993; 82:22–28.

128. Disteche CM, Brannan CI, Larsen A, Adler DA, Schorderet DG, Gearing D, Copeland NG, Jenkins NA, Park LS. The human pseudoautosomal GM-CSF receptor and subunit gene is autosomal in mouse. Nature Genet 1992; 1:333–336.

129. Chan JY, Slamon DJ, Nimer SD, Golde DW, Gasson JC. Regulation of expression of human granulocyte/macrophage colony-stimulating factor. Proc Natl Acad Sci USA 1986; 83:8669–8673.

130 Bickel M, Cohen RB, Pluznik DH. Post-transcriptional regulation of granulocyte–macrophage colony-stimulating factor synthesis in murine T cells. J Immunol 1990; 145:840–845.

131. Thorens B, Mermod JJ, Vassalli P. Phagocytosis and inflammatory stimuli induce GM-CSF mRNA in macrophages through posttranscriptional regulation. Cell 1987; 48:671–679.

132. Akahane K, Cohen RB, Bickel M, Pluznik DH. IL-1 alpha induces granulocyte–macrophage colony-stimulating factor gene expression in murine B lymphocyte cell lines via mRNA stabilization. J Immunol 1991; 146:4190–4196.

133. Kaushansky K. Control of granulocyte–macrophage colony-stimulating factor production in normal endothelial cells by positive and negative regulatory elements. J Immunol 1989; 143:2525–2529.

134. Razanajaona D, Maroc C, Lopez M, Mannoni P, Gabert J. Shift from posttranscriptional to predominant transcriptional control of the expression of the GM-CSF gene during activation of human Jurkat cells. Cell Growth Differ 1992; 3:299–305.

135. Ernst TJ, Ritchie AR, O'Rourke R, Griffin JD. Colony-stimulating factor gene expression in human acute myeloblastic leukemia cells is posttranscriptionally regulated. Leukemia 1989; 3:620–625.

136. Hahn S, Wodnar FA, Nair AP, Moroni C. *Ras* oncogenes amplify lymphokine (interleukin 3, granulocyte–macrophage colony-stimulating factor) induction by calcium ionophore. Oncogene 1991; 6:2327–2332.

137. Gough NM, Nicola NA. Granulocyte–macrophage colony stimulating factor. In: Dexter TM, Garland JM, Testa NG, eds. Colony-Stimulating Factors, New York: Marcel Dekker, 190:111–153.

138. Shaw G, Kamen R. A conserved AU sequence from the 3′ untranslated region of GM-CSF mRNA mediates selective mRNA degradation. Cell 1986; 46:659–667.

139. Iwai Y, Akahane K, Pluznik DH, Cohen RB. Ca^{2+} ionophore A23187-dependent stabilization of granulocyte–macrophage colony-stimulating factor messenger RNA in murine thymoma EL-4 cells is mediated through two distinct regions in the 3′-untranslated region. J Immunol 1993; 150:4386–4394.

140. Iwai Y, Bickel M, Pluznik DH, Cohen RB. Identification of sequences within the murine granulocyte–macrophage colony-stimulating factor mRNA 3′-untranslated region that mediate mRNA stabilization induced by mitogen treatment of EL-4 thymoma cells. J Biol Chem 1991; 266:17959–17965.

141. Akashi M, Shaw G, Gross M, Saito M, Koeffler HP. Role of AUUU sequences in stabilization of granulocyte–macrophage colony-stimulating factor RNA in stimulated cells. Blood 1991; 78:2005–2012.

142. Kelso A, Gough N. Expression of hemopoietic growth factor genes in murine T lymphocytes. In: Webb DR, Goeddel DV, eds. The Lymphokines, New York: Academic Press, 1987:209–238.

143. Shannon MF, Gamble JR, Vadas MA. Nuclear proteins interacting with the promoter region of the human granulocyte/macrophage colony-stimulating factor gene. Proc Natl Acad Sci USA 1988; 85:674–678.

144. Nimer SD, Morita EA, Martis MJ, Wachsman W, Gasson JC. Characterization of the human granulocyte–macrophage colony-stimulating factor promoter region by genetic analysis: correlation with DNase I footprinting. Mol Cell Biol 1988; 8:1979–1984.

145. Nimer S, Fraser J, Richards J, Lynch M, Gasson J. The repeated sequence CATT(A/T) is required for granulocyte–macrophage colony-stimulating factor promoter activity. Mol Cell Biol 1990; 10:6084–6088.

146. Nishida J, Yoshida M, Arai K, Yokota T. Definition of a GC-rich motif as regulatory sequence of the

human IL-3 gene: coordinate regulation of the IL-3 gene by CLE2/GC box of the GM-CSF gene in T cell activation. Int Immunol 1991; 3:245–254.

147. Miyatake S, Seiki M, Yoshida M, Arai K. T-cell activation signals and human T-cell leukemia virus type I-encoded p40x protein activate the mouse granulocyte–macrophage colony-stimulating factor gene through a common DNA element. Mol Cell Biol 1988; 8:5581–5587.

148. Shannon MF, Pell LM, Lenardo MJ, Kuczek ES, Occhiodoro FS, Dunn SM, Vadas MA. A novel tumor necrosis factor-responsive transcription factor which recognizes a regulatory element in hemopoietic growth factor genes. Mol Cell Biol 1990; 10:2950–2959.

149. Tsuboi A, Sugimoto K, Yodoi J, Miyatake S, Arai K, Arai N. A nuclear factor NF-GM2 that interacts with a regulatory region of the GM-CSF gene essential for its induction in responses to T-cell activation: purification from human T-cell leukemia line Jurkat cells and similarity to NF-kappa B. Int Immunol 1991; 3:807–817.

150. Sugimoto K, Tsuboi A, Miyatake S, Arai K, Arai N. Inducible and non-inducible factors co-operatively activate the GM-CSF promoter by interacting with two adjacent DNA motifs. Int Immunol 1990; 2:787–794.

151. Miyatake S, Shlomai J, Arai K, Arai N. Characterization of the mouse granulocyte–macrophage colony-stimulating factor (GM-CSF) gene promoter: nuclear factors that interact with an element shared by three lymphokine genes—those for GM-CSF, interleukin-4 (IL-4), and IL-5. Mol Cell Biol 1991; 11:5894–5901.

152. Kuczek ES, Shannon MF, Pell LM, Vadas MA. A granulocyte–colony-stimulating factor gene promoter element responsive to inflammatory mediators is functionally distinct from an identical sequence in the granulocyte–macrophage colony-stimulating factor gene. J Immunol 1991; 146:2426–2433.

153. Schreck R, Baeuerle PA. NF-kappa B as inducible transcriptional activator of the granulocyte–macrophage colony-stimulating factor gene. Mol Cell Biol 1990; 10:1281–1286.

154. Nomiyama H, Hieshima K, Hirokawa K, Hattori T, Takatsuki K, Miura R. Characterization of cytokine LD78 gene promoters: positive and negative transcriptional factors bind to a negative regulatory element common to LD78, interleukin-3, and granulocyte–macrophage colony-stimulating factor gene promoters. Mol Cell Biol 1993; 13:2787–2801.

155. Burgess AW, Metcalf D, Sparrow LG, Simpson RJ, Nice EC. Granulocyte/macrophage colony-stimulating factor from mouse lung conditioned medium. Purification of multiple forms and radioiodination. Biochem J 1986; 235:805–814.

156. Zupo S, Perussia B, Baldi L, Corcione A, Dono M, Ferrarini M, Pistoia V. Production of granulocyte–macrophage colony-stimulating factor but not IL-3 by normal and neoplastic human B lymphocytes. J Immunol 1992; 148:1423–1430.

157. Pluznik DH, Bickel M, Mergenhagen SE. B lymphocyte derived hematopoietic growth factors. Immunol Invest 1989; 18:103–116.

158. Fibbe WE, Kluck PM, Duinkerken N, Voogt PJ, Willemze R, Falkenburg JH. Factors influencing release of granulocyte–macrophage colony-stimulating activity from human mononuclear phagocytes. Eur J Haematol 1988; 41:352–358.

159. Rich IN. A role for the macrophage in normal hemopoiesis. I. Functional capacity of bone-marrow-derived macrophages to release hemopoietic growth factors. Exp Hematol 1986; 14:738–745.

160. Demetri GD, Zenzie BW, Rheinwald JG, Griffin JD. Expression of colony-stimulating factor genes by normal human mesothelial cells and human malignant mesothelioma cells lines in vitro. Blood 1989; 74:940–946.

161. Kupper TS, Lee F, Coleman D, Chodakewitz J, Flood P, Horowitz M. Keratinocyte derived T-cell growth factor (KTGF) is identical to granulocyte macrophage colony stimulating factor (GM-CSF). J Invest Dermatol 1988; 91:185–188.

162. Wodnar Filipowicz A, Heusser CH, Moroni C. Production of the haemopoietic growth factors GM-CSF and interleukin-3 by mast cells in response to IgE receptor-mediated activation. Nature 1989; 339:150–152.

163. Weir EC, Insogna KL, Horowitz MC. Osteoblast-like cells secrete granulocyte–macrophage colony-

stimulating factor in response to parathyroid hormone and lipopolysaccharide. Endocrinology 1989; 124:899–904.

164. Robertson SA, Mayrhofer G, Seamark RF. Uterine epithelial cells synthesize granulocyte–macrophage colony-stimulating factor and interleukin-6 in pregnant and nonpregnant mice. Biol Reprod 1992; 46:1069–1079.

165. Agro A, Jordana M, Chan KH, Cox G, Richards C, Stepien H, Stanisz AM. Synoviocyte derived granulocyte macrophage colony stimulating factor mediates the survival of human lymphocytes. J Rheumatol 1992; 19:1065–1069.

166. Plaut M, Pierce JH, Watson CJ, Hanley HJ, Nordan RP, Paul WE. Mast cell lines produce lymphokines in response to cross-linkage of Fc epsilon RI or to calcium ionophores. Nature 1989; 339:64–67.

167. Yamato K, El HZ, Kuo JF, Koeffler HP. Granulocyte–macrophage colony-stimulating factor: signals for its mRNA accumulation. Blood 1989; 74:1314–1320.

168. Tobler A, Marti HP, Gimmi C, Cachelin AB, Saurer S, Fey MF. Dexamethasone and 1,25-dihydroxyvitamin D3, but not cyclosporine A, inhibit production of granulocyte–macrophage colony-stimulating factor in human fibroblasts. Blood 1991; 77:1912–1918.

169. Koury MJ, Pragnell IB. Retroviruses induce granulocyte–macrophage colony stimulating activity in fibroblasts. Nature 1982; 299:638–640.

170. Koury MJ, Balmain A, Pragnell IB. Induction of granulocyte–macrophage colony-stimulating activity in mouse skin by inflammatory agents and tumor promoters. EMBO J 1983; 2:1877–1882.

171. Kaushansky K, Lin N, Adamson JW. Interleukin 1 stimulates fibroblasts to synthesize granulocyte–macrophage and granulocyte colony-stimulating factors. Mechanism for the hematopoietic response to inflammation. J Clin Invest 1988; 81:92–97.

172. Munker R, Gasson J, Ogawa M, Koeffler HP. Recombinant human TNF induces production of granulocyte-monocyte colony-stimulating factor. Nature 1986; 323:79–82.

173. Zucali JR, Dinarello CA, Oblon DJ, Gross MA, Anderson L, Weiner RS. Interleukin 1 stimulates fibroblasts to produce granulocyte–macrophage colony-stimulating activity and prostaglandin E2. J Clin Invest 1986; 77:1857–1863.

174. Szczylik C, Skorski T, Ku DH, Nicolaides NC, Wen SC, Rudnicka L, Bonati A, Malaguarnera L, Calabretta B. Regulation of proliferation and cytokine expression of bone marrow fibroblasts: role of c-*myb*. J Exp Med 1993; 178:997–1005.

175. Mano H, Nishida J, Usuki K, Maru Y, Kobayashi Y, Hirai H, Okabe T, Urabe A, Takaku F. Constitutive expression of the granulocyte–macrophage colony-stimulating factor gene in human solid tumors. Jpn J Cancer Res 1987; 78:1041–1043.

176. Pisa P, Halapi E, Pisa EK, Gerdin E, Hising C, Bucht A, Gerdin B, Kiessling R. Selective expression of interleukin 10, interferon gamma, and granulocyte–macrophage colony-stimulating factor in ovarian cancer biopsies. Proc Natl Acad Sci USA 1992; 89:7708–7712.

177. Kittler EL, McGrath H, Temeles D, Crittenden RB, Kister VK, Quesenberry PJ. Biologic significance of constitutive and subliminal growth factor production by bone marrow stroma. Blood 1992; 79:3168–3178.

178. Rajavashisth TB, Andalibi A, Territo MC, Berliner JA, Navab M, Fogelman AM, Lusis AJ. Induction of endothelial cell expression of granulocyte and macrophage colony-stimulating factors by modified low-density lipoproteins. Nature 1990; 344:254–257.

179. Broudy VC, Kaushansky K, Harlan JM, Adamson JW. Interleukin 1 stimulates human endothelial cells to produce granulocyte–macrophage colony-stimulating factor and granulocyte colony-stimulating factor. J Immunol 1987; 139:464–468.

180. Broudy VC, Kaushansky K, Segal GM, Harlan JM, Adamson JW. Tumor necrosis factor type alpha stimulates human endothelial cells to produce granulocyte/macrophage colony-stimulating factor. Proc Natl Acad Sci USA 1986; 83:7467–7471.

181. Seelentag WK, Mermod JJ, Montesano R, Vassalli P. Additive effects of interleukin 1 and tumour necrosis factor-alpha on the accumulation of the three granulocyte and macrophage colony-stimulating factor mRNAs in human endothelial cells. EMBO J 1987; 6:2261–2265.

182. Bagby GJ, Dinarello CA, Wallace P, Wagner C, Hefeneider S, McCall E. Interleukin 1 stimulates granulocyte macrophage colony-stimulating activity release by vascular endothelial cells. J Clin Invest 1986; 78:1316–1323.

183. Sieff CA, Tsai S, Faller DV. Interleukin 1 induces cultured human endothelial cell production of granulocyte–macrophage colony-stimulating factor. J Clin Invest 1987; 79:48–51.

184. Zsebo KM, Yuschenkoff VN, Schiffer S, Chang D, McCall E, Dinarello CA, Brown MA, Altrock B, Bagby GJ. Vascular endothelial cells and granulopoiesis: interleukin-1 stimulates release of G-CSF and GM-CSF. Blood 1988; 71:99–103.

185. Malone DG, Pierce JH, Falko JP, Metcalfe DD. Production of granulocyte–macrophage colony-stimulating factor by primary cultures of unstimulated rat microvascular endothelial cells. Blood 1988; 71:684–689.

186. Akahane K, Pluznik DH. Interferon-gamma destabilizes interleukin-1-induced granulocyte–macrophage colony-stimulating factor mRNA in murine vascular endothelial cells. Exp Hematol 1993; 21:878–884.

187. Kelso A, Troutt AB, Maraskovsky E, Gough NM, Morris L, Pech MH, Thomson JA. Heterogeneity in lymphokine profiles of CD4$^+$ and CD8$^+$ T cells and clones activated in vivo and in vitro. Immunol Rev 1991; 123:85–114.

188. Imboden JB, Weiss A, Stobo JB. Transmembrane signalling by the T3-antigen receptor complex. Immunol Today 1985; 6:328–331.

189. Kelso A, Gough NM. Differential inhibition by cyclosporin A reveals two pathways for activation of lymphokine synthesis in T cells. Growth Factors 1989; 1:165–177.

190. Hamilton TA, Adams DO. Molecular mechanisms of signal transduction in macrophages. Immunol Today 1987; 8:151–158.

191. Tannenbaum CS, Koerner TJ, Jansen MM, Hamilton TA. Characterization of lipopolysaccharide-induced macrophage gene expression. J Immunol 1988; 140:3640–3645.

192. Nicola NA. Why do hemopoietic growth factor receptors interact with each other? Immunol Today 1987; 8:134–140.

193. DiPersio JF, Abboud CN. Activation of neutrophils by granulocyte–macrophage colony-stimulating factor. Immunol Ser 1992; 57:457–484.

194. Walker F, Burgess AW. Specific binding of radioiodinated granulocyte–macrophage colony-stimulating factor to hemopoietic cells. EMBO J 1985; 4:933–939.

195. Park LS, Friend D, Gillis S, Urdal DL. Characterization of the cell surface receptor for granulocyte–macrophage colony-stimulating factor. J Biol Chem 1986; 261:4177–4183.

196. Walker F, Burgess AW. Internalisation and recycling of the granulocyte–macrophage colony-stimulating factor (GM-CSF) receptor on a murine myelomonocytic leukemia. J Cell Physiol 1987; 130:255–261.

197. Metcalf D. The Wellcome Foundation Lecture, 1986. The molecular control of normal and leukaemic granulocytes and macrophages. Proc R Soc Lond Biol 1987; 230:389–423.

198. Metcalf D. The Molecular Control of Blood Cells. Cambridge MA: Harvard University Press, 1988.

199. Gasson JC. Molecular physiology of granulocyte–macrophage colony-stimulating factor. Blood 1991; 77:1131–1145.

200. Boise LH, Gonzalez GM, Postema CE, Ding L, Lindsten T, Turka LA, Mao X, Nunez G, Thompson CB. bcl-x, a bcl-2-related gene that functions as a dominant regulator of apoptotic cell death. Cell 1993; 74:597–608.

201. Oltvai ZN, Milliman CL, Korsmeyer SJ. Bcl-2 heterodimerizes in vivo with a conserved homolog, Bax, that accelerates programmed cell death. Cell 1993; 74:609–619.

202. Vaux DL, Cory S, Adams JM. Bcl-2 gene promotes haemopoietic cell survival and cooperates with c-myc to immortalize pre-B cells. Nature 1988; 335:440–442.

203. Dancey JT, Deubelbeiss KA, Harker LA, Finch CA. Neutrophil kinetics in man. J Clin Invest 1976; 58:705.

204. Colotta F, Re F, Polentarutti N, Sozzani S, Mantovani A. Modulation of granulocyte survival and programmed cell death by cytokines and bacterial products. Blood 1992; 80:2012–2020.

205. Begley CG, Lopez AF, Nicola NA, Warren DJ, Vadas MA, Sanderson CJ, Metcalf D. Purified colony-stimulating factors enhance the survival of human neutrophils and eosinophils in vitro: a rapid and sensitive microassay for colony-stimulating factors. Blood 1986; 68:162–166.

206. Yamaguchi M, Hirai K, Morita Y, Takaishi T, Ohta K, Suzuki S, Motoyoshi K, Kawanami O, Ito K. Hemopoietic growth factors regulate the survival of human basophils in vitro. Int Arch Allergy Appl Immunol 1992; 97:322–329.

207. Metcalf D, Merchav S. Effects of GM-CSF deprivation on precursors of granulocytes and macrophages. J Cell Physiol 1982; 112:411–418.

208. Nicola NA, Metcalf D. Analysis of purified fetal liver hemopoietic progenitor cells in liquid culture. J Cell Physiol 1982; 112:257–264.

209. Williams GT, Smith CA, Spooncer E, Dexter TM, Taylor DR. Haemopoietic colony stimulating factors promote cell survival by suppressing apoptosis. Nature 1990; 343:76–79.

210. Nunez G, London L, Hockenbery D, Alexander M, McKearn JP, Korsmeyer SJ. Deregulated *Bcl-2* gene expression selectively prolongs survival of growth factor-deprived hemopoietic cell lines. J Immunol 1990; 144:3602–3610.

211. Burgess AW, Nicola NA, Johnson GR, Nice EC. Colony-forming cell proliferation: a rapid and sensitive assay system for murine granulocyte and macrophage colony-stimulating factors. Blood 1982; 60:1219–1223.

212. Lotem J, Sachs L. Hematopoietic cytokines inhibit apoptosis induced by transforming growth factor beta 1 and cancer chemotherapy compounds in myeloid leukemic cells. Blood 1992; 80:1750–1757.

213. Bodine DM, Crosier PS, Clark SC. Effects of hematopoietic growth factors on the survival of primitive stem cells in liquid suspension culture. Blood 1991; 78:914–920.

214. Rajotte D, Haddad P, Haman A, Cragoe EJ, Hoang T. Role of protein kinase C and the Na^+/H^+ antiporter in suppression of apoptosis by granulocyte macrophage colony-stimulating factor and interleukin-3. J Biol Chem 1992; 267:9980–9987.

215. Metcalf D. The Colony Stimulating Factors. Amsterdam: Elsevier, 1984.

216. Hogge DE, Sutherland HJ, Cashman JD, Lansdorp PM, Humphries RK, Eaves CJ. Cytokines acting early in human haematopoiesis. In: Brenner MK, ed. Cytokines and Growth Factors. London: Bailliere Tindall, 1994:49–63.

217. Fan K, Ruan Q, Sensenbrenner L, Chen BD. Up-regulation of granulocyte–macrophage colony-stimulating factor (GM-CSF) receptors in murine peritoneal exudate macrophages by both GM-CSF and IL-3. J Immunol 1992; 149:96–102.

218. Metcalf D. Clonal analysis of proliferation and differentiation of paired daughter cells: action of granulocyte–macrophage colony-stimulating factor on granulocyte–macrophage precursors. Proc Natl Acad Sci USA 1980; 77:5327–5330.

219. Lardon F, Van BD, Snoeck HW, Peetermans ME. Quantitative cell-cycle progression analysis of the first three successive cell cycles of granulocyte colony-stimulating factor and/or granulocyte–macrophage colony-stimulating factor-stimulated human $CD34^+$ bone marrow cells in relation to their colony formation. Blood 1993; 81:3211–3216.

220. Koike K, Ogawa M, Ihle JN, Miyake T, Shimizu T, Miyajima A, Yokota T, Akai K. Recombinant murine granulocyte–macrophage (GM) colony-stimulating factor supports formation of GM and multipotential blast cell colonies in culture: comparison with the effects of interleukin-3. J Cell Physiol 1987; 131:458–464.

221. Robinson BE, McGrath HE, Quesenberry PJ. Recombinant murine granulocyte macrophage colony-stimulating factor has megakaryocyte colony-stimulating activity and augments megakaryocyte colony stimulation by interleukin 3. J Clin Invest 1987; 79:1648–1652.

222. Metcalf D, Burgess AW, Johnson GR, Nicola NA, Nice EC, DeLamarter J, Thatcher DR, Mermod JJ. In vitro actions on hemopoietic cells of recombinant murine GM-CSF purified after production in *Escherichia coli*: comparison with purified native GM-CSF. J Cell Physiol 1986; 128:421–431.

223. Mazur EM, Cohen JL, Wong GG, Clark SC. Modest stimulatory effect of recombinant human GM-CSF on colony growth from peripheral blood human megakaryocyte progenitor cells. Exp Hematol 1987; 15:1128–1133.

224. Migliaccio AR, Bruno M, Migliaccio G. Evidence for direct action of human biosynthetic (recombinant) GM-CSF on erythroid progenitors in serum-free culture. Blood 1987; 70:1867–1871.

225. Dexter TM, Garland J, Scott D, Scolnick E, Metcalf D. Growth of factor-dependent hemopoietic precursor cell lines. J Exp Med 1980; 152:1036–1047.

226. Morgan C, Pollard JW, Stanley ER. Isolation and characterization of a cloned growth factor dependent macrophage cell line, BAC1.2F5. J Cell Physiol 1987; 130:420–427.

227. Hara K, Suda T, Suda J, Eguchi M, Ihle JN, Nagata S, Miura Y, Saito M. Bipotential murine hemopoietic cell line (NFS-60) that is responsive to IL-3, GM-CSF, G-CSF, and erythropoietin. Exp Hematol 1988; 16:256–261.

228. Valtieri M, Santoli D, Caracciolo D, Kreider BL, Altmann SW, Tweardy DJ, Gemperlein I, Mavilio F, Lange B, Rovera G. Establishment and characterization of an undifferentiated human T leukemia cell line which requires granulocyte–macrophage colony stimulatory factor for growth. J Immunol 1987; 138:4042–4050.

229. Kupper T, Flood P, Coleman D, Horowitz M. Growth of an interleukin 2/interleukin 4-dependent T cell line induced by granulocyte–macrophage colony-stimulating factor (GM-CSF). J Immunol 1987; 138:4288–4292.

230. Begley CG, Metcalf D, Nicola NA. Primary human myeloid leukemia cells: comparative responsiveness to proliferative stimulation by GM-CSF or G-CSF and membrane expression of CSF receptors. Leukemia 1987; 1:1–8.

231. Asano Y, Shibuya T, Okamura S, Yamaga S, Otsuka T, Niho Y. Effect of human recombinant granulocyte/macrophage colony-stimulating factor on clonogenic leukemic blast cells. Cancer Res 1987; 47:5647–5648.

232. Vellenga E, Young DC, Wagner K, Wiper D, Ostapovicz D, Griffin JD. The effects of GM-CSF and G-CSF in promoting growth of clonogenic cells in acute myeloblastic leukemia. Blood 1987; 69:1771–1776.

233. Kelleher C, Miyauchi J, Wong G, Clark S, Minden MD, McCulloch EA. Synergism between recombinant growth factors, GM-CSF and G-CSF, acting on the blast cells of acute myeloblastic leukemia [published erratum appears in Blood 1987; 70:339]. Blood 1987; 69:1498–1503.

234. Mitjavila MT, Villeval JL, Cramer P, Henri A, Gasson J, Krystal G, Tulliez M, Berger R, Breton GJ, Vainchenker W. Effects of granulocyte–macrophage colony-stimulating factor and erythropoietin on leukemic erythroid colony formation in human early erythroblastic leukemias. Blood 1987; 70:965–973.

235. McNiece IK, Stewart FM, Deacon DM, Quesenberry PJ. Synergistic interactions between hematopoietic growth factors as detected by in vitro mouse bone marrow colony formation. Exp Hematol 1988; 16:383–388.

236. Metcalf D, Nicola NA, Gough NM, Elliott M, McArthur G, Li M. Synergistic suppression: Anomalous inhibition of factor-dependent hemopoietic cell proliferation by combination of two colony stimulating factors. Proc Natl Acad Sci USA 1992; 89:2819–2823.

237. Bot FJ, van Eijk L, Schipper P, Backx B, Lowenberg B. Synergistic effects between GM-CSF and G-CSF or M-CSF on highly enriched human marrow progenitor cells. Leukemia 1990; 4:325–328.

238. Caracciolo D, Shirsat N, Wong GG, Lange B, Clark S, Rovera G. Recombinant human macrophage colony-stimulating factor (M-CSF) requires subliminal concentrations of granulocyte/macrophage (GM)-CSF for optimal stimulation of human macrophage colony formation in vitro. J Exp Med 1987; 166:1851–1860.

239. McNiece IK, Langley KE, Zsebo KM. Recombinant human stem cell factor synergises with GM-CSF, G-CSF, IL-3 and EPO to stimulate human progenitor cells of the myeloid and erythroid lineages. Exp Hematol 1991; 19:226–231.

240. Broxmeyer HE, Sherry B, Cooper S, Lu L, Maze R, Beckmann MP, Cerami A, Ralph P. Comparative analysis of the human macrophage inflammatory protein family of cytokines (chemokines) on proliferation of human myeloid progenitor cells. Interacting effects involving suppression, synergistic suppression, and blocking of suppression. J Immunol 1993; 150:3448–3458.

241. Metcalf D, Nicola NA. The clonal proliferation of normal mouse hemopoietic cells: enhancement and suppression by colony-stimulating factor combinations. Blood 1992; 79:2861–2866.

242. Gliniak BC. Park LS, Rohrschneider LR. A GM-colony-stimulating factor (CSF) activated ribonuclease system transregulates M-CSF receptor expression in the murine FDC-P1/MAC myeloid cell line. Mol Biol Cell 1992; 3:535–544.

243. Metcalf D. Hemopoietic regulators and leukemia development: a personal retrospective. Adv Cancer Res 1994; pp. 41–91.

244. Fairbairn LJ, Cowling GJ, Reipert BM, Dexter TM. Suppression of apoptosis allows differentiation and development of a multipotent hemopoietic cell line in the absence of added growth factors. Cell 1993; 74:823–832.

245. Metcalf D, Burgess AW. Clonal analysis of progenitor cell commitment of granulocyte or macrophage production. J Cell Physiol 1982; 111:275–283.

246. Metcalf D, Johnson GR, Burgess AW. Direct stimulation by purified GM-CSF of the proliferation of multipotential and erythroid precursor cells. Blood 1980; 55:138–147.

247. Inaba K, Steinman RM, Pack MW, Aya H, Inaba M, Sudo T, Wolpe S, Schuler G. Identification of proliferating dendritic cell precursors in mouse blood. J Exp Med 1992; 175:1157–1167.

248. Inaba K, Inaba M, Deguchi M, Hagi K, Yasumizu R, Ikehara S, Maramatsu S, Steinman RM. Granulocutes, macrophages, and dendritic cells arise from a common major histocompatibility complex class II-negative progenitor in mouse bone marrow. Proc Natl Acad Sci USA 1993; 90:3038–3042.

249. Caux C, Dezutter DC, Schmitt D, Banchereau J. GM-CSF and TNF-alpha cooperate in the generation of dendritic Langerhans cells. Nature 1992; 360:258–261.

250. Clark EA, Grabstein KH, Shu GL. Cultured human follicular dendritic cells. Growth characteristics and interactions with B lymphocytes. J Immunol 1992; 148:3327–3335.

251. Reid CD, Stackpoole A, Meager A, Tikerpae J. Interactions of tumor necrosis factor with granulocyte–macrophage colony-stimulating factor and other cytokines in the regulation of dendritic cell growth in vitro from early bipotent CD34[+] progenitors in human bone marrow. J Immunol 1992; 149:2681–2688.

252. Fischer HG, Nitzgen B, Germann T, Degitz K, Daubener W, Hadding U. Differentiation driven by granulocyte–macrophage colony-stimulating factor endows microglia with interferon-gamma-independent antigen presentation function. J Neuroimmunol 1993; 42:87–92.

253. Povolny BT, Lee MY. The role of recombinant human M-CSF, IL-3, GM-CSF and calcitriol in clonal development of osteoclast precursors in primate bone marrow. Exp Hematol 1993; 21:532–537.

254. Lotem J, Sachs L. In vivo control of differentiation of myeloid leukemic cells by recombinant granulocyte–macrophage colony-stimulating factor and interleukin 3. Blood 1988; 71:375–382.

255. Metcalf D. Clonal analysis of the action of GM-CSF on the proliferation and differentiation of myelomonocytic leukemic cells. Int J Cancer 1979; 24:616–623.

256. Tomonaga M, Golde DW, Gasson JC. Biosynthetic (recombinant) human granulocyte–macrophage colony-stimulating factor: effect on normal bone marrow and leukemia cell lines. Blood 1986; 67:31–36.

257. Begley CG, Metcalf D, Nicola NA. Purified colony stimulating factors (G-CSF and GM-CSF) induce differentiation in human HL60 leukemic cells with suppression of clonogenicity. Int J Cancer 1987; 39:99–105.

258. Collins HL, Bancroft GJ. Cytokine enhancement of complement-dependent phagocytosis by macrophages: synergy of tumor necrosis factor-alpha and granulocyte–macrophage colony-stimulating factor for phagocytosis of *Cryptococcus neoformans*. Eur J Immunol 1992; 22:1447–1454.

259. Vadas MA, Nicola NA, Metcalf D. Activation of antibody-dependent cell-mediated cytotoxicity of human neutrophils and eosinophils by separate colony-stimulating factors. J Immunol 1983; 130:795–799.

260. Dranoff G, Jaffee E, Lazenby A, Golumbek P, Levitsky H, Brose K, Jackson V, Hamada H, Pardoll D, Mulligan RC. Vaccination with irradiated tumor cells engineered to secrete murine granulocyte–

macrophage colony-stimulating factor stimulates potent, specific, and long-lasting anti-tumor immunity. Proc Natl Acad Sci USA 1993; 90:3539–3543.

261. Grabstein KH, Urdal DL, Tushinski RJ, Mochizuki DY, Price VL, Cantrell MA, Gillis S, Conlon PJ. Induction of macrophage tumoricidal activity by granulocyte–macrophage colony-stimulating factor. Science 1986; 232:506–508.

262. Burnett D, Chamba A, Stockley RA, Murphy TF, Hill SL. Effects of recombinant GM-CSF and IgA opsonisation on neutrophil phagocytosis of latex beads coated with P6 outer membrane protein from *Haemophilus influenzae*. Thorax 1993; 48:638–642.

263. Fleischmann J, Golde DW, Weisbart RH, Gasson JC. Granulocyte–macrophage colony-stimulating factor enhances phagocytosis of bacteria by human neutrophils. Blood 1986; 68:708–711.

264. Kohn EC, Hollister GH, DiPersio JD, Wahl S, Liotta LA, Schiffmann E. Granulocyte–macrophage colony-stimulating factor induces human melanoma-cell migration. Int J Cancer 1993; 53:968–972.

265. Vaillant P, Muller V, Martinet Y, Martinet N. Human granulocyte- and granulocyte–macrophage-colony stimulating factors are chemotactic and "competence" growth factors for human mesenchymal cells. Biochem Biophys Res Commun 1993; 192:879–885.

266. Arnaout MA, Wang EA, Clark SC, Sieff CA. Human recombinant granulocyte–macrophage colony-stimulating factor increases cell-to-cell adhesion and surface expression of adhesion-promoting surface glycoproteins on mature granulocytes. J Clin Invest 1986; 78:597–601.

267. Yong KL, Linch DC. Granulocyte–macrophage-colony-stimulating factor differentially regulates neutrophil migration across IL-11-activated and nonactivated human endothelium. J Immunol 1993; 150:2449–2456.

268. Weisbart RH, Golde DW, Clark SC, Wong GG, Gasson JC. Human granulocyte–macrophage colony-stimulating factor is a neutrophil activator. Nature 1985; 314:361–363.

269. Wang JM, Colella S, Allavena P, Mantovani A. Chemotactic activity of human recombinant granulocyte–macrophage colony-stimulating factor. Immunology 1987; 60:439–444.

270. Rossman MD, Ruiz P, Comber P, Gomez F, Rottem M, Schreiber AD. Modulation of macrophage Fc gamma receptors by rGM-CSF. Exp Hematol 1993; 21:177–183.

271. Gosselin EJ, Wardwell K, Rigby WF, Guyre PM. Induction of MHC class II on human polymorphonuclear neutrophils by granulocyte/macrophage colony-stimulating factor, IFN-gamma, and IL-3. J Immunol 1993; 151:1482–1490.

272. Grabbe S, Gallo RL, Lindgren A, Granstein RD. Deficient antigen presentation by Langerhans cells from athymic (nu/nu) mice. Restoration with thymic transplantation or administration of cytokines. J Immunol 1993; 151:3430–3439.

273. Witmer-Pack MD, Olivier W, Valinsky J, Schuler G, Steinman RM. Granulocyte/macrophage colony-stimulating factor is essential for the viability and function of cultured murine epidermal Langerhans cells. J Exp Med 1987; 166:1484–1498.

274. Cannistra SA, Rambaldi A, Spriggs DR, Herrmann F, Kufe D, Griffin JD. Human granulocyte–macrophage colony-stimulating factor induces expression of the tumor necrosis factor gene by the U937 cell line and by normal human monocytes. J Clin Invest 1987; 79:1720–1728.

275. Horiguchi J, Warren MK, Kufe D. Expression of the macrophage-specific colony-stimulating factor in human monocytes treated with granulocyte–macrophage colony-stimulating factor. Blood 1987; 69:1259–1261.

276. Denzlinger C, Tetzloff W, Gerhartz HH, Pokorny R, Sagebiel S, Haberl C, Wilmanns W. Differential activation of the endogenous leukotriene biosynthesis by two different preparations of granulocyte–macrophage colony-stimulating factor in healthy volunteers. Blood 1993; 81:2007–2013.

277. Denzlinger C, Walther J, Wilmanns W, Gerhartz HH. Interleukin-3 enhances the endogenous leukotriene production [letter]. Blood 1993; 81:2466–2468.

278. Takahashi GW, Andrews D, Lilly MB, Singer JW, Alderson MR. Effect of granulocyte–macrophage colony-stimulating factor and interleukin-3 on interleukin-8 production by human neutrophils and monocytes. Blood 1993; 81:357–364.

279. Lindemann A, Riedel D, Oster W, Meuer SC, Blohm D, Mertelsmann RH, Herrmann F. Granulo-

cyte/macrophage colony-stimulating factor induces interleukin 1 production by human polymorphonuclear neutrophils. J Immunol 1988; 140:837–839.

280. Silberstein DS, Owen WF, Gasson JC, DiPersio JF, Golde DW, Bina JC, Soberman R, Austen KF, David JR. Enhancement of human eosinophil cytotoxicity and leukotriene synthesis by biosynthetic (recombinant) granulocyte–macrophage colony-stimulating factor. J Immunol 1986; 137:3290–3294.

281. Amann U, Bender A, Heidenreich S, Schmidt A, Nain M, Gemsa D. Activation of mononuclear phagocytes by granulocyte–macrophage colony-stimulating factor. In: van Furth R, ed. Hemopoietic Growth Factors and Mononuclear Phagocytes, Basel: Karger, 1993:21–28.

282. Hogasen AK, Hestdal K, Abrahamsen TG. Granulocyte–macrophage colony-stimulating factor, but not macrophage colony-stimulating factor, suppresses basal and lipopolysaccharide-stimulated complement factor production in human monocytes. J Immunol 1993; 151:3215–3224.

283. Hogasen AK, Abrahamsen TG. Increased C3 production in human monocytes after stimulation with *Candida albicans* is suppressed by granulocyte–macrophage colony-stimulating factor. Infect Immun 1993; 61:1779–1785.

284. Sullivan GW, Carper HT, Mandell GL. The effect of three human recombinant hematopoietic growth factors (granulocyte–macrophage colony-stimulating factor, granulocyte colony-stimulating factor, and interleukin-3) on phagocyte oxidative activity. Blood 1993; 81:1863–1870.

285. van der Bruggen T, Kok PT, Raaijmakers JA, Verhoeven AJ, Kessels RG, Lammers JW, Koenderman L. Cytokine priming of the respiratory burst in human eosinophils is Ca^{2+} independent and accompanied by induction of tyrosine kinase activity. J Leukoc Biol 1993; 53:347–353.

286. Bussolino F, Wang JM, Defilippi P, Turrini F, Sanavio F, Edgell CJ, Aglietta M, Arese P, Mantovani A. Granulocyte- and granulocyte–macrophage-colony stimulating factors induce human endothelial cells to migrate and proliferate. Nature 1989; 337:471–473.

287. De Nichilo MO, Burns GF. Granulocyte macrophage and macrophage colony-stimulating factors differentially regulate alpha v beta integrin expression on cultured human macrophages. Proc Natl Acad Sci USA 1993; 90:2517–2521.

288. Stanley E, Lieschke GJ, Grail D, Metcalf D, Hodgson G, Gall JAM, Maher DW, Cebon J, Sinickas V, Dunn AR. Granulocyte/macrophage colony-stimulating factor-deficient mice show no major perturbation of hematopoiesis but develop a characteristic pulmonary pathology. Proc Natl Acad Sci USA 1994; 91:5592–5596.

289. Dranoff G, Crawford AD, Sadelain M, Ream B, Rashid A, Bronson RT, Dickersin GR, Bachurski CJ, Mark EL, Whitsett JA, Mulligan RC. Involvement of granulocyte–macrophage colony-stimulating factor in pulmonary hemostasis. Science 1994; 264:713–716.

290. Lieschke GJ, Stanley E, Grail D, Hodgson G, Sinickas V, Gall JAM, Sinclair RA, Dunn AR. Mice lacking both macrophage- and granulocyte–macrophage colony-stimulating factor have macrophages and coexistent osteopetrosis and severe lung disease. Blood 1994; 84:27–35.

291. Bilyk N, Holt PG. Inhibition of the immunosuppressive activity of resident pulmonary alveolar macrophages by granulocyte/macrophage colony-stimulating factor. J Exp Med 1993; 177:1773–1777.

292. Holt PG, Oliver J, Bilyk N, McMenamin C, McMenamin PG, Kraal G, Thepen T. Downregulation of the antigen presenting cell function(s) of pulmonary dendritic cells in vivo by resident alveolar macrophages. J Exp Med 1993; 177:397–407.

293. Zimmer WE, Chew FS. Pulmonary alveolar proteinosis [clinical conference]. AJR 1993; 161:26.

294. Pech N, Hermine O, Goldwasser E. Further study of internal autocrine regulation of multipotent hematopoietic cells. Blood 1993; 82:1502–1506.

295. Cheers C, Haigh AM, Kelso A, Metcalf D, Stanley ER, Young AM. Production of colony-stimulating factors (CSFs) during infection: separate determinations of macrophage-, granulocyte-, granulocyte–macrophage-, and multi-CSFs. Infect Immun 1988; 56:247–251.

296. Cebon J, Layton JE, Maher D, Morstyn G. Endogenous haemopoietic growth factors in neutropenia and infection. Br J Haematol 1994; 86:265–274.

297. Metcalf D, Begley CG, Williamson DJ, Nice EC, De LJ, Mermod JJ, Thatcher D, Schmidt A.

Hemopoietic responses in mice injected with purified recombinant murine GM-CSF. Exp Hematol 1987; 15:1–9.

298. Morrissey PJ, Bressler L, Charrier K, Alpert A. Response of resident murine peritoneal macrophages to in vivo administration of granulocyte–macrophage colony-stimulating factor. J Immunol 1988; 140:1910–1915.

299. Mayer P, Lam C, Obenaus H, Liehl E, Besemer J. Recombinant human GM-CSF induces leukocytosis and activates peripheral blood polymorphonuclear neutrophils in nonhuman primates. Blood 1987; 70:206–213.

300. Donahue RE, Wang EA, Stone DK, Kamen R, Wong GG, Sehgal PK, Nathan DG, Clark SC. Stimulation of haematopoiesis in primates by continuous infusion of recombinant human GM-CSF. Nature 1986; 321:872–875.

301. Johnson GR, Gonda TJ, Metcalf D, Hariharan IK, Cory S. A lethal myeloproliferative syndrome in mice transplanted with bone marrow cells infected with a retrovirus expressing granulocyte–macrophage colony stimulating factor. EMBO J 1989; 8:441–448.

302. Lang RA, Metcalf D, Cuthbertson RA, et al. Transgenic mice expressing a hemopoietic growth factor gene (GM-CSF) develop accumulations of macrophages, blindness, and a fatal syndrome of tissue damage. Cell 1987; 51:675–686.

303. Metcalf D, Elliott MJ, Nicola NA. The excess numbers of peritoneal macrophages in granulocyte–macrophage colony-stimulating factor transgenic mice are generated by local proliferation. J Exp Med 1992; 175:877–884.

304. Tran HT, Metcalf D, Cheers C. Anti-bacterial activity of peritoneal cells from transgenic mice producing high levels of GM-CSF. Immunology 1990; 71:377–382.

305. Elliott MJ, Faulkner JB, Stanton H, Hamilton JA, Metcalf D. Plasminogen activator in granulocyte–macrophage-CSF transgenic mice. J Immunol 1992; 149:3678–3681.

306. Sisson SD, Dinarello CA. Production of interleukin-1 alpha, interleukin-1 beta and tumor necrosis factor by human mononuclear cells stimulated with granulocyte–macrophage colony-stimulating factor [see comments]. Blood 1988; 72:1368–1374.

307. Lang RA, Cuthbertson RA, Dunn AR. TNF alpha, IL-1 alpha and bFGF are implicated in the complex disease of GM-CSF transgenic mice. Growth Factors 1992; 6:131–138.

308. Sporn MB, Roberts AB. Autocrine growth factors and cancer. Nature 1985; 313:745–747.

309. Moore MA, Spitzer G, Metcalf D, Penington DG. Monocyte production of colony stimulating factor in familial cyclic neutropenia. Br J Haematol 1974; 27:47–55.

310. Freedman MH, Grunberger T, Correa P, Axelrad AA, Dube ID, Cohen A. Autocrine and paracrine growth control by granulocyte–monocyte colony-stimulating factor of acute lymphoblastic leukemia cells. Blood 1993; 81:3068–3075.

311. Young DC, Griffin JD. Autocrine secretion of GM-CSF in acute myeloblastic leukemia. Blood 1986; 68:1178–1181.

312. Cheng GY, Kelleher CA, Miyauchi J, Wang C, Wong G, Clark SC, McCulloch EA, Minden MD. Structure and expression of genes of GM-CSF and G-CSF in blast cells from patients with acute myeloblastic leukemia. Blood 1988; 71:204–208.

313. Rogers SY, Bradbury D, Kozlowski R, Russell NH. Evidence for internal autocrine regulation of growth in acute myeloblastic leukemia cells. Exp Hematol 1994; 22:593–598.

314. Lowenberg B, van Putten WLJ, Touw IP, Delwel R, Santini V. Autonomous proliferation of leukemic cells in vitro as a determinant of prognosis in adult acute myeloid leukemia. N Engl J Med 1993; 328:614–619.

315. Dührsen U. In vitro growth patterns and autocrine production of hemopoietic colony stimulating factors: analysis of leukemic populations arising in irradiated mice from cells of an injected factor-dependent continuous cell line. Leukemia 1988; 2:334–342.

316. Dührsen U, Metcalf D. A model system for leukemic transformation of immortalized hemopoietic cells in irradiated recipient mice. Leukemia 1988; 2:329–333.

317. Dührsen U, Stahl J, Gough NM. In vivo transformation of factor-dependent hemopoietic cells:

role of intracisternal A-particle transposition for growth factor gene activation. EMBO J 1990; 9:1087–1096.

318. Stocking C, Loliger C, Kawai M, Suciu S, Gough N, Ostertag W. Identification of genes involved in growth autonomy of hematopoietic cells by analysis of factor-independent mutants. Cell 1988; 53:869–879.

319. Lang RA, Metcalf D, Gough NM, Dunn AR, Gonda TJ. Expression of a hemopoietic growth factor cDNA in a factor-dependent cell line results in autonomous growth and tumorigenicity. Cell 1985; 43:531–542.

320. Metcalf D, Rasko JEJ. Leukemic transformation of immortalized FDC-P1 cells engrafted in GM-CSF transgenic mice. Leukemia 1993; 7:878–886.

321. Wheeler EF, Rettenmier CW, Look AT, Sherr CJ. The v-*fms* oncogene induces factor independence and tumorigenicity in CSF-1 dependent macrophage cell line. Nature 1986; 324:377–380.

322. Yoshimura A, Longmore G, Lodish HF. Point mutation in the exoplasmic domain of the erythropoietin receptor resulting in hormone-independent activation and tumorigenicity. Nature 1990; 348:647–649.

323. Longmore GD, Lodish HF. An activating mutation in the murine erythropoietin receptor induces erythroleukemia in mice: a cytokine receptor superfamily oncogene. Cell 1991; 67:1089–1102.

324. Gough NM. Leukemogenic potential of the hemopoietic growth factor GM-CSF and its receptor. In: Brugge J, Curran T, Harlow E, McCormick F, eds. Origins of Human Cancer. Cold Spring Harbor NY: Cold Spring Harbor Press, 1991:485–492.

325. Moore MA. Review: Stratton Lecture 1990. Clinical implications of positive and negative hematopoietic stem cell regulators. Blood 1991; 78:1–19.

326. Dai CH, Krantz SB, Dessypris EN, Means RJ, Horn ST, Gilbert HS. Polycythemia vera. II. Hypersensitivity of bone marrow erythroid, granulocyte–macrophage, and megakaryocyte progenitor cells to interleukin-3 and granulocyte–macrophage colony-stimulating factor. Blood 1992; 80:891–899.

327. Kyas U, Pietsch T, Welte K. Expression of receptors for granulocyte colony-stimulating factor on neutrophils from patients with severe congenital neutropenia and cyclic neutropenia. Blood 1992; 79:1144–1147.

328. Schrezenmeier H, Raghavachar A, Heimpel H. Granulocyte–macrophage colony-stimulating factor in the sera of patients with aplastic anemia. Clin Invest 1993; 71:102–108.

329. Zwierzina H, Schollenberger S, Herold M, Schmalzl F, Besemer J. Endogenous serum levels and surface receptor expression of GM-CSF and IL-3 in patients with myelodysplastic syndromes. Leuk Res 1992; 16:1181–1186.

330. Fiehn C, Wermann M, Pezzutto A, Hufner M, Heilig B. Plasma GM-CSF concentrations in rheumatoid arthritis, systemic lupus erythematosus and spondyloarthropathy. Z Rheumatol 1992; 51:121–126.

331. Re MC, Zauli G, Furlini G, Giovannini M, Ranieri S, Ramazzotti E, Vignoli M, La PM. GM-CSF production by CD4[+] T-lymphocytes is selectively impaired during the course of HIV-1 infection. A possible indication of a preferential lesion of a specific subset of peripheral blood CD4[+] T-lymphocytes. Microbiologica 1992; 15:265–270.

332. Agostini C, Trentin L, Zambello R, Bulian P, Caenazzo C, Cipriani A, Cadrobbi P, Garbisa S, Semenzato G. Release of granulocyte–macrophage colony-stimulating factor by alveolar macrophages in the lung of HIV-1 infected patients. A mechanism accounting for macrophage and neutrophil accumulation. J Immunol 1992; 149:3379–3385.

333. Marini M, Avoni E, Hollemborg J, Mattoli S. Cytokine mRNA profile and cell activation in bronchoalveolar lavage fluid from nonatopic patients with symptomatic asthma. Chest 1992; 102:661–669.

334. Warringa RA, Mengelers HJ, Kuijper PH, Raaijmakers JA, Bruijnzeel PL, Koenderman L. In vivo priming of platelet-activating factor-induced eosinophil chemotaxis in allergic asthmatic individuals. Blood 1992; 79:1836–1841.

335. Massey W, Friedman B, Kato M, Cooper P, Kagey-Sobotka A, Lichtenstein LM, Schleimer RP.

Appearance of granulocyte–macrophage colony-stimulating factor activity at allergen-challenged cutaneous late-phase reaction sites. J Immunol 1993; 150:1084–1092.

336. Neta R, Oppenheim JJ. Cytokines in therapy of radiation injury. Blood 1988; 72:1093–1095.

337. Neta R, Oppenheim JJ, Douches SD. Interdependence of the radioprotective effects of human recombinant interleukin 1 alpha, tumor necrosis factor alpha, granulocyte colony-stimulating factor, and murine recombinant granulocyte–macrophage colony-stimulating factor. J Immunol 1988; 140:108–111.

338. Muench MO, Moore MA. Accelerated recovery of peripheral blood cell counts in mice transplanted with in vitro cytokine-expanded hematopoietic progenitors. Exp Hematol 1992; 20:611–618.

339. Blazar BR, Widmer MB, Soderling CC, Urdal DL, Gillis S, Robison LL, Vallera DA. Augmentation of donor bone marrow engraftment in histoincompatible murine recipients by granulocyte/macrophage colony-stimulating factor. Blood 1988; 71:320–328.

340. Naparstek E, Ohana M, Greenberger JS, Slavin S. Continuous intravenous administration of rmGM-CSF enhances immune as well as hematopoietic reconstitution following syngeneic bone marrow transplantation in mice. Exp Hematol 1993; 21:131–137.

341. Monroy RL, Skelly RR, MacVittie TJ, Davis TA, Sauber JJ, Clark SC, Donahue RE. The effect of recombinant GM-CSF on the recovery of monkeys transplanted with autologous bone marrow. Blood 1987; 70:1696–1699.

342. Nienhuis AW, Donahue RE, Karlsson S, Clark SC, Agricola B, Antinoff N, Pierce JE, Turner P, Anderson WF, Nathan DG. Recombinant human granulocyte–macrophage colony-stimulating factor (GM-CSF) shortens the period of neutropenia after autologous bone marrow transplantation in a primate model. J Clin Invest 1987; 80:573–577.

343. Lieschke GJ, Burgess AW. Granulocyte colony-stimulating factor and granulocyte–macrophage colony-stimulating factor (second of two parts). N Engl J Med 1992; 327:99–106.

344. Vadhan RS, Broxmeyer HE, Hittelman WN, et al. Abrogating chemotherapy-induced myelosuppression by recombinant granulocyte–macrophage colony-stimulating factor in patients with sarcoma: protection at the progenitor cell level. J Clin Oncol 1992; 10:1266–1277.

345. Williams DE, Dunn JT, Park LS, Frieden EA, Seiler FR, Farese AM, Macvittie TJ. A GM-CSF/IL-3 fusion protein promotes neutrophil and platelet recovery in sublethally irradiated rhesus monkeys. Biotechnol Ther 1993; 4:17–29.

346. Nemunaitis J, Buckner CD, Appelbaum FR, et al. Phase I/II trial of recombinant human granulocyte-macrophage colony-stimulating factor following allogeneic bone marrow transplantation. Blood 1991; 77:2065–2071.

347. Powles R, Smith C, Milan S, Treleaven J, Millar J, McElwain T, Gordon-Smith E, Milliken S, Tiley C. Human recombinant GM-CSF in allogeneic bone-marrow transplantation for leukaemia: double-blind, placebo-controlled trial. Lancet 1990; 336:1417–1420.

348. Khwaja A, Linch DC, Goldstone AH, et al. Recombinant human granulocyte–macrophage colony-stimulating factor after autologous bone marrow transplantation for malignant lymphoma: a British National Lymphoma Investigation double-blind, placebo-controlled trial. Br J Haematol 1992; 82:317–323.

349. Campos L, Bastion Y, Roubi N, Felman P, Espinouse D, Dumontet C, Debost M, Tremisi JP, Coiffier B. Peripheral blood stem cells harvested after chemotherapy and GM-CSF for treatment intensification in patients with advanced lymphoproliferative diseases. Leukemia 1993; 7:1409–1415.

350. Serke S, Kirsch A, Huhn D. Mobilization of circulating haemopoietic cells (blood stem cells) by GM-CSF in patients with malignancies. Infection 1992; 20(suppl 2):S100–S102.

351. Scadden DT. The use of GM-CSF in AIDS. Infection 1992; 20(suppl 2):S103–S106.

352. Roilides E, Pizzo PA. Biologicals and hematopoietic cytokines in prevention or treatment of infections in immunocompromised hosts. Hematol Oncol Clin North Am 1993; 7:841–864.

353. Charak BS, Agah R, Mazumder A. Granulocyte–macrophage colony-stimulating factor-induced antibody-dependent cellular cytotoxicity in bone marrow macrophages: application in bone marrow transplantation. Blood 1993; 81:3474–3479.

354. van Furth R. Hemopoietic Growth Factors and Mononuclear Phagocytes. Symposium, Lugarno: Karger, Basel, 1993.

355. Rubbia-Brandt L, Sappino AP, Gabbiani G. Locally applied GM-CSF induces the accumulation of alpha-sm actin containing myofibroblasts. Virchows Archiv B Cell Pathol 1991; 60:73–82.

356. Lieschke GJ, Burgess AW. Granulocyte colony-stimulating factor and granulocyte-macrophage colony-stimulating factor (first of two parts). N Engl J Med 1992; 327:28–35.

357. Demetri GD, Antman KH. Granulocyte–macrophage colony-stimulating factor (GM-CSF): preclinical and clinical investigations. Semin Oncol 1992; 19:362–385.

358. Antman KS, Griffin JD, Elias A, Socinski MA, Ryan L, Cannistra SA, Oette D, Whitley M, Frei E, Schnipper LE. Effect of recombinant human granulocyte–macrophage colony-stimulating factor on chemotherapy-induced myelosuppression. N Engl J Med 1988; 319:593–598.

359. Nemunaitis J, Singer JW, Buckner CD, Hill R, Storb R, Thomas ED, Appelbaum FR. Use of recombinant human granulocyte–macrophage colony-stimulating factor in autologous marrow transplantation for lymphoid malignancies. Blood 1988; 72:834–836.

360. Gorin NC, Coiffier B, Hayat M, Fouillard L, Kuentz M, Flesch M, Colombat P, Boivin P, Slavin S, Philip T. Recombinant human granulocyte–macrophage colony-stimulating factor after high-dose chemotherapy and autologous bone marrow transplantation with unpurged and purged marrow in non-Hodgkin's lymphoma: a double-blind placebo-controlled trial. Blood 1992; 80:1149–1157.

361. Nemunaitis J, Rabinowe SN, Singer JW, et al. Recombinant granulocyte–macrophage colony-stimulating factor after autologous bone marrow transplantation for lymphoid cancer. N Engl J Med 1991; 324:1773–1778.

362. Link H, Boogaerts MA, Carella AM, et al. A controlled trial of recombinant human granulocyte–macrophage colony-stimulating factor after total body irradiation, high-dose chemotherapy, and autologous bone marrow transplantation for acute lymphoblastic leukemia or malignant lymphoma. Blood 1992; 80:2188–2195.

363. Gulati SC, Bennett CL. Granulocyte–macrophage colony-stimulating factor (GM-CSF) as adjunct therapy in relapsed Hodgkin disease. Ann Intern Med 1992; 116:177–182.

364. Glaspy JA, Jakway J. Cost considerations in therapy with myeloid growth factors. Am J Hosp Pharm 1993; 50(suppl.3):S19–S26.

365. Weber RJ. Pharmacoeconomic issues in the use of granulocyte–macrophage colony-stimulating factor for bone marrow transplantation or chemotherapy-induced neutropenia. Clin Ther 1993; 15:180–191.

366. Neidhart JA, Mangalik A, Stidley CA, Tebich SL, Sarmiento LE, Pfile JE, Oette DH, Oldham FB. Dosing regimen of granulocyte–macrophage colony-stimulating factor to support dose-intensive chemotherapy. J Clin Oncol 1992; 10:1460–1469.

367. Ragnhammar P, Fagerberg J, Frodin JE, Hjelm AL, Lindemalm C, Magnusson I, Masucci G, Mellstedt H. Effect of monoclonal antibody 17-1A and GM-CSF in patients with advanced colorectal carcinoma—long-lasting, complete remissions can be induced. Int J Cancer 1993; 53:751–758.

368. Dranoff G, Jaffee E, Lazenby A, Columbek P, Levitsky H, Brose K, Jackson V, Hamada H, Pardoll D, Muligan RC. Vaccination with irradiated tumor cells engineered to secrete murine granulocyte–macrophage colony-stimulating factor stimulates potent, specific, and long-lasting anti-tumor immunity. Proc Natl Acad Sci USA 1993; 90:3539–3543.

369. Tao MH, Levy R. Idiotype/granulocyte–macrophage colony-stimulating factor fusion protein as a vaccine for B-cell lymphoma. Nature 1993; 362:755–758.

370. Edmonson JH, Hartmann LC, Long HJ, Colon OG, Fitch TR, Jefferies JA, Braich TA, Maples WJ. Granulocyte–macrophage colony-stimulating factor. Preliminary observations on the influences of dose, schedule, and route of administration in patients receiving cyclophosphamide and carboplatin. Cancer 1991; 70:2529–2539.

371. Cebon J, Lieschke GJ, Bury RW, Morstyn G. The dissociation of GM-CSF efficacy from toxicity according to route of administration: a pharmacodynamic study. Br J Haematol 1992; 80:144–150.

372. Sheridan WP, Fox RM. Haemopoietic growth factors and cancer therapy. Med J Aust 1993; 158:514–516.

373. Lieschke GJ, Maher D, Cebon J, et al. Effects of bacterially synthesized recombinant human granulocyte–macrophage colony-stimulating factor in patients with advanced malignancy. Ann Intern Med 1989; 110:357–364.

374. Mehregan DR, Fransway AF, Edmonson JH, Leiferman KM. Cutaneous reactions to granulocyte–monocyte colony-stimulating factor. Arch Dermatol 1992; 128:1055–1059.

375. Stern AC, Jones TC. The side-effect profile of GM-CSF. Infection 1992; 20(suppl 2):S124–S127.

376. Lieschke GJ, Cebon J, Morstyn G. Characterization of the clinical effects after the first dose of bacterially synthesized recombinant human granulocyte–macrophage colony-stimulating factor. Blood 1989; 74:2634–2643.

377. De Vries EGE, Willemse PH, Biesma B, Stern AC, Limburg PC, Vellenga E. Flare-up of rheumatoid arthritis during GM-CSF treatment after chemotherapy [letter]. Lancet 1991; 338:517–518.

378. Stehle B, Weiss C, Ho AD, Hunstein W. Serum levels of tumor necrosis factor alpha in patients treated with granulocyte–macrophage colony-stimulating factor. Blood 1990; 75:1895–1896.

379. Hazenberg BP, Van LM, Van RM, Stern AC, Vellenga E. Correction of granulocytopenia in Felty's syndrome by granulocyte–macrophage colony-stimulating factor. Simultaneous induction of inter-leukin-6 release and flare-up of the arthritis. Blood 1989; 74:2769–2770.

380. Hoekman K, Von BvdFB, Wagstaff J, Drexhage HA, Pinedo HM. Reversible thyroid dysfunction during treatment with GM-CSF. Lancet 1991; 338:541–542.

381. Berney T, Shibata T, Merino R, Chicheportiche Y, Kindler V, Vassalli P, Izui S. Murine autoimmune hemolytic anemia resulting from Fc gamma receptor-mediated erythrophagocytosis: protection by erythropoietin but not by interleukin-3, and aggravation by granulocyte–macrophage colony-stimu-lating factor. Blood 1992; 79:2960–2964.

382. Logothetis CJ, Dexeus FH, Sella A, Amato RJ, Kilbourn RG, Finn L, Gutterman JU. Escalated therapy for refractory urothelial tumors: methotrexate–vinblastine–doxorubicin–cisplatin plus un-glycosylated recombinant human granulocyte–macrophage colony-stimulating factor. J Natl Cancer Inst 1990; 82:667–672.

383. Wilson PA, Ayscue LH, Jones GR, Bentley SA. Bone marrow histiocytic proliferation in association with colony-stimulating factor therapy. Am J Clin Pathol 1993; 99:311–313.

384. Lang E, Cibull ML, Gallicchio VS, Henslee-Downey PJ, Davey DD, Messino MJ, Harder EJ. Proliferation of abnormal bone marrow histiocytes, an undesired effect of granulocyte macrophage-colony-stimulating factor therapy in a patient with Hurler's syndrome undergoing bone marrow transplantation. Am J Hematol 1992; 41:280–284.

385. Stephens LC, Haire WD, Schmit-Pokorny K, Kessinger A, Kotulak G. Granulocyte macrophage colony stimulating factor: high incidence of apheresis catheter thrombosis during peripheral stem cell collection. Bone Marrow Transplant 1993; 11:51–54.

386. Buchner T, Hiddemann W, Koenigsmann M, et al. Recombinant human granulocyte–macrophage colony-stimulating factor after chemotherapy in patients with acute myeloid leukemia at higher age or after relapse. Blood 1991; 78:1190–1197.

387. Willemze R, van DLN, Zwierzina H. A randomized phase-I/II multicenter study of recombinant human granulocyte–macrophage colony-stimulating factor (GM-CSF) therapy for patients with myelodysplastic syndromes and a relatively low risk of acute leukemia. EORTC Leukemia Cooper-ative Group [published erratum appears in Ann Hematol 1992; 64:312]. Ann Hematol 1992; 64:173–180.

388. Ganser A, Seipelt G, Eder M, Geissler G, Ottmann OG, Hess U, Hoelzer D. Treatment of myelo-dysplastic syndromes with cytokines and cytotoxic drugs. Semin Oncol 1992; 19(suppl 4):95–101.

389. Economopoulos T, Papageorgiou E, Stathakis N, Asprou N, Karmas P, Dervenoulas J, Bouronikou H, Chalevelakis G, Raptis S. Treatment of myelodysplastic syndromes with human granulocytic–macrophage colony stimulating factor (GM-CSF) or GM-CSF combined with low-dose cytosine arabinoside. Eur J Haematol 1992; 49:138–142.

390. Verhoef G, Van DBH, Boogaerts M. Cytogenetic effects on cells derived from patients with myelodysplastic syndromes during treatment with hemopoietic growth factors. Leukemia 1992; 6:766–769.

391. Gradishar WJ, Le BM, O'Laughlin R, Vardiman JW, Larson RA. Clinical and cytogenetic responses to granulocyte–macrophage colony-stimulating factor in therapy-related myelodysplasia. Blood 1992; 80:2463–2470.

392. Osborne CS, Vadas MA, Cockerill PN. Transcriptional regulation of mouse granulocyte–macrophage colony-stimulating factor/IL-3 locus. J Immunol 1995; 155:226–235.

393. Brown MA, Harrison-Smith M, DeLuca E, Begley CG, Gough NM. No evidence for GM-CSF receptor alpha chain gene mutation in AML-M2 leukemias which have lost a sex chromosome. Leukemia 1994; 8:1774–1779.

394. Barahmand-pour F, Meinke A, Kieslinger M, Eilers A, Decker T. A role for STAT family transcription factors in myeloid differentiation. Cur Top Microbiol Immunol 1996; 211:121–128.

395. Taglietti M. Vaccine adjuvancy: a new potential area of development for GM-CSF. Adv Exp Med Biol 1995; 378:565–569.

11

Granulocyte Colony-Stimulating Factor

Andrew W. Roberts and Nicos A. Nicola
The Walter and Eliza Hall Institute of Medical Research, Melbourne, Victoria, Australia

I. INTRODUCTION

Granulocyte colony-stimulating factor (G-CSF) is a member of the family of secreted glycoproteins that regulate the survival, proliferation, differentiation, and function of hematopoietic cells. In the 17 years since its original functional description, in 1979, in the human (1) and in 1980 in the mouse (2), G-CSF has been purified, cloned, extensively evaluated in preclinical and clinical trials, and licensed for routine used in the management of hematopoietic disorders and malignant diseases. This sequence of events has proceeded at a remarkable pace, on a background of rapidly evolving understanding of both hematopoiesis in general and intercellular signaling. This chapter will focus on the basic biology of G-CSF, with particular emphasis on its structure, function, and physiology.

II. ISOLATION OF G-CSF

What is now known as human G-CSF was originally separated from granulocyte–macrophage (GM)-CSF activity in human placental-conditioned medium by hydrophobic chromatography and called CSF-β (1). It was described as a relatively selective stimulus for day-7 granulocyte colony-forming cells from human bone marrow. Murine G-CSF was first described by Burgess and Metcalf, as an activity present in the serum of endotoxin-injected mice, that induced differentiation in the murine myelomonocytic leukemia cell line WEHI-3B D$^+$ (2). Nicola et al. subsequently purified a glycoprotein with a relative molecular mass (Mr) 25,000 from the conditioned medium from lungs of mice injected with endotoxin, using a bioassay based on this differentiation-inducing activity (3). This glycoprotein stimulated specific neutrophil granulocyte colony formation in agar cultures of normal bone marrow cells and, hence, was designated granulocyte colony-stimulating factor.

The human functional homologue of murine G-CSF was identified as CSF-β by showing that these two activities were completely species cross-reactive in all assays, including the WEHI 3B D$^+$ assay (4). Human (h)G-CSF was also purified to homogeneity from the conditioned media of carcinoma cell lines that constitutively produced colony-stimulating activity (5,6). Following the partial determination of the amino acid sequence of purified hG-CSF, oligonucleotide probes were designed and Nagata et al. and Souza et al. isolated the cDNA for hG-CSF from cDNA libraries constructed from CHU-2 and 5637 tumor cell lines, respectively (7,8). The relation between the murine and human functional homologues was confirmed when the mouse (m)G-CSF cDNA was isolated by cross-hybridization with the human cDNA from a cDNA library constructed from a murine fibrosarcoma cell line that constitutively secreted G-CSF (9).

III. GENE AND PROTEIN STRUCTURE

A. Structure of the Gene and Its Transcripts

The gene for human G-CSF exists in single copy and is located on the q21-22 region of chromosome 17 (10). In mice, the gene has been mapped to the distal half of the syntenic murine chromosome 11 (11). Both genes consist of five exons and four introns, with similar spacings, and are about 2500 nucleotides in length (Fig. 1). The coding regions of the exons are highly conserved, with 69% identity between mouse and human. Similarly, the 300 base pairs (bp) upstream from the transcription initiation site are conserved and they incorporate key promoter sequences. However, there is little sequence homology between the introns.

   The promoter and enhancer elements of the G-CSF gene and their DNA-binding proteins have not been fully defined. For the murine G-CSF gene, three regulatory sequences have been described and candidate transcription factors have been identified (see Fig. 1). Nishizawa et al. (13,14) described three cis-regulatory elements 192, 116, and 87 bp upstream from the transcription initiation site—G-CSF promoter elements (GPE) 1, 2, and 3, respectively. GPE-1 consists of a decanucleotide sequence and the adjacent downstream 23-bp sequence. Although the proximal sequence appears to be unique to the G-CSF gene, the decanucleotide is highly conserved in sequence within the promoter regions of the G-CSF, GM-CSF, and interleukin (IL)-3 genes and has been variously named the CSF box, CK-1 (cytokine-1), or CLE-1 (conserved lymphokine element-1) (13,15,16). Several nuclear factors that bind either the decanucleotide or the adjacent region have been identified, but their precise roles in regulation

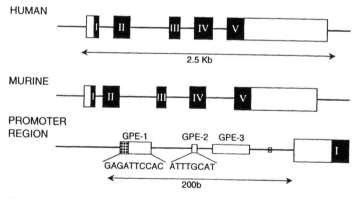

Figure 1   The chromosomal gene structures for human and murine G-CSFs and an expanded view of the 5′-flanking region, indicating the promoter elements. Exons are indicated by wide boxes, with the coding regions shaded black. Introns are indicated by lines. In the promoter region, the boxes represent G-CSF promoter elements (GPE). The dotted area of GPE-1 indicates the "CSF box"; the "TATA box" is indicated by the small dotted box. (Modified from Ref. 12.)

of gene expression remain unclear (15,17,18). GPE-2 is an octamer sequence, which is also found in the promoter regions of immunoglobulin and histone genes, where it functions as the binding site for octamer transcription factors (19). GPE-3 is a poorly characterized region of approximately 30 bp, with no homology to known regulatory elements of other genes.

The human G-CSF gene codes for two different mRNAs that are generated by alternative use of two 5′ splice donor sequences in the second intron (10). These mRNAs code for precursor proteins of 204 and 207 amino acids. The NH₂-terminal 30 amino acids comprise an hydrophilic leader sequence typical of secreted proteins and which, when cleaved, results in mature molecules of 174 and 177 amino acids in length.

B. Protein Sequence

The amino acid sequences for human and murine G-CSF are represented in Figure 2. Native human and murine G-CSF exist as glycoproteins, with the threonine at amino acid position 133 being *O*-glycosylated in the human molecule. This glycosylation appears to stabilize the molecule and protects it from aggregation at neutral pH, but does not influence receptor binding or the specific activity of G-CSF in vitro (21). Bovine and canine G-CSFs have also been cloned, and their amino acid sequences share greater than 70% homology with human and murine

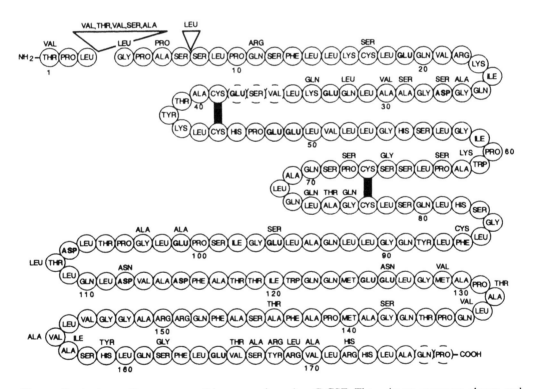

Figure 2 Amino acid sequences of human and murine G-CSF. The primary sequence shown and numbering system are for human G-CSF. Insertions in the mouse sequence are shown by triangles, and deletions are shown by dashed circles. Substitutions in the mouse sequence are shown above the human sequence. Amino acids in position 36–38 occur in only one type of human cDNA and not in the mouse. The connectivity of the disulfide bonds is shown by heavy bars. (From Ref. 20.)

G-CSFs (22). The G-CSF amino acid sequence is distantly related to that of IL-6, with apparent conservation of the two disulfide bonds.

C. Tertiary Structure

The three-dimensional structure of recombinant hG-CSF expressed in *Escherichia coli* has been solved by X-ray crystallography (23). G-CSF is an antiparallel, four-α-helical bundle with a left-handed twist (Fig. 3); 104 of the 175 (174 plus the f-met residue) amino acids are incorporated in this bundle. In addition to the four major helices, there is a shorter helical section within the long connecting loop between helices A and B. Of particular note, two disulfide bonds, Cys36–Cys42 and Cys64–Cys74 are located at opposite ends of the long A–B connecting loop. The up-up–down-down pattern of connection of the four major helices with the two long crossover loops is a common feature of a distinct structural class of growth factors, which includes growth hormone (24), GM-CSF (25), interferon-alpha (IFN-α) (26), IL-2 (27,28), and IL-4 (29). Although members of this class share the same basic architecture, there is little sequence similarity (23). This class is also predicted to include IL-6 and leukemia inhibitory factor (LIF).

D. Structure–Function Relations

Structure–function studies performed before the determination of the three-dimensional structure of G-CSF had confirmed the importance of conformational integrity for biological activity. Disruption of either, or both, disulfide bonds interferes with folding and markedly reduces biological activity (30). Similarly, mutants created with deletions or tandem repeats in the protein structure between residue 18 and the COOH-terminus, which would be expected to substantially

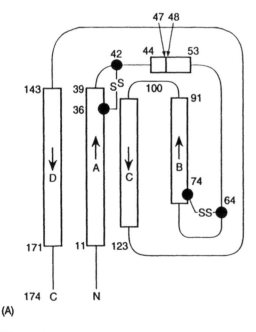

(A)

Figure 3 (A) Connectivity diagram of rhG-CSF: The lengths of the secondary structural elements are drawn in proportion to the number of residues. (B) Ribbon diagram of rhG-CSF: The main bundle helices A,B,C,D are labeled near their NH₂-termini. The positions of the NH₂- and COOH-termini, the two disulfide bonds, and Glu-46 are marked. (Modified from Ref. 23.)

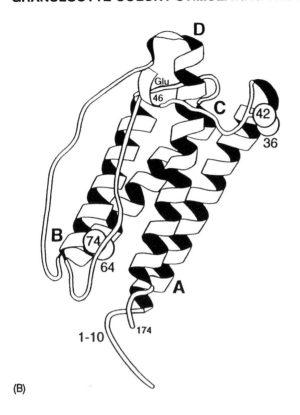

(B)

alter the tertiary structure, all result in loss of biological activity (31). Of note, deletion of the COOH-terminus (residues 165–174) results in loss of activity. This feature is shared by other members of the structural class: growth hormone (32), GM-CSF (33), IL-4 (34), and IL-6 (35). In contrast, deletion of the first 11 amino acids does not affect biological activity, and mutations of Cys-17 have been described with increased biological activity (31) and stability (36).

Point mutations at Leu-35 and Glu-46 (Glu-47 in r-met-G-CSF) (22) result in loss of function. The importance of the region flanked by these residues to biological activity is reflected by two further observations. (1) The 177-amino acid form of G-CSF contains an additional three residues between Leu-35 and Cys-36, and shows a marked reduction in activity. (2) Layton et al. (37) generated neutralizing antibodies to G-CSF, and their binding was localized to residues 20–46. Consideration of the three-dimensional structure reveals that this region comprises the COOH-terminus of helix A and the segment of the AB loop containing the first disulfide bond and part of the minor helix. It seems likely that this region, in conjunction with residues 165–174 on the adjacent D helix, compose a receptor-binding site (22).

IV. BIOLOGY OF G-CSF

Although regulation of neutrophil production is its most striking action, the biology of G-CSF is considerably more diverse than the single function implied by its name. As a framework for discussing these functions, the in vitro actions of G-CSF will be addressed initially, followed by reviews on the production of G-CSF, the function and distribution of its receptor, and the effects of G-CSF treatment in whole animals. The chapter will conclude with a summary of the evidence for key in vivo roles for G-CSF as an hematopoietic regulator.

A. In Vitro Actions

Proliferation

The first action of G-CSF on normal hematopoietic cells to be recognized was its ability to stimulate the formation of small numbers of neutrophil colonies from cultures of unfractionated human and murine bone marrow cells in agar (1,2,3,38). The proliferative activity on progenitor cells of the neutrophil lineage has been confirmed to be a specific and direct action of G-CSF (38,39).

However, the proliferative activity of G-CSF in not restricted to progenitor cells committed to the neutrophil lineage. When cultures of murine bone marrow are initiated with G-CSF and emergent clones transferred to secondary cultures stimulated by others CSFs, G-CSF clearly promotes the survival or initial proliferation of macrophage, eosinophil, and erythroid colony-forming cells (38). Indirect evidence implicates G-CSF in triggering the entry of dormant progenitors into the cell cycle. In agar culture, the addition of G-CSF significantly decreases the time interval between plating the cells and the emergence of blast cell colonies, when compared with IL-3 alone (40). The activity of G-CSF on the proliferation of primitive as well as committed progenitor cells has also been observed in other settings, including long-term suspension cultures using stromal cells engineered to produce G-CSF (41) and stroma-free liquid suspension cultures (42). G-CSF is a key component in cytokine combinations used in liquid suspension cultures designed to maximize the output of mature and directly clonogenic cells from input populations of purified primitive progenitors or unfractionated marrow (43). The foregoing evidence clearly indicates that G-CSF acts in a synergistic fashion with other growth factors to promote the proliferation of primitive progenitor cells. In keeping with these latter observations of the biological effects of G-CSF as a culture stimulus, G-CSF receptors have been detected by autoradiography on purified murine progenitor cell subpopulations, including the population highly enriched for lymphohematopoietic stem cells (44).

Differentiation

The proliferative action of G-CSF on normal neutrophil precursors appears to be invariably coupled with differentiation, at least after several cell divisions have occurred. In these cells, G-CSF is sufficient for proliferation and differentiation. In contrast, G-CSF is insufficient for either differentiation or sustained proliferation of primitive stem cells and progenitor cells of nonneutrophil lineages. This dissociation of differentiation and proliferation can also be demonstrated in some immortalized leukemia cell lines and in some primary myeloid leukemias. A prime example of the differentiation-inducing activity is the effect of G-CSF on the morphology of WEHI-3B D[+] colony formation in agar. G-CSF induces terminal differentiation of some blast cells in these colonies, causing the normally tight colony to develop a halo of neutrophils and macrophages (2), and inducing a marked reduction in the clonogenicity of cells within G-CSF-stimulated colonies (45).

Among human myeloid cell lines, G-CSF is able to induce differentiation with some loss of proliferative capacity in cultures of the HL60 promyelocytic leukemia cell line (46). When primary human myeloid leukemias have been analyzed for their response to G-CSF in vitro, the results have not been uniform. Primary myeloid leukemia cells are absolutely dependent on colony-stimulating factors for survival and growth in vitro, making analysis of the effects of G-CSF in isolation difficult to dissect. Cells from most myeloid leukemias proliferate in the presence of G-CSF, and differentiation has been observed in some instances (8,47–52). Analogous to the findings using murine and HL60 cell lines, some myeloid leukemias appear to undergo clonal extinction in response to G-CSF; however, this is not a consistent finding. Many myeloid leukemias show minimal differentiation in both short-term and long-term cultures (48) and, within individual primary leukemic populations, many blast cells fail to express G-CSF receptors (47).

Actions on Mature Neutrophils

The presence of G-CSF enhances the survival time of purified human blood neutrophils in vitro (53). A broad range of neutrophil functions are also enhanced by G-CSF. In particular, G-CSF increases the cellular processes involved in defense against infections by priming neutrophils to be more responsive to a variety of stimuli. In vitro and in vivo, G-CSF enhances superoxide anion production by neutrophils in response to various stimuli (54,55). G-CSF increases both phagocytic and bactericidal activities of neutrophils against opsonized *Staphylococcus aureus*, but not against *Candida albicans* blastoconidia (56).

G-CSF increases the binding of the chemotactic peptide, *N*-formyl-Met-Leu-Phe (fMLP), to neutrophils (57) and also increases the in vitro responses to this stimulating agent. G-CSF upregulates the expression of C3b receptors (58), the generation of superoxide anions (59,60), the release of arachidonic acid (60), and the phagocytosis of opsonized microspheres (58) by neutrophils exposed to fMLP, as compared with controls exposed to fMLP alone. The ability of neutrophils to mediate antibody-dependent cell-mediated cytotoxicity in vitro against antibody-coated murine thymoma cells (61) and leukemia and lymphoma cell lines (62) is also increased by G-CSF. G-CSF (and GM-CSF) can induce a change from low- to high-affinity Fc receptors for IgA on neutrophils, facilitating IgA-mediated phagocytosis (63). Finally, G-CSF possesses chemotactic activity for neutrophils and monocytes and induces neutrophil adhesion (64,59).

Actions on Nonhematopoietic Cells

Granulocyte colony-stimulating factor may have effects outside the hematopoietic system. Bussolino et al. have reported that G-CSF and GM-CSF promote the in vitro proliferation and migration of human umbilical vein endothelial cells, without modulating other endothelial cell functions related to hemostasis and inflammation (65,66). Furthermore, G-CSF has angiogenic activity in an in vivo model of new vessel formation in rabbit corneas (66). However, other authors have reported no effects on either primary or passaged endothelial cells (67).

Occasional cell lines from nonmyeloid malignancies respond in vitro to G-CSF. The growth of several small-cell lung carcinoma cell lines (60) and colonic adenocarcinoma lines (68) can be promoted by G-CSF. However, there have been no reports of proliferative responses by primary malignant cells.

B. Production of G-CSF: Cellular Sources, Stimuli, and Regulation

Conditioned media from almost all murine tissues cultured in vitro contain G-CSF. The levels of G-CSF in these media can be substantially increased by the injection of endotoxin into mice 3 hr before harvesting the tissues (69). These initial observations highlight the key features of cellular production of G-CSF. First, the capacity to produce G-CSF is distributed throughout all organs, and second, G-CSF production is highly inducible. Subsequent investigations using purified primary cells and cell lines have revealed the specific cellular sources of G-CSF to be monocytes and macrophages (70–73), endothelial cells (74,75), fibroblasts and related mesenchymal cells (76–81), and bone marrow stromal cells (82–84).

It remains contentious as to whether any of these cell types truly constitutively produce G-CSF. The bulk of information on G-CSF production comes from experiments demonstrating induction of expression of mRNA or protein by a variety of stimuli typically associated with an inflammatory response (Table 1).

These in vitro cell- and tissue-based observations gel nicely with in vivo observations of G-CSF production in whole animals. Serum levels of G-CSF are normally undetectable, or are at very low levels, in normal mice and humans (86–88). However, in response to significant infection, serum levels are markedly elevated. With a murine model, Cheers et al. (89) demonstrated a pronounced increase in serum G-CSF levels 48 hr after the inoculation of *Listeria monocytogenes*. In a susceptible strain of mice, this culminated in fatal infection within 72 hr. Circulating G-CSF levels continued to rise until death. In mice genetically resistant to listeria infection, G-CSF levels reached a peak at 48 hr and, subsequently, declined with the resolution of the infection. This pattern of circulating G-CSF levels peaking in the acute phase of infection and falling with recovery has been confirmed in humans (87).

Although the production of G-CSF is clearly highly responsive to infective or inflammatory insults, little is known about the baseline production of G-CSF by primary cells or, until recently, whether such a process exists. As will be discussed later, the role of G-CSF in regulating steady-state hematopoiesis is only now beginning to be understood. No conclusive evidence of constitutive expression and secretion of G-CSF protein by unstimulated primary cells has been reported. Established human bone marrow stromal layers in long-term culture and various transformed stromal cell lines may apparently constitutively produce G-CSF, as determined by bioassay of conditioned media, or by detection of mRNA (83,84,90,91). However, the manipulations required to establish these cultures prevent the exclusion of other stimulatory factors or events driving G-CSF expression. The kinetics and regulation of G-CSF production by bone marrow stromal cells during steady-state hematopoiesis remain unknown.

Table 1 Stimuli for the Cellular Production of G-CSF

Stimulus	Cell type	Ref.
Endotoxin	Whole animal and tissues	69
	Monocytes and macrophages	71,72
Interleukin-1	Endothelial cells	74,75
	Fibroblasts	
	Dermal	76
	Synovial	78
	Stromal cells	
	Cell lines	82,83
	Long-term culture stroma	84,90
	Astroglia	79
	Chondrocytes	81
	Arterial smooth-muscle cells	80
TNF-α	Endothelial cells	74
	Fibroblasts: adult lung	77
	Astroglia	85
	Arterial smooth-muscle cells	80
M-CSF	Monocytes and macrophages	70
Products of activated T cells (IL-3, IFN-γ)	Monocytes and macrophages	70,73

The regulation of G-CSF gene expression at the molecular level appears to be complex and may involve both transcriptional and posttranscriptional mechanisms. As described earlier in this chapter, three cis-acting promoter sequences are required for induction of G-CSF production by monocytes stimulated with lipopolysaccharide (14). The components of one of these regions, GPE-1, have been studied by several groups in some detail, and transcription factors capable of binding various sequences in this region have been described. In transfected fibroblasts treated with tumor necrosis factor-alpha (TNF-α) and IL-1, the CSF box (or CK-1) sequence directs increased transcription from a heterologous promoter (15,92). The same investigators also reported that a nuclear factor, NF-GMa, binds to this CK-1 sequence in response to the same stimuli and that mutations that affect NF-GMa binding correlate with diminished transcriptional activity (15). Other nuclear proteins may also be involved in the regulation of G-CSF gene expression. A leucine zipper protein, GPE-1–binding protein (GPE-1BP), binds GPE-1 in the proximal nonconsensus region, but does not appear to independently regulate transcription. Three other leucine zipper proteins that interact with GPE-1BP and possibly bind (and regulate) GPE-1 have also been identified (17,18).

In model systems characterized by constitutive transcription of the G-CSF gene, evidence for posttranscriptional regulation has also been documented. In such cells, mRNA has a short half-life, which increases transiently after exposure to TNF-α, IL-1, cyclohexamide, or lipopoly-saccharide (LP5), resulting in accumulation of G-CSF message (93–96). Further investigations are required to elucidate the relative importances of the transcriptional and posttranscriptional mechanisms for regulation of G-CSF gene expression. Given the wide variety of cell types known to be capable of producing G-CSF and the diverse data already derived from various model systems, it seems possible that distinct mechanisms operate in different cell types and in response to different inductive signals.

C. G-CSF Receptor

Granulocyte colony-stimulating factor interacts with target cells through a specific transmembrane receptor. Studies using ^{125}I-labeled G-CSF have demonstrated that G-CSF receptors are largely confined to the hematopoietic system, and within this system to be predominantly expressed on mature neutrophils and their morphologically recognizable precursors (4,8,97). In these studies, essentially all cells within the neutrophil lineage displayed receptors, whereas only a proportion of monocytic cells displayed G-CSF receptors. Table 2 summarizes the results of quantitative studies of receptor numbers on normal murine and human hematopoietic cells. The G-CSF receptors expressed on normal hemopoietic cells and cell lines bind G-CSF with an affinity of 60–300 pM. Primary human acute and chronic myeloid leukemia cells also expressed receptors for G-CSF in similar numbers (broad range 100–3000 per cell) and with binding affinities similar to normal myeloid cells (8,47,98). Receptors have also been found on two normal nonhematopoietic cell types: human endothelial cells (65) and human placental and trophoblastic cells (99).

The kinetics of G-CSF binding to its receptor and subsequent cellular processing of the ligand–receptor complex at physiological temperature have been described in detail (100,101). Ligand binding is characterized by rapid kinetic association rates and slow dissociation rates. Most of the G-CSF bound to receptors is internalized, and G-CSF accumulates intracellularly. Occupied G-CSF receptors are internalized five to ten times more rapidly than unoccupied receptors so that incubation with G-CSF results in a net loss of receptors from the cell surface.

The G-CSF receptor belongs to the class of cytokine receptors known as the hematopoietin receptor family. This family of receptors is characterized by a conserved 200-amino–acid

Table 2   G-CSF Receptors on Normal
Hematopoietic Cells

	Mean number of receptors per cell	
	Murine	Human
Bone marrow		
Blast cells	160	300
Promyelocytes	180	500
Metamyelocytes	120	400
Neutrophils	300	300
Promonocytes	60	100
Monocytes	30	50
Eosinophils	0	0
Lymphocytes	0	0
Nucleated erythroid cells	0	0
Peripheral blood		
Neutrophils		500–3000

Source: Modified from Ref. 20.

extracellular domain (with a predicted structure of two barrels, each containing seven strands; 24, 102) and a cytoplasmic domain devoid of apparent enzymic activity. The human and murine G-CSF receptors consist of 812 and 813 amino acids, respectively (103–105). Figure 4 is a stylized representation of the receptor molecule indicating the important structural and functional domains. As with the erythropoietin receptor and growth hormone receptor, ligand binding to the extracellular domains is thought to induce homodimerization, a necessary prelude to signal transduction. At the amino acid sequence level, however, the G-CSF receptor is most closely related to several of the members of the hematopoietin receptor family, which heterodimerize in response to ligand-binding: leukemia inhibitory factor (LIF) receptor and GP130 (the subunit of the IL-6, LIF, CNTF, oncostatin-M, and IL-11 receptors; 106). In keeping with other members of the family, the G-CSF receptor does not have a catalytic domain, but, nevertheless, induces tyrosine phosphorylation of multiple cellular substrates (107). It is presumed that, following ligand binding, signal transduction proceeds by the association and activation of cytoplasmic kinases. Indeed, recent data have suggested that JAK-1, which also associates with other hematopoietin domain receptors, is associated with the G-CSF receptor, and that its tyrosine kinase activity is activated by receptor dimerization (108). Nicholson et al. (108) detected tyrosine phosphorylation of both the G-CSF receptor and JAK-1 within 2 min of exposure of Chinese hamster ovary (CHO) cells transfected with human G-CSF receptor cDNA to G-CSF. By analogy with the action of JAK-related kinases in interferon signaling, it is likely that JAK-1 activation leads to phosphorylation of cytoplasmic transcription factors that then translocate to the nucleus and activate gene transcription (109).

Recent evidence strongly implicates the G-CSF receptor as one potential source of diversity in cellular responses to G-CSF. Structure–function studies have demonstrated that separate regions of the cytoplasmic domain of the G-CSF receptor are responsible for the signals driving the distinct responses of proliferation and differentiation. Independent studies by Fukunaga et al. (110) and Ziegler et al. (106), using murine cell lines transfected with deletion-mutant and

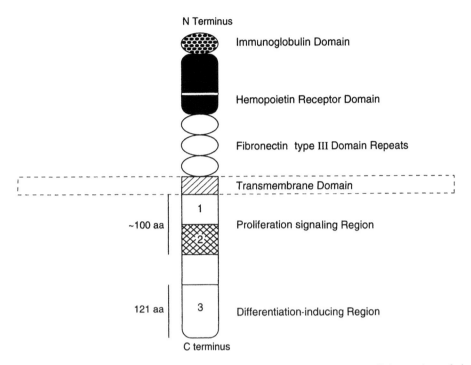

Figure 4 Stylized representation of the G-CSF receptor: The extracellular region of the molecule comprises an NH_2-terminal immunoglobulin-like domain; a typical hematopoietin receptor family domain, containing the WSXWS motif (solid white bar); and three fibronectin type III domain repeats. In the intracytoplasmic domain, areas 1 and 2 are required for transduction of a maximum proliferative signal. Within the 57 residues of area 1 are sequences absolutely required for the proliferative signal. Area 3 contains sequences necessary, but not sufficient, for the delivery of a differentiative signal.

chimeric-mutant receptors, have indicated that the region responsible for the transduction of the growth signal lies within the membrane-proximal 100 amino acids of the cytoplasmic tail of the G-CSF receptor. The latter investigators reported that a region that is homologous to the conserved box-1 domains in other receptors and that lies within the first 57 residues was absolutely required and sufficient to deliver a proliferative signal in transfected cells. However, a maximal proliferative response required that other sequences in the first 96 amino acids be present. Fukunaga et al. (110) found the membrane-proximal 76 amino acids of the cytoplasmic domain was sufficient for growth signal transduction, whereas both the membrane-proximal and the COOH-terminal regions of the receptor were required to transduce a differentiation signal, as measured by induction of expression of the myeloperoxidase gene.

   The relevance of these studies in artificial cell–receptor systems to the physiology of primary myeloid cells normally expressing G-CSF receptors has been partly validated by a recent report describing a naturally occurring mutation of the G-CSF receptor in a patient with Kostmann's syndrome. Kostmann's syndrome, or severe congenital neutropenia, is a rare disease characterized by arrested maturation of myeloid precursors in the bone marrow and, consequently, decreased numbers of neutrophils in the peripheral blood. The cause of most cases of Kostmann's syndrome is unknown and most patients respond favorably to G-CSF therapy (111). However, in one instance, a nonsense mutation in the G-CSF receptor gene has been identified, and in vitro studies indicate a causal role in the patient's neutropenia (112). The mutant gene encodes a COOH-terminally truncated receptor. When transfected, the mutant receptor can transduce a proliferative, but not a differentiative signal, analogous to the COOH-terminal deletion mutants previously described. The mutation appears to have been a somatic mutation in an abnormal clone of progenitor cells. Because the other allele of the gene appears normal, it is postulated that the mutant G-CSF receptor protein has a dominant, negative effect on the wild-type G-CSF receptor (112,113). Abnormalities of the G-CSF receptor are unlikely to explain most cases of Kostmann's syndrome. A study of the transmembrane and cytosolic portions of the receptor by polymerase chain reaction (PCR) and single-strand conformational polymorphism analysis in cells from six patients with Kostmann's syndrome failed to detect major structural abnormalities (114).

D. In Vivo Biology of G-CSF Treatment

This section summarizes key aspects of the effects of G-CSF treatment on the biology of whole animals. There is already a large and ever-increasing body of literature addressing this field. Consequently, the references selected are not intended to be comprehensive, but to be representative examples illustrating major findings. Preclinical data are emphasized; the effects of G-CSF in human trials are examined in detail in Chapter 24.

Normal Animals

In keeping with the in vitro actions of G-CSF in promoting the proliferation and maturation of neutrophil precursors, the most pronounced effect of treatment with G-CSF is the development of a striking neutrophil leukocytosis. This was initially demonstrated in Syrian hamsters (115) and mice (116,117) and, subsequently, in dogs (118), primates (119), and humans (120–122). The magnitude of the neutrophilic response is dose-dependent (119,121,123). In mice, G-CSF induces a 2.5-fold rise in neutrophil numbers in the peripheral blood within 2–3 hr of a single injection (116). The same response is observed in humans, although it is preceded by a transitory neutropenia in the first hour after injection (121). Neutrophilia persists throughout the duration of G-CSF treatment, although plateau levels are reached after 3–5 days. The peak levels observed represent more than tenfold rises in hamsters (115), mice (116,123), primates (119), and humans (121).

At least two mechanisms of action appear to be involved in the induction of the peripheral blood neutrophil leukocytosis by G-CSF treatment. The rapid rise in neutrophil numbers within hours of injection is best explained by release of mature neutrophils from the large storage pool in the bone marrow. Subsequently, the persistent neutrophilia is sustained by an increased production of neutrophils. Kinetic studies in mice and humans have measured the rate of production of mature neutrophils during G-CSF therapy and also their half-life in the circulation (124–126). The half-life of neutrophils in the blood was unaffected by G-CSF, but the transit time from the marrow into the blood for newly produced neutrophils was reduced from 4–5 days to 1–2 days. Production was estimated to be amplified by approximately 14.5–20 times over baseline in mice, and 9 times over baseline in humans. The authors calculated this to be equivalent to an extra three to four amplification divisions in the mitotic neutrophil–precursor compartment (neutrophil–CFC and their morphologically recognizable progeny: myeloblasts, promyelocytes, and myelocytes). In keeping with these findings, in vivo administration of G-CSF has been demonstrated to increase the proportion of progenitor cells, myeloblasts, promyelocytes, and myelocytes in S phase (124,125,127).

In contrast with the uniform peripheral blood responses across species, the effects of G-CSF treatment on bone marrow cellularity and splenic hematopoiesis vary between species. In mice, total marrow cellularity falls, and there is a dramatic redistribution of hematopoiesis. The numbers of committed progenitor cells and CFU-S in the bone marrow fall, and splenic hematopoiesis increases substantially (116,128–130). This effect is particularly marked for erythropoiesis. In hamsters (115), primates (119), and humans (121), bone marrow cellularity increases, and redistribution to the spleen has not been identified as a significant feature of the response to G-CSF. This presumably reflects intrinsic physiological differences in hematopoietic responses to stress between species. However, extramedullary hematopoiesis may occur to a limited extent in humans and has been observed in monkeys. Splenomegaly is a common finding in patients with severe chronic neutropenias treated with G-CSF on a long-term basis (131).

An important and unpredicted response to G-CSF in vivo is the mobilization of progenitor cells into the peripheral blood. This was first documented in humans (132) and, subsequently, in mice (128–130). The effect is dramatic and exceeds the neutrophilic response in relative magnitude. Increases of 40- and 570-fold have been documented in circulating progenitor cell numbers, and the effect is not restricted to progenitor cells of the neutrophil lineage. Erythroid, megakaryocyte, eosinophil, and mixed CFC are all increased. The presence of pluripotent lymphohematopoietic stem cells among these circulating cells has been documented by serial transplantation in mice (133).

The mechanism by which progenitor cell numbers are increased in the blood is unknown. As elevated levels of circulating progenitor cells are also seen after treatment with several other hematopoietic growth factors and during recovery from chemotherapy-induced marrow aplasia, it is likely that this effect of G-CSF is indirect. However, mobilization is not simply a process of random release from the marrow. The circulating progenitor cell population induced by G-CSF is a highly selected population: megakaryocyte progenitor cells are markedly more overrepresented in the blood than in the marrow (130), and cycling cells are excluded (134). In contrast to the rapid development of neutrophilia, the increase in progenitor cell numbers in the blood takes 3–4 days of G-CSF treatment to develop in humans, reaching a peak after 5–7 days and then persisting at a lower, but elevated, level while G-CSF injections continue (135,136).

Neutropenic Animals

Preclinical studies with G-CSF showed accelerated neutrophil recovery after cyclophosphamide treatment in mice (116) and primates (119) and after irradiation in mice (137) and dogs (138). Furthermore, in mice (139) and hamsters (140), treatment with G-CSF alone significantly reduces the mortality associated with induced bacterial infections. Matsumoto et al. (141) also documented the profoundly protective effect of G-CSF given prophylactically in neutropenic mice subsequently infected with a range of different bacteria and candida.

Animal Models of Infection

Granulocyte colony-stimulating factor has also been investigated in nonneutropenic animal models of infection. The underlying premise examined in these experiments was that augmentation of the functional activity of preexisting neutrophils (as seen in vitro), along with an increase in total neutrophil number, could prevent or ameliorate bacterial infections. G-CSF treatment dramatically improved the outcome of induced *Pseudomonas aeruginosa* infections in mice (139,142). In a rat model of neonatal group B streptococcal infection, G-CSF alone did not reduce the mortality rate; however, in combination with the appropriate antibiotics, survival at 72 hr was increased from 4% in control rats to 91% (143).

Animal Models of Leukemia

The inhibitory and differentiative effects of G-CSF on some myeloid leukemia cell lines and primary leukemias, observed in vitro, prompted investigation of G-CSF as an antileukemia agent in vivo. Several studies have demonstrated beneficial effects of daily G-CSF injections in mice implanted with autonomous murine myeloid leukemic cell lines (144,145); however, the benefits were modest and amounted to a short prolongation of mean survival times in uniformly fatal diseases. These models are clearly limited in terms of relevance to primary human leukemias; nevertheless, they suggest G-CSF is unlikely to possess significant clinical antileukemic activity as a single agent.

Chronic G-CSF Excess and Miscellaneous Effects

The biology of long-term exposure to high circulating levels of G-CSF has been studied in mice using retroviral expression of the gene in bone marrow cells transplanted into irradiated recipients (146). This model essentially confirmed the principal findings of short-term experiments using injected G-CSF: neutrophilic granulocytosis was observed in all hematopoietic tissues, with marked increases in splenic and peripheral blood hemopoiesis, and neutrophilic infiltration was prominent in the liver and lungs. This hyperplasia was nonneoplastic and apparently benign. In vivo, G-CSF also rapidly and selectively increases, approximately twofold, the bloodflow to the bone marrow, but not to the bone, in rats within 8 hr of a single injection (147).

E. Physiology of G-CSF

The precise physiological roles for G-CSF have yet to be defined. Although the actions of G-CSF have been well characterized in in vitro systems, and the effects of pharmacological doses of G-CSF have been described in whole animals and humans, little is known about the function of G-CSF in homeostasis in whole animals.

When one considers its most specific action in vitro—the production of mature neutrophils from precursors—G-CSF could play a role in the regulation of hematopoiesis in two conceptually distinct settings. Clearly, G-CSF could be an important positive regulator of neutrophil numbers during resting or baseline hematopoiesis. If this were true, then deficiency of G-CSF would be expected to result in deficiencies in neutrophil numbers or function. Alternatively, G-CSF could be required only as a positive regulator in stress or emergency situations (e.g., marrow aplasia or infection). This latter model would predict that the absence of G-CSF does not lead to perturbations of neutrophil numbers or function at rest, but that the response to a challenge is diminished. This simplistic division of potential physiological roles for G-CSF does not imply that the roles are mutually exclusive. It also ignores the in vitro evidence of the effects of G-CSF on nonhematopoietic tissues.

Indirect evidence does exist for important roles for G-CSF in the regulation of neutrophil production in both steady-state and stress hematopoiesis. Hammond and co-workers (148) have developed a model of severe chronic neutropenia in dogs by inducing deficiency with neutralizing antibodies to G-CSF. After stimulating an initial neutrophilia, the administration of recombinant human G-CSF immunizes normal dogs, which then develop severe neutropenia lasting for up to 4 months. The mechanism of neutropenia is the neutralization of native canine G-CSF by cross-reactive antibodies induced by the human G-CSF. The circulating and bone marrow levels of cells of other lineages are apparently unaffected. This model of induced "G-CSF deficiency" argues strongly for a physiological role for G-CSF in baseline hematopoiesis.

The most convincing evidence for a physiological role for G-CSF in the maintenance of steady-state neutrophil production is the creation of G-CSF–deficient mice by homologous recombination, resulting in gene disruption. Lieschke et al. (149) have confirmed that G-CSF–deficient mice have reduced levels of mature neutrophils in the blood and bone marrow and a reduction in neutrophil progenitor cells in the bone marrow. These observations suggest that, although G-CSF is necessary for the maintenance of normal neutrophil production, it is not absolutely required for the generation of mature neutrophils. This is consistent with the apparent redundancy of actions among the hematopoietic regulators evident in vitro and, perhaps, reflects the complexity and subtlety of the system in vivo (150).

In stress hematopoiesis, a role for G-CSF can be inferred from several observations. First, serum levels of G-CSF are markedly elevated in response to significant infection (87–89) or neutropenia (88,151). In infection, this is accompanied by the development of neutrophilia. Second, the administration of pharmacological doses of G-CSF to patients with chemotherapy-induced marrow aplasia markedly accelerates the recovery of circulating neutrophil numbers (121,152,153).

Finally, a principal physiological role of G-CSF would appear to be as a key mediator of the neutrophil response to microbial infection. G-CSF–deficient mice have a markedly impaired ability to control infection with *Listeria*, demonstrating a high mortality rate and excessive microabscess formation, compared with normal controls (149). During infection, G-CSF acts not only as a systemic hormone to increase neutrophil production, but also as a locally produced and locally acting positive regulator of neutrophil function.

ACKNOWLEDGMENT

The authors' work is supported by the National Health and Medical Research Council of Australia, Grant CA 22556 from the United States National Institutes of Health, and the Cooperative Research Centre Program of the Australian Government.

REFERENCES

1. Nicola NA, Metcalf D, Johnson GR, Burgess AW. Separation of functionally distinct human granulocyte–macrophage colony-stimulating factors. Blood 1979; 54:614–627.
2. Burgess AW, Metcalf D. Characterization of a serum factor stimulating the differentiation of myelomonocytic leukemic cells. Int J Cancer 1980; 39:647–654.
3. Nicola NA, Metcalf D, Matsumoto M, Johnson GR. Purification of a factor inducing differentiation in murine myelomonocytic leukemia cells: identification as granulocyte colony-stimulating factor (G-CSF). J Biol Chem 1983; 258:9017–9023.
4. Nicola NA, Begley CG, Metcalf D. Identification of the human analogue of a regulator that induces differentiation in murine leukaemic cells. Nature 1985; 314:625–628.
5. Welte K, Platzer E, Lu L, Gabrilove JL, Levi E, Mertelsmann R, Moore MAS. Purification and biochemical characterization of human pluripotent hematopoietic colony-stimulating factor. Proc Natl Acad Sci USA 1985; 82:1526–1530.
6. Nomura H, Imazeki I, Oheda M, Kubota N, Tamura M, Ono M, Ueyama Y, Asano S. Purification and characterization of human granulocyte colony-stimulating factor (G-CSF). EMBO J 1986; 5:871–876.
7. Nagata S, Tsuchiya M, Asano S, Kaziro Y, Yamazaki T, Yamamoto O, Hirata Y, Kubota N, Oheda M, Nomura H, Ono M. Molecular cloning and expression of cDNA for human granulocyte colony-stimulating factor. Nature 1986; 319:415–418.
8. Souza LM, Boone TC, Gabrilove J, Lai PH, Zsebo KM, Murdock DC, Chazin VR, Bruszewski J, Lu H, Chen KK, Barendt J, Platzer E, Moore MAS, Mertelsman R, Welte K. Recombinant human granulocyte colony-stimulating factor: effects on normal and leukemic myeloid cells. Science 1986; 232:61–65.
9. Tsuchiya M, Asano S, Kaziro Y, Nagata S. Isolation and characterization of the cDNA for murine granulocyte colony-stimulating factor. Proc Natl Acad Sci USA 1986; 83:7633–7637.
10. Nagata S, Tsuchiya M, Asano S, Yamamoto O, Hirata Y, Kubota N, Oheda M, Nomura H, Yamazaki

T. The chromosomal gene structure and two mRNAs for human granulocyte colony-stimulating factor. EMBO J 1986; 5:575–581.

11. Tsuchiya M, Kaziro Y, Nagata S. The chromosomal gene structure for murine granulocyte colony-stimulating factor. Eur J Biochem 1987; 165:7–12.

12. Nagata S. Gene structure and function of granulocyte colony-stimulating factor. Bioessays 1989; 10:113–117.

13. Nishizawa M, Tsuchiya M, Watanabe-Fukunaga R, Nagata S. Multiple elements in the promoter of granulocyte colony-stimulating factor gene regulate its constitutive expression in human carcinoma cells. J Biol Chem 1990; 265:5897–5902.

14. Nishizawa M, Nagata S. Regulatory elements responsible for inducible expression of the granulocyte colony-stimulating factor gene in macrophages. Mol Cell Biol 1990; 10:2002–2011.

15. Kuczek ES, Shannon MF, Pell LM, Vadas MA. A granulocyte–colony-stimulating factor gene promoter element responsive to inflammatory mediators is functionally distinct from an identical sequence in the granulocyte–macrophage colony-stimulating factor gene. J Immunol 1991; 146:2426–2433.

16. Miyatake S, Seiki M, Yoshida M, Arai K-I. T-cell activation signals and human T-cell leukemia virus type 1-encoded p40x protein activate the mouse granulocyte–macrophage colony-stimulating factor gene through a common DNA element. Mol Cell Biol 1988; 8:5581–5587.

17. Nishizawa M, Wakabayashi-Ito N, Nagata S. Molecular cloning of cDNA and a chromosomal gene encoding GPE-1BP, a nuclear protein which binds to granulocyte colony-stimulating factor promoter element 1. FEBS Lett 1991; 282:95–97.

18. Nishizawa M, Nagata S. cDNA clones encoding leucine-zipper proteins which interact with G-CSF gene promoter element 1-binding protein. FEBS Lett 1992; 299:36–38.

19. Muller M, Ruppert S, Schaffner W, Matthias P. A cloned octamer transcription factor stimulates transcription from lymphoid-specific promoters in non-B cells. Nature 1988; 336:544–551.

20. Nicola NA. Granulocyte colony-stimulating factor. In: Dexter TM, Garland JM, Testa NG, eds. Colony-Stimulating Factors, Molecular and Cellular Biology. New York: Marcel Dekker, 1990:77–109.

21. Oheda M, Nasegawa M, Hattori K, Kuboniwa H, Kojima T, Orita T, Tomonou K, Yamazaki T, Ochi N. O-Linked sugar chain of human granulocyte colony-stimulating factor protects it against polymerization and denaturation allowing it to retain its biological activity. J Biol Chem 1990; 265:11432–11435.

22. Osslund T, Boone T. Biochemistry and structure of filgrastim (r-metHuG-CSF). In: Morstyn G, Dexter TM, eds. Filgrastim (r-metHug-CSF) in Clinical Practice. New York: Marcel Dekker, 1994:23–31.

23. Hill CP, Osslund TD, Eisenberg D. The structure of granulocyte colony-stimulating factor and its relationship to other growth factors. Proc Natl Acad Sci USA 1993; 90:5167–5171.

24. de Vos AM, Ultsch M, Kossiakoff AA. Human growth hormone and extracellular domain of its receptor: crystal structure of the complex. Science 1992; 255:306–312.

25. Walter MR, Cook WJ, Ealick SE, Nagabhushan TL, Trotta PP, Bugg CE. Three-dimensional structure of recombinant human granulocyte-macrophage colony-stimulating factor. J Mol Biol 1992; 224:1075–1085.

26. Senda T, Shimazu T, Matsuda S, Kawano G, Shimizu H, Nakamura KT, Mitsui Y. Three-dimensional crystal structure of recombinant murine interferon-alpha. EMBO J 1992; 11:3193–3201.

27. McKay DB. Unraveling the structure of IL-2. Science 1992; 257:412–413.

28. Bazan JF. Unraveling the structure of IL-2. Science 1992; 257:410–412.

29. Powers R, Garrett DS, March CJ, Frieden EA, Gronenborn AM, Clore GM. Three-dimensional solution structure of human interleukin-4 by multidimensional heteronuclear magnetic resonance spectroscopy. Science 1992; 256:1673–1677.

30. Lu HS, Clogston CL, Narhi LO, Merewether LA, Pearl WR, Boone TC. Folding and oxidation of recombinant human granulocyte colony-stimulating factor produced in *Escherichia coli*. Characterization of the disulfide-reduced intermediates and cysteine–serine analogs. J Biol Chem 1992; 267:8770–8777.

31. Kuga T, Komatsu Y, Yamasaki M, Sekine S, Miyaji H, Nishi T, Sato M, Yokoo T, Asano M, Okabe M, Morimoto M, Itoh S. Mutagenesis of human granulocyte colony stimulating factor. Biochem Biophys Res Commun 1989; 159:103–111.

32. Cunningham BC, Wells JA. High-resolution epitope mapping of hGH-receptor interactions by alanine-scanning mutagenesis. Science 1989; 244:1081–1084.

33. Clark-Lewis I, Lopez AF, To LB, Vadas MA, Schrader JW, Hood LE, Kent SBH. Structure–function studies of human granulocyte–macrophage colony-stimulating factor. Identification of residues required for activity. J Immunol 1988; 141:881–889.

34. Le HV, Seelig GF, Syto R, Ramanathan L, Windsor WT, Borkowski D, Trotta PP. Selective proteolytic cleavage of recombinant human interleukin-4. Evidence for a critical role of the C-terminus. Biochemistry 1991; 30:9576–9582.

35. Brakenhoff JPJ, Hart M, De Groot ER, Di Padova F, Aarden LA. Structure–function analysis of IL-6. Epitope mapping of neutralizing monoclonal antibodies with amino- and carboxyl-terminal deletion mutants. J Immunol 1990; 145:561–568.

36. Ishikawa M, Iijima H, Satake-Ishikawa R, Tsumura H, Iwamatsu A, Kadoya T, Shimada Y, Fukamachi H, Kobayashi K, Matsuki S. The substitution of cysteine 17 of recombinant human G-CSF with alanine greatly enhanced its stability. Cell Struct Funct 1992; 17:61–65.

37. Layton JE, Morstyn G, Fabri LJ, Reid GE, Burgess AW, Simpson RJ, Nice EC. Identification of a functional domain of human granulocyte colony-stimulating factor using neutralizing monoclonal antibodies. J Biol Chem 1991; 266:23815–23823.

38. Metcalf D, Nicola NA. Proliferative effects of purified granulocyte colony-stimulating factor (G-CSF) on normal mouse hemopoietic cells. J Cell Physiol 1983; 116:198–206.

39. Strife NA, Lambek C, Wisniewski D, Gulati S, Gasson JC, Golde DW, Welte K, Gabrilove JL, Clarkson B. Activities of four purified growth factors on highly enriched human hematopoietic progenitor cells. Blood 1987; 1508–1523.

40. Leary AG, Zeng HQ, Clark SC, Ogawa M. Growth factor requirements for survival in G_0 and entry into the cell cycle of primitive human hemopoietic progenitors. Proc Natl Acad Sci USA 1992; 89:4013–4017.

41. Sutherland HJ, Eaves CJ, Lansdorp PM, Thacker JD, Hogge DE. Differential regulation of primitive human hematopoietic cells in long-term cultures maintained on genetically engineered murine stromal cells. Blood 1991; 78:666–672.

42. Bodine DM, Crosier PS, Clark SC. Effects of hematopoietic growth factors on the survival of primitive stem cells in liquid suspension culture. Blood 1991; 78:914–920.

43. Haylock D, To LB, Dowse TL, Juttner CA, Simmons PA. Ex vivo expansion and maturation of peripheral blood CD34[+] cells into the myeloid lineage. Blood 1992; 80:1405–1412.

44. McKinstry WJ, Li C-L, Rasko JEJ, Nicola NA, Johnson GR, Metcalf D. Cytokine receptor expression on hematopoietic stem and progenitor cells. Blood (in press).

45. Metcalf D. Regulator-induced suppression of myelomonocytic leukemic cells: clonal analysis of early cellular events. Int J Cancer 1982; 30:203–210.

46. Begley CG, Metcalf D, Nicola NA. Purified colony-stimulating factors (G-CSF and GM-CSF) induce differentiation in human HL60 leukemic cells with suppression of clonogenicity. Int J Cancer 1987; 39:99–105.

47. Begley CG, Metcalf D, Nicola NA. Primary human myeloid leukemia cells: comparative responsiveness to proliferative stimulation by GM-CSF or G-CSF and membrane expression of CSF receptors. Leukemia 1987; 1:1–8.

48. Vellenga E, Ostapovicz D, O'Rourke B, Griffin JD. Effects of recombinant IL-3, GM-CSF, and G-CSF on proliferation of leukemic clonogenic cells in short-term and long-term cultures. Leukemia 1987; 1:584–589.

49. Kelleher C, Miyauchi J, Wong G, Clark S, Minden MD, McCulloch EA. Synergism between recombinant growth factors, GM-CSF and G-CSF, acting on the blast cells of acute myeloblastic leukemia. Blood 1987; 69:1498–1503.

50. Miyauchi J, Kelleher C, Yang Y-C, Wong GG, Clark SC, Minden MD, Minkin S, McCulloch EA.

The effects of three recombinant growth factors, IL-3, GM-CSF and G-CSF, on the blast cells of acute myeloblastic leukemia maintained in short-term suspension culture. Blood 1987; 70:657–663.

51. Asano Y, Shibuya T, Okamura S, Yamaga S, Otsuka T, Niho Y. Effect of human recombinant granulocyte/macrophage colony-stimulating factor and native granulocyte colony-stimulating factor on clonogenic leukemic blast cells. Cancer Res 1987; 47:5647–5648.

52. Nara N, Tohda S, Suzuki T, Nagata K, Yamashita Y, Imai Y, Morio T, Besho M, Shiuya A, Adachi Y. Terminal differentiation to mature neutrophils and eosinophils in suspension culture of the blast progenitors in acute myeloblastic leukemia. Hematol Pathol 1990; 4:125–134.

53. Begley CG, Lopez AF, Nicola NA, Warren DJ, Vadas MA, Sanderson CJ, Metcalf D. Purified colony-stimulating factors enhance the survival of human neutrophils and eosinophils in vitro: a rapid and sensitive microassay for colony-stimulating factors. Blood 1986; 68:162–166.

54. Kitagawa S, Yuo A, Souza LM, Saito M, Miura Y, Takaku F. Recombinant human granulocyte colony-stimulating factor enhances superoxide release in human granulocytes stimulated by the chemotactic peptide. Biochem Biophys Res Commun 1987; 144:1143–1146.

55. Lindemann A, Herrmann F, Oster W, Haffner G, Meyenburg W, Souza LM, Mertelsmann R. Hematological effects of recombinant human granulocyte colony-stimulating factor in patients with malignancy. Blood 1989; 74:2644–2651.

56. Roilides E, Walsh TJ, Pizzo PA, Rubin M. Granulocyte colony-stimulating factor enhances the phagocytic and bactericidal activity of normal and defective human neutrophils. J Infect Dis 1991; 163:579–583.

57. Platzer E, Welte K, Gabrilove J, Lu L, Harris P, Mertelsmann R, Moore MAS. Biological activities of a human pluripotent hemopoietic colony stimulating factor on normal and leukemic cells. J Exp Med 1985; 162:1788–1801.

58. Ogle JD, Noel JG, Sramkoski RM, Ogle CK, Alexander JW. The effects of cytokines, platelet activating factor, and arachidonate metabolites on C3b receptor (CR1, CD35) expression and phagocytosis by neutrophils. Cytokine 1990; 2:447–455.

59. Yuo A, Kitagawa S, Ohsaka A, Saito M, Takaku F. Stimulation and priming by granulocyte colony-stimulating factor and granulocyte–macrophage colony-stimulating factor: qualitative and quantitative differences. Biochem Biophys Res Commun 1990; 171:491–497.

60. Avalos BR, Gasson JC, Hedvat C, Quan SG, Baldwin CG, Weisbart RH, Williams RE, Golde DW, DiPersio JF. Human granulocyte colony-stimulating factor: biological activities and receptor characterization on hematopoietic cells and small cell lung cancer cell lines. Blood 1990; 75:851–857.

61. Vadas MA, Nicola NA, Metcalf D. Activation of antibody-dependent cell-mediated cytotoxicity of human neutrophils and eosinophils by separate colony-stimulating factors. J Immunol 1983; 130:795–799.

62. Platzer E, Oez S, Welte K, Sendler A, Gabrilove JL, Mertelsmann R, Moore MA, Kalden JR. Human pluripotent hemopoietic colony stimulating factor: activities on human and murine cells. Immunobiology 1986; 172:185–193.

63. Weibart KH, Kacena A, Schuh A, Golde DW. GM-CSF induces human neutrophil IgA-mediated phagocytosis by an IgA Fc receptor activation mechanism. Nature 1988; 332:647–648.

64. Wang JM, Chen ZG, Colella S, Bonilla MA, Welte K, Bordignon C, Mantovani A. Chemotactic activity of recombinant human granulocyte colony-stimulating factor. Blood 1988; 72:1456–1460.

65. Bussolino F, Wang JM, Defilippi P, Turrini F, Sanavio F, Edgell CJ, Aglietta M, Arese P, Mantovani A. Granulocyte- and granulocyte–macrophage colony-stimulating factors induce human endothelial cells to migrate and proliferate. Nature 1989; 337:471–473.

66. Bussolino F, Ziche M, Wang JM, Alessi D, Morbidelli L, Cremona O, Bosia A, Marchisio PC, Mantovani A. In vitro and in vivo activation of endothelial cells by colony-stimulating factors. J Clin Invest 1991; 87:986–995.

67. Yong K, Cohen H, Khwaja A, Jones HM, Linch DC. Lack of effect of granulocyte–macrophage and granulocyte colony-stimulating factors on cultured endothelial cells. Blood 1991; 77:1675–1680.

68. Berdel WE, Danhauser RS, Steinhauser G, Winton EF. Various human hematopoietic growth factors

(interleukin-3, GM-CSF, G-CSF) stimulate clonal growth of nonhematopoietic tumor cells. Blood 1989; 73:80–83.

69. Nicola NA, Metcalf D. Biochemical properties of differentiation factors for murine myelomonocytic leukemia cells in organ conditioned media. Separation from colony-stimulating factors. J Cell Physiol 1981; 112:257–264.

70. Metcalf D, Nicola NA. Synthesis by mouse peritoneal cells of G-CSF, the differentiation inducer for myeloid leukemia cells: stimulation by endotoxin, M-CSF and multi-CSF. Leuk Res 1985; 9:35–50.

71. Vadas MA, Lopez AF. Regulation of granulocyte function by colony-stimulating factors and monoclonal antibodies. Lymphokines 1985; 12:179–200.

72. Sieff CA, Niemeyer CM, Faller DF. The production of hematopoietic growth factors by endothelial accessory cells. Blood Cells 1987; 13:65–74.

73. Rambaldi A, Young DC, Griffin JD. Expression of the M-CSF (CSF-1) gene by human monocytes. Blood 1987; 69:1409–1413.

74. Seelentag WK, Mermod JJ, Montesano R, Vassalli P. Additive effects of interleukin 1 and tumour necrosis factor on the accumulation of the three granulocyte and macrophage colony-stimulating factor mRNAs in human endothelial cells. EMBO J 1987; 6:2261–2265.

75. Zsebo KM, Yuschenkoff VN, Schiffer S, Chang D, McCall E, Dinarello CA, Brown MA, Altrock B, Bagby GC. Vascular endothelial cells and granulopoiesis: interleukin-1 stimulates release of G-CSF and GM-CSF. Blood 1988; 71:99–103.

76. Kaushansky K, Lin N, Adamson JW. Interleukin 1 stimulates fibroblasts to synthesize granulocyte-macrophage and granulocyte colony-stimulating factors. J Clin Invest 1988; 81:92–97.

77. Koeffler HP, Gasson J, Ranyard J, Souza L, Shepard M, Munker R. Recombinant human TNF stimulates production of granulocyte colony-stimulating factor. Blood 1987; 70:55–59.

78. Leizer T, Cebon J, Layton JE, Hamilton JA. Cytokine regulation of colony-stimulating factor production in cultured human synovial fibroblasts: I. Induction of GM-CSF and G-CSF production by interleukin-1 and tumour necrosis factor. Blood 1990; 76:1989–1996.

79. Tweardy DJ, Mott PL, Glazer EW. Monokine modulation of human astroglial cell production of granulocyte colony-stimulating factor and granulocyte–macrophage colony-stimulating factor. I. Effects of IL-1 and IL-1. J Immunol 1990; 144:2233–2241.

80. Zoellner H, Filonzi EL, Stanton HR, Layton JE, Hamilton JA. Human arterial smooth muscle cells synthesize granulocyte colony-stimulating factor in response to interleukin-1 alpha and tumor necrosis factor-alpha. Blood 1992; 80:2805–2810.

81. Campbell IK, Novak U, Cebon J, Layton JE, Hamilton JA. Human articular cartilage and chondro-cytes produce hemopoietic colony-stimulating factors in culture in response to IL-1. J Immunol 1991; 147:1238–1246.

82. Rennick D, Yang G, Gemmell L, Lee F. Control of hemopoiesis by a bone marrow stromal cell clone: lipopolysaccharide- and interleukin-1-inducible production of colony-stimulating factors. Blood 1987; 69:682–691.

83. Yang Y-C, Tsai S, Wong GG, Clark SC. Interleukin-1 regulation of hematopoietic growth factor production by human stromal fibroblasts. J Cell Physiol 1988; 134:292–296.

84. Kittler ELW, McGrath H, Temeles D, Crittenden RB, Kister VK, Quesenberry PJ. Biological significance of constitutive and subliminal growth factor production by bone marrow stroma. Blood 1992; 79:3168–3178.

85. Tweardy DJ, Glazer EW, Mott PL, Anderson K. Modulation by tumor necrosis factor-α of human astroglial cell production of granulocyte–macrophage colony-stimulating factor (GM-CSF) and granulocyte colony-stimulating factor (G-CSF). J Neuroimmunol 1991; 32:269–278.

86. Watari K, Asano S, Shirafuji N, Kodo H, Ozawa K, Takaku F, Shin-ichi K. Serum granulocyte colony-stimulating factor levels in healthy volunteers and patients with various disorders as estimated by enzyme immunoassay. Blood 1989; 73:117–122.

87. Kawakami M, Tsutsumi H, Kumakawa T, Abe H, Hirai M, Kurosaw S, Mori M, Fukushima M. Levels of serum granulocyte colony-stimulating factor in patients with infections. Blood 1990; 76:1962–1964.

88. Cebon J, Layton JE, Maher D, Morstyn G. Endogenous haemopoietic growth factors in neutropenia and infection. Br J Haematol 1994; 86:265–274.

89. Cheers C, Haigh AM, Kelso A, Metcalf D, Stanley ER, Young AM. Production of colony-stimulating factors (CSFs) during infection: separate determinations of macrophage-, granulocyte-, granulocyte–macrophage-, and multi-CSFs. Infect Immun 1988; 56:247–251.

90. Fibbe WE, van Damme J, Billiau A, Goselink HM, Voogt PJ, van Eeden G, Ralph P, Altrock BW, Falkenburg JHF. Interleukin 1 induces human marrow stromal cells in long-term culture to produce granulocyte colony-stimulating factor and macrophage colony-stimulating factor. Blood 1988; 71:430–435.

91. Ciccutina FM, Martin M, Salvaris E, Ashman L, Begley CG, Novotny J, Maher D, Boyd AW. Support of human cord blood progenitor cells on human stromal cell lines transformed by SV40 large T antigen under the influence of an inducible (metallothionein) promoter. Blood 1982; 80:102–112.

92. Shannon MF, Pell LM, Lenardo MJ, Kuczek ES, Occhiodoro FS, Dunn SM, Vadas MA. A novel TNF-α responsive transcription factor which recognizes a regulatory element in haemopoietic growth factor genes. Mol Cell Biol 1990; 10:2950–2959.

93. Koeffler HP, Gasson J, Tobler A. Transcriptional and posttranscriptional modulation of myeloid colony-stimulating factor expression by tumor necrosis factor and other agents. Mol Cell Biol 1988; 8:3432–3438.

94. Ernst TJ, Ritchie AR, Demetri GD, Griffin JD. Regulation of granulocyte- and monocyte-colony stimulating factor mRNA levels in human blood monocytes is mediated primarily at a post-transcriptional level. J Biol Chem 1989; 264:5700–5703.

95. Demetri G, Ernst T, Pratt E, Zenzie B, Rheinwald J, Griffin J. Expression of *ras* oncogenes in cultured human cells alters the transcriptional and posttranscriptional regulation of cytokine genes. J Clin Invest 1990; 86:1261–1269.

96. Demetri GD, Graffin JD. Granulocyte colony-stimulating factor and its receptor. Blood 1991; 78:2791–2808.

97. Nicola NA, Metcalf D. Binding of ^{125}I-labeled granulocyte colony-stimulating factor to normal murine hemopoietic cells. J Cell Physiol 1985; 124:313–321.

98. Begley CG, Metcalf D, Nicola NA. Binding characteristics and proliferative action of purified granulocyte colony-stimulating factor (G-CSF) on normal and leukemic human promyelocytes. Exp Hematol 1988; 16:71–79.

99. Uzumaki H, Okabe Y, Sasaki N, Hagiwara K, Takaku F, Tobita M, Yasukawa K, Ho S, Umezawa Y. Identification and characterization of receptors for granulocyte colony-stimulating factor on human placenta and trophoblastic cells. Proc Natl Acad Sci USA 1989; 86:9323–9326.

100. Nicola NA, Peterson L, Hilton DJ, Metcalf D. Cellular processing of murine colony-stimulating factor (multi-CSF, GM-CSF, G-CSF) receptors by normal hemopoietic cells and cell lines. Growth Factors 1988; 1:41–49.

101. Nicola NA. Kinetic aspects of the interaction of colony-stimulating factors with cellular receptors. In: Gale RP, Golde DW, eds. Recent Advances in Leukemia and Lymphoma. New York: Alan R. Liss, 1987:215–228.

102. Bazan JF. Structural design and molecular evolution of a cytokine receptor superfamily. Proc Natl Acad Sci USA 1990; 87:6934–6938.

103. Fukunaga R, Seto Y, Mizushim S, Nagata S. Three different mRNAs encoding human granulocyte colony-stimulating factor receptor. Proc Natl Acad Sci USA 1990; 87:8702–8706.

104. Fukunaga R, Ishizaka-Ikeda E, Seto Y, Nagata S. Expression cloning of a receptor for murine granulocyte colony-stimulating factor. Cell 1990; 61:341–350.

105. Larson A, David T, Curtis BM, Gimpel S, Sims JE, Cosman D, Park L, Sorensen E, March CJ, Smith CA. Expression cloning of a human granulocyte colony-stimulating factor receptor: a structural mosaic of hemopoietin receptor, immunoglobulin, and fibronectin domains. J Exp Med 1990; 172:1559–1570.

106. Ziegler SF, Bird TA, Morella KK, Mosley B, Gearing DP, Baumann H. Distinct regions of the human

granulocyte colony-stimulating factor receptor cytoplasmic domain are required for proliferation and gene induction. Mol Cell Biol 1993; 13:2384–2390.

107. Isford RJ, Ihle JN. Multiple hematopoietic growth factors signal through tyrosine phosphorylation. Growth Factors 1990; 2:213–220.

108. Nicholson SE, Oates AC, Harpur AG, Ziemiecki A, Wilks AF, Layton JE. Tyrosine kinase JAK1 is associated with the granulocyte-colony-stimulating factor receptor and both become tyrosine-phosphorylated after receptor activation. Proc Natl Acad Sci USA 1994; 91:2985–2988.

109. Ihle JN, Witthuhn B, Tang B, Yi T, Quelle FW. Cytokine receptors and signal transduction. Baillière's Clin Haematol 1994; 7:17–48.

110. Fukunaga R, Ishizaka-Ikeda E, Nagata S. Growth and differentiation signals mediated by different regions in the cytoplasmic domain of granulocyte colony-stimulating factor receptor. Cell 1993; 74:1079–1087.

111. Welte K, Zeidler C, Reiter A, Müller W, Oderwald E, Souza L, Riehm H. Differential effects of granulocyte–macrophage colony-stimulating factor and granulocyte colony-stimulating factor in children with severe congenital neutropenia. Blood 1990; 75:1056–1063.

112. Dong F, Hoefsloot LH, Schelen AM, Broeders LCAM, Meijer Y, Veeman AJP, Touw IP, Löwenberg B. Identification of a nonsense mutation in the granulocyte-colony-stimulating factor receptor in severe congenital neutropenia. Proc Natl Acad Sci USA 1994; 91:4480–4484.

113. D'Andrea AD. Cytokine receptors in congenital hematopoietic disease. N Engl J Med 1994; 330:839–846.

114. Guba SC, Sartor CA, Hutchinson R, Boxer LA, Emerson SG. Granulocyte colony-stimulating factor (G-CSF) production and G-CSF receptor structure in patients with congenital neutropenia. Blood 1994; 83:1486–1492.

115. Cohen AM, Zsebo KM, Inoue H, Hines D, Boone TC, Chazin VR, Tsai L, Ritch T, Souza LM. In vivo stimulation of granulopoiesis by recombinant human granulocyte colony-stimulating factor. Proc Natl Acad Sci USA 1987; 84:2484–2488.

116. Tamura M, Hattori K, Nomura H, Oheda M, Kubota N. Induction of neutrophilic granulocytosis in mice by administration of purified human native granulocyte colony-stimulating factor (G-CSF). Biochem Biophys Res Commun 1987; 142:454–460.

117. Broxmeyer HE, Williams DE, Cooper S, Hangoc G, Ralph P. Recombinant human granulocyte colony-stimulating factor and recombinant human macrophage colony-stimulating factor synergize in vivo to enhance proliferation of granulocyte–macrophage, erythroid, and multipotential progenitor cells in mice. J Cell Biochem 1988; 38:127–136.

118. Lothrop CJ, Warren DJ, Souza LM, Jones JB, Moore MA. Correction of canine cyclic hematopoiesis with recombinant human granulocyte colony-stimulating factor. Blood 1988; 72:1324–1328.

119. Welte K, Bonilla MA, Gillio AP, Boone TC, Potter GK, Gabrilove JL, Moore MAS, O'Reilly RJ, Souza LM. Recombinant human granulocyte colony-stimulating factor: effects on hematopoiesis in normal and cyclophosphamide treated primates. J Exp Med 1987; 165:941–948.

120. Bronchud MH, Scarffe JH, Thatcher N, Crowther D, Souza LM, Alton NK, Testa NG, Dexter TM. Phase I/II study of recombinant human granulocyte colony-stimulating factor in patients receiving intensive chemotherapy for small cell lung cancer. Br J Cancer 1987; 56:809–813.

121. Morstyn G, Campbell L, Souza LM, Alton NK, Keech J, Green M, Sheridan W, Metcalf D, Fox R. Effect of granulocyte colony-stimulating factor on neutropenia induced by cytotoxic chemotherapy. Lancet 1988; 1:667–672.

122. Gabrilove JL, Jakubowski A, Fain K, Grous J, Scher H, Sternberg C, Yagoda A, Clarkson B, Bonilla MA, Oettgen HF, Alton K, Boone T, Altrock B, Welte K, Souza L. Phase I study of granulocyte colony-stimulating factor in patients with transitional cell carcinoma of the urothelium. J Clin Invest 1988; 82:1454–1461.

123. Podja Z, Molineux G, Dexter TM. Hemopoietic effects of short-term in vivo treatment of mice with various doses of rhG-CSF. Exp Hematol 1990; 18:27–31.

124. Lord BI, Bronchud MH, Owens S, Chang J, Howell A, Souza L, Dexter TM. The kinetics of human

granulopoiesis following treatment with granulocyte colony-stimulating factor in vivo. Proc Natl Acad Sci USA 1989; 86:9499–9503.

125. Lord BI, Molineux G, Podja Z, Souza LM, Mermod JJ, Dexter TM. Myeloid cell kinetics in mice treated with recombinant interleukin-3, granulocyte colony-stimulating factor (CSF), or granulocyte–macrophage CSF in vivo. Blood 1991; 77:2154–2159.

126. Uchida T, Yamagiwa A. Kinetics of rG-CSF-induced neutrophilia in mice. Exp Hematol 1992; 20:152–155.

127. Broxmeyer HE, Benninger L, Shreyaskumar RP, Benjamin RS, Vadhan-Raj S. Kinetic response of human marrow myeloid progenitor cells to in vivo treatment of patients with granulocyte colony-stimulating factor is different from the response to treatment with granulocyte–macrophage colony-stimulating factor. Exp Hematol 1994; 22:100–102.

128. Molineux G, Podja Z, Dexter TM. A comparison of hematopoiesis in normal and splenectomized mice treated with granulocyte colony-stimulating factor. Blood 1990; 75:563–569.

129. Bungart B, Loeffler M, Goris H, Dontje B, Diehl V, Nijhof W. Differential effects of recombinant human granulocyte colony-stimulating factor (rhG-CSF) on stem cells in marrow, spleen, and peripheral blood in mice. Br J Haematol 1990; 76:174–179.

130. Roberts AW, Metcalf D. Granulocyte colony-stimulating factor induces selective elevations of progenitor cells in the peripheral blood of mice. Exp Hematol 1994; 22:1156–1163.

131. Dale DC, Bonilla MA, Davis MW, Nakanishi AM, Hammond WP, Kurtzberg J, Wang W, Jakubowski A, Winton E, Lalezari P, Robinson W, Glaspy JA, Emerson S, Gabrilove J, Vincent M, Boxer LA. A randomized controlled phase III trial of recombinant human granulocyte colony-stimulating factor (filgrastim) for treatment of severe chronic neutropenia. Blood 1993; 81:2496–2502.

132. Dührsen U, Villeval JL, Boyd J, Kannourakis G, Morstyn G, Metcalf D. Effects of recombinant human granulocyte colony-stimulating factor on hematopoietic progenitor cells in cancer patients. Blood 1988; 72:20744–2081.

133. Molineux G, Podja Z, Hampson IN, Lord BI, Dexter TM. Transplantation potential of peripheral blood stem cells induced by granulocyte colony-stimulating factor. Blood 1990; 76:2153–2158.

134. Roberts AW, Metcalf D. Noncycling state of peripheral blood progenitor cells mobilized by granulocyte colony-stimulating factor and other cytokines. Blood 1995; 86:1600–1605.

135. DeLuca E, Sheridan WP, Watson D, Szer J, Begley CG. Prior chemotherapy does not prevent effective mobilisation by G-CSF of peripheral blood progenitor cells. Br J Cancer 1992; 66:893–899.

136. Grigg AP, Roberts AW, Raunow H, Houghton S, Layton JE, Boyd AW, McGrath KM, Maher D. Optimising dose and scheduling of filgrastim (granulocyte colony-stimulating factor) for mobilization and collection of peripheral blood progenitor cells from normal volunteers. Blood 1995; 86:2162–2170.

137. Tanikawa S, Nose M, Aoki Y, Tsuneoka K, Shikita M, Nara N. Effects of recombinant human granulocyte colony-stimulating factor on the hematological recovery and survival of irradiated mice. Blood 1990; 76:445–449.

138. Macvittie TJ, Monroy RL, Parchen ML, Souza LM. Therapeutic use of recombinant human G-CSF (rhG-CSF) in a canine model of sublethal and lethal whole body irradiation. Int J Rad Biol 1990; 57:723–736.

139. Yasuda H, Ajiki Y, Shimozato T, Kasahara M, Kawada H, Iwata M, Shimizu K. Therapeutic efficacy of granulocyte colony-stimulating factor alone and in combination with antibiotics against *Pseudomonas aeruginosa* infections in mice. Infect Immun 1990; 58:2502–2509.

140. Cohen AM, Hines DK, Korach ES, Ratzkin BJ. In vivo activation of neutrophil function in hamsters by recombinant human granulocyte colony-stimulating factor. Infect Immun 1988; 56:2861–2865.

141. Matsumoto M, Matsubara S, Matsuno T, Tamura M, Hattori K, Nomura H, Ono M, Yokota T. Protective effect of human granulocyte colony-stimulating factor on microbial infection in neutropenic mice. Infect Immun 1987; 55:2715–2720.

142. Silver GM, Gamelli RL, O'Reilly M. The beneficial effect of granulocyte colony-stimulating factor (G-CSF) in combination with gentamicin on survival after *Pseudomonas* burn would infection. Surgery 1989; 106:452–456.

143. Cairo MS, Mauss D, Kommareddy S, Norris K, Van de Ven C, Modanlou H. Prophylactic or simultaneous administration of recombinant human granulocyte colony stimulating factor in the treatment of group B streptococcal sepsis in neonatal rats. Pediatr Res 1990; 27:612–616.

144. Tamura M, Hattori K, Ono M, Hata S, Hayata I, Asano S, Bessho M, Hirashima K. Effects of recombinant human granulocyte colony-stimulating factor (rG-CSF) on murine myeloid leukemia: stimulation of leukemic cells in vitro and inhibition of development of leukemia in vivo. Leukemia 1989; 3:853–858.

145. Tamura M, Orita T, Oh-eda M, Hasegawa M, Nomura H, Maekawa T, Abe T, Ono M. Reduction of leukemogenic potential of malignant murine leukemic cells by in vivo treatment with recombinant human granulocyte colony-stimulating factor. Leuk Res 1993; 17:593–600.

146. Chang JM, Metcalf D, Gonda TJ, Johnson GR. Long-term exposure to retrovirally expressed granulocyte colony-stimulating factor induces a nonneoplastic granulocytic and progenitor cell hyperplasia without tissue damage in mice. J Clin Invest 1989; 84:1488–1496.

147. Iversen PO, Nicolaysen G, Benestad HB. The leukopoietic cytokine granulocyte colony-stimulating factor increases blood flow to rat bone marrow. Exp Hematol 1993; 21:231–235.

148. Hammond WP, Csiba E, Canin A, Hockman H, Souza LM, Layton JE, Dale DC. Chronic neutropenia. A new canine model induced by human granulocyte colony-stimulating factor. J Clin Invest 1991; 87:704–710.

149. Lieschke GJ, Grail D, Hodgson G, Metcalf D, Stanley E, Cheers C, Fowler KJ, Basu S, Zhan YF, Dunn AR. Mice lacking granulocyte colony-stimulating factor have chronic neutropenia, granulocyte and macrophage progenitor cell deficiency and impaired neutrophil mobilization. Blood 1994; 84:1737–1746.

150. Metcalf D. Hematopoietic regulators: redundancy or subtlety. Blood 1993; 82:3515–3523.

151. Kawano Y, Takaue Y, Saito S, Sato J, Shimizu T, Suzue T, Hirao A, Okamoto Y, Abe T, Watanabe T, Kuroda Y, Kimura F, Motoyoshi K, Asano S. Granulocyte colony-stimulating factor (CSF), macrophage-CSF, granulocyte–macrophage CSF, interleukin-3, and interleukin-6 levels in sera from children undergoing blood stem cell autografts. Blood 1993; 81:856–860.

152. Gabrilove JL, Jakubowski A, Scher H, Sternberg C, Wong G, Grous J, Yagoda A, Fain K, Moore MAS, Clarkson B, Oettge HF, Alton K, Welte K, Souza L. Effect of granulocyte colony-stimulating factor on neutropenia and associated morbidity due to chemotherapy for transitional-cell carcinoma of the urothelium. N Engl J Med 1988; 319:1414–1422.

153. Sheridan WP, Morstyn G, Wolf M, Dodds A, Lusk J, Maher D, Layton JE, Green MD, Souza L, Fox RM. Granulocyte colony-stimulating factor and neutrophil recovery after high-dose chemotherapy and autologous bone marrow transplantation. Lancet 1989; 2:891–895.

12

Human Interleukin-3
An Overview

Yu-Chung Yang
Indiana University School of Medicine, Indianapolis, Indiana

I. INTRODUCTION

The term *interleukin-3* (IL-3) was first proposed by Ihle and co-workers to define an activity in the conditioned media of activated T cells that was capable of inducing the expression of 20α-hydroxysteroid dehydrogenase in cultures of splenic lymphocytes from nude mice (1). Subsequent studies have indicated that this cytokine has a broad spectrum of biological activities and is equivalent to other biological activities, including multicolony-stimulating factor (multi-CSF), mast cell growth factor, P cell-stimulating factor, burst-promoting activity, WEHI-3 growth factor, as well as others (2). The cloning of the human homologue turned out to be more difficult than anticipated owing to the low sequence homology between the human and rodent species. The failure to identify the human counterpart has led to the speculation that the gene encoding human IL-3 was lost during evolution, and its functions were replaced by granulocyte–macrophage colony-stimulating factor (GM-CSF; 3). In 1986, the cDNA encoding gibbon IL-3 was first cloned by functional expression cloning based on the ability of the cytokine to stimulate the proliferation of a chronic myelogeneic leukemic (CML) cell line (4). Although mouse and human IL-3 share very low sequence homology, subsequent molecular and biological characterization of this cytokine has indicated that human IL-3, similar to its murine counterpart, represents a multifunctional growth factor possessing many different biological activities (5). In vitro, in vivo, and clinical studies were soon carried out over the following 10 years since its discovery. Although IL-3 is believed to be a multifunctional growth factor with therapeutic potentials, as with many other hematopoietic growth factors, its clinical efficacy should be interpreted with caution and may be further improved by the combinations with other hematopoietic growth factors.

The expression of IL-3 has been detected mainly in human T cells and natural killer (NK) cells. Because of its close proximity to other growth factors on the long arm of human chromosome 5, the coordinated expression of these cytokine genes has been proposed. Recent discovery of the sharing of the common signal transducer among IL-3, GM-CSF, and IL-5 has provided, at least in part, the explanation for the biological redundancy of these cytokines. With the advances in the control mechanisms of gene expression and signal transduction for different cytokines, it is now possible to better define the physiological role of IL-3 within the complicated cytokine network. This information will also be useful in the future design of combined cytokine therapy. Because many reviews dealing with similar topics are available (5–13), this review will focus only on human IL-3 owing to its clinical potentials and its species specificity. The goal is to summarize many observations from different investigators and update the most recent studies on the molecular biology, signal transduction, in vitro–in vivo biological characterization, and clinical studies of human IL-3.

II. MOLECULAR AND STRUCTURAL CHARACTERIZATION OF HUMAN IL-3

A. Human IL-3 cDNA

The human IL-3 (hIL-3) cDNA contains a long open-reading frame of 456 nucleotides that encodes a 152-amino acid polypeptide with a 19-amino acid signal peptide and a calculated relative molecular mass of 14.6 kDa (4). The amino-terminus starts with the dipeptide Ala-Pro (residues 20–21), which is found at the amino (NH_2)-termini of many cytokines (14). The cDNA encodes two cysteine residues (residues 35 and 103) and two N-linked glycosylation sites. The 3′-noncoding region contains several copies of the ATTTA sequence, which may be important for controlling the stability of the mRNA (15). Murine and human IL-3 coding sequences show 45% identity at the nucleotide level and 29% identity at the amino acid level (Fig. 1) (4).

B. Human IL-3 Protein

The predicted structure of hIL-3, as with many other cytokines, consists of four α-helix bundles and two β-sheets (Fig. 2) (16). Mutagenesis has been carried out to identify the critical amino acid residues responsible for protein folding and biological activities. Structure–function relation studies of hIL-3, by deletion or substitution variants and interspecies chimera in conjunction with epitope mapping of specific neutralizing monoclonal antibodies, have shown that two noncontiguous helical domains located near the NH_2-terminus and the COOH-terminus of the molecule are responsible for binding to the receptor (Fig. 3) (17–19). The two active sites (NH_2-terminal

```
Human IL-3        MSRLPVLLLLQLLVRPGLQAPMTQTTSLKTSW-VNCSNMIDEIITHLKQPPLP
                   *    **** * ; **** ;  : : :   *** ; *** ;*   * *
Mouse IL-3        MVLASSTTSIHTMLLLLLMLFHLGLQASISGRDTHRLTRTLNCSSIVKEIIGKL---PEP
                                                            #

Human IL-3        LLDFNNLNGEDQDILMENNLRRPNLEAFNRAVKSL--QNASAIESILKNLLPCLPLATAA
                   :   ::   *    ** **  *          ;* * *   *  ***
Mouse IL-3        --ELKT--DDEGPSLRNKSFRRVNLSKFVESQGEVDPEDRYVIKSNLQKLNCCLPTSAND
                                                                       ^#

Human IL-3        PTRHPIHIKDGDWNEFRRKLTFYLKTLENAQAQQTTLSLAIF
                   : *: *  ;**;** **   *  ;   *:
Mouse IL-3        SALPGVFIR--DLDDFRKKLRFYMVHLNDLETVLTSRPPQPASGSVSPNRGTVEC
                                                                       ^
```

Figure 1 Comparison of the amino acid sequences of the human and mouse IL-3s: Identical residues (*) and conservative changes (:) between the two sequences are as indicated. The positions of conserved cysteine residues (#) and the two cysteines found only in mouse sequences (^) are also shown. (Modified from Ref. 4.)

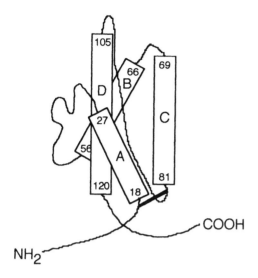

Figure 2 Predicted four-helix bundle tertiary structure of human IL-3: Residue numbers correspond to the boundaries of four helices that are represented by rectangles. The thick line represents the disulfide bond present in the molecule. (Modified from Ref. 18.)

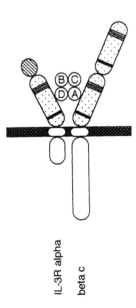

Figure 3 Predicted interactions between human IL-3 and the human IL-3 receptor. (Modified from Ref. 12.)

residues Pro-33 and Leu-34; COOH-terminal residue Leu-111) are in close proximity, owing to the specific helical structure. The study by Lopez et al. (20) shows the generation of a hIL-3 analogue with increased biological and binding activities and supports the model that the COOH-terminus of hIL-3 interacts with the α-chain of the hIL-3 receptor. In addition, Barry et al. (21) have shown that two contiguous residues (Asp-21 and Glu-22) in the NH_2-terminal region of hIL-3 mediate binding to the two different chains of the hIL-3 receptor and emphasize the functional significance of the conserved glutamic acid residue in the first helix of the IL-3, GM-CSF, and IL-5 cytokine subfamily.

C. Human IL-3 Genomic DNA

The human IL-3 gene has been mapped to the long arm of human chromosome 5 at 5q31-33 (22). The gene is 3.2 kilobases (kb) in length and consists of five exons and four introns (23). The 5′-flanking region contains several potential transcriptional control elements, including AP-1, AP-2, CK-1, CK-2, CREB, and NFAT-1 sites. There are at least six copies of ATTTA sequence clustered in one segment of the hIL-3 3′-flanking region. The role of some of these transcriptional or posttranscriptional control elements in the regulation of hIL-3 gene expression will be addressed in the following section. Interestingly, many cytokine genes, including GM-CSF, IL-3, IL-4, IL-5, IL-9, IL-13, and ECGF, have been mapped to the nearby region of human chromosome 5q (24). Kluck et al. (25) have determined the order of the human IL-3, IL-4, IL-5, IL-9, GM-CSF, and MCSFR genes to be 5cen-IL3/GMCSF-IL5-IL4-IL9-MCSFR+++-qter. Interestingly, the genes for human IL-3 and GM-CSF are only 9 kb apart (26), and the IL-4 and IL-5 genes are separated by 90–140 kb (24). The close proximity and the similar gene organization of GM-CSF and IL-3 have prompted the speculation that these two genes may have derived through gene duplication (26). Equally interesting is the finding that the gene for the IL-3 receptor α-chain has been mapped to the X-Y pseudoautosomal region at bands Xp22.3 and Yp11.3 near the gene for the α-subunit of GM-CSF receptor (27).

D. Species Specificity

Sequence comparison of the IL-3-coding regions from different species, including human, gibbon, chimpanzee, tamarin, marmoset, rhesus monkey, ovine, rat, and mouse, has shown that only a few regions are conserved during mammalian evolution, which are likely to be associated with functional domains of the IL-3 protein (28). Interestingly, two of the four cysteine residues of murine IL-3 that are important for its biological activities are well conserved in the hIL-3 molecule (29). The relative affinity of recombinant human IL-3 binding to normal rhesus monkey bone marrow cells is 25- to 50-fold less than that of homologous IL-3, which explained the species specificity of human IL-3 observed when tested in *Macaca* species (30). In contrast, only a small difference was found between human and rhesus monkey IL-3 in relative-binding affinity for receptors on human AML cells, which confirmed that the species specificity of IL-3 is largely unidirectional. This unidirectional species specificity has also been observed for other cytokines such as IL-9 and the c-*kit* ligand.

III. REGULATION OF THE EXPRESSION OF HUMAN IL-3 IN T LYMPHOCYTES

A. Sources of Human IL-3

The human IL-3 transcript has been detected in human T cells and NK cells (31–33). Other sources included monocytes, epithelial cells, and epidermal keratinocytes (34,35). The size of the hIL-3 transcript is 0.8 kb in most of the cell types analyzed. The expression of hIL-3 by peripheral mononuclear cells has also been assessed in patients with human immunodeficiency virus (HIV) infection (36), aplastic anemia (37), rheumatoid arthritis (38), T-cell lymphoma (39), chronic hyperplastic sinusitis (40), multiple sclerosis (41), and in those undergoing allogeneic or autologous bone marrow transplantation. With the exception of eosinophilia associated with chronic hyperplastic sinusitis or T-cell lymphoma and a low percentage of rheumatoid arthritis, the level of hIL-3 expression decreases in these patients. When serum IL-3 levels following autologous or allogeneic bone marrow transplantation were measured to evaluate the physiological role of hIL-3 in the engraftment process (42–45), removal of T lymphocytes from donor marrow or acceleration of engraftment by use of stem cells or growth factors appeared to blunt the endogenous release of IL-3, whereas use of antithymocyte globulin posttransplant increased IL-3 release. Measurable amounts of IL-3 are produced in a significant subgroup of patients suffering from extensive chronic graft-versus-host disease (GVHD).

B. Control Mechanisms of Human IL-3 Gene Expression in T Lymphocytes

Northern blot analysis, nuclear run-on assay, and mRNA stability measurement have been employed to dissect the control mechanisms of the hIL-3 gene expression in human T lymphocytes. The studies have demonstrated that hIL-3 gene is constitutively transcribed but no IL-3 mRNA can be detected in unstimulated T lymphocytes. The expression of the hIL-3 gene can be induced by 12-O-tetradecanoylphorbol-13-acetate (TPA) and phytohemagglutinin (PHA) through a transient increase in both gene transcription and mRNA stabilization (46).

Many investigators have shown that both positive and negative regulatory elements appear to control expression of the human IL-3 gene in activated T cells (47,48). Transient transfection studies have localized two positive control sequences and an interposed repressor sequence within 315 nucleotides upstream from the transcriptional start site (47–49). The proximal regulatory region is specific to IL-3 and is essential for efficient transcription. Molecular and biochemical characterization of this region has shown that it binds an inducible T-cell–specific factor and Oct-1. The characteristics of this binding site are very similar to the activating ARRE-1 site in the IL-2 promoter (50). The transcriptional activation of this region can be enhanced by a second, more distal activating sequence consisting of AP-1- and Ef1-binding sites, which appear to also confer certain specificity (51). Between the two activators lies a transcriptional silencer, which is a potent repressor in the absence of the AP-1 site (47). In vivo footprinting studies have further mapped a core sequence TGTGGTTT (IF-1IL3) that is not found in other cytokine promoters, but is conserved in the IL-3 promoter of several species and is similar to certain viral and T-cell–specific cellular enhancers. Although IF-1IL3 is necessary for hIL-3 expression, a truncated IL-3 promoter, with an intact IF-1IL3 site, but no other activator sites, is transcriptionally silent (52).

Nishida et al. (53) have identified a GC-rich region (–76 to –47) as essential for basal transcriptional activity and transactivation by human T-cell Lymphotropic virus-1 (HTLV-I) encoded *Tax* protein. A novel cDNA encoding a zinc finger protein, DB1, which binds to one of the GC boxes in this region, has recently been isolated in addition to EGR1 and EGR2 (54). The study suggests that DB1 is responsible for the basal transcriptional activity and can cooperate with *Tax* to activate IL-3 promoter while EGR1 has a role in inducible transcription of the gene (54). The role of *Tax* protein in IL-3 gene expression has also been analyzed in HTLV-infected and uninfected cells. The results indicated that, although *Tax* can transactivate IL-3 gene in HTLV-uninfected cells, the expression of IL-3 in HTLV-infected cells is not detectable. It appears that CK-1 and CK-2 sites and their binding proteins participate in the repression of IL-3 gene expression in HTLV-infected T cells (55).

C. Coordinated Expression of Cytokine Genes

It is speculated that the expression of many cytokine genes mapped on human chromosome 5q can be coordinately controlled through common cis- and trans-transcriptional elements. Oct-1, a member of the POU family of transcriptional factors, binds to regions of the GM-CSF, IL-3, and IL-5 promoters and contributes to basal T-cell gene transcription (56).

In addition, two novel T-cell–specific proteins (45 and 43 kDa) can bind to the GM-CSF, IL-3, and IL-5 promoters and are involved in transcriptional enhancement of GM-CSF and IL-3 genes in stimulated T cell (56). The mechanisms of coordinated expression are more complicated because the selective expression of certain cytokines can be explained either by activation of a selective intracellular signaling pathway, or by selective activation of a T-cell subset or the combination of both (57).

IV. IL-3 RECEPTOR AND SIGNAL TRANSDUCTION

A. Structure of Human IL-3 Receptor

The receptor for hIL-3 (hIL-3R) consists of two subunits, the 70-kDa ligand-binding subunit (IL-3Rα) (58) and a 120-kDa signal transducer (βc) (see Fig. 3) (59). This 120-kDa protein is shared by two other cytokines, GM-CSF and IL-5, each using a different ligand-binding subunit (Fig. 4) (12,60–63). The sharing of common signal transducer partly explains the biological redundancy, cross-competition for ligand-receptor interactions (64,65), and the overlapping signaling events (66,67) of these three cytokines mapped onto human chromosome 5q. Neither subunit contains intrinsic tyrosine kinase activity, but mutagenesis studies of the β-chain cytoplasmic domain has defined certain regions that are important for signal transduction (68). One domain between Arg-456 and Phe-487 appears to be essential for proliferation, and the second domain between Val-518 and Asp-544 enhances the response to GM-CSF, but is not absolutely required for proliferation. Interestingly, the region between Val-518 and Leu-626, is responsible for major tyrosine phosphorylation of 95- and 60-kDa proteins, indicating that βc-mediated tyrosine phosphorylation of these proteins may not be involved in proliferation. However, the β-517 mutant lacking residues downstream from Val-518 transmitted a herbimycin-sensitive proliferation signal, suggesting that β-517 still activates a tyrosine kinase(s) (68). The area from Arg-456 to Val-518 is further required for JAK2 activation (69) and the induction of c-*myc* and *pim-1* (70), and Leu-626 to Ser-763 for activation of Ras, Raf-1, MAPK, PI3K, and p70S6 kinase as well as induction of c-*fos* and c-*jun* (70). The former is sensitive to tyrosine kinase inhibitors herbimycin and genistein, whereas the later is only partially sensitive to herbimycin and is resistant to genistein (71). Mutagenesis of the IL-3Rα has indicated that deletion of the cytoplasmic domain abolishes tyrosine phosphorylation of βc, JAK2, and Shc (72).

Interestingly, a mutant form of human βc has been isolated from growth factor-independent cells that arose spontaneously after infection of a murine factor-dependent hematopoietic cell line FDCP-1 with a retroviral vector expressing human βc. Analysis of this βc mutation shows that a small (37-amino acid) duplication of extracellular sequence that includes two conserved sequence motifs is sufficient to confer ligand-independent growth of these cells and lead to tumorigenicity (73).

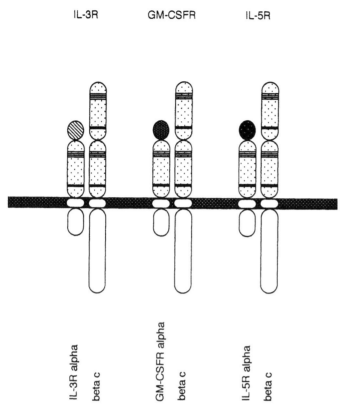

Figure 4 The IL-3, GM-CSF, IL-5 receptor family: The shaded region containing lines (thin lines represent cysteine residues and the thick line represents the WSXWS motif) represents the conserved domain found in the hematopoietic growth factor receptor family, and the circle represents the immunoglobulin domain. (Modified from Ref. 63.)

B. Human IL-3–Mediated Signal Transduction

Many signaling studies have been carried out with human GM-CSF, which shares βc with human IL-3. In addition, because human βc has significant homology with its murine counterpart, it, therefore, is assumed that many of the signaling pathways are common for both IL-3 and GM-CSF in both human and mouse systems. Activation of protein tyrosine phosphorylation (74) and primary response gene expression (75) is important in growth factor-mediated signal transduction. The following summarizes signaling macromolecules involved in human IL-3 signaling that have been the subject of many recent reviews (12,60–63,70,76,77).

Involvement of Protein Kinases and Other Signaling Molecules

The JAK Family Tyrosine Kinases.   The JAK family tyrosine kinases are involved in the signal transduction pathways mediated by several growth factors (78,79). Quelle et al. (69) have shown, by immunoprecipitation and immunoblotting, that JAK2 is tyrosine phosphorylated following stimulation with hGM-CSF. Also tyrosine kinase JAK2 physically associates with βc, and the in vitro kinase activity of JAK2 is greatly enhanced following stimulation with hGM-CSF. Mutagenesis has further demonstrated that only the membrane-proximal 62 amino acids of βc cytoplasmic domain are required for JAK2 activation. Signal transducers and activators of transcription (STAT) family of transcriptional factors have been shown, initially in the interferon system, to be the substrates of JAK family kinases. Stat 91 and related proteins are cytoplasmic proteins that were originally identified as interferon-activated transcriptional factors (80). Larner et al. (81) have shown that IL-3 and GM-CSF induce tyrosine phosphorylation of a transcriptional factor distinct from Stat 91. Recent study, using a factor-dependent cell line Mo7E, further suggested that these two cytokines may activate different STATs (82).

Src Family Tyrosine Kinases.   The Src family of tyrosine kinases participate in the signal transduction pathways of many different cytokines, especially in the IL-2 system. It has recently been shown that the phenotypic changes induced by IL-2 and IL-3 in an immature T-lymphocyte leukemia are associated with regulated expression of IL-2R β-chain and of protein tyrosine kinases lck and lyn (83). p53/p56 lyn is highly expressed in IL-3–dependent cells and the expression level decreases with growth in IL-2. In addition, p53/56 lyn kinase participates in early IL-3–initiated signaling events, at least in some human leukemic cell lines (84). Also both IL-3 and GM-CSF induce tyrosine phosphorylation and kinase activity of the c-*fps/fes* protooncogene product (p92c-*fec*) in the human erythroleukemia cell line TF-1, which requires GM-CSF or IL-3 for growth. In addition, GM-CSF induces physical association between p92c-*fes* and the β-chain of the GM-CSF receptor (85).

Ras/Raf/MAPK.   The activation of Ras (86,87), followed by the activation of the kinase cascade of Raf-1 (88,89), MAPK kinase, and MAPK (90–92) that leads to the induction of c-*fos* and c-*jun* has been well documented in the human IL-3 system. In addition, many adaptor molecules, such as Shc (93), Grb2, and the c-*vav* (94) oncogene have also been linked to this pathway. These adaptor molecules, which contain SH2 or SH3 domains, are tyrosine phosphorylated following IL-3 stimulation. Vav is an SH2/SH3-containing, hematopoietic-specific molecule with RasGTP-exchange activity (95). Interestingly, p95*vav* is tyrosine phosphorylated by GM-CSF, IL-3, steel factor, and p210*bcr/abl*, and is associated with JAK kinases in myeloid cells (96). Mutagenesis analysis of human GM-CSF receptor βc has localized COOH-terminal amino acid 626–763 to be essential for Shc phosphorylation, activation of Ras, Raf, and MAPK, and induction of c-*fos* and c-*jun* (70). IL-3 can induce MAP kinase-activated protein kinase 2 activity and phosphorylate the human small heat-shock protein Hsp27 on serine residues in vitro. Protein phosphatase A_2 abolished IL-3-induced serine phosphorylation of Hsp27 (97). Phosphorylation of Hsp27, however, appears to be independent of cell proliferation.

Protein Kinase C.   Human IL-3 induces the translocation of protein kinase C (PKC) from cytosol to the cell membrane of MO7E cells (98) and human platelets (99). This translocation is accompanied by rapid accumulation of diacylglycerol (DAG) in the absence of an increase in intracellular calcium. IL-3 did not induce phosphatidic acid accumulation, and the IL-3–induced DAG accumulation can be blocked by a phospholipase C inhibitor. Genistein and herbimycin inhibited both IL-3–induced protein kinase C translocation and the accumulation of DAG, suggesting that IL-3–induced tyrosine phosphorylation may result in activation of a phosphatidylcholine phospholipase C and protein kinase C (98). In monocytes, however, human IL-3–induced c-*jun* expression is independent of tyrosine kinase and involves PKC (100). Bryostatin 1, a PKC-activating agent, modulates proliferation and lineage commitment of hematopoietic progenitors stimulated by IL-3 (101).

p70S6 Kinase.   Human IL-3 activates p70S6 kinase, which phosphorylates the 40S ribosomal protein S6 in vitro but, unlike pp90*rsk*, is not regulated by MAPK. The region between amino acids 626 and 763 of hIL-3 receptor βc is required for p70S6 kinase activation (70).

PI3K.   Phosphatidylinositol 3-kinase (PI3K) is composed of an 85-kDa regulatory subunit (containing two SH2 domains and one SH3 domain) and a 110-kDa catalytic subunit that phosphorylates the D3 position of phosphatidylinositol (PI) or its derivatives. p85 interacts with receptors for IL-2, IL-4, and erythropoietin (EPO). In the human IL-3 system, the region between amino acids 626 and 763 of the hIL-3R βc is required for PI3K activation (70).

Activation of Primary and Secondary Response Genes

Human IL-3 stimulation resulted in the induction of c-*myc*, c-*fos*, c-*jun*, and *pim-1* (102). Mutagenesis of IL-3R βc has indicated that a membrane-proximal sequence (amino acids 456–518) is required for the induction of c-*myc* and *pim-1*, whereas the residues 626–763 are required for the induction of c-*fos* and c-*jun*. In addition to c-*myc*, c-*fos*, c-*jun*, and *pim-1*, IL-3 rapidly and transiently induces expression of early growth response gene 1 (*egr-1*) in the human factor-dependent cell line TF-1. Mutational analysis of the cytokine-responsive region of the *egr-1* promoter revealed that both the cAMP response and serum response elements are required for induction by GM-CSF and IL-3. The study also suggests that the *egr-1* protein may further stimulate transcription of the *egr-1* gene in response to GM-CSF as a secondary event (103). In the more downstream events, no change has been detected in the level of expression of early (c-*myc*), mid (ornithine decarboxylase) or mid-late G_1 (p53, c-*myb*) cell cycle genes after restoration of IL-3 in deprived cells. The fact that only late G_1–S-phase genes (proliferating cell nuclear antigen [PCNA], thymidine kinase [TK], and histone H3) are modulated by IL-3 suggests that this factor may control human cell proliferation by acting at the G_1–S boundary (104).

V. BIOLOGICAL ACTIVITIES OF HUMAN IL-3

A. In Vitro Biological Activities

Activities on Primitive and Committed Hematopoietic Progenitor Cells

Human IL-3 supports colony formation by normal hematopoietic progenitors, including multilineage (CFU-GEMM), granulocyte–macrophage (CFU-GM), granulocyte (CFU-G), macrophage (CFU-M), megakaryocyte (CFU-Mk), eosinophil (CFU-eo), basophil (CFU-b) and, in the presence of EPO, both early (BFU-E) and late (CFU-E) erythroid colonies (105–110). This broad spectrum of activities demonstrates that IL-3 can interact with early progenitor cells that are common to all the major myeloid lineages as well as many of the later cells committed to development along the individual lineages. The blast cell assay has shown that IL-3 synergizes with blast cell growth factors such as IL-1, IL-6, IL-11, and G-CSF, to support IL-3– or IL-4–dependent proliferation of primitive progenitors, and part of this synergism is to shorten the G_0 period of the early hematopoietic progenitors (111–113). In this system, IL-3 supported the formation of various types of single-lineage as well as multilineage colonies by CD34$^+$ bone marrow cells in the presence of EPO. In contrast, hGM-CSF supported the formation of single-lineage colonies and only a small number of multilineage colonies. IL-3 also supported the formation of blast cell colonies with variable, but high-replating capability compared with GM-CSF. Many studies subsequently support the concept that human IL-3 and GM-CSF have overlapping, but distinct, hematopoietic activities and suggest a potential role for the clinical application of combined IL-3–GM-CSF therapy. The effects of IL-3 on later progenitors will be discussed individually.

Growth Promotion of Myeloid Leukemic Cells

The initial identification and cloning of human IL-3 was based on its ability to stimulate the proliferation of CML cells. Subsequent studies have shown the responsiveness of primary blast cells from patients with acute myeloblastic leukemia (AML; 114–119). Many different studies show that combinations of hematopoietic growth factors containing IL-3 induce AML cell sensitivity to cycle-specific (such as cytarabine [Ara-C]) and nonspecific drugs (such as daunorubicin), and to the alkylating agents (such as 4-hydroxyperoxycyclophosphamide [4-HC]) (118,120–122). The effects of IL-3 on AML cells suggested the potential therapeutic use of combining hematopoietic growth factors, especially IL-3, with chemotherapy in the AML treatment. The colony growth of leukemic colony-forming unit (L-CFU) obtained from patients with primary AML stimulated with recombinant human IL-3 is significantly potentiated by tumor necrosis factor-alpha (TNF-α) as a result of induction of secondary hematopoietic cytokines by TNF-α (123). Pretreatment of AML cells with hIL-3 increased their susceptibility to lympho-kine-activated killing (LAK) possibly mediated through the expression of cell adhesion molecule LFA-3 (CD58; 124).

Megakaryocyte Colony-Stimulating Activity

Interleukin-3 is a potent stimulator of the early events of human megakaryocyte progenitor cell development, promoting predominantly mitosis and early megakaryocytic differentiation (125–127). The effect of IL-3 on megakaryocytopoiesis has also been implicated in patients with myeloproliferative disorders (MPD; 128). The CFU-Mk of MPD patients with thrombocytosis are hypersensitive to IL-3. Although Meg-CSF activity of IL-3 has been the basis of many clinical trials, the results suggest the importance of combined therapy with different growth factors discussed later.

Erythroid Potentiating Activity

Interleukin-3 can stimulate primitive multilineage hematopoietic progenitors as well as precursors at various stages of erythroid differentiation (106,110,129). The effects of IL-3 on erythroid progenitors have also been suggested in patients with polycythemia vera (PV) (130,131) and Diamond-Blackfan anemia (DBA; 132). Burst-forming unit—erythroid (BFU-E) and CFU-E progenitor cells from patients with PV are much more sensitive to IL-3 and may contribute to the pathogenesis of PV. IL-3 increases the number and size of BFU-E colonies from DBA bone marrow, suggesting its potential therapeutic use in this particular setting (see later discussion).

Eosinophil Differentiation Activity

Eosinophil functions can be modulated by several cytokines, such as IL-3 (133), GM-CSF, and IL-5. IL-3 and GM-CSF support early and intermediate stages of differentiation of eosino-philic progenitor cells. IL-5, on the other hand, supports terminal proliferation and maturation of eosinophilic progenitor cells. IL-3 appears to maintain the viability of eosinophils in vitro, to augment the calcium ionophore-induced generation of LTC4, and to induce eosinophils to become hypodense cells. This effect may be partly mediated by a change in the affinity of the complement receptor type 3 on the eosinophils with opsonized particles (134). Growth factors, such as transforming growth factor-β (TGF-β) promote IL-3-dependent differentiation of human basophils, but inhibit IL-3-dependent differentiation of human eosinophils. Interferon (IFN)-α and IFN-γ, on the other hand, down-regulate IL-3-dependent formation of both basophils and eosinophils (135).

Basophil Differentiation Activity

Interleukin-3 can activate human basophils from peripheral blood, bone marrow, fetal liver, and cord blood to release histamine. It induces an increase in histamine production by human bone marrow, fetal liver, and cord blood cells (136,137). The α-chain of Mac-1 may be involved in the priming effect of IL-3 to increase histamine release (138). IL-3 can alter free arachidonic acid generation in C5a-stimulated human basophils, possibly through the transient cytosolic calcium ion response (139). However, IL-3 does not alter intracellular free calcium concentration in basophils, whereas C5a induces a transient rise independent of IL-3 pretreatment, indicating that the priming effect of IL-3 cannot be explained by alterations in calcium changes (140). Both IL-4 and nerve growth factor (NGF) can function as cofactors with IL-3 in the support of histamine production in human umbilical cord blood cells (141).

Biological Effects with Lymphoid Cells

Interleukin-3 potentiates the human immune response by stimulation of B-cell differentiation, and its effect is dependent on the target B-cell population, and its stage of activation or maturation (142,143). Leukemic B-cell precursors express functional receptors for human IL-3 (144,145). More than one-half of adults with non-Hodgkin's B-cell lymphoma present with low-grade follicular lymphoma. IL-3 either promoted or inhibited the in vitro proliferation of follicular tumor cells, suggesting that therapies directed against IL-3, its receptor, or the T cells that produce IL-3 may be effective in treating follicular lymphoma (146).

IL-3 induces short-term proliferation of unfractionated peripheral blood lymphocytes (PBL) and lectin-stimulated T cells. It also potentiates the proliferative responses and the long-term growth of both unstimulated and mitogen-activated lymphocytes in the presence of IL-2 (147). IL-3 stimulation of human T lymphocytes may be related to its ability to enhance monocyte functions (148; see following discussion). Also, IL-2 plus IL-3 phenotypically induce not only NK-like LAK cells, but also T cell-like LAK cells (149).

Monocyte-Activating Activity

Interleukin-3 modulates not only blood progenitor cell activity, but also alters the function of mature phagocytes. It serves as a survival factor for monocytes, allowing them to remain functional in culture for longer periods (150). Also, in the presence of dexamethasone, IL-3 can induce high levels of histocompatibility antigen (HLA) expression on human monocytes, indicating that they may play a role in inflammatory response in vivo (151). When alveolar macrophages and blood monocytes from hIL-3–treated patients were studied, both monocytes and aveolar macrophages were activated by IL-3 to secrete TNF-α, IL-1β, and IL-6 (152). Compared with IFN-γ and TNF-α, IL-3 is more efficient in priming cultured human peripheral blood monocytes for enhanced stimulation of respiratory burst activity (153).

Effects on Endothelial Cells

Human umbilical vein endothelial cells (HUVEC) constitutively express functional high-affinity receptors for IL-3. Functional activation of endothelial cells by IL-3 was shown by the expression of endothelial–leukocyte adhesion molecule 1 (ELAM-1) and enhanced adhesion of neutrophils and CD4$^+$ T cells to HUVEC (154). Tumor necrosis factor-alpha (TNF-α), IL-1 β, or lipopolysaccharide can induce the expression of IL-3 receptors, and IL-3 binding to TNF-α–activated HUVEC further enhanced IL-8 production, E-selectin expression, and neutrophil transmigration. The selective induction of a functional IL-3 receptor on endothelial cells suggests that IL-3 may have an important role in chronic inflammation and in allergic diseases (155).

Growth of Nonhematopoietic Tumor Cells

Interleukin-3 has proliferative effect on certain human transitional cell carcinoma cell lines (156). IL-3 and GM-CSF can stimulate the proliferation of the small-cell lung cancer cell lines H69 and N417 (157). These findings suggest that hematopoietic growth factors can stimulate the growth of some malignant nonhematopoietic cells in vitro. Therefore, it is important to test the responsiveness of different solid tumors when applying IL-3 for hematopoietic recovery after chemotherapy.

B. In Vivo Biological Activities

Donahue et al. (158) first showed that IL-3, in comparison with GM-CSF, elicited a delayed and relatively modest leukocytosis in primates. However, IL-3 greatly potentiated the responsiveness of the animal to subsequent administration of a low dose of GM-CSF. These earlier results suggested that IL-3 can expand an early cell population in vivo that requires the action of a late-acting factor, such as GM-CSF, to complete its development. Administration of hIL-3 in primates (158,159) induced a delayed and modest leukocytosis, mainly caused by an increase in neutrophils, lymphocytes, eosinophils, and atypical basophils, and occasionally, an increase in reticulocytes and platelets. The ability of IL-3 to expand the pool of circulating stem cells is reflected by the increases in CFU-GM, BFU-E, CFU-mix, and CFU-Mk (160). IL-3 also induced proliferation of inflammatory cells and keratinocytes in vivo (161). Administration of GM-CSF to IL-3–pretreated primates resulted in further hematopoietic enhancement mediated by IL-3 or GM-CSF alone. Megakaryocytes from IL-3–treated monkeys had many ultra structural characteristics indicative of increased maturation (162). The changes in mean DNA content and megakaryocyte size were larger in monkeys treated with sequential IL-3 and then GM-CSF and with GM-CSF alone than in simultaneously treated monkeys (163). The combined administration of IL-3 and GM-CSF in normal primates suggested that a sequential protocol of IL-3 followed by GM-CSF would be more effective than that of GM-CSF alone in producing neutrophils.

In addition to combinations of GM-CSF and IL-3, optimal stimulation of hematopoiesis has also been attempted with combinations of IL-3 and IL-6 or EPO. Geissler et al. (164) have shown that administration of IL-3 alone increased levels only of CFU-Mk without raising the platelet counts in primates. IL-6, on the other hand, did not increase CFU-Mk levels, but significantly raised platelet counts in the same model. When primates were pretreated with IL-3 to expand CFU-Mk, the thrombopoietic effect of rhIL-6 was synergistically enhanced. The sequential administration of IL-3 and IL-6, therefore, may present a powerful strategy to stimulate thrombopoiesis (164). The cooperative effects of IL-3 and EPO on in vivo erythropoiesis and F-cell formation have also been examined in baboons and macaques (165). IL-3 administration alone resulted in increases in CFU-E and BFU-E in both normal and anemic animals. In parallel to the increase in peripheral reticulocytes, IL-3 increased the frequency of F reticulocytes in these animals. Sequential administration of IL-3 and EPO resulted in an expansion of erythroid progenitors and an increase in reticulocytes to an extent higher than that of EPO alone. The combination, however, did not further increase the rate of F reticulocytes mediated by EPO alone. These results suggested that IL-3 enhances the effect of EPO on erythropoiesis, but the combination of the two growth factors did not lead to a preferential and significant enhancement of HbF production (165).

A severe combined immunodeficient (SCID)-hu mouse model implanted with human fetal bone has been used to assess the effects of hIL-3. IL-3 induced significant increases of eosinophilic, granulocyte, and BFU-E activity. Pretreatment with IL-3 followed by EPO enhanced EPO-induced human erythropoiesis significantly. No synergistic effects on myelopoiesis were observed using sequential treatment with IL-3 followed by G-CSF in such a setting (166). Another similar approach was to transplant gene-transduced human CD34$^+$ progenitor cells into immunodeficient (bnx) mice together with primary human bone marrow stromal cells engineered to produce human IL-3 (167). The IL-3–secreting stroma, but not control stroma, supported human hematopoiesis from the cotransplanted human bone marrow CD34$^+$ progenitors.

VI. PRECLINICAL AND CLINICAL STUDIES

A. Preclinical Studies

Gillio et al. (168) have examined the effects of IL-3 on hematopoietic recovery after 5-fluoro-uracil (5-FU) or cyclophosphamide treatment in primates. The administration of rhIL-3 following intensive myelosuppressive therapy dramatically enhances myeloid recovery and ablates the predicted period of prolonged severe neutropenia. rhIL-3 was administered subcutaneously or intravenously to primates for 14 days starting 24 hr after the end of chemotherapy. Cyclophosphamide or 5-FU was administered intravenously on two consecutive days. The addition of IL-3 resulted in a higher neutrophil nadir count and reduction of the period of severe neutropenia compared with controls.

In sublethally irradiated primates, coadministration of IL-3 and GM-CSF shortens the period of neutropenia and antibiotic support (169). Similarly, the average period of severe thrombocytopenia, which necessitated platelet transfusion in the control animals, was also reduced when IL-3 and GM-CSF were coadministered. The sequential administration of IL-3 followed by GM-CSF had no greater effect on PMN production than GM-CSF alone and was less effective than IL-3 alone in reducing thrombocytopenia. It appears that IL-3 may be a potential therapeutic agent for thrombocytopenia and neutropenia following radiation- or drug-induced marrow aplasia.

B. Clinical Trials

Numerous clinical trials have shown the beneficial effect of hIL-3 in myelosuppressive conditions after cytotoxic therapy, in posttransplant period, in bone marrow transplantation procedures, and in neutropenic conditions as part of other clinical settings. Some of the human studies are summarized as follows:

Human Studies with IL-3 with or without Chemotherapy

In the initial phase I/II studies, rhIL-3 functioned as a multilineage hematopoietin in vivo in patients with normal bone marrow function and in patients with secondary bone marrow failure (170). IL-3 treatment resulted in a dose-dependent increase in platelet counts as well as a substantial increase in the number of circulating neutrophils, eosinophils, monocytes, and lymphocytes in patients with advanced malignancies, but normal hematopoiesis. In patients with secondary hematopoietic failure caused by prolonged chemo- or radiotherapy or bone marrow infiltration by tumor cells, IL-3 treatment leads to a clinically significant restoration of hematopoiesis, especially of thrombopoiesis and granulopoiesis (171–174).

Several subsequent studies have been carried out with specific groups of patients. Patients with myelodesplastic syndromes (MDS) frequently present with anemia, leukopenia, and thrombocytopenia owing to defective maturation of bone marrow cells. In several phase I/II trials, treatment with IL-3 has resulted in increases in neurophil, platelet, leukocyte, and reticulocyte counts in certain patients (175–177). The ability of monocytes to secrete secondary cytokines is impaired in MDS patients, but can be restored by in vivo administration of GM-CSF and IL-3 (178). TNF-α not only is induced during IL-3 therapy in MDS patients, but also this elevation might be associated with a poor platelet response to therapy (179). The patients with MDS who underwent IL-3 treatment had elevated levels of sIL-2R. The increased expression could be a primary event because of involvement of lymphocytes in the malignant clone, or a secondary alteration of the cytokine network caused by chronic neutropenia (180). Long-term treatment with low-dose IL-3 stimulates megakaryocytopoiesis with increase of platelet counts, but additional later-acting factors probably will be required to augment neutrophil and erythrocyte counts. A sequential administration of IL-3 and GM-CSF has been attempted in this group of patients, the results showed that, although neutrophil counts improved in most patients, the effect on red blood cells and platelets is minimal.

The human studies with IL-3 on patients with Diamond-Blackfan anemia (DBA) (181–183) suggested that 10–20% of patients responded to hIL-3 therapy and that the response can persist after its withdrawal (184). hIL-3 increased total WBC count, secondary to increases in neutrophils, eosinophils, and lymphocytes. Another study indicated the response of DBA patients to IL-3 in vivo is heterogeneous and cannot be predicted from in vitro studies (185). The absence of a corrective effect of IL-3 in these patients with DBA indicates that a deficiency of the cytokine may not be central in the pathogenesis of the disorder (186). The experience with aplastic anemia has been moderate and usually no lasting effects were obtained when treating with IL-3 (187,188).

Human Studies with IL-3 and Chemotherapy

Several clinical trials have been performed to evaluate whether rhIL-3 reduced chemotherapy-induced neutropenia and thrombocytopenia. The initial studies indicated that sequential administration of IL-3 and GM-CSF following standard-dose combination chemotherapy with etoposide, ifosfamide, and cisplatin in patients with advanced malignancies resulted in WBC recovery in all patients and an accelerated platelet recovery in some (189,190). In this setting, IL-3 plus GM-CSF recruits peripheral blood progenitor cells, which may partly shorten the hematopoietic recovery after high-dose chemotherapy. The effects of IL-3 on bone marrow cell proliferation and differentiation have been observed in patients following high-dose cyclophosphamide cancer therapy (191). In this setting, IL-3 increased bone marrow-derived megakaryocyte number, proliferative activity of erythroid cells, and fibrosis of bone marrow cells.

The IL-3 treatment of patients with small-cell lung carcinoma following the second cycle of chemotherapy, which consisted of either cyclophosphamide, doxorubicin, and etoposide (CDE), or of vincristine, ifosfamide, mesna, and carboplatin (VIMP), has resulted in accelerated recovery of leukocytes, neutrophils, and platelets, as well as an increase in monocytes and eosinophils (192). In addition, IL-3 augmented plasma levels of TNF-α and IL-6. In another study, patients received IL-3 either before or after chemotherapy consisting of carboplatin, etoposide, and epirubicin (CVE). During the prechemotherapy course, there was a dose-dependent platelet and neutrophil rise. IL-3 infusion following the second cycle of CVE in these patients appears to reduce chemotherapy-induced myelosuppression, but does not alter tumor response or patient survival rates (193).

In patients with advanced ovarian cancer treated with carboplatin–cyclophosphamide (194), IL-3 increased the recovery of leukocyte, neutrophil, basophil, eosinophil, monocyte, lymphocyte, and platelet counts. Less frequent postponement of chemotherapy for insufficient bone marrow recovery was necessary after cycles during which IL-3 was administered. hIL-3 also increased plasma levels of TNF-α, C-reactive protein, and serum amyloid A. In this setting, IL-3 administration appears to reduce chemotherapy-induced myelotoxicity. IL-3 has also been administered to previously untreated breast cancer patients following high-dose therapy with cyclophosphamide (195). IL-3 significantly accelerated granulocyte, platelet, and reticulocyte recovery. In addition, no platelet transfusions and fewer erythrocyte transfusions were required in IL-3–treated patients, indicating that IL-3 may represent a well-tolerated cytokine, clinically useful for accelerating trilineage hematopoietic recovery following severely myelotoxic treatments such as high-dose cyclophosphamide. In patients with relapsed malignant lymphoma undergoing DHAP chemotherapy (cisplatin, cyterabine, dexamethasone), administration of IL-3 proved to be safe and resulted in an increase in eosinophils (196).

Human Studies with hIL-3 and Bone Marrow Transplantation

In patients with delayed engraftment after autologous bone marrow transplantation (ABMT) for hematological malignancies, IL-3 appears to be of limited benefit, especially for those with very low levels of bone marrow progenitors (197). However, sequential administration of hIL-3 and hGM-CSF in patients after ABMT for both Hodgkin's and non-Hodgkin's lymphoma has resulted in fewer days of platelet transfusions than were seen in historical control groups using GM-CSF, G-CSF, or IL-3 alone. In addition, there were fewer days of red blood cell transfusions compared with no cytokines or hGM-CSF alone. Therefore, sequential administration of IL-3 and GM-CSF after ABMT is safe, generally well-tolerated, and results in rapid recovery of multilineage hematopoiesis (198).

Human Studies with IL-3 and Peripheral Stem or Progenitor Cell Harvesting

Many hematopoietic growth factors are increasingly used for mobilization of stem and progenitor cells in the peripheral blood. Several studies (199–202) have demonstrated the efficacy of sequential administration of IL-3 followed by GM-CSF or G-CSF in mobilizing hematopoietic stem and progenitor cells with the capacity of reconstituting marrow functions following high-dose chemotherapy.

VII. CONCLUSIONS

Molecular cloning of human IL-3 and its cognate receptor has advanced our understanding of a growth factor that was thought not to exist in humans many years ago. The IL-3 gene has been mapped to the long arm of human chromosome 5 (5q) where the genes for many other growth factors and growth factor receptors are localized. It is clear from the transcriptional studies that different positive- and negative-control elements are involved in the regulation of IL-3 gene expression. Although the search for common control elements responsible for the coordinated expression of cytokine genes mapped to 5q is far from over, it appears that similar transcriptional control mechanisms are employed to activate certain cytokine genes in a cell–type-specific manner. The elucidation of the control mechanisms of cytokine gene expression remains an important task in our understanding of the host response to different inflammatory stimuli.

IL-3 is a multifunctional cytokine biologically related to GM-CSF and IL-5, which utilize different ligand-binding subunits, but share a common signal transducer. It has become apparent from recent studies in the gp130 family (203) and the IL-2 receptor γ-chain family (204), that many different cytokines use heterodimeric receptors to achieve biological specificity and redundancy. It is also clear that many signaling events occur following growth factor stimulation, and cross talk among different signaling pathways may determine the ultimate signals transduced into the nucleus. The understanding of signal transduction pathways mediated by IL-3 and related cytokines may help define the common and unique biological properties of these growth factors in different cell types. It has also become apparent from the in vitro, in vivo, and clinical studies that many hematopoietic growth factors function in a synergistic manner. IL-3, which stimulates early hematopoietic progenitor cells, can be used to prime the actions of other lineage-specific growth factors, such as GM-CSF, IL-6, or EPO, to elicit more dramatic hematopoietic responses. The mechanisms of synegy among different growth factors are largely unknown. The understanding of the signaling pathways involved in the synergistic effects of these cytokines will allow us to better design future combination therapy. In vivo administration of IL-3 in animals and humans has demonstrated the beneficial effects of this cytokine. Further clinical studies with IL-3 alone and in combination with other biological and chemotherapeutic agents will establish the role of this cytokine in the treatment of various diseases.

ACKNOWLEDGMENTS

The cloning of human IL-3 was accomplished while the author was at the Genetics Institute. The author would like to express the highest gratitude to Dr. Steve Clark for continuous support and encouragement, Aggie Ciarletta, Patty Temple, Sharlotte Kovacic, Bob Donahue, Gordon Wong, and many others in the hematopoiesis group for technical support and helpful discussions. Without them, the discovery of human IL-3 would not have been possible.

REFERENCES

1. Ihle JN, Pepersack L, Rebar L. Regulation of T cell differentiation: in vitro induction of 20 alpha hydroxysteroid dehydrogenase in splenic lymphocytes from athymic mice by a unique lymphokine. J Immunol 1981; 126:2184–2189.

2. Ihle JN, Keller J, Oroszlan S, et al. Biological properties of homogeneous interleukin-3. I. Demonstration of WEHI-3 growth factor activity, mast cell growth factor activity, P-cell stimulating factor activity, colony stimulating factor activity and histamine producing factor activity. J Immunol 1983; 131:282–287.

3. Emerson SG, Sieff CA, Wang EA, Wong GG, Clark SC, Nathan DG. Purification of fetal hematopoietic progenitors and demonstration of recombinant multipotential colony-stimulating activity. J Clin Invest 1985; 76:1286–1290.

4. Yang YC, Ciarletta AB, Temple PA, Chung MP, Kovacic S, Witek-Giannotti JS, Leary AC, Kriz R, Donahue RE, Wong GG, Clark SC. Human IL-3 (multi-CSF): identification by expression cloning of a novel hematopoietic growth factor related to murine IL-3. Cell 1986; 47:3–10.

5. Yang YC, Clark SC. Interleukin-3: molecular biology and biologic activities. Hematol Oncol Clin North Am 1989; 3:441–452.

6. Morris CF, Young IG, Hapel AJ. Molecular and cellular biology of interleukin-3. Immunol Ser 1990; 49:177–214.

7. Ihle JN. Interleukin-3 and hematopoiesis. Chem Immunol 1992; 51:65–106.

8. Ganser A. Clinical results with recombinant human interleukin-3. Cancer Invest 1993; 11:212–218.

9. Oster W, Schulz G. Interleukin 3: biological and clinical effects. Int J Cell Cloning 1991; 9:5–23.

10. Lindemann A, Mertelsmann R. Interleukin-3: structure and function. Cancer Invest 1993; 11:609–623.

11. de Vries EG, van Gameren MM, Willemse PH. Recombinant human interleukin 3 in clinical oncology. Stem Cells 1993; 11:72–80.

12. Miyajima A, Mui AL, Ogorochi T, Sakamaki K. Receptors for granulocyte–macrophage colony-stimulating factor, interleukin-3, and interleukin-5. Blood 1993; 82:1960–1974.

13. Urdal DL, Price V, Sassenfeld HM, Cosman D, Gillis S, Park LS. Molecular characterization of colony-stimulating factors and their receptors: human interleukin-3. Ann NY Acad Sci 1989; 554:167–176.

14. Schrader JW, Ziltener HJ, Leslie KB. Structural homologies among the hemopoietins. Proc Natl Acad Sci USA 1986; 83:2458–2462.

15. Shaw G, Kamen R. A conserved AU sequence from the 3′ untranslated region of GM-CSF mRNA mediates selective mRNA degradation. Cell 1986; 46:659–667.

16. Bazan JF. Neuropoietic cytokines in the hematopoietic fold. Neuron 1991; 7:197–208.

17. Kaushansky K, Shoemaker SG, Broudy VC, et al. Structure–function relationships of interleukin-3. An analysis based on the function and binding characteristics of a series of interspecies chimera of gibbon and murine interleukin-3. J Clin Invest 1992; 90:1879–1888.

18. Lokker NA, Zenke G, Strittmatter U, Fagg B, Movva NR. Structure–activity relationship study of human interleukin-3: role of the C-terminal region for biological activity. EMBO J 1991; 10:2125–2131.

19. Lokker NA, Movva NR, Strittmatter U, Fagg B, Zenke G. Structure–activity relationship study of human interleukin-3. Identification of residues required for biological activity by site-directed mutagenesis. J Biol Chem 1991; 266:10624–10631.

20. Lopez AF, Shannon MF, Barry S, Phillips JA, Cambareri B, Dottore M, Simmons P, Vadas MA. A human interleukin 3 analog with increased biological and binding activities. Proc Natl Acad Sci USA 1992; 89:11842–11846.

21. Barry SC, Bagley CJ, Phillips J, et al. Two contiguous residues in human interleukin-3, Asp21 and Glu22, selectively interact with the alpha- and beta-chains of its receptor and participate in function. J Biol Chem 1994; 269:8488–8492.

22. Le Beau MM, Epstein ND, O'Brien SJ, Nienhuis AW, Yang YC, Clark SC, Rowley JD. The interleukin 3 gene is located on human chromosome 5 and is deleted in myeloid leukemias with a deletion of 5q. Proc Natl Acad Sci USA 1987; 84:5913–5917.

23. Yang YC, Clark SC. Human interleukin 3: analysis of the gene and its role in the regulation of hematopoiesis. Int J Cell Cloning 1990; 1:121–128.

24. van Leeuwen BH, Martinson ME, Webb GC, Young IG. Molecular organization of the cytokine gene cluster, involving the human IL-3, IL-4, IL-5, and GM-CSF genes, on human chromosome 5. Blood 1989; 73:1142–1148.

25. Kluck PM, Weigant J, Raap AK, Vrolijk H, Tanke HJ, Willemze R, Landegent JE. Order of human hematopoietic growth factor and receptor genes on the long arm of chromosome 5, as determined by fluorescence in situ hybridization. Ann Hematol 1993; 66:15–20.

26. Yang YC, Kovacic S, Kriz R, Wolf S, Clark SC, Wellems TE, Nienhuis A, Epstein N. The human genes for GM-CSF and IL-3 are closely linked in tandem on chromosome 5. Blood 1988; 71:958–961.

27. Milatovich A, Kitamura T, Miyajima A, Francke U. Gene for the alpha-subunit of the human interleukin-3 receptor (IL3RA) localized to the X-Y pseudoautosomal region. Am J Hum Genet 1993; 53:1146–1153.

28. Burger H, Wagemaker G, Leunissen JA, Dorssers LC. Molecular evolution of interleukin-3. J Mol Evol 1994; 39:255–267.

29. Clark-Lewis I, Hood LE, Kent SB. Role of disulfide bridges in determining the biological activity of interleukin 3. Proc Natl Acad Sci USA 1988; 85:7897–7901.

30. van Gils FC, Budel LM, Burger H, van Leen RW, Lowenberg B, Wagemaker G. Interleukin-3 (IL-3) receptors on rhesus monkey bone marrow cells: species specificity of human IL-3, binding characteristics, and lack of competition with GM-CSF. Exp Hematol 1994; 22:248–255.

31. Otsuka T, Miyajima A, Brown N, et al. Isolation and characterization of an expressible cDNA encoding human IL-3. Induction of IL-3 mRNA in human T cell clones. J Immunol 1988; 140:2288–2295.

32. Wimperis JZ, Niemeyer CM, Sieff CA, Mathey-Prevot B, Nathan DG, Arceci RJ. Granulocyte–macrophage colony-stimulating factor and interleukin-3 mRNAs are produced by a small fraction of blood mononuclear cells. Blood 1989; 74:1525–1530.

33. Niemeyer CM, Sieff CA, Mathey-Prevot B, Wimperis JZ, Bierer BE, Clark SC, Nathan DG. Expression of human interleukin-3 (multi-CSF) is restricted to human lymphocytes and T-cell tumor lines. Blood 1989; 73:945–951.

34. Dalloul AH, Arock M, Fourcade C, Hatzfeld A, Bertho JM, Debre P, Mossalayi MD. Human thymic epithelial cells produce interleukin-3. Blood 1991; 77:69–74.

35. Dalloul AH, Arock M, Fourcade C, Beranger JY, Jaffray P, Debre P, Mossalayi MD. Epidermal keratinocyte-derived basophil promoting activity. Role of interleukin 3 and soluble CD23. J Clin Invest 1992; 90:1242–1247.

36. Re MC, Zauli G, Furlini G, Ranieri S, La Placa M. Progressive and selective impairment of IL-3 and IL-4 production by peripheral blood CD4[+] T-lymphocytes during the course of HIV-1 infection. Viral Immunol 1992; 5:185–194.

37. Kawano Y, Takaue Y, Hirao A, et al. Production of interleukin 3 and granulocyte–macrophage

colony-stimulating factor from stimulated blood mononuclear cells in patients with aplastic anemia. Exp Hematol 1992; 20:1125–1128.

38. Waalen K, Sioud M, Natvig JB, Forre O. Spontaneous in vivo gene transcription of interleukin-2, interleukin-3, interleukin-4, interleukin-6, interferon-gamma, interleukin-2 receptor (CD25) and proto-oncogene c-*myc* by rheumatoid synovial T lymphocytes. Scand J Immunol 1992; 36:865–873.

39. Fermand JP, Mitjavila MT, Le Couedic JP, Tsapis A, Berger R, Modigliani R, Seligmann M, Brouet JC, Vainchenker W. Role of granulocyte–macrophage colony-stimulating factor, interleukin-3 and interleukin-5 in the eosinophilia associated with T cell lymphoma. Br J Haematol 1993; 83:359–364.

40. Hamilos DL, Leung DY, Wood R, et al. Chronic hyperplastic sinusitis: association of tissue eosinophilia with mRNA expression of granulocyte–macrophage colony-stimulating factor and interleukin-3. J Allergy Clin Immunol 1993; 92:39–48.

41. Huberman M, Shalit F, Roth-Deri I, Gutman B, Kott E, Sredni B. Decreased IL-3 production by peripheral blood mononuclear cells in patients with multiple sclerosis. J Neurol Sci 1993; 118:79–82.

42. Valent P, Sillaber KC, Scherrer R, et al. Detection of circulating endogenous interleukin-3 in extensive chronic graft-versus-host disease. Bone Marrow Transplant 1992; 9:331–336.

43. Atkinson K, Seymour R, Altavilla N, Cooley M, Biggs J. Cytokine activity after allogeneic bone marrow transplantation. IV. Production of mRNA for IL-3 and GM-CSF by mitogen-stimulated circulating mononuclear cells. Bone Marrow Transplant 1992; 9:175–183.

44. Mangan KF, Mullaney MT, Barrientos TD, Kernan NA. Serum interleukin-3 levels following autologous or allogeneic bone marrow transplantation: effects of T-cell depletion, blood stem cell infusion, and hematopoietic growth factor treatment. Blood 1993; 81:1915–1922.

45. Bruserud O, Ehninger G, Hamann W, Pawelec G. Secretion of IL-2, IL-3, IL-4, IL-6 and GM-CSF by CD4[+] and CD8[+] TCR alpha beta[+] T-cell clones derived early after allogeneic bone marrow transplantation. Scand J Immunol 1993; 38:65–74.

46. Ryan GR, Milton SE, Lopez AF, Bardy PG, Vadas MA, Shannon MF. Human interleukin-3 mRNA accumulation is controlled at both the transcriptional and posttranscriptional level. Blood 1991; 77:1195–1202.

47. Mathey-Prevot B, Andrews NC, Murphy HS, Kreissman SG, Nathan DG. Positive and negative elements regulate human interleukin 3 expression. Proc Natl Acad Sci USA 1990; 87:5046–5050.

48. Shoemaker SG, Hromas R, Kaushansky K. Transcriptional regulation of interleukin 3 gene expression in T lymphocytes. Proc Natl Acad Sci USA 1990; 87:9650–9654.

49. Park JH, Kaushansky K, Levitt L. Transcriptional regulation of interleukin 3 (IL3) in primary human T lymphocytes. Role of AP-1- and octamer-binding proteins in control of IL3 gene expression. J Biol Chem 1993; 268:6299–6308.

50. Davies K, TePas EC, Nathan DG, Mathey-Prevot B. Interleukin-3 expression by activated T cells involves an inducible, T-cell-specific factor and an octamer binding protein. Blood 1993; 81:928–934.

51. Gottschalk LR, Giannola DM, Emerson SG. Molecular regulation of the human IL-3 gene: inducible T cell-restricted expression requires intact AP-1 and Elf-1 nuclear protein binding sites. J Exp Med 1993; 178:1681–1692.

52. Cameron S, Taylor DS, TePas EC, Speck NA, Mathey-Prevot B. Identification of a critical regulatory site in the human interleukin-3 promoter by in vivo footprinting. Blood 1994; 83:2851–2859.

53. Nishida J, Yoshida M, Arai K, Yokota T. Definition of a GC-rich motif as regulatory sequence of the human IL-3 gene: coordinate regulation of the IL-3 gene by CLE2/GC box of the GM-CSF gene in T cell activation. Int Immunol 1991; 3:245–254.

54. Koyano-Nakagawa N, Nishida J, Baldwin D, Arai K, Yokota T. Molecular cloning of a novel human cDNA encoding a zinc finger protein that binds to the interleukin-3 promoter. Mol Cell Biol 1994; 14:5099–5107.

55. Wolin M, Kornuc M, Hong C, Shin SK, Lee F, Lau R, Nimer S. Differential effect of HTLV infection and HTLV *Tax* on interleukin 3 expression. Oncogene 1993; 8:1905–1911.

56. Kaushansky K, Shoemaker SG, O'Rork CA, McCarty JM. Coordinate regulation of multiple human

lymphokine genes by Oct-1 and potentially novel 45 and 43 kDa polypeptides. J Immunol 1994; 152:1812–1820.

57. Van Straaten JF, Dokter WH, Stulp BK, Vellenga E. The regulation of interleukin 5 and interleukin 3 gene expression in human T cells. Cytokine 1994; 6:229–234.

58. Kitamura T, Sato N, Arai K, Miyajima A. Expression cloning of the human IL-3 receptor cDNA reveals a shared beta subunit for the human IL-3 and GM-CSF receptors. Cell 1991; 66:1165–1174.

59. Hayashida K, Kitamura T, Gorman DM, Yokota T, Miyajima A. Molecular cloning of a second subunit of the human GM-CSF receptor: reconstitution of a high-affinity GM-CSF receptor. Proc Natl Acad Sci USA 1990; 87:9655–9659.

60. Miyajima A. Molecular structure of the IL-3, GM-CSF and IL-5 receptors. Int J Cell Cloning 1992; 10:126–134.

61. Sato N, Miyajima A. Multimeric cytokine receptors: common versus specific functions. Curr Opin Cell Biol 1994; 6:174–179.

62. Mui AL, Miyajima A. Interleukin-3 and granulocyte–macrophage colony-stimulating factor receptor signal transduction. Proc Soc Exp Biol Med 1994; 206:284–288.

63. Mui AL, Miyajima A. Cytokine receptors and signal transduction. Prog Growth Factor Res 1994; 5:15–35.

64. Lopez AF, Eglinton JM, Gillis D, Park LS, Clark S, Vadas MA. Reciprocal inhibition of binding between interleukin 3 and granulocyte–macrophage colony-stimulating factor to human eosinophils. Proc Natl Acad Sci USA 1989; 86:7022–7026.

65. Park LS, Friend D, Price V, Anderson D, Singer J, Prickett KS, Urdal DL. Heterogeneity in human interleukin-3 receptors. A subclass that binds human granulocyte/macrophage colony stimulating factor. J Biol Chem 1989; 264:5420–5427.

66. Linnekin D, Farrar WL. Signal transduction of human interleukin 3 and granulocyte–macrophage colony-stimulating factor through serine and tyrosine phosphorylation. Biochem J 1990; 271:317–324.

67. Kanakura Y, Druker B, Cannistra SA, Furukawa Y, Torimoto Y, Griffin JD. Signal transduction of the human granulocyte–macrophage colony-stimulating factor and interleukin-3 receptors involves tyrosine phosphorylation of a common set of cytoplasmic proteins. Blood 1990; 76:706–715.

68. Sakamaki K, Miyajima I, Kitamura T, Miyajima A. Critical cytoplasmic domains of the common beta subunit of the human GM-CSF, IL-3 and IL-5 receptors for growth signal transduction and tyrosine phosphorylation. EMBO J 1992; 11:3541–3549.

69. Quelle FW, Sato N, Witthuhn BA, Inhorn RC, Eder M, Miyajima A, Griffin JD, Ihle JN. JAK2 associates with the beta-c chain of the receptor for granulocyte-macrophage colony-stimulating factor, and its activation requires the membrane-proximal region. Mol Cell Biol 1994; 14:4335–4341.

70. Sato N, Sakamaki K, Terada N, Arai K, Miyajima A. Signal transduction by the high-affinity GM-CSF receptor: two distinct cytoplasmic regions of the common beta subunit responsible for different signaling. EMBO J 1993; 12:4181–4189.

71. Watanabe S, Muto A, Yokota T, Miyajima A, Arai K. Differential regulation of early response genes and cell proliferation through the human granulocyte macrophage colony-stimulating factor receptor: selective activation of the c-*fos* promoter by genistein. Mol Biol Cell 1993; 4:983–992.

72. Doshi PD, Rapoport AP, Venepalli JR, Zhang R, Leung M, DiPersio JF. Specific regions of cytoplasmic domains of the human GM-CSF and IL-3 receptor alpha subunits are essential for ligand-induced transmembrane signaling. Blood 1994; 84:224a.

73. D'Andrea R, Rayner J, Moretti P, Lopez A, Goodall GJ, Gonda TJ, Vadas M. A mutation of the common receptor subunit for interleukin-3 (IL-3), granulocyte–macrophage colony-stimulating factor, and IL-5 that leads to ligand independence and tumorigenicity. Blood 1994; 83:2802–2808.

74. Cantely LC, Auger KR, Carpenter C, Duckworth B, Graziani A, Kapeller R, Soltoff S. Oncogenes and signal transduction. Cell 1991; 64:281–302.

75. Herschman HR. Primary response genes induced by growth factors and tumor promoters. Annu Rev Biochem 1991; 60:281–319.

76. Miyajima A, Hara T, Kitamura T. Common subunits of cytokine receptors and the functional redundancy of cytokines. Trends Biochem Sci 1992; 17:378–382.

77. Miyajima A, Kitamura T, Harada N, Yokota T, Arai K. Cytokine receptors and signal transduction. Annu Rev Immunol 1992; 10:295–331.

78. Ihle JN. The Janus kinase family and signaling through members of the cytokine receptor superfamily. Proc Soc Exp Biol Med 1994; 206:268–272.

79. Ihle JN, Witthuhn BA, Quelle FW, Yamamoto K, Thierfelder WE, Kreider B, Silvennoinen O. Signaling by the cytokine receptor superfamily: JAKs and STATs. Trends Biochem Sci 1994; 19:222–227.

80. Darnell JE Jr, Kerr IM, Stark GR. Jak–Stat pathways and transcriptional activation in response to IFNs and other extracellular signalling proteins. Science 1994; 264:1415–1421.

81. Larner AC, David M, Feldman GM, Igarashi K, Hackett RH, Webb DS, Sweitzer SM, Petricoin E, Finbloom DS. Tyrosine phosphorylation of DNA binding proteins by multiple cytokines [see comments]. Science 1993; 261:1730–1733.

82. Frank D, Salgia R, Bhatt A, Griffin JD, Greenberg M. Identification of a 93 kDa STAT protein phosphorylated following treatment by GM-CSF or IL-3 in hematopoietic cells. Blood 1994; 84:224a.

83. O'Connor R, Torigoe T, Reed JC, Santoli D. Phenotypic changes induced by interleukin-2 (IL-2) and IL-3 in an immature T-lymphocytic leukemia are associated with regulated expression of IL-2 receptor beta chain and of protein tyrosine kinases LCK and LYN. Blood 1992; 80:1017–1025.

84. Torigoe T, O'Connor R, Santoli D, Reed JC. Interleukin-3 regulates the activity of the LYN protein-tyrosine kinase in myeloid-committed leukemic cell lines. Blood 1992; 80:617–624.

85. Hanazono Y, Chiba S, Sasaki K, Mano H, Miyajima A, Arai K, Yazaki Y, Hirai H. c-*fps/fes* protein-tyrosine kinase is implicated in a signaling pathway triggered by granulocyte–macrophage colony-stimulating factor and interleukin-3. EMBO J 1993; 12:1641–1646.

86. Duronio V, Welhaam MJ, Abraham S, Dryden P, Schrader JW. p21ras activation via hemopoietin receptors and c-*kit* requires tyrosine kinase activity but not tyrosine phosphorylation of p21*ras* GTPase-activating protein. Proc Natl Acad Sci USA 1992; 89:1587.

87. Satoh T, Nakafuku M, Miyajima A, Kaziro Y. Involvement of *ras*p21 protein in signal-transduction pathways from interleukin 2, interleukin 3 and granulocyte/macrophage colony-stimulating factor, but not from interleukin 4. Proc Natl Acad Sci USA 1991; 88:3314.

88. Kanakura Y, Druker B, Wood KW, Mamon HJ, Okuda K, Roberts TM, Griffin JD. Granulocyte–macrophage colony-stimulating factor and interleukin-3 induce rapid phosphorylation and activation of the proto-oncogene *Raf-1* in a human factor-dependent myeloid cell line. Blood 1991; 77:243–248.

89. Carroll MP, Clark-Lewis I, Rapp UR, May WS. Interleukin-3 and granulocyte-macrophage colony-stimulating factor mediate rapid phosphorylation and activation of cytosolic c-*raf.* J Biol Chem 1990; 265:19812–19817.

90. Okuda K, Sanghera JS, Pelech SL, Kanakura Y, Hallek M, Griffin JD, Druker BJ. Granulocyte–macrophage colony-stimulating factor, interleukin-3, and steel factor induce rapid tyrosine phosphorylation of p42 and p44 MAP kinase. Blood 1992; 79:2880–2887.

91. Raines MA, Golde DW, Daeipour M, Nel AE. Granulocyte–macrophage colony-stimulating factor activates microtubule-associated protein 2 kinase in neutrophils via a tyrosine kinase-dependent pathway. Blood 1992; 79:3350–3354.

92. Whelham MJ, Duronio V, Sanghera JS, Pelech SL, Schrader JW. Multiple hematopoietic growth factors stimulate activation of mitogen-activated protein kinase family members. J Immunol 1992; 149:1683–1693.

93. Matsuguchi T, Salgia R, Hallek M, Eder M, Druker B, Ernst TJ, Griffin JD. Shc phosphorylation in myeloid cells is regulated by granulocyte macrophage colony-stimulating factor, interleukin-3, and steel factor and is constitutively increased by p210*BCR/ABL.* J Biol Chem 1994; 269:5016–5021.

94. Mui A, Cutler R, Alai M, Bustelo X, Barbacid M, Krystal G. Steel factor and interleukin-3 stimulate the tyrosine phosphorylation of p95*vav* in hemopoietic cell lines. Exp Hematol 1992; 20:752.

95. Gulbins E, Coggeshall K, Baier G, Katzav S, Burn P, Altman A. Tyrosine kinase stimulated guanine nucleotide exchange activity of *vav* in T cell activation. Science 1993; 260:822–825.

96. Matsuguchi T, Inhorn RC, Carlesso N, Griffin JD. p95*vav* is tyrosine phosphorylated by GM-CSF, IL-3, steel factor and p210*bcr/abl*, and is associated with JAK kinases in myeloid cells. Blood 1994; 84:369a.

97. Ahlers A, Engel K, Sott C, Gaestel M, Herrmann F, Brach MA. Interleukin-3 and granulocyte–macrophage colony-stimulating factor induce activation of the MAPKAP kinase 2 resulting in in vitro serine phosphorylation of the small heat shock protein (Hsp 27). Blood 1994; 83:1791–1798.

98. Rao P, Mufson RA. Human interleukin-3 stimulates a phosphatidylcholine specific phospholipase C and protein kinase C translocation. Cancer Res 1994; 54:777–783.

99. Cook PP, Chen J, Ways DK. Interleukin-3 induces translocation and down-regulation of protein kinase C in human platelets. Biochem Biophys Res Commun 1992; 185:670–675.

100. Mufson RA, Szabo J, Eckert D. Human IL-3 induction of c-*jun* in normal monocytes is independent of tyrosine kinase and involves protein kinase C. J Immunol 1992; 148:1129–1135.

101. Li F, Grant S, Pettit GR, McCrady CW. Bryostatin 1 modulates the proliferation and lineage commitment of human myeloid progenitor cells exposed to recombinant interleukin-3 and recombinant granulocyte–macrophage colony-stimulating factor. Blood 1992; 80:2495–2502.

102. Lilly M, Le T, Holland P, Hendrickson SL. Sustained expression of the *pim-1* kinase is specifically induced in myeloid cells by cytokines whose receptors are structurally related. Oncogene 1992; 7:727–732.

103. Sakamoto KM, Fraser JK, Lee HJ, Lehman E, Gasson JC. Granulocyte–macrophage colony-stimulating factor and interleukin-3 signaling pathways converge on the CREB-binding site in the human *egr-1* promoter. Mol Cell Biol 1994; 14:5975–5985.

104. Avanzi GC, Porcu P, Brizzi MF, Ghigo D, Bosia A, Pegoraro L. Interleukin 3-dependent proliferation of the human Mo-7e cell line is supported by discrete activation of late G_1 genes. Cancer Res 1991; 51:1741–1743.

105. Messner HA, Yamasaki K, Jamal N, Minden MM, Yang YC, Wong GG, Clark SC. Growth of human hemopoietic colonies in response to recombinant gibbon interleukin 3: comparison with human recombinant granulocyte and granulocyte–macrophage colony-stimulating factor. Proc Natl Acad Sci USA 1987; 84:6765–6769.

106. Sieff CA, Niemeyer CM, Nathan DG, Ekern SC, Bieber FR, Yang YC, Wong G, Clark SC. Stimulation of human hematopoietic colony formation by recombinant gibbon multi-colony-stimulating factor or interleukin 3. J Clin Invest 1987; 80:818–823.

107. Lopez AF, To LB, Yang YC, Gamble JR, Shannon MF, Burns GF, Dyson PG, Juttner CA, Clark S, and Vadas MA. Stimulation of proliferation, differentiation, and function of human cells by primate interleukin 3. Proc Natl Acad Sci USA 1987; 84:2761–2765.

108. Lopez AF, Dyson PG, To LB, Elliott MJ, Milton SE, Russell JA, Juttner CA, Yang YC, Clark SC, Vadas MA. Recombinant human interleukin-3 stimulation of hematopoiesis in humans: loss of responsiveness with differentiation in the neutrophilic myeloid series. Blood 1988; 72:1797–1804.

109. Emerson SG, Yang YC, Clark SC, Long MW. Human recombinant granulocyte–macrophage colony stimulating factor and interleukin 3 have overlapping but distinct hematopoietic activities. J Clin Invest 1988; 82:1282–1287.

110. Sonoda Y, Yang YC, Wong GG, Clark SC, Ogawa M. Erythroid burst-promoting activity of purified recombinant human GM-CSF and interleukin-3: studies with anti-GM-CSF and anti-IL-3 sera and studies in serum-free cultures. Blood 1988; 72:1381–1386.

111. Leary AG, Hirai Y, Kishimoto T, Clark SC, Ogawa M. Survival of hemopoietic progenitors in the G_0 period of the cell cycle does not require early hemopoietic regulators. Proc Natl Acad Sci USA 1989; 86:4535–4538.

112. Leary AG, Ikebuchi K, Hirai Y, Wong GG, Yang YC, Clark SC, Ogawa M. Synergism between

interleukin-6 and interleukin-3 in supporting proliferation of human hematopoietic stem cells: comparison with interleukin-1 alpha. Blood 1988; 71:1759–1763.

113. Sonoda Y, Yang YC, Wong GG, Clark SC, Ogawa M. Analysis in serum-free culture of the targets of recombinant human hemopoietic growth factors: interleukin 3 and granulocyte/macrophage-colony-stimulating factor are specific for early developmental stages. Proc Natl Acad Sci USA 1988; 85:4360–4364.

114. Miyauchi J, Kelleher CA, Yang YC, Wong GG, Clark SC, Minden MD, Minkin S, McCulloch EA. The effects of three recombinant growth factors, IL-3, GM-CSF, and G-CSF, on the blast cells of acute myeloblastic leukemia maintained in short-term suspension culture. Blood 1987; 70:657–663.

115. Miyauchi J, Kelleher CA, Wong GG, Yang YC, Clark SC, Minkin S, Minden MD, McCulloch EA. The effects of combinations of the recombinant growth factors GM-CSF, G-CSF, IL-3, and CSF-1 on leukemic blast cells in suspension culture. Leukemia 1988; 2:382–387.

116. Budel LM, Touw IP, Delwel R, Clark SC, Lowenberg B. Interleukin-3 and granulocyte–monocyte colony-stimulating factor receptors on human acute myelocytic leukemia cells and relationship to the proliferative response. Blood 1989; 74:565–571.

117. Salem M, Delwel R, Mahmoud LA, Clark S, Elbasousy EM, Lowenberg B. Maturation of human acute myeloid leukaemia in vitro: the response to five recombinant haematopoietic factors in a serum-free system. Br J Haematol 1989; 71:363–370.

118. Tafuri A, Lemoli RM, Chen R, Gulati SC, Clarkson BD, Andreeff M. Combination of hematopoietic growth factors containing IL-3 induce acute myeloid leukemia cell sensitization to cycle specific and cycle non-specific drugs. Leukemia 1994; 8:749–757.

119. Saeland S, Caux C, Favre C, et al. Effects of recombinant human interleukin-3 on CD34-enriched normal hematopoietic progenitors and on myeloblastic leukemia cells. Blood 1988; 72:1580–1588.

120. Hiddemann W, Kiehl M, Zuhlsdorf M, Busemann C, Schleyer E, Wormann B, Buchner T. Granulocyte–macrophage colony-stimulating factor and interleukin-3 enhance the incorporation of cytosine arabinoside into the DNA of leukemic blasts and the cytotoxic effect on clonogenic cells from patients with acute myeloid leukemia. Semin Oncol 1992; 19:31–37.

121. Bhalla K, Bullock G, Lutzky J, Holladay C, Ibrado AM, Jasiok M, Singh S. Effect of combined treatment with interleukin-3 and interleukin-6 on 4-hydroperoxycyclophosphamide-mediated reduction of glutathione levels and cytotoxicity in normal and leukemic bone marrow progenitor cells. Leukemia 1992; 6:814–819.

122. te Boekhorst PA, Lowenberg B, Vlastuin M, Sonneveld P. Enhanced chemosensitivity of clonogenic blasts from patients with acute myeloid leukemia by G-CSF, IL-3 or GM-CSF stimulation. Leukemia 1993; 7:1191–1198.

123. Brach MA, Gruss HJ, Asano Y, de Vos S, Ludwig WD, Mertelsmann R, Herrmann F. Synergy of interleukin 3 and tumor necrosis factor alpha in stimulating clonal growth of acute myelogenous leukemia blasts is the result of induction of secondary hematopoietic cytokines by tumor necrosis factor alpha. Cancer Res 1992; 52:2197–2201.

124. Cesano A, Lista P, Bellone G, Geuna M, Brizzi MF, Rossi PR, Pegoraro L, Oberholtzer E, Matera L. Effect of human interleukin 3 on the susceptibility of fresh leukemia cells to interleukin-2-induced lymphokine activated killing activity. Leukemia 1992; 6:567–573.

125. Mazur EM, Cohen IL, Bogart L, Mufson RA, Gesner TG, Yang YC, Clark SC. Recombinant gibbon interleukin-3 megakaryocyte colony growth in vitro from human peripheral blood progenitor cells. J Cell Physiol 1988; 136:439.

126. Briddell RA, Brandt JE, Leemhuis TB, Hoffman R. Role of cytokines in sustaining long-term human megakaryocytopoiesis in vitro. Blood 1992; 79:332–337.

127. Teramura M, Katahira J, Hoshino S, Motoji T, Oshimi K, Mizoguchi H. Clonal growth of human megakaryocyte progenitors in serum-free cultures: effect of recombinant human interleukin 3. Exp Hematol 1988; 16:843–848.

128. Kobayashi S, Teramura M, Hoshino S, Motoji T, Oshimi K, Mizoguchi H. Circulating megakaryo-

cyte progenitors in myeloproliferative disorders are hypersensitive to interleukin-3. Br J Haematol 1993; 83:539–544.

129. Migliaccio G, Migliaccio AR, Visser JWM. Synergism between erythropoietin and interleukin-3 in the induction of hematopoietic stem cell proliferation and erythroid burst colony formation. Blood 1988; 72:944–951.

130. de Wolf JT, Beentjes JA, Esselink MT, Smit JW, Halie RM, Clark SC, Vellenga E. In polycythemia vera human interleukin 3 and granulocyte–macrophage colony-stimulating factor enhance erythroid colony growth in the absence of erythropoietin. Exp Hematol 1989; 17:981–983.

131. Dai CH, Krantz SB, Dessypris EN, Means R Jr, Horn ST, Gilbert HS. Polycythemia vera. II. Hypersensitivity of bone marrow erythroid, granulocyte–macrophage, and megakaryocyte progenitor cells to interleukin-3 and granulocyte–macrophage colony-stimulating factor. Blood 1992; 80:891–899.

132. Halperin DS, Estrov Z, Freedman MH. Diamond-Blackfan anemia: promotion of marrow erythropoiesis in vitro by recombinant interleukin-3. Blood 1989; 73:1168–1174.

133. Rothenberg ME, Owen W Jr, Silberstein DS, Woods J, Soberman RJ, Austen KF, Stevens RL. Human eosinophils have prolonged survival, enhanced functional properties, and become hypodense when exposed to human interleukin 3. J Clin Invest 1988; 81:1986–1992.

134. Blom M, Tool AT, Kok PT, Koenderman L, Roos D, Verhoeven AJ. Granulocyte–macrophage colony-stimulating factor, interleukin-3 (IL-3), and IL-5 greatly enhance the interaction of human eosinophils with opsonized particles by changing the affinity of complement receptor type 3. Blood 1994; 83:2978–2984.

135. Sillaber C, Geissler K, Scherrer R, Kaltenbrunner R, Bettelheim P, Lechner K, Valent P. Type beta transforming growth factors promote interleukin-3 (IL-3)-dependent differentiation of human basophils but inhibit IL-3-dependent differentiation of human eosinophils. Blood 1992; 80:634–641.

136. Valent P, Besemer J, Muhm M, Majdic O, Lechner K, Bettelheim P. Interleukin 3 activates human blood basophils via high-affinity binding sites. Proc Natl Acad Sci USA 1989; 86:5542–5546.

137. Minkowski M, Label B, Arnould A, Dy M. Interleukin 3 induces histamine synthesis in the human hemopoietic system. Exp Hematol 1990; 18:1158–1163.

138. Watanabe A, Tominaga T, Tsuji J, Yanagihara Y, Koda A. Effects of anti-CD11a, anti-CD11b and anti-CD18 on histamine release from human basophils primed with IL-3. Int Arch Allergy Immunol 1992; 98:308–310.

139. MacGlashan D Jr, Hubbard WC. IL-3 alters free arachidonic acid generation in C5a-stimulated human basophils. J Immunol 1993; 151:6358–6369.

140. Krieger M, von Tscharner V, Dahinden CA. Signal transduction for interleukin-3–dependent leukotriene synthesis in normal human basophils: opposing role of tyrosine kinase and protein kinase. Eur J Immunol 1992; 22:2917–2913.

141. Richard A, McColl SR, Pelletier G. Interleukin-4 and nerve growth factor can act as cofactors for interleukin-3–induced histamine production in human umbilical cord blood cells in serum-free culture. Br J Haematol 1992; 81:6–11.

142. Tadmori W, Feingersh D, Clark SC, Choi YS. Human recombinant IL-3 stimulates B cell differentiation. J Immunol 1989; 142:1950–1955.

143. Xia X, Li L, Choi YS. Human recombinant IL-3 is a growth factor for normal B cells. J Immunol 1992; 148:491–497.

144. Uckun FM, Gesner TG, Song CW, Myers DE, Mufson A. Leukemic B-cell precursors express functional receptors for human interleukin-3. Blood 1989; 73:533–542.

145. Wormann B, Gesner TG, Mufson RA, LeBien TW. Proliferative effect of interleukin-3 on normal and leukemic human B cell precursors. Leukemia 1989; 3:399–404.

146. Clayberger C, Luna-Fineman S, Lee JE, Pillai A, Campbell M, Levy R, Krensky AM. Interleukin 3 is a growth factor for human follicular B cell lymphoma. J Exp Med 1992; 175:371–376.

147. Santoli D, Clark SC, Kreider BL, Maslin PA, Rovera G. Amplication of IL-2-drive T cell proliferation

by recombinant human IL-3 and granulocyte-macrophage colony-stimulating factor. J Immunol 1988; 141:519–526.

148. Mossalayi MD, Dalloul AH, Bertho JM, Lecron JC, Debre P. The role of monocytes in IL-3 enhancement of human T cell proliferation. Int Immunol 1990; 2:495–499.

149. Okuno K, Ohnishi H, Shilayama Y, Hirohata T, Ozaki M, Yasutomi M. Augmentation of human lymphokine-activated killer cell activity in splenic lymphocytes by the combination of low-dose interleukin 2 plus interleukin 3. Mol Biother 1992; 4:83–86.

150. Young DA, Lowe LD, Clark SC. Comparison of the effects of IL-3, granulocyte–macrophage colony-stimulating factor, and macrophage colony-stimulating factor in supporting monocyte differentiation in culture. Analysis of macrophage antibody-dependent cellular cytotoxicity. J Immunol 1990; 145:607–615.

151. Sadeghi R, Feldmann M, Hawrylowicz C. Upregulation of HLA class II, but not intercellular adhesion molecule 1 (ICAM-1) by granulocyte–macrophage colony stimulating factor (GM-CSF) or interleukin-3 (IL-3) in synergy with dexamethasone. Eur Cytokine Netw 1992; 3:373–380.

152. Thomassen MJ, Antal JM, Connors MJ, McLain D, Sandstrom K, Meeker DP, Budd GT, Levitt D, Bukowski RM. Immunomodulatory effects of recombinant interleukin-3 treatment on human alveolar macrophages and monocytes. J Immunother 1993; 14:43–50.

153. Jendrossek V, Buth S, Stetter C, Gahr M. Modulation of human monocyte superoxide production by recombinant interleukin-3. Agents Actions 1992; 37:127–133.

154. Brizzi MF, Garbarino G, Rossi PR, Pagliardi GL, Arduino C, Avanzi GC, Pegaroro L. Interleukin 3 stimulates proliferation and triggers endothelial–leukocyte adhesion molecule 1 gene activation of human endothelial cells. J Clin Invest 1993; 91:2887–2892.

155. Korpelainen EI, Gamble JR, Smith WB, Goodall GJ, Qiyu S, Woodcock JM, Dottore M, Vadas MA, Lopez AF. The receptor for interleukin 3 is selectively induced in human endothelial cells by tumor necrosis factor alpha and potentiates interleukin 8 secretion and neutrophil transmigration [see comments]. Proc Natl Acad Sci USA 1993; 90:11137–11141.

156. Block T, Schmid F, Geffken B, Treiber U, Busch R, Hartung R. Modulation of in vitro cell growth of human and murine urothelial tumor cell lines under the influence of interleukin-3, granulocyte-, macrophage- and granulocyte-colony-stimulating factor. Urol Res 1992; 20:289–292.

157. Pedrazzoli P, Bacciocchi G, Bergamaschi G, et al. Effects of granulocyte–macrophage colony-stimulating factor and interleukin-3 on small cell lung cancer cells. Cancer Invest 1994; 12:283–288.

158. Donahue RE, Seehra J, Metzger M, et al. Human IL-3 and GM-CSF act synergistically in stimulating hematopoiesis in primates. Science 1988; 241:1820–1823.

159. Mayer P, Valent P, Schmidt G, Liehl E, Bettelheim P. The in vivo effects of recombinant human interleukin-3: demonstration of basophil differentiation factor, histamine-producing activity, and priming of GM-CSF-responsive progenitors in nonhuman primates. Blood 1989; 74:613–621.

160. Geissler K, Valent P, Mayer P, Liehl E, Hinterberger W. Lechner K, Bettelheim P. Recombinant human interleukin-3 expands the pool of circulating hematopoietic progenitor cells in primates—synergism with recombinant granulocyte/macrophage colony-stimulating factor. Blood 1990; 75:2305–2310.

161. Volc-Platzer B, Valent P, Radaszkiewicz T, Mayer P, Bettelheim P, Wolff K. Recombinant human interleukin 3 induces proliferation of inflammatory cells and keratinocytes in vivo. Lab Invest 1991; 64:557–566.

162. Dvorak AM, Monahan-Earley RA, Estrella P, Kissell S, Donahue RE. Ultrastructure of monkey peripheral blood basophils stimulated to develop in vivo by recombinant human interleukin 3. Lab Invest 1989; 61:677–690.

163. Stahl CP, Winton EF, Monroe MC, Haff E, Holman RC, Myers L, Liehl E, Evatt BL. Differential effects of sequential, simultaneous, and single agent interleukin-3 and granulocyte–macrophage colony-stimulating factor on megakaryocyte maturation and platelet response in primates. Blood 1992; 80:2479–2485.

164. Geissler K, Valent P, Bettelheim P, Sillaber C, Wagner B, Kyrle P, Hinterberger W, Lechner K, Liehl

E, Mayer P. In vivo synergism of recombinant human interleukin-3 and recombinant human interleukin-6 on thrombopoiesis in primates. Blood 1992; 79:1155–1160.

165. Umemura T, al-Khatti A, Donahue RE, Papayannopoulou T, Stamatoyannopoulos G. Effects of interleukin-3 and erythropoietin on in vivo erythropoiesis and F-cell formation in primates. Blood 1989; 74:1571–1576.

166. Kyoizumi S, Murray LJ, Namikawa R. Preclinical analysis of cytokine therapy in the SCID-hu mouse. Blood 1993; 81:1479–1488.

167. Nolta JA, Hanley MB, Kohn DB. Sustained human hematopoiesis in immunodeficient mice by cotransplantation of marrow stroma expression human interleukin-3: analysis of gene transduction of long-lived progenitors. Blood 1994; 83:3041–3051.

168. Gillio AP, Gasparetto C, Laver J, Abboud M, Bonilla MA, Garnick MB, O'Reilly RJ. Effects of interleukin-3 on hematopoietic recovery after 5-fluorouracil or cyclophosphamide treatment of cynomolgus primates. J Clin Invest 1990; 85:1560–1565.

169. Farese AM, Williams DE, Seiler FR, MacVittie TJ. Combination protocols of cytkine therapy with interleukin-3 and granulocyte–macrophage colony-stimulating factor in a primate model of radiation-induced marrow aplasia. Blood 1993; 82:3012–3018.

170. Ganser A, Lindemann A, Seipelt G, Ottmann OG, Herrmann F, Eder M, Frisch J, Schulz G, Mertelsmann R, Hoelzer D. Effects of recombinant human interleukin-3 in patients with normal hematopoiesis and in patients with bone marrow failure. Blood 1990; 76:666–676.

171. Ottmann OG, Ganser A, Seipelt G, Eder M, Schulz G, Hoelzer D. Effects of recombinant human interleukin-3 on human hematopoietic progenitor and precursor cells in vivo. Blood 1990; 76:1494–1502.

172. Falk S, Seipelt G, Ganser A, Ottmann OG, Hoelzer D, Stutte HJ, Hubner K. Bone marrow findings after treatment with recombinant human interleukin-3. Am J Clin Pathol 1991; 95:355–362.

173. Ganser A, Ottmann OG, Seipelt G, et al. Results of phase I/II trials with recombinant human interleukin-3. Biotechnol Ther 1991; 2:313–325.

174. Kurzrock R, Talpaz M, Estrov Z, Rosenblum MG Gutterman JU. Phase I study of recombinant human interleukin-3 in patients with bone marrow failure. J Clin Oncol 1991; 9:1241–1250.

175. Ganser A, Ottmann OG, Seipelt G, et al. Effect of long-term treatment with recombinant human interleukin-3 in patients with myelodysplastic syndromes. Leukemia 1993; 7:696–701.

176. Ganser A, Ottmann OG, Hoelzer D. Interleukin-3 in the treatment of myelodysplastic syndromes [editorial]. Int J Clin Lab Res 1992; 22:125–128.

177. Ganser A, Seipelt G, Lindemann A, et al. Effects of recombinant human interleukin-3 in patients with myelodysplastic syndromes. Blood 1990; 76:455–462.

178. Maurer AB, Ganser A, Buhl R, Seipelt G, Ottmann OG, Mentzel U, Geissler RG, Hoelzer D. Restoration of impaired cytokine secretion from monocytes of patients with myelodysplastic syndromes after in vivo treatment with GM-CSF or IL-3. Leukemia 1993; 7:1728–1733.

179. Seipelt G, Ganser A, Duranceyk H, Maurer A, Ottmann OG, Hoelzer D. Induction of TNF-alpha in patients with myelodysplastic syndromes undergoing treatment with interleukin-3. Br J Haematol 1993; 84:749–751.

180. Seipelt G, Ganser A, Duranceyk H, Maurer A, Ottmann OG, Hoelzer D. Induction of soluble IL-2 receptor in patients with myelodysplastic syndromes undergoing high-dose interleukin-3 treatment. Ann Hematol 1994; 68:167–170.

181. Dunbar CE, Smith DA, Kimball J, Garrison L, Nienhuis AW, Young NS. Treatment of Diamond-Blackfan anaemia with haematopoietic growth factors, granulocyte–macrophage colony stimulating factor and interleukin 3: sustained remissions following IL-3 [see comments]. Br J Haematol 1991; 79:316–321.

182. Gillio AP, Faulkner LB, Alter BP, Reilly L, Klafter R, Heller G, Young DC, Lipton JM, Moore MA, O'Reilly RJ. Successful treatment of Diamond-Blackfan anemia with interleukin 3. Stem Cells 1993; 2:123–130.

183. Gillio AP, Faulkner LB, Alter BP, Reilly L, Klafter R, Heller G, Young DC, Lipton JM, Moore MA,

O'Reilly RJ. Treatment of Diamond-Blackfan anemia with recombinant human interleukin-3 [see comments]. Blood 1993; 82:744–751.

184. Bastion Y, Bordigoni P, Debre M, et al. Sustained response after recombinant interleukin-3 in Diamond Blackfan anemia [letter; comment]. Blood 1994; 83:617–618.

185. Olivieri NF, Feig SA, Valentino L, Berriman AM, Shore R, Freedman MH. Failure of recombinant human interleukin-3 therapy to induce erythropoiesis in patients with refractory Diamond-Blackfan anemia. Blood 1994; 83:2444–2450.

186. Freedman MH. Erythropoiesis in Diamond-Blackfan anemia and the role of interleukin 3 and steel factor. Stem Cells 1993; 2:98–104.

187. Nimer SD, Paquette RL, Ireland P, Resta D, Young D, Golde DW. A phase I/II study of interleukin-3 in patients with aplastic anemia and myelodysplasia. Exp Hematol 1994; 22:875–880.

188. Ganser A, Lindemann A, Seipelt G, et al. Effects of recombinant human interleukin-3 in aplastic anemia. Blood 1990; 76:1287–1292.

189. Brugger W, Frisch J, Schulz G, Pressler K, Mertelsmann R, Kanz L. Sequential administration of interleukin-3 and granulocyte–macrophage colony-stimulating factor following standard-dose combination chemotherapy with etoposide, ifosfamide, and cisplatin. J Clin Oncol 1992; 10:1452–1459.

190. Brugger W, Bross K, Frisch J, Dern P, Weber B, Mertelsmann R, Kanz L. Mobilization of peripheral blood progenitor cells by sequential administration of interleukin-3 and granulocyte–macrophage colony-stimulating factor following polychemotherapy with etoposide, ifosfamide, and cisplatin [see comments]. Blood 1992; 79:1193–2000.

191. Orazi A, Cattoretti G, Schiro R, Siena S, Bregni M, Di Nicola M, Gianni AM. Recombinant human interleukin-3 and recombinant human granulocyte–macrophage colony-stimulating factor administered in vivo after high-dose cyclophosphamide cancer chemotherapy: effect on hematopoiesis and microenvironment in human bone marrow. Blood 1992; 79:2610–2619.

192. Postmus PE, Gietema JA, Damsma O, Biesma B, Limburg PC, Vellenga E, de Vries EG. Effects of recombinant human interleukin-3 in patients with relapsed small-cell lung cancer treated with chemotherapy: a dose-finding study. J Clin Oncol 1992; 10:1131–1140.

193. D'Hondt V, Weynants P, Humblet Y, et al. Dose-dependent interleukin-3 stimulation of thrombopoiesis and neutropoiesis in patients with small-cell lung carcinoma before and following chemotherapy: a placebo-controlled randomized phase Ib study [see comments]. J Clin Oncol 1993; 11:2063–2071.

194. Biesma B, Willemse PH, Mulder NH, Sleijfer DT, Gietema JA, Mull R, Limburg PC, Bouma J, Vellenga E, de Vries EG. Effects of interleukin-3 after chemotherapy for advanced ovarian cancer. Blood 1992; 80:1141–1148.

195. Gianni AM, Siena S, Bregni M, DiNicola M, Peccatori F, Magni M, Ravagnani F, Sklenar I, Bonadonna G. Recombinant human interleukin-3 hastens trilineage hematopoietic recovery following high-dose (7 g/m^2) cyclophosphamide cancer therapy. Ann Oncol 1993; 4:759–766.

196. Raemaekers JM, van Imhoff GW, Verdonck LF, Hessels JA, Fibbe WE. The tolerability of continuous intravenous infusion of interleukin-3 after DHAP chemotherapy in patients with relapsed malignant lymphoma. A phase-I study. Ann Hematol 1993; 67:175–181.

197. Crump M, Couture F, Kovacs M, Saragosa R, McCrae J, Brandwein J, Huebsch L, Beauregard-Zollinger L, Keating A. Interleukin-3 followed by GM-CSF for delayed engraftment after autologous bone marrow transplantation. Exp Hematol 1993; 21:405–410.

198. Fay JW, Lazarus H, Herzig R, et al. Sequential administration of recombinant human interleukin-3 and granulocyte–macrophage colony-stimulating factor after autologous bone marrow transplantation for malignant lymphoma: a phase I/II multicenter study. Blood 1994; 84:2151–2157.

199. D'Hondt V, Guillaume T, Humblet Y, Doyen C, Chatelain B, Feyens AM, Staquet P, Osselaer JC, Mull B, Symann M. Tolerance of sequential or simultaneous administration of IL-3 and G-CSF in improving peripheral blood stem cells harvesting following multi-agent chemotherapy: a pilot study. Bone Marrow Transplant 1994; 13:261–264.

200. Henon PR, Becker M. Cytokine enhancement of peripheral blood stem cells. Stem Cells 1993; 2:65–71.

201. Hocker P, Geissler K, Kurz M, Wagner A, Gerhartl K. Potentiation of GM-CSF or G-CSF induced mobilization of circulating progenitor cells by pretreatment with IL-3 and harvest by apheresis. Int J Artif Organs 1993; 5:25–29.

202. Haas R, Ehrhardt R, Witt B, Goldschmidt H, Hohaus S, Pforsich M, Ehrlich H, Farber L, Hunstein W. Autografting with peripheral blood stem cells mobilized by sequential interleukin-3/granulocyte-macrophage colony-stimulating factor following high-dose chemotherapy in non-Hodgkin's lymphoma. Bone Marrow Transplant 1993; 12:643–649.

203. Kishimoto T, Taga T, Akira S. Cytokine signal transduction. Cell 1994; 76:253–262.

204. Noguchi M, Yi H, Rosenblatt HM, Filipovich AH, Adelstein S, Modi WS, McBride OW, Leonard WJ. Interleukin-2 receptor gamma chain mutation results in X-linked severe combined immunodeficiency in humans. Cells 1993; 73:147–157.

13

Macrophage Colony-Stimulating Factor

Malcolm A. S. Moore
Memorial Sloan-Kettering Cancer Center, New York, New York

I. INTRODUCTION

Over 25 years ago, a colony-stimulating factor (CSF) was identified in murine serum (1), human urine (2), and in murine cell-conditioned media (3). What distinguished this CSF from what were subsequently identified as granulocyte colony-stimulating factor (G-CSF), granulocyte–monocyte colony-stimulating factor (GM-CSF), or multi-CSF (IL-3) was its restricted ability to stimulate the development of only macrophage colonies. This M-CSF, or CSF-1, as it was first termed (reviewed in Ref. 4), was also independently identified in other assays and termed macrophage–granulocyte inducer-1M (5) or macrophage growth factor (MGF), owing to its ability to stimulate peritoneal macrophage proliferation (6).

Highly purified M-CSF was obtained from mouse L–cell-conditioned medium as a glycoprotein of 70 kDa composed of two 35-kDa polypeptide chains, with disulfide bonds (7,8). Purification was facilitated by a sensitive bioassay based upon macrophage colony formation by single murine marrow progenitors in agar culture (9). One unit of M-CSF was defined as the amount of factor required to stimulate a single murine macrophage colony in the linear part of the dose–response curve, representing 0.44 fmol of murine M-CSF (10). This bioassay allowed detection of M-CSF, in mouse or human sera, urine, or conditioned media, at concentration as low as 1 pM, with maximum stimulation at 250 pM. With this assay, it was possible to purify M-CSF from mouse L–cell-conditioned medium (7,10,11) or human urine (12) by sequential chromatographic and molecular-sieving procedures, or subsequently, by immunoaffinity chromatography with immobilized antibodies to M-CSF (11,12).

II. MOLECULAR BIOLOGY OF M-CSF

A. Transcription, Translation, and Posttranslational Modification

Degenerate oligonucleotide probes predicted from the NH_2-terminal sequence of M-CSF purified from mouse L–cell-conditioned medium or human urine were used to isolate partial human CSF-1 genome clones. The nucleotide sequences of these clones were then used to isolate

M-CSF clones from a human pancreatic tumor cell line cDNA library. Kawasaki et al. (13) isolated a 1.6-kb clone encoding a leader sequence of 32 amino acids followed by 256 amino acids. This species is referred to as the *short* or α-form of M-CSF. In addition to this species, longer, polyadenylated RNA species of up to 4 kb were isolated, including a "long" or β-form of 554 amino acids (14) and a γ-form of 438 amino acids (15; Fig 1). Because Southern blotting analysis indicated the presence of only a single M-CSF gene, the mRNAs of different lengths are presumed to be products of alternative splicing (13).

Probes from the human M-CSFα cDNA were used to isolate M-CSF cDNA from mouse L cells, resulting in isolation of a 4-kb and a biologically active 2.3-kb clone, both encoding a precursor of 552 amino acids, including the signal peptide (16,17). Human and murine M-CSFs have a 69.5% homology at the amino acid level, with the greatest similarity in the biologically active region between residues 1 and 149 (80.5%) and least between 150 and 447 (64%). The various species of human M-CSF are identical at the NH_2- and COOH-terminal regions, but M-CSFβ and M-CSFγ contain inserts at residue 149, as a result of alternative splicing. The fully active form of M-CSF is produced using a construct with as few as the first 150 amino acids (3). The COOH-terminus is not required for biological activity, but has a hydrophobic transmembrane domain, followed by positively charged residues and a cytoplasmic tail characteristic of membrane proteins. The COOH-terminus is believed to anchor some forms of M-CSF in an active cell-associated form, with the native secreted forms of M-CSFα being generated by proteolytic cleavage at or near amino acid 158.

M-CSFα has two possible N-linked glycosylation sites of the Asn–X–Ser/Thr variety and M-CSFβ has four. The first two sites common to both species of CSF are probably responsible for the large amount of glycosylation and heterogeneity seen in the native protein. A high molecular mass form of M-CSF (> 200 kDa) has been reported to be produced by Chinese hamster ovary (CHO) cells transfected with 4-kDa M-CSF cDNA (18) and by a human osteoblastic cell line (19). This species is a proteoglycan, with an alternatively spliced core protein of 45 kDa, and binds in a dose-dependent manner to collagen-coated wells. Immunoblot analysis of urea-extracted bone M-CSF has revealed the presence of this proteoglycan form in bone matrix, suggesting its possible importance in bone metabolism (19).

The 18 cysteines in the 49-kDa homodimeric species form three intermolecular and three intramolecular disulfide bonds. The close proximity of four cysteines (Cys = 139, = 146, = 157, and = 159) result in a tight core complex, rendering the protein resistant to proteolytic digestion (20). The X-ray crystallographic structure of M-CSFα has been determined at 0.25 nM (21). The dimer consists of two bundles of four α-helices laid end-to-end with an interchain disulfide bond. Individual monomers of M-CSF show a close structural similarity to GM-CSF and human growth hormone, suggesting that receptor-binding determinants for all three cytokines may be similar, with the exposed surface of helix D or the residues on the side of the bundle on helices A and C most likely to be implicated in receptor binding.

The M-CSFα is synthesized as a membrane-bound precursor, which undergoes cotranslational glycosylation and assembly into disulfide-linked 64-kDa homodimers containing mannose-rich asparagine-*N*-linked oligosaccharide chains. Modification of N-linked carbohydrates from high-mannose forms to complex oligosaccharide chains during intracellular transport yields a 68-kDa surface dimer, immobilized through the chain predicted hydrophobic anchor (3). Killed, fixed 3T3 cells expressing this form of CSF are capable of stimulating M-CSF-dependent cell proliferation, indicating functional activity of the ligand by direct cell–cell interaction (22). The large species of M-CSF is synthesized as an integral transmembrane glycoprotein of 70 kDa and undergoes rapid disulfide-mediated assembly into homodimers. However, the homodimer acquires both N- and O-linked sugars during transport and is cleaved from the membrane within

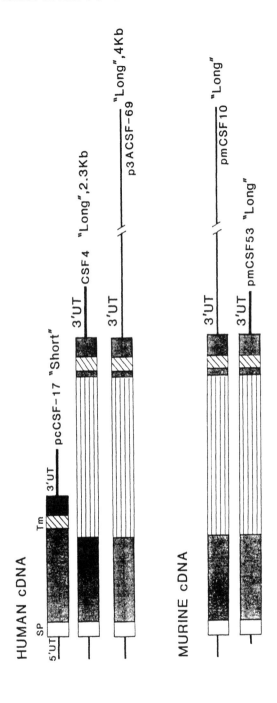

Figure 1 Structure of the human and murine cDNA clones. The postulated 32-amino-acid signal peptide (SP) is indicated by the clear block. The coding sequences in both the short [pcCSF-17(7)] and long [CSF4(8), p3ACSF-69(9), pMCSF10 and pMCSF53 (12)] protein forms of M-CSF are indicated by the dark areas. The coding sequence unique to the long protein forms is indicated by the blocks with horizontal lines, and the transmembrane domain (Tm) is depicted by boxes with slanted lines. The short 3′ untranslated (UT) sequences found in pcCSF-17, CSF4, and pmCSF53 are shown by thick lines and the 2-kb 3′ UT regions in p3ACSF-69 and pmCSF10 are depicted by the long, thin lines. (From Ref. 27a.)

the intracellular secretory compartment before its release from the cells. Plasma membrane-bound forms of this species are not detected, and the proteolyzed homodimer is efficiently secreted into the medium as an 86-kDa glycoprotein. In contrast, the smaller precursor (M-CSFα) is relatively resistant to proteolysis and is only inefficiently released from the plasma membrane through the action of extracellular proteases that cleave the molecule at an alternative site. In the presence of phorbol ester, stimulation of protein kinase C (PKC) results in rapid cleavage of this membrane-bound form, possibly by activating endogenous cellular protease production (23). The soluble proteolytically cleaved M-CSF species is biologically active, further confirming that residues 1–158 are sufficient for biological activity. The potential of varying the balance between membrane-associated and soluble forms of bioactive M-CSF in different cell types or under different activation conditions provides the potential for endocrine, paracrine and juxtacrine interactions between M-CSF and its receptor. Additionally, the binding of membrane-anchored M-CSF to its receptor mediates specific cytoadhesion between stromal cells and M-CSF receptor-bearing hematopoietic cells (24).

B. Chromosomal Organization of the Human M-CSF Gene

The human M-CSF gene is located on chromosome 1 bands p13-p21, rather than the long arm of chromosome 5, as initially reported (25). This is consistent with murine studies that localize the murine M-CSF gene to mouse chromosome 3 at band 3F3, closely linked to a region bearing several genes that are also expressed on human chromosome 1 (26).

The M-CSF gene spans 21 kb and comprises ten exons and nine introns. The latter range in size from 0.3 to 2.7 kb, and the exons vary from 52 base pairs (bp) to more than 2 kb. The majority of exons are relatively small, with the exception of exon 6, which is either 131 or 1025 bp, depending on which splice acceptor site is used (27).

III. M-CSF PRODUCTION AND GENE EXPRESSION

Macrophage colony-stimulating factor is present in normal plasma at levels of 100–200 units(U) /ml. Fluctuations in levels of circulating M-CSF reflect a balance between production and excretion with M-CSF receptors on Kupffer cells in the liver, and splenic macrophages responsible for significant clearance by specifically binding M-CSF, which is subsequently endocytosed and destroyed intracellularly (28).

Most tissues produce M-CSF transcripts of several sizes, and this ubiquitous expression may reflect production by a restricted population of cell types common to many organs (e.g., macrophages, fibroblasts, mesothelial cells, or endothelial cells). Table 1 is a list of normal cell types identified as constitutive or inducible sources of M-CSF.

A. M-CSF Production by Monocytes and Macrophages

Circulating monocytes, isolated by centrifugal elutriation, do not constitutively express M-CSF mRNA or protein, but these are up-regulated following adherence to a plastic surface (29). Low levels of prostaglandin E (PGE) or other arachidonic acid metabolites produced by monocytes may limit M-CSF gene expression (30). Lipopolysaccharide (LPS) is a potent inducer of monocyte M-CSF production, and its activity is enhanced by cyclooxygenase inhibition and reversed by the action of prostaglandin E_2 (31). Interleukin (IL)-4 and glucocorticoids also suppress LPS-induced M-CSF production. Phorbol esters are strong inducers of monocyte M-CSF mRNA and protein production, inducing predominantly transcripts of the 4-kb species (32–34). GM-CSF is a strong inducer of M-CSF message in monocytes, and both tumor necrosis

Table 1 Sources of M-CSF

Cell type	Inducer	Ref.
Monocyte	Adherence	29
Monocyte	LPS	31
Monocyte	Phorbol esters	32–34
Monocyte	GM-CSF	31,34,35
Monocyte	TNF-α	31,36
Monocyte	IFN-γ	31,36
Monocyte	IL-2	37,38
Monocyte	Anti-CD45,CD44,LFA-2,MAb	39
Monocyte	1,25(OH)$_2$ vitamin-D$_3$	71
Macrophage	None	42,43,45
Macrophage	LPS	42,43,45
T Lymphocyte	T-cell mitogens	47–49
T Lymphocyte	Antigen + IL-2	47–49
T Lymphocyte	TNF, IL-1	49
T Lymphocyte	Anti-CD3,CD2,CD28,MAB	30
Epithelial cells–keratinocytes	None	52
Epithelial cells–keratinocytes	LPS	52
Epithelial cells–nasal	None	53
Epithelial cells–nasal	IL-1,TNF,bacterial products	53
Epithelial cells–thymic	None	54
Epithelial cells–thymic	IL-1,IFN-γ	54
Fibroblasts–various strains	IL-1,TNF-α,IL-1	7,57
Fibroblasts–mesenchymal	TNF-α	58
Fibroblasts–synovial	IL-1,TNF-α,IFN-γ,IL-4	59
Fibroblasts–hematopoietic tissue	Constitutive	55,56
Fibroblasts–hematopoietic tissue	IL-1,TNF-α	55,56
Articular cartilage/chondrocytes	IL-1,TNF-α	60
Vascular endothelium	IL-1,TNF-α,LPS	61,63
Vascular endothelium	Oxidized LDLs	62
Vascular smooth muscle	IL-1,TNF-α,LPS	66
Myogenic cells	Proliferation	67
Peritoneal mesothelium	IL-1	69
Renal mesangial cells	IL-1,TNF-α	70
Placenta–trophoblast decidual stroma	None or IL-1	72–83
Endometrial gland cells, villous mesenchyme	None or IL-1	72–83

factor-alpha (TNF-α) and interferon gamma (IFN-γ) enhanced M-CSF message levels induced by GM-CSF, but only TNF-α synergized with GM-CSF in the induction of M-CSF protein (31,34,35). Thus, M-CSF message expression induced by cytokines does not always correlate with M-CSF protein secretion. IFN-γ (33) and TNF-α (36) have been reported to induce monocyte M-CSF directly, but adherent activation of the monocyte may be necessary for this effect. IL-2, at concentrations higher than 0.05 ng/ml up-regulated IL-2 receptor α-chain on monocytes, and both directly and indirectly enhanced production of M-CSF (37). Transcriptional activation of the M-CSF gene by IL-2 is preceded by enhanced binding activity of the transcriptional factor NF-κB to its recognition sequence in the 5$'$ regulatory enhancer region of the M-CSF gene (38).

Human monocyte cell surface molecules CD45, CD44, and LFA-3 can act as physiological triggers for the autocrine production of M-CSF (39). The binding of monoclonal antibody MAb to these cell surface antigens induced M-CSF message and a small amount of protein. IL-1 acted as a second signal strongly augmenting M-CSF secretion in this system, partly by increasing gene transcription, with no effect on M-CSF message half-life (40). Both IL-4 and IL-10 strongly inhibited M-CSF secretion by anti-CD45/IL-1β-induced monocytes by repressing M-CSF gene transcription. This contrasts with the observation of Wieser et al. (41), who demonstrated that IL-4 induced M-CSF secretion by human monocytes. This discrepancy may be due to the differences in isolation methods, because Gruber et al. (40) used centrifugal elutriation to isolate monocytes, whereas Wieser et al. (41) used repeated plastic adherence, which serves to activate monocyte function. Following prolonged culture at high cell density or in the presence of M-CSF, monocytes or their progenitors differentiate into macrophages in suspension culture (42) or in Dexter-type long-term marrow cultures (43). Such macrophages exhibit basal constitutive expression of M-CSF message, and LPS induces protein production, whereas increasing levels of exogenous M-CSF down-modulate CSF production.

The ubiquity of M-CSF transcripts in such tissues as liver, lung, brain, and spleen (14,44) partly reflects the presence of significant numbers of phagocytic mononuclear cells, such as Kupffer cells in liver, alveolar macrophages in lung, microglial cells in brain, and littoral cells or fixed tissue macrophages in lymphoid organs. The number and function of pulmonary macrophages are critical to lung homeostasis, and there is a significant correlation between the numbers of macrophages and the levels of M-CSF in bronchoalveolar lavage fluid (45). Constitutive M-CSF mRNA expression was found in alveolar macrophages, and the M-CSF was further up-regulated by LPS in macrophages from individuals who smoked, but not from nonsmokers.

The detection of both M-CSF and c-*fms* expression by cells of the monocyte lineage has suggested that M-CSF may act by an autocrine mechanism and it can induce the expression of its own gene by activation, and by translocation of protein kinase C and an associated expression of NF-κB protein in nuclear extracts (46).

B. M-CSF Production by T Cells

Activated, but not resting T cells or thymocytes, can be induced to transcribe M-CSF cDNA within 92–120 hr of stimulation by a variety of T-cell mitogens (47–49), antigen plus IL-2 in the case of alloreactive T-cell clones (47–49), TNF, IL-1 (49), and MAb directed against CD3 (49). The MAb against the T-cell adherence molecule CD28 provides a costimulatory signal in combination with either CD2 or CD3 MAb to produce transcription of the M-CSF gene (50). The expressed M-CSF mRNA is predominantly 4 kb, with lesser expression of the 2-kb form (48), and this is followed by bioactive M-CSF production. Malignant T cells also produce M-CSF on induction (49), and the T helper cell line HUT 102 constitutively produces M-CSF, probably because viral proteins expressed by integrated HTLV-1 function as trans-acting transcriptional activators inducing constitutive M-CSF and GM-CSF production (51).

C. M-CSF Production by Epithelial Cells

Various epithelial cell types can be induced to produce M-CSF, implicating this cytokine in various immune or inflammatory events occuring at epithelial surfaces. Murine and human keratinocytes produce several immunoreactive cytokines, including constitutive expression of M-CSF mRNA and protein (52). Furthermore, M-CSF production is increased by bacterial LPS, indicating a role in regulation of the cutaneous immune response.

Constitutive M-CSF production has been reported in primary nasal epithelial cells isolated from the upper airways and continuous epithelial cells lines from normal persons and from patients with cystic fibrosis (53). Epithelial cell M-CSF production was enhanced by substances likely to be present in the inflamed lungs of patients with cystic fibrosis; namely, IL-1, TNF-α, and bacterial products. Dexamethasone does not suppress this M-CSF production, in contrast with its action on endothellial IL-8, GM-CSF, and IL-6 release (53). Thymic stromal cells and isolated thymic epithelial cell cultures show constitutive M-CSF production and up-regulated M-CSF levels when induced with IL-1 or IFN-γ (54). In contrast IL-4 did not block this induction in macrophages.

D. M-CSF Production by Fibroblasts

Following the early observations that various normal and transformed fibroblast lines produced M-CSF (7), numerous studies attest to the ability of fibroblasts from diverse sources to produce M-CSF on induction by inflammatory mediators. Stromal cells from marrow (55) and SV40-transformed human fetal spleen and marrow fibroblasts produce M-CSF mRNA constitutively (56). The acute-phase response mediators IL-1, TNF, and IL-6, which are abundantly produced by activated monocytes, enhance levels of M-CSF in fibroblasts by both transcriptional and posttranscriptional mechanisms (57). Mesenchymal cell or fibroblast production of M-CSF (and GM-CSF, IL-8, and IL-6) is induced by TNF-α, and glucocorticoids repress the induction of most cytokines, but not M-CSF (58). IL-1, TNF-α, IFN-γ, IL-4, and dexamethasone increase production of M-CSF above constitutive levels in human synovial fibroblast-like cells, suggesting that these may be the source of M-CSF production in the joints of patients with inflammatory arthritis (59). Articular cartilage and chondrocyte monolayers also produce M-CSF in response to IL-1 or TNF-α within 2 hr of exposure (60), suggesting that another source of M-CSF is implicated in chronic rheumatoid disease. In contrast with synovial fibroblasts, chrondrocyte M-CSF production was inhibited by anti-inflammatory corticosteroids, but not by cyclo-oxygenase inhibitors (60).

E. M-CSF Production by Endothelial Cells

Vascular endothelium produces M-CSF upon IL-1, TNF-α, or LPS stimulation (61). The pathological implication of this observation is emerging from studies of the mechanisms involved in the development of atherosclerotic plaques. Oxidized lipoproteins have been identified in such plaques and in early lesions. Treatment of endothelial cells with modified low-density lipoprotein obtained by mild iron oxidation results in rapid induction of M-CSF, GM-CSF, and G-CSF (62). Rosenfeld et al. (63) reported that inflammatory cytokines induced M-CSF mRNA accumulation and protein synthesis in endothelial cells and smooth-muscle cells of atheromatous plaques in humans and rabbits. Infiltration of monocytes into the vascular wall and their transformation into lipid-laden foam cells characterizes early atherogenesis and is probably a consequence of local endothelial and smooth-muscle cell M-CSF production. Foam cells isolated from plaques expressed M-CSF mRNA and protein, contributing to the local accumulation of M-CSF which stimulated survival and proliferation of macrophages and activation of specific macrophage functions, such as expression of acetyl–low-density lipoprotein (scavenger) receptors and secretion of apolipoprotein E (63,64).

F. M-CSF Production by Miscellaneous Tissues

Macrophage colony-stimulating factor may play an important intrathecal immunoregulatory role in neoplastic and infectious diseases of the central nervous system (CNS). Elevated levels of M-CSF have been detected in the cerebrospinal fluid of patients with brain tumors, bacterial meningitis, and acquired immunodeficiency syndrome (AIDS) dementia complex (65). Low levels of M-CSF mRNA and protein can be detected in fetal astrocytes and microglial cells, and both increase after LPS stimulation (66). IL-1 and TNF-α significantly increased production by astrocytes, but not in microglial cells. The latter cells proliferated and differentiated in response to M-CSF and showed marked suppression of class II major histocompactibility complex (MHC) antigen expression, even following IFN-γ induction (66). This suggests the possibility that *in situ* M-CSF production may regulate normal glial development and may contribute to the immunological status of the CNS through down-regulation of class II MHC expression.

The normal and neoplastic proliferation of muscle cells may also be influenced by M-CSF, because the 4-kb mRNA transcripts are expressed during proliferation of myogenic cells and are down-regulated after the cells differentiate to myotubes (67). This down-regulation was at the posttranscriptional level, in contrast to regulation of M-CSF expression in phorbol ester-treated monocytes, which is at the transcriptional level (68).

Human peritoneal mesothelial cells constitutively produce many cytokines, including M-CSF, and IL-1 up-regulates expression of these genes (69). Mesangial cells of the renal glomeruli produce high levels of M-CSF upon IL-1 or TNF-α stimulation (70). This observation supports the view that mesangial cells have an effector role in modulating immune inflammatory responses in glomeruli. Release of cytokines may activate not only infiltrating inflammatory cells, but also mesangial cells themselves through an autocrine pathway.

M-CSF can also be produced by various placental tissues (see section V.), constitutively by leukemia and lymphoma cells, and by a variety of solid tumors (see later discussions).

IV. THE M-CSF RECEPTOR (M-CSFR), DOWN-REGULATION, AND TRANSMODULATION

A. Expression, Down-Regulation, and Transmodulation

High-affinity ($K_d \cong 10^{-13}$ M) M-CSF-binding sites are expressed at levels of 1×10^4–1×10^5 per cell on mature monocytes, tissue macrophages, osteoclasts (3), and placental trophoblasts (84), and at low levels on normal or neoplastic B lymphocytes (85). During maturation of cells in the macrophage lineage, the number of receptors increases with maturation, but the proliferative response to M-CSF declines.

The M-CSFR has a half-life of 2–3 hr at the cell surface in the absence of ligand binding, but is rapidly (15 min) internalized and degraded on ligand binding. This down-modulation requires receptor kinase activity (86). The receptors are initially replaced from an intracellular pool and later by de novo synthesis. IFN-γ, LPS, and the phorbol ester PMA produce a diminution of M-CSF-binding sites on monocytes and macrophages, and this "transmodulation" occurs in the presence or absence of M-CSF (87,88). This form of transmodulation involves activation of protein kinase C, which does not phosphorylate M-CSFR, but activates a protease that cleaves the receptor near its transmembrane segment, releasing the intact ligand-binding domain from the cell (86). This mechanism serves to desensitize the activated macrophage, switching off M-CSFR–mediated proliferation and chemotaxis. A further indirect mechanism of receptor trans-down–modulation involves IFN-γ or LPS stimulation of macrophage TNF-α production (88). TNF-α itself trans-down–modulates M-CSFR on bone marrow-derived macrophages (89) and on murine peritoneal macrophages (90–92). A greater than 90% decrease in M-CSF binding was observed within 15 min of exposure of macrophages to TNF-α, and this reduction in binding was reversed by prolonged (8–12 hr) incubation at 37°C, even in the presence of TNF, in a process requiring protein synthesis (90).

The M-CSFRs on murine bone marrow cells are also trans-down–modulated by IL-3 and GM-CSF at 37°C after 30 min, but not by G-CSF (93). This hierarchical down-modulation was investigated in a murine factor-dependent cell line responsive to IL-3 and GM-CSF and engineered to respond to M-CSF following c-*fms* transduction (94). Both IL-3 and GM-CSF extinguished cell surface expression of M-CSFR within 6–12 hr at both the protein and mRNA level, even in the presence of M-CSF. There was a GM-CSF concentration-dependent reduction in processed cytoplasmic c-*fms* transcripts (but not in the 6.5-kb–unprocessed transcript). It would appear that c-*fms* is regulated posttranslationally by modulation of mRNA stabilization and accumulation in the cytoplasm, mediated by labile protein-stabilizing and -nonstabilizing factors (94).

B. Molecular Biology of c-*fms*

The M-CSFR is encoded by the protooncogene c-*fms* (95). When a COOH-terminally truncated and mutated cDNA copy of the c-*fms* gene was transduced as the v-*fms* oncogene in feline sarcoma virus, it transformed fibroblasts and induced tumor formation (reviewed in Ref. 96). Biologically active c-*fms* cDNAs, obtained from human placental (97) and murine pre-B-cell libraries (98), show sequence similarity to each other and to the feline v-*fms* oncogene product.

The M-CSFR is a member of a family of growth factor receptors that exhibit ligand-induced tyrosine-specific protein kinase activity (96). The receptor is an integral membrane glycoprotein of 972 amino acids, including a signal peptide. The extracellular NH_2-terminal segment is 512 amino acids, with five immunoglobulin-like domains, separated by a 23-amino acid transmembrane domain from the 435-amino acid intracellular domain that includes all sequences necessary for tyrosine kinase activity (96; Fig. 2).

The v-*fms*–coded polypeptides are synthesized as integral transmembrane glycoproteins, oriented with their NH_2-terminal ligand-binding domains in the endoplasmic reticulum cisternae, and their COOH-terminal ligand-binding domains in the cytoplasm. The polypeptides are glycosylated, acquiring N-linked oligosaccharide chains in their NH_2-terminal domains (96). Transport of the chains to the cell surface is associated with sugar chain modification and increases in molecular mass (99).

M-CSFR is most closely related to the A- and B- type of platelet-derived growth factor (PDGF) receptor and the receptor encoded by the c-*kit* protooncogene, with all of these sharing a distinct pattern of cysteine spacing within their extracellular domains that includes sequences characteristic of the immunoglobulin superfamily (96). The enzymatic domains of this receptor family are interrupted by hydrophilic spacer sequences of 74–104 amino acids that are unique to each receptor and that may function to provide substrate recognition. The M-CSFR gene maps in a closely linked tandem array with the B-type PDGFR gene on human chromosome 5 at band 5q33-3 (100). The close sequence homology between c-*kit*, c-*fms*, and PDGFR is maximal through amino acids 563–678 and 752–906, with residues 589–594 being a consensus sequence for ATP binding, as well as the ATP-binding site itself at Lys-616 (101). There is no homology with other tyrosine kinases in the COOH-terminal tail of M-CSFR at amino acids 907–972, and this region acts to negatively regulate receptor activity (96).

C. Receptor Signal Transduction

Ligand-induced kinase activation of the M-CSFR occurs by receptor oligomerization, without propagation of conformational changes through the transmembrane domain, which itself may play a role in receptor oligomerization (102). The M-CSFR mediates its pleiotropic effects through the coupling of its ligand-activated tyrosine kinase (TK) to multiple intracellular effector

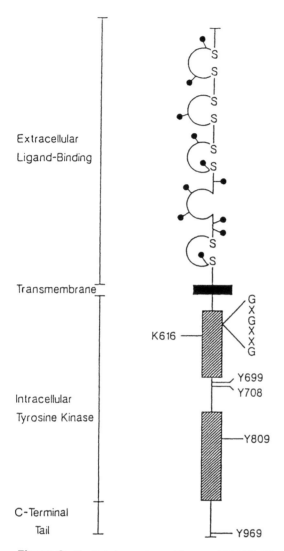

Figure 2 Predicted structure of human CSF-1R: The extracellular ligand-binding portion is organized
into five immunoglobulin-like domains, four of which are stabilized by disulfide bonds (S-S). Positions of
asparagine-linked oligoasaccharide chains are indicated. The intracellular kinase domain (stippled bars) is
interrupted by spacer sequences containing two predicted sites of receptor phosphorylation at tyrosines (Y)
699 and 708. A third phosphorylation site (Y809) occurs in the distal core consensus sequences. The
membrane-proximal kinase segment includes a glycine-rich signature sequence (G-X-G-X-X-G), characteristic of kinases in general. This is followed by the ATP-binding site at lysine (K) 616. The COOH-
terminal tail contains a single tyrosine residue (Y969), the removal of which up-regulates receptor kinase
activity. (From Ref. 96.)

proteins, the combined actions of which determine the magnitude and specificity of the
biological response (Table 2).

M-CSF–induced tyrosine phosphorylation of GTPase-activating protein and its associated
cellular proteins p62 and p196 induce mitogenic signals through p21 *ras* activation (103). The
activated M-CSFR physically associates with phosphatidylinositol [(PI)3′-kinase and induces

Table 2 M-CSF-Induced Molecular Responses

Response	Cell type	Ref.
PI-3'kinase phosphorylation	Fibroblast (c-*fms*)	104,120,121
PI-3'kinase phosphorylation	Macrophage line	104,120
PI-3'kinase phosphorylation	Macrophage	104
GTP-binding/GTPase activity	Monocytes	103,122
GTP-binding/GTPase activity	Fibroblast (c-*fms*)	103,104
p21 *ras* phosphorylation	Fibroblast (c-*fms*)	103,110,111
p26 "GAP"-associated protein tyrosine phosphorylation	Fibroblast (c-*fms*)	103,104
DAG turnover	Macrophage	115,117
DAG turnover	Monocytes	114
Phospholipase-A_2 activation	Monocytes	115
PKC translocation	Monocytes	114
RAF-1 kinase	Macrophage line	123
cAMP↓	Macrophage	116,117
Na^+,H^+ exchange	Progenitors	124
Na^+,H^+ exchange	Macrophage	118,125
Na^+,H^+ exchange	Monocytes	122
Na^+,H^+-ATPase	Macrophage	125
Na^+,H^+-ATPase	Monocytes	122
c-*fos* mRNA	Macrophage	126,127
c-*fos* mRNA	Macrophage line	128
c-*fos* mRNA	Monocyte	115
c-*fos* mRNA	Fibroblast (c-*fms*)	121
c-*myc* mRNA	Macrophage	126,127
c-*myc* mRNA	Macrophage line	128
c-*myc* mRNA	Fibroblast (c-*fms*)	115
c-*jun* mRNA	Monocyte	112,115
c-*jun* mRNA	Fibroblast (c-*fms*)	121
p21*ras* mRNA	Fibroblast (c-*fms*)	110,111
Src, Fyn, Yes activation	Fibroblast (c-*fms*)	108
hck mRNA	Monocyte	112
Fibronectin-Rβ mRNA	Monocyte	112
β-Actin mRNA	Monocyte	112
Plasminogen activator mRNA/protein	Macrophage	117,129
G_1 cyclin mRNA/protein	Macrophage line	130
Glucose uptake	Macrophage	131
Pinocytosis	Macrophage	132
Arachidonate and PGE production	Monocytes	115

DAG, 1,2-diacylglycerol; PI, phosphatidylinositol; GAP, GTPase-activating protein; PKC, protein kinase C
Source: Ref. 117.

phosphorylation of its GTPase-activating protein (104). Deletion mutants of the kinase insert (KI) domain confirm the role of PI 3'-kinase in mutagenesis (104,105). Replacement of Leu-301 in the receptor by certain amino acids induces ligand-independent transforming activity in NIH 3T3 cells, associated with constitutive receptor kinase activity and increased phosphorylation of cellular substrates on tyrosine (106). Mutations of the autophosphorylation site at Tyr-809 block

M-CSF mitogenic action and induction of c-*myc* mRNA, but do not impair ligand stimulation of tyrosine kinase activity, receptor binding to PT 3′-kinase or induction of c-*fos* and *Jun* B (107). These studies suggest a receptor-mediated bifurcation of intracellular transduction pathways during the immediate early response, for enforced expression of an endogenous c-*myc* gene restored the proliferative response to M-CSF. M-CSF stimulation also results in activation of three Src family kinases—Src, Fyn, and Yes—all associated with the ligand-activated receptor and probably playing roles in the mitogenic response (108). Their interaction did not involve the KI domain of the receptor, but mutation of the receptor autophosphorylation site at Tyr-809 reduced both their binding and enzymatic activation.

Phosphorylation and activation of RAF-1 kinase is among the early events in M-CSF signal transduction, and oncogenically activated v-*raf* induces M-CSF–independent macrophage proliferation in the absence of autocrine growth factor production (108). These studies indicate that *RAF-1* and not MAP-kinase is a key component of the M-CSF mitogenic signal transduction pathway and possibly involves transcriptional activation of genes by *AP-1* (109). The p21 *ras* activation is required for the proliferative response to M-CSF (110). The p21 *ras* signals stimulate transcription from promoter elements containing overlapping binding sites for Fos/Jun and Ets-related proteins (111). Mitogenic signaling by both M-CSF and *ras* is suppressed by the *Ets-2* DNA-binding domain and restored by *myc* overexpression.

In monocytes, a group of coordinately induced early-response genes are induced by M-CSF within 15–30 min (112). This includes c-*jun*, fibronectin receptor, actin, and a myelomonocytic-specific tyrosine kinase *hck* gene. The rapid and transient increase in c-*jun* mRNA occurs by increase in transcription rate and prolongation of the half-life of transcripts and also by rapid induction of the *jun-B* gene (113). The c-*fos* mRNA levels are also increased through control at both the transcriptional and posttranscriptional level following induction of the *fos-B* gene (113).

Enhanced 1,2-diacylglycerol (DAG) turnover may represent an important early signal for macrophage proliferation. Elevated DAG turnover, possibly by phosphatidylinositol 4,5-biphosphate-specific phospholipase C-mediated phosphatidylcholine (PC) hydrolysis, has been described in human monocytes in response to M-CSF (114). Increased formation of DAG and PC hydrolysis is associated with M-CSF stimulation of a cytoplasmic phospholipase-A_2 activity, in part by posttranscriptional stabilization of the phospholipase transcripts (115). Furthermore, M-CSF treatment leads to phosphorylation of the phospholipase protein, indicating regulation at multiple levels. Stimulation of the phospholipase-A_2 pathway results in increased arachidonate and prostaglandin synthesis by monocytes and macrophages.

M-CSF inhibits cAMP accumulation, overcoming the macrophage proliferation–inhibition that is mediated by elevated cAMP levels (116,117). This may be achieved by inhibitory adenylate cyclase or increasing cAMP phosphodiesterase activity (50,116). cAMP elevation inhibits M-CSF–induced plasminogen activator enzyme production in bone marrow-derived macrophages (117).

Stimulation of the amiloride-sensitive Na^+/H^+ exchanger is a ubiquitous early response to CSF stimulation. The inhibition of CSF-stimulated proliferation in the presence of HCO_3 suggests that the Na^+ influx, rather than any pH increase, may be the more important consequence of Na^+/H^+ exchange activation (117,118). This influx also activates the Na^+/K^+-ATPase, which is responsible for maintaining intracellular Na^+ levels (117). Although Na^+ cycle activity appears necessary for M-CSF-stimulated bone marrow-derived macrophage proliferation, this is not so for a variety of other early M-CSF responses (118), indicating that Na^+ cycle activation itself is unlikely to be a general mediator of M-CSF action.

Gene transfer technology indicates that the M-CSFR delivers a proliferative signal when expressed in certain IL-3- or GM-CSF–dependent cell lines and murine fetal liver progenitors

(119). These studies show that, when activated by M-CSF, c-*fms*, a receptor normally exhibiting lineage-restricted expression, can deliver a proliferative and differentiative signal to progenitors of several lineages (e.g., erythroid and megakaryocytic); however, the failure of M-CSF to elicit a response in c-*fms*–overexpressing granulocytes also suggests lineage differences in signaling pathways (119).

V. BIOLOGICAL ACTIVITIES OF M-CSF

A. Action on Progenitor Cells

Macrophage colony-stimulating factor is a specific proliferative stimulus for monocyte–macrophage progenitors and mature macrophages in various tissues (133). M-CSF and TNF markedly synergize in enhancing proliferation of bone marrow macrophages (134). Because macrophages can be induced to produce M-CSF and TNF and respond to these factors by proliferation, an autostimulatory loop exists. Inhibition of macrophage proliferation in a selective manner by prostaglandins of the E series provides a negative-feedback, and both M-CSF and TNF induce macrophage prostaglandin production (135,136). M-CSF stimulates progenitors that are unipotential, bipotential, and multipotential (137). Highly purified murine progenitor populations are stimulated by M-CSF as well as by GM-CSF and IL-3, with a frequency indicating that many progenitors have receptors for all three growth factors (137). Furthermore, cultures of marrow that were initiated with M-CSF with delayed addition of GM-CSF, formed granulocyte–macrophage colonies indistinguishable from those initiated with GM-CSF (138). This indicates the ability of M-CSF to sustain a progenitor population with the potential to form granulocytes. Multipotential progenitors or true stem cells respond to M-CSF, but only in synergistic interaction with IL-1, IL-6, or IL-3 (139–142).

Recombinant human M-CSF is 10- to 100-fold less active in stimulating human macrophage colony formation than mouse macrophage colony formation in agar culture (143). This paradox can be partly explained by culture conditions, because in agarose cultures, particularly under low oxygen tension, rhM-CSF was only two to four times more active on mouse than human marrow (143). Rosenfeld et al. (144) reported that both native urinary and rhM-CSF stimulated human macrophage colonies in a dose-dependent manner after treatment of the marrow with L-phenylalanine methyl ester, which efficiently removes monocytes and myeloid cells that appear to selectively inhibit macrophage colony formation. rhM-CSF also acts directly on purified human CD34$^+$ cells in a serum-free fibrin clot system (145).

B. Action of CSFs on Monocytes–Macrophages

The brief survival of blood monocytes can be influenced by agents that prevent or delay their apoptotic death within a few hours of their entry into the circulation, or agents that promote differentiation into more long-lived macrophages. M-CSF prevents the death of monocytes in culture and promotes their differentiation into macrophages (146). Upper airway structural cell-derived cytokines, including M-CSF, also support human monocyte survival and may account for accumulation of inflammatory cells that play a pathological role (147).

M-CSF induces or augments the production of a variety of cytokines by monocytes or macrophages. It stimulates the synthesis and release of IL-1β in monocytes (148) and macrophages (149), TNF-α in monocytes (150), macrophages (151), and Kupffer cells (152); and interferon (150,153), and G-CSF (154,155) in monocytes and macrophages. The extent to which unstimulated or unprimed monocytes can respond directly to M-CSF alone to produce these cytokines has been questioned, because elutriation-purified monocytes, as opposed to adherence-separated monocytes, do not produce IL-1, IL-6, TNF-α, PGE$_2$ (156), or interferon (151) after stimulation with M-CSF. The secretion of urokinase-type plasminogen activator activity and plasminogen activator inhibitor by human monocytes after M-CSF stimulation has possible consequences for this enzyme system and M-CSF in plasminogen activation and cell migration (156,157). M-CSF also potentiates the production of thromboplastin (158) and the release of reactive oxygen products from phagocytic mononuclear cells (159), particularly following treatment with receptor-mediated agonists, such as N-formyl-methianyl-leucyl-phenylalanine (fMLP), or fMLP concanavin A, (160) or pathogens (161). Nitric oxide is also produced by endothelial cells and macrophages from endotoxemic rats following synergistic stimulation with M-CSF and LPS or IFN-γ (162). Monocyte-derived macrophages cultured with M-CSF normally acquire a characteristic morphology and phenotype that differs from macrophages developing in the presence of GM-CSF or IL-3 (163,164). The cells reenter cell cycle and change from accessory cells for mitogen-induced lymphocyte activation to suppressor cells (164). This is associated with expression of FcγR III (CD16), the development of antibody-dependent cell-mediated cytotoxicity (ADCC), and antibody-independent cytotoxicity, peaking at 8–12 days (164). IFN-γ blocks the development of some of these attributes; however, it enhances accessory cell function and oxidative burst activity (163).

M-CSF enhances Kupffer cell and macrophage phagocytosis of microspheres and LPS (165) and synergizes with phorbol esters to activate glucose transport into macrophages (166). The third complement component (C3) is an important factor in the host defense mechanism in which monocytes and macrophages participate as the primary phagocytes and as the principal extrahepatic source of C3. M-CSF enhances C3 production, C3R expression, and C3R-mediated phagocytes (167).

Activation of monocytes or macrophages by M-CSF can protect the cell from the lytic effects of vesicular stomatitis infection (168) and enhances amebicidal capacity against *Entamoeba histolytica* in conjunction with increased reactive oxygen production (160). In combination with IFN-γ, M-CSF is very effective in inhibiting *Mycobacterium avium* growth in human alveolar macrophages (169) and in activating intramacrophage killing of *Leishmania* (170). In the absence of IFN-γ, M-CSF increased the in vitro growth of a virulent strain of *M. avium* in monocytes (171). M-CSF also stimulates macrophage killing of *Candida albicans* (172) and inhibits intracellular growth of the yeast *Histoplasma capsulatum* (173).

M-CSF activates macrophage antibody-independent tumor cell cytolysis (174,175). Cytolysis may be mediated by different effector mechanisms, for in one study, M-CSF stimulated tumor cytolytic capacity in monocytes and Kupffer cells within 72–96 hr, without enhancement of reactive oxygen production or induction of IL-1 and TNF (176). In contrast, Curley et al. (152) observed that M-CSF enhanced murine Kupffer cell cytotoxicity against a TNF-sensitive tumor cell line by producing a dose-dependent increase in TNF release. Human macrophages cultured with a combination of IL-2 and M-CSF show high tumoricidal activity against human and murine leukemic cell lines, and the effector mechanisms involve the production of cytotoxic molecules (177). Synergy between M-CSF and IFN-γ has also been reported to induce a tumor cytotoxic phenotype that is developmentally regulated by IFN-γ (163). Monocytes and macrophages exposed to M-CSF for up to 48 hr exhibit enhanced FcRγ expression and effectively mediate antibody-dependent cell-mediated cytotoxicity (ADCC; 178–180). With antibodies to disialogangliosides expressed on neuroectodermal tumor targets, M-CSF-induced monocyte ADCC, and significant enhancement was seen with IFN-γ (179). This ADCC killing of tumor cells involves phagocytosis of intact cells as the principal mechanism of antitumor cytotoxicity (180).

Enhanced human immunodeficiency virus (HIV) replication, as measured by p24 release and levels of reverse transcriptase, has been reported in CD34$^+$CD4$^-$ human hematopoietic precursors (181) and in monocytes (182) treated with M-CSF. If these in vitro observations extend to the HIV-infected patient, then the variable levels of M-CSF in tissue or blood may determine both the susceptibility of macrophage to virus infection and the extent of virus replication in infected cells.

C. The Role of M-CSF in Osteoclast Development

Osteopetrotic (op/op) mice are characterized by impaired mononuclear phagocyte development involving both macrophages and osteoclasts. The mice are toothless, bone density is increased, and marrow cavities are occluded, with marrow cellularity 10% and marrow macrophage content 1% that of control mice (183,184). Osteopetrotic mice are characterized by an autosomal recessive inactivating mutation in the M-CSF gene, resulting in absence of M-CSF (185,186). Administration of M-CSF, particularly at early stages of development, corrected the osteopetrosis, toothless phenotype, osteoclast deficit, and defects in splenic and femoral macrophages (but did not correct deficiencies of pleural and peritoneal cavity macrophages and reduced female fertility that may be controlled by more local M-CSF action; 187).

Splenic red pulp macrophages are reduced to 60% of normal, and splenic marginal zone macrophages are absent in op/op mice (188). The former recover rapidly after M-CSF treatment, whereas the latter recover slowly and are still subnormal after 12 weeks of treatment. This suggests that local production of M-CSF is important for marginal zone macrophages involved in immune response against T-independent type 2 antigens. More recent data suggest that op/op mice undergo an age-related hematopoietic recovery and resolution of osteopetrosis, suggesting that the hematopoietic system has the capacity to use alternative mechanisms to compensate for loss of M-CSF (189).

A marrow stromal cell line developed from newborn op/op mice proved capable of supporting hematopoiesis, but not differentiation of macrophages and osteoclasts (190). Osteoclasts appeared only when marrow cells were cultured in contact with the stromal line in the presence of M-CSF and dihydroxy vitamin-D$_3$ [$1\alpha25(OH)_2D_3$] (190). Colonies of tartrate-resistant–acid phosphatase-positive, calcitonin receptor-positive osteoblasts have been reported to develop from osteoblast progenitors in semisolid cultures of CD34$^+$ cell stimulated with a combination of M-CSF, IL-3, and calcitriol (191). The recent isolation of a factor that selectively stimulates hematopoietic progenitors to form colonies composed of cells exhibiting osteoclast functional features and tartrate-resistant acid phosphatase has raised questions about the role of M-CSF in early osteoclast development (192,193). Coculture of marrow on op/op and control marrow revealed that significantly higher numbers of osteoclast progenitors (CFU-O) were found in op/op than in control cultures, when assayed with osteoclast colony-stimulating factor (193). It is possible that the osteoclast CSF supports proliferation of osteoclast progenitors, whereas M-CSF plays a role in the later development and maturation of the progenitors and possibly blocks apoptosis. Certainly, M-CSF dramatically stimulates osteoclast motility, spreading, and survival in vitro, and prevents apoptosis (194). It does not act as a regulator of bone turnover, because it does not stimulate bone resorption and, indeed, may suppress resorption (194).

In human malignant osteopetrosis, serum levels of bioactive and immunoreactive M-CSF were not depressed, suggesting that deficiency of M-CSF is unlikely to be a major contributor to the etiologic basis for most children with congenital osteopetrosis (196).

D. The Role of M-CSF in Atherogenesis

Atherogenesis is characterized by the proliferation of smooth-muscle cells, macrophages, and lymphocytes in the artery wall, and the accumulation of cholesterol and cholesterol esters in the surrounding connective tissue matrix and associated cells. Macrophages serve as the principal inflammatory cell in the atheromatous plaque microenvironment, and their localization and function are probably determined by M-CSF produced locally by endothelial and smooth-muscle cells (61–63) following activation with inflammatory cytokines or modified low-density lipoprotein (LDL; 62). Lipid-laden macrophages (foam cells) developing in plaques also produce M-CSF and contribute to survival, proliferation, and activation of specific macrophage functions. Among the latter are the expression of acetyl-LDL (scavenger) receptors (63) and the secretion of apolipoprotein E (64). The scavenger receptors mediate the uptake of modified lipoprotein–cholesterol and subsequent cholesterol ester accumulation and foam cell formation (197). M-CSF markedly and selectively increases scavenger receptor synthesis, and posttranslationally, the receptor becomes stabilized and shifted to a predominantly surface distribution (198). In addition, the receptor has a novel function as a divalent cation independent adhesion molecule, promoting the retention of mononuclear phagocytes and subsequent foam cell formation (198). Increased monocyte trapping during atherogenesis has also been attributed to M-CSF stimulation of monocyte chemotactic protein-1 expression by vascular endothelium, producing increased monocyte adhesion (199). M-CSF also specifically induces the expression of the α- versus β 5-integrin receptor on macrophages, promoting binding to the vitronectin component of the extracellular matrix in adhesion plaques (200). The membrane-anchored form of M-CSF expressed on activated endothelial cells can also mediate specific cytoadhesion to c-*fms*-bearing monocytes, macrophages, and foam cells (201). The uptake of acetylated LDL and oxidized LDL by macrophages is enhanced by M-CSF (202,203), but so also is the efflux of cholesterol from cholesterol-laden macrophages in the arterial wall, which would serve to reduce the rate of atherogenesis (202). M-CSF also regulates the activities of macrophage neutral and acidic cholesteryl ester hydrolases, enhancing net hydrolysis of acidic cholesterol ester and possibly reducing the rate of atherogenesis (204). Systemic M-CSF treatment in humans has been reported to reduce serum cholesterol (see last section on clinical studies). In both normal rabbits and Watanabe hyperlipidemic rabbits deficient in LDL receptors, M-CSF reduced cholesterol levels by one-third (205). M-CSF stimulated clearance of lipoproteins containing apolipoprotein B-100 by both an LDL receptor-dependent and receptor-independent pathway.

E. The Role of M-CSF in Pregnancy

The impaired fertility of female op/op mice (187) and the marked pregnancy-associated increase in serum levels of M-CSF, peaking by 28-weeks gestation (79,206), has suggested a role for M-CSF in blastocyst attachment, trophoblast outgrowth, implantation, and proliferation of placental tissue (206). M-CSF is first produced by oocytes and their vestment, the corona cumulus complex (207). In the first trimester, M-CSF is produced by cytotrophoblasts lining the villus core (72) and by syncytiotrophoblasts (75). At this stage, it induces the differentiation of trophoblasts, causing cytotrophoblastic fusion and formation of typical syncytiotrophoblasts producing chorionic gonadotropin and placental lactogen (77). In the second trimester, it is expressed in villous mesenchyme, decidual stroma, tissue macrophages (Hofbauer cells), endometrial gland cells, and trophoblasts (77,78,81). In the third trimester, M-CSF is produced by cells lining the villous vessels (72). IL-1 produced by decidualized endometrium during pregnancy may play an important role in inducing placental villous mesenchyme M-CSF (83).

The c-*fms* mRNA is expressed in extravillous cytotrophoblast columns, fetal Hofbauer cells, decidual macrophages at implantation sites, mesenchyme of chorionic villi, and glandular epithelial cells (80,82). Secretory endometrium expresses low levels of c-*fms* mRNA relative to high levels in endometrial cancer and intermediate levels in proliferative or hyperplastic endometrium (208,209). Sex steroids, secreted by corpus luteum or placenta, influence endometrial and placental growth and differentiation by local production of M-CSF and its receptor in the endometrium (210).

F. Role of c-*fma* Expression and M-CSF in Hematological Malignancy

In vitro clonogenic cells from patients with acute myeloid leukemia (AML), chronic myeloid leukemia (CML), and myelodysplastic syndrome (MDS) require hematopoietic growth factors for their in vitro survival and proliferation (211–214). Although G-CSF, GM-CSF, and IL-3 activate DNA synthesis in 80% of AML cases, M-CSF stimulates proliferation in approximately 50% of cases (214). No clear relation between cytokine response and the French–American–British (FAB) classification of leukemias has been reported. M-CSF often favors terminal differentiation of AML blast cells to macrophage–monocyte-like adherent cells incapable of growth. In studies using leukemic cell lines, differentiation and proliferative senescence were obtained by treatment with M-CSF in combination with $1,25(OH)_2D_3$ (215,216) or IFN-γ (217). When M-CSF was added to long-term cultures of AML bone marrow, it inhibited leukemic hematopoiesis in six of seven cases and stimulated putative normal hematopoiesis (218). This effect, either direct or indirect, could be a valuable purging procedure and may be attributed to M-CSF stimulation of TNF-α and prostaglandin E production in the cultures (219). In 7-day suspension cultures of AML marrow from 15 patients, CFU-leukemic (CFU-L) were either suppressed or stimulated by M-CSF, reflecting the underlying heterogeneity of AML even with the same FAB classification (220). The possibility of an autocrine role for M-CSF in some leukemias is suggested by a cell–density-dependent development of "spontaneous" leukemic colonies, owing to endogenous production of CSF species (211–214). In many cases, leukemic blasts produce various cytokines, including M-CSF, but rarely constitutively. IL-1 and TNF-α effectively induce CSFs by a paracrine or autocrine mechanism (214).

The oncogene c-*fms* is expressed in AML, yet although receptor overexpression has been implicated as one mechanism of ligand-independent activation of the M-CSF receptor (96), this has not been reported in AML. An etiologic role of M-CSF might be expected in the acute and chronic leukemias involving the monocytic lineage. Fifty percent of 25 patients with acute myelomonocytic leukemia had c-*fms* transcripts, and only a few cases had coexpression of M-CSF and c-*fms*, but with no gene rearrangement (221). Furthermore, the autocrine M-CSF production and response seen in normal tissue macrophage does not lead to leukemia and is rarely seen in leukemia. Even though serum M-CSF levels are significantly elevated in 83% of 316 patients with AML, MDS, lymphoma, and myeloma, the high levels did not correlate with phenotype, but rather with disease activity or tumor burden (222).

Experimental evidence indicates that certain mutations in c-*fms* cause leukemia in mice, and substitution of serine at codon 301 of the human c-*fms* gene may render it transforming in 3T3 cells (96). There is no proof of such a mechanism in human AML or MDS. The frequency of c-*fms* point mutations in hematological malignancy has ranged from absent to moderately high in different surveys. At one extreme, point mutations at codon 969 were not detected in 40 MDS or leukemia samples (223). In another study of 110 MDS and AML patients, 12% had mutations at codon 969 and 1.8% in codon 301, with mutations most common in chronic myelomonocytic leukemia (20%) and FAB type 4 AML (23%) (224). Tobal et al. (225) reported mutations at codon 301 or 969 in 8 of 41 patients with AML or MDS. One of 51 normal subjects has also been reported with a mutation in codon 969, indicating the alterations at this point do not automatically lead to MDS or AML (224). This observation implies either a low degree of polymorphism or a second promoting event necessary for overt transformation. Random mutagenesis of c-*fms* has revealed multiple sites for activating mutations within the extracellular domain that are worthy of investigation in AML or MDS (226). An obvious role for M-CSF is seen in the majority of lymphoid malignancies, even though c-*fms* mRNA is expressed in some B-cell leukemia–lymphomas (85). In Hodgkin's disease, M-CSF and c-*fms* are expressed in cell lines derived from the lymphoma (227), including lines with the morphology of Reed-Sternberg cells (228).

G. Expression of M-CSF and c-*fms* in Solid Tumors

High levels of c-*fms* mRNA have been observed in 56–100% of ovarian and endometrial cancers and elevated levels correlated with high histological grade and advanced clinical presentation (229–231). Half of the invasive adenocarcinomas of the ovary and endometrial cancer coexpressed c-*fms* and M-CSF (229,231). Constitutive production of M-CSF has been reported in cultures of normal ovarian epithelium at levels comparable with ovarian cancer cell lines (234). However, the coexpression of c-*fms* on M-CSF may establish an autocrine loop that may play a role in progression to the metastatic state. In an endometrial adenocarcinoma cell line, M-CSF expression was increased two to three times following treatment with synthetic progestin, and this induction was antagonized by the antiprogestin mefepristome (RU 486; 235). In this study, agar colony formation by the tumor cell line was inhibited by M-CSF, and glycogen granules accumulated in the cells, suggesting that M-CSF suppressed growth and enhanced differentiation. In contrast, another study that used the same Ishikawa cell line was unable to demonstrate increased M-CSF production following progesterone or estrogen treatment, and a tumor proliferative response to M-CSF was observed (233). In neither study did hormone treatment influence c-*fms* mRNA levels.

Serum levels of M-CSF have been reported as markedly elevated (> 100 ng/ml versus control levels of 1.2 ng/ml) in patients with endometrial cancer with active or recurrent disease (232). Elevated serum M-CSF also characterizes most of the clinically apparent epithelial ovarian cancers. Sixty-eight percent of sera from 69 patients with ovarian cancer had 2.5 ng/ml, or higher, of M-CSF versus 2% of 80 normal donors, and 31% of 134 patients with behign disease (236,237) reported a control level of 750 U/ml M-CSF, with an upper limit of 1056 U/ml in 634 healthy donors and a significant elevation to 1460 U/ml in 96% of a group of 69 patients with ovarian cancer. Sixty-one of these cancer patients had levels exceeding the upper limit of normal, and only 7% of 55 patients with benign ovarian tumors exceeded 1.56 U/ml. In ovarian cancer patients undergoing a "second-look" laparotomy, patients who were tumor-positive had higher levels than control or tumor-negative patients (238). Neither measurement of M-CSF alone nor serum levels of the CA125 tumor-associated antigen can be used unequivocally as diagnostic of ovarian cancer, but in a retrospective study measuring three markers, M-CSF, CA125, and OVX1, at least one of the markers was elevated in 98% of 46 patients with stage i ovarian cancer, versus 11% in 204 healthy women and 51% of patients with a benign pelvic mass (239). Serum M-CSF levels were also elevated in 27 patients with stages I–IV lung cancer, with levels of 515–10,000 U/ml, exceeding the control range of 300–500 U/ml (240). Serial measurements correlated with change in clinical status, and levels were normal in benign pulmonary disease, with the exception of idiopathic pulmonary fibrosis.

Coexpression of c-*fms* and M-CSF was seen in 36% of 200 nonselected cases of breast cancer associated with invasive ductal adenocarcinomas (241), in 33% of 17 unselected breast cancer patients (242), and in 95% of 50 patients with invasive breast cancer, but not in situ preinvasive cancer (243). Endogenous M-CSF increased the invasiveness of *fms*-positive breast cancer in a cell line model (244), and this may be related to M-CSF induction of urokinase-type plasminogen activator (UPA; 241). In this context, UPA is a serine esterase that converts plasminogen to the active protease plasmin, which activates procollagenase, leading to local degradation of interstitial tissue and subsequent invasion. M-CSF increased the invasiveness of a c-*fms*-positive lung carcinoma cell line by increasing its production of UPA (245).

Molecular abnormalities involving c-*fms* have been observed in association with certain solid tumors. Tumor-specific loss of heterozygosity at the c-*fms* locus on chromosome 5 was reported in 11 of 22 cases of colorectal cancer (246). A four- and tenfold amplification of c-*fms* was observed in 31 cases of bone and soft-tissue sarcoma, specifically in 2 of 3 cases of liposarcoma, suggesting a causal relation (247).

VI. PRECLINICAL IN VIVO STUDIES OF M-CSF

The physiological significance of M-CSF can be addressed in vivo by either blocking or "knocking out" CSF production, or by elevating M-CSF levels by administration of exogenous M-CSF, either acutely or chronically. The former approach has utilized the op/op mouse with a mutation inactivating M-CSF. The consequence of this has been discussed earlier. The eventual recovery from defective production of monocytes, macrophages, and osteoclasts in these mice suggests that there is redundancy in the cytokine network regulating differentiation in these lineages (183–189). The ability of GM-CSF and, to a lesser extent (and possibly by an indirect route) IL-3, to substitute for M-CSF in macrophage development would account for persisting monocyte–macrophage production in op/op mice. However, in mice deficient in both GM-CSF and M-CSF (generated by interbreeding GM-CSF–deficient mice produced by gene targeting and op/op mice), circulating levels of monocytes are comparable with those in M-CSF-deficient mice, and the lungs contain numerous phagocytically active macrophages (248). Similar to M-CSF–deficient mice, they have osteopetrosis and are toothless, and as do GM-CSF–deficient mice, they have a characteristic, but more severe, alveolar proteinosis-like lung abnormality. It is possible that IL-3 is compensating for this combined CSF deficiency, or that some other mechanism not involving cytokines is involved, for example, $1,25(OH)_2$ vitamin D_3, which can be mitogenic for macrophages.

Initial in vivo studies of M-CSF were carried out with purified natural material, and only minimal effects on hematopoiesis were observed (249). In normal mice, rats, rabbits, and primates, 10–300 µg/kg per day of rhM-CSF for 7–14 days leads to an increase in circulating monocytes, promonocytes, and large vacuolated macrophage-like cells that make up 30–70% of circulating nucleated cells (250–252). Thrombocytopenia and anemia were also noted. A single dose of M-CSF in mice 7 days after cyclophosphamide administration, when CFU-GEMM, CFU-GM, and BFU-E were only slowly cycling, stimulated these cells into cycle within 3 hr, but probably by an indirect mechanism (253). Increased progenitor cell cycling was also seen with two M-CSF injections at 0 and 12 hr, but a third injection at 24 hr significantly suppressed cell cycling (251). Increases in progenitor cell numbers and responsiveness have been reported in some studies (252,254), but not in others (249,251). Macrophage numbers increase in the peritoneum, liver, and spleen. Monocyte ADCC and tumoricidal capacity were also enhanced by in vivo M-CSF therapy, and this correlated with a significant reduction in metastasis in an experimental melanoma tumor system (254). In cynomolgus monkeys, 50–100 µg/kg per day of rhM-CSF for 14 days produced a fivefold increase in activated monocytes exhibiting enhanced direct and ADCC activity (255). The serum half-life of rhM-CSF was 2–6 hr in rats and 6–4 hr in primates with doses of 100 µg/kg (256). Clearance was dose-dependent, with a minimum $t_{1/2}$ of 23 min, increasing to 10 hr at doses approaching saturation (1000 µg/kg). After M-CSF induced tenfold increases in circulating monocytes, M-CSF clearance increased significantly, with the $t_{1/2}$ falling to 5 min in rats and 2 hr in primates. The extensive distribution of macrophages expressing large numbers of M-CSFRs and their increase after M-CSF therapy provides a mechanism by which M-CSF regulates its own plasma concentration by receptor-mediated binding, internalization, and degradation of the molecule.

VII. CLINICAL STUDIES OF M-CSF

Phase I clinical trials of rhM-CSF have demonstrated that treatment leads to the expansion of circulating mononuclear phagocytes (257–259). Serum cholesterol levels were reduced at all dose levels, and dose-limiting thrombocytopenia is observed at higher dose levels (257,258,260). Phase II trials of an M-CSF, purified from human urine, in 141 patients with cancer, after chemotherapy, and in 30 patients after marrow transplantation revealed that this source of M-CSF accelerated production of neutrophils and platelets (261). In contrast, in a randomized phase I–II study of urine-derived M-CSF, after chemotherapy with carboplatin and etoposide, in 26 patients with small-cell lung cancer, no differences were seen between M-CSF-treated or untreated groups in chemotherapy dose, survival, WBC nadir, duration of leukocytopenia, number of red cells, number of platelet transfusions, number of infections, or antibiotic use (262). Administration of rhM-CSF has proved effective in greatly reducing the mortality from invasive fungal infections, when given as an adjunct to standard antifungal treatment in patients undergoing bone marrow transplantation (260).

A phase I clinical trial of rhM-CSF with rIFN-γ has been undertaken in 36 patients with advanced malignancies (263). The rationale being that IFN-γ in vivo induces nonspecific monocyte tumor cytotoxicity at doses well below the maximally tolerated dose (264), and the same dose also stimulates mononuclear phagocyte ADCC (265). The minimal activity of IFN-γ on human neoplasia may be because its antitumor activity requires a critical number of mononuclear phagocytes, which could be provided by M-CSF, that also synergizes with IFN-γ in providing macrophage activation and ADCC. Peripheral monocytosis was observed at all M-CSF dose levels exceeding 40 μg/kg per day, and the maximally tolerated dose of M-CSF was 120 μg/kg per day M-CSF and 0.1 mg/m^2 per day IFN-γ. The induction of monocytes coincided with the onset of thrombocytopenia, and unexpected pulmonary and hepatic toxicities were observed at the higher tested dose. At the conclusion of therapy with 100 μg/kg per day M-CSF and 0.1 mg/m^2 per day IFN-γ, 78% of peripheral blood monocytes expressed low-affinity FcγIIIR (CD16), whereas CD14 was expressed by only 36% of the cells—a phenotype associated with cellular activation. A partial clinical response was noted in a patients with renal cell carcinoma, and minor clinical responses were observed in patients with lymphoma, renal cell carcinoma, and thymoma.

The administration of M-CSF for 21 days following autologous bone marrow transplantation in 20 patients with malignant lymphoma resulted in no significant difference in neutrophil recovery and suggestive evidence of accelerated platelet recovery with M-CSF, but relatively large marrow inocula were required to see this effect (266). In a 2-year follow-up of a randomized, double-blind, placebo-controlled phase III trial of urinary M-CSF, after allogeneic or syngeneic bone marrow transplantation in acute lymphocytic or nonlymphocytic leukemia, M-CSF improved the recovery of neutrophils and survival time in the first 120 days, without influencing the occurrence of leukemic relapse or graft-versus-host disease (267). In one patient with acute monocytic leukemia, a combination of M-CSF and low-dose cytarabine (cytosine arabinoside) produced a complete remission that persisted for 5 months (268). This observation suggests that this combination may be useful in treating some patients with acute monocytic leukemia.

The major demonstrated effects of rhM-CSF include an increase in the number and the activation of monocytes and macrophages, the ability to enhance antibody-dependent cellular cytotoxicity, enhancement of macrophage microbial phagocytic and killing activity, cholesterol lowering, and platelet lowering. The range of potential indications for the use of rhM-CSF makes the further clinical development of this molecule both a challenge and an opportunity. Investigation of combination therapies with other cytokines or monoclonal antibodies will be an important aspect of future investigations.

REFERENCES

1. Robinson W, Metcalf D, Bradley TR. Stimulation by normal and leukemic mouse sera of colony formation in vitro by mouse bone marrow cells. J Cell Physiol 1967; 69:83–92.

2. Stanley ER, McNeill TA, Chan SH. Antibody production to the factor in human urine stimulating colony formation in vitro by bone marrow cells. Br J Haematol 1970; 18:585–590.

3. Sherr CJ, Stanley ER. Colony stimulating factor 1 (macrophage colony-stimulating factor). In: Sporn MB, Roberts AB, eds. Peptide Growth Factors and Their Receptors I. Berlin: Springer Verlag, 1990:667–698.

4. Bradley TR, Sumner MA. Stimulation of mouse bone marrow colony growth in vitro by conditioned medium. Aust J Exp Biol Med Sci 1968; 46:607–618.

5. Sachs L. The molecular control of blood cell development. Science 1987; 238:1374–1379.

6. Virolainen M, Defendi V. Dependence of macrophage growth in vitro upon interaction with other cell types. Wistar Inst Symp Monogr 1967; 7:67–85.

7. Stanley ER, Heard PM. Factors regulating macrophage production and growth. Purification and some properties of the colony stimulating factor from medium conditioned by mouse L cells. J Biol Chem 1977; 252:4305–4312.

8. Das SK, Stanley ER. Structure–function studies of a colony stimulating factor (CSF-1). J Biol Chem 1982; 257:13679–13684.

9. Metcalf D. Studies on colony formation in vitro by mouse bone marrow cells. II. Action of colony stimulating factor. J Cell Physiol 1970; 76:89–99.

10. Stanley ER, Guilbert LJ. Methods for the purification, assay, characterization and target cell binding of a colony stimulating factor (CSF-1). J Immunol Methods 1981; 42:253–284.

11. Waheed A, Shadduck RK. Purification and properties of L cell-derived colony stimulating factor. J Lab Clin Med 1979; 94:180–194.

12. Das SK, Stanley ER, Guilbert LJ, Forman LW. Human colony stimulating factor (CSF-1) radioimmunoassay: resolution of three subclasses of human colony stimulating factors. Blood 1981; 58:630–641.

13. Kawasaki ES, Ladner MB, Wang AM, Van Arsdell J, Warren MK, Coyne MY, Schweickart VL, Lee MT, Wilson KJ, Boosman A, Stanley ER, Ralph P, Mark DF. Molecular cloning of a complementary DNA encoding human macrophage-specific colony stimulating factor (CSF-1). Science 1985; 230:291–296.

14. Wong GG, Temple PA, Leary AC, Witek-Giannotti JS, Yang Y-C, Ciarletta AB, Chung M, Murtha P, Kriz R, Kaufman RJ, Ferenz CR, Sibley BS, Turner KJ, Hewick RM, Clark SC, Yanai N, Yokota H, Yamada M, Saito M, Motoyoshi K, Takaku F. Human CSF-1: molecular cloning and expression of 4 kb cDNA encoding the human urinary protein. Science 1987; 235:1504–1508.

15. Ceretti DP, Wignall J, Anderson D, Tushninski RJ, Gallis BM, Stata M, Gillis S, Urdal DL, Cosman D. Human macrophage-colony stimulating factor: alternative RNA and protein processing from a single gene.

16. DeLamarter JF, Hession C, Semon D, Gough NM, Rothenbuhler R, Mermod J-J. Nucleotide sequence of a cDNA encoding murine CSF-1 (macrophage-CSF). Nucleic Acids Res 1987; 15:2389–2390.

17. Ladner MB, Martin GA, Noble JA, Wittman VP, Shadle PJ, Warrens MK, McGrogan M, Stanley

ER. cDNA cloning and expression of murine CSF-1 from L929 cells. Proc Natl Acad Sci USA 1988; 85:6706–6710.

18. Suzu S, Ohtsuki T, Yanai N, Takatsu Z, Kawashima T, Takaku F, Nagata N, Motoyoshi K. Identification of a high molecular weight macrophage colony stimulating factor as a glycosaminoglycan-containing species. J Biol Chem 1992; 267:4345–4348.

19. Ohtsuki T, Suzu S, Hatake K, Nagata N, Miura Y, Motoyoshi K. A proteoglycan form of macrophage colony-stimulating factor that binds to bone-derived collagens and can be extracted from bone matrix. Biochem Biophys Res Commun 1993; 190:215–222.

20. Glocker MO, Arbogast B, Schreurs J, Deinzer ML. Assignment of the inter- and intramolecular disulfide linkages in recombinant human macrophage colony stimulating factor using fast atom bombardment mass spectrometry. Biochemistry 1993; 32:482–488.

21. Pandit J, Bohn A, Jancarik J, Halenbeck R, Koths K, Kim S-H. Three-dimensional structure of dimeric human recombinant macrophage colony stimulating factor. Science 1992; 258:1358–1362.

22. Stein J, Borzillo GV, Rettenmier CW. Direct stimulation of cells expressing receptors for macrophage colony stimulating factor (CSF-1) by a plasma membrane-bound precursor of human CSF-1. Blood 1990; 76:1308–1314.

23. Stein J, Rettenmier CW. Proteolytic processing of a plasma membrane-bound precursor to human macrophage colony stimulating factor (CSF-1) is accelerated by phorbol ester. Oncogene 1991; 6:601–605.

24. Uemura N, Ozawa K, Takahashi K. Binding of membrane-anchored macrophage colony stimulating factor (M-CSF) to its receptor mediates specific adhesion between stromal cells and M-CSF receptor-bearing hematopoietic cells. Blood 1993; 82:2634–2640.

25. Morris SW, Valentine MB, Shapiro DN, Sublett JE, Deaven LL, Foust JT, Roberts WM, Ceretti DP, Look AT. Reassignment of the human CSF-1 gene to chromosome 1p13-p21. Blood 1991; 78:2013–2020.

26. Gisselbrecht S, Sola B, Fichelson S, Bordeaux D, Tambourin P, Mattei MG, Simon D, Guenet JL. The murine M-CSF gene is localized on chromosome 3. Blood 1989; 73:1742–1748.

27. Kawasaki ES, Ladner MB. Molecular biology of macrophage colony stimulating factor. Immunol Ser 1990; 49:155–176.

27a. Kawasaki ES, Ladner MB. Molecular biology of macrophage colony-stimulation factor. In: Dexter TM, Garland JM, Testra NG, eds. Colony-Stimulating Factor: Molecular and Cell Biology. New York: Marcel Dekker, 1990:155–176.

28. Bartocci A, Mastrogiannis DS, Migliorati G, Stockert RJ, Wolkoff AW, Stanler ER. Macrophages specifically regulate the concentration of their own growth factor in the circulation. Proc Natl Acad Sci USA 1987; 84:6179–6183.

29. Haskill S, Warren MK, Becker S, Ladner MB, Johnson C, Eierman D, Ralph P, Mark DF. Adherence induces CSF-1 gene expression in monocytes. J Leukoc Biol 1987; 42:359.

30. Lee MT, Kaushansky K, Ralph P, Ladner MB. Differential expression of M-CSF, G-CSF, and GM-CSF by human monocytes. J Leukoc Biol 1990; 47:275–282.

31. Hamilton JA. Coordinate and noncoordinate colony stimulating factor formation by human monocytes. J Leukoc Biol 1994; 55:355–361.

32. Horiguchi J, Warren MK, Ralph P, Kufe D. Expression of the macrophage specific colony stimulating factor (CSF-1) during human monocytic differentiation. Biochem Biophys Res Commun 1986; 141:924–930.

33. Rambaldi A, Young DC, Griffin JD. Expression of the M-CSF (CSF-1) gene by human monocytes. Blood 1987; 69:1409–1413.

34. Gruber MF, Gerrard TL. Production of macrophage colony stimulating factor (M-CSF) by human monocytes is differentially regulated by GM-CSF, TNF-alpha, and IFN-gamma. Cell Immunol 1992; 142:361–369.

35. Horiguchi J, Warren MK, Kufe D. Expression of the macrophage-specific colony stimulating factor in human monocytes treated with granulocyte–macrophage colony stimulating factor. Blood 1987; 69:1259–1261.

36. Oster W, Lindemann A, Horn S, Mertelsmann R, Hermann F. Tumor necrosis factor (TNF)-alpha but not TNF-beta induces secretion of colony stimulating factor for macrophages (CSF-1) by human monocytes. Blood 1987; 70:1700–1703.

37. Misago M, Tsukada J, Ogawa R, Kikjuchi M, Hanamura T, Chiba S, Oda S, Morimoto I, Eto S. Enhancing effects of IL-2 on M-CSF production by human peripheral blood monocytes. Int J Hematol 1993; 58:43–51.

38. Brach MA, Arnold C, Kiehntopf M, Gruss HJ, Herrmann F. Transcriptional activation of the macrophage colony stimulating factor gene by IL-2 is associated with secretion of bioactive macrophage colony stimulating factor protein by monocytes and involves activation of the transcription factor NF-kappa B. J Immunol 1993; 150:5535–5543.

39. Gruber MF, Webb DSA, Gerrard TL. Stimulation of human monocytes via CD45, CD44, and LFA-3 triggers macrophage colony stimulating factor production synergism with lipopolysaccharide and IL-1β. J Immunol 1992; 148:1113.

40. Gruber MF, Williams CC, Gerrard TL. Macrophage colony stimulating factor expression by anti-CD45 stimulated human monocytes is transcriptionally up-regulated by IL-1β and inhibited by IL-4 and IL-10. J Immunol 1994; 152:1354–1361.

41. Wieser M, Bonifer R, Oster W, Lindemann A, Mertelsmann R, Herrmann F. Interleukin-4 induces secretion of CSF for granulocytes and CSF for macrophages by peripheral blood monocytes. Blood 1989; 73:1105–1109.

42. Scheibenbogen C, Andreesen R. Developmental regulation of the cytokine repertoire in human macrophages: IL-1, IL-6, TNF-alpha, and M-CSF. J Leukoc Biol 1991; 50:35–42.

43. Temeles DS, McGrath HE, Kittler ELW, Shadduck RK, Kister VK, Crittenden RB, Turner BL, Quesenberry PJ. Cytokine expression from bone marrow derived macrophages. Exp Hematol 1993; 21:388–393.

44. Rajavashisth TB, Eng R, Shadduck RK, Waheed A, Ben-Avram CM, Shively JE, Lusis AJ. Cloning and tissue-specific expression of mouse macrophage colony stimulating factor mRNA. Proc Natl Acad Sci USA 1987; 84:1157–1161.

45. Rose RM, Kobzik L, Filderman AE, Vermeulen MW, Dushay K, Donahue RE. Characterization of colony stimulating factor activity in the human respiratory tract. Comparison of healthy smokers and nonsmokers. Am Rev Respir Dis 1992; 145:394–399.

46. Brach MA, Henschler R, Mertelsmann RH, Herrmann F. Regulation of M-CSF expression by M-CSF: role of protein kinase C and transcription factor NF kappa B. Pathobiology 1991; 59:283–288.

47. Cerdan C, Courcoul M, Razanajaona D, Pierres A, Maroc N, Lopez M, Mannoni P, Mawas C, Olive D, Birg F. Activated but not resting T cells or thymocytes express colony-stimulating factor 1 mRNA without co-expressing c-fms mRNA. Eur J Emmunol 1990; 20:331–335.

48. Hallet MM, Praloran V, Vie H, Peyrat MA, Wong G, Witek-Giannotti J, Soulillou JP, Moreau JF. Macrophage colony-stimulating factor (CSF-1) gene expression in human T-lymphocyte clones. Blood 1991; 77:780–786.

49. Praloran V, Gascan H, Papin S, Chevalier S, Trossaert M, Boursier MC. Inducible production of macrophage colony-stimulating factor (CSF-1) by malignant and normal human T cells. Leukemia 1990; 4:411–414.

50. Cerdan C, Razanajaona D, Martin Y, Courcoul M, Pavon C, Mawas C, Olive D, Birg F. Contributions of the CD2 and CD28 T lymphocyte activation pathways to the regulation of the expression of the colony-stimulating factor (CSF-1) gene. J Immunol 1992; 149:373–379.

51. Kawasaki C, Okamura S, Omori F, Shimoda K, Hayashi S, Kondo S, Yamada M, Niho Y. Both granulocyte–macrophage colony-stimulating factor and monocytic colony-stimulating factor are produced by the human T-cell line, HUT 102. Exp Hematol 1990; 18:1090–1093.

52. Chodakewitz JA, Lacy J, Edwards SE, Birchall N, Coleman DL. Macrophage colony-stimulating factor production by murine and human keratinocytes. Enhancement by bacterial lipopolysaccharide. J Immunol 1990; 144:2190–2196.

53. Bedard M, McClure CD, Schiller NL, Francoeur C, Cantin A, Denis M. Release of interleukin-8,

interleukin-6, and colony-stimulating factors by upper airway epithelial cells: implications for cystic fibrosis. Am J Respir Cell Mol Biol 1993; 9:455–462.

54. Galy AH, Spits H, Hamilton JA. Regulation of M-CSF production by cultured human thymic epithelial cells. Lymphokine Cytokine Res 1993; 12:265–270.

55. Lanotte M, Metcalf D, Dexter TM. Production of monocyte/macrophage colony-stimulating factor by preadipocyte cell lines derived from murine bone marrow stroma. J Cell Physiol 1982; 112:123–127.

56. Goldstein NL, Moore MAS, Allen C, Tackney C. A human fetal spleen cell line, immortalized with SV40 T-antigen, will support the growth of CD34$^+$ long-term culture-initiating cells. Mol Cell Differ 1993; 1:301–321.

57. Mantovani L, Henschler R, Brach MA, Mertelsmann RH, Herrmann F. Regulation of gene expression of macrophage-colony-stimulating factor in human fibroblasts by the acute phase response mediators interleukin (IL)-1 beta, tumor necrosis factor-alpha and IL-6. FEBS Lett 1991; 280:97–102.

58. Tobler A, Meier R, Seitz M, Dewald B, Baggiolini M, Fey MF. Glucocorticoids downregulate gene expression of GM-CSF, NAP-1/IL-8, and IL-6, but not of M-CSF in human fibroblasts. Blood 1992; 79:45–51.

59. Hamilton JA, Filonzi EL, Ianches G. Regulation of macrophage colony-stimulating factor (M-CSF) production in cultured human synovial fibroblasts. Growth Factors 1993; 9:157–165.

60. Campbell IK, Ianches G, Hamilton JA. Production of macrophage colony-stimulating factor (M-CSF) by human articular cartilage and chondrocytes. Modulation by interleukin-1 and tumor necrosis factor alpha. Biochim Biophys Acta 1993; 1182:57–63.

61. Seelentag WK, Mermod JJ, Montesano R, Vassalli P. Additive effects of interleukin 1 and tumor necrosis factor-alpha on the accumulation of the three granulocyte and macrophage colony-stimulating factor mRNAs in human endothelial cells. EMBO J 1987; 6:2261–2265.

62. Rajavashisth TB, Andalibi A, Territo MC, Berliner JA, Naab M, Fogelman AM, Lusis AJ. Induction of endothelial cell expression of granulocyte and macrophage colony-stimulating factors by modified low-density lipoproteins. Nature 1990; 344:254–257.

63. Rosenfeld ME, Yla-Herttuala S, Lipton BA, Ord VA, Witztum JL, Steinberg D. Macrophage colony-stimulating factor mRNA and protein in atherosclerotic lesions of rabbits and humans. Am J Pathol 1992; 140:291–300.

64. Clinton SK, Underwood R, Hayes L, Sherman ML, Kufe DW, Libby P. Macrophage colony-stimulating factor gene expression in vascular cells and in experimental and human atherosclerosis. Am J Pathol 1992; 140:301–306.

65. Gallo P, Pagni S, Giometto B, Piccinno MG, Bozza F, Argentiero V, Tavolato B. Macrophage-colony-stimulating factor (M-CSF) in the cerebrospinal fluid. J Neuroimmunol 1990; 29:105–112.

66. Lee SC, Liu W, Roth P, Dickson DW, Berman JW, Brosnan CF. Macrophage colony-stimulating factor in human fetal astrocytes and microglia. Differential regulation by cytokines and lipopolysaccharide, and modulation of class II MHC on microglia. J Immunol 1993; 150:594–604.

67. Borycki A, Lenormund J, Guillier M, Leibovitch SA. Isolation and characterization of a cDNA clone encoding for rat CSF-1 gene. Post-transcriptional repression occurs in myogenic differentiation. Biochim Biophys Acta 1993; 1174:143–152.

68. Horiguchi J, Sariban E, Kufe D. Transcriptional and post-transcriptional regulation of CSF-1 gene expression in human monocytes. Mol Cell Biol 1988; 8:3951–3954.

69. Lanfrancone L, Boraschi D, Ghiara P, Falini B, Grignani F, Peri G, Mantovani A, Pelicci PG. Human peritoneal mesothelial cells produce many cytokines (granulocyte colony-stimulating factor [CSF], granulocyte–monocyte-CSF, macrophage-CSF, interleukin-1 [IL-1], and IL-6) and are activated and stimulated to grow by IL-1. Blood 1992; 80:2835–2842.

70. Zoja C, Wang JM, Bettoni S, et al. Interleukin-1 beta and tumor necrosis factor-alpha induce gene expression and production of leukocyte chemotactic factors, colony-stimulating factors, and interleukin-6 in human mesangial cells. Am J Pathol 1991; 138:991–1003.

71. Kaneki M, Inoue S, Hosoi T, Mizuno Y, Akedo Y, Ikegami A, Nakamura T, Shiraki M, Ito H, Suzu S, Motoyoshi K, Ouchi Y, Orimo H. Effects of 1α, 25-dihydroxyvitamin D$_3$ on macrophage

colony-stimulating factor production and proliferation of human monocytic cells. Blood 1994; 83:2285–2293.

72. Daiter E, Pampfer S, Yeung YG, Barad D, Stanler ER, Pollard JW. Expression of colony-stimulating factor-1 in the human uterus and placenta. J Clin Endocrinol Metab 1992; 74:850–858.

73. Guilbert L, Robertson SA, Wegmann TG. The trophoblast as an integral component of a macrophage–cytokine network. Immunol Cell Biol 1993; 71:49–57.

74. Harty JR, Kauma SW. Interleukin-1 beta stimulates colony-stimulating factor-1 production in placental villous core mesenchymal cells. J Clin Endocrinol Metab 1992; 75:947–950.

75. Haynes MK, Jackson LG, Tuan RS, Shepley KJ, Smith JB. Cytokine production in first trimester chorionic villi: detection of mRNAs and protein products in situ. Cell Immunol 1993; 151:300–308.

76. Kauma SW. Interleukin-1 beta stimulates colony-stimulating factor-1 production in human term placenta. J Clin Endocrinol Metab 1993; 76:701–703.

77. Saito S, Motoyoshi K, Saito M, Kato Y, Enomoto M, Nishikawa K, Morii T, Ichijo M. Localization and production of human macrophage colony-stimulating factor (hM-CSF) in human placental and decidual tissues. Lymphokine Cytokine Res 1993; 12:101–107.

78. Saito M, Saito S, Nakagawa T, Ichijo M, Motoyoshi K. Origin of macrophage colony-stimulating factor (M-CSF) and granulocyte colony-stimulating factor (G-CSF) in amniotic fluid. Asia Oceania J Obstetr Gynaecol 1992; 18:355–361.

79. Saito S, Motoyoshi K, Ichijo M, Saito M, Takaku F. High serum human macrophage colony-stimulating factor level during pregnancy. Int J Hematol 1992; 55:219–225.

80. Pampfer S, Daiter E, Barad D, Pollard JW. Expression of the colony-stimulating factor-1 receptor (c-*fms* proto-oncogene product) in the human uterus and placenta. Biol Reprod 1992; 46:48–57.

81. Kanzaki H, Yui J, Iwai M, Imai K, Kariya M, Hatayama H, Mori T, Guilbert LJ, Wegmann TG. The expression and localization of mRNA for colony-stimulating factor (CSF-1) in human term placenta. Hum Reprod 1992; 7:563–567.

82. Jokhi PP, Chumbley G, King A, Gardner L, Loke YW. Expression of the colony-stimulating factor-1 receptor (c-*fms* product) by cells at the human uteroplacental interface. Lab Invest 1993; 68:308–320.

83. Haynes MK, Jackson LG, Tuan RS, Shepley KJ, Smith JB. Cytokine production in first trimester chorionic villi: detection of mRNAs and protein products in situ. Cell Immunol 1993; 151:300–308.

84. Bartocci A, Pollard JW, Stanley ER. Regulation of colony-stimulating factor-1 during pregnancy. J Exp Med 1986; 164:956–961.

85. Baker AH, Ridge SA, Hoy T, Cachia PG, Culligan D, Baines P, Whittaker JA, Jacobs A, Padua RA. Expression of the colony-stimulating factor-1 receptor in B lymphocytes. Oncogene 1993; 8:371–378.

86. Downing JR, Roussel MF, Sherr CJ. Ligand and protein kinase C downmodulate the colony-stimulating factor-1 receptor by independent mechanisms. Mol Cell Biol 1989; 9:2890–2896.

87. Wheeler EF, Rettenmier CW, Look AT, Sherr CJ. The v-*fms* oncogene induces factor independence and tumorigenicity in CSF-1 dependent macrophage cell line. Nature 1986; 324:377–380.

88. Baccarini M, Buscher D, Dello Sbarba P, Stanley ER. Receptor cross-talk: protein kinase C dependent down-modulation of the colony-stimulating factor-1 receptor upon activation with interferon-γ/lipopolysaccharide. In: Freund, Link, Schmidt, Welte, eds. Cytokines in Hemopoiesis, Oncology, and AIDS II. Berlin: Springer-Verlag, 1992:285–290.

89. Branch RB, Turner R, Guilbert LJ. Synergistic stimulation of macrophage proliferation by the monokines tumor necrosis factor-alpha and colony-stimulating factor-1. Blood 1989; 73:307–311.

90. Shieh J-H, Peterson RHF, Warren DJ, Moore MAS. Modulation of colony-stimulating factor-1 receptors on macrophages by tumor necrosis factor. J Immunol 1989; 143:2534–2539.

91. Shieh J-H, Moore MAS. Hematopoietic growth factor receptors. Cytotechnology 1989; 2:269–286.

92. Moore MAS. Clinical implications of positive and negative hematopoietic stem cell regulators [review: Stratton Lecture 1990]. Blood 1991; 78:1–19.

93. Walker F, Nicola NA, Metcalf D, Burgess AW. Hierarchical down-modulation of hemopoietic growth factor receptors. Cell 1985; 43:269–276.

94. Gliniak NC, Rohrschneider LR. Expression of the M-CSF receptor is controlled posttranscriptionally by the dominant actions of GM-CSF or multi-CSF. Cell 1990; 63:1073–1083.

95. Sherr CJ, Rettenmier CW, Sacca R, Roussel MF, Look AT, Stanley ER. The c-*fms* proto-oncogene product is related to the receptor for the mononuclear phagocyte growth factor CSF-1. Cell 1985; 41:665–676.

96. Sherr CJ. Colony-stimulating factor-1 receptor. Blood 1990; 75:1–12.

97. Coussens L, Van Beveren C, Smith D, Chen E, Mitchell RL, Isacke CM, Verma IM, Ullrich A. Structural alteration of viral homologue of receptor protooncogene *fms* at carboxyl terminus. Nature 1986; 320:277–280.

98. Rothwell VM, Rohrschneider LR. Murine c-*fms* cDNA; cloning, sequence analysis, and retroviral expression. Oncogene Res 1987; 1:311–324.

99. Rettenmier CW, Sacca R, Furman WL, Roussel MF, Holt JT, Nienhuis AW, Stanley ER, Sherr CJ. Expression of the human c-*fms* proto-oncogene product (colony-stimulating factor-1 receptor) on peripheral blood mononuclear cells and choriocarcinoma cell lines. J Clin Invest 1986; 77:1740–1746.

100. Roberts M, Look AT, Roussel MF, Sherr CJ. Tandem linkage of human CSF-1 receptor (c-*fms*) and PDGF receptor genes. Cell 1988; 55:655–661.

101. Downing JR, Rettenmier CW, Sherr CJ. Ligand-induced tyrosine kinase activity of the colony-stimulating factor-1 receptor in a murine macrophage cell line. Mol Cell Biol 1988; 8:1795–1799.

102. Lee AW, Nienhuis AW. Mechanism of kinase activation in the receptor for colony-stimulating factor-1. Proc Natl Acad Sci USA 1990; 87:7270–7274.

103. Heidara MA, Molloy CJ, Pangelinan M, Choudhury GG, Wang LM, Fleming TP, Sakaguchi AY, Pierce JH. Activation of the colony-stimulating factor-1 receptor leads to the rapid tyrosine phosphorylation of GTPase-activating protein and activation of cellular p21*ras*. Oncogene 1992; 7:147–152.

104. Reedijk M, Liu XQ, Pawson T. Interactions of phosphatidylinositol kinase, GTPase-activating protein (GAP), and GAP-associated proteins with the colony-stimulating factor 1 receptor. Mol Cell Biol 1990; 10:5601–5608.

105. Choudhury GG, Wang LM, Pierce J, Harvey SA, Sakaguchi AY. A mutational analysis of phosphatidylinositol-3-kinase activation by human colony-stimulating factor-1 receptor. J Biol Chem 1991; 266:8068–8072.

106. Roussel MF, Downing JR, Sherr CJ. Transforming activities of human CSF-1 receptors with different point mutations at codon 301 in their extracellular domains. Oncogene 1990; 5:25–30.

107. Roussel MF, Cleveland JL, Shurtleff SA, Sherr CJ. *myc* rescue of a mutant CSF-1 receptor impaired in mitogenic signalling. Nature 1991; 353:361–363.

108. Courtneidge SA, Dhand R, Pilat D, Twamley GM, Waterfield MD, Roussel MF. Activation of Src family kinases by colony-stimulating factor-1, and their association with its receptor. EMBO J 1993; 12:943–950.

109. Buscher D, Dello Sbarba P, Hipskind RA, et al. v-*raf* confers CSF-1 independent growth to a macrophage cell line and leads to immediate early gene expression without MAP-kinase activation. Oncogene 1993; 8:3323–3332.

110. Bortner DM, Ulivi M, Roussel MF, Ostrowski MC. The carboxy-terminal catalytic domain of the GTPase-activating protein inhibits nuclear signal transduction and morphological transformation mediated by the CSF-1 receptor. Genes Dev 1991; 5:1777–1785.

111. Langer SJ, Bortner DM, Roussel MF, Sherr CJ, Ostrowski MC. Mitogenic signaling by colony-stimulating factor 1 and *ras* is suppressed by the *ets-2* DNA-binding domain and restored by *myc* overexpression. Mol Cell Biol 1992; 12:5355–5362.

112. Mufson RA. Induction of immediate early response genes by macrophage colony-stimulating factor in normal human monocytes. J Immunol 1990; 145:2333–2339.

113. Nakamura T, Datta R, Kharbanda S, Kufe D. Regulation of *jun* and *fos* gene expression in human monocytes by the macrophage colony-stimulating factor. Cell Growth differ 1991; 2:267–272.

114. Imamura K, Dianoux A, Nakamura T, Kufe D. Colony-stimulating factor 1 activates protein kinase c in human monocytes. EMBO J 1990; 9:2423–2429.

115. Nakamura T, Lin LL, Kharbanda S, Knopf J, Kufe D. Macrophage colony-stimulating factor activates phosphatidylcholine hydrolysis by cytoplasmic phospholipase A_2. EMBO J 1992; 11:4917–4922.

116. Vairo G, Argyriou S, Bordum A-M, Gonda TJ, Gragoe EJ, Hamilton JA. Na^+/H^+ exchange involvement in colony-stimulating factor-1-stimulated macrophage proliferation. J Biol Chem 1990; 265:2692–2701.

117. Vairo G, Hamilton JA. Signalling through CSF receptors. Immunol Today 1991; 12:362–369.

118. Vairo G, Argyriou S, Bordum A-M, Gonda TJ, Cragoe EJ, Hamilton JA. Na^+/H^+ exchange involvement in colony-stimulating factor1 stimulated macrophage proliferation. J Biol Chem 1990; 265:16929–16939.

119. McArthur GA, Rohrschneider LR, Johnson. Induced expression of c-*fms* in normal hematopoietic cells shows evidence for both conservation and lineage restriction of signal transduction in response to macrophage colony-stimulating factor. Blood 1994; 83:972–981.

120. Varticovski L, Druker B, Morrison D, Cantley L, Roberts T. The colony-stimulating factor-1 receptor associates with and activates phosphatidylinositol-3 kinase. Nature 1989; 342:699–702.

121. Roussel MF, Shurtleff SA, Downing JR, Sherr CJ. A point mutation at tyrosine 809 in the human colony-stimulating factor 1 receptor impairs mitogenesis without abrogating tyrosine kinase activity, association with phosphatidylinositol-3-kinase, or induction of c-*fos* and *jun B* genes. Proc Natl Acad Sci USA 1990; 87:6738–6742.

122. Imamura K, Kufe D. Colony-stimulating factor 1-induced Na^+ influx into human monocytes involves activation of a pertussis toxin-sensitive GTP-binding protein. J Biol Chem 1988; 263:14093–14098.

123. Baccarini M, Sabatini DM, App H, Rapp UR, Stanley ER. Colony-stimulating factor-1 (CSF-1) stimulates temperature dependent phosphorylation and activation of the RAF-1 proto-oncogene product. EMBO J 1990; 9:3649–3657.

124. Cook N, Dexter TM, Lord BI, Cragoe EJ, Whetton AD. Identification of a common signal associated with cellular proliferation stimulated by four haemopoietic growth factors in a highly enriched population of granulocyte/macrophage colony-forming cells. EMBO J 1989; 8:2967–2974.

125. Vairo G, Hamilton JA. Activation and proliferation signals in murine macrophages: stimulation of Na^+,K^+-ATPase activity by hematopoietic growth factors and other agents. J Cell Physiol 1988; 134:13–24.

126. Hamilton JA, Veis N, Bordun A-M, Cairo G, Gonda TJ, Phillips WA. Activation and proliferation signals in murine macrophages: relationships among c-*fos* and c-*myc* expression, phosphoinositol hydrolysis, superoxide formation and DNA synthesis. J Cell Physiol 1989; 141:618–626.

127. Muller R, Curran T, Muller D, Guilbert L. Induction of c-*fos* during myelomonocytic differentiation and macrophage proliferation. Nature 1985; 314:546–548.

128. Orlofsky A, Stanley ER. CSF-1-induced gene expression in macrophages: dissociation from the mitogenic response. EMBO J 1987; 6:2947–2952.

129. Hamilton JA, Vairo G, Knight KR, Cocks BG. Activation and proliferation signals in murine macrophages. Biochemical signals controlling the regulation of macrophage urokinase-type plasminogen activator activity by colony-stimulating factors and other agents. Blood 1991; 77:616–627.

130. Matsushime J, Roussel MF, Ashmun RA, Sherr CJ. Colony-stimulating factor 1 regulates novel cyclins during the G_1 phase of the cell cycle. Cell 1991; 65:701–713.

131. Hamilton JA, Vairo G, Lingelbach SR. Activation and proliferation signals in murine macrophages: stimulation of glucose uptake by hematopoietic growth factors and other agents. J Cell Physiol 1988; 134:405–412.

132. Racoosin EL, Swanson JA. Macrophage colony-stimulating factor (rM-CSF) stimulates pinocytosis in bone marrow-derived macrophages. J Exp Med 1989; 170:1635–1648.

133. Stanley ER, Heard PM. Factors regulating macrophage production and growth. Purification and some properties of the colony-stimulating factor from medium conditioned by mouse L cells. J Biol Chem 1977; 252:4305.

134. Branch DR, Turner AR, Guilbert LJ. Synergistic stimulation of macrophage proliferation by the monokines tumor necrosis factor-alpha and colony-stimulating factor 1. Blood 1989; 73:307–311.

135. Kurland JI, Broxmeyer HE, Pelus LM, et al. Role for monocyte–macrophage-derived colony-stimulating factor and prostaglandin E in the positive and negative feedback control of myeloid stem cell proliferation. Blood 1978; 52:388–407.

136. Kurland JI, Pelus LM, Ralph P, et al. Induction of prostaglandin E synthesis in normal and neoplastic macrophages: role for colony-stimulating factor(s) distinct from effects on myeloid progenitor cell proliferation. Proc Natl Acad Sci USA 1979; 76:2326–2330.

137. Whetton AD, Dexter TM. Myeloid haemopoietic growth factors. Biochim Biophys Acta 1989; 989:111–132.

138. Rothstein G, Rhondeau SM, Peters CA, et al. Stimulation of neutrophil production in CSF-1 responsive clones. Blood 1988; 72:898–902.

139. McNiece IK, Stewart FM, Deacon DM, Bockman RS, Moore MAS. Synergistic interactions between hematopoietic growth factors as detected by in vitro mouse bone marrow colony formation. Exp Hematol 1988; 16:383–388.

140. Moore MAS. The use of colony-stimulating factors in combinations. Curr Opin Biotechnol 1991; 2:854.

141. Moore MAS, Warren DJ, Souza L. Synergistic interaction between interleukin-1 and CSFs in hematopoiesis. In: Gale RP, Golde DW, eds. UCLA Symposium on Leukemia, Recent Advances in Leukemia and Lymphoma. New York: Alan R Liss, 1987:445–456.

142. Moore MAS. Combination biotherapy: synergistic, additive and concatenate interactions between CSFs and interleukin in hematopoiesis. In: Fortner JG, Rhoads JE, eds. Accomplishments in Cancer Research 1987. Philadelphia: JB Lippincott, 1988:335–350.

143. Broxmeyer HE, Cooper S, Lu L, Miller ME, Langefeld CD, Ralph P. Enhanced stimulation of human bone marrow macrophage colony formation in vitro by recombinant human macrophage colony-stimulating factor in agarose medium and at low oxygen tension. Blood 1990; 76:323–329.

144. Rosenfeld CS, Evans C, Shadduck RK. Human macrophage colony-stimulating factor induces macrophage colonies after L-phenylalanine methylester treatment of human marrow. Blood 1990; 76:1783–1787.

145. Sato N, Sawada K, Kannoji M, Tarumi T, Sakai N, Ieko M, Sakurama S, Nakagawa S, Yasukouchi T, Krantz SB. Purification of human marrow progenitor cells and demonstration of the direct action of macrophage colony-stimulating factor on colony-forming unit-macrophage. Blood 1991; 78:967–974.

146. Lopez M, Martinache C, Canepa S. Autologous lymphocytes prevent the death of monocytes in culture and promote, as do GM-CSF, IL3-, and M-CSF, their differentiation into macrophages. J Immunol Methods 1993; 159:29–38.

147. Xing Z, Ohtoshi T, Ralph P. Human upper airway structural cell-derived cytokines support human peripheral blood monocyte survival: a potential mechanism for monocyte/macrophage accumulation in the tissue. Am J Respir Cell Mol Biol 1992; 6:212–218.

148. Oster W, Brach MA, Gruss HJ, Mertelsmann R, Herrmann F. Interleukin-1 beta (IL-1 beta) expression in human blood mononuclear phagocytes is differentially regulated by granulocyte–macrophage colony-stimulating factor (GM-CSF), M-CSF, and IL-3. Blood 1992; 79:1260–1265.

149. Moore RN, Oppenheim JJ, Farrar JJ, Carter CS Jr, Waheed A, Shadduck RK. Production of lymphocyte-activating factor (interleukin-1) by macrophages activated with colony-stimulating factors. J Immunol 1980; 125:1302–1305.

150. Warren MK, Ralph P. Macrophage growth factor CSF-1 stimulates human monocyte production of interferon, tumor necrosis factor and colony stimulating activity. J Immunol 1986; 137:2281–2285.

151. James SL, Cook KW, Lazdins JK. Activation of human monocyte-derived macrophages to kill schistosomula of *Schistosoma mansoni* in vitro. J Immunol 1990; 145:2686–2690.

152. Curley SA, Roh MS, Kleinerman E, Klostergaard J. Human recombinant macrophage colony-stimulating factor activates murine Kupffer cells to a cytotoxic state. Lymphokine Res 1990; 9:355–363.

153. Moore RN, Larsen HS, Horohov DW, Rouse BT. Endogenous regulation of macrophage proliferative expansion by colony-stimulating factor-induced interferon. Science 1984; 223:178–180.

154. Motoyoshi K, Suda T, Kusumoto K, Takaku F, Miura Y. Granulocyte–macrophage colony stimulating and binding activities of purified human urinary colony-stimulating factor to murine and human bone marrow cells. Blood 1982; 60:1378–1386.

155. Metalf D, Nicola A. Synthesis by mouse peritoneal cells of G-CSF, the differentiation inducer for myeloid leukemia cells: stimulation by endotoxin, M-CSF and multi-CSF. Leuk Res 1985; 9:35–50.

156. Hamilton JA, Whitty GA, Stanton H, Meager A. Effects of macrophage–colony-stimulating factor on human monocytes: induction of expression of urokinase-type plasminogen activator, but not of secreted prostaglandin E2, interleukin-6, interleukin-1, or tumor necrosis factor-alpha. J Leukoc Biol 1993; 53:707–714.

157. Hamilton JA, Whitty GA, Stanton H, Wojta J, Gallichio M, McGrath K, Ianches B. Macrophage colony-stimulating factor and granulocyte–macrophage colony-stimulating factor stimulate the synthesis of plasminogen-activator inhibitors by human monocytes. Blood 1993; 82:3616–3621.

158. Lyberg T, Stanley ER, Prydz H. Colony stimulating factor-1 induces thromboplastin activity in murine macrophages and human monocytes. J Cell Physiol 1987; 132:367–370.

159. Wing EJ, Ampel NM, Waheed A, Shadduck RK. Macrophage colony-stimulating factor (M-CSF) enhances the capacity of murine macrophages to secrete oxygen reduction products. J Immunol 1985; 135:2052–3056.

160. Garcia Lloret MI, Rocha Ramirez LM, Ramirez A, Santos Preciado JI. Macrophage colony-stimulating factor enhances the respiratory burst of human monocytes in response to *Entamoeba histolytica*. Arch Med Res 1992; 23:139–141.

161. Yuo A, Kitagawa S, Motoyoshi K, Azuma E, Saito M, Takaku F. Rapid priming of human monocytes by human hematopoietic growth factors: granulocyte–macrophage colony-stimulating factor (CSF), macrophage-CSF, and interleukin-3 selectively enhance superoxide release triggered by receptor-mediated agonists. Blood 1992; 79:1553–1557.

162. Feder LS, Laskin DL. Regulation of hepatic endothelial cell and macrophage proliferation and nitric oxide production by GM-CSF, M-CSF, and IL-1β following acute endotoxemia. J Leuk Biol 1994; 55:507–513.

163. Munn DH, Armstrong E. Cytokine regulation of human monocyte differentiation in vitro: the tumor-cytotoxic phenotype induced by macrophage colony-stimulating factor is developmentally regulated by gamma-interferon. Cancer Res 1993; 53:2603–2613.

164. Young DA, Lowe LD, Clark SC. Comparison of the effects of IL-3, granulocyte–macrophage colony-stimulating factor, and macrophage colony-stimulating factor in supporting monocyte differentiation in culture. Analysis of macrophage antibody-dependent cellular cytotoxicity. J Immunol 1990; 145:607–615.

165. Urano F, Imoto M, Fukuda Y, Koyama K, Nakano I. Effects of macrophage colony-stimulating factor on the proliferation and the function of Kupffer cells. Arzneimittelforschung 1993; 43:804–807.

166. Rist RJ, Jones GE, Naftalin RJ. Synergistic activation of 2-deoxy-D-glucose uptake in rat and murine peritoneal macrophages by human macrophage colony-stimulating factor-stimulated coupling between transport and hexokinase activity and phorbol-dependent stimulation of pentose phosphate-shunt activity. Biochem J 1990; 265:243–249.

167. Andoh A, Fujiyama Y, Kitoh K, Niwakawa M, Hodohara K, Bamba T, Hosoda S. Macrophage colony-stimulating factor (M-CSF) enhances complement component C3 production by human monocytes/macrophages. Int J Hematol 1993; 57:53–59.

168. Lee MT, Warren MK. CSF-1 induced resistance to viral infection in murine macrophages. J Immunol 1987; 138:3019–3022.

169. Rose RM, Fuglestad JM, Remington L. Growth inhibition of *Mycobacterium avium* complex in human alveolar macrophages by the combination of recombinant macrophage colony-stimulating factor and interferon-gamma. Am J Respir Cell Mol Biol 1991; 4:248–254.

170. Ho JL, Reed SG, Wick EA, Giordano M. Granulocyte–macrophage and macrophage colony-

stimulating factors activate intramacrophage killing of *Leishmania mexicana amazonensis.* J Infect Dis 1990; 162:224–230.

171. Denis M. Killing of *Mycobacerium tuberculosis* within human monocytes: activation by cytokines and calcitriol. Clin Exp Immunol 1991; 84:200–206.

172. Karbassi A, Becker JM, Foster JS, Moore RN. Enhanced killing of *Candida albicans* by murine macrophages treated with macrophage colony-stimulating factor: evidence for augmented expression of mannose receptors. J Immunol 1987; 139:417–421.

173. Newman SL, Gootee L. Colony-stimulating factors activate human macrophages to inhibit intracellular growth of *Histoplasma capsulatum* yeasts. Infect Immun 1992; 60:4593–4597.

174. Wing EJ, Waheed A, Shadduck RK, Nagle LS, Stephenson K. Effect of colony-stimulating factor on murine macrophages. Induction of antitumor activity. J Clin Invest 1982; 69:270–276.

175. Ralph P, Nakoinz I. Stimulation of macrophage tumoricidal activity by the growth and differentiation factor CSF-1. Cell Immunol 1987; 105:270–279.

176. Thomassen MN, Barna NP, Widemann HP, Ahmad M. Modulation of human alveolar macrophage tumoricidal activity by recombinant macrophage colony-stimulating factor. J Biol Response Modif 1990; 9:87–91.

177. Nishimura Y, Higashi N, Tsuji T, Higuchi M, Osawa T. Activation of human monocytes by interleukin-2 and various cytokines. J Immunother 1992; 12:90–97.

178. Suzu S, Yanai N, Saito M, Kawashima T, Kasahara T, Saito M, Takaku F, Motoyoshi K. Enhancement of the antibody-dependent tumorcidal activity of human monocytes by human monocytic colony-stimulating factor. Jpn J Cancer Res 1990; 81:79–84.

179. Baldwin GC, Chung GY, Kaslander C, Esmail T, Reisfeld RA, Golde DW. Colony-stimulating factor enhancement of myeloid effector cell cytotoxicity towards neuroectodermal tumour cells. Br J Haematol 1993; 83:545–553.

180. Munn DH, Cheung NK. Phagocytosis of tumor cells by human monocytes cultured in recombinant macrophage colony-stimulating factor. J Exp Med 1990; 172:231–237.

181. Kitano K, Abbound CN, Ryan DH, Quan SG, Baldwin GC, Golde DW. Macrophage-active colony-stimulating factors enhance human immunodeficiency virus type 1 infection in bone marrow stem cells. Blood 1991; 77:1699–1705.

182. Kalter DC, Nakamura M, Turpin JA, Baca LM, Hoover DL, Dieffenbach C, Ralph P, Gendelman JE, Meltzer MS. Enhanced HIV replication in macrophage colony-stimulating factor-treated monocytes. J Immunol 1991; 146:298–306.

183. Wiktor-Jedrezjczak W, Ahmed A, Szczylik C, Skelly RR. Hematological characterization of congenital osteopetrosis in op/op mouse. J Exp Med 1982; 156:1516–1527.

184. Wiktor-Jedrezjczak W, Szczylik C, Rataczak MZ, Ahmed A. Congenital murine osteopetrosis inherited with osteosclerotic (*oc*) gene: hematological characterization. Exp Hematol 1986; 14:819–825.

185. Wiktor-Jedrezjczak W, Bartocci A, Ferrante AW Jr, Ahmed-Ansari A, Sell KW, Pollard JW, Stanley ER. Total absence of colony-stimulating factor 1 in the macrophage-deficient osteopetrotic (op/op) mouse. Proc Natl Acad Sci USA 1990; 87:4828–4832.

186. Yoshida H, Hayashi S, Kuniscada T, Ogawa M, Nishikawa S, Okamura H, Sudo T, Shultz LD, Nishikawa SI. The murine mutation osteopetrosis is in the coding region of the macrophage colony-stimulating factor gene. Nature 1990; 345:442–443.

187. Wiktor-Jedrezjczak W, Urbanowska E, Aukerman SL, Pollard JW, Stanley ER, Ralph P, Ansari AA, Sell KW, Szperl M. Correction by CSF-1 of defects in the osteopetrotic op/op mouse suggests local developmental, and humoral requirements for this growth factor. Exp Hematol 1991; 19:1049–1054.

188. Takahashi K, Umeda S, Schultz LD, Hayashi S-I, Nishikawa S-I. Effect of macrophage colony-stimulating factor (M-CSF) on the development, differentiation, and maturation of marginal metallophilic macrophages and marginal zone macrophages in the spleen of osteopetrosis (*op*) mutant mice lacking functional M-CSF activity. J Leukoc Biol 1994; 55:581–588.

189. Begg SK, Radley JM, Pollard JW, Chisholm OT, Stanley ER, Bertoncello I. Delayed hematopoietic development in osteopetrotic (op/op) mice. J Exp Med 1993; 177:237–242.

190. Kodama H, Nose M, Niida S, Ohagame Y, Abe M, Kumagawa M, Suda T. Essential role of macrophage colony-stimulating factor in the osteoclast differentiation supported by stromal cells. J Exp Med 1991; 173:1291–1294.

191. Povolny BT, Lee MY. The role of recombinant human M-CSF, IL-3, GM-CSF and calcitriol in clonal development of osteoclast precursors in primate bone marrow. Exp Hematol 1993; 21:532–537.

192. Lee MY, Eyre DR, Osborne WRA. Isolation of a murine osteoclast colony-stimulating factor. Proc Natl Acad Sci USA 1991; 88:8500–8504.

193. Lee TH, Fevold KL, Muguruma Y, et al. Relative roles of osteoclast colony-stimulating factor and macrophage colony-stimulating factor in the course of osteoclast development. Exp Hematol 1994; 22:66–73.

194. Fuller K, Owens JM, Jagger CJ, Wilson A, Moses R, Chambers TJ. Macrophage colony-stimulating factor stimulates survival and chemotactic behavior in isolated osteoclasts. J Exp Med 1993; 178:1733–1744.

195. Bertolini DR, Strassmann G. Differential activity of granulocyte-macrophage and macrophage colony-stimulating factors on bone resorption in fetal rat long bone organ cultures. Cytokine 1991; 3:421–427.

196. Orchard PJ, Dahl N, Aukerman SL, Blazer BR, Key LL. Circulating macrophage colony-stimulating factor is not reduced in malignant osteopetrosis. Exp Hematol 1992; 20:103–105.

197. Krieger M, Acton S, Ashkenas J, Pearson A, Penman M, Resnick D. Molecular flypaper, host defense, and atherosclerosis. J Biol Chem 1993; 268:4569–4574.

198. de Villiers WJS, Fraser IP, Hughes DA, Doyle AG, Gordon S. Macrophage-colony-stimulating factor selectively enhances macrophage scavenger receptor expression and function. J Exp Med 1994; 180:705–709.

199. Shyy YJ, Wickham LL, Hagan JP, Hsieh HJ, Hu YL, Telian SH, Valente AJ, Sung KL, Chien S. Human monocyte colony-stimulating factor stimulates the gene expression of monocyte chemotactic protein-1 and increases the adhesion of monocytes to endothelial monolayers. J Clin Invest 1993; 92:1745–1751.

200. De Nichilo MO, Burns GF. Granulocyte–macrophage and macrophage colony-stimulating factors differentially regulate alpha v beta integrin expression on cultured human macrophages. Proc Natl Acad Sci USA 1993; 90:2517–2521.

201. Uemura N, Ozawa K, Takahashi K, et al. Binding of membrane-anchored macrophage colony-stimulating factor (M-CSF) to its receptor mediates specific adhesion between stromal cells and M-CSF receptor-bearing hematopoietic cells. Blood 1993; 82:2634–2640.

202. Yamada N, Ishibashi S, Shimano H, et al. Role of monocyte colony-stimulating factor in foam cell generation. Proc Soc Exp Biol Med 1992; 20:240–244.

203. Ishibashi S, Inaba T, Shimano H, Harada K, Inoue I, Mokuno H, Mori N, Gotoda T, Takaku F, Yamada N. Monocyte colony-stimulating factor enhances uptake and degradation of acetylated low density lipoproteins and cholesterol esterification in human monocyte-derived macrophages. J Biol Chem 1990; 265:14109–14117.

204. Inaba T, Shimano H, Gotoda T, Harada K, Shimada M, Kawamura M, Yazaki Y, Yamada N. Macrophage colony-stimulating factor regulates both activities of neutral and acidic cholesteryl ester hydrolases in human monocyte-derived macrophages. J Clin Invest 1993; 92:750–757.

205. Shimano H, Yamada N, Motoyoshi K, Matsumoto A, Ishibashi S, Mori N, Takaku F. Plasma cholesterol-lowering activity of monocyte colony-stimulating factor (M-CSF). Ann NY Acad Sci 1990; 587:362–370.

206. Yong K, Salooja N, Donahue RE, Hegde U, Linch DC. Human macrophage colony-stimulating factor levels are elevated in pregnancy and in immune thrombocytopoenia. Blood 1992; 80:2897–2902.

207. Zolti M, Ben-Rafael Z, Meirom R, Shemesh M, Bider D, Mashiach S, Apte RN. Cytokine involvement in oocytes and early embryos. Fertil Steril 1991; 56:265–272.

208. Leiserowitz GS, Harris SA, Subramaniam M, Keeney GL, Podratz KC, Spelsberg TC. The proto-oncogene c-*fms* is overexpressed in endometrial cancer. Gynecol Oncol 1993; 49:190–196.

209. Kauma SW. Interleukin-1 beta stimulates colony-stimulating factor-1 production in human term placenta. J Clin Endocrinol Metab 1993; 76:701–703.

210. Azuma C, Saji F, Kimura T, Tokugawa Y, Takemura M, Samejima Y, Tanizawa O. Steroid hormones induce macrophage colony-stimulating factor (MCSF) and MCSF receptor mRNAs in the human endometrium. J Mol Endocrinol 1990; 5:103–108.

211. Moore MAS, Williams N, Metcalf D. In vitro colony formation by normal and leukemic human hemopoietic cells: characterization of the colony-forming cells. J Natl Cancer Inst 1973; 50:603.

212. Moore MAS, Spitzer G, Williams N, Metcalf D, Buckley J. Agar culture studies in 127 cases of untreated acute leukemia. The prognostic value of reclassification of leukemia according to in vitro growth characteristics. Blood 1974; 44:1–6.

213. Moore MAS. Marrow culture—a new approach to classification of leukemias. Blood cells 1975; 1:149–158.

214. Lowenberg B, Touw IP. Hematopoietic growth factors and their receptors in acute leukemia. Blood 1993; 81:281–292.

215. Valtieri M, Boccoli G, Testa U, Barletta C, Peschle C. Two step differentiation of AML-193 leukemic line: terminal maturation is induced by positive interaction of retinoic acid with granulocyte colony-stimulating factor (CSF) and vitamin-D3 with monocyte CSF. Blood 1991; 77:1804–1812.

216. Kato H, Adachi K, Suzuki M, Tanimoto M, Saito H. Macrophage colony-stimulating factor stimulates growth progression of the G_1-phase fraction and induces monocytic differentiation of the G_2/M phase fraction in human myeloid leukemia cells. Exp Hematol 1993; 21:1597–1604.

217. Vassiliadis S, Guilbert LJ. CSF-1 and immune interferon synergize to induce differentiation of human leukemic cells bearing CSF-1 receptors. Leuk Res 1991; 15:943–952.

218. Mayani H, Shen SY, Guilbert LJ, Clark SC, Sych I, Janowska-Wieczorek A. Effect of rhCSF-1 on human hemopoiesis in long-term cultures from patients with acute myelogenous leukemia. Leukemia 1991; 5:8–13.

219. Mayani HL, Guilbert LJ, Sych I, Janowska-Wieczorek A. Production of tumor necrosis factor-alpha in human long-term marrow cultures from normal subjects and patients with acute myelogenous leukemia: effect of recombinant macrophage colony-stimulating factor. Leukemia 1992; 6:1148–1154.

220. Ferrero D, Carlesso N, Pregno P, Gallo E, Pileri A. Self-renewal inhibition of acute myeloid leukemia clonogenic cells by biological inducers of differentiation. Leukemia 1992; 6:100–106.

221. Parwaresch MR, Kreipe H, Felgner J, Heidorn K, Jaquet K, Bodewadt-Radzun S, Radzun HJ. M-CSF and M-CSF-receptor gene expression in acute myelomonocytic leukemias. Leuk Res 1990; 14:27–37.

222. Stanley ER. Increased circulating colony-stimulating factor-1 in patients with preleukemia, leukemia, and lymphoid malignancies. Blood 1991; 77:796–803.

223. Jaquet K, Kreipe H, Felgner J, Radzun HJ, Parwaresch MR. Restriction map modification for the detection of point mutations within the human M-CSF receptor gene [abstr]. J Cancer Res Clin Oncol 1991; 117(suppl 3):S117.

224. Ridge SA, Worwood M, Oscier D, Jacobs A, Padua RA. FMS mutations in myelodsplastic, leukemic, and normal subjects. Proc Natl Acad of Sci USA. 1990; 87:1377–1380.

225. Tobal K, Pagliuca A, Bhatt B, Bailey N, Layton DM, Mufti GJ. Mutation of the human FMS gene (M-CSF receptor) in myelodysplastic syndromes and acute myeloid leukemia. Leukemia 1990; 4:486–489.

226. Van Daalen Wetters T, Hawkins SA, Roussel MF, Sherr CJ. Random mutagenesis of CSF-1 receptor (FMS) reveals multiple sites for activating mutations within the extracellular domain. EMBO J 1992; 11:551–556.

227. Paietta E, Racevskis J, Stanley ER, Andreeff M, Papenhausen P, Wiernik PH. Expression of the macrophage growth factor, CSF-1 and its receptor c-*fms* by a Hodgkin's disease-derived cell line and its variants. Cancer Res 1991; 51:5690–5696.

228. Hsu PL, Lin YC, Hsu SM. Expression of macrophage colony-stimulating factor (M-CSF) in two

Hodkin's Reed-Sternberg (H-RS) cell lines, HDLM-1 and KM-H2, and in H-RS cells in tissues. Int J Hematol 1991; 54:315–326.

229. Kacinski BM, Carter D, Mittal K, Yee LD, Scata KA, Donofrio L, Chambers SK, Wang KI, Yang-Feng T, Rohrschneider LR. Ovarian adenocarcinomas express *fms*-complementary transcripts and *fms* antigen, often with coexpression of CSF-1. Am J Pathol 1990; 137:135–147.

230. Bast RC Jr, Boyer CM, Jacobs I, Xu FJ, Wu S, Wiener J, Kohler M, Berchuck A. Cell growth regulation in epithlial ovarian cancer [review]. Cancer 1993; 71:1597–1601.

231. Baiocchi G, Kavanagh JJ, Talpaz M, Wharton JT, Gutterman JU, Kurzrock R. Expression of the macrophage colony-stimulating factor and its receptor in gynecologic malignancies. Cancer 1991; 67:990–996.

232. Kacinski BM, Chambers SK, Stanley ER, et al. The cytokine CSF-1 (M-CSF) expressed by endometrial carcinomas in vivo and in vitro, may also be a circulating tumor marker of neoplastic disease activity in endometrial carcinoma patients. Int J Radiat Oncol Biol Phys 1991; 19:619–626.

233. Crostall JD, Pollard JW, Carey F, Forder RA, White JO. Colony stimulating factor-1 stimulates Ishikawa cell proliferation and lipocortin II synthesis. J Steroid Biochem Mol Biol 1992; 42:121–129.

234. Lidor YJ, Xu FJ, Marinez-Maza O, Olt GJ, Marks JR, Berchuch A, Ramakrishnan S, Berek JS, Bast RC Jr. Constitutive production of macrophage colony-stimulating factor and interleukin-6 by human ovarian surface epithlial cells. Exp Cell Res 1993; 207:332–339.

235. Kimura T, Azuma C, Saji F, Tokugawa Y, Takemura M, Miki M, Ono M, Tanizawa O. The biological effects of macrophage-colony-stimulating factor induced by progestin on growth and differentiation of endometrial adenocarcinoma cells. Int J Cancer 1991; 49:229–233.

236. Xu FJ, Ramakrishnan S, Daly L, Soper JT, Bernuck A, Clarke-Pearson D, Bast RC Jr. Increased serum levels of macrophage colony-stimulating factor in ovarian cancer. Am J Obstet Gynecol 1991; 165:1356–1362.

237. Suzuki M, Ohwada M, Aida I, Tamada T, Hanamura T, Nagatomo M. Macrophage colony-stimulating factor as a tumor marker for epithelial ovarian cancer. Obstet Gynecol 1993; 82:946–950.

238. Elg SA, Yu Y, Carson LF, Adcock LL, Twigg LB, Prem KA, Ramakrishnan S. Serum levels of macrophage colony-stimulating factor in patients with ovarian cancer undergoing second-look laparotomy. Am J Obstet Gynecol 1992; 166:134–137.

239. Woolas RP, Xu FJ, Jacobs IJ, Yu YH, Daly L, Berchuck A, Soper JT, Clarke-Pearson DL, Oram DH, Bast RC Jr. Elevation of multiple serum markers in patients with stage I ovarian cancer. J Natl Cancer Inst 1993; 85:1748–1751.

240. Filderman AE, Stanley ER, Kacinski BM. Circulating macrophage-colony stimulating factor as a tumor marker for lung cancer [abstr]. 1990; World Conference on Lung Health. May 20–24, 1990:34.

241. Scholl SM, Tang T, Beuvon F, Palud C, Lidereau R. Pouillart P. Expression of colony-stimulating factor-1 and its receptor (the protein product of c-*fms*) in invasive breast tumor cells [abstr]. Pisa Symposia in Oncology, Breast Cancer: From Biology to Therapy. October 19–21, 1992, Pisa, Italy, p.44.

242. Kacinski BM, Scata KA, Carter D, Yee LD, Sapi E, King BL, Chambers SK, Jones MA, Pirro MH, Stanley ER. FMS (CSF-1 receptor) and CSF-1 transcripts and protein are expressed by human breast carcinomas in vivo and vitro. Oncogene 1991; 6:941–952.

243. Tang R, Beuvon F, Ojeda M, Mosseri V, Pouilart P, Scholl S. M-CSF (monocyte colony stimulating factor) and M-CSF receptor expression by breast tumour cells: M-CSF mediated recruitment of tumour infiltrating monocytes? J Cell Biochem 1992; 50:350–356.

244. Filderman AE, Bruckner A, Kacinski BM, Deng N, Remold HG. Macrophage colony-stimulating factor (CSF-1) enhances invasiveness in CSF-1 receptor-positive carcinoma cell lines. Cancer Res 1992; 52:3661–3666.

245. Bruckner A, Kirchheimer J, Filderman A, Remold H. CSF-1 upregulates urokinase-dependent invasion in lung tumor cells [abstr]. FASEB J 1991; 5:A1462.

246. Gope R, Christensen M, Christensen MA, Thorson A, Smyrk T, Lynch HT, Boman BM, Gope ML.

Loss of heterozygosity at the colony-stimulating factor-1 receptor locus on chromosome 5 in human colorectal cancers [abstr]. Proc Annu Meet Am Assoc Cancer Res 1990; 31:A1847.

247. Castresana JS, Barrios C, Ruiz J, Gomez L, Kreicbergs A. Sporadic amplification of the c-*fms* proto-oncogene in human musculoskeletal sarcomas. Anticancer Res 1993; 13:807–810.

248. Lieschke GJ, Stanley E, Grail D, Hodgson G, Sinickas V, Gall JAM, Sinclair RA, Dunn AR. Mice lacking both macrophage-and granulocyte-macrophage colony-stimulating factor have macrophages and coexistent osteopetrosis and severe lung disease. Blood 1994; 84:27–35.

249. Shadduck RK, Waheen A, Boegel F. The effect of colony stimulating factor-1 in vivo. Blood Cells 1987; 13:49–63.

250. Hume DA, Pavli P, Donahue RE, et al. The effect of human recombinant macrophage colony-stimulating factor (CSF-1) on the murine mononuclear phagocyte system in vivo. J. Immunol 1988; 141:3405–3409.

251. Chikkappa G, Broxmeyer HE, Cooper S. Effect in vivo of multiple injections of purified murine and recombinant human macrophage colony-stimulating factor to mice. Cancer Res 1989; 49:3558–3561.

252. Garnick MB, Stoudemire J, Donahue RE. In vitro and in vivo biological effects of recombinant human macrophage colony-stimulating factor (rhm-CSF). Proc Annu Meet Soc Clin Immunol 1989; 8:A717.

253. Broxmeyer HE, Williams DE, Cooper S. The influence in vivo of murine colony-stimulating factor-1 on myeloid progenitor cells in mice recovering from sublethal dosages of cyclophosphamide. Blood 1987; 69:913–918.

254. Hume DA, Donahue RE, Fidler IJ. The therapeutic effect of human recombinant macrophage colony-stimulating factor (CSF-1) in experimental murine metastatic melanoma. Lymphokine Res 1989; 8:69–77.

255. Munn DH, Garnick MB, Cheung N-KV. Effects of parenteral recombinant human macrophage colony-stimulating factor on monocyte number, phenotype, and antitumor cytotoxity in nonhuman primates. Blood 1990; 75:2042–2050.

256. Stoudemire J, Metzger M, Timony G. Pharmacokinetics of recombinant human macrophage colony-stimulating factor (rhM-SCF) in primates and rodents. Proc Annu Meet AACR 1989; 30:A2139.

257. Sanda MG, Yang JC, Topalian SL, Groves ES, Childs A, Belfort R, deSmet MD, Schwartzentruber DJ, White DE, Lotze MT, Rosenberg SA. Intravenous administration of recombinant human macrophage colony stimulating factor to patients with metastic cancer: a phase I study. J Clin Oncol 1992; 10:1643–1649.

258. Zamkoff KW, Hudson J, Groves ES, Childs A, Konrad M, Rudolph AR. A phase I trial of recombinant human macrophage colony-stimulating factor by rapid intravenous infusion in patients with refractory malignancy. J Immunother 1992; 11:103–110.

259. Redman BG, Flaherty L, Chou TH, Kraut M, Martino S, Simon M, Valdivieso M, Groves E. Phase I trial of recombinant macrophage colony stimulating factor by rapid intravenous infusion in patients with cancer. J Immunother 1992; 12:50–54.

260. Nemunaitis J, Mayers JD, Buckner CD, Shannon-Dorcy K, Mori H, Shulman H, Bianco JA, Higano CS, Groves E, Storb R, Hansen J, Applebaum ER, Singer JW. Phase I trial of recombinant human macrophage colony-stimulating factor in patients with invasive fungal infections. Blood 1991; 78:907–913.

261. Motoyoshi K, Takaku F. Human monocytic colony-stimulating factor (HM-CSF): phase I/II clinical studies. In: Mertelsmann R, Herrmann F, eds. Hematopoietic Growth Factors in Clinical Applications. New York: Marcel Dekker, 1992:161–175.

262. Van Hoef ME, Radford JA, Jones A, Smith IE, Earl HM, Thatcher N. Phase I–II study of monocyte colony-stimulating factor (p-100; Alpha Pharmaceutic Corporation) following treatment with carboplatin and etoposide in small cell lung cancer (SCLC) [abstr]. Ann Oncol 1992; 3(suppl 5):6.

263. Weiner LM, Li W, Holmes M, Catalano RB, Dovnarsky M, Padavic K, Alpaugh RK. Phase I trial of recombinant macrophage colony-stimulating factor and recombinant interferon: toxity, monocytosis, and clinical effects. Cancer Res 1994; 54:4084–4090.

264. Weiner LM, Steplewki A, Koprowski H, Litwin S, Comis RL. Divergent dose-related effects of interferon therapy on in vitro antibody-dependent cellular and nonspecific cytotoxicity by human peripheral blood monocytes. Cancer Res 1988; 48:1042–1046.

265. Weiner LM, Moldofsky P, Gatenby R, O'Dwyer J, O'Brien J, Litwin S, Comis RL. Antibody delivery and effector cell activation in a phase II trial of recombinant interferon and murine monoclonal antibody CO17-1A in advanced colorectal carcinoma. Cancer Res 1988; 48:2568–2573.

266. Khwaja A, Yong K, Jones HM, Chopra R, McMillan AK, Goldstone AH, Patterson KG, Matheson C, Ruthven K, Abramson SB. The effect of macrophage colony-stimulating factor on haemopoietic recovery after autologous bone marrow transplantation. Br J Haematol 1992; 81:288–295.

267. Masaoka T, Shibata H, Ohno R, Katch S, Harada M, Motoyoshi K, Takaku F, Sakuma A. Double-blind test of human urinary macrophage colony-stimulating factor for allogeneic and syngeneic bone marrow transplantation: effectiveness of treatment and 2-year follow-up for relapse of leukaemia. Br J Haematol 1990; 76:501–505.

268. Kitano K, Kobayashi H, Maeyama H, Miyabayashi H, Furuta S. Treatment of acute monoblastic leukaemia by combination of recombinant human macrophage colony-stimulating factor and low dose of Ara-C. Br J Haematol 1993; 85:176–178.

14

Erythropoietin

Stephanie S. Watowich*

The Whitehead Institute for Biomedical Research, Cambridge, Massachusetts

I. INTRODUCTION

Erythropoietin (EPO) is a serum glycoprotein hormone that supports the survival, proliferation, and maturation of committed erythroid progenitor cells into circulating red blood cells. EPO acts primarily on a subset of hematopoietic precursor cells, in contrast with many of the cytokines, and this specificity has made it an attractive system to study hematopoietic regulation. Despite this specificity, however, the role that EPO plays in controlling cell proliferation and differentiation, and the molecular events of EPO-mediated signal transduction, are only beginning to be unraveled.

This chapter will focus on recent advances made in understanding the expression, structure and biological functions of EPO and the EPO receptor (EPOR), particularly at the molecular level. Other reviews on these topics are available (66,144).

II. EPO STRUCTURAL AND FUNCTIONAL RELATIONSHIPS

A. Purification, Cloning, and Sequence Analysis

Erythropoietin was first purified from the urine of patients with severe aplastic anemia in 1977 (86) and was found to be a glycoprotein with a molecular mass of 34,000 Da. The human EPO gene was cloned by the use of degenerate oligonucleotide probes, based on limited amino acid sequence information from the purified hormone (57,70). The cDNA, which encoded a biologically active product, was isolated from a fetal liver library, demonstrating that EPO is expressed in human fetal liver. Sequence analysis of the cDNA predicts an open-reading frame of 193 amino acids, the first 27 amino acids constitute the signal sequence and the last 166 form the mature protein (57). The gene is present as a single copy in humans and is composed of five

Current affiliation: The University of Texas M. D. Anderson Cancer Center, Houston, Texas.

exons spread over a little more than 3 kilobase pairs (kb; 57,70). The EPO genes and cDNAs have been cloned from a variety of species, and regions of identity in both the flanking sequences and the coding sequences have been identified (46,136). These regions may play important roles in conserved functions, such as receptor binding or transcriptional regulation.

B. Carbohydrate Analysis

Carbohydrate modifications on EPO account for approximately 30% of the weight of the hormone. Human EPO contains three asparagine (N)-linked carbohydrates on residues 24, 38, and 83, and one O-linked carbohydrate on serine-126, and this pattern of glycosylation is identical in both urinary EPO and recombinant EPO (68,121). The N-linked oligosaccharide addition sites are highly conserved among different species, whereas the O-linked site is not found in all species (136).

Studies that have investigated the function of the carbohydrate side chains on the recombinant hormone by using different approaches have yielded complementary results. Initial studies, using the glycoslyation inhibitor tunicamycin, suggested that the oligosaccarides are required for efficient secretion and biological activity (94). Removal of any one of the N-linked chains alone by site-directed mutagenesis had no effect on EPO secretion or in vitro biological activity, as measured by erythroid colony formation or ^{59}Fe-uptake by erythroid precursor cells (32). Double mutants, missing glycosylation sites at positions 24 and 38 or positions 24 and 83, were secreted normally and had full in vitro biological activity, but the mutant missing sites 38 and 83 was secretion-defective. As expected, removal of all three N-linked sugars by mutagenesis also yielded a secretion-defective molecule (32). The inactive forms of EPO had altered conformations, as judged by radioimmunoassay, suggesting that glycosylation at residue 38 or 83 is important for proper folding and secretion of EPO (32). By contrast, enzymatic cleavage of the mature, secreted hormone, to remove N-linked carbohydrates, had no effect on its in vitro biological activity (54,120,126). Together, these results suggest that the presence of at least one of the N-linked carbohydrates, at position 38 or 83, is important for the folding and secretion of EPO; however, once fully folded and secreted from the cell, the de-N-glycosylated hormone is fully active in vitro. Desialated EPO competes for binding to the EPOR better than the fully glycosylated hormone, suggesting that the carbohydrate chains on EPO may be positioned close to the receptor-binding surface (89).

EPO can be efficiently secreted with full in vitro activity without addition of the O-linked oligosaccharide, as judged by synthesis in a glycosylation-defective cell line, enzymatic cleavage, or site-directed mutagenesis of serine-126 (32,54,131). However, two point mutations of residue 126, either an alanine or glycine substitution, strongly inhibited the biological activity of EPO (32,34). These mutations appear to induce a conformational change in EPO (32). In vivo, the O-linked oligosaccharide is dispensable for full activity; the N-linked carbohydrate side chains are, however, essential for full activity (32,54,131). This is thought to be due to rapid clearance of the desialated hormone from the circulation by the liver (32,45,54,131).

C. Role of Disulfide Bonds

Human EPO contains two intramolecular disulfide bonds, between residues 7 and 161 and between residues 29 and 33 (68). The former disulfide bridge, which links the NH$_2$- and COOH-terminal portions of EPO, is highly conserved across species. The small disulfide loop (C29–C33) is not highly conserved; this disulfide bridge is predicted to be in the loop region between α-helices A and B (10,136). Reduction and alkylation of the sulfhydryl groups rendered EPO completely inactive, demonstrating that the integrity of at least one of the disulfide bonds is required for hormone function (128). Site-directed mutagenesis of the cysteine residues in the small disulfide loop (C29–C33) in human EPO, or complete deletion of this loop, demonstrated that it is not required for secretion and in vitro activity. The fully conserved loop (C7–C161) is necessary for full activity of EPO, indicating that this loop may be important in stabilizing the tertiary structure of EPO (10).

D. Structure–Function Studies

A variety of techniques have been used to probe EPO structure–function relations, particularly concerning hormone–receptor interactions. Monoclonal antibodies (MAbs) have been raised against EPO; some of these antibodies compete with the receptor for hormone binding and have neutralizing affects on EPO-stimulated proliferation (24,41). A recent study, which used epitope mapping of neutralizing antibodies, suggested that the receptor-binding site may consist of a tertiary structure, which is composed of peptides from various regions in the primary structure of the hormone, rather than a single polypeptide stretch in the primary structure (41).

Computer-assisted molecular modeling, using secondary structure predictions, suggested that the cytokine molecules, including EPO, are structurally similar, despite the fact that they share little amino acid identity (3). These molecules are predicted to be four α-helical bundles, with the α-helices in an antiparallel arrangement (3). EPO appears to be more closely related to growth hormone (GH) and prolactin (Pr) than to other cytokines and is predicted to have helical-loop arrangements similar to them, with long loops between helices A and B and between C and D, and a short B–C loop (3). Determination of the GH structure by X-ray crystallography demonstrated that the molecule is a four-helical bundle, in good agreement with the proposed structure (31). In support of the structural model for EPO, circular dichroism studies have indicated that at least 50% of the molecule is composed of α-helical structures (68). Deletional mutagenesis of the EPO cDNA, to test the predicted structure, is also compatible with the model. Deletion of predicted loop regions had no effect on EPO secretion or in vitro proliferative activity; however, deletion of predicted helical regions, which would be more likely to destabilize the tertiary structure of the hormone, resulted in secretion-defective molecules (10).

III. EXPRESSION OF EPO

A. Sites of EPO Production

An elegant control system, which is responsive to the oxygen-carrying capacity of the animal, regulates the synthesis and secretion of EPO. In the mouse, the fetus produces EPO primarily in the liver. By late gestation the liver production declines and production in the kidney increases. In the adult, the kidney is the major site of EPO production (58,64,110). Severe anemia, cobalt treatment, or hypoxia, caused by low ambient O_2, boosts the production of EPO, resulting in up to a 1000-fold increase in the serum EPO concentration, from 100–200 pg/ml to ~ 100 ng/ml (8,11,65,109,110). In the adult animal this results from an increased synthesis of EPO in the kidney and some synthesis in the liver (approximately 10–20% of the total EPO production; 8,11). In the absence of EPO synthesis by the kidneys, as seen in anephric patients, some EPO production is maintained by the liver (66).

The EPO mRNA is synthesized by interstitial cells in the renal cortex under normal and anemic conditions (64,65,67,110). With induction of anemia more cells in this region are recruited to produce EPO and, in fact, the hematocrit is exponentially related to the number of EPO-synthesizing cells in the kidney, the total renal EPO mRNA levels, and the serum concentration of EPO (65). In situ hybridization experiments showed that the number of EPO-producing cells is very low under normal physiological conditions. The number of EPO-producing cells increases under conditions of mild or moderate anemia, and the cells are found in clusters. Under severely anemic conditions, the majority of inner cortex cells, and some cells in the outer medulla or subcapsular cortex, are producing EPO (65). From these studies, it has been proposed that the kidney cells capable of producing EPO are either in an "on" or "off" state in terms of EPO synthesis. The microcirculation or the local O_2 consumption may be responsible for triggering the selection of specific clusters of cells (65).

B. EPO Gene Regulation

The steady-state levels of EPO mRNA increase over 100-fold in the kidney and liver with hypoxia or anemia (11,110). Run-on transcription experiments in nuclei isolated from rat kidney cells, or in nuclei isolated from the hepatoma cell line Hep3B, have shown that this increase is partly due to an increase in the transcriptional rate of the EPO gene (48,109), demonstrating that these cells both sense and respond to hypoxia. EPO mRNA is detectably increased by approximately 1 hr of physiological stress and reaches maximal steady-state levels by 2–4 hr (11,110). Protein synthesis is required for EPO gene induction, suggesting that a labile protein is involved in the oxygen-sensing pathway (47). Because the kinetics of EPO mRNA synthesis are similar following induction of hypoxia or anemia and because ongoing protein synthesis is required for both inducers, they may stimulate EPO transcription by a common signaling pathway (47). EPO mRNA has a relatively short half-life ($t_{1/2} \sim 2$ hr), which provides a mechanism for a rapid return to the basal state of EPO synthesis in normal growth conditions (40,48,110).

A major regulator of EPO gene induction is the capacity of the blood to deliver oxygen, represented by the hematocrit in anemia and oxygen loading on hemoglobin under hypoxic conditions. Exactly how the oxygen levels are sensed by EPO-producing cells in the kidney and liver is unclear, but several lines of research have been aimed at addressing this question. To probe the nature of the cellular oxygen-sensing mechanism, other inducers of EPO mRNA synthesis were sought; exposure to cobalt, nickel, or manganese also stimulated EPO expression (8,47). EPO induction by hypoxia, cobalt, or nickel was blocked by inhibitors of heme biosynthesis in Hep3B cells (47). Together these results suggested that a heme-containing protein may be involved in the EPO-signaling pathway. Because cobalt, nickel, or manganese can substitute for iron in the heme moiety, they may lock the heme-containing protein in a deoxy-conformation, similar to their effects on hemoglobin, thereby inducing EPO transcription by the same pathway as hypoxia (47). Supportive evidence for this model comes from the fact that all inducers require ongoing protein synthesis and that hypoxia and nickel together do not have additive effects. In addition, carbon monoxide inhibited EPO production under hypoxic conditions, but not in the presence of cobalt or nickel, consistent with known properties of heme binding (47).

C. EPO Promoter Analysis

Several groups have performed extensive analysis of regulatory control regions in the flanking portions of the EPO gene, as well as identifying potential transcription factors, as another approach toward understanding oxygen-sensing and EPO gene regulation. These studies have been carried out primarily in hepatoma cell lines, such as Hep3B, which show inducible EPO expression in culture. Oxygen-responsive sequences have been mapped to regions in both the 5′- and 3′-flanking sequences (5,9,75,100,112). In particular, an enhancer element of about 50 nucleotides (nt), located approximately 120 nt 3′ of the polyadenylation site of the human EPO gene, is responsive to hypoxia and cobalt induction (5,9,100,114). Mutational analysis identified three functional regions within the enhancer, two of which are required for hypoxia-inducible activity (114). This enhancer functions in Hep3B cells, but not in several other cell types tested, suggesting that cell-type specific factors are also involved in hypoxia induction (5,100). The enhancer region is highly conserved between human and mouse EPO genes, indicating that it plays an essential regulatory role (46).

A variety of techniques, including DNase I footprinting, methylation interference, and electrophoretic mobility shift assays, have demonstrated the existence of proteins in Hep3B cell extracts that specifically bind the 3′ enhancer element (6,9,114). One of these proteins, hypoxia-inducible factor-1 (HIF-1), has hypoxia-inducible DNA-binding activity. HIF-1 binds within the first 18 nt of the 3′ enhancer element and this inducible DNA-binding activity is blocked in the presence of protein synthesis inhibitors (6,114), suggesting that it is involved in the oxygen-sensing pathway (Fig. 1). HIF-1-inducible–binding activity has been found in a wide variety of cell lines, in which the EPO gene is not responsive to hypoxia, indicating that it may also be involved in the inducible expression of other oxygen-sensitive genes (6,77,129).

EPO mRNA levels are very low in the kidney under normal physiological conditions, and EPO transcription was not detected in nuclei isolated from normal rat kidney (109). Results of in situ hybridization on kidney preparations suggest that cells in the kidney are either on or off in terms of their EPO transcription, with the number of cells producing EPO increasing with the severity of anemia, rather than an induction of EPO mRNA synthesis by cells already producing EPO (65). In contrast, a low level of EPO transcription is detectable in Hep3B cells grown under normal oxygen levels, and the transcription rate of the EPO gene is stimulated under stress conditions (48). Although the former experiments were performed with primary kidney cell

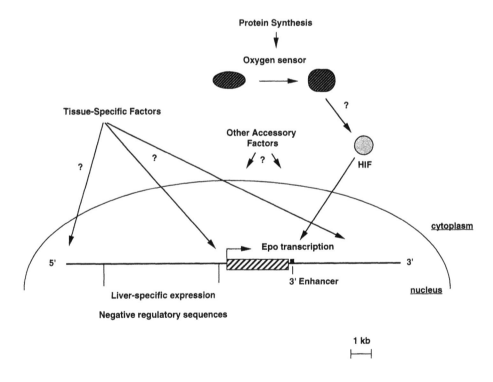

Figure 1 Pathways of EPO gene regulation: A decrease in blood oxygen levels is sensed by cells in the kidney or liver, which respond by inducing EPO transcription. Positive regulatory regions of the EPO gene are found in both the 5′- and 3′-flanking sequences, and these include tissue-specific elements as well as a hypoxia-responsive enhancer element. Negative-regulatory sequences, which maintain the EPO gene in an inactive form in nonhepatic and nonrenal tissues, map to the 5′-flanking sequences. Protein synthesis is required for EPO gene induction, including the activation of HIF DNA binding to the 3′-enhancer element.

preparations and the latter were performed in an established hepatoma cell line, together these results suggest that there may be tissue-specific differences in the transcriptional control of EPO.

Studies in transgenic mice also indicated that EPO transcription is regulated in a tissue-specific manner. Transgenic mice were generated containing the human EPO gene plus 0.7 kb of 3'-flanking sequences and 0.4 or 6.0 kb of 5'-flanking sequences. Mice with the 0.4 kb of 5'-flanking sequences had EPO expression in the liver, kidney, and a variety of other tissues. EPO expression was inducible in liver, but not in kidney (111,113). These results demonstrate that the 0.4-kb 5'-flanking region is missing regulatory sequences for inducible kidney expression and for negative control of EPO expression in other tissues. Transgenic mice with 6.0 kb of 5'-flanking sequences had detectable EPO expression only in the liver under hypoxic conditions. Thus, a negative control region of EPO transcription is localized between 0.4 and 6.0 kb of 5'-flanking sequences. Sequences that control kidney-inducible expression map beyond 6 kb of the 5' end and 0.7 kb of the 3' end of the gene, whereas sequences controlling liver-inducible expression map within 0.4 kb of the 5' end and 0.7 kb of the 3' end (111; see Fig. 1). Blanchard and co-workers (9) have found that a 53-bp region in the promoter acts synergistically with the 3'-enhancer element to stimulate hypoxia-inducible EPO transcription in transiently transfected Hep3B cells, thus further delineating liver-specific cis-acting elements. Cis-acting elements for regulated kidney expression have yet to be defined, most likely because of the lack of an adequate model cell line.

IV. BIOLOGICAL ACTIVITIES OF EPO

The principal functions of EPO are to regulate the proliferation, survival, and differentiation of erythroid precursor cells in the adult bone marrow and in the fetal liver, thereby regulating the oxygen-carrying capacity of the animal by effects on the red blood cell mass (37,66). In vivo, administration of recombinant EPO leads to an increased production of circulating red blood cells (38,42,49) and dysregulated expression of EPO in vivo gives rise to polycythemia in mice (111,113). In vitro, EPO is required for erythroid colony formation and hemoglobin synthesis (50,51,56,119,122). It is generally accepted that EPO is a late-acting, lineage-specific hematopoietic growth factor, which supports the proliferation and maturation of progenitor cells committed to the erythroid lineage (66,96).

Although the effects of EPO on in vivo and in vitro erythroid production have been clearly documented, the precise molecular events required for erythroid development, and the role that EPO plays in these processes, remain unclear. EPO appears to have direct effects on at least three processes involved in erythroid cell production: progenitor cell survival, proliferation, and differentiation. These processes are under intensive investigation in a variety of in vitro model systems. Erythroid progenitors have been isolated from fetal livers of normal or anemic mice, or from the spleens of mice infected with the anemia-inducing strain of spleen focus-forming virus (FVA cells). These two systems have the advantage of being primary cultures, but the disadvantage of being heterogeneous populations of cells, and the FVA cells are virally infected. Various cell lines have also been used to study EPO-mediated effects. Although these lines provide the advantage of being homogeneous populations, enabling molecular and biochemical events to be more clearly defined, they are established cell lines and thus may not provide an accurate reflection of in vivo events.

A. EPO as a Proliferation-Inducing Factor

Direct evidence that EPO can act as a mitogen comes from a variety of in vitro systems. EPO is absolutely required for the proliferation of isolated erythroid progenitor cells in culture. Cells at the colony-forming unit–erythroid (CFU-E) stage of development are highly responsive to EPO; they give rise to erythroid colonies in semisolid media by day 2 (mouse) or by day 7 (human) in culture in the presence of EPO (119,122). The burst-forming unit–erythroid (BFU-E) is an earlier committed progenitor, which requires the presence of stem cell factor and interleukin-3 (IL-3) or granulocyte–macrophage colony-stimulating factor (GM-CSF), as well as EPO, to complete the differentiation program. These cells generate erythroid colonies by day 8 (mouse) or day 15 (human) in culture (36,50,51,104).

EPO also stimulates the proliferation of several erythroleukemic cell lines, such as HCD57 and J2E cell lines (13,118). The EPOR cDNA has been ectopically expressed in IL-3–dependent cell lines and this expression enables the cells to grow in the presence of either IL-3 or EPO, demonstrating that EPO and the EPOR can also stimulate cell proliferation in nonerythroid hematopoietic cell lines (69,82). These cell lines have been used to identify some of the biochemical events involved in EPO-mediated proliferation (discussed in more detail later).

B. EPO as a Survival Factor

Studies done in vitro on purified FVA cells have clearly shown that the presence of EPO in these cultures is required for inhibition of apoptosis, or programmed cell death (62). EPO appears to retard or reduce the extent of DNA cleavage in progenitor cultures, enabling more cells to terminally differentiate (62). Apoptosis may be a normal component of erythropoiesis in vivo (59,62,145). According to this hypothesis, the circulating levels of EPO under normal physiological conditions are able to support the survival of only a small fraction of the available progenitors. This surviving population is, however, large enough to provide a sufficient level of erythrocytes under normal conditions. Under anemic or hypoxic conditions, with the increase in serum EPO concentration, the survival and proliferation of numerous erythroid progenitors is sustained, leading to an increase in the number of circulating red cells (62). In support of this model, Kelley and co-workers (59) have found that apoptosis and terminal erythroid differentiation are temporally and morphologically distinct processes, with the number of isolated progenitor cells undergoing apoptosis in culture directly related to EPO concentration. Whether or not EPO can directly signal cell survival, without concomitantly stimulating cell proliferation, remains to be determined.

C. The Role of EPO in Erythroid Differentiation

Committed erythroid progenitor cells, isolated from adult bone marrow or fetal liver, require EPO for colony formation and hemoglobinization in vitro (119); the steps involved in commitment to the erythroid lineage are unclear, but they do not appear to require EPO (96). An important question now under investigation is whether the intracellular signals EPO sends to sustain erythroid differentiation are similar to, or different from, the signals used to sustain erythroid survival and proliferation. This question has been difficult to answer in primary cell cultures, in which the processes of proliferation and differentiation are difficult to uncouple. Studies in established cell lines indicate that induction of proliferation and of differentiation may be distinct processes, with EPO able to activate each one separately (13,95). Several cell lines exist that do not require EPO for their survival or proliferation, but do respond to EPO by undergoing terminal differentiation (13,60,95,135). These systems will be very useful in defining the biochemical events involved in EPO-induced erythroid differentiation.

D. EPO as a Mitogen in Nonerythroid Cell Types

EPO receptors are expressed on various nonerythroid tissues, indicating that EPO may play a role in the development of these cell types. EPO has an effect on thrombopoiesis in rodents and on megakaryocyte colony formation in vitro (39,81,127). Megakaryocytes express the EPOR on their cell surface (43), and they appear to be responsive to the mitogenic effects of an ectopically expressed, constitutively activated EPOR (73). However, whether or not EPO plays a significant role in the development of megakaryocytes in vivo is unclear. Vascular endothelial cells express the EPOR, and EPO acts as a mitogen and positive chemotactic agent for these cells in culture (1). EPO appears to have neurotrophic effects on cholinergic neurons in vivo, and EPOR expression has been detected in the brain and in two neuronal cell lines, PC12 and SN6, although the physiological role of this expression is unclear (61,76,138).

E. Role of Maternal EPO in Fetal Erythropoiesis

Erythropoietin receptors are expressed on the placenta, suggesting that they may be involved in maternal-to-fetal transfer of EPO (63). A small amount (7–10%) of maternal-to-fetal transfer of EPO has been detected in some experiments (63), whereas other studies do not confirm placental transfer (146). This discrepancy may be partly explained by the different animal models and assays used in each study. Clearly, more work is needed to resolve this issue. Maternal and fetal erythropoiesis appear to be independent, although the physiological state of the mother can influence erythropoiesis in the fetus (146).

V. CLINICAL USES FOR EPO

There have been many uses for recombinant EPO in the clinic, mainly in treating patients with end-stage renal failure or in treating anemias arising in patients with chronic disease. Because patients with renal failure do not produce adequate levels of EPO, administration of EPO has alleviated the requirement for frequent blood transfusions (38). Recombinant EPO has been used to treat AIDS patients, who often develop anemia with the progression of the disease. AIDS patients undergoing zidovudine (AZT) therapy, which is a bone marrow-suppressive therapy, are also treated with EPO to replenish some of the circulating red blood cell numbers (42). Similarly, bone marrow suppression in cancer patients undergoing chemotherapy is treated with EPO, and these treatments have minimized the frequency of necessary blood transfusions.

VI. EPOR STRUCTURE

A. Isolation and Analysis of the EPOR cDNA

The EPOR cDNA was isolated by expression cloning from a murine erythroleukemia cell line library (23). Sequence analysis of the cDNA predicted the synthesis of a polypeptide of near 55 kDa, with a single membrane-spanning domain, oriented with the NH_2-terminus outside the cell and the COOH-terminus inside (23). Expression of the EPOR cDNA in IL-3-dependent cell lines, such as BA/F3 or 32D, rendered these cells EPO-responsive, demonstrating that the cloned receptor is functional in terms of stimulating cell proliferation (69,82).

B. Cytokine Receptor Family

Initial analysis of the EPOR sequence revealed that it shares homology with the β-chain of the interleukin-2 receptor (IL-2R), in both the extracellular and cytoplasmic domains (22). Subsequent isolation of the cDNAs for several cytokine receptors has shown that these receptors consitute a new gene family (4,19). Members of the cytokine receptor family are type I membrane-spanning proteins, with structurally similar exoplasmic domains, including four similarly spaced cysteine residues and a strikingly homologous sequence motif, WSXWS. These receptors have variable-slzed cytoplasmic domains, lacking kinase-related sequences. Two regions of homology (box 1 and box 2) have been identified in the membrane-proximal region of the cytoplasmic domains of several members of the family, including the EPOR (Fig. 2). These regions appear to play a crucial role in mediating cell proliferation (90).

The crystal structure of the exoplasmic region of the human growth hormone receptor (GH-R), complexed with GH, was determined recently and this structure has provided a paradigm for cytokine receptor structure (31). The GH-R forms a homodimer, complexed to a single ligand molecule (20,31). The exoplasmic region of each monomer of the receptor is composed of two subdomains; each subdomain is a seven-stranded β-barrel structure (31). The highly conserved cysteine residues are found buried within the interior of the first subdomain

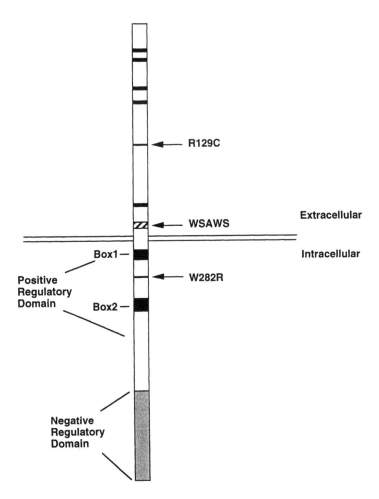

Figure 2 Structural features of the EPOR. The schematic diagram shows the relative positions of the four conserved cysteine residues (initial four thick, black bars from the NH$_2$-terminal end), the fifth extracellular cysteine residue (thick black bar close to transmembrane region), the conserved WSXWS motif (hatched box), the position of the activating point mutation (R129C) in the extracellular domain and the inactivating point mutation (W282R) in the cytoplasmic domain, and the positive (including the location of the Box 1 and Box 2 motifs) and negative-signaling domains in the cytoplasmic region.

and are likely to contribute to the stability of that region. Residues that interact with the ligand have been identified in both subdomains of the receptor. A membrane-proximal region, in the second subdomain of the receptor, is involved in stabilizing intersubunit contacts between the receptor monomers (the dimer interface region; 31). The conserved WSXWS motif (YGEFS in GH-R) is partially solvent-exposed and is located away from all binding interfaces; this motif may contribute to the stability of the second subdomain. In addition, the YGEFS motif is part of a unique pattern of alternating polar and charged surface residues, which could provide an interface for the binding of accessory proteins (31).

C. EPO- and EPOR-Binding Characteristics

Equilibrium-binding experiments using iodinated EPO on erythroid progenitor cells, erythroleukemia cell lines, or transfected cell lines stably expressing the EPOR cDNA, demonstrate that the EPOR is expressed at low levels on the cell surface (between 100 and 2000 sites per cell; 12,107,123). Some studies have reported a single affinity ($K_d \sim 200$–800 pM) for iodinated EPO (79,85,123), whereas others have reported two ligand-binding affinities ($K_d \sim 30$–90 pM and 200–600 pM) of the receptor (23,107,108). Recently, Nagao and co-workers (92) described the conversion of the two receptor affinities into one affinity in the presence of the glycosylation inhibitor tunicamycin. As the binding experiments cited here have been performed in different cell types and with different preparations of EPO, the discrepancy over the number of receptor affinities remains unresolved, and further experiments are necessary to clarify this issue.

D. Metabolism of the EPOR

The biosynthesis of the EPOR has been studied in a variety of hematopoietic cell lines stably transfected with the EPOR cDNA, as well as in cells that synthesize endogenous receptors. The receptor is synthesized primarily as a 64-kDa species, containing one N-linked oligosaccharide, sensitive to endoglycosidase H (endo H); a minor amount of the newly synthesized receptor (less than 10%) remains unglycosylated after synthesis. The endo H-sensitive form of the receptor is poorly processed to an endo H-resistant form (93,140). The extent of processing varies in different cell lines tested, but it is rarely more than 50%, indicating that a large portion of the newly made receptors are retained in the endoplasmic reticulum (ER). Similarly, endogenous receptors, synthesized in murine fetal liver cells or HCD57 cells, are poorly processed to endo H-resistant forms (132,138). Newly made EPORs are found in a complex with the ER chaperone protein GRP78/BiP, indicating that a portion of the receptors may be slowly or improperly folded (91). The fraction of Golgi-processed EPORs that are transported to the cell surface is unknown; however, under steady-state conditions in BA/F3 cells expressing the EPOR, less than 1% of the total cell-associated receptors are on the surface, as judged by proteinase K treatment of intact cells (140).

The EPOR may be poorly processed as a consequence of inefficient folding and transport along the secretory pathway. Alternatively, it may require the presence of another molecule, perhaps another subunit with which it oligomerizes, to be transported. Once the receptor is at the cell surface, it is rapidly internalized in the presence of ligand (25,107).

E. Cell Surface Organization of the EPOR

Erythroid precursor cells and murine erythroleukemic (MEL) cell lines express specific-binding sites for iodinated EPO (78,79,89,107,108). Binding and cross-linking studies, using iodinated EPO, have been performed on a variety of cell types to determine the molecular structure of the EPOR. Two polypeptides of about 85 kDa and 100 kDa were identified on FVA cells, and limited peptide mapping suggested that they are related (105,108). Conflicting studies have reported that the polypeptides that cross-link to iodinated EPO are or are not immunoreactive with anti-EPOR serum (80,85), thus the nature of these species remains unclear. In addition, another polypeptide of about 95 kDa was detected by binding and cross-linking experiments; this component appears to be specific to hematopoietic cells and is immunologically unrelated to the cloned EPOR (85). Unlike many other members of the cytokine receptor family, a second subunit for the EPOR has not yet been identified, although some indirect evidence suggests there may be a second component of the activated receptor–ligand complex (16–33).

Purification of the cell surface form of the receptor, using biotinylated EPO, detected a species of about 72 kDa, indicating that this form of the receptor is more extensively processed than the Golgi form (141). Studies in EPO-deprived murine erythroleukemia cells (HCD57 cells) demonstrated that a higher molecular weight form of the EPOR (~78 kDa) was down-regulated by EPO addition (106). Because the appearance of the higher molecular weight form correlated with an increase in the number of surface-binding sites, and its disappearance correlated with binding and internalization of EPO, it is likely to be the cell surface form of the receptor (106). The increased molecular weight has been attributed to tyrosine phosphorylation or to increased glycosylation of the receptor (84,106,141). More work is necessary to delineate the specific modifications and molecular structure of the cell surface form of the receptor.

VII. EPOR STRUCTURE–FUNCTION STUDIES

A. Conserved Exoplasmic Domain Motifs

The mature (signal sequence-cleaved) murine EPOR contains 225 amino acids in the extracellular domain, including five cysteine residues at positions 28, 38, 66, 82, and 180 (23). The first four cysteine residues are highly conserved among members of the cytokine receptor family; by analogy with the disulfide-bonding pattern of the human growth hormone receptor (GH-R), it is likely that Cys-28 and Cys-38 are disulfide-bonded, as are Cys-66 and Cys-82 (44). The remaining cysteine at position 180 appears to be present in the reduced state under normal conditions (134).

The WSXWS motif is strikingly conserved among members of the cytokine receptor family, yet its functional role has been difficult to determine. Mutagenesis studies have suggested that these residues play a role in distinct processes, such as ligand binding, receptor internalization, receptor folding and transport, and signal transduction (87,101,143). In an attempt to resolve this issue, saturation mutagenesis was performed on the EPOR WSAWS motif (55). Each residue in the motif was changed, individually, to all other possible amino acids. The positions held by the tryptophan residues tolerated substitution poorly; only receptors with phenylalanine or tyrosine substitutions were expressed at the cell surface (at levels 10- to 100-fold less than the wild-type receptor). A slightly higher number of substitutions were tolerated in the positions normally held by serine residues, whereas the middle position, alanine in the EPOR, tolerated the greatest number of substitutions. The mutants could be separated into two classes on the basis of their ability to stimulate cell proliferation. The proliferation-inactive mutants were retained in the ER, as judged by sensitivity to endo H, and had no detectable surface expression. Interestingly, all mutants that were active in proliferation assays bound EPO with an affinity similar to that of the wild-type receptor, although the surface expression of these mutants varied. Thus, the motif appears to play an important structural role, specifically contributing to efficient folding and surface transport of the receptor. These data also indicate that the WSXWS motif is not directly involved in ligand binding, receptor internalization, or signal transduction (55).

B. An Activating Mutation in the EPOR

A constitutively active (ligand-independent) EPOR was isolated by a retroviral transduction system (142). This mutant has provided insight into the signal transductory mechanism that the wild-type EPOR is likely to use. The constitutively active receptor (R129C) has a single amino acid substitution in the exoplasmic domain, converting an arginine residue at position 129 to a cysteine (142). Mutation of residue 129 indicated that the presence of cysteine, and not the loss of arginine, is required for constitutive activity. The R129C mutant forms disulfide-linked homodimers shortly after synthesis, in the absence of ligand, which are transported to and expressed on the cell surface. These disulfide-linked dimers display a single affinity for iodinated EPO, similar to that of the wild-type receptor (134). The R129C receptor is able to support both the proliferation and differentiation of isolated erythroid progenitor cells in vitro, in the absence of EPO, indicating that it signals in a manner similar to that of the ligand-occupied wild-type receptor (99). Together, these results, along with the ligand-induced homodimeric structure of the GH-R, suggested that ligand-induced receptor homodimerization may be an initial step in EPOR activation (134).

C. Dimerization and Activation of the EPOR

To test the role of dimerization in EPOR activation, a series of amino acid substitutions were made in the region of the EPOR that aligned with the GH-R dimer interface. New constitutively active receptors were generated by substitution of the glutamate residues at position 132 or position 133 with cysteine. These mutants formed disulfide-linked dimers in the absence of ligand, similar to R129C. Other cysteine substitutions in the predicted dimer interface region (at positions R130 or A131) had no effect on the activity of the receptor; these mutants were strictly EPO-dependent and did not form disulfide-linked dimers (133). The absolute correlation between constitutive activation and disulfide-linked dimerization of the EPOR strongly suggests that the wild-type receptor is activated by homodimerization. In addition, these results demonstrate that residues 129, 132, and 133 of the EPOR are in proximity during receptor synthesis and folding, and are likely to form a portion of the wild-type EPOR dimer interface (133).

A further indication that receptor homo-oligomerization is an initial step in EPO-mediated signal transduction was demonstrated by the identification of dominant inhibitory mutants of the EPOR. Cytoplasmic truncation mutants, deleted for all or part of their cytoplasmic domain and that are inactive on their own, were coexpressed with the wild-type receptor in EPO-dependent hematopoietic cells. Overexpression of these mutants inhibited EPO-stimulated cell proliferation, suggesting that ligand-induced heterodimerization occurred between wild-type and mutant EPORs on the cell surface, inhibiting activation of the intracellular-signaling cascade (2,133).

D. EPOR Cytoplasmic Domain

The cytoplasmic domain of the murine EPOR contains 236 amino acids. Analysis of cytoplasmic domain mutants has suggested regions that may be involved in specific receptor-mediated functions. A hypersensitive form of the receptor, isolated by retroviral transduction, is missing the COOH-terminal 42 amino acids. This mutant receptor enabled cells to grow in one-tenth the normal concentration of EPO, without altering receptor affinity for ligand or receptor cell surface numbers (25,142). Similarly, a deletion mutant of the EPOR, missing 91 amino acids of the cytoplasmic domain, is hypersensitive to growth in EPO (25). Thus, the deleted portion, spanning residues 382–483 of the mature EPOR sequence, defines a negative-regulatory domain (25,142; see Fig. 2). This region of the receptor may be part of the binding site for a negative-regulatory molecule, such as the hematopoietic cell phosphatase (HCP; 139; Fig. 3).

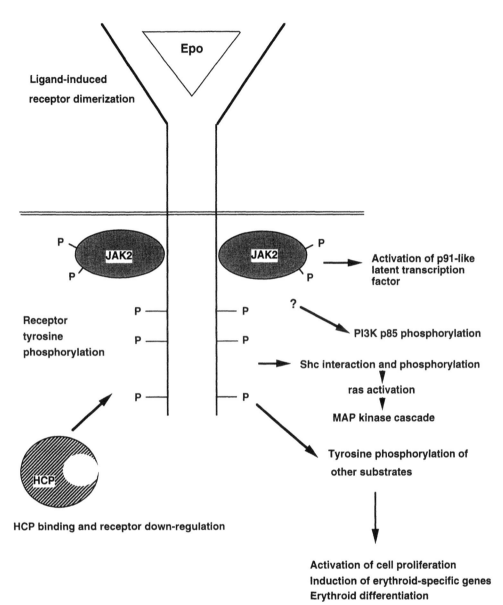

Figure 3 A model of the EPOR-signaling complex. Ligand-induced receptor dimerization is an initial step in EPOR activation. Dimerization of the receptor stimulates protein tyrosine kinase activity, including the associated JAK2 kinase. Tyrosine phosphorylation (P) of the receptor and substrate molecules then occurs. Phosphorylation of the receptor stimulates interactions between the receptor and SH2-containing effector molecules, including P13K and the adapter protein Shc, which may link the activated receptor to $p21^{ras}$ and the MAP kinase pathway. The activated JAK2 kinase may directly phosphorylate latent cytoplasmically localized transcription factors (e.g., STAT5), resulting in their translocation to the nucleus and activation of specific genes. Other molecules may also be involved in eliciting proliferation- and differentiation-specific signals. Receptor inactivation may occur following receptor tyrosine phosphorylation, due to an interaction between the receptor and the SH2-domain containing HCP, leading to receptor dephosphorylation.

An EPOR deletion mutant missing 133 amino acids from the COOH-terminus (truncated at residue 350) retained the ability to stimulate EPO-mediated proliferation. Deletion of 221 residues from the COOH-terminus (truncated at residue 262) abrogated EPO-responsive proliferation, thereby defining a region between amino acids 262 and 350 of the mature EPOR that is essential for EPO-mediated proliferation (25). Other studies have confirmed that the residues in this region are essential for EPOR function in proliferation assays (84,102). A mutant EPOR containing substitution of a highly conserved tryptophan residue in this region (residue 282) to arginine, is also nonfunctional in cell proliferation assays (83). Thus, the membrane-proximal domain of the EPOR may contain binding sites for positive-regulatory molecules (see Figs. 2 and 3).

VIII. EPOR SIGNAL TRANSDUCTION

The cytoplasmic domain of the EPOR does not contain kinase-related sequences, yet tyrosine phosphorylation of cellular proteins, including the receptor, occurs on EPO binding (26,35,71,84). Tyrosine phosphorylation appears to be necessary for EPO-mediated proliferation as mutant EPORs, which are inactive in proliferation assays, are also unable to mediate tyrosine phosphorylation of cellular substrates (84,102). Recent reports have demonstrated a physical association between the JAK2 kinase and the EPOR. The unoccupied receptor is associated with JAK2, however EPO binding stimulated tyrosine phosphorylation of JAK2 and its autophosphorylation activity (137). The membrane-proximal subdomain of the EPOR, which is required for cell proliferation, appears to mediate the association with JAK2 (84,137). These results suggest that JAK2 may couple EPO–EPOR binding to tyrosine phosphorylation and cell proliferation, possibly by direct phosphorylation and activation of latent transcription factors, similar to the role of related kinases in the signaling pathway of interferons (see Fig. 3).

The EPOR also appears to associate with the p85 subunit of phosphatidylinositol 3-kinase (PI3K; 28,53). There are conflicting reports over whether the association between PI3K p85 and the EPOR is independent or dependent on ligand binding and whether phosphorylation of PI3K p85 is stimulated by EPO (28,53). However, different cell types were used in these studies, which may explain why different results were obtained. PI3K p85 subunit appears to interact with the EPOR through its SH2 domains (28). One potential site of interaction is in a membrane-proximal region of the receptor that includes tyrosine-343 (53); sequences in the membrane-distal region may also mediate PI3K association.

Several components of the *ras*-signaling pathway are activated by EPO, suggesting that this pathway may also be involved in EPOR-mediated signaling (see Fig. 3). EPO stimulates an increase in the activity of $p21^{ras}$, tyrosine phosphorylation of p120GAP, and inhibition of GAP activity (125). Tyrosine phosphorylation of the adapter protein $p52^{shc}$ is stimulated by EPO and the phosphorylated protein appears to interact with the EPOR in some cell types (27). Shc may function to couple tyrosine kinases to downstream effectors in the *ras* pathway, such as Grb2 and the *ras* guanine nucleotide exchange factor *sos* (21,74,97,103,115). EPO also stimulates the tyrosine and serine phosphorylation of *Raf1* and its in vitro kinase activity, and inhibition of *Raf1* expression by antisense oligonucleotides inhibited EPO-stimulated proliferation (14).

Several other genes also appear to be responsive to EPO, including c-*myc* and c-*myb* (15,98,116,117). Expression of these genes has been linked to erythroid differentiation, although the exact role they play in mediating this process is as yet unclear (15,18,88,124). Expression of erythroid lineage-specific genes are also stimulated by EPO, including α- and β-globin and *GATA-1* (13,16). In addition, the protooncogene c-*fps/fes* is phosphorylated after EPO stimulation and has an increased in vitro kinase activity (52). Much current work is aimed at dissecting steps in the signal transduction pathway of EPO and the EPOR to understand the role these various gene products play in controlling erythroid proliferation and differentiation.

IX. EPOR AND DISEASE

A. Murine Erythroleukemias

Two types of virally mediated erythroleukemia have been associated with the murine EPOR. One of these, Friend disease, is characterized by an initial phase of polyclonal erythroblastosis in mice, followed by the development of acute leukemia (7). The Friend virus complex is composed of both a replication-competent helper component, Friend murine leukemia virus, and a replication-defective pathogenic component, spleen focus-forming virus (SFFV). The envelope gene encoded by SFFV, *gp55,* is a recombinant–deletion membrane glycoprotein that is directly involved in leukemogenesis; no other oncogenic sequences have been found in the SFFV genome. The gp55 protein binds to and stimulates the proliferative activity of the EPOR in the absence of ligand (69). This association between the EPOR and gp55 is likely to be the principal cause of the initial polyclonal erythroblastosis of Friend disease. Subsequent genetic alterations in the proliferating erythroid precursors generate leukemic clones that are prominent in the terminal phases of the disease (7).

Introduction of the constitutively activated (R129C) EPOR into mice, using a viral vector in which the envelope-coding sequences of SFFV were replaced with the R129C-coding sequences, induced erythrocytosis and leukemia in adult mice (72). Similar to Friend disease, the initial stage of this disease was characterized by erythrocytosis, primarily owing to the constitutive activation of the ectopically expressed EPOR. Genetic mutations in the proliferating precursors gave rise to leukemic clonal cell lines (72).

B. Human Diseases Associated with the EPOR

Study of the murine receptor has revealed that it can be activated for proliferation by point mutation, partial deletion, or by interaction with the gp55 protein of SFFV. These activation mechanisms can lead to disease in mice, suggesting that mutations in the human receptor may be involved in some human hematopoietic disorders. Two human erythroleukemic cell lines, UT7 and TF-1, display an elevated number of EPO-binding sites and have rearrangements and amplification of the EPOR gene, possibly implicating receptor mutations in the development of erythroleukemia (17,130).

Some members of a large Finnish family have a benign erythrocytosis disorder, which is transmitted in an autosomal dominant inheritance pattern. Cultured bone marrow erythroid progenitors from affected family members are hypersensitive to EPO, suggesting that a mutation in the EPO-signaling pathway may contribute to the observed erythrocytosis. Sequence analysis of genomic DNA isolated from affected individuals demonstrated that they carry a mutation in one allele of the EPOR that introduces a premature stop codon in the cytoplasmic tail of the receptor, deleting the last 70 amino acids (29,30). Thus, it appears that the negative-regulatory domain of the EPOR, mapped by in vitro deletional mutagenesis, may play an important regulatory role in vivo, as mutation of this region contributes to a mild, dominant, pathological condition (30). It will be of interest to determine if the dominant phenotype is also observed in in vitro cultured cells, coexpressing the hypersensitive EPOR and the wild-type receptor. Such a system may help elucidate the biochemical nature of the hypersensitive phenotype.

X. SUMMARY

Molecular analyses of EPO and the EPOR have led to many advances in understanding the regulation, structure, and function of these proteins. Further work will be aimed at characterizing the cellular pathways that activate EPO transcription and expression and the pathways that are involved in EPOR expression during erythroid differentiation. In addition, future studies will likely clarify the biochemical events that occur following ligand binding, and their role in mediating erythroid cell proliferation and differentiation. The identification of human and murine diseases linked to the EPOR indicates that a complete understanding of EPO signal transduction may help to elucidate potential modes of therapy.

NOTE ADDED IN PROOF

Since the submission of this manuscript, a number of important advances have been made with regard to EPO and EPOR. Most significant are the generation and careful analysis of genetically engineered mice with homozygous deletions in the EPO or EPOR genes (147,148) and the determination of the three-dimensional structure of human EPOR extracellular region complexed to an agonist peptide that is unrelated to EPO (149). For a current review containing information on EPOR signal transduction, see the work of Ihle (150).

ACKNOWLEDGMENTS

I would like to thank Harvey F. Lodish and Gregory D. Longmore for advise and many helpful discussions. I also thank Hong Wu and Stanley J. Watowich for their critical review of the manuscript and for discussions.

REFERENCES

1. Anagnostou A, Lee ES, Kessimian N, Levinson R, Steiner M. Erythropoietin has a mitogenic and positive chemotactic effect on endothelial cells. Proc Natl Acad Sci USA 1990; 87:5978–5982.
2. Barber DL, DeMartino JC, Showers MO, D'Andrea AD. A dominant negative erythropoietin (EPO) receptor inhibits EPO-dependent growth and blocks F-gp55-dependent transformation. Mol Cell Biol 1994; 14:2257–2265.
3. Bazan JF. Haemopoietic receptors and helical cytokines. Immunol Today 1990; 11:350–354.
4. Bazan JF. Structural design and molecular evolution of a cytokine receptor superfamily. Proc Natl Acad Sci USA 1990; 87:6934–6938.
5. Beck I, Ramirez S, Weinmann R, Caro J. Enhancer element at the 3'-flanking region controls transcriptional response to hypoxia in the human erythropoietin gene. J Biol Chem 1991; 266:15563–15566.
6. Beck I, Weinmann R, Caro J. Characterization of hypoxia-responsive enhancer in the human erythropoietin gene shows presence of hypoxia-inducible 120-Kd nuclear DNA-binding protein in erythropoietin-producing and nonproducing cells. Blood 1993; 82:704–711.
7. Ben-David Y, Bernstein A. Friend virus-induced erythroleukemia and the multistage nature of cancer. Cell 1991; 66:831–834.
8. Beru N, McDonald N, Lacombe C, Goldwasser E. Expression of the erythropoietin gene. Mol Cell Biol 1986; 6:2571–2575.
9. Blanchard KL, Acquaviva AM, Galson L, Bunn HF. Hypoxic induction of the human erythropoietin gene: cooperation between the promoter and enhancer, each of which contains steroid receptor response elements. Mol Cell Biol 1992; 12:5373–5385.
10. Boisse JP, Lee WR, Presnell SR, Cohen FE, Bunn HF. Erythropoietin structure–function relationships. Mutant proteins that test a model of tertiary structure. J Biol Chem 1993; 268:15983–15993.
11. Bondurant MC, Koury MJ. Anemia induces accumulation of erythropoietin mRNA in the kidney and liver. Mol Cell Biol 1986; 6:2731–2733.
12. Broudy VC, Lin N, Brice M, Nakamoto B. Papayannopoulou T. Erythropoietin receptor characteristics on primary human erythroid cells. Blood 1991; 77:2583–2590.

13. Busfield SJ, Klinken SP. Erythropoietin-induced stimulation of differentiation and proliferation in J2E cells is not mimicked by chemical induction. Blood 1992; 80:412–419.

14. Carroll MP, Spivak JL, McMahon M, Weich N, Rapp UR, May WS. Erythropoietin induces Raf-1 activation and Raf-1 is required for erythropoietin-mediated proliferation. J Biol Chem 1991; 266:14964–14969.

15. Chern Y, Spangler R, Choi H-S, Sytkowski AJ. Erythropoietin activates the receptor in both Rauscher and Friend murine erythroleukemia cells. J Biol Chem 1991; 266:2009–2012.

16. Chiba T, Nagata Y, Kishi A, Sakamaki K, Miyajima A, Yamamoto M, Engel JD, Todokoro K. Induction of erythroid-specific gene expression in lymphoid cells. Proc Natl Acad Sci USA 1993; 90:11593–11597.

17. Chretien S, Moreau-Gachelin F, Apiou F, Courtois G, Mayeux P, Dutrillaux B, Cartron JP, Gisselbrecht S, Lacombe C. Putative oncogenic role of the erythropoietin receptor in murine and human erythroleukemia cells. Blood 1994; 83:1813–1821.

18. Clarke MF, Kukowska-Latallo JF, Westin E, Smith M, Prochownik EV. Constitutive expression of a c-*myb* cDNA blocks Friend murine erythroleukemia cell differentiation. Mol Cell Biol 1988; 8:884–892.

19. Cosman D, Lyman SD, Idzerda L, Beckmann MP, Park LS, Goodwin RG, March CJ. A new cytokine receptor superfamily. Trends Biochem Sci 1990; 15:265–270.

20. Cunningham, BC, Ultsch M, De Vos AM, Mulkerrin MG, Clauser KR, Wells JA. Dimerization of the extracellular domain of the human growth hormone receptor by a single hormone molecule. Science 1991; 254:821–825.

21. Cutler RL, Liu L, Damen JE, Krystal G. Multiple cytokines induce the tyrosine phosphorylation of Shc and its association with Grb2 in hemopoietic cells. J Biol Chem 1993; 268:21463–21465.

22. D'Andrea AD, Fasman GD, Lodish HF. Erythropoietin receptor and interleukin-2 receptor beta chain: a new receptor family. Cell 1989; 58:1023–1024.

23. D'Andrea AD, Lodish HF, Wong GG. Expression cloning of the murine erythropoietin receptor. Cell 1989; 57:277–285.

24. D'Andrea AD, Szklut PJ, Lodish HF, Alderman EM. Inhibition of receptor binding and neutralization of bioactivity of antierythropoietin monoclonal antibodies. Blood 1990; 75:874–880.

25. D'Andrea AD, Yoshimura A, Youssoufian H, Zon LI, Koo JW, Lodish HF. The cytoplasmic region of the erythropoietin receptor contains nonoverlapping positive and negative growth-regulatory domains. Mol Cell Biol 1991; 11:1980–1987.

26. Damen J, Mui AL, Hughes P, Humphries K, Krystal G. Erythropoietin-induced tyrosine phosphorylations in a high erythropoietin receptor-expressing lymphoid cell line. Blood 1992; 80:1923–1932.

27. Damen JE, Liu L, Cutler RL, Krystal G. Erythropoietin stimulates the tyrosine phosphorylation of Shc and its association with Grb2 and a 145-Kd tyrosine phosphorylated protein. Blood 1993; 82:2296–2303.

28. Damen JE, Mui AL, Puil L, Pawson T, Krystal G. Phosphatidylinositol 3-kinase associates, via its Src homology 2 domains, with the activated erythropoietin receptor. Blood 1993; 81:3204–3210.

29. de la Chapelle A, Sistonen P, Lehvaslaiho H, Ikkala E, Juvonen E. Familial erythrocytosis genetically linked to erythropoietin receptor gene. Lancet 1993; 341:82–84.

30. de la Chapelle A, Traskelin AL, Juvonen E. Truncated erythropoietin receptor causes dominantly inherited benign human erythrocytosis. Proc Natl Acad Sci USA 1993; 90:4495–4499.

31. De Vos AM, Ultsch M, Kossiakoff AA. Human growth hormone and extracellular domain of its receptor: crystal structure of the complex. Science 1992; 255:306–311.

32. Delorme E, Lorenzini T, Giffin J, Martin F, Jacobsen F, Boone T, Elliott S. Role of glycosylation on the secretion and biological activity of erythropoietin. Biochemistry 1992; 31:9871–9876.

33. Dong YJ, Goldwasser E. Evidence for an accessory component that increases the affinity of the erythropoietin receptor. Exp Hematol 1993; 21:483–486.

34. Dube S, Fisher JW, Powell JS. Glycosylation at specific sites of erythropoietin is essential for biosynthesis, secretion, and biological function. J Biol Chem 1988; 263:17516–17521.

35. Dusanter-Fourt I, Casadevall N, Lacombe C, Muller O, Billat C, Fischer S, Mayeux P. Erythropoietin

induces the tyrosine phosphorylation of its own receptor in human erythropoietin-responsive cells. J Biol Chem 1992; 267:10670–10675.

36. Emerson SG, Sieff CA, Wang EA, Wong GG, Clark SC, Nathan DG. Purification of fetal hemato-poietic progenitors and demonstration of recombinant multipotential colony-stimulating activity. J Clin Invest 1985; 76:1286–1290.

37. Erslev A. Humoral regulation of red cell production. Blood 1953; 8:349–357.

38. Eschbach JW, Kelly MR, Haley NR, Abels RI, Adamson JW. Treatment of the anemia of progressive renal failure with recombinant human erythropoietin. N Engl J Med 1989; 321:158–163.

39. Evatt BL, Spivak JL, Levin J. Relationships between thrombopoiesis and erythropoiesis: with studies of the effects of preparations of thrombopoietin and erythropoietin. Blood 1976; 48:547–558.

40. Fandrey J, Bunn HF. In vivo and in vitro regulation of erythropoietin mRNA: measurement by competitive polymerase chain reaction. Blood 1993; 81:617–623.

41. Fibi MR, Aslan M, Hintz-Obertreis P, Pauly JU, Gerken M, Luben G, Lauffer L, Siebold B, Stuber W, Nau G, Zettlmeiss G, Harthus H-P. Human erythropoietin-specific sites of monoclonal antibody-mediated neutralization. Blood 1993; 81:670–675.

42. Fischl M, Galpin JE, Levine JD, Groopman JE, Henry DH, Kennedy P, Miles S, Robbins W, Starrett B, Zalusky R, Abels RI, Tsai HC, Rudnick SA. Recombinant human erythropoietin for patients with AIDS treated with zidovudine [see coments]. N Engl J Med 1990; 322:1488–1493.

43. Fraser JK, Tan AS, Lin FK, Berridge MV. Expression of specific high-affinity binding sites for erythropoietin on rat and mouse megakaryocytes. Exp Hematol 1989; 17:10–16.

44. Fuh G, Mulkerrin MG, Bass S, McFarland N, Brochier M, Bourell JH, Light DR, Wells JA. The human growth hormone receptor. Secretion from *Escherichia coli* and disulfide bonding pattern of the extracellular binding domain. J Biol Chem 1990; 265:3111–3115.

45. Fukuda MN, Sasaki H, Lopez L, Fukuda M. Survival of recombinant erythropoietin in the circulation: the role of carbohydrates. Blood 1989; 73:84–89.

46. Galson DL, Tan CC, Ratcliffe PJ, Bunn HF. Comparison of the human and mouse erythropoietin genes shows extensive homology in the flanking regions. Blood 1993; 82:3321–3326.

47. Goldberg MA, Dunning SP, Bunn HF. Regulation of the erythropoietin gene: evidence that the oxygen sensor is a heme protein. Science 1988; 242:1412–1415.

48. Goldberg MA, Gaut CC, Bunn HF. Erythropoietin mRNA levels are governed by both the rate of gene transcription and posttranscriptional events. Blood 1991; 77:271–277.

49. Goodnough LT, Rudnick S, Price TH, Ballas SK, Collins ML, Crowley JP, Kosmin M, Kruskall MS, Lenes BA, Menitove JE, Silberstein LE, Smith KJ, Wallas CH, Abels R. Von Tress M. Increased preoperative collection of autologous blood with recombinant human erythropoietin therapy. N Engl J Med 1989; 321:1163–1168.

50. Gregory CJ, Eaves AC. Human marrow cells capable of erythropoietic differentiation in vitro: definition of three erythroid colony responses. Blood 1977; 49:855–864.

51. Gregory CJ, Eaves AC. Three stages of erythropoietic progenitor cell differentiation distinguished by a number of physical and biologic properties. Blood 1978; 51:527–537.

52. Hanazono Y, Chiba S, Sasaki K, Mano H, Yazaki Y, Hirai H. Erythropoietin induces tyrosine phosphorylation and kinase activity of the c-*fps/fes* proto-oncogene product in human erythropoie-tin-responsive cells. Blood 1993; 81:3193–3196.

53. He TC, Zhuang H, Jiang N, Waterfield MD, Wojchowski DM. Association of the p85 regulatory subunit of phosphatidylinositol 3-kinase with an essential erythropoietin receptor subdomain. Blood 1993; 82:3530–3538.

54. Higuchi M, Oh-eda M, Kuboniwa H, Tomonoh K, Shimonaka Y, Ochi N. Role of sugar chains in the expression of the biological activity of human erythropoietin. J Biol Chem 1992; 267:7703–7709.

55. Hilton DJ, Watowich SS, Katz L, Lodish HF. Saturation mutagenesis of the WSXWS motif of the erythropoietin receptor. J Biol Chem 1996; 271:4699–4708.

56. Iscove NN, Sieber F, Winterhalter KH. Erythroid colony formation in cultures of mouse and human bone marrow: analysis of the requirement for erythropoietin by gel filtration and affinity chromatog-raphy on agarose–concanavalin A. J Cell Physiol 1974; 83:309–320.

57. Jacobs K, Shoemaker C, Rudersdorf R, Neill SD, Kaufman RJ, Mufson A, Seehra J, Jones SS, Hewick R, Fritsch EF, Kawakita M, Shimizu T, Miyake T. Isolation and characterization of genomic and cDNA clones of human erythropoietin. Nature 1985; 313:806–810.

58. Jacobson LO, Goldwasser E, Fried W, Plzak L. Role of the kidney in erythropoiesis. Nature 1957; 179:633–634.

59. Kelley LL, Koury MJ, Bondurant MC, Koury ST, Sawyer ST, Wickrema A. Survival or death of individual proerythroblasts results from differing erythropoietin sensitivities: a mechanism for controlled rates of erythrocyte production. Blood 1993; 82:2340–2352.

60. Klinken SP, Nicola NA, Johnson GR. In vitro-derived leukemic erythroid cell lines induced by a *raf*- and *myc*-containing retrovirus differentiate in response to erythropoietin. Proc Natl Acad Sci USA 1988; 85:8506–8510.

61. Konishi Y, Chui D-H, Hirose H, Kunishita T, Tabira T. Trophic effect of erythropoietin and other hematopoietic factors on central cholinergic neurons in vitro and in vivo. Brain Res 1993; 609:29–35.

62. Koury MH, Bondurant MC. Erythropoietin retards DNA breakdown and prevents programmed death in erythroid progenitor cells. Science 1990; 248:378–381.

63. Koury MJ, Bondurant MC, Graber SE, Sawyer ST. Erythropoietin messenger RNA levels in developing mice and transfer of ^{125}I-erythropoietin by the placenta. J Clin Invest 1988; 82:154–159.

64. Koury ST, Bondurant MC, Koury MJ. Localization of erythropoietin synthesizing cells in murine kidneys by in situ hybridization. Blood 1988; 71:524–527.

65. Koury ST, Koury MJ, Bondurant MC, Caro J, Graber SE. Quantitation of erythropoietin-producing cells in kidneys of mice by in situ hybridization: correlation with hematocrit, renal erythropoietin mRNA, and serum erythropoietin concentration. Blood 1989; 74:645–651.

66. Krantz SB. Erythropoietin. Blood 1991; 77:419–434.

67. Lacombe C, Da Silva JL, Bruneval P, Fournier JG, Wendling F, Casadevall N, Camilleri JP, Bariety J, Varet B, Tambourin P. Peritubular cells are the site of erythropoietin synthesis in the murine hypoxic kidney. J Clin Invest 1988; 81:620–623.

68. Lai PH, Everett R, Wang FF, Arakawa T, Goldwasser E. Structural characterization of human erythropoietin. J Biol Chem 1986; 261:3116–3121.

69. Li J-P, D'Andrea AD, Lodish HF, Baltimore D. Activation of cell growth by binding of Friend spleen focus-forming virus gp55 glycoprotein to the erythropoietin receptor. Nature 1990; 343:762–764.

70. Lin FK, Suggs S, Lin CH, Browne JK, Smalling R, Egrie JC, Chen KK, Fox GM, Martin F, Stabinsky Z, Badrawi SM, Lai P-H, Goldwasser E. Cloning and expression of the human erythropoietin gene. Proc Natl Acad Sci USA 1985; 82:7580–7584.

71. Linnekin D, Evans GA, D'Andrea A, Farrar WL. Association of the erythropoietin receptor with protein tyrosine kinase activity. Proc Natl Acad Sci USA 1992; 89:6237–6241.

72. Longmore GD, Lodish HF. An activating mutation in the murine erythropoietin receptor induces erythroleukemia in mice: a cytokine receptor superfamily oncogene. Cell 1991; 67:1089–1102.

73. Longmore GD, Pharr P, Neumann D, Lodish HF. Both megakaryocytopoiesis and erythropoiesis are induced in mice infected with a retrovirus expressing an oncogenic erythropoietin receptor. Blood 1993; 82:2386–2395.

74. Lowenstein EJ, Daly RJ, Batzer AG, Li W, Margolis B, Lammers R, Ullrich A, Skolnik EY, Bar-Sagi D, Schlessinger J. The SH2 and SH3 domain-containing protein GRB2 links receptor tyrosine kinases to *ras* signaling. Cell 1992; 70:431–442.

75. Madan A, Curtin PT. A 24-base-pair sequence 3' to the human erythropoietin gene contains a hypoxia-responsive transcriptional enhancer. Proc Natl Acad Sci USA 1993; 90:3928–3932.

76. Masuda S, Nagao M, Takahata K, Konishi Y, Gallyas J, Tabira T, Sasaki R. Functional erythropoietin receptor of the cells with neuronal characteristics. J Biol Chem 1993; 268:11208–11216.

77. Maxwell PH, Pugh CW, Ratcliffe PJ. Inducible operation of the erythropoietin 3' enhancer in multiple cell lines: evidence for a widespread oxygen-sensing mechanism. Proc Natl Acad Sci USA 1993; 90:2423–2427.

78. Mayeux P, Billat C, Jacquot R. The erythropoietin receptor of rat erythroid progenitor cells. J Biol Chem 1987; 262:13986–13990.

79. Mayeux P, Billat C, Jacquot R. Murine erythroleukaemia cells (Friend cells) possess high-affinity binding sites for erythropoietin. FEBS Lett 1987; 211:229–233.

80. Mayeux P, Lacombe C, Casadevall N, Chretien S, Dusanter I, Gisselbrecht S. Structure of the murine erythropoietin receptor complex. Characterization of the erythropoietin cross-linked proteins. J Biol Chem 1991; 266:23380–23385.

81. McLeod DL, Shreeve MM, Axelrad AA. Induction of megakaryocyte colonies with platelet formation in vitro. Nature 1976; 261:492–494.

82. Migliaccio AR, Migliaccio G, D'Andrea AD, Baiocchi M, Crotta S, Nicolis S, Ottolenghi S, Adamson JW. Response to erythropoietin in erythroid subclones of the factor-dependent cell line 32D is determined by translocation of the erythropoietin receptor to the cell surface. Proc Natl Acad Sci USA 1991; 88:11086–11090.

83. Miura O, Cleveland JL, Ihle JN. Inactivation of erythropoietin receptor function by point mutations in a region having homology with other cytokine receptors. Mol Cell Biol 1993; 13:1788–1795.

84. Miura O, D'Andrea AD, Kabat D, Ihle JN. Induction of tyrosine phosphorylation by the erythropoietin receptor correlates with mitogenesis. Mol Cell Biol 1991; 11:4895–4902.

85. Miura O, Ihle JN. Subunit structure of the erythropoietin receptor analyzed by ^{125}I-Epo cross-linking in cells expressing wild-type or mutant receptors. Blood 1993; 81:1739–1744.

86. Miyake T, Kung CK, Goldwasser E. Purification of human erythropoietin. J Biol Chem 1977; 252:5558–5564.

87. Miyazaki T, Maruyama M, Yamada G, Hatakeyama M, Taniguchi T. The integrity of the conserved "WS motif" common to IL-2 and other cytokine receptors is essential for ligand binding and signal transduction. EMBO J 1991; 10:3191–3197.

88. Mucenski ML, McLain K, Kier AB, Swerdlow SH, Schreiner CM, Miller TA, Pietryga DW, Scott J, Potter SS. A functional c-*myb* gene is required for normal murine fetal hepatic hematopoiesis. Cell 1991; 65:677–689.

89. Mufson RA, Gesner TG. Binding and internalization of recombinant human erythropoietin in murine erythroid precursor cells. Blood 1987; 69:1485–1490.

90. Murakami M, Narazaki M, Hibi M, Yawata H, Yasukawa K, Hamaguchi M, Taga T, Kishimoto T. Critical cytoplasmic region of the interleukin 6 signal transducer gp130 is conserved in the cytokine receptor family. Proc Natl Acad Sci USA 1991; 88:11349–11353.

91. Murray PJ, Watowich SS, Lodish HF, Young RA, Hilton DJ. Epitope tagging of the human endoplasmic reticulum HSP70 protein, BiP, to facilitate analysis of BiP-substrate interactions. Anal Biochem 1995; 229:170–179.

92. Nagao M, Matsumoto S, Masuda S, Sasaki R. Effect of tunicamycin treatment on ligand binding to the erythropoietin receptor: conversion from two classes of binding sites to a single class. Blood 1993; 81:2503–2510.

93. Neumann D, Wikstrom L, Watowich SS, Lodish HF. Intermediates in degradation of the erythropoietin receptor accumulate and are degraded in lysosomes. J Biol Chem 1993; 268:13639–13649.

94. Nielsen OJ, Schuster SJ, Kaufman R, Erslev AJ, Caro J. Regulation of erythropoietin production in a human hepatoblastoma cell line. Blood 1987; 70:1904–1909.

95. Noguchi T, Fukumoto H, Mishina Y, Obinata M. Differentiation of erythroid progenitor (CFU-E) cells from mouse fetal liver cells and murine erythroleukemia (TSA8) cells without proliferation. Moll Cell Biol 1988; 8:2604–2609.

96. Ogawa M. Differentiation and proliferation of hematopoietic stem cells. Blood 1993; 81:2844–2853.

97. Olivier JP, Raabe T, Henkemeyer M, Dickson B, Mbamalu G, Margolis B, Schlessinger J, Hafen E, Pawson T. A *Drosophila* SH2-SH3 adaptor protein implicated in coupling the sevenless tyrosine kinase to an activator of *Ras* guanine nucleotide exchange, Sos. Cell 1993; 73:179–191.

98. Patel HR, Choi HS, Sytkowski AJ. Activation of two discrete signaling pathways by erythropoietin. J Biol Chem 1992; 267:21300–21302.

99. Pharr PN, Hankins D, Hofbauer A, Lodish HF, Longmore GD. Expression of a constitutively active

erythropoietin receptor in primary hematopoietic progenitors abrogates erythropoietin dependence and enhances erythroid colony-forming unit, erythroid burst-forming unit, and granulocyte/macrophage progenitor growth. Proc Natl Acad Sci USA 1993; 90:938–942.

100. Pugh CW, Tan CC, Jones RW, Ratcliffe PJ. Functional analysis of an oxygen-regulated transcriptional enhancer lying 3′ to the mouse erythropoietin gene. Proc Natl Acad Sci USA 1991; 88:10553–10557.

101. Quelle DE, Quelle FW, Wojchowski DM. Mutations in the WSAWSE and cytosolic domains of the erythropoietin receptor affect signal transduction and ligand binding and internalization. Mol Cell Biol 1992; 12:4553–4561.

102. Quelle DE, Wojchowski DM. Localized cytosolic domains of the erythropoietin receptor regulate growth signaling and down-modulate responsiveness to granulocyte–macrophage colony-stimulating factor. Proc Natl Acad Sci USA 1991; 88:4801–4805.

103. Rozakis-Adcock M, McGlade J, Mbamalu G, Pelicci G, Daly R, Li W, Batzer A, Thomas S, Brugge J, Pelicci PG, Schlessinger J, Pawson T. Association of the Shc and Grb/2Sem5 SH2-containing proteins is implicated in activation of the Ras pathway by tyrosine kinases. Nature 1992; 360:689–692.

104. Sawada K, Krantz SB, Dai CH, Koury ST, Horn ST, Glick AD, Civin CI. Purification of human blood burst-forming units–erythroid and demonstration of the evolution of erythropoietin receptors. J Cell Physiol 1990; 142:219–230.

105. Sawyer ST. The two proteins of the erythropoietin receptor are structurally similar. J Biol Chem 1989; 264:13343–13347.

106. Sawyer ST, Hankins WD. The functional form of the erythropoietin receptor is a 78-kDa protein: correlation with cell surface expression, endocytosis, and phosphorylation. Proc Natl Acad Sci USA 1993; 90:6849–6853.

107. Sawyer ST, Krantz SB, Goldwasser E. Binding and receptor-mediated endocytosis of erythropoietin in Friend virus-infected erythroid cells. J Biol Chem 1987; 262:5554–5562.

108. Sawyer ST, Krantz SB, Luna J. Identification of the receptor for erythropoietin by cross-linking to Friend virus-infected erythroid cells. Proc Natl Acad Sci USA 1987; 84:3690–3694.

109. Schuster SJ, Badiavas EV, Costa-Giomi P, Weinmann R, Erslev AJ, Caro J. Stimulation of erythropoietin gene transcription during hypoxia and cobalt exposure. Blood 1989; 73:13–16.

110. Schuster SJ, Wilson JH, Erslev AJ, Caro J. Physiologic regulation and tissue localization of renal erythropoietin messenger RNA. Blood 1987; 70:316–318.

111. Semenza GL, Dureza RC, Traystman MD, Gearhart JD, Antonarakis SE. Human erythropoietin gene expression in transgenic mice: multiple transcription initiation sites and cis-acting regulatory elements. Mol Cell Biol 1990; 10:930–938.

112. Semenza GL, Nejfelt MK, Chi SM, Antonarakis SE. Hypoxia-inducible nuclear factors bind to an enhancer element located 3′ to the human erythropoietin gene. Proc Natl Acad Sci USA 1991; 88:5680–5684.

113. Semenza GL, Traystman MD, Gearhart JD, Antonarakis SE. Polycythemia in transgenic mice expressing the human erythropoietin gene. Proc Natl Acad Sci USA 1989; 86:2301–2305.

114. Semenza GL, Wang GL. A nuclear factor induced by hypoxia via de novo protein synthesis binds to the human erythropoietin gene enhancer at a site required for transcriptional activation. Mol Cell Biol 1992; 12:5447–5454.

115. Simon MA, Dodson GS, Rubin GM. An SH3-SH2-SH3 protein is required for p21Ras1 activation and binds to sevenless and Sos proteins in vitro. Cell 1993; 73:169–177.

116. Spangler R, Bailey SC, Sytkowski AJ. Erythropoietin increases c-*myc* mRNA by a protein kinase C-dependent pathway. J Biol Chem 1991; 266:681–684.

117. Spangler R, Sytkowski AJ. c-*myc* is an erythropoietin early response gene in normal erythroid cells: evidence for a protein kinase C-mediated signal. Blood 1992; 79:52–57.

118. Spivak JL, Pham T, Isaacs M, Hankins WD. Erythropoietin is both a mitogen and a survival factor. Blood 1991; 77:1228–1233.

119. Stephenson JR, Axelrad AA, McLeod DL, Shreeve MM. Induction of colonies of hemoglobin-synthesizing cells by erythropoietin in vitro. Proc Natl Acad Sci USA 1971; 68:1542–1546.

120. Sytkowski AJ, Feldman L, Zurbuch DJ. Biological activity and structural stability of N-deglyco-sylated recombinant human erythropoietin. Biochem Biophys Res Commun 1991; 176:698–704.

121. Takeuchi M, Takasaki S, Miyazaki H, Kato T, Hoshi S, Kochibe N, Kobata A. Comparative study of the asparagine-linked sugar chains of human erythropoietins purified from urine and the culture medium of recombinant Chinese hamster ovary cells. J Biol Chem 1988; 263:3657–3663.

122. Tepperman AD, Curtis JE, McCulloch A. Erythropoietic colonies in cultures of human marrow. Blood 1974; 44:659–669.

123. Todokoro K, Kanazawa S, Amanuma H, Ikawa Y. Characterization of erythropoietin receptor on erythropoietin-unresponsive mouse erythroleukemia cells. Biochim Biophys Acta 1988; 943:326–330.

124. Todokoro K, Watson RJ, Higo H, Amanuma H, Kuramochi S, Yanagisawa H, Ikawa Y. Down-regulation of c-*myb* gene expression is a prerequisite for erythropoietin-induced erythroid differentiation. Proc Natl Acad Sci USA 1988; 85:8900–8904.

125. Torti M, Marti KB, Altschuler D, Yamamoto K, Lapetina EG. Erythropoietin induces p21*ras* activation and p120GAP tyrosine phosphorylation in human erythroleukemia cells. J Biol Chem 1992; 267:8293–8298.

126. Tsuda E, Kawanishi G, Ueda M, Masuda S, Sasaki R. The role of carbohydrate in recombinant human erythropoietin. Eur J Biochem 1990; 188:405–411.

127. Vainchenker W, Bouguet J, Guichard J, Breton-Gorius J. Megakaryocyte colony formation from human bone marrow precursors. Blood 1979; 54:940–945.

128. Wang FF, Kung CK, Goldwasser E. Some chemical properties of human erythropoietin. Endocrinology 1985; 116:2286–2292.

129. Wang GL, Semenza GL. General involvement of hypoxia-inducible factor 1 in transcriptional response to hypoxia. Proc Natl Acad Sci USA 1993; 90:4304–4308.

130. Ward JC, Harris KW, Penny LA, Forget BG, Kitamura T, Winkelmann JC. A structurally abnormal erythropoietin receptor gene in a human erythroleukemia cell line. Exp Hematol 1992; 20:371–373.

131. Wasley LC, Timony G, Murtha P, Stoudemire J, Dorner AJ, Caro J, Krieger M, Kaufman RJ. The importance of N- and O-linked oligosaccharides for the biosynthesis and in vitro and in vivo biologic activities of erythropoietin. Blood 1991; 77:2624–2632.

132. Watowich SS, Hilton DJ. Unpublished results.

133. Watowich SS, Hilton DJ, Lodish HF. Activation and inhibition of erythropoietin receptor function: role of receptor dimerization. Mol Cell Biol 1994; 14:3535–3549.

134. Watowich SS, Yoshimura A, Longmore GD, Hilton DJ, Yoshimura Y, Lodish HF. Homodimerization and constitutive activation of the erythropoietin receptor. Proc Natl Acad Sci USA 1992; 89:2140–2144.

135. Weiss TL, Barker ME, Selleck SE, Wintroub BU. Erythropoietin binding and induced differentiation of Rauscher erythroleukemia cell line red 5-1.5. J Biol Chem 1989; 264:1804–1810.

136. Wen D, Boissel JP, Tracy TE, Gruninger RH, Mulcahy LS, Czelusniak J, Goodman M, Bunn HF. Erythropoietin structure–function relationships: high degree of sequence homology among mammals. Blood 1993; 82:1507–1516.

137. Witthuhn BA, Quelle FW, Silvennoinen O, Yi T, Tang B, Miura O, Ihle JN. JAK2 associates with the erythropoietin receptor and is tyrosine phosphorylated and activated following stimulation with erythropoietin. Cell 1993; 74:227–236.

138. Wu H. Unpublished results.

139. Yi T, Mui AL, Krystal G, Ihle JN. Hematopoietic cell phosphatase associates with the interleukin-3 (IL-3) receptor beta chain and down-regulates IL-3-induced tyrosine phosphorylation and mitogenesis. Mol Cell Biol 1993; 13:7577–7586.

140. Yoshimura A, D'Andrea AD, Lodish HF. Friend spleen focus-forming virus glycoprotein gp55 interacts with the erythropoietin receptor in the endoplasmic reticulum and affects receptor metabolism. Proc Natl Acad Sci USA 1990; 87:4139–4143.

141. Yoshimura A, Lodish HF. In vitro phosphorylation of the erythropoietin receptor and an associated protein, pp130. Mol Cell Biol 1992; 12:706–715.

142. Yoshimura A, Longmore G, Lodish HF. Point mutation in the exoplasmic domain of the erythropoietin receptor resulting in hormone-independent activation and tumorigenicity. Nature 1990; 348:647–649.

143. Yoshimura A, Zimmers T, Neumann D, Longmore G, Yoshimura Y, Lodish HF. Mutations in the Trp-Ser-X-Trp-Ser motif of the erythropoietin receptor abolish processing, ligand binding, and activation of the receptor. J Biol Chem 1992; 267:11619–11625.

144. Youssoufian H, Longmore G, Neumann D, Yoshimura A, Lodish HF. Structure, function, and activation of the erythropoietin receptor. Blood 1993; 81:2223–2236.

145. Yu H, Bauer B, Lipke GK, Phillips RL, Van Zant G. Apoptosis and hematopoiesis in murine fetal liver. Blood 1993; 81:373–384.

146. Zanjani ED, Pixley JS, Slotnick N, Mac Kintosh FR, Ekhterae D, Clemons G. Erythropoietin does not cross the placenta into the fetus. Pathobiology 1993; 61:211–215.

147. Wu et al. Cell 1995; 83:59–67.

148. Lin et al. Genes Dev 1996; 10:154–164.

149. Livnah et al. Science 1996; 273:464.

150. Ihle. Nature 1995; 377:591–594.

15

The Role of Interleukin-4 in Normal and Malignant Human Hematopoiesis

Sem Saeland and Jacques Banchereau
Schering-Plough, Dardilly, France

I. INTRODUCTION

Interleukin 4 (IL-4) was identified in 1982 for its ability to induce the proliferation of murine B lymphocytes costimulated with anti-IgM (1) and for its capacity to induce lipopolysaccharide (LPS)-activated B cells to produce IgG1 (2). In 1986, a cDNA encoding human IL-4 was isolated (3). IL-4 is a member of a cytokine family including IL-3, IL-5, IL-13, and granulocyte–macrophage colony-stimulating factor (GM-CSF).

Although first described as a stimulatory factor for B cells, IL-4 was thereafter found to promote the growth of mast cells and T cells, indicating that its cellular targets encompass several hematopoietic lineages. Subsequently, IL-4 was recognized to modulate the development of various hematopoietic progenitor cells, thereby adding to the rapidly expanding list of cytokines that regulate hematopoiesis. The actions of IL-4 are not limited to the progenitor cells, as they also extend to the stromal cell supportive microenvironment of the bone marrow. Thus, it is likely that IL-4 plays a fundamental role in hematopoiesis, which will be developed in the present chapter. Herein, we will mainly refer to studies performed with human IL-4. Additional references on murine IL-4 can be found in several reviews (4–7). Clinical trials are presently being performed aimed at determining the possible therapeutic use of IL-4 as an antitumor agent.

II. MOLECULAR ASPECTS OF IL-4

The principal characteristics of IL-4 are summarized in Table 1. Molecular cloning of human IL-4 cDNA revealed a single open-reading frame of 153 amino acids, yielding a secreted glycoprotein of 129 amino acids (3). Expression of the recombinant protein in mammalian cells demonstrates three variants with apparent M_r values of 15, 18, and 19 kDa. The microheterogeneity of recombinant human (rHu) IL-4 appears to be related to the nature of the N-linked oligosaccharides in the 18- and 19-kDa variants. Human IL-4 contains three disulfide

Table 1 Properties of Human IL-4 and IL-4 Receptor (CDW124)

Property	IL-4	IL-4 receptor
Protein		
Precursor protein: amino acids (aa)	153	825
Mature protein: aa	129	800
Extracellular domain: aa		207
Transmembrane domain: aa		24
Cytoplasmic domain: aa		569
Molecular weight (kDa)	15–19	140
Glycosylation sites	2N	6N
Disulfide bonds	3 (6Cys)	
Intracytoplasmic Tyr		5
Gene		
Chromosome	5q23q31	16p12.1–p11.2
Exons	4	
Introns	3	
Cell sources	T cells (T_H2)	Ubiquitous
	NK cells	
	Basophils/mast cells	
	Stromal cells (?)	

bridges between C-3 and C-127, C-4 and C-65, and C-46 and C-99. X-ray diffraction of IL-4 crystals (8) and magnetic resonance spectroscopy of IL-4 in solution (9) indicate that IL-4 is a left-handed four–α-helices bundle, with short stretches of β-sheets. The four α-helices are situated between the residues 9 and 21, 45 and 64, 74 and 96, and 113 and 129, and the mini-antiparallel β-sheets are between residues 32 and 34 and 110 and 112. The structure of IL-4 bears close resemblance to GM-CSF, M-CSF, and growth hormone. The human IL-4 gene, composed of four exons and three introns, is localized on the long arm of chromosome 5 on bands q23-31 together with genes of other related cytokines including IL-3, IL-5, IL-9, IL-13, and GM-CSF (10). The promoter of the human IL-4 gene in the 5′-flanking region contains a unique repetitive motif (P-sequence) (11) that binds the T-cell–specific activation-inducible factor NF(P) (12,13). Another site of the IL-4 promoter binds the ubiquitous and constitutive nuclear factor NF-Y. Furthermore, the IL-4 promoter contains two OAP-40 sites that bind factors that may consist of dimers of *jun*-C or *jun*-D. The human IL-4 promoter displays a negative regulatory element containing two protein-binding sites (NRE-I and NRE-II) and interacts with the T-cell–specific NEG-1 and the ubiquitous NEG-2 nuclear factors (14). The silencer suppresses the activity of an enhancer element in the human IL-4 promoter (14). IL-4 production is also controlled posttranscriptionally through stabilization of IL-4 mRNA.

Interleukin-4 displays 20% homology at the amino acid level, with the more recently identified human IL-13 (15,16).

III. IL-4 RECEPTOR

Interleukin-4 binds to high-affinity (K_d, 100-pM) receptors, which are expressed in low number (a few hundred to a few thousand) on virtually every cell type tested, including T and B lymphocytes, monocytes, granulocytes, fibroblasts, epithelial, and endothelial cells (17,18).

Human and murine IL-4 are species-specific relative to receptor binding (17). Studies with radiolabeled IL-4 showed that the IL-4 receptor includes molecular species of M_r 130–140, 70–75, and 65 kDa (17,19,20), with the 70-kDa polypeptide representing a breakdown product of p140 (21). The 140-kDa glycoprotein has been purified and binds IL-4 with a high affinity, close to that of the receptor expressed on intact cells. A cDNA coding for the human 140-kDa IL-4–binding protein (p140) has been isolated, and encodes an open-reading frame of 825 amino acids that includes a 25-amino acid signal sequence (22,23) (see Table 1). p140 displays a 207-amino acid extracellular domain that contains the characteristic features (four conserved cysteines and a WSXWS box) of the class-1 cytokine receptor superfamily (23,24), a 24-amino acid transmembrane domain, and a long 569-amino acid intracellular portion. The intracellular domain of p140 (Fig. 1) contains five tyrosine residues (at positions 497, 603, 631, 713, and 821), a membrane-proximal box-1 (25) sequence (residues 262–269), a conserved acidic portion (EAPV-X3-EEEEE), and an I4R motif, surrounding tyrosine-497 and shared with the insulin and insulin-like growth factor 1 (IGF-1) receptors. The IL-2 receptor γc-chain (M_r 64 kDa) represents a common component to the receptors for IL-2, IL-4, IL-7, IL-9, and IL-15 (26–30). The γc-chain increases the affinity of IL-4 for p140 by two- to three-fold. The wide use of this common γc-chain, the gene of which is located on the X chromosome, explains the severe X-SCID immunodeficiency characterized by a genetic alteration of the γc-gene (31).

Interleukin-4 receptor expression is regulated by both the transcription rate and the stabilization of p140 mRNA. The IL-4 receptor appears to be differentially regulated according to cell type, as its expression is affected by IL-4 and Ca^{2+}-dependent pathways in T cells (19), but not in monocytic cells. Following internalization, IL-4 is degraded, most likely in lysosomes. Preliminary studies suggest a model of receptor–ligand interaction involving the formation of a ternary complex consisting of two molecules of the extracellular portion of the receptor and one molecule of IL-4.

Finally, a distinct 65- to 75-kDa low-affinity (K_d, 600-nM) IL-4 binding protein has been described. This low-affinity receptor was detected both at the cell surface and as a soluble protein, the partial NH_2-terminal amino acid sequence of which is unrelated to p140 (32).

IV. BIOCHEMICAL PATHWAYS OF IL-4 ACTION

The signaling mechanisms through the IL-4 receptor has recently been reviewed (33). Cross-linking studies have shown that the p140 chain of the IL-4 receptor associates with the common γc-chain following binding of IL-4, suggesting that such heterodimerization is important for biological activity (28). The cytoplasmic region of the human γc-chain is essential for IL-4–induced growth signal transduction in murine cells transfected with the human p140 IL-4-binding chain. After binding to its receptor, IL-4 induces tyrosine phosphorylation of multiple proteins, including species of 170, 140, 130, 110–120, 100, and 92 kDa (34–36). Because neither the cytoplasmic domain of p140 nor that of γc contain consensus sequences established for catalytic activity, it appeared likely that binding of IL-4 would recruit nonreceptor tyrosine kinases. Indeed, analogous to signaling through several other cytokine receptors, JAK family kinases are mobilized by IL-4 (see Fig. 1). Thus, JAK-1 (130 kDa) and JAK-3 (120 kDa) are tyrosine phosphorylated and activated following IL-4 binding (36,37). JAK-1 complexes with 4PS (170 kDa) (36), which rapidly becomes tyrosine-phosphorylated in response to IL-4 (34). IL-13 shares with IL-4 the capacity to phosphorylate 4PS (33), thereby explaining shared biological activities of the two cytokines. The 4PS substrate is closely related to insulin receptor substrate-1 (IRS-1) (38), which, through tyrosine-phosphorylated sites, serves as a docking protein for signaling molecules that contain a src-homology 2 (SH2) domain. The I4R motif

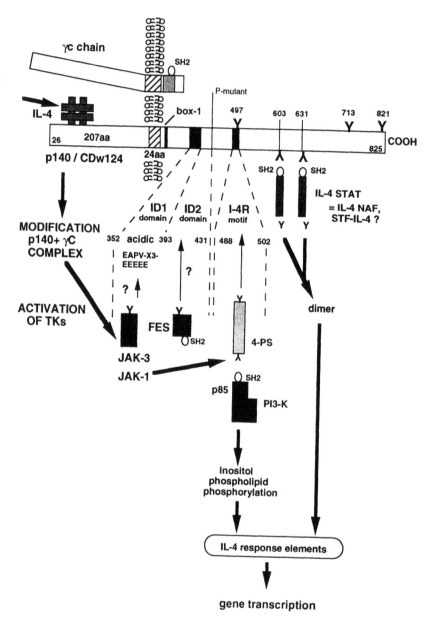

Figure 1   IL-4 signaling through the p140/γc IL-4 receptor complex: Binding of IL-4 activates tyrosine kinase (TK) activity (JAK-1, JAK-3, FES), probably as a result of conformational change in the p140/γc complex. The 4-PS substrate becomes tyrosine (Y)-phosphorylated, allowing activation of PI3-kinase. In addition, phosphorylation occurs on Y-residues of the p140 receptor, allowing recruitment of IL-4 STAT transcription factor and activation of IL-4-responsive genes. Association between phosphotyrosine groups and SH2-domains play a key role in IL-4 signal transduction. The ID1/ID2 domains and the I-4R motif of p140 have been described (39,40). Numbers correspond to amino acids starting from residue 26 (aa 1–25, signal peptide).

surrounding tyrosine-497 in the cytoplasmic region of the p140 IL-4-binding chain bears homology to sequences in the insulin and IGF-1 receptors, suggesting that it has been conserved among receptors that link to the IRS-1–4PS pathway (39). Of interest, expression of the γc-chain is required for IL-4-induced phosphorylation of 4PS (28). Phosphorylated 4PS associates with the SH2-containing p85 subunit of phosphotidylinositol 3-(PI_3)-kinase (34), thus activating a pathway likely associated with cell proliferation. However, 4PS-independent signals may also lead to proliferation, as suggested from studies with an IL-4 receptor mutant lacking tyrosine-497 (33) and with the P-mutant lacking the entire COOH-terminal portion from residue 432 (40).

Recently, JAK-1 was shown to complex with the IL-4 receptor following ligand binding (36), thereby likely playing a role in the previously observed tyrosine phosphorylation of the 140-kDa IL-4 receptor (34,35). In addition to the JAK family kinases, IL-4 binding induces tyrosine phosphorylation of a 92-kDa protein, identical with or closely related to the FES tyrosine kinase (41). p92 FES associates with the p140 IL-4 receptor, within a domain between residues 352 and 431 of the receptor (see Fig. 1), and association is enhanced on ligand binding (41). The stretch of 41 amino acids between residues 352 and 392 is required to signal growth through the human IL-4 receptor in murine Ba/F3 transfectants, probably through the acidic EAPV-X3-EEEEE domain (40). These results would be consistent with the existence of binding sites for p92 FES and, possibly, also for the JAK-family kinases in this portion of the receptor.

Recently, it has been recognized that tyrosine phosphorylation, initiated by engagement of cytokine receptors, can recruit factors, termed *signal transducers and activators of transcription* (STATs), that directly translocate to the nucleus, leading to activation of cytokine-responsive genes. In this context, IL-4 activates IL-4 STAT, a 100-kDa DNA-binding protein, containing contiguous SH2 and SH3 domains, and recognizing an IL-4 response element upstream from the human FcγRI gene (42). IL-4 STAT binds to either, or both, sequences surrounding phosphorylated tyrosines at positions 603 and 631 in the p140 IL-4 receptor (see Fig. 1). Following binding of IL-4 to its receptor, IL-4 STAT becomes tyrosine phosphorylated by a as yet undefined tyrosine kinase, allowing dimerization and subsequent release from the receptor (42). Importantly, the IL-4 STAT pathway has not yet been linked to signaling cell proliferation. Other tyrosine-phosphorylated DNA-binding factors induced by IL-4 have been described, that may be identical with IL-4 STAT. Thus, IL-4 NAF binds to the interferon (IFN)-γ response region of the human Fcγl gene, to the human FcεRIIb promoter, and to the murine Cγl promoter, all of which contain an inverted GAA repeat structure (43). The factor STF-IL-4 binds to an IFN-γ activation site (GAS element) of the IRF-1 promoter, and to a sequence at the initiation site of immunoglobulin germline ε transcription (44). Furthermore, a factor has been reported that binds to an IL-4 responsive element upstream from the lε exon of the human Cε gene. Of interest, IL-4-induced DNA-binding factors recognize IFN-γ response sites, which allows the two cytokines to modulate transcription of the same genes, thereby providing an explanation for the fact that IL-4 and IFN-γ may have either antagonistic or analogous biological effects.

Interleukin-4 thus mediates its actions through at least two distinct pathways (PI_3-kinase and STAT) linked to the p140 IL-4 receptor, in line with earlier speculations that more than one pathway is involved in IL-4 signaling. The involvement of other pathways is likely, but is less clearly defined. IL-4 inhibits transcription of the c-*fos* and c-*jun* genes in activated human monocytes and U937 cells (45), and the effect on c-*fos* expression is partly controlled by the lipoxygenase-directed pathway of arachidonic acid metabolism. The contribution of phosphoinositide metabolism to IL-4 signaling is controversial. Thus, IL-4 increases inositol 1,4,5-triphosphate (IP_3) and intracellular Ca^{2+} levels in human (46), but not murine (47,48) B cells. In support of an involvement the phosphoinositide pathway, IL-4 induces a redistribution of protein kinase C (PKC) from the cytosol to the nucleus in human monocytes (49). In addition,

IL-4 may recruit the adenylate cyclase pathway, because it increases the levels of cAMP in human B lymphocytes (46) and natural killer (NK) cells (50). IL-4 can increase the levels of cGMP in IFN-γ–activated human monocytes, likely as a result of enhanced nitric oxide (NO)-synthase activity and NO release (51). Two groups have reported the phosphorylation of a membrane-bound 42- to 44-kDa protein in response to IL-4 in B cells (48). Finally, IL-4–induced proliferation of human myeloid cell lines involves dephosphorylation of an 80-kDa protein, suggesting involvement of a tyrosine phosphatase (52).

The collective information thus indicates that IL-4 can signal through multiple pathways and suggests that the signaling used is dependent on cell type and cell activation status.

V. CELLULAR SOURCES OF IL-4

Unlike IL-1, IL-6 and IL-10, which are produced by many different cell types, IL-4, similar to IL-2 and IFN-γ, is secreted by restricted cell types.

A. T Lymphocytes

T cells constitute a major source of IL-4. Thus, T_H2 cells secrete IL-4 and IL-5, and lead to a preferential stimulation of humoral immunity. In contrast, T_H1 cells, which produce IL-2 and IFN-γ, lead to a preferential stimulation of cellular immunity.

An important feature of T_H1 and T_H2 cells is the ability of one subset to regulate the activities of the other. It occurs at the level of the effector cells triggered by these subsets, as indicated by the inhibitory effects of IFN-γ on IL-4–induced B-cell activation, or those of IL-4 on IL-2–induced T- and B-lymphocyte proliferation. It also occurs directly at the level of these subsets, for the products of one subset can antagonize the activation of the other: IFN-γ inhibits proliferation of T_H2 cells, whereas IL-4 and IL-10 inhibit cytokine production by T_H1 cells (53,54).

The mechanisms controlling the development of T_H1 and T_H2 cells are becoming clearer. The different functional subsets of T_H cells arise postthymically from naive precursors (T_{Hp}) as a consequence of antigenic activation. T_{Hp} cells are short-lived $CD4^+$ cells that secrete IL-2, but not other cytokines. After activation by antigen, T_{Hp} cells develop into effector cells (T_H0) that can be induced to secrete very high titers of IL-3, IL-4, IL-5, GM-CSF, and IFN-γ. In contact with IL-4, activated T cells mature into T_H2 cells, whereas IFN-γ and IL-12 allow the generation of T_H1 cells (55–59).

More recent evidence suggests that $CD8^+$ T cells may also be subdivided into T_H1 and T_H2 subpopulations. In humans, IL-4–producing $CD8^+$ T cells have been observed in diseases such as leprosy (60) and, more recently, the acquired immunodeficiency syndrome (AIDS) (61). The emergence of such cells may contribute to the disease, possibly because of reduced overall cytolytic activity against microbe-infected cells.

B. NK Cells

A minor population of mouse $CD4^+$, NK 1.1^+ T cells produces large amounts of IL-4 in response to in vivo challenge with anti-CD3 (62). This T-cell population, which represents an important lymphokine-producing cell of the thymus and bone marrow (63), does not require exogenous IL-4 priming to secrete IL-4.

C. Mast Cells–Basophils

Interleukin-4 is also produced by basophils and mast cells, activated by cross-linkage of $FcεR_I$ and $FcγR_{II}$ (64–66). Nasal biopsy specimens from patients with allergic rhinitis, and bronchial biopsy specimens from patients with allergic asthma, display mast cells with intracellular IL-4 (67). Interestingly, anti-IgE induces maximum release of IL-4 within 1 hr, suggesting that it is present within the cells in a preformed state similar to that reported for tumor necrosis factor-α (TNF-α).

Recent data, however, imply that basophil–mast cell-derived IL-4 is not required to prime T_H2 cells to express IL-4 (68).

D. Stromal Cells

Murine fibroblastic bone marrow stromal cells can produce IL-4 constitutively, and this function is up-regulated by exposure to IL-1 (69). Heterogeneous primary stromal cell populations, as well as cell lines of mouse origin, have been demonstrated to produce IL-4 (69,70). However, no IL-4 transcripts were detected in primary cultured bone marrow stromal cells that produced various cytokines, including IL-7, M-CSF, c-*kit* ligand, IGF-1, and leukemia inhibitory factor (LIF). These findings suggest that IL-4-producing stromal cell lines reflect a relatively rare subset of the stromal cell population in the bone marrow.

VI.   EFFECTS OF IL-4 ON IMMUNE EFFECTOR CELLS

A.   B Lymphocytes

Interleukin-4 increases the volume of resting B cells (71), induces hyperexpression of major histocompatibility complex (MHC) class II and CD23 (the low-affinity receptor for IgE-FcϵRII and a ligand for CD21) on normal and leukemic B lymphocytes (72). Two CD23 cDNAs, FcϵRIIα and FcϵRIIβ, have been isolated that differ in the first few amino acids of the cytoplasmic tail (73). In normal B cells, FcϵRIIα, but not FcϵRIIβ, is expressed spontaneously at low levels, and IL-4 enhances expression of FcϵRIIα and induces expression of FcϵRIIβ. Both IFN-α and IFN-γ block the IL-4-dependent increase of CD23 on B cells. IL-4 up-regulates expression of CD40, the ligand of which is expressed on activated T cells and plays a key role in B-cell proliferation and differentiation (74). IL-4 also up-regulates expression of surface IgM (75) and B7-1/CD80 (76,77), a member of the immunoglobulin superfamily and the B-cell counterstructure of T-cell CD28 and CTLA4. All these effects of IL-4 on B cells are suggestive of a role of IL-4 in the enhancement of antigen-presenting capacity of B cells towards T cells. Furthermore, IL-4 induces activated B cells to produce IL-6 and TNF (78), cytokines that play an important role in the activation and clonal expansion of activated T cells. Thus, IL-4 favors T-cell–B-cell interactions, concurring to enhanced antibody production.

IL-4 allows the proliferation of B cells activated through their antigen receptor (79,80). Addition of IL-4 to anti–CD40-activated B cells results in a strong B-cell proliferation, allowing the generation of factor-dependent long-term normal B-cell lines (81). Paradoxically, IL-4 inhibits the IL-2-induced proliferation of B cells activated through their antigen receptor (82,83). This effect may be due to the preferential sequestering by IL-4 of the γc-chain, constituting also the IL-2 receptor.

Early studies have shown that murine IL-4 can induce murine B cells to produce IgG1 (2) and IgE (84). Addition of CD4$^+$ T cells and monocytes (85), anti-CD40 monoclonal antibody (86,87), or hydrocortisone to human B lymphocytes allow them to secrete IgE in response of IL-4. The IL-4-dependent production of IgE is the consequence of IL-4-induced isotype switching (88–92). Human IL-4 also appears to direct switching toward the production of IgG4 (93). In vivo experiments in mice have confirmed the crucial role of IL-4 and interferons in the regulation of IgE production (94). Inactivation of the IL-4 gene in mice through gene-targeting is associated with normal T- and B-cell development, but with a strong reduction of seric IgG1 and disappearance of IgE (95). Conversely, IL-4 transgenic mice have increased IgE and an allergic-like disease, with ocular lesions infiltrated with mast cells and eosinophils (96).

IL-13 has IL-4-like effects on B cells and on monocytes. The IL-4- and IL-13-binding proteins are distinct entities, but which can be physically associated as a receptor complex (97,98).

B. T Lymphocytes and NK Cells

The T-cell growth-promoting effects of IL-4 were initially discovered on continuous T-cell lines (99). IL-4 can act on activated normal $CD4^+$ and $CD8^+$ T cells (100,101) in an IL-2-independent fashion. Studies with antisense oligonucleotides showed that IL-4 and IL-2 are autocrine growth factors of T_H2 and T_H1 T-cell clones, respectively (102). Combinations of IL-2 and IL-4 result in a proliferation of activated T cells that is greater than that obtained with each cytokine alone. However, IL-4 blocks the IL-2-induced proliferation of naive $CD4^+/CD45RA^+$ T cells (103).

IL-4 is also able to inhibit the production of IFN-γ by activated T cells (53) and, in doing so, favors the generation of T cells of the T_H2 type (see foregoing).

IL-4 turns on the expression of CD8 on $CD4^+$ T-cell clones and neonatal T cells (104). The induced $CD4^+CD8^+$ neonatal cells subsequently differentiate to express only CD8.

Analogous to its effect of B cells, IL-4 can induce the expression of CD23 on activated T lymphocytes (105).

When added to mixed leukocyte cultures, IL-4 increases antigen-specific cytotoxic activity against allogeneic stimulator cells (106–108). It also enhances the development of virus-specific cytotoxic T cells (109). The endogenous production of IL-4 in response to viral challenge is likely to play an important role in the generation of antigen-specific cytotoxic cells. T-cell clones, cultured in the presence of IL-4, display higher cytolytic ability than when cultured in IL-2. Administration of soluble IL-4 receptor to mice inhibits an allogenic response in vivo and enhances cardiac allograft survival, thus suggesting in vivo relevance for IL-4 in the development of cytotoxic T cells. Finally, IL-4 increases the adhesiveness of T cells for endothelial cells (110).

$CD3^-$ NK cells proliferate in response to IL-4, provided that the cells have received a mitogenic stimulus (50,108,111,112). However, IL-4 inhibits IL-2–dependent proliferation of these cells and, thereby, blocks IL-2-dependent generation of human LAK cells. Interestingly, IL-13 does not act on either T cells or NK cells, and these cells do not bind IL-13 even after activation.

C. Myelomonocytic Cells

Mononuclear Phagocytes

Interleukin-4 enhances the antigen-presenting capacity of monocytes–macrophages through an increased expression of MHC class II molecules (113), and down-regulates expression of the differentiation antigen CD14, a receptor for LPS (114). In addition, IL-4 up-regulates expression of the low-affinity IgE receptor–CD23 (114), but down-regulates expression of the three types of IgG receptors (115), resulting in inhibition of antibody-dependent cytotoxicity. Adherent blood mononuclear cells, cultured in the presence of both GM-CSF, and IL-4 mature into dendritic cells that can efficiently present soluble antigen to specific T-cell clones (116). IL-4 activates tumoricidal activity of macrophages and induces the formation of giant multinucleated cells (117). An important property of IL-4 is its ability to block spontaneous and induced production of the proinflammatory cytokines IL-1, IL-6, IL-8, and TNF-α (113,114,118), while stimulating the production of IL-1 receptor antagonist (IL-1RA) (119). IL-4 also inhibits the secretion of interstitial collagenase and 92-kDa type IV collagenase (120,121), thereby reducing the ability of macrophages to degrade extracellular matrix. IL-4 has a profound inhibitory effect on the release of superoxide (122) and the secretion of prostaglandin E_2 (PGE_2) (118). In addition, IL-4 inhibits IFN-γ–induced expression of the proinflammatory α-chemokine IP-10 in murine macrophages. Taken together these properties suggest a powerful anti-inflammatory effect of IL-4. Accordingly, in vivo administration of IL-4 inhibited the development of an antigen-specific T-cell–mediated inflammatory response in a hapten-induced model of contact sensitivity (123). The anti-inflammatory effects of IL-4 have recently been postulated to be due to up-regulation of expression and activity of cell membrane aminopeptidase-N (CD13) (124), which inactivates proinflammatory peptides.

IL-4 inhibits the killing by macrophages of various parasites, such as *Leishmania* (125) and asexual forms of *Plasmodium falciparum* (126).

Finally, complete resistance to human immunodeficiency virus (HIV-1) infection was observed in monocytes that had been treated with IL-4 or IL-13 (127,128).

Granulocytes

Eosinophils express reduced levels of FcγR when cultured with IL-4 (129). The down-regulation of FcγR results in a decreased secretion of glucuronidase and arylsulfatase in response to IgG-coated beads and in reduced antibody-dependent killing of schistosomulas (129). Of importance, IL-4 enhances the transendothelial migration of eosinophils, an effect related to up-regulation of adhesion molecules, such as VCAM-1, on endothelial cells (130–132). Infiltration of eosinophils contributes to the antitumor effects of IL-4 in vivo (133), and is a feature of IL-4 transgenic mice (96). IL-4 also displays an anti-inflammatory effect on neutrophils by enhancing respiratory burst and phagocytic properties (134), and by inhibiting their secretion of IL-8 (135). IL-4 enhances expression and secretion of membrane and soluble type II IL-1 receptor by neutrophils (136). This may also contribute to IL-4's anti-inflammatory effects by blocking IL-1's effects.

Finally, IL-4 induces the adherence of basophils, but not of neutrophils, to endothelial cells (132).

VII. IL-4 AND NORMAL MYELOPOIESIS

Several studies indicate that IL-4 acts as a cofactor in modulating myelopoiesis (5) (see Table 2; also Fig. 2, color plate).

A. Primitive Progenitors

Interleukin-4 in combination with erythropoietin (EPO) supports the development of small numbers of erythroid-mixed colonies from murine bone marrow (137). Studies using fluorouracil (5-FU)-treated mice to select early quiescent progenitors demonstrated that this effect of IL-4 is further enhanced by IL-6, as measured by both formation of multilineage blast colonies (138) or macroscopic (> 0.5 mm) HPP-CFU colonies (139).

The effects of IL-4 on early progenitors do not appear to be borne by the most primitive cells detectable, but rather, by their immediate progeny. Thus, day 2 post-5-FU cells (more primitive) are refractory to IL-4, whereas their day 4 post-5-FU descendants are responsive (138,139).

No effect of IL-4 was observed on mixed erythroid colony formation from T cell and monocyte-depleted human bone marrow in the presence of EPO (140). However, a recent study

Table 2 Effects of IL-4 on Normal and Malignant Hematopoietic Progenitors

Cells[a]	Proliferation	Cofactors	Antagonists	Gene expression	Maturation
Normal					
Multipotent	+	c-*kit* ligand, IL-3			
Neutrophil	+	G-CSF	IFN-γ		
Basophil/mast	+	IL-3		c-*kit* ↓	
	−	c-*kit* ligand			
Eosinophil	+	IL-3			
Monocyte/ macrophage	−		IFN-γ		
Erythroid	+	Epo, IL-3			
Thrombocyte	−(?)				
B lymphoid	−				+(?)
T lymphoid	+/−				+
Malignant					
AML	+	G-CSF, IL-1, IL-6 (IL-3, GM-CSF)		c-*kit* ↓ CD14↓ CD23↑	
	− (Particularly FAB M4 and M5)	G-CSF			+ After CD23 ligation
CML	−	G-CSF	G-CSF		
BCP-ALL	−			CD23↑ CD20↑	

[a]AML, acute myeloblastic leukemia; CML, chronic myelogenous leukemia; BCP-ALL, B-lineage acute lymphoblastic leukemia.

indicates that IL-4 does enhance the output of human CFU-GM in stromal cell-dependent long-term cultures, a likely result of synergy between IL-4 and stem cell factor (SCF)–c-*kit* ligand (141).

B. Progenitors of the Granulocytic Series

Neutrophilic Progenitors

Interleukin-4 acts as a cofactor for the generation of neutrophilic granulocytes in the presence of G-CSF, but not of IL-3 or GM-CSF. This cofactor effect of IL-4 was noted in early murine studies (137,142) and, subsequently, extended to human granulocytic precursors (140,143,144). This potentiating effect of IL-4 is antagonized by IFN-γ (145), thereby extending the well-established counteracting properties of IL-4–IFN-γ to hematopoietic progenitor cells. Early claims that the effect of IL-4 is directly exerted on the granulocytic progenitors, rather than indirectly mediated by accessory cells, have been confirmed by single-cell cultures of human bone marrow CD34+ cells (145). The synergy between IL-4 and G-CSF is borne by CD34+ HLA-DR++ cells, through recruitment by IL-4 of a higher number of G-CSF–responsive progenitors and their subsequent enhanced proliferation in response to IL-4 plus G-CSF, as compared with G-CSF alone, resulting in both more colonies and in an increase in the size of colonies (146). Of interest, CD34+ HLA-DR-low cells, which are more primitive than the HLA-DR++ progenitors (147), are refractory to IL-4 plus G-CSF, but become responsive following preincubation with IL-4 (146). These data support the notion of a sequential mechanism for the IL-4–G-CSF synergy, and raise the possibility that IL-4 may up-regulate the expression of G-CSF receptors.

Basophils–Mast Cell Progenitors

Historically, the capacity of IL-4 to enhance the proliferation of murine factor-dependent mast cell lines was the first indication that IL-4 acted on myeloid cells. Subsequently, IL-4 was shown to enhance the size of mast cell colonies from murine bone marrow Thy-1–low progenitors in the presence of IL-3, although this effect was noted only after a considerable (21 days) culture period (142). A further synergy, resulting in increased numbers of mucosal-like mast cell colonies, was noted when mesenteric lymph node cells from *Nippostrongylus brasiliensis*-infected mice were cultured in a combination of IL-4, IL-3, and IL-10 (5). It is likely that IL-4 is an important component in the process leading to mastocytosis in parasitic infections. IL-4 also supports the formation of colonies from murine peritoneal connective tissue-type mast cells (148,149). An important question will be to define the earliest stage of IL-4 responsiveness during mast cell development. Mature mast cells are longlived and are capable of proliferation (148); thus, it cannot be excluded that early mast cell progenitors are triggered by IL-3 alone, and it is their progeny that subsequently become responsive to IL-4, thereby explaining the delay required to observe the IL-4 effect.

In humans, IL-4 results in enhanced generation of basophils from T- and monocyte-depleted cord blood cells cultured in the presence of IL-3 (150). The cells thus generated have a morphologic appearance and a cytochemical composition (alcian–blue-positive, safranin–O-negative) compatible with blood basophils, but not mast cells.

IL-4 down-regulates c-*kit* gene expression in human bone marrow progenitors and in the immature mast cell line HMC-1 (151). In line with the observed down-regulation of c-*kit*, IL-4 has been reported to inhibit SCF–c-*kit* ligand-dependent human mast cell development (152). These results suggest that IL-4 has agonistic or antagonistic effects on the generation of basophils–mast cells, depending on its association either with IL-3 or c-*kit* ligand. However, expression of the c-*kit* gene in murine progenitors and mast cells has been reported to be down-regulated by IL-3, GM-CSF, or EPO, but IL-4 failed to display such an activity (153).

Eosinophilic Granulocyte Progenitors

Divergent results have been obtained for the capacity of IL-4 to recruit eosinophilic progenitors. IL-4 has been described as enhancing the generation of eosinophils from T- and monocyte-depleted human cord blood cells in the presence of IL-3 (150). In contrast, in another study performed on nonadherent human bone marrow mononuclear cells, no significant effect of IL-4 was observed on eosinophilic colony formation in response to IL-3 (144). Further studies will be required to evaluate whether these differences may be related to the tissue origin of the progenitors, or to the presence of T cells.

IL-4 does clearly play a role in the eosinophilic pathway, as demonstrated by in vivo murine experiments in which both IL-4 transgenic animals and nude mice injected with IL-4-producing tumor cells exhibit tissue infiltration of eosinophils (96,154). However, the exact contribution of IL-4 to the development of eosinophilic progenitors remains to be clarified.

C. Generation of Mononuclear Phagocytes

Early experiments revealed suppressive effects of IL-4 on macrophage colony formation by murine bone marrow cells cultured in IL-3 (142). This inhibitory effect of IL-4 was confirmed on human progenitors, is direct, and is independent of the stimulation signal provided (i.e., M-CSF or GM-CSF) (140,155). Analogous to its action on neutrophil progenitors, IFN-γ also counteracts the effects of IL-4 on macrophage development. Thus, IFN-γ can overcome the inhibitory effect of IL-4 through direct action on macrophage progenitors (145). Single-cell cultures indicate that IL-4 inhibits human macrophage development at the level of CD34$^+$ HLA-DR^{++} progenitors, but not at the earlier CD34$^+$ HLA-DR-low stage (146). The mechanisms by which IL-4 inhibits macrophage development remain to be determined.

IL-4 can also suppress IL-3-dependent development of progenitors not restricted to macrophage development, as noted by reduction in numbers of colony-forming units (CFU)-GM and mixed erythroid colonies (142,144). Recent studies suggest that these effects of IL-4 are indirect (146); thus, they could be related to previously described IL-4-mediated production of inhibitory factors by stromal cell components, including macrophages (143,156). Recent studies indicate that the capacity of IL-4 to suppress monocyte–macrophage development permits the development of dendritic cells from peripheral blood progenitors or precursors in the presence of GM-CSF (157).

D. Erythropoiesis

Early studies noted that IL-4 enhances both murine (142,156) and human (143) erythroid development in the presence of EPO. This effect is visualized essentially by increased burst-promoting activity (BFU-E development). The stimulating effect of IL-4 on erythroid development is strongly potentiated by IL-6 (139), a cytokine that has no burst-promoting activity of its own. In contrast with the foregoing reports, however, other investigators have either failed to detect a potentiating effect of IL-4, or have observed an inhibitory effect on erythropoiesis (140,158,159).

E. Thrombopoiesis

Interleukin-4 induces megakaryocyte colony formation from murine bone marrow cultured in the presence of IL-1 or EPO (137). This effect, however, may be indirect through release of megakaryocyte-stimulating activity by accessory cells. This possibility could explain the lack of costimulatory effect of IL-4 and EPO reported in a subsequent study. Indeed, recent findings indicate that IL-4 can be a negative regulator of thrombopoiesis, by inhibiting the development of both pure and mixed megakaryocyte colonies from human CD34$^+$ HLA-DR$^+$ progenitors cultured in IL-3 plus EPO (158).

VIII. IL-4 AND NORMAL LYMPHOPOIESIS

A. B Lymphopoiesis

Proliferation of early B-cell progenitors is enhanced in long-term Whitlock-Witte type murine bone marrow cultures in the presence of IL-4 (160). Thus, IL-4 favors the expansion of early B-lineage cells that bear DJ$_H$ rearrangements, but do not yet express B220 antigen. This effect likely results from IL-4–mediated induction of a potentiating factor by stromal cells. Other studies have demonstrated a direct growth-promoting effect of IL-4 on murine pro-B cell clones (germline immunoglobulin gene configuration) with the capacity to differentiate either into B cells or myeloid cells (161,162). However, the expression of CD5 by such pro–B-cell clones could suggest that they are not representative of the majority of early B-cell progenitors. In contrast with these stimulatory effects, IL-4 can also suppress B lymphopoiesis. In the mouse, IL-4 inhibits the development of B220$^+$ B-cell precursors (142,160), and both the emergence (163) and proliferation (142) of pre-B cells (expressing cytoplasmic μ-chain) in stroma-dependent cultures. Thus, murine studies would suggest that negative regulatory activities of IL-4 are mainly exerted at later developmental stages in B lymphopoiesis. Several groups have extended the growth-inhibitory properties of IL-4 to human bone marrow B-cell progenitors (164–166). In apparent contrast with murine data, primitive human lymphoid progenitors (CD34[+] cells which form TdT[+] colonies that include early B-lineage cells) are also inhibited by IL-4 (166). It has been suggested that IL-4 is involved in a network with other cytokines, such as IL-1α and TNF-α, in orienting the bone marrow microenvironment toward supporting myeloid development at the expense of B lymphopoiesis (166) (see Fig. 3, color plate). The growth-inhibitory effect of IL-4 on human B-cell precursors is shared by IL-13 (98).

In contrast with mature B cells, IL-4 does not appear to induce CD23 expression in human B-cell precursors (167; our unpublished observations). This does not, however, represent an intrinsic failure to express CD23, as ligation of surface CD40 antigen results in induction of CD23 in B-cell precursors (168).

IL-4 plays a role in the final maturation step during bone marrow B lymphopoiesis, by enhancing the output of sIgM[+] cells in murine cultures (69,70). This effect likely results from enhanced cell survival, rather than from induction of maturation (169). In humans, there is controversy over whether IL-4 induces an increase in the numbers of sIgM[+] cells generated from precursors (164), or only enhances sIgM expression of sIgM[+] cells that have emerged independently of IL-4 (165,167). Finally, human bone marrow B-cell precursors have been described to differentiate into immunoglobulin-secreting cells when cultured with IL-4 in association with activated T cells (170). Thus, although IL-4 clearly functions in positively regulating the final stages of B lymphopoiesis, the exact contribution of direct versus indirect pathways remains to be clarified.

In view of the agonistic and antagonistic effects of IL-4 on primary B-cell production, it is of importance that local microenvironments are likely to be quite different as the B-cell progenitors migrate from the periphery to the center of the bone marrow during their development.

B. T Lymphopoiesis

Both murine (171) and human (101,172) thymocytes proliferate in response to IL-4 and a costimulant such as phorbol esters (phorbol myristate acetate, PMA). In the mouse, IL-4 induces the proliferation of the most immature CD4[-]CD8[-] subset, which also has the ability to produce IL-4. The intermediate CD8[+]CD4[+] subset virtually fails to proliferate under these conditions, whereas the most mature CD4[+]CD8[-] and CD4[-]CD8[+] subsets also proliferate in response to IL-4 and PMA. More recently, IL-4 has been shown to protect murine thymocytes from dexamethasone-induced apoptosis (173). This effect is borne primarily by CD4[-]CD8[-] and CD4[+]CD8[-] cells. Fetal thymocytes also respond to IL-4, and in situ hybridization studies (174) have demonstrated the transcription of the IL-4 gene in fetal thymus. In addition to its effects on cell proliferation and survival, IL-4 can induce the maturation of thymocytes through induction of mature T-cell antigens (CD3, CD5, T-cell receptor [TCR]) and a loss of CD1. Concomitantly, the CD4[+]CD8[+] subset disappears and CD4[-]CD8[-], CD4[+]CD8[-], CD4[-]CD8[+] cells are generated (175). These data indicate that IL-4 plays an important role in T-cell ontogeny. Paradoxically, IL-4 transgenic mice have involuted thymuses (96). In accordance with this latter finding, IL-4 inhibits early T-cell development in fetal thymus organ culture, probably during the differentiation of CD4[-]CD8[-] cells into CD4[+]CD8[+] thymocytes. The generation of TCRαβ thymocytes appears to be more impaired than that of TCRγδ thymocytes (176). Likewise, human IL-4 induces a preferential differentiation of TCRγ/δ pre-T cells (172).

IX. IL-4 AND MALIGNANCIES OF IMMATURE MYELOID AND LYMPHOID CELLS

A. IL-4 and Myeloid Leukemia

Acute myeloblastic leukemia (AML) cells display as many IL-4 receptors as normal myeloid cells (177). IL-4 can stimulate DNA replication and clonogenic expansion in AML cells isolated from patient bone marrow or peripheral blood (144,177–179). Stimulation of growth concerns both CD34$^+$- and CD34$^-$-negative AML cases (179), and does not appear to be mediated through GM-CSF or G-CSF production (180). In contrast, IL-4 can inhibit DNA replication of AML cells (177,179), an effect likely due to interference with IL-1–induced growth or with production of stimulatory factors, such as GM-CSF, IL-6, and TNF-α (180,181). In this context, IL-4 inhibits production of IL-1 both in normal and malignant myeloid cells (118,182,183). IL-4 also down-regulates expression of c-*kit* (151) and the CD14 LPS receptor (181) in AML cells. In addition, and as on other cell types, AML cells up-regulate CD23 expression in response to IL-4.

IL-4 displays highly divergent effects on AML cells in association with other exogenously added cytokines. IL-4 potentiates the stimulatory effect of IL-1 and IL-6 on AML growth (179,180). It can either agonize or antagonize the effects of IL-3, GM-CSF, or G-CSF on AML cells (144,178,179,184). The agonistic effects observed between IL-4 and G-CSF (144,179,184) may be related to the clear synergism between these two factors for normal neutrophil development (see foregoing). Of interest, IL-4 suppresses DNA replication and the clonogenicity of AML cells with myelomonocytic differentiation (M4 and M5 FAB-classification types) (185), analogous to its effect on normal monocytic development (see foregoing). Suppression of growth does not appear to be due to inhibition of autocrine IL-6 production (181,185), although the finding that IL-4 reduces IL-1β mRNA levels in the promonocytic cell line U937 may favor an indirect mechanism of action (182). IL-4 inhibits growth of U937 cells (186), and G-CSF synergizes with IL-4 for suppression of clonogenicity (186). Interestingly, ligation of surface membrane CD23 following IL-4 treatment triggers differentiation of U937 cells and can also induce macrophage-like features in fresh AML cells. A suppressive effect of IL-4 has also been reported in chronic myelomonocytic leukemia (187). IL-4 has divergent effects on factor-dependent differentiation of myeloid leukemic cells to macrophages. Thus, IL-4 can inhibit differentiation of the murine M1 and human HL60 cell lines (188,189). However, macrophage-like differentiation of M1 cells and of the murine WEHI-3B JCS cell line can be enhanced by IL-4 when induced, respectively, by 1α,25-dihydroxyvitamin D$_3$ and TNF-α (188,190). The divergent responses of M1 cells have been related to the capacity of IL-4 to either antagonize or agonize down-regulation of c-*myc* protooncogene expression induced by these factors (191).

Finally, IL-4 suppresses CSF-dependent CFU-GM colony formation in chronic myelogenous leukemia (CML), an effect apparently related to inhibition of IL-1β and IL-6 production (183). Either suppressive or enhancing effects on CML CFU-GM growth were exerted by IL-4 in combination with G-CSF (183).

Multiple underlying parameters thus appear to contribute to the divergent effects of IL-4 on growth and differentiation of myeloid leukemia cells, with the most important variables likely to be the endogenous cytokine production pattern and the maturation stage of the leukemic cells.

B. IL-4 and Acute Lymphoblastic Leukemia

B-lineage acute lymphocytic leukemia (ALL) cells express IL-4 receptors, although the proportion of freshly isolated cells that bind IL-4 is heterogeneous, compared with leukemic cell lines (167). IL-4 has not been reported to have growth-stimulatory activity on leukemic B-cell precursors (BCP). Rather, and as also noted on normal BCP (see foregoing), several studies indicate that IL-4 inhibits growth and survival of BCP-ALL cells. Thus, DNA replication is suppressed both in Ph[1] positive [t(9;22)] (192) and Ph[1]-negative ALL (98,165), although not all cases are responsive. The effect of IL-4 on DNA replication is directly exerted on the leukemic cells in IL-4-sensitive cell lines, and does not appear to be due to production of inhibitory factors, such as IFN-α, -β, -γ, or TNF-α (192). In vitro viability of freshly isolated ALL cells can be maintained when cultured on bone marrow stromal layers (193). Under such conditions, IL-4 has a cytotoxic effect in most (16/21) of the ALL cases tested (194). Cytotoxicity is due to induction of apoptosis in G_0/G_1 phases of cell cycle, and is believed to be a direct effect of IL-4 (194). The molecular basis of the inhibitory action of IL-4 in BCP-ALL is currently unknown. Accordingly, it will also be important to determine whether the IL-4-refractory cases lack expression of IL-4 receptors. IL-4 can induce CD23 (167) and CD20 (165) in a proportion of BCP-ALL cases. In addition, IL-4-mediated inhibition of proliferation of the REH cell line was recently reported to be paralleled by an increased proportion of sIgM[+] cells (195), although it is unclear whether this effect represents cell maturation or expansion of an sIgM[+] subset. Given the documented inhibitory effects of IL-4 on leukemic cell growth and survival, this cytokine appears as a candidate for clinical trials in cases of high-risk BCP-ALL. IL-4 was cytotoxic for all the cases analyzed (n = 9) that correspond to poor-prognosis ALL that was associated with either t(9;22) or t(4;11) translocations (194). Recently, administration of human IL-4 (200 µg/kg per day) was reported to suppress human ALL xenografts in scid mice (196). Finally, IL-13 shares with IL-4 the capacity to inhibit BCP-ALL growth (98). Dichotomy was observed in the pattern of IL-13 versus IL-4 responsiveness in several cases (98).

X. IL-4 AND THE BONE MARROW STROMAL MICROENVIRONMENT

Stromal cells in the bone marrow represent a heterogeneous cell population that comprises mainly fibroblast-like cells and macrophages. This cellular network constitutes an obligate microenvironment for hematopoietic development, through the production of a vast array of regulatory factors. Of importance, stromal cells display a high degree of functional plasticity, as a result of their multiple interactions with soluble and cell membrane molecules. As an example of such plasticity, murine stromal cells are both responsive to IL-4 and can produce IL-4. Thus, stromal cell lines display IL-4 receptors, and IL-4 induces fibroblastic-like stromal cells to produce soluble activity that inhibits pre–B-cell formation from murine bone marrow (163). The latter effect may be related to the reported capacity of IL-4 to induce production of G-CSF and M-CSF from the fibroblastic 3T3 cell line (197), as these factors reduce stroma-dependent B lymphopoiesis. IL-4 can thus contribute to shaping a myeloid-supportive microenvironment (see Fig. 3, color plate), a feature shared with IL-1 and TNF (198). In a recent study, IL-4 inhibited IL-1α-induced production of myeloid factors (G-CSF, GM-CSF, IL-1β) by stromal cells, suggesting that a predominant effect of IL-4 might be to favor early, rather than late, myelopoiesis (141). Of interest, IL-4 up-regulates stromal cell expression of surface-membrane VCAM-1 (199), a molecule that mediates adhesion of hematopoietic progenitors (200). Accordingly, preincubation of stromal cells with IL-4 results in increased adherence of CD34[+] progenitor cells (141).

In addition to their responsiveness to IL-4, bone marrow stromal cells can produce IL-4. This has been demonstrated both in heterogeneous populations and in fibroblastic-like murine stromal cell lines (69,70). Stromal cell production of IL-4 is strongly up-regulated by IL-1 (69), likely reflecting a contribution of bone marrow macrophages. However, the IL-4-producing stromal cells probably represent only a particular subset or physiological state, as indicated by the lack of detectable IL-4 transcripts in LPS-stimulated Whitlock-Witte stromal cultures. Finally, the relative contribution of stromal cell versus T-cell–derived IL-4 in determining the levels of this cytokine that will be available to hematopoietic progenitors in the bone marrow microenvironment remains to be investigated.

XI. CLINICAL ASPECTS OF IL-4

Several aspects of the biology of IL-4 appear relevant for potential clinical applications. First, the key role of IL-4 in positively regulating IgE production suggests that antagonists of IL-4 could be useful in the treatment of certain allergic disorders. In this context, spontaneous IgE synthesis in vitro by lymphocytes from patients with atopic dermatitis is partially inhibited by anti-IL-4 antibodies (201), and anti-IL-4 antibodies inhibit polyclonal and antigen-specific primary and secondary IgE responses. IL-4 antagonists may also be of value in parasitology. Thus, anti–IL-4 suppresses the eosinophilia, hyper-IgE, and intestinal mastocytosis found in helminth infections, but not IgG1 and the protective immunity to the infection. Anti–IL-4 also reduces granuloma size in *Schistosoma mansoni* infection (202). The strong anti-inflammatory potential of IL-4 represents another area in which IL-4 may have clinical significance. This effect of IL-4 has now been demonstrated in vivo in rodent models, for IL-4 suppressed the chronic phase of arthritis induced by streptococcal cell wall fragments (203), and protected against IgG immune complex-induced lung injury (204). For more comprehensive information on IL-4 within the fields of allergy and inflammation, the reader is invited to consult a previous review (205).

A. IL-4 and Neoplasia

Preclinical evidence for antitumor effects of IL-4 has been demonstrated both in murine models and in human in vitro studies, although the cytotoxic or antiproliferative effects induced in response to IL-4 are highly variable. Tumorigenic plasmacytoma cell lines expressing IL-4 as a result of IL-4 gene transfection fail to grow in nude mice, and IL-4-producing tumors confer immunity against rechallenge with untransfected tumor (154). Eosinophil- and macrophage-dependent cytotoxicity appears to be an important component of the antitumor activity of IL-4 (133). However, development of specific tumor resistance depends mainly on the recruitment of cytotoxic CD8$^+$ T cells (206,207). In this context, IL-4 augments the generation of antigen-specific cytotoxic T cells in vitro (108,109) and promotes in vitro proliferation and cytolytic activity of antigen-specific tumor-infiltrating lymphocytes cytotoxic for human melanoma cells (106,208). However, the foregoing effects must be weighed against evidence suggesting that some of the immunomodulatory properties of IL-4 might favor tumor growth. Thus, IL-4 inhibits the ability of IL-2 to induce NK cells to develop lymphokine-activated killer (LAK) activity (108,112,209). Furthermore, IL-4 can inhibit in vitro tumoricidal activity of human monocytes (113,210).

In addition to indirect effects, IL-4 also displays direct antitumor activity both against solid tissue and hematological malignancies. Thus, IL-4 inhibits in vitro proliferation of human melanoma (211), gastric carcinoma (212), renal carcinoma (211,213), and colon and breast carcinoma (214) cell lines. IL-4 also reduces the proliferative activity of some human non–small-cell lung carcinoma (NSCLC) cell lines in vitro as well as in nude mice (215). Several types of human hematological malignancies are responsive to IL-4. Thus, IL-4 inhibits in vitro cell proliferation and can induce apoptotic cell death in a number of cases of B-lineage ALL (165,192–194). Antitumor effects are also observed in B-CLL, in which IL-4 counteracts the growth-promoting activity of IL-2 (216,217). IL-4 can provide an antiproliferative signal to non-Hodgkin's B lymphoma (NHML) cells (218,219). In addition, inhibition of in vitro growth occurs in several multiple myeloma samples through IL-4–mediated suppression of IL-6 expression (218,220). Antitumor effects of IL-4 have also been identified in myeloid malignancies. Thus, in vitro growth of AML cells is inhibited by IL-4 in a proportion of the samples analyzed (177,179–181). The tumor-suppressive effects of IL-4 may appear of particular interest in the context of AML with myelomonocytic differentiation (185). In addition, IL-4 inhibits GM-CFU colony formation in CML (183).

However, in contrast with the foregoing antitumor effects, growth-stimulatory effects of IL-4 have been observed in vitro in some cases of NHML (218,219), and multiple myeloma (218). Furthermore, IL-4 can stimulate DNA replication in hairy cell leukemia (221), and has been reported to be an autocrine growth factor for Reed-Sternberg cells.

B. Clinical Studies

The direct and indirect antitumor activities of IL-4 observed in vitro and in animal models have resulted in the initiation of phase I and phase II clinical studies to evaluate the effects of IL-4 in various advanced and refractory malignancies.

Doses and Schedule

Several phase I studies have been conducted with both recombinant *Escherichia coli*-produced human IL-4 (Schering-Plough Research Institute) and recombinant yeast-derived human IL-4 (Sterling Winthrop). The studies with *E. coli*-derived IL-4 were conducted with daily intravenous infusion in doses of 0.25–5 µg/kg per day, or daily subcutaneous dosing with 0.25–5 µg/kg per day in a single dose (217,222). Studies were also conducted with yeast-derived IL-4, with intravenous bolus injections every 8 hr at 10–15 µg/kg per dose, as an intravenous bolus, followed by intravenous infusion and daily subcutaneous injections in doses up to 400 µg/m^2 per day, or as single daily subcutaneous injections (223,224).

In addition, phase II studies have been initiated, with *E. coli*-derived IL-4 administered subcutaneously (0.25 µg/kg and 1.0 µg/kg, three times a week), and yeast-derived IL-4 as thrice-daily bolus intravenous injections of 800 µg/m^2 (225).

Pharmacokinetics of IL-4

Following intravenous bolus injection IL-4 is rapidly cleared from the circulation (total plasma clearance of 109 to 240 ml/min), with a mean half-life of 19 ± 8.7 min (224). This short half-life is consistent with that seen for many other cytokines. The volume of distribution is small (V_d between 3.4 and 7.2 L), suggesting that IL-4 is retained in the circulating blood volume and immediate extravascular space (224). Following intravenous bolus, IL-4 displays linear pharmacokinetics over a 40- to 400-µg/m^2–dose range (the latter value corresponding approximately to 10 µg/kg), with a maximum observed plasma concentration of 100–200 ng/ml at a 400-µg/m^2 dose. Consistent with a slow absorption rate, detectable circulating concentrations of IL-4 persisted for up to 8 hr following subcutaneous administration. Thus, although bioavailability is less than 100% following subcutaneous injection (224), this mode of administration may be preferable to achieve sustained concentrations of IL-4.

Toxicity in Clinical Studies

Low doses of *E. coli*-derived human IL-4 are relatively well-tolerated, although generally mild and reversible toxicity was observed independently of the mode of administration. In the dose range of 0.5 µg/kg per day up to 5 µg/kg per day, the most common side effects (10% or more of patients) include headache, flulike symptoms, and fever (< 39°C). In addition, sweats, chills, anorexia, epigastric discomfort, nausea or vomiting, diarrhea, lethargy, pedal and periorbital edema, hypotension, fluid retention, capillary leak syndrome, and pruritus, have been reported. These side effects only rarely interfere with prolonged IL-4 therapy. Dose-limiting toxicity of IL-4 is reached at 5 µg/kg per day, as cases of WHO grade-3 arthralgia and headache, were observed in two of three patients treated at this dose (222). Collective results suggest that the maximum tolerated dose (MTD) of *E. coli*-derived IL-4 administered subcutaneously is approximately 2 µg/kg per day (226). Studies performed at higher doses (10–20 µg/kg per day) of yeast-derived human IL-4 have reported cases of severe toxicity, including life-threatening gastrointestinal hemorrhage (224) and a fatal myocardial infarction, which revealed an inflammatory infiltrate that included a high proportion of neutrophils and mast cells (227).

IL-4 shows minimal clinical myelosuppression. Low-dose treatment has been associated with modest elevations of neutrophil, platelet, and lymphocyte counts and induces no detectable bone marrow toxicity (222). However, some cases of non–dose-related WHO grade 1–2 anemia have been reported (224). Relatively little effect on serum glucose was noted, despite the appearance of hypoglycemia in some animal studies of rHuIL-4.

Immunological and Antitumor Activity

Only limited data are available on immunological effects of rHuIL-4 in vivo. An increase in plasma IL-1 receptor antagonist (IL-1RA) and soluble CD23 levels were reported following IL-4 administration (223,225). In addition, increased HLA class II expression has been reported on monocytes (223,228), and the percentage of monocytes expressing CD16 antigen was decreased following IL-4 treatment (223). Of interest, given the known in vitro activity of IL-4 on IgE production, there was neither evidence of elevation in serum IgE levels following high-dose IL-4 administration (225), nor during long-term low-dose administration (Rybak ME, Schering-Plough Research Institute, personal communication). No changes in LAK or NK activity were noted, but slightly increased proliferative responses of PBMC to mitogens were observed in some patients (222,228). Patients treated with low-dose IL-4 displayed no consistent trend relative to lymphocyte counts, although individual patients had evidence of elevations in CD4$^+$ and CD8$^+$ cells, which did not affect the CD4/CD8 ratio. Finally, anti–IL-4 antibody could not be detected in patients treated with high-dose IL-4 (225).

Limited antitumor activity was observed in phase I studies, although positive clinical responses have been noted in non-Hodgkin's lymphoma (224), Hodgkin's disease, CLL (217), multiple myeloma (222), and NSCLC.

Phase II trials have been initiated with IL-4 in advanced malignancies. When using low dose *E. coli*-derived IL-4 administered subcutaneously (0.25 µg/kg or 1.0 µg/kg, three times a week), antitumor activity with an apparent dose–response was seen in 7 of 32 patients suffering from advanced NSCLC. In addition, activity has been noted in Hodgkin's disease (226). In another study, however, using high-dose yeast-derived IL-4 injected as a thrice-daily intravenous bolus of 600–800 µg/m^2, no meaningful antitumor activity was found in patients with metastatic melanoma or advanced renal cancer (225). Although preliminary, these data suggest that IL-4 has potential as an antitumor agent when administered over a more prolonged course, rather than short intravenous bolus. This is consistent with the rapid clearance and short half-life of IL-4.

Finally, phase I studies using combination IL-4–IL-2 therapy have been initiated in patients with refractory malignancies (229,230). Antitumor responses were observed with a combination of low doses of IL-2 and IL-4, with complete resolution of pulmonary metastatic disease from renal carcinoma (229). In another study, one partial response was observed among 27 patients evaluated after IL-4–IL-2 combination therapy.

XII. CONCLUDING REMARKS

In vitro studies have now shown that IL-4 can act on many cell types and that it can act at various stages of maturation of a given cell type. Both normal and malignant myeloid and lymphoid human hematopoietic progenitors are directly or indirectly susceptible to IL-4. The biological effects of IL-4 on a given cell type depend on the surronding cells and cytokines. Indeed, IL-4 can play an important biological role in an indirect fashion, as it modulates cytokine production by T, B, and NK cells, monocytes–macrophages, endothelial cells, and fibroblasts. Thus, the plasticity of the bone marrow microenvironment is likely to be determinant in shaping the response of progenitor cells to IL-4. The relative in situ contribution of various cell types (T cells, stromal cells, basophil–mast cell precursors?) to IL-4 production within the bone marrow remains to be determined. In vivo murine studies and in vitro analysis of human cells have demonstrated that IL-4 displays both indirect and direct antitumor activity. Such properties have encouraged the initiation of clinical trials with IL-4. When administered in doses of 1–2 μg/kg per day to adult patients with advanced cancer, IL-4 is safe and well-tolerated. Current data suggest some antitumor activity; however, the optimal dose and schedule to achieve desired immunomodulatory and antitumor effects in vivo remain to be determined. Furthermore, in vitro studies point to the importance of patient selection relative to administration of IL-4 in hematological malignancies. Future protocols in experimental cancer therapy will likely include associations of IL-4 with cytokines (i.e., IL-2) or with conventional chemotherapy. In addition to its potential role in oncology, the biology of IL-4 suggests that it may be appropriate for evaluation in a variety of diseases in which its immunomodulatory and anti-inflammatory properties would be of clinical benefit. In particular, the inhibitory effects of IL-4 on the production of cytokines give a strong rationale for the use of IL-4 in chronic inflammatory diseases.

The close future will see many clinical trials with IL-4 itself or drugs affecting IL-4 production or IL-4 effects, such as interferons. In the long-term, structural studies on IL-4, IL-4 receptor, and their complex, and studies on the intracellular pathways specifically activated by IL-4 should permit the design of chemical agents that will either inhibit or mimic part of or all the biological effects of IL-4.

ACKNOWLEDGMENTS

We would like to thank Dr. M. E. Rybak for critically reviewing the clinical aspects of this manuscript. Also, we are most grateful to Ms. N. Courbière and Ms. S. Bonnet-Arnaud for excellent editorial help.

REFERENCES

1. Howard M, Farrar J, Hilfiker M, Johnson B, Takatsu K, Hamaoka T, Paul WE. Identification of a T-cell derived B cell growth factor distinct from interleukin-2. J Exp Med 1982; 155:914–921.
2. Isacksson PC, Pure E, Vitetta S, Krammer PH. T cell-derived B cell differentiation factor(s). Effect on the isotype switch of murine B cells. J Exp Med 1982; 155:734–748.
3. Yokota T, Otsuka T, Mosmann T, Banchereau J, Defrance T, Blanchard D, de Vries JE, Lee F, Arai K. Isolation and characterization of a human interleukin cDNA clone, homologous to mouse B-cell stimulatory factor 1, that expresses B-cell-stimulatory activities. Proc Natl Acad Sci USA 1986; 83:5894–5898.
4. Paul WE. Interleukin-4: a prototypic immunoregulatory lymphokine. Blood 1991; 77:1859–1870.
5. Rennick, DM, Moore JG, Thompson-Snipes L. IL-4 and hematopoiesis. In: Spits H, ed. IL-4: Structure and Function. London: CRC Press, 1992:151–168.

6. Spits H. IL-4: Structure and Function. London: CRC Press, 1992.

7. O'Garra A, Spits H. The immunobiology of interleukin 4. Res Immunol 1993; 144:567–643.

8. Walter MR, Cook WJ, Zhao BG, Cameron RP, Ealick SE, Walter RL, Reichert P, Nagabhushan TL, Trotta PP, Bugg CE. Crystal structure of recombinant human interleukin-4. J Biol Chem 1992; 267:20371–20376.

9. Powers R, Garrett DS, March CJ, Frieden EA, Gronenborn AM, Clore GM. Three-dimensional solution structure of human interleukin-4 by multidimensional heteronuclear magnetic resonance spectroscopy. Science 1992; 256:1673–1677.

10. Morgan JG, Dolganov GM, Robbins SE, Hinton LM, Lovett M. The selective isolation of novel cDNAs encoded by the regions surrounding the human interleukin 4 and 5 genes. Nucleic Acids Res 1992; 20:5173–5179.

11. Abe E, De Waal Malefyt R, Matsuda I, Arai K, Arai N. An 11-base-pair DNA sequence motif apparently unique to the human interleukin 4 gene confers responsiveness to T-cell activation signals. Proc Natl Acad Sci USA 1992; 89:2864–2868.

12. Todd MD, Grusby MJ, Lederer JA, Lacy E, Lichtman AH, Glimcher LH. Transcription of the interleukin 4 gene is regulated by multiple promoter elements. J Exp Med 1993; 177:1663–1674.

13. Szabo SJ, Gold JS, Murphy TL, Murphy KM. Identification of cis-acting regulatory elements controlling interleukin-4 gene expression in T cells. Mol Cell Biol 1993; 13:4793–4805.

14. Li-Weber M, Krafft H, Krammer PH. A novel enhancer element in the human IL-4 promoter is suppressed by a position-independent silencer. J Immunol 1993; 151:1371–1382.

15. McKenzie ANJ, Culpepper JA, de Waal Malefyt R, Brière F, Punnonen J, Aversa G, Sato A, Dang W, Cocks BG, Menon S, de Vries JE, Banchereau J, Zurawski G. Interleukin-13, a novel T cell-derived cytokine that regulates human monocyte and B cell function. Proc Natl Acad Sci USA 1993; 90:3735–3739.

16. Minty A, Chalon P, Derocq J-M, Dumont X, Guillemot J-C, Kaghad M, Labit C, Leplatois P, Liauzun P, Miloux B, Minty C, Casellas P, Loison G, Lupker J, Shire D, Ferrara P, Caput D. Interleukin-13 is a new human lymphokine regulating inflammatory and immune responses. Nature 1993; 362:248–250.

17. Park LS, Friend D, Sassenfeld HM, Urdal DL. Characterization of the human B cell stimulatory factor 1 receptor. J Exp Med 1987; 166:476–488.

18. Cabrillat H, Galizzi JP, Djossou O, Arai N, Yokota T, Arai K, Banchereau J. High affinity binding of human interleukin 4 to cell lines. Biochem Biophys Res Commun 1987; 149:995–1001.

19. Galizzi J-P, Zuber CE, Cabrillat H, Djossou O, Banchereau J. Internalization of human interleukin 4 and transient down-regulation of its receptor in the CD23-inducible Jijoye cells. J Biol Chem 1989; 264:6984–6989.

20. Foxwell BMJ, Woerly G, Ryffel B. Identification of interleukin-4 receptor-associated proteins and expression of both high and low affinity binding on human lymphoid cells. Eur J Immunol 1989; 19:1637–1641.

21. Keegan AD, Beckmann MP, Park LS, Paul WE. The IL-4 receptor: Biochemical characterization of IL-4-binding molecules in a T cell line expressing large numbers of receptors. J Immunol 1991; 146:2272–2279.

22. Galizzi JP, Zuber CE, Harada N, Gorman DM, Djossou O, Kastelein R, Banchereau J, Howard M, Miyajima A. Molecular cloning of a cDNA encoding the human interleukin-4 receptor. Int Immunol 1990; 2:669–675.

23. Idzerda RL, March CJ, Mosley B, Lyman SD, Vanden Bos T, Gimpel SD, Din WS, Grabstein KH, Widmer MB, Park LS, Cosman D, Beckmann MP. Human interleukin 4 receptor confers biological responsiveness and defines a novel receptor superfamily. J Exp Med 1990; 171:861–873.

24. Miyajima A, Kitamura T, Harada N, Yokota T, Arai K. Cytokine receptors and signal transduction. Annu Rev Immunol 1992; 10:295–331.

25. Murakami M, Narazaki M, Hibi M, Yawata H, Yasukawa K, Hamaguchi M, Taga T, Kishimoto T. A critical cytoplasmic region of the interleukin 6 signal transducer gp130 is conserved in the cytokine receptor family. Proc Natl Acad Sci USA 1991; 88:11349–11353.

26. Kondo M, Takeshita T, Ishii N, Nakamura M, Watanabe S, Arai K-A, Sugamura K. Sharing of the interleukin-2 (IL-2) receptor γ chain between receptors for IL-2 and IL-4. Science 1993; 262:1874–1877.

27. Noguchi M, Nakamura Y, Russel SM, Ziegler SF, Tsang M, Cao X, Leonard WJ. Interleukin-2 receptor γ chain: a functional component of the interleukin-7 receptor. Science 1993; 262:1877–1880.

28. Russel SM, Keegan AD, Harada N, Nakamura Y, Noguchi M, Leland P, Friedmann MC, Miyajima A, Puri RK, Paul WE, Leonard WJ. Interleukin-2 receptor γ chain: a functional component of the interleukin-4 receptor. Science 1993; 262:1880–1883.

29. Russell SM, Johnston JA, Noguchi M, Kawamura M, Bacon CM, Friedmann M, Berg M, McVicar DW, Witthuhn BA, Silvennoinen O, Goldman AS, Schmalstieg FC, Ihle JN, O'Shea JJ, Leonard WJ. Interaction of IL-2Rβ and γc chains with Jak1 and Jak3: implications for XSCID and XCID. Science 1994; 266:1042.

30. Giri JG, Ahdieh M, Eisenman J, Shanebeck K, Grabstein K, Kumaki S, Namen A, Park LS, Cosman D, Anderson D. Utilization of the β and γ chains of the IL-2 receptor by the novel cytokine IL-15. EMBO J 1994; 13:2822–2830.

31. Noguchi M, Yi H, Rosenblatt HM, Filipovich AH, Adelstein S, Modi WS, McBride OW, Leonard WJ. Interleukin-2 receptor γ chain mutation results in X-linked severe combined immunodeficiency in humans. Cell 1993; 73:147–157.

32. Fanslow WC, Spriggs MK, Rauch CT, Clifford KN, Macduff BM, Ziegler SF, Schooley KA, Mohler KM, March CJ, Armitage RJ. identification of a distinct low-affinity receptor for human interleukin-4 on pre-B cells. Blood 1993; 81:2998–3005.

33. Keegan AD, Nelms K, Wang LM, Pierce JH, Paul WE. Interleukin-4 receptor: signaling mechanisms. Immunol Today 1994; 15:423–432.

34. Wang L-M, Keegan AD, Paul WE, Heidaran MA, Gutkind JS, Pierce JH. IL-4 activates a distinct signal transduction cascade from IL-3 in factor-dependent myeloid cells. EMBO J 1992; 11:4899–4908.

35. Izuhara K, Harada N. Interleukin-4 (IL-4) induces protein tyrosine phosphorylation of the IL-4 receptor and association of phosphatidylinositol 3-kinase to the IL-4 receptor in a mouse T cell line, HT2. J Biol Chem 1993; 268:13097–13102.

36. Yin T, Tsang ML-S, Yang Y-C. JAK1 kinase forms complexes with interleukin-4 receptor and 4PS/insulin receptor substrate-1-like protein and is activated by interleukin-4 and interleukin-9 in T lymphocytes. J Mol Biol 1994; 269:26614–26617.

37. Witthuhn BA, Silvennoinen O, Miura O, Lai KS, Cwik C, Liu ET, Ihle JN. Involvement of the Jak-3 Janus kinase in signalling by interleukin 2 and 4 in lymphoid and myeloid cells. Nature 1994; 370:153–157.

38. Wang L-M, Keegan AD, Li W, Lienhard GE, Pacini S, Gutkind JS, Myers MG, Sun X-J, White MF, Aaronson SA, Paul WE, Pierce JH. Common elements in interleukin-4 and insulin signaling pathways in factor-dependent hematopoietic cells. Proc Natl Acad Sci USA 1993; 90:4032–4036.

39. Keegan AD, Nelms K, White M, Wang LM, Pierce JH, Paul WE. An IL-4 receptor region containing an insulin receptor motif is important for IL-4-mediated IRS-1 phosphorylation and cell growth. Cell 1994; 76:811–820.

40. Harada N, Yang G, Miyajima A, Howard M. Identification of an essential region for growth signal transduction in the cytoplasmic domain of the human interleukin-4 receptor. J Biol Chem 1992; 267:22752–22758.

41. Izuhara K, Feldman RA, Greer P, Harada N. Interaction of the c-fes protooncogene product with the interleukin-4 receptor. J Biol Chem 1994; 269:18623–18629.

42. Hou J, Schindler U, Henzel WJ, Ho TC, Brasseur M, McKnight SL. An interleukin-4-induced transcription factor: IL-4 STAT. Science 1994; 265:1701–1706.

43. Kotanides H, Reich NC. Requirement of tyrosine phosphorylation for rapid activation of a DNA binding factor by IL-4. Science 1993; 262:1265–1267.

44. Schindler C, Kashleva H, Pernis A, Pine R, Rothman P. STF-IL-4: a novel IL-4-induced signal transducing factor. EMBO J 1994; 13:1350–1356.

45. Dokter WHA, Esselink MT, Halie MR, Vellenga E. Interleukin-4 inhibits the lipopolysaccharide-induced expression of c-*jun* and c-*fos* messenger RNA and activator protein-1 binding activity in human monocytes. Blood 1993; 81:337–343.

46. Finney M, Guy GR, Michell RH, Gordon J, Dugas B, Rigley KP, Callard RE. Interleukin 4 activates human B lymphocytes via transient inositol lipid hydrolysis and delayed cyclic adenosine monophosphate generation. Eur J Immunol 1990; 20:151–156.

47. Mizuguchi J, Beaven MA, O'Hara J, Paul WE. BSF-1 action on resting B-cells does not require elevation of inositol phospholipid metabolism or increased (Ca^{2+}). J Immunol 1986, 137:2215–2219.

48. Justement L, Chen ZZ, Harris LK, Ransom JT, Sandoval VS, Smith C, Rennick D, Roehm N, Cambier J. BSF-1 induces membrane protein phosphorylation but not phosphoinositide metabolism, Ca^{2+} mobilization, protein kinase C translocation, or membrane depolarization in resting murine B lymphocytes. J Immunol 1986; 137:3664–3670.

49. Arruda S, Ho JL. IL-4 receptor signal transduction in human monocytes is associated with protein kinase C translocation. J Immunol 1992; 149:1258–1264.

50. Blay JY, Branellec D, Robinet E, Dugas B, Gay F, Chouaib S. Involvment of cAMP in the IL-4 inhibitory effect on IL-2 induced LAK generation. J Clin Invest 1990; 85:1909–1913.

51. Kolb JP, Paul-Eugène N, Damais C, Yamaoka K, Drapier JC, Dugas B. Interleukin-4 stimulates cGMP production by IFN-γ-activated human monocytes. J Biol Chem 1994; 269:9811–9816.

52. Mire-Sluis AR, Thorpe R. Interleukin-4 proliferative signal transduction involves the activation of a tyrosine-specific phosphatase and the dephosphorylation of an 80-kDa protein. J Biol Chem 1991; 266:18113–18118.

53. Peleman R, Wu J, Fargeas C, Delespesse G. Recombinant interleukin-4 suppresses the production of interferon γ by human mononuclear cells. J Exp Med 1989; 170:1751–1756.

54. Vieira P, de Waal-Malefyt R, Dang MN, Johnson KE, Kastelein R, Fiorentino DF, de Vries JE, Roncarolo MG, Mosmann TR, Moore KW. Isolation and expression of human cytokine synthesis inhibitory factor (CSIF/IL-10) cDNA clones: homology to Epstein-Barr virus open reading frame BCRF1. Proc Natl Acad Sci USA 1991; 88:1172–1176.

55. Coffman RL, Varkila K, Scott P, Chatelain R. Role of cytokines in the differentiation of CD4+ T cell subsets in vivo. Immunol Rev 1991; 123:189–207.

56. Maggi E, Parronchi P, Manetti R, Simonelli C, Piccinni M-P, Rugiu FS, De Carli M, Ricci M, Romagnani S. Reciprocal regulatory effects of IFN-γ and IL-4 on the in vitro development of human Th1 and Th2 clones. J Immunol 1992; 148:2142–2147.

57. Swain SL, Bradley LM, Croft M, Tonkonogy S, Atkins G, Weinberg AD, Duncan DD, Hedrick SM, Dutton RW, Huston G. Helper T-cell subsets: phenotype, function and the role of lymphokines in regulating their development. Immunol Rev 1991; 123:115–144.

58. Seder RA, Paul WE, Davis MM, Fazekas de St. Groth BF. The presence of interleukin 4 during in vitro priming determines the lymphokine-producing potential of CD4+ T cells from T cell receptor transgenic mice. J Exp Med 1992; 176:1091–1098.

59. Manetti R, Gerosa F, Giudizi MG, Biagiotti R, Parronchi P, Piccini MP, Sampognaro S, Maggi E, Romagnani S, Trinchieri G. Interleukin 12 induces stable priming for interferon gamma (IFN-gamma) production during differentiation of human T helper (Th) cells and transient IFN-gamma production in established Th2 cell clones. J Exp Med 1994; 179:1273–1283.

60. Salgame P, Abrams JS, Clayberger C, Goldstein H, Convit J, Modlin RL, Bloom BR. Differing lymphokine profiles of functional subsets of human CD4 and CD8 T cell clones. Science 1991; 254:279–282.

61. Maggi E, Giudizi MG, Biagiotti R, Annunziato F, Manetti R, Piccinni M-P, Parronchi P, Sampognaro S, Giannarini L, Zuccati G, Romagnani S. Th2-like CD8+ T cells showing B cell helper function and reduced cytolytic activity in human immunodeficiency virus type 1 infection. J Exp Med 1994; 180:489–495.

62. Yoshimoto T, Paul WE. CD4pos, NK1.1pos T cells promptly produce interleukin 4 in response to in vivo challenge with anti-CD3. J Exp Med 1994; 179:1285–1295.

63. Arase H, Arase N, Nakagawa K, Good RA, Onoe K. NK1.1⁺ CD4⁺ CD8⁺ thymocytes with specific lymphokine secretion. Eur J Immunol 1993; 23:307–310.

64. Brown MA, Pierce JH, Watson CJ, Falco J, Ihle JN, Paul WE. B cell stimulatory factor-1/interleukin-4 mRNA is expressed by normal and transformed mast cells. Cell 1987; 50:809–818.

65. Brunner T, Heusser CH, Dahinden CA. Human peripheral blood basophils primed by interleukin-3 (IL-3) produce IL-4 in response to immunoglobulin E receptor stimulation. J Exp Med 1993; 177:605–611.

66. Piccinni MP, Macchia D, Parronchi P, Giudizi MG, Bani D, Alterini R, Grossi A, Ricci M, Maggi E, Romagnani S. Human bone marrow non-B, non-T cells produce IL-4 in response to cross-linkage of Fcε and Fcγ receptors. Proc Natl Acad Sci USA 1991; 88:8656–8660.

67. Bradding P, Feather IH, Howarth PH, Mueller R, Roberts JA, Britten K, Bews JPA, Hunt TG, Okayama Y, Heusser CH. Interleukin 4 is localized to and released by human mast cells. J Exp Med 1992; 176:1381–1386.

68. Schmitz J, Thiel A, Kühn R, Rajewsky K, Müller W, Assenmacher M, Radbruch A. Induction of Interleukin 4 (IL-4) expression in T helper (Th) cells is not dependent on IL-4 from non-Th cells. J Exp Med 1994; 179:1349–1353.

69. King AG, Wierda D, Landreth KS. Bone marrow stromal cell regulation of B-lymphopoiesis. I. The role of macrophages, IL-1, and IL-4 in pre-B cell maturation. J Immunol 1988; 141:2016–2026.

70. Kinashi T, Inaba K, Tsubata T, Tashiro K, Palacios R, Honjo T. Differentiation of an interleukin 3-dependent precursor B-cell clone into immunoglobulin-producing cells in vitro. Proc Natl Acad Sci USA 1988; 85:4473–4477.

71. Vallé A, Zuber CE, Defrance T, Djossou O, De Rie M, Banchereau J. Activation of human B lymphocytes through CD40 and interleukin 4. Eur J Immunol 1989; 19:1463–1467.

72. Defrance T, Aubry JP, Rousset F, Vanbervliet B, Bonnefoy JY, Arai N, Takebe Y, Yokota T, Lee F, Arai K, De Vries J, Banchereau J. Human recombinant interleukin 4 induces Fcε receptors (CD23) on normal human B lymphocytes. J Exp Med 1987; 165:1459–1467.

73. Yokota A, Kikutani H, Tanaka T, Sato R, Barsumian EL, Suemura M, Kishimoto T. Two species of human Fc epsilon receptor II (Fc epsilon RII/CD23): tissue-specific and IL-4-specific regulation of gene expression. Cell 1988; 55:611–618.

74. Banchereau J, Bazan F, Blanchard D, Brière F, Galizzi J-P, van Kooten C, Liu Y-J, Rousset F, Saeland S. The CD40 antigen and its ligand. Annu Rev Immunol 1994; 12:881–922.

75. Shields JG, Armitage RJ, Jamieson BN, Beverley PCL, Callard RE. Increased expression of surface IgM but not IgD or IgG on human B cells in response to IL-4. Immunology 1989; 66:224–227.

76. Vallé A, Aubry J-P, Durand I, Banchereau J. IL-4 and IL-2 upregulate the expression of antigen B7, the B cell counterstructure to T cell CD28: an amplification mechanism for T–B cell interactions. Int Immunol 1991; 3:229–235.

77. Ranheim EA, Kipps TJ. Activated T cells induce expression of B7/BB1 on normal or leukemic B cells through a CD40-dependent signal. J Exp Med 1993; 177:925–935.

78. Smeland EB, Blomhoff HK, Funderud S, Shalaby MR, Espevik T. Interleukin 4 induces selective production of interleukin 6 from normal human B lymphocytes. J Exp Med 1989; 170:1463–1468.

79. Defrance T, Vanbervliet B, Aubry JP, Takebe Y, Arai N, Miyajima A, Yokota T, Lee T, Arai K, de Vries JE, Banchereau J. B cell growth-promoting activity of recombinant human interleukin-4. J Immunol 1987; 139:1135–1141.

80. Llorente L, Mitjavila F, Crevon MC, Galanaud P. Dual effects of interleukin 4 on antigen-activated human B cells: induction of proliferation and inhibition of interleukin 2-dependent differentiation. Eur J Immunol 1990; 20:1887–1892.

81. Banchereau J, de Paoli P, Vallé A, Garcia E, Rousset F. Long term human B cell lines dependent on interleukin 4 and antibody to CD40. Science 1991; 251:70–72.

82. Defrance T, Vanbervliet B, Aubry JP, Banchereau J. Interleukin 4 inhibits the proliferation but not the differentiation of activated human B cells in response to interleukin 2. J Exp Med 1988; 168:1321–1337.

83. Jelinek DF, Lipsky PE. Inhibitory influence of IL-4 on human B cell responsiveness. J Immunol 1988; 141:164–173.

84. Coffman RL, Ohara J, Bond MW, Carty J, Zlotnick A, Paul WE. B cell stimulatory factor-1 enhances the IgE response of lipopolysaccharide-activated B cells. J Immunol 1986; 136:4538–4541.

85. Pène J, Rousset F, Brière F, Chrétien I, Bonnefoy JY, Spits H, Yokota T, Arai N, Arai KI, Bancherreau J, De Vries JE. IgE production by normal human lymphocytes is induced by interleukin 4 and suppressed by interferons gamma and alpha and prostaglandin E$_2$. Proc Natl Acad Sci USA 1988; 85:6880–6884.

86. Jabara HH, Fu SM, Geha RS, Vercelli D. CD40 and IgE: synergism between anti-CD40 monoclonal antibody and interleukin 4 in the induction of IgE synthesis by highly purified human B cells. J Exp Med 1990; 172:1861–1864.

87. Rousset F, Garcia E, Bancherreau J. Cytokine-induced proliferation and immunoglobulin production of human B lymphocytes triggered through their CD40 antigen. J Exp Med 1991; 173:705–710.

88. Shapira SK, Jabara HH, Thienes CP, Ahern DJ, Vercelli D, Gould HJ, Geha RS. Deletional switch recombination occurs in interleukin-4 induced isotype switching to IgE expression by human B cells. Proc Natl Acad Sci USA 1991; 88:7528–7532.

89. Shapira SK, Vercelli D, Jabara HH, Fu SM, Geha RS. Molecular analysis of the induction of immunoglobulin E synthesis in human B cells by interleukin 4 and engagement of CD40 antigen. J Exp Med 1992; 175:289–292.

90. Mills FC, Thyphronitis G, Finkelman FD, Max EE. Immunoglobulin μ–ε isotype switch in IL-4 treated human B lymphoblastoid cells; evidence for a sequential switch. J Immunol 1992; 149:1075–1085.

91. Jabara HH, Loh R, Ramesh N, Vercelli D, Geha RS. Sequential switching from μ to epsilon via gamma4 in human B cells stimulated with IL-4 and hydrocortisone. J Immunol 1993; 151:4528–4533.

92. Zhang K, Mills FC, Saxon A. Switch circles from IL-4-directed epsilon class switching from human B lymphocytes. Evidence for direct, sequential, and multiple step sequential switch from μ to ε Ig heavy chain gene. J Immunol 1994; 152:3427–3435.

93. Gascan H, Gauchat J-F, Roncarolo M-G, Yssel H, Spits H, de Vries JE. Human B cell clones can be induced to proliferate and to switch to IgE and IgG4 synthesis by interleukin 4 and a signal provided by activated CD4$^+$ T cell clones. J Exp Med 1991; 173:747–750.

94. Finkelman FD, Holmes J, Katona IM, Urban JF, Beckmann MP, Schooley KA, Coffman RL, Mosmann TR, Paul WE. Lymphokine control of in vivo immunoglobulin isotype selection. Annu Rev Immunol 1990; 8:303–333.

95. Kühn R, Rajewsky K, Müller W. Generation and analysis of interleukin-4 deficient mice. Science 1991; 254:707–710.

96. Tepper RI, Levinson DA, Stanger BZ, Campos-Torres J, Abbas AK, Leder P. IL-4 induces allergic-like inflammatory disease and alters T cell development in transgenic mice. Cell 1990; 62:457–467.

97. Zurawski G, de Vries JE. Interleukin 13, an interleukin 4-like cytokine that acts on monocytes and B cells, but not on T cells. Immunol Today 1994; 15:19–26.

98. Renard N, Duvert V, Bancherreau J, Saeland S. Interleukin-13 inhibits the proliferation of normal and leukemic human B-cell precursors. Blood 1994; 84:2253–2260.

99. Mosmann TR, Bond MW, Coffman RL, Ohara J, Paul WE. T-cell and mast cell lines respond to B-cell stimulatory factor 1. Proc Natl Acad Sci USA 1986; 83:5654–5658.

100. Hu-Li J, Shevach EM, Mizuguchi J, Ohara J, Mosmann T, Paul WE. B cell stimulatory factor 1 (interleukin 4) is a potent costimulant for normal resting T lymphocytes. J Exp Med 1987; 165:157–172.

101. Spits H, Yssel H, Takebe Y, Arai N, Yokota T, Lee F, Arai K, Bancherreau J, de Vries JE. Recombinant interleukin-4 promotes the growth of human T cells. J Immunol 1987; 135:1142–1147.

102. Harel-Bellan A, Durum S, Muegge K, Abbas AK, Farrar WL. Specific inhibition of lymphokine biosynthesis and autocrine growth using antisense oligonucleotides in Th1 and Th2 helper T cell clones. J Exp Med 1988; 168:2309–2318.

103. Gaya A, Alsinet E, Martorell J, Places L, De La Calle O, Yagüe J, Vives J. Inhibitory effect of IL-4 on the sepharose–CD3-induced proliferation of the CD4CD45RO human T cell subset. Int Immunol 1990; 2:685–689.

104. Paliard X, de Waal Malefijt R, de Vries JE, Spits H. Interleukin 4 can mediate the induction of CD8 on human CD4+ T cell clones. Nature 1988; 335:642–644.

105. Prinz JC, Baur X, Mazur G, Rieber EP. Allergen-directed expression of Fc receptors for IgE (CD23) on human T lymphocytes is modulated by interleukin-4 and interferon-γ. Eur J Immunol 1990; 20:1259–1264.

106. Widmer MB, Acres RB, Sassenfeld HM, Grabstein KH. Regulation of cytolytic cell populations from human peripheral blood by B cell stimulatory factor 1 (interleukin 4). J Exp Med 1987; 166:1447–1455.

107. Widmer MB, Grabstein KH. Regulation of cytolytic T lymphocyte generation by B-cell stimulatory factor. Nature 1987; 326:795–798.

108. Spits H, Yssel H, Paliard X, Kastelein R, Figdor D, de Vries JE. Interleukin-4 inhibits interleukin-2 mediated induction of human lymphokine activated killer cells, but not the generation of antigen specific cytotoxic T lymphocytes in mixed leucocyte cultures. J Immunol 1988; 141:29–36.

109. Horohov DW, Crim JA, Smith PL, Siegel JP. IL-4 (B cell-stimulatory factor 1) regulates multiple aspects of influenza virus-specific cell-mediated immunity. J Immunol 1988; 141:4217–4223.

110. Thornhill MH, Kyan-Aung U, Haskard DO. IL-4 increases human endothelial cell adhesiveness for T cells but not for neutrophils. J Immunol 1990; 144:3060–3065.

111. Nagler A, Lanier LL, Phillips JH. The effects of IL-4 on human natural killer cells. A potent regulator of IL-2 activation and proliferation. J Immunol 1988; 141:2349–2351.

112. Kawakami Y, Custer MC, Rosenberg SA, Lotze MT. IL-4 regulates IL-2 induction of lymphokine-activated killer activity from human lymphocytes. J Immunol 1989; 142:3452–3461.

113. Te Velde AA, Klomp JPG, Yard BA, de Vries JE, Figdor CG. Modulation of phenotypic and functional properties of human peripheral blood monocytes by IL-4. J Immunol 1988; 140:1548–1554.

114. Vercelli D, Jabara HH, Lee W, Woodland N, Geha RS, Leung DYM. Human recombinant interleukin 4 induces Fcε R2/CD23 on normal human monocytes. J Exp Med 1988; 167:1406–1416.

115. Te Velde AA, Rousset F, Péronne C, de Vries JE, Figdor C. IL-4 induced expression of FcεRIIb and soluble FcεRIIb release from human monocytes is downregulated by IFNα and IFNγ. J Immunol 1990; 144:3052–3059.

116. Sallusto F, Lanzavecchia A. Efficient presentation of soluble antigen by cultured human dendritic cells is maintained by granulocyte/macrophage colony-stimulating factor plus interleukin 4 and downregulated by tumor necrosis factor alpha. J Exp Med 1994; 179:1109–1118.

117. McInnes R, Rennick DM. Interleukin 4 induces cultured monocytes/macrophages to form giant multinucleated cells. J Exp Med 1988; 167:598–611.

118. Hart PH, Vitti GF, Burgess DR, Whitty GA, Piccoli DS, Hamilton JA. Potential antiinflammatory effects of interleukin 4: suppression of human monocyte tumor necrosis factor α, interleukin 1, and prostaglandin E2. Proc Natl Acad Sci USA 1989; 86:3803–3807.

119. Vannier E, Miller LC, Dinarello CA. Coordinated anti-inflammatory effects of interleukin 4: IL-4 suppresses IL-1 production but up-regulates gene expression and synthesis of interleukin 1 receptor antagonist. Proc Natl Acad Sci USA 1992; 89:4076–4080.

120. Corcoran ML, Stetler-Stevenson WG, Brown PD, Wahl LM. Interleukin 4 inhibition of prostaglandin E2 synthesis blocks interstitial collagenase and 92-kDa type IV collagenase/gelatinase production by human monocytes. J Biol Chem 1992; 267:515–519.

121. Lacraz S, Nicod L, Galve-de-Rochemonteix B, Baumberger C, Dayer J-M, Welgus HG. Suppression of metalloproteinase biosynthesis in human alveolar macrophages by interleukin-4. J Clin Invest 1992; 90:382–388.

122. Abramson SL, Gallin JI. IL-4 inhibits superoxide production by human mononuclear phagocytes. J Immunol 1990; 144:625–630.

123. Gautam SC, Chikkala NF, Hamilton TA. Anti-inflammatory action of IL-4. J Immunol 1992; 148:1411–1415.

124. van Hal PTW, Hopstaken-Broos JPM, Prins A, Favaloro EJ, Huijbens RJF, Hilvering C, Figdor CG, Hoogsteden HC. Potential indirect anti-inflammatory effects of IL-4. Stimulation of human monocytes, macrophages, and endothelial cells by IL-4 increases aminopeptidase-N activity (CD13; EC 3.4.11.2). J Immunol 1994; 153:2718–2728.

125. Lehn M, Weiser WY, Engelhorn S, Gillis S, Remold HG. IL-4 inhibits H_2O_2 production and antileishmanial capacity of human cultured monocytes mediated by IFN-γ. J Immunol 1989; 143:3020–3024.

126. Kumaratilake LM, Ferrante A. IL-4 inhibits macrophage-mediated killing of *Plasmodium falciparum* in vitro. J Immunol 1992; 149:194–199.

127. Schuitemaker H, Kootstra NA, Koppelman MHGM, Bruisten SM, Huisman HG, Tersmette M, Miedema F. Proliferation-dependent HIV-1 infection of monocytes occurs during differentiation into macrophages. J Clin Invest 1992; 89:1154–1160.

128. Montaner LJ, Doyle AG, Collin M, Herbein G, Illei P, James W, Minty A, Caput D, Ferrara P, Gordon S. Interleukin 13 inhibits human immunodeficiency virus type 1 production in primary blood-derived human macrophages in vitro. J Exp Med 1993; 178:743–747.

129. Baskar P, Silberstein DS, Pincus SH. Inhibition of IgG-triggered human eosinophil function by IL-4. J Immunol 1990; 144:2321–2326.

130. Thornhill MH, Kaskard DO. IL-4 regulates endothelial cell activation by IL-1, tumor necrosis factor, or IFN-γ. J Immunol 1990; 145:865–872.

131. Moser R, Fehr J, Bruijnzeel PLB. IL-4 controls the selective endothelium-driven transmigration of eosinophils from allergic individuals. J Immunol 1992; 149:1432–1438.

132. Schleimer RP, Sterbinsky SA, Kaiser J, Bickel CA, Klunk DA, Tomioka K, Newman W, Luscinskas FW, Gimbrone MA, McIntyre BW, Bochner BS. IL-4 induces adherence of human eosinophils and basophils but not neutrophils to endothelium. Association with expression of VCAM-1. J Immunol 1992; 148:1086–1092.

133. Tepper RI, Coffman RL, Leder P. An eosinophil-dependent mechanism for the antitumor effect of interleukin-4. Science 1992; 257:548–551.

134. Boey H, Rosenbaum R, Castracane J, Borish L. Interleukin-4 is a neutrophil activator. J Allergy Clin Immunol 1989; 83:978–984.

135. Wertheim WA, Kunkel SL, Standiford TJ, Burdick MD, Becker FS, Wilke CA, Gilbert AR, Strieter RM. Regulation of neutrophil-derived IL-8: the role of prostaglandin E_2, dexamethasone, and IL-4. J Immunol 1993; 151:2166–2175.

136. Colotta F, Re F, Muzio M, Bertini R, Polentarutti N, Sironi M, Giri JG, Dower SK, Sims JE, Mantovani A. Interleukin-1 type II receptor: as a decoy target for IL-1 that is regulated by IL-4. Science 1993; 261:472–475.

137. Peschel C, Paul WE, Ohara J, Green I. Effects of B cell stimulatory factor-1/interleukin 4 on hematopoietic progenitor cells. Blood 1987; 70:254–263.

138. Kishi K, Ihle JN, Urdal DL, Ogawa M. Murine B-cell stimulatory factor-1 (BSF-1)/interleukin-4 (IL-4) is a multilineage colony-stimulating factor that acts directly on primitive hemopoietic progenitors. J Cell Physiol 1989; 139:463–468.

139. Rennick D, Jackson J, Moulds C, Lee F, Yang G. IL-3 and stromal cell-derived factor synergistically stimulate the growth of pre-B cell lines cloned from long-term lymphoid bone marrow cultures. J Immunol 1989; 142:161–166.

140. Sonoda Y, Okuda T, Yokota S, Maekawa T, Shizumi Y, Nishigaki H, Misawa S, Fujii H, Abe T. Actions of human interleukin-4/B-cell stimulatory factor-1 proliferation and differentiation of enriched hematopoietic progenitor cells in culture. Blood 1990; 75:1615–1621.

141. Keller U, Aman MJ, Derigs G, Huber C, Peschel C. Human interleukin-4 enhances stromal cell-dependent hematopoiesis: costimulation with stem cell factor. Blood 1994; 84:2189–2196.

142. Rennick D, Yang G, Muller-Sieburg C, Smith C, Arai N, Takabe Y, Gemmell L. Interleukin 4 (B-cell stimulatory factor 1) can enhance or antagonize the factor-dependent growth of hemopoietic progenitor cells. Proc Natl Acad Sci USA 1987; 84:6889–6893.

143. Broxmeyer HE, Lu L, Cooper S, Tushinski R, Mochizuki D, Rubin BY, Gillis S, William DE.

Synergistic effects of purified recombinant human and murine B cell growth factor-1/IL-4 on colony formation in vitro by hematopoietic progenitor cells. J Immunol 1988; 141:3852–3862.

144. Vellenga E, De Wolf JT, Beentjes JAM, Esselink MT, Smit JW, Haee MR. Divergent effects of interleukin 4 (IL-4) on the granulocyte colony stimulating factor and IL-3 supported myeloid colony formation from normal and leukemic bone marrow cells. Blood 1990; 75:633–637.

145. Snoeck H-W, Lardon F, Lenjou M, Nys G, Van Bockstaele DR, Peetermans ME. Interferon-γ and interleukin-4 reciprocally regulate the production of monocytes/macrophages and neutrophils through a direct effect on committed monopotential bone marrow progenitor cells. Eur J Immunol 1993; 23:1072–1077.

146. Snoeck H-W, Lardon F, Lenjou M, Nys G, Van Bockstaele DR, Peetermans ME. Differential regulation of the expression of CD38 and human leukocyte antigen-DR on CD34$^+$ hematopoietic progenitor cells by interleukin-4 and interferon-γ. Exp Hematol 1993; 21:1480–1486.

147. Brandt JE, Baird N, Lu L, Srour E, Hoffman R. Characterization of a human hematopoietic progenitor cell capable of forming blast cell containing colonies in vitro. J Clin Invest 1988; 82:1017–1027.

148. Hamaguchi Y, Kanakura Y, Fusita J, Takeoa SI, Nakano T, Tarui S, Honjo T, Kitamura Y. Interleukin 4 as an essential factor for in vitro clonal growth of murine connective tissue-type mast cells. J Exp Med 1987; 165:268–273.

149. Tsuji K, Nakahata T, Takagi M, Kobayashi T, Ishiguro A, Kikuchi T, Naganuma K, Koike K, Miyajima A, Arai K, Akabane T. Effects of interleukin 3 and interleukin 4 on the development of connective tissue type mast cells: interleukin 3 supports their survival and interleukin 4 triggers and supports their proliferation synergistically with interleukin 3. Blood 1990; 75:421–427.

150. Favre C, Saeland S, Caux C, Duvert V, De Vries JE. Interleukin 4 has basophilic and eosinophilic cell growth-promoting activity on cord blood cells. Blood 1990; 75:67–73.

151. Sillaber C, Strobl H, Bevec D, Ashman LK, Butterfield JH, Lechner K, Maurer D, Bettelheim P, Valent P. IL-4 regulates c-kit proto-oncogene product expression in human mast and myeloid progenitor cells. J Immunol 1991; 147:4224–4228.

152. Sillaber C, Sperr WR, Agis H, Geissler K, Spanblöchl E, Ashman LK, Bettelheim P, Lechner K. IL-4 inhibits SCF-dependent growth of human mast cells and downregulates c-kit expression. Blood 1992; 80(suppl):92a.

153. Welham MJ, Schrader JW. Modulation of c-kit mRNA and protein by hemopoietic growth factors. Mol Cell Biol 1991; 11:2901–2904.

154. Tepper RI, Pattengale PK, Leder P. Murine interleukin-4 displays potent anti-tumor activity in vivo. Cell 1989; 57:503–512.

155. Jansen JH, Wientjens G-JHM, Fibbe WE, Willemze R, Kluin-Nelemans HC. Inhibition of human macrophage colony formation by interleukin 4. J Exp Med 1989; 170:577–582.

156. Peschel C, Green I, Paul WE. Interleukin-4 induces a substance in bone marrow stromal cells that reversibly inhibits factor-dependent and factor-independent cell proliferation. Blood 1989; 73:1130–1141.

157. Romani N, Gruner S, Brang D, Kämpgen E, Lenz A, Trockenbacher B, Konwalinka G, Fritsch PO, Steinman RM, Schuler G. Proliferating dendritic cell progenitors in human blood. J Exp Med 1994; 180:83–93.

158. Sonoda Y, Kuzuyama Y, Tanaka S, Yokota S, Maekawa T, Clark SC, Abe T. Human interleukin-4 inhibits proliferation of megakaryocyte progenitor cells in culture. Blood 1993; 81:624–630.

159. de Wolf JTM, Beentjes JAM, Esselink MT, Smit JW, Halie MR, Vellenga E. Interleukin-4 suppresses the interleukin-3 dependent erythroid colony formation from normal human bone marrow cells. Br J Haematol 1990; 74:246–250.

160. Peschel C, Green I, Paul WE. Preferential proliferation of immature B lineage cells in long-term stromal cell-dependent cultures with IL-4. J Immunol 1989; 142:1558–1568.

161. Sideras P, Palacios R. Bone marrow pro-T and pro-B lymphocyte clones express functional receptors for interleukin (IL)3 and IL-4/BSF-1 and nonfunctional receptors for IL2. Eur J Immunol 1987; 17:217–221.

162. Kinashi T, Tashiro K, Inaba K, Takeda T, Palacios R, Honjo T. An interleukin-4-dependent precursor clone is an intermediate of the differentiation pathway from an interleukin-3-dependent precursor clone into myeloid cells as well as B lymphocytes. Int Immunol 1989; 1:11–19.

163. Billips L, Petitte D, Landreth KS. Bone marrow stromal cell regulation of B lymphopoiesis: interleukin-1 (IL-1) and IL-4 regulate stromal cell support of pre-B cell production in vitro. Blood 1990; 75:611–619.

164. Hofman FM, Brock M, Taylor CR, Lyons B. IL-4 regulates differentiation and proliferation of human precursor B cells. J Immunol 1988; 141:1185–1190.

165. Pandrau D, Saeland S, Duvert V, Durand I, Manel AM, Zabot MT, Philippe N, Banchereau J. Interleukin-4 inhibits in vitro proliferation of leukemic and normal human B cell precursors. J Clin Invest 1992; 90:1697–1706.

166. Ryan DH, Nuccie BL, Ritterman I, Liesveld JL, Abboud CN. Cytokine regulation of early human lymphopoiesis. J Immunol 1994; 152:5250–5258.

167. Law C-L, Armitage RJ, Villablanca JG, LeBien TW. Expression of interleukin-4 receptors on early human B-lineage cells. Blood 1991; 78:703–710.

168. Saeland S, Duvert V, Moreau I, Banchereau J. Human B cell precursors proliferate and express CD23 after CD40 ligation. J Exp Med 1993; 178:113–120.

169. Simons A, Zharhary D. The role of IL-4 in the generation of B lymphocytes in the bone marrow. J Immunol 1989; 143:2540–2545.

170. Punnonen J, Aversa G, de Vries JE. Human pre-B cells differentiate into Ig-secreting plasma cells in the presence of interleukin-4 and activated CD4$^+$ T cells or their membranes. Blood 1993; 82:2781–2789.

171. Zlotnik A, Ramsom J, Franck G, Fischer M. Howard M. Interleukin 4 is a growth factor for activated thymocytes: possible role in T cell ontogeny. Proc Natl Acad Sci USA 1987; 84:3856–3860.

172. Barcena A, Toribio ML, Pezzi L, Martinez-A C. A role for interleukin 4 in the differentiation of mature T cell receptor γ/δ$^+$ cells from human intrathymic T cell precursors. J Exp Med 1990; 172:439–446.

173. Migliorati G, Nicoletti I, Pagliacci MC, D'Adamio L, Riccardi C. Interleukin-4 protects double-negative and CD4 single-positive thymocytes from dexamethasone-induced apoptosis. Blood 1993; 81:1352–1358.

174. Sideras P, Funa K, Zalcberg-Quintana I, Xanthopoulos KG, Kisielow P, Palacios R. Analysis by in situ hybridization of cells expressing mRNA for interleukin 4 in the developing thymus and in peripheral lymphocytes from mice. Proc Natl Acad Sci USA 1988; 85:218–221.

175. Ueno Y, Boone T, Uittenbogaart CH. Selective stimulation of human thymocyte subpopulations by recombinant IL-4 and IL-3. Cell Immunol 1989; 118:382–393.

176. Plum J, De Smedt M, Leclercq G, Tison B. Inhibitory effect of murine recombinant IL-4 on thymocyte development in fetal thymus organ cultures. J Immunol 1990; 145:1066–1073.

177. Wagteveld AJ, van Zanten AK, Esselink MT, Halie MR, Vellenga E. Expression and regulation of IL-4 receptors on human monocytes and acute myeloblastic leukemic cells. Leukemia 1991; 5:782–788.

178. Miyauchi J, Clark SC, Tsunematsu Y, Shimizu K, Park J-W, Ogawa T, Toyama K. Interleukin-4 as a growth regulator of clonogenic cells in acute myelogenous leukemia in suspension culture. Leukemia 1991; 5:108–115.

179. Akashi K, Harada M, Shibuya T, Eto T, Takamatsu Y, Teshima T, Niho Y. Effects of interleukin-4 and interleukin-6 on the proliferation of CD34$^+$ and CD34$^-$ blasts from acute myelogenous leukemia. Blood 1991; 78:197–204.

180. Wagteveld AJ, Esselink MT, Limburg P, Halie MR, Vellenga E. The effects of IL-1β and IL-4 on the proliferation and endogenous secretion of growth factors by acute myeloblastic leukemic cells. Leukemia 1992; 6:1020–1024.

181. Jansen JH, Fibbe WE, Wientjens GJHM, Van Damme J, Landegent JE, Willemze R, Kluin-Nelemans JC. Interleukin 4 down-regulates the expression of CD14 and the production of interleukin 6 in acute myeloid leukemia cells. Lymph Cytol Res 1991; 10:457–461.

182. Noma T, Nakakubo H, Hama K, Honjo T. Multiple effects of human recombinant interleukin 4 on human myeloid monocyte cell lines. Immunol Lett 1989; 21:323–328.

183. Estrov Z, Markowitz AB, Kurzrock R, Wetzler M, Kantarjian HM, Ferrajoli A, Gutterman JU, Talpaz M. Suppression of chronic myelogenous leukemia colony growth by interleukin-4. Leukemia 1993; 7:214–220.

184. Imai Y, Tohda S, Nagata K, Suzuki T, Nara N, Aoki N. Effects of recombinant interleukin 4 on the growth and differentiation of blast progenitors stimulated with G-CSF, GM-CSF and IL-3 from acute myeloblastic leukaemia patients. Br J Haematol 1991; 78:173–179.

185. Jansen JH, Fibbe WE, Wientjens GJHM, Willemze R, Kluin-Nelemans JC. Inhibitory effect of interleukin-4 on the proliferation of acute myeloid leukemia cells with myelo-monocytic differentiation (AML-M4/M5); the role of interleukin-6. Leukemia 1993; 7:643–645.

186. Maekawa T, Sonoda Y, Kuzuyama Y, Inazawa J, Kimura S, Nakamichi K, Abe T. Synergistic suppression of the clonogenicity of U937 leukemic cells by combinations of recombinant human interleukin 4 and granulocyte colony-stimulating factor. Exp Hematol 1992; 20:1201–1207.

187. Akashi K, Shibuya T, Harada M, Takamatsu Y, Uike N, Eto T, Niho Y. Interleukin 4 suppresses the spontaneous growth of chronic myelomonocytic leukemia cells. J Clin Invest 1991; 88:223–230.

188. Kasukabe T, Okabe-Kado J, Honma Y, Hozumi M. Interleukin-4 inhibits the differentiation of mouse myeloid leukemia M1 cells induced by dexamethasone, D-factor/leukemia inhibitory factor and interleukin-6, but not 1α,25-hydroxyvitamin D3. FEBS Lett 1991; 291:181–184.

189. Yamashita U, Tanaka Y, Shirakawa F. Suppressive effect of interleukin-4 on the differentiation of M1 and HL60 myeloid leukemic cells. J Leukoc Biol 1993; 54:133–137.

190. Leung KN, Mak NK, Fung MC, Hapel AJ. Synergistic effect of IL-4 and TNF-α in the induction of monocytic differentiation of a mouse myeloid leukaemic cell line (WEHI-3B JCS). Immunology 1994; 81:65–72.

191. Kasukabe T, Okabe-Kado J, Hozumi M, Honma Y. Inhibition by interleukin 4 of leukemia inhibitory factor-, interleukin 6-, and dexamethasone-induced differentiation of mouse myeloid leukemia cells: role of c-*myc* and *jun* B proto-oncogenes. Cancer Res 1994; 54:592–597.

192. Okabe M, Kuni-eda Y, Sugiwura T, Tanaka M, Miyagishima T, Saiki I, Minagawa T, Kurosawa M, Itaya T, Miyazaki T. Inhibitory effect of interleukin-4 on the in vitro growth of Ph[1]-positive acute lymphoblastic leukemia cells. Blood 1991; 78:1574–1580.

193. Manabe A, Coustan-Smith E, Behm FG, Raimondi SC, Campana D. Bone marrow-derived stromal cells prevent apoptotic cell death in B-lineage acute lymphoblastic leukemia. Blood 1992; 79:2370–2377.

194. Manabe A, Coustan-Smith E, Kumagai M-A, Behm FG, Raimondi SC, Pui C-H, Campana D. Interleukin-4 induces programmed cell death (apoptosis) in cases of high-risk acute lymphoblastic leukemia. Blood 1994; 83:1731–1737.

195. Ouaaz F, Mentz F, Mossalayi MD, Schmitt C, Michel A, Debré P, Guillosson JJ, Merle-Béral H, Arock M. Interleukin 4 inhibits the proliferation and promotes the maturation of human leukemic pre B cells. Eur J Haematol 1993; 51:276–281.

196. Mitchell P, Clutterbuck R, Powles R, De Lord C, Titley J, Millar J. Interleukin-4 (IL-4) improves survival and suppresses human ALL xenografts in SCID mice, confirming in vitro growth inhibition. Blood 1994; 84(suppl):48a.

197. Tushinski RJ, Larsen A, Park LS, Spoor E. Interleukin 4 alone or in combination with interleukin 1 stimulates 3T3 fibroblasts to produce colony-stimulating factors. Exp Hematol 1991; 19:238–244.

198. Lovhaug D, Pelus LM, Nordlie EM, Boyum A, Moore MAS. Monocyte-conditioned medium and interleukin 1 induce granulocyte–macrophage colony-stimulating factor production in the adherent layer of murine bone marrow cultures. Exp Hematol 1986; 14:1037.

199. Dittel BN, McCarthy JB, Wayner EA, LeBien TW. Regulation of human B-cell precursor adhesion to bone marrow stromal cells by cytokines that exert opposing effects on the expression of vascular cell adhesion molecule-1 (VCAM-1). Blood 1993; 81:2272–2282.

200. Teixido J, Hemler ME, Greenberger JS, Anklesaria P. Role of β1 and β2 integrins in the adhesion of human CD34[hi] stem cells to bone marrow stroma. J Clin Invest 1992; 90:358–367.

201. Vollenweider S, Saurat J-H, Röcken M, Hauser C. Evidence suggesting involvement of interleukin-4 (IL-4) production in spontaneous in vitro IgE synthesis in patients with atopic dermatitis. J Allergy Clin Immunol 1991; 87:1088–1095.

202. Chensue SW, Terebuh PD, Warmington KS, Hershey SD, Evanoff HL, Kunkel SL, Higashi GI. Role of IL-4 and IFN-γ in *Schistosoma mansoni* EGG-induced hypersensitivity granuloma formation. Orchestration, relative contribution, and relationship to macrophage function. J Immunol 1992; 148:900–906.

203. Allen JB, Wong HL, Costa GL, Bienkowski MJ, Wahl SM. Suppression of monocyte function and differential regulation of IL-1 and IL-1ra by IL-4 contribute to resolution of experimental arthritis. J Immunol 1993; 151:4344–4351.

204. Mulligan MS, Jones ML, Vaporciyan AA, Howard MC, Ward PA. Protective effects of IL-4 and IL-10 against immune complex-induced lung injury. J Immunol 1993; 151:5666–5674.

205. Banchereau J. Interleukin 4. Trends Exp Clin Med 1994; 4:514–531.

206. Golumbek PT, Lazenby AJ, Levitsky HI, Jaffee LM, Karasuyama H, Baker M, Pardoll DM. Treatment of established renal cancer by tumor cells engineered to secrete interleukin-4. Science 1991; 254:713–716.

207. Bosco MC, Giovarelli M, Forni M, Modesti A, Scarpa G, Masuelli L, Forni G. Low doses of interleukin 4 injected perilymphatically in tumor-bearing mice inhibit the growth of poorly and apparently nonimmunogenic tumors and induce a tumor-specific immune memory. J Immunol 1990; 145:3136–3143.

208. Kawakami Y, Rosenberg SA, Lotze MT. Interleukin 4 promotes the growth of tumor-infiltrating lymphocytes cytotoxic for human autologous melanoma. J Exp Med 1988; 168:2183–2191.

209. Brooks B, Rees RC. Human recombinant IL-4 suppresses the induction of human IL-2 induced lymphokine activated killer (LAK) activity. Clin Exp Immunol 1988; 74:162–165.

210. Hudson MM, Markowitz AB, Gutterman JU, Knowles RD, Snyder JS, Kleinerman ES. Effect of recombinant human interleukin 4 on human monocyte activity. Cancer Res 1990; 50:3154–3158.

211. Hoon DSB, Okun E, Banez M, Irie RF, Morton DL. Interleukin-4 alone and with γ-interferon or α-tumor necrosis factor inhibits cell growth and modulates cell surface antigens on human renal cell carcinomas. Cancer Res 1991; 51:2002–2008.

212. Morisaki T, Yuzuki DH, Lin RT, Foshag LJ, Morton DL, Hoon DSB. Interleukin 4 receptor expression and growth inhibition of gastric carcinoma cells by interleukin 4. Cancer Res 1992; 52:6059–6065.

213. Obiri NI, Hillman GG, Haas GP, Sud S, Puri RK. Expression of high affinity interleukin-4 receptors on human renal cell carcinoma cells and inhibition of tumor cell growth in vitro by interleukin-4. J Clin Invest 1993; 91:88–93.

214. Toi M, Bicknell R, Harris AL. Inhibition of colon and breast carcinoma cell growth by interleukin-4. Cancer Res 1992; 52:275–279.

215. Topp MS, Koenigsmann M, Mire-Sluis A, Oberberg D, Eitelbach F, Von Marschall Z, Notter M, Reufi B, Stein H, Thiel E, Berdel WE. Recombinant human interleukin-4 inhibits growth of some human lung tumor cell lines in vitro and in vivo. Blood 1993; 82:2837–2844.

216. Karray S, Defrance T, Merle-Béral H, Banchereau J, Debré P, Galanaud P. Interleukin 4 counteracts the interleukin 2-induced proliferation of monoclonal B cells. J Exp Med 1988; 168:85–94.

217. Maher D, Boyd A, McKendrick J, Begley G, Lieschke M, Green M, Fox R, Rallings M, Bonnem E, Morstyn G. Rapid response of B-cell malignancies induced by interleukin-4 (IL-4). Blood 1990; 76(suppl):152a.

218. Taylor CW, Grogan TM, Salmon SE. Effects of interleukin-4 on the in vitro growth of human lymphoid and plasma cell neoplasms. Blood 1990; 75:1114–1118.

219. Defrance T, Fluckiger AC, Rossi JF, Magaud JP, Sotto JJ, Banchereau J. Anti-proliferative effects of IL-4 on freshly isolated non Hodgkin malignant B lymphoma cells. Blood 1992; 79:990–996.

220. Herrmann F, Andreeff M, Gruss H-J, Brach MA, Lübbert M, Mertelsmann R. Interleukin-4 inhibits growth of multiple myelomas by suppressing interleukin-6 expression. Blood 1991; 78:2070–2074.

221. Barut BA, Cochran MK, O'Hara C, Anderson X. IL-4 and B cell tumours response patterns of hairy cell leukemia to B-cell mitogens and growth factors. Blood 1990; 76:2091–2097.

222. Gilleece MH, Scarffe JH, Ghosh A. Recombinant human interleukin-4 (IL-4) given as daily subcutaneous injections—a phase I dose toxicity trial. Br J Cancer 1992; 66:204–210.

223. Atkins MB, Vachino G, Tilg HG, Karp DD, Robert NJ, Kappler K, Mier JW. Phase I evaluation of thrice-daily intravenous bolus interleukin-4 in patients with refractory malignancy. J Clin Oncol 1992; 10:1802–1809.

224. Prendiville J, Thatcher N, Lind M, McIntosh R, Ghosh A, Stern P, Crowther D. Recombinant human interleukin-4 (RhuIL-4) administered by the intravenous and subcutaneous routes in patients with advanced cancer—a phase I toxicity study and pharmacokinetic analysis. Eur J Cancer 1993; 29A:1700–1707.

225. Margolin K, Aronson FR, Sznol M, Atkins MB, Gucalp R, Fisher RI, Sunderland M, Doroshow JH, Ernest ML, Mier JW, Dutcher JP, Gaynor ER, Weiss GR. Phase II studies of recombinant human interleukin-4 in advanced renal cancer and malignant melanoma. J Immunother 1994; 15:147–153.

226. Banchereau J, Rybak ME. Interleukin 4. In: Thomson A, ed. The Cytokine Handbook, 2nd ed. London: Academic Press, 1994:99–126.

227. Trehu EG, Isner JM, Mier JW, Karm DD, Atkins MB. Possible myocardial toxicity associated with interleukin-4 therapy. J Immunother 1993; 14:348–351.

228. Ghosh AK, Smith NK, Prendiville J, Thatcher N, Crowther D, Stern PL. A phase I study of recombinant human interleukin-4 administered by the intravenous and subcutaneous route in patients with advanced cancer: immunological studies. Eur Cytokine Netw 1993; 4:205–211.

229. Lotze MT. Role of IL4 in the antitumor response. In: Spits H, ed. IL4: Structure and function. New York: CRC Press, 1992:238–258.

230. Sosman JA, Ellis T, Bodner B, Kefer C, Fisher RI. A phase Ia/Ib trial of continuous infusion (ci)-interleukin-4 alone and following interleukin-2 in cancer patients. Proc Annu Meet Am Assoc Cancer Res 1992; 33:347.

16
Hematopoietic Effects of Interleukin-1

James R. Zucali
University of Florida, Gainesville, Florida

Claire M. Dubois
University of Sherbrooke, Sherbrooke, Quebec, Canada

Joost J. Oppenheim
National Cancer Institute, Frederick Cancer Research and Development Center, Frederick, Maryland

I. INTRODUCTION

Interleukin-1 (IL-1), notwithstanding its name, is a major hematopoietic growth factor (HGF). Historically, IL-1 was discovered by infectious disease researchers engaged in studies of endogenous pyrogen (EP; 1), and by immunologists investigating macrophage-derived lymphocyte-activating factors (LAF; 2). In the course of these early studies, it became clear that IL-1 can be produced by virtually all nucleated cell types and has pleiotropic-activating effects on many cell types (reviewed in Refs. 3–5). The capacity of IL-1 to protect mice from the lethal hematopoietic suppression, caused by high doses of radiation, provided the first evidence of the potent in vivo hematopoietic activity of IL-1 (6). Subsequently, two laboratories concomitantly discovered that hematopoietin 1 activity was also attributable to IL-1 (7,8). Hematopoietin 1 was defined as a synergistic costimulant with colony-stimulating factor 1 (CSF-1) of early progenitor cells.

Although the hematopoietic and myeloprotective effects of IL-1 have been copiously documented over the past 8 years, the basis for these effects of IL-1 are less well defined. The pleiotropic effects of IL-1 include activities with potential hematopoietic consequences that enhance the cycling of bone marrow (BM) cells (9). Although IL-1 increases the pool of BM stem cell precursors with hematopoietic reconstitutive capacity, it has not been shown to have any direct effects on bone marrow hematopoietic progenitor cells. IL-1 has abundant indirect effects by inducing stromal cell-derived hematopoietic growth factors, such as granulocyte—macrophage colony-stimulating factor (GM-CSF), G-CSF, M-CSF, and IL-3. It also induces the generation of cells that express more receptors for hematopoietic growth factors, such as M-CSF,

GM-CSF, and stem cell factor (SCF). IL-1 may protect hematopoietic cells by inducing radical scavengers, such as manganese superoxide dismutase (MnSOD), that presumably counteract the damaging effects of radiation-induced oxygen radicals. In this chapter we will discuss the wide variety of prohematopoietic effects of IL-1 and will cover the clinical applications of IL-1 in the treatment of bone marrow suppression and transplantation.

II. INTERLEUKIN-1 STIMULATES HEMATOPOIETIC PROGENITOR CELLS

In 1985, Bartelmez and Stanley (10) reported on a factor, found in the conditioned medium of the human bladder carcinoma cell line 5637, that exhibited dramatic synergism with CSF-1, and termed the factor hemopoietin-1. In the presence of CSF-1, hemopoietin-1 stimulated the proliferation of cells that were at earlier stages of development more than those that responded to CSF-1 and interleukin-3 (IL-3), either alone or in combination. Hemopoietin-1 plus IL-3 generated cells that could be subcultured at least twice, whereas cells from colonies induced by IL-3 alone could not be subcultured (11). Thus, hemopoietin-1 affected primitive multipotent target cells, termed high-proliferative potential colony-forming cells (HPP-CFC). In addition, hemopoietin-1 stimulated an increase in the expression of CSF-1 receptors on cultured early hematopoietic cells. In 1987, Mochizuchi et al. (7) completely characterized hemopoietin-1 from the 5637 tumor cell line and determined that it was identical with interleukin-1α (IL-1α). Together with Zsebo et al. (12), they also demonstrated that IL-1α had the same synergistic activity formerly ascribed to hemopoietin-1 and that IL-1α has a survival-enhancing and proliferation-inducing effect on primitive hematopoietic stem cells.

In vitro, IL-1α synergizes with CSF-1, GM-CSF, IL-3 and G-CSF in stimulating the production of HPP-CFC in primary cultures and in expanding hematopoietic progenitors with both high- and low-proliferative potential in short-term suspension cultures (10–12). In 1987, Moore and Warren (8) demonstrated that IL-1α and IL-3 could stimulate the in vitro proliferation and differentiation of bone marrow cells (BMC) from 5-fluorouracil (5-FU)-treated mice. BMCs from mice that were treated with 5-FU and then placed in suspension culture for 7 days with IL-1α and IL-3 showed a 100-fold expansion of both HPP- and LPP-CFU and a 1000-fold expansion of G-CSF-responsive CFU-C. Although 5-FU was profoundly immunosuppressive, IL-1α synergized with G-CSF to reduce the resultant neutropenia and to restore neutrophil counts. Therefore, IL-1 and G-CSF in combination were considered potentially useful in the treatment of radiation- or chemotherapy-induced myelosuppression.

Iscove and colleagues (13,14), demonstrated that pluripotent hematopoietic precursors from mouse bone marrow increased 8–12 times over a 4-day period when cultured in the presence of IL-1 and IL-3. They defined a population of precursors of multilineage CFCs (pre-CFC-multi) on the basis of this cell's ability to generate CFC multiprogeny within suspension cultures. A distinguishing hallmark of this primitive pre-CFC multicell population was its responsiveness to IL-1, which was absolutely necessary in addition to IL-3 for maximum yield.

The addition of IL-1 to purified murine bone marrow cultures colony-forming units (CFU)-GM promoted survival of the CFU-GM cells in the absence of colony-stimulating factors. This survival occurred over 2–4 days, with no change in the pattern of CFU-GM differentiation (15). IL-1, by itself, did not enhance the proliferation of human CFU-GM or burst-forming units–erythroid (BFU-E), but did promote the survival of these progenitors, and the survival effects could be neutralized by anti–IL-1 antibody (16). Taken together, these results implicate IL-1 in having a direct effect on the prolonged survival of CFU-GM and BFU-E cells.

In recent years, peripheral blood as well as ex vivo expanded cell populations have been used as sources of hematopoietic progenitor cells for transplantation. Fibbe et al. (7) reported that a single dose of IL-1 injected into mice mobilized both progenitor cells as well as repopulating cells into the peripheral blood. They further demonstrated that transplantation of peripheral blood cells obtained from IL-1-treated donor animals results in the long-term survival and long-term repopulation of the hematopoietic system of irradiated recipient mice. A single injection of IL-1 induces a shift in hematopoietic repopulating cells into the peripheral blood, and these cells can be used to rescue and permanently repopulate the BM of lethally irradiated recipients.

Muench and Moore (18) found that mice transplanted with BMCs that were expanded with IL-1 and IL-3 in vitro, shortened the period of cytopenia and accelerated the recovery of peripheral blood total leukocyte, neutrophil, red blood cell, and platelets numbers. The ex vivo expansion of BM progenitors also reduced the numbers of transplanted hematopoietic cells required for survival in lethally irradiated animals without a decrease in the proliferative capacity of the earliest hematopoietic progenitors (18). Expansion of BM progenitors in culture without loss of proliferative potential may be of clinical use in transplantation and gene therapy.

Studies were performed by Kobayashi et al. (19) to evaluate whether the effects of IL-1 on human hematopoietic progenitors were direct or indirect. These investigators determined that the synergistic effects of IL-1 and IL-3 in liquid culture occurred even when they used selected $CD34^+$ cells. Although anti–GM-CSF could partially block the synergistic effects observed with IL-1 and IL-3, limiting dilution assays of suspension cultures of $CD34^+$ cells indicated that IL-1 directly increased the number of colony-forming cells. Their results, therefore, suggested that IL-1 has unique direct and indirect effects on expansion of enriched human hematopoietic progenitors.

More recently, with the discovery of the ligand for the c-*kit* receptor as an important regulator of hematopoiesis, this growth factor has been studied for its ability to expand hematopoietic progenitors (20–23). Alone, the c-*kit* ligand (KL) is a stimulator of mast cell growth and has only a limited ability to stimulate hematopoiesis. However, it is a powerful synergistic factor with other growth factors. Bone marrow transplantation with IL-1- and KL-expanded progenitors was most effective in accelerating the recovery of peripheral blood leukocytes, platelets, and erythrocytes in lethally irradiated mice (20). Expansion of bone marrow cells ex vivo by IL-1 and KL also reduced the number of transplanted cells needed to provide radioprotection. All mice survived when receiving as few as 5000 IL-1- plus KL-cultured bone marrow cells, whereas it took a 200-fold greater number of fresh bone marrow cells to bring about complete survival (21). Additional studies by Han et al. (22) demonstraed that IL-1 plus IL-3 plus KL increased the number of HPP-CFC, CFUs, as well as cells capable of reconstituting lethally irradiated mice, and that reconstitution occurred earlier than when transplanting fresh bone marrow cells. Consequently, BMC can be incubated with various HGF in vitro, with improved reconstitutive capabilities.

Finally, to provide sufficient numbers of peripheral blood progenitor cells (PBPC) for reconstitution after high-dose chemotherapy and transplantation, Brugger and colleagues (23) investigated the ability of combinations of hematopoietic growth factors to expand the number of clonogenic PBPCs ex vivo. Optimum expansion of enriched peripheral blood $CD34^+$ progenitors occurred ex vivo in the presence of a combination of KL, IL-1, IL-3, IL-6, interferon gamma (IFN-γ) and erythropoietin. Expanded progenitor cells generated a 250-fold increase in multilineage colonies, as compared with preexpansion values, and the absolute number of early hematopoietic progenitor cells ($CD34^+/DR^-$;$CD34^+/38^-$) as well as the number of 4-HC-resistant progenitors within the expanded population increased significantly.

III. THE CONTRIBUTION OF THE CYTOKINE–RECEPTOR NETWORK TO HEMATOPOIESIS

Most of the hematopoietic effects of IL-1 are indirect and presumably based on the stimulation of IL-1 receptor-expressing accessory cells to produce hematopoietic growth factors (Fig. 1). IL-1 stimulates production of the colony-stimulating factors G-CSF, M-CSF, GM-CSF, as well as IL-6 and IL-11 by a variety of cell types, including BM stromal cells, macrophages, fibroblasts, and endothelial cells (24–28). This increase in production is based on increased gene transcription or, as in the case of GM-CSF and M-CSF, through stabilization of mRNA (25).

A. IL-1 Regulates Receptor Expression for CSFs In Vitro and In Vivo

In 1985, Bartelmez and Stanley demonstrated that hemopoietin-1 (IL-1) increased the average CSF-1R density on maturing monocytes–macrophages by about tenfold (10). This observation led to the idea that IL-1 may synergize with CSFs by up-regulating the expression of CSF receptors on progenitor cells. Subsequently, Jacobsen and colleagues (1992), using a progenitor-enriched BMC population, have shown that IL-1 up-regulates GM-CSF receptors on progenitor cells in vitro in a dose- and time-dependent manner (29). This effect preceded the enhanced proliferative response, morphological changes, and differentiation induced by GM-CSF and IL-1, suggesting that it is mechanistically important. Similarly, administration of IL-1 to mice up-regulated the expression of GM-CSF, IL-3, and CSF receptors on progenitor cells, reaching a maximum 10–24 hr after treatment (30). This enhanced HGF receptor expression by IL-1

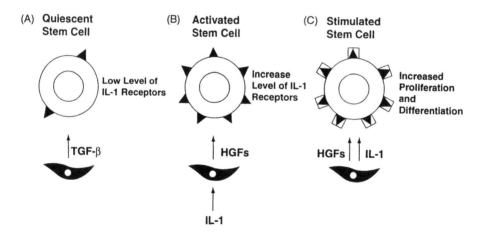

TGF-β = Transforming Growth Factor Beta
IL-1   = Interleukin-1
HGFs = Hematopoietic Growth Factors

Figure 1 Proposed role of IL-1 and IL-1 receptor in hematopoiesis. (A) Quiescent hematopoietic progenitors are maintained in G_1 phase through the action of TGF-β and other inhibitors. These cells expressed low levels of receptors for IL-1 and other hematopoietic growth factors. (B) IL-1 injection would result in a rapid production of hematopoietic growth factors which, in turn, would increase IL-1 receptor and HGFs receptor expression on early progenitors as well as IL-1 receptors on myeloid progenitors. (C) This would prime the primitive and more-committed progenitors to the proliferative effects of IL-1-induced increased local concentrations of hematopoietic growth factor.

correlated with increased IL-3- and GM-CSF-induced growth of CFU-c, CFU-Mix, and HPP-CFC. Therefore, IL-1 promotion of HPP colony formation in vitro and granulopoiesis in vivo might be partly based on stimulation of the expression of CSF receptors on hematopoietic progenitors that render them more responsive to circulating CSFs.

B. Expression of IL-1 Receptors on Hematopoietic Progenitors

Two distinct types of IL-1 receptors, both members of the immunoglobulin superfamily, have been identified (31,32). The 80-kDa receptor (IL-1R-I) is expressed on T cells, keratinocytes, hepatocytes, endothelial cells, fibroblasts, chondrocytes, and synovial-lining cells. The binding of IL-1 to these receptors results in signal transduction (33) and phosphorylation of threo-nine/serine residues (34). The second IL-1 receptor kDa 65 (IL-1R-II) has been cloned and detected on B cells, monocytes, and mature neutrophils. However, IL-1R-II has only a very short intracellular cytoplasmic tail and has not yet been shown to transduce signals (32,35). In fact, the extracytoplasmic domain of type II IL-1R is actively shed by cells and functions as an IL-1 inhibitor (35).

Interleukin-1 by itself does not directly promote the growth of hematopoietic progenitors in soft-agar colony assays. On the other hand, results from single-cell assay, using purified progenitor populations, accessory cell depletion studies, and serum-free cultures, suggest that IL-1 can act on its own and directly affect primitive progenitor cells (16,36–39). Given these findings, these progenitor cells must express specific IL-1-binding sites and transducing proteins. With use of radiolabeled IL-1α, the density of IL-1 receptors expressed on freshly isolated BMCs was demonstrated to be between 50 and 100 receptors per cell, with an affinity for Il-1α of 2×10^{-10} M (40,41). The observed receptor levels were similar to those on freshly isolated, mature lymphoid and myeloid cells. Further studies of fluorescence-activated cell sorter (FACS)-purified BMCs, labeled with granulocyte-specific monoclonal antibody, revealed that most IL-1–specific binding was expressed on maturing myeloid cells (42). This was confirmed by results from autoradiographic analysis showing that most of the labeled cells belonged to the morphologically identifiable granulocytic series, whereas only 7% of the early or blast cells were IL-1R-positives (43). The IL-1-binding protein expressed on BMCs and progenitor-enriched cells had an apparent molecular weight of 65 kDa and was immunologically distinct from type I IL-1R, indicating that bone marrow cells expressed most type II IL-1R (40).

C. Regulation of IL-1 Receptors by Hematopoietic Growth Factors

The level of receptor expression also regulates the biological effect of hematopoietic growth factors. In fact, it has been shown by two independent laboratories that IL-1 administration to mice resulted in a marked up-regulation of IL-1R expression on myeloid progenitor cells (42,44). This up-regulation occurred through an indirect mechanism involving both production of HGFs, such as GM-CSF, G-CSF, IL-8, and IL-6, by type I IL-1R-expressing stromal cells, as well as production of glucocorticoid by the adrenal gland (43). This was supported by results from studies of BMCs in liquid culture where type II IL-1R was up-regulated by IL-3, IL-6, G-CSF, and GM-CSF, but not by IL-1 itself (40). These studies demonstrated that IL-3 induced IL-1R on undifferentiated blast cells, whereas G-CSF affected mostly the more committed myeloid precursors. These findings can be related to the synergistic effect of IL-1 on the expansion of several progenitor cell compartments. Taken together IL-1 induced HGF production in vivo through the type I IL-1R, followed by the up-regulation of type II IL-1R on BMCs by the endogenously produced HGFs. The relevance of up-regulation of the nonsignaling IL-1RII to the hematopoietic in vivo effects of IL-1 remains unclear.

The IL-1 receptors expressed on hematopoietic progenitor cells can also be modulated by negative regulators of hematopoiesis, such as transforming growth factor-β (TGF-β). TGF-β is a potent inhibitor of type II IL-1R expressed on several murine myeloid progenitor cell lines, as well as bone marrow cells from 5-FU-treated mice bone (enriched for HPP-CFC; 41). Such trans-down-modulation of type II IL-1R expression possibly occurs through mRNA destabilization, as it occurs for c-*kit* (45), the receptor for stem cell factor (SCF). Local production of TGF-β is involved in the regulation of the cycling of stem cells and maintains these cells in a quiescent state (46). However, the inhibition of type II IL-1R cannot account for this effect of TGF-β. Perhaps, TGF-β decreased the putative low number of type I IL-1R on BM stem cells or suppressed the production of other HGFs and their receptors.

Primitive hematopoietic progenitor cells express low levels of growth factor receptors, as indicated by direct-binding studies (29,47), and they need multiple growth factors to be induced to growth (48). Consequently, the inhibition of expression of IL-1 receptors on progenitor cells, together with the down-modulation of c-*kit* and CSF receptors by TGF-β (45,49) provides an additional mechanism (other than the inhibition of cell cycle-related genes (50)), by which TGF-β would maintain progenitors in a non- or low-responsive state by reducing the levels of receptors for HGF.

D. Role of IL-1R Type I and Type II in Hematopoietic Cell Growth

The contribution of type I and type II IL-1R to hematopoiesis has been assessed using blocking monoclonal antibodies (MAbs). Neta and colleagues have shown that treatment of murine stromal cells with type I-specific antibody (35F5 MAb) inhibited IL-1-induced production of IL-6 in vitro (51). Similar involvement of type 1R in IL-1-mediated cytokine production has been observed in human neutrophils and monocytes (35,39). By using a single-cell assay with purified progenitors, Hestdal and colleagues have evaluated the mechanism of IL-1-stimulated growth and differentiation of progenitor cells. The synergistic effect of IL-1 on IL-3-responsive Lin-Thy-1[+] progenitors was indirect (presumably mediated by IL-6), because it could be blocked by the 35F5 antibody, suggesting that this was a type I IL-1R-mediated event (38). On the other hand, the direct synergistic effect of IL-1 on GM-CSF-induced myeloid colony formation was direct because it could be only partially inhibited by anti-type II receptor antibody. In vivo administration of 35F5 MAb blocked the radioprotective and chemoprotective effects of IL-1 (52). This may be related to the observed blockage by the 35F5 MAb of IL-1–induced circulating CSFs, as well as the in vivo inhibition of IL-1–mediated up-regulation of both IL-1 and CSF receptors on BMCs (51,52). Taken together, this evidence shows that the in vivo effects and some of the in vitro hematopoietic effects of IL-1 are mediated by type I IL-1Rs. Although type I IL-1Rs have been too few to be detected on BMC, a very low number of IL-1RI may suffice to transduce signals. Alternatively, the existence of yet another unidentified signal-transducing IL-1 receptor may account for the hematopoietic effects of IL-1.

Hematopoietic cell proliferation and differentiation are based on a complex cascade of cellular events involving coordinate action of multiple growth factors and receptors. In at least one subset of early progenitor cells, receptors for IL-1 and other HGFs, such as IL-3 and SCF, appear to be present in low numbers, presumably sufficient to permit a survival response, but below the threshold level required for the proliferative response. For example, the number of c-*kit* receptors was dose-limiting for the proliferative, but not for the survival response of SCF (53). In this context, the capacity for exogenous (probably endogenous as well) IL-1 to expand early progenitor cell growth might be partly due to its unique ability to sequentially induce the expression of receptors for HGF. This would render the primitive, as well as the more committed progenitor cells, more responsive to the proliferative effects of local, induced, and exogenous growth factors and may explain the unexpected multiplicity of effects of in vivo IL-1.

IV. INTERLEUKIN-1 EFFECTS IN PROTECTION FROM CHEMOTHERAPY

Although increasing the dose of chemotherapy is emerging as a crucial determinant for the success in cytotoxic cancer therapy, myelosuppression often results as a detrimental side effect. The administration of cytokines, either before or after cytotoxic therapy is being used to overcome this myelosuppression. Several studies describe the beneficial effects of IL-1 in protecting mice from chemotherapy. In vivo, IL-1 significantly hastened granulocyte recovery, following treatment with 200 mg/kg cyclophosphamide, by enhancing both the proliferation and maturation of myeloid precursors (54–56). IL-1 also rescued mice if it was administered either 20 hr before or 48 hr after lethal doses of cyclosphosphamide (57,58). This IL-1 protective activity correlated with increased colony-forming frequency and total colony number, as well as an increased cellularity in the BM and peripheral blood. The continuous administration of IL-1 to mice for 7 days by subcutaneous implanted Alzet osmotic minipumps markedly stimulated granulopoiesis and protected against depletion of myeloid and erythroid cells of mice treated with a single injection of 20 or 30 mg/kg doxorubicin (59). Recombinant human IL-1 has also been used to protect normal and tumor-bearing mice from the acute lethal effects of cyclophosphamide, mafosfamide, and 5-flourouracil by accelerating hematopoiesis in these mice (60). Pretreatment for 7 days with IL-1 protected 70–100% of mice from lethal doses of cyclophosphamide, mafosfamide, and 5-FU; whereas, single or multiple posttreatment doses of IL-1 were not chemoprotective. Pretreatment increased the acute LD_{90} of cyclophosphamide from 380 to higher than 500 mg/kg giving a dose-modifying effect of at least 1.25. In addition, mice bearing murine renal cancer were also protected from the acute toxic effects of cyclophosphamide by IL-1 pretreatment, resulting in greater cyclophosphamide antitumor effect.

Interleukin-1 in combination with G-CSF accelerates the recovery of hematopoietic progenitors and decreases the severity and duration of neutropenia associated with the administration of 5-FU to mice (8). Female mice, bearing spontaneous breast tumors, if treated with IL-1 alone or in combination with recombinant human G-CSF or GM-SCF with 5-FU, showed significantly inhibited tumor growth and improved survival when compared with 5-FU alone (61). IL-1 and G-CSF or GM-CSF also increased the maximally tolerated dose of 5-FU, accelerated recovery of blood neutrophil counts, and reduced the duration of significant neutropenia and loss of body weight. Similarly, primates treated with 5-FU had a significantly shortened time to achieve an absolute neutrophil count over 500/µL after receiving 2- and 7-day courses of recombinant human IL-1B when compared with controls (62). These results paved the way for clinical trials of IL-1 in patients with advanced cancers.

Pretreatment of bone marrow cells with IL-1 results in the protection of primitive hematopoietic progenitor cells from the lethal effects of the chemotherapeutic agent 4-hydroperoxy-cyclophosphamide (4-HC), a derivative of cyclophosphamide (63,64). The only other tested cytokine that showed similar protection from 4-HC was tumor necrosis factor (TNF) and IL-1 and TNF synergize with each other in their protective effects. Although all types of colonies in culture are protected by IL-1 and TNF from 4-HC, the most affected are the undifferentiated blast colony-forming cells. These cells grow quickly into large colonies and demonstrate a high-replating ability (63,64).

Interleukin-1 and TNF protect hematopoietic cells capable of repopulating irradiated long-term bone marrow stromal cultures from 4-HC (65). Irradiated long-term one marrow cultures reconstituted with hematopoietic cells pretreated with IL-1 and TNF before 4-HC give rise to greater numbers of colony-forming cells at 4–5 weeks of culture within both the nonadherent and adherent cells populations of the long-term cultures, when compared with controls. Further studies (Fig. 2) demonstrate that bone marrow cells pretreated with IL-1 before 4-HC were able to bring about 30-day survival of 90% of irradiated animals, compared with 40% survival in animals transplanted with cells not pretreated with IL-1 (66). Taken together, these results suggest that IL-1 and TNF not only protect human long-term culture-initiating cells, but also protect cells that are capable of short-term reconstitution in lethally irradiated mice.

For the protection of IL-1 and TNF to be clinically relevant, the protection from 4-HC toxicity should be selective for normal hematopoietic progenitor cells and should not protect malignant cells. Clonal analysis and [^{3}H]thymidine uptake were used to evaluate the protective effects of IL-1 and TNF on various leukemic cell lines and primary leukemic blasts from the peripheral blood of six AML patients (63). Prior incubation with IL-1 or IL-1 stimulated bone marrow-conditioned medium did not protect the leukemic cell lines HL-60, K562, KG1, KG1a, U937, and THP-1, nor the primary acute myeloid leukemic cells from the lethal effects of 4-HC. In addition, mixing experiments with normal bone marrow cells and leukemic cells also showed no protection for the leukemic cells, while still protecting normal bone marrow colony growth (Table 1). Thus, IL-1 appears to selectively protect early human hematopoietic progenitors from 4-HC while providing not in vitro protection for leukemic cells.

The possibility of compromising stem cell function is compounded when using hematopoietic growth factor therapy combined with repeated cycles of cycle-specific or proliferation-dependent chemotherapy. This was documented by Hornung and Longo (67), who reported that treatment of mice with repeated cycles of high-dose cyclophosphamide, followed by restorative therapy with either G-CSF or GM-CSF led to a permanent impairment of stem cell function, as measured by serial transplantation of allotype-marked marrow into lethally irradiated mice. The impairment did not compromise the hematopoietic restoration of the primary mice, but the results

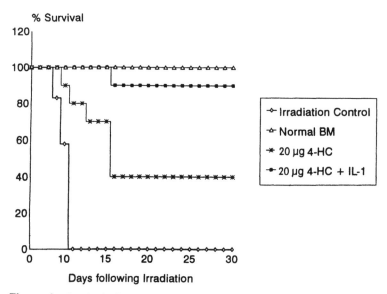

Figure 2 Survival of mice injected with bone marrow cells treated for 20 hr with or without 100 ng/ml IL-1 before 30-min exposure to 4-HC.

Table 1 Effect of 20-hr Preincubation with IL-1 plus TNF Before
4-HC on the Number of Colony-Forming Cells Obtained from
Normal Human Bone Marrow Cells Containing 17% AML Cells

| | Total colonies | | | |
| | Control | | +IL-1 + TNF | |
	–4-HC[a]	+4-HC[b]	–4-HC[a]	+4-HC[b]
Normal CFCs	118 ± 24	4 ± 2	125 ± 30	138 ± 15
Leukemic CFCs	56 ± 6	0	34 ± 6	0

[a]Colonies per 2×10^4 cells plated.
[b]Colonies per 5×10^5 cells plated.

raise troubling questions about the short-term advantages of CSF therapy in permitting chemotherapy dose and schedule intensification, if irreversible stem cell depletion were to occur in the
clinical situation. Strategies may have to be devised to protect against stem cell damage and
Hornung and Longo demonstrated that administration of a single dose of IL-1, 24 hr before each
cycle of cyclophosphamide, prevented the stem cell depletion with subsequent CSF treatments.

Interleukin-1 pretreatment apparently promotes stem cell proliferation and expands the stem
cell pool, resulting in more stem cells that survive the chemotherapeutic insult. The mechanisms
by which IL-1 may protect the stem cell compartment from chemotherapy are only beginning to
be studied. IL-1 may initiate a cytokine cascade providing elevated levels of HGFs and their
receptors that promote accelerated regeneration of stem cells and progenitor cells surviving
chemotherapy. Alternatively, stem cell drug resistance could theoretically involve increased
enzymatic inactivation, decreased drug influx, or increased drug efflux and increased DNA
repair. Stem cells are relatively more resistant to 4-HC than normal mature hematopoietic or
malignant cells (68,69). These observations were the basis for the use of 4-HC as a purging agent
in autologous bone marrow transplantation studies. Kasten et al. (70) reported that CD34[+] cells
contain the highest levels of aldehyde dehydrogenase (ALDH), an enzyme responsible for the
inactivation of 4-HC. Cytokine modulation of this detoxifying enzyme system has been reported
by Moreb et al. (71), who showed that in vitro preincubation with IL-1 for 20 hr was required
for the protection of early hematopoietic progenitors from 4-HC. Cycloheximide totally abolished this protection, as did diethylaminobenzaldehyde (DEAB), a specific inhibitor of aldehyde
dehydrogenase (ALDH). These studies indicate that the up-regulation of ALDH by IL-1 may
play a role in the protection of hematopoietic progenitor cells from 4-HC. These authors were
also able to show that enhanced transcription or stabilization of the ALDH mRNA may be
directly induced by IL-1, providing additional evidence for the importance of ALDH in the
protection from 4-HC (66). That DEAB had no effect on IL-1 protection when phenylketo-
phosphamide, another cyclophosphamide derivative, was used as the chemotherapeutic agent
suggests that other mechanisms of protection may also be induced by IL-1. IL-1 also protects
normal hematopoietic progenitors from the toxic effects of melphalan, another alkylating agent
used in cancer chemotherapy (72). Buthionine sulfoximide, a specific inhibitor of glutathione,
used at the time of IL-1 preincubation, completely abolished this protection, implicating
glutathione or its dependent enzymes as possible mechanisms in the protection of cells from
melphalan (Table 2).

Table 2 Effect of IL-1 on Protection from Melphalan

Melphalan added (μM)	Total colonies		
	−IL-1	+IL-1	+IL-1 +BSO
20	31 ± 6	105 ± 12	12 ± 4
40	0.3 ± 0.3	20 ± 2	4 ± 1
50	0	8 ± 1	0
60	0	4 ± 1	0

V. RADIOPROTECTIVE EFFECTS OF INTERLEUKIN-1

It was shown in the 1950s and 1960s that prior administration of immunostimulatory adjuvants, such as endotoxin (LPS), resulted in the survival and recovery of mice exposed to lethal doses of radiation (73). Lethally irradiated mice die from the damaging consequences of radiation to their lymphoid and hematopoietic tissues, resulting in anemia and decreased resistance to opportunistic infections. Since 1986 it has been established that the radioprotective effects of LPS can be entirely equaled or bettered by the combination of IL-1 and TNF-α (6,74). Consequently, although many cytokines are induced by the in vivo administration of LPS, IL-1 and TNF-α potentially account for all radiorestorative effects of LPS. In contrast, the other hematopoietic growth factors, such as the CSFs, that are induced by LPS have the capacity to promote recovery of hematopoiesis from sublethal, but not from lethal, doses of irradiation (75).

Preadministration of IL-1 can protect mice from $LD_{100/30}$ (100% lethal by 30 days) doses of radiation with a dose-reducing factor (DRF) of about 1.25 (the ratio at which 50% of the treated to untreated mice die). When administered after radiation IL-1 promotes the recovery of up to 100% of $LD_{95/30}$ sublethally irradiated mice (76). The radioprotective effects of these cytokines are chiefly due to the recovery of suppressed hematopoietic tissues, although IL-1 is also reported to benefit the recovery of pulmonary and gastrointestinal tissues from radiation injury (77). Administration of neutralizing anti–IL-1 receptor (35F5) and anti–TNF-α antibodies reduced the radioresistance of normal mice and also completely blocked the radioprotective effect of endotoxin (78). This established that IL-1 and TNF account for the radioprotective effects of LPS. In addition to inhibiting the radioprotective effect of IL-1, anti–IL-1R also blocked radioprotection by TNF-α. Similarly anti–TNF-α reduced radioprotection by TNF-α and also IL-1, suggesting that synergistic cooperative interactions between IL-1 and TNF-α are obligatory for radioprotection (78).

The radioprotective effect of IL-1 is presumably related to the fact that Il-1 is a hematopoietin (7). IL-1 (hematopoietin-1) acts as a necessary costimulant with colony-stimulating factors (CSFs) of high-proliferative potential (HPP) primitive hematopoietic progenitor colony-forming cells (CFC; 11,12). That the hematopoietic effects of IL-1 are important is supported by the fact that prior treatment with IL-1 synergizes with allogeneic BM transplants in protecting mice from lethal doses of radiation (79). This has been further substantiated by data showing that bone marrow cells from male mice pretreated with IL-1 in vitro increased the survival of irradiated female recipient mice (80). These recipients had increased male cells in their bone marrow, spleen, and thymus for up to 3 months posttransplant. Furthermore, serial transplantation studies revealed that male cells could be recovered only from recipients that had received

IL-1 treatment, but not from untreated male cells. Thus, IL-1 pretreatment could protect both short- and long-term repopulating cells from an irradiation insult (80).

A. The Role of the Cytokine Network in Radioprotection by IL-1

Both IL-1 and TNF-α are potent inducers of the cytokine cascade; that is, they promote the production of other cytokines, including GM-CSF, G-CSF, and IL-6 (81), as well as one another (82). Although GM-CSF, G-CSF, and IL-6 by themselves do not prevent the lethal effects of irradiation, GM-CSF, G-CSF, or IL-6 in conjunction with suboptimal doses of IL-1, synergize to prevent radiation lethality (74). Thus, the "radioprotective" effects of IL-1 and TNF are partly mediated by the rapid in vivo induction of CSEs and other cytokines (74).

The radioprotective effects of IL-1 is also based on induction of the expression of functional cytokine receptors. For example, administration of IL-1 to mice up-regulates not only the production of G-CSF and GM-CSF, but also the expression of type II IL-1R by tenfold as well as receptor for GM-CSF on bone marrow cells (BMC). Consequently, mice pretreated with IL-1 that develop more binding sites for IL-1, GM-CSF, and IL-3 also respond with greater HPP colony formation to combinations of IL-1 and G-CSF. The increase in the functional capacity of the BM progenitor cells correlated directly with increases in the expression of nonsignaling type II IL-1R. However, because type II IL-1R does not transduce signals, we presume that parallel effects on the low numbers of type I IL-1R may account for the greater functional effects of IL-1 on cells expressing more type II IL-1R. Thus, myelorestorative effects of IL-1 and TNF appear to at least partly depend on both the induction of hematopoietic growth factors and their receptors.

Radiation injury itself induces the production of proinflammatory cytokines, including IL-1 and TNF (83–85). These observations may also account for the fact that anti–IL-1R antibodies radiosensitize mice, even if given after radiation, by blocking the IL-1 that is produced following irradiation (78), whereas shorter-lived IL-1RA does not have such an inhibitory effect (unpublished observation).

B. The Relation of IL-1 and Stem Cell Factor in Radioprotection

Another hematopoietic growth factor, stem cell factor (SCF), has considerable radioprotective activity in mice (30). Preadministration of SCF synergizes with IL-1 to promote the survival of mice exposed to lethal doses of radiation. Furthermore, SCF is required for the radioprotective effects of IL-1 and TNF, because antibody to murine SCF reduces the radioprotective effects of these cytokines. Administration of antibody to murine (Mu)SCF also increased the lethal effects of radiation in normal mice and acted as a radiosensitizer. Conversely, anti–IL-1 reduced the radioprotective effect of SCF. However, neither IL-1 nor TNF can induce the production of SCF, or the reverse. SCF also failed to induce MnSOD, IL-6, CSF, and acute-phase proteins. In fact, BMC from mice treated with IL-1 bind more ^{125}I-SCF and also show increased expression of mRNA for c-*kit* (30). FACS analysis has shown that IL-1 induces increased SCF-binding sites, a subset enriched in c-*kit* plus BMCs. These observations suggest that the interaction of c-*kit* ligand (SCF) with other radioprotective cytokines may be based on an increase in the number of BMCs that express receptors for SCF (c-*kit*). Thus, although IL-1 and SCF do not induce one another, they are required to interact to achieve radioprotection. Whether SCF accounts for all the hematopoietic effects of IL-1 remains uncertain.

C. The Contribution of MnSOD to Radioprotection by IL-1

Agents that scavenge oxygen radicals are potential radioprotectors. IL-1 induces circulating acute-phase proteins, several of which (e.g., ceruloplasmin and metallothionein) have scavenger activity (86). IL-1 as well as TNF selectively induced the prominent expression of a p25 mitochondrial protein, which was subsequently shown to have sequence identity with MnSOD (87). IL-1 and TNF do not induce other intracellular antioxidant enzymes such as Cu/Zn SOD, glutathione peroxidase, glutathione S-transferase or catalase. These enzymes, are all constitutively and widely expressed by mammalian cells, but only mitochondrial MnSOD is regulated by cytokines (88). Induction of MnSOD activity and mRNA by IL-1 and TNF-α has been observed in human melanoma (A375) cell line cells, human skin fibroblasts, BMCs, and peripheral blood mononuclear cells (88,89). Furthermore, BMCs from IL-1-treated mice show dose-dependent increases in MnSOD mRNA (89). The expression of high basal levels of intracellular MnSOD mRNA in a variety of cell lines correlates with greater radioresistance (89). Consequently, inducers of MnSOD may prevent cell damage by radiation, agents, or drugs that generate oxygen radicals.

The effects of exposure to various types of oxidative stress have been assessed using cell lines transfected with vectors that express sense or antisense mRNA for MnSOD (90). Cell lines transfected with sense MnSOD overexpress mRNA for MnSOD, whereas those transfected with antisense expressed only minimal MnSOD mRNA. As predicted, IL-1-stimulated sense MnSOD-expressing cells produced more MnSOD protein, whereas production of MnSOD in IL-1-stimulated antisense transfected cells was reduced. Cell fractionation studies showed that the overexpressed MnSOD was localized in the mitochondrial fraction, but not in the cytoplasmic fraction. Overexpression of MnSOD by A375 human melanoma cells was associated with increased resistance to the cytostatic and cytotoxic effects of IL-1 and TNF. Conversely, reduction in endogenous MnSOD through the expression of antisense MnSOD mRNA was associated with increased susceptibility to the cytostatic and cytotoxic effects of IL-1 and TNF-α. Thus, the induction of MnSOD by pretreatment with lower doses of IL-1 and TNF can protect cells from radical damage induced by subsequent higher doses of these cytokines. Consequently, induction of MnSOD may contribute to the "desensitizing" or "tachyphylactic" effects of repeated exposure of cells to a cytokine.

Exposure of these transfected cells to a panel of anticancer drugs showed that MnSOD-overproducing clones of hamster kidney CHO cells were more resistant to mitomycin C and doxorubicin, both of which generate oxygen radicals, but were still equally susceptible to chemotherapeutic agents for which the mechanisms of cytotoxicity is not dependent on oxygen radicals (e.g., melphalan, vincristine, 5-fluorouracil, and cisplatin [cis-diammine dichloroplatinum] as shown in Table 3; 90). As predicted, CHO clones that overexpress MnSOD also survived better following in vitro radiation exposure (Fig. 3). The basis for this apparent increase in radioresistance can be attributed to the protective scavenging effects of MnSOD because there is a shift in the D_0 (e.g., 37% survival level) of cells transfected with MnSOD. The D_0 of transfected cells was increased from 1.5 Gy to about 3 Gy and 3.6 Gy, respectively, indicating that an increased number of cells were more resistant to radiation damage. In addition, K562 cells transfected with MnSOD in the antisense orientation display increased sensitivity to irradiation, whereas the reverse is seen with A375 human melanoma cell lines transfected with MnSOD, compared with their parental or with vector-transfected controls (91). However, these in vitro results are merely correlative, and it remains to be established whether MnSOD contributes to in vivo radioprotection.

VI. HEMATOPOIETIC EFFECTS OF IL-1 IN HUMANS

Interleukin-1 has considerable stimulatory effects on hematopoiesis in cancer patients. This effect of IL-1 is potentially useful in counteracting the bone marrow suppressive effects of

Table 3 Relation of Level of MnSOD Activity of Cell Line to Sensitivity to Anticancer Drugs

Cell lines	MnSOD activity (units/mg)	IC_{50} (μM)						
		Dox	Mit	Mel	Vin	5-Fu	Cis
cVec (control)	4.3 ± 0.3	0.80 ± 0.12 (1.00)	1.10 ± 0.35 (1.00)	2.84 ± 0.34 (1.00)	0.15 ± 0.04 (1.00)	0.93 ± 0.10 (1.00)	2.05 ± 0.02 (1.00)
cSOD-C1	10.1 ± 0.7	1.52 ± 0.53^a (1.91)	5.78 ± 1.06^c (5.25)	2.95 ± 0.13 (1.04)	0.14 ± 0.02 (0.93)	0.99 ± 0.15 (1.06)	2.00 ± 0.09 (0.98)
cSOD-C2	16.9 ± 1.1	2.10 ± 0.04^b (2.60)	9.80 ± 0.65^c (8.91)	2.78 ± 0.93 (0.98)	0.14 ± 0.05 (0.93)	0.93 ± 0.20 (1.00)	2.15 ± 0.54 (1.05)

Exponentially growing cells were added to 96-well microtiter plates and incubated in the absence or presence of various doses of doxorubicin (Dox), mitomycin (Mit), melphalan (Mel), vincristine (Vin), 5-fluorouracil (5-Fu), and cisplatin (Cis). After 96 hr of incubation, MTT assay was carried out. IC_{50} ($\pm$ SE) values were calculated from the survival curves (data not shown). The resistance ratio ( ) is the IC_{50} of transfectant with human MnSOD cDNA divided by the IC_{50} of cVec, respectively. MnSOD activity of whole cell lysate was determined. All values represent mean $\pm$ SE. Significantly different from control values: $p < 0.05$ (a); $p < 0.01$ (b); $p < 0.001$ (c).

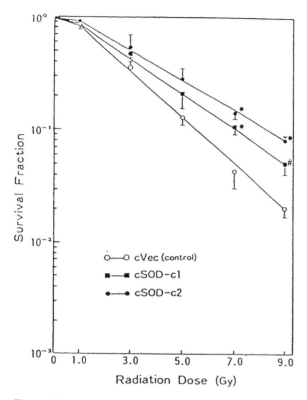

Figure 3 Clonogenic radiation survival curves for CHO cells transfected with vector alone or human MnSOD cDNA. Cell suspensions were irradiated in isotonic phosphate-buffered saline (PBS; pH 7.4) in Corning centrifuge tubes (15 ml) with a Theratron 80 ^{60}Co unit; a lateral field in a specially constructed Lucite block holder was used. Immediately after irradiation, cells were plated in triplicate at 100–100,000 cells per 60-mm dish. After 10 days, the plates were fixed, stained in the solution of 50% methanol, 15% acetic acid, and 0.2% of coomassie brilliant blue R, and colonies containing > 50 cells were scored. Survival curves were plotted as the log of the survival fraction of cells versus the radiation dose. Experimental points are plotted as the mean ± SE. Significantly different from control value; $p < 0.001(*)$, $p < 0.05(\#)$.

chemotherapy or radiotherapy. Even at the relatively low doses tolerated by humans, IL-1 resulted in significant increases in peripheral blood granulocyte counts and bone marrow cellularity (92). The induction of G-CSF production by bone marrow stromal cells, as well as increases in GM-CSF and SCF receptor expression by progenitor cells, probably contribute to the hematopoietic effects of IL-1 (30,47). The cancer patients being treated with IL-1 also exhibited an increase in the number and ploidy of BM megakaryocytes and a delayed increase of 70% in their platelet counts that occurred 1–2 weeks after IL-1 therapy. The previously noted increase in the IL-1-induced IL-6 levels may have mediated the increase in platelets (92,93). Several other clinical studies revealed the hematopoietic effects of IL-1 in humans. A phase I trial of recombinant human IL-1β alone and in combination with myelosuppressive doses of 5-fluorouracil in patients with gastrointestinal cancer has also been reported (94). An IL-1β dose-dependent neutrophil leukocytosis as well as increases in platelet counts extending through 23 days were observed after IL-1β administration. Fewer neutropenia days were noted after 5-FU plus IL-1β when compared with 5-FU alone. IL-1, if given within 24 hr after chemotherapy with carboplatin, shortens the period of thrombocytopenia (95).

IL-1α was used after an autologous BM transplant in a group of lymphoma patients. Patients who received IL-1 at doses higher than or equal to 3.0 μg/m^2 per day achieved neutrophil counts higher than 500 an average of 9 days sooner than control patients. There was a trend toward earlier platelet recovery and fewer days spent in the hospital. Significant elevations in G-CSF and IL-6 were detected in the plasma of the IL-1-treated patients (96). IL-1 treatment also is reported to shorten the period of granulocytopenia following bone marrow transplantation, when given before bone marrow ablation by chemotherapy, presumably by promoting the engraftment of the transplant (97). More recently, a phase 1 study of recombinant IL-1β in patients undergoing BM transplantation for acute myelogenous leukemia was reported (98). Recombinant human IL-1β was administered by 30-min intravenous infusions once a day, beginning on the day of bone marrow transplantation and continuing for 5 days. The number of days required to achieve an absolute neutrophil count higher than 500 in patients who received IL-1β was significantly fewer than historic controls. This appeared to correlate with a reduced incidence of infection as well. Survival was also improved. IL-1 treatment also shortens the period of granulocytopenia following BM transplantation, when given before BM ablation by chemotherapy. This presumably occurs by promoting the engraftment of the BM transplant. The radioprotective effects of IL-1β were studied in a group of colon cancer patients before a dose of ^{131}I-monoclonal antibody. In patients who had been treated with IL-1, exposure to SCF or PIXY-321 in culture increased BFU-E, CFU-GM, CFU-GEMM, and HPP-CFC activity in the peripheral blood, without any increase in the number of CD34$^+$ cells detected by flow cytometry. There was less marrow toxicity in the patients receiving the radiolabeled monoclonal antibody. They concluded that IL-1β was a potent primer of early marrow progenitors and has some radioprotective effects (99).

VII. CONCLUSIONS

The identification of hemopoietin-1 as being identical with IL-1α significantly extended the known biological activity of this pleiotropic cytokine. The action of IL-1 in hematopoiesis involves not only its ability to induce HGF production and release, but also its ability to stimulate early stem cells and to synergize with various HGFs to stimulate hematopoietic cell production. The biological effects of IL-1 on hematopoietic cells are mediated through two distinct, specific cell surface receptors with the 80-kDa species thought to be the most important for any direct effects on hematopoiesis. IL-1 also upregulates CSF receptors on hematopoietic progenitors and, thus, may promote hematopoiesis indirectly through this mechanism. As a result, IL-1 acts in synergy with other cytokines, including G-CSF, GM-CSF, IL-3, IL-6, and the c-*kit* ligand or SCF, to stimulate the proliferation of primitive murine and human precursors with functional or phenotypic features of stem cells.

Pretreatment of mice with IL-1 provides a dose-dependent protection from the lethal effects of both ionizing radiation and chemotherapy. The radioprotective effects of IL-1 may be due to the in vivo induction of HGF or other cytokines; the induction of increased expression of functional cytokine receptors; and the induction by IL-1 of oxygen–radical-scavenging molecules such as MnSOD. Similarly, the promotion of drug resistance by IL-1 has also recently come under study and the mechanisms responsible for this resistance are only beginning to be investigated. The documented ability of IL-1 to promote drug resistance would be of little practical value if this cytokine similarly protected malignant cells. However, the data so far suggest that IL-1 is selective in its protection, at least in vitro with myeloid leukemic cells (64) and in vivo in a murine breast tumor model (61) and a renal tumor model (60). Use of IL-1 in vitro as a purging protocol employing chemotherapy before autologous BM transplantation may prove clinically relevant if relapse rates are shown to be due to transplantation of tumor cells present within the autograft. Two enzymes, MnSOD and ALDH are induced by IL-1 and have been implicated as possible mechanisms for the observed protection of hematopoietic cells from irradiation and 4-HC. However, it would not be surprising that more than one mechanism may account for the IL-1–induced protection from different therapeutic modalities.

ACKNOWLEDGMENT

We are grateful for the critical review of this manuscript by Drs. Jonathan Keller and Ruth Neta and the secretarial assistance of Ms. Roberta Unger.

REFERENCES

1. Atkins E. Fever: historical aspects. In: Bomford R, Henderson B, (eds.) Interleukin I, Inflammation and Disease. New York: Elsevier Science, 1989:3–15.
2. Gery I, Gershon RH, Waksman BH. Potentiation of cultured mouse thymocyte responses by factors released by peripheral leucocytes. J Immunol 1971; 107:1778–1786.
3. Oppenheim JJ, Gery I. From lymphodrek to interleukin 1. Immunol Today 1993; 14:232–234.
4. Dinarello CA. Interleukin 1. In Thompson AW, ed. The Cytokine Handbook. New York: Academic Press, 1991:47–82.
5. Neta R, Oppenheim JJ. Interleukin-1: can we exploit Jekyll and subjugate Hyde. In: DeVita VT, Hellman S, and Rosenberg SA, eds. Biologic Therapy of Cancer Updates, vol. 2. Philadelphia; JB Lippincott, 1992; 1–11.
6. Neta R, Douches S, Oppenheim JJ. IL-1 is a radioprotector. J Immunol 1986; 136:2483.
7. Mochizuki DY, Eisenman JR, Conlon PJ, Larsen AD, Tushinski RJ. Interleukin 1 regulates hematopoietic activity, a role previously ascribed to hemopoietin 1. Proc Natl Acad Sci USA 1987; 84:5267–5271.
8. Moore MAS, Warren DJ. Synergy of interleukin-1 and granulocyte colony-stimulating factor: in vivo stimulation of stem cell recovery and hematopoietic regeneration following 5-fluorouracil treatment of mice. Proc Natl Acad Sci USA 1987; 84:7134–7138.
9. Neta R, Sztein MB, Oppenheim JJ, Gillis S, Douches SD. The in vivo effecs of IL-1. I. Bone marrow cells are induced to cycle after administration of IL-1. J Immunol 1987; 139:1861.
10. Bartelmez SH, Stanley ER. Synergism between hemopoietic factors (HGFs) detected by their effects on cells bearing receptors for specific HGF: assay for hemopoietin-1. J Cell Physiol 1985; 122:370–378.
11. Stanley ER, Bartocci A, Patinkin D, Rosendaal M, Bradley TR. Regulation of very primitive, multipotent hemopoietic cells by hemopoietin-1. Cell 1986; 45:667–674.
12. Zsebo KM, Wypych J, Yuschenkoff VN, Lu H, Hunt P, Dukes PP, Langley KE. Effects of hematopoietin-1 and interleukin 1 activities on early hematopoietic cells of the bone marrow. Blood 1988; 71:962–968.
13. Iscove NN, Shaw AR, Keller G. Net increase of pluripotential hematopoietic precursors in suspension culture in response to IL-1 and IL-3. J Immunol 1989; 142:2332–2337.
14. Iscove NN, Yan X-Q. Precursors (pre-CFC_{multi}) of multilineage hematopoietic colony-forming cells quantitated in vitro. J Immunol 1990; 145:190–195.
15. Williams DE, Broxmeyer HE. Interleukin-1α enhances the in vitro survival of purified murine granulocyte–macrophage progenitor cells in the absence of colony-stimulating factors. Blood 1988; 72:1608–1615.
16. Hangoc G, Williams DE, Falkenburg JHF, Broxmeyer HE. Influence of IL-1α and -1β on the survival of human bone marrow cells responding to hematopoietic colony-stimulating factors. J Immunol 1989; 142:4329–4334.
17. Fibbe WE, Hamilton MS, Laterveer LL, Kibbelaar RE, Falkenburg JHF, Visser JWM, Willemze R.

Sustained engraftment of mice transplanted with IL 1-primed blood-derived stem cells. J Immunol 1992; 148:417–421.

18. Muench MO, Moore MAS. Accelerated recovery of peripheral blood cell counts in mice transplanted with in vitro cytokine-expanded hematopoietic progenitors. Exp Hematol 1992; 20:611–618.

19. Kobayashi M, Imamura M, Gotohda Y, Maeda S, Iwasaki H, Sakurada K, Kasai M, Hapel AJ, Miyazaki T. Synergistic effects of interleukin-1β and interleukin-3 on the expansion of human hematopoietic progenitor cells in liquid cultures. Blood 1991; 78:1947–1953.

20. Muench MO, Schneider JG, Moore MAS. Interactions among colony-stimulating factors, IL-1β, IL-6, and kit-ligand in the regulation of primitive murine hematopoietic cells. Exp Hematol 1992; 20:339–349.

21. Muench MO, Firpo MT, Moore MAS. Bone marrow transplantation with interleukin-1 plus kit-ligand ex vivo expanded bone marrow accelerates hematopoietic reconstitution in mice without the loss of stem cell lineage and proliferative potential. Blood 1993; 81:3463–3473.

22. Han M, Kobayashi M, Imamura M, Hashino S, Kobayashi H, Maeda S, Iwasaki H, Fujii Y, Musashi M, Sakurada K, Kasai M, Hapel AJ, Miyazaki T. In vitro expansion of murine hematopoietic progenitor cells in liquid cultures for bone marrow transplantation: effects of stem cell factors. Int J Hematol 1993; 57:113–120.

23. Brugger W, Mocklin W, Heimfeld S, Berenson RJ, Mertelsmann R, Kanz L. Ex vivo expansion of enriched peripheral blood CD34$^+$ progenitor cells by stem cell factor, interleukin-1β, IL-6, IL-3, interferon-γ, and erythropoietin. Blood 1993; 81:2579–2584.

24. Zucali JR, Dinarello CA, Oblon DJ, Gross MA, Anderson L, Weiner RS. Interleukin 1 stimulates fibroblasts to produce granulocyte–macrophage colony-stimulating activity and prostaglandin E2. J Clin Invest 1986; 77:1857.

25. Bagby GC, Dinarello CA, Wallace P, Wagner C, Hefeneider S, McCall E. Interleukin 1 stimulates granulocyte–macrophage colony-stimulating activity release by vascular endothelial cells. J Clin Invest 1986; 78:1316.

26. Fibbe WE, Van Damme J, Billiau A, Voogt PJ, Duinkerken N, Kluck PMC, Falkenburg JHF. Interleukin 1 (22K factor) induces release of granulocyte–macrophage colony-stimulating activity from human mononuclear phagocytes. Blood 1986; 68:1316.

27. Kaushanski K, Lin N, Adamson JW. Interleukin 1 stimulates fibroblasts to synthesize granulocyte–macrophage and granulocyte colony stimulating factors. J Clin Invest 1988; 81:92.

28. Maier R, Ganu V, Lotz M. Interleukin 11, an inducible cytokine in human articular chondrocytes and synoviocytes, stimulates the production of the tissue inhibitor of metalloproteinases. J Biol Chem 1993; 268:21527.

29. Jacobsen SE, Ruscetti FW, Dubois CM, Wine J, Keller JR. Induction of colony stimulating factor receptor expression on hematopoietic progenitors. New model for growth factor synergism. Blood 1992; 80:678–687.

30. Neta R, Oppenheim JJ, Wang JM, Snapper CM, Morman MA, Dubois CM. Synergy of IL 1 and c-kit ligand (KL) in radioprotection of mice correlates with IL 1 upregulation of mRNA and protein expression for c-kit on bone marrow cells. J Immunol (in press).

31. Sims JE, March CJ, Cosman D. et al. cDNA expression cloning of the IL 1 receptor, a member of the immunoglobulin superfamily. Science 1988; 241:585.

32. McMahan CJ, Slack JL, Mosley B, Cosman D, Lupton SD, Brunton LL, Grubin CE, Wignall JM, Jenkins NA, Brannan CI, Copeland NG, Huebner K, Croce CM, Cannizzarro LA, Benjamin D, Dower SK, Spriggs MK, Sims JE. A novel IL-1 receptor, cloned from B cells by mammalian expression, is expressed in many cell types. EMBO J 1991; 10:2821–2832.

33. Sims JE, Gayle MA, Slack JL, Alderson MR, Bird TA, Giri JG, Colotta F, Re F, Mantovani A, Shanebeck K, Grabstein RH, Dower SK. Interleukin 1 signaling occurs exclusively via the type I receptor. Proc Natl Acad Sci USA 1993; 90:6155:6159.

34. O'Neill LAJ, Bird TA, Skatvala J. How does IL 1 activate cells? Immunol Today 1990; 11:392–394.

35. Colotta F, Re F, Muzio M, Bertini R, Polentarutti N, Sironi M, Giri JG, Dower SK, Sims JE, Mantovani

A. Interleukin-1 type II receptor: a decoy traget for IL-1 that is regulated by IL-4. Science 1993; 261:472–475.

36. Moore MAS. Clinical implications of positive and negative hematopoietic stem cell regulators. Blood 1991; 78:1–19.

37. Williams DE, Broxmeyer HE. Interleukin-1 alpha enhances the in vitro survival of purified murine granulocyte–macrophage progenitor cells in the absence of colony-stimulating factors. Blood 1988; 72:1608–1615.

38. Hestdal K, Ruscetti FW, Ghizzonite R, Ortiz M, Gooya J, Longo DL, Keller J. Interleukin-1 directly and indirectly promotes hematopoietic cell growth through type I interleukin-1 receptor. Blood (in press).

39. Snour EF, Brandt JE, Leemhuis T, Ballas CB, Hoffman R. Relationship between cytokine-dependent cell cycle progression and MHC class II antigen expression by human $CD34^+$ HLA-DR bone marrow cells. J Immunol 1992; 148:815–820.

40. Dubois CM, Ruscetti FW, Jacobsen SEW, Oppenheim JJ, Keller JR. Hematopoietic growth factors up-regulate the p65 type II IL-1 receptor on bone marrow progenitor cells in vitro. Blood 1992; 80:600–608.

41. Dubois CM, Ruscetti FW, Palaszynski EW, Falk LA, Oppenheim JJ, Keller JR. Transforming growth factor β is a potent inhibitor of interleukin-1 (IL-1) receptor expression proposed mechanism of inhibition of IL-1 action. J Exp Med 1990; 172:737–744.

42. Dubois CM, Ruscetti FW, Keller JR, Oppenheim JJ, Hestdal K, Chizzonite R, Neta R. In vivo IL-1 administration indirectly promotes type II IL-1R expression on hematopoietic bone marrow cells: novel mechanism for the hematopoietic effects of IL-1. Blood 1991; 78:2841–2847.

43. Dubois CM, Neta R, Keller JR, Jacobsen SEW, Oppenheim JJ, Ruscetti FW. Hematopoietic growth factors and glucocorticoids synergize to mimic the effects of IL-1 on granulocyte differentiation and IL-1 receptor induction on bone marrow cells in vivo. Exp Hematol 1993; 21:303–310.

44. Shieh JH, Peterson RHF, Moore MAS. IL-1 modulation of cytokine receptors on bone marrow cells. In vitro and in vivo studies. J Immunol 1991; 147:1273–1278.

45. Dubois CM, Ortiz M, Gooya JM, Ruscetti FW, Grondin F, Stankova J, Keller J. Transforming growth factor beta regulates c-*kit* message stability and cell-surface protein expression in hematopoietic progenitors. Blood 1994; 83:3138–3143.

46. Hatzfeld J, Li ML, Brown EL, Sookdeo H, Levesque JP, O'Toole T, Gurney C, Clark SC, Hartzfeld A. Release of early human hematopoietic progenitors from quiescence by antisense transforming growth factor β1 or Rβ oligonucleotides. J Exp Med 1991; 174:925–929.

47. Dubois CM, Ruscetti FW, Jacobsen SEW, Oppenheim JJ, Keller JR. Hematopoietic growth factors up-regulate the p65 type II IL-1 receptor on bone marrow progenitor cells in vitro. Blood 1992; 80:600.

48. Bartelmez S, Bradley TR, Bertoncello T. Interleukin-1 plus interleukin-3 plus colony-stimulating factor-1 are essential for clonal proliferation of primitive myeloid bone marrow cells. Exp Hematol 1991; 17:240.

49. Jacobsen SEW, Ruscetti FW, Dubois GM, Lee J, Boone TC, Keller JR. Transforming growth factor-β transmodulates the expression of colony stimulating factor receptors on murine hematopoietic progenitor cell lines. Blood 1991; 77:1706–1716.

50. Ewen ME, Sluss HK, Whitehous LL, Livingston DM. TGFβ inhibition of Cdk4 synthesis is linked to cell cycle arrest. Cell 1993; 74:1009–1020.

51. Hestdal K, Jacobsen SEW, Ruscetti FW, Dubois CM, Longo DL, Chizzonite R, Oppenheim JJ, Keller J. In vivo effect of interleukin-1α on hematopoiesis: role of colony-stimulating factor receptor modulation. Blood 1992; 80:2486–2494.

52. Neta R, Vogel SW, Plocinsky JM, Tare NS, Benjamin W, Chizzonite R, Pilcher M. In vivo modulation with anti IL-1 receptor antibody 35F5 of the response to IL 1. The relationship of radioprotection, CSF and IL-6. Blood 1993; 76:57.

53. Nelson SY, Pack I, Besmer P. Role of *kit* ligand in proliferation and suppression of apoptosis in

mast cells: basis for radiosensitivity of white spotting and steel mutant mice. J Exp Med 1994; 179:1777–1787.

54. Stork L, Barczuk L, Kissinger M, Robinson WA. Interleukin-1 accelerates murine granulocyte recovery following treatment with cyclophosphamide. Blood 1989; 73:938–944.

55. Fibbe WE, van der Meer JWM, Falkenburg JHF, Hamilton MS, Kluin PM, Dinarello CA. A single low dose of human recombinant interleukin 1 accelerates recovery of neutrophils in mice with cyclophosphamide-induced neutropenia. Exp Hematol 1989; 17:805–808.

56. Benjamin WR, Tare NS, Hayes TJ, Becker JM, Anderson TD. Regulation of hemopoiesis in myelo-suppressed mice by human recombinant IL-1α. J Immunol 1989; 142:792–799.

57. Castelli MP, Black PL, Schneider M, Pennington R, Abe F, Talmadge JE. Protective, restorative, and therapeutic properties of recombinant human IL-1- in rodent models. J Immunol 1988; 140:3830–3837.

58. Ido M, Harada M, Furuichi H, Matsuoka N, Nakano K, Sohmura Y. Interleukin 1-induced sequential myelorestoration: dynamic relation between granulopoiesis and progenitor cell recovery in myelo-suppressed mice. Exp Hematol 1992; 20:161–166.

59. Eppstein DA, Kurahara CG, Bruno NA, Terrell TG. Prevention of doxorubicin-induced hematotoxicity in mice by interleukin 1. Cancer Res 1989; 49:3955–3960.

60. Futami H, Jansen R, Macphee MJ, Keller J, McCormick K, Longo DL, Oppenheim JJ, Ruscetti FW, Wiltrout RH. Chemoprotective effects of recombinant human IL-1α in cyclophosphamide-treated normal and tumor-bearing mice. J Immunol 1990; 145:4121–4130.

61. Moore MAS, Stolfi RL, Martin DS. Hematologic effects of interleukin-1β, granulocyte colony-stimulating factor, and granulocyte-macrophage colony-stimulating factor in tumor-bearing mice treated with fluorouracil. J Natl Cancer Inst 1990; 82:1031–1037.

62. Gasparetto C, Laver J, Abboud M, Gillio A, Smith C, O'Reilly RJ, Moore MAS. Effects of interleukin-1 on hematopoietic progenitors: evidence of stimulatory and inhibitory activities in a primate model. Blood 1989; 74:547–550.

63. Moreb J, Zucali JR, Gross MA, Weiner RS. Protective effects of IL-1 on human hematopoietic progenitor cells treated in vitro with 4-hydroperoxycyclophosphamide. J Immunol 1989; 142:1937–1942.

64. Moreb J, Zucali JR, Rueth S. The effects of tumor necrosis factor-α on early human hematopoietic progenitor cells treated with 4-hydroperoxycyclophosphamide. Blood 1990; 76:681–689.

65. Zucali JR, Moreb J, Bain C. Protection of cells capable of reconstituting long-term bone marrow stromal cultures from 4-hydroperoxycyclophosphamide by interleukin-1 and tumor necrosis factor-α. Exp Hematol 1992; 20:969–973.

66. Zucali JR, Moreb J. Protective effects of TNF and IL-1. In: Guigon M, et al, eds. The Negative Regulation of Hematopoiesis, Colloque INSERM/John Libbey Eurotext 1993; 229:429–435.

67. Hornung RL, Longo DL. Hematopoietic stem cell depletion by restorative growth factor regimens during repeated high dose cyclophosphamide therapy. Blood 1992; 80:77.

68. Kohn FRW, Sladek NE. Aldehyde dehydrogenase activity as the basis for the relative insensitivity of murine pluripotent hematopoietic stem cells to oxazaphosphorines. Biochem Pharmacol 1985; 34:3465–3471.

69. Sahovic EA, Colvin MO, Hilton J, Ogawa M. Role of aldehyde dehydrogenase in survival of progenitors for murine blast cell colonies after treatment with 4-hydroperoxycyclophosphamide in vitro. Cancer Res 1988; 48:1223–1226.

70. Kasten MB, Schlaffer E, Russo JE, Colvin MO, Civin CI, Hilton J. Direct demonstration of elevated aldehyde dehydrogenase in human hematopoietic progenitor cells. Blood 1990; 75:1947–1950.

71. Moreb J, Zucali JR Zhang Y, Colvin MO, Gross MA. Role of aldehyde dehydrogenase in the protection of hematopoietic progenitor cells from 4-hydroperoxycyclophosphamide by interleukin-1β and tumor necrosis factor. Cancer Res 1992; 52:1770–1774.

72. Zucali JR, Zhang C, Pincus R, Moreb J. Protection of normal hematopoietic progenitors from L-phenylalanine mustard by interleukin-1. Exp Hematol (in press).

73. Ainsworth EJ, Hatch MH. Decreased x-ray mortality in endotoxin-treated mice. Radiat Res 1958; 9:96.

74. Neta R, Oppenheim JJ, Douches S. Interdependence of the radioprotective effects of human recombinant IL 1, TNF, G-CSF and murine GM-CSF. J Immunol 1988; 140:108.

75. Neta R, Oppenheim JJ, Douches SD, Giclas PC, Imbra RJ, Karin M. Radioprotection with IL 1. Comparison with other cytokines. Prog Immunol 1986; 6:900.

76. Neta R, Oppenheim JJ. Cytokines in therapy of radiation injury. Blood 1988; 72:1093–1095.

77. Hancock SL, Chung RT, Cox RS, Kallman RF. Interleukin 1β initially sensitizes and subsequently protects murine intestinal stem cells exposed to photon radiation. Cancer Res 1991; 51:2280–2285.

78. Neta R, Oppenheim JJ, Schreiber RD, Chizzonite R, Ledney GD, MacVittie TJ. Role of cytokines (IL-1, TNF and TGFβ) in natural and LPS enhanced radioresistance. J Exp Med 1991; 173:1177.

79. Oppenheim JJ, Neta R, Tiberghien P, Gress R, Kenny JJ, Longo DL. IL 1 enhances survival of lethally irradiated mice treated with allogeneic bone marrow cells. Blood 1989; 74:2257.

80. Zucali JR, Moreb J, Gibbons W, Alderman J, Suresh A, Zhang Y, Shelby B. Radioprotection of hematopoietic stem cells by IL-1. Exp Hematol 1994; 22:130–135.

81. Neta R, Sayers TJ, Oppenheim JJ. Relationship of tumor necrosis factor to interleukins. In: Aggarwal BB, Vilcek J, eds. Tumor Necrosis Factors. New York: Marcel Dekker, 1992:499.

82. Bernheim HA, Beutler B, Cerami A, Figari IS, Palladino JA Jr, O'Connor JV. Tumor necrosis factor (cachectin) is an endogenous pyrogen and induces interleukin-1. J Exp Med 1986; 163:1433.

83. Sherman ML, Datta R, Hallahan ED, Weischelbaum RR, Kufe DW. Regulation of TNF gene expression by ionizing radiation in human myeloid leukemia cells and peripheral blood monocytes. J Clin Invest 1991; 87:1794.

84. Woloschak GE, Chang-Liu CM, Jones PS, Jones CA. Modulation of gene expression in Syrian hamster embryo cells following ionizing radiation. Cancer Res 1990; 50:339.

85. Witta L, Fuks Z, Haimovitz-Friedman A, Vlodavsky I, Goodman DS, Eldor A. Effects of irradiation on the release of growth factors from cultured bovine, porcine and human endothelial cells. Cancer Res 1989; 49:5066.

86. Thornally PJ, Vajak M. Possible role of methionein in protection against radiation-induced oxidative stress kinetics and mechanism of its reaction with superoxide and hydroxyl radicals. J Immunol 1985; (in press).

87. Masuda A, Longo DL, Kobayaski Y, Appella E, Oppenheim JJ, Matsushima K. Induction of MnSOD by IL 1. FASEB J 1988; 2:3087.

88. Wong GHW, Goeddel DV. Induction of MnSOD by TNF: possible protective mechanism. Science 1988; 242:941.

89. Eastgate J, Moreb J, Nick HS, Suzuki K, Taniguchi N, Zucali JR. A role for manganese superoxide dismutase in radioprotection of hematopoietic stem cells by IL 1. Blood 1993; 81:639–646.

90. Hirose K, Longo DL, Oppenheim JJ, Matsushima K. Overexpression of mitochondrial manganese superoxide dismutase promotes the survival of tumor cells exposed to interleukin 1, tumor necrosis factor, selected anti-cancer drugs and ionizing radiation. FASEB J 1993; 7:361–368.

91. Suresh A, Jung F, Mareb J, Zucali JR. Role of MnSOD in radio-protection using gene transfer studies. Cancer Gene Ther 1994; 1:1–7.

92. Smith JW II, Urba WJ, Curti BD, Elwood LJ, Steis RG, Janik JE, Sharfman WH, Miller LL, Fenton RG, Conlon KC, Sznol M, Creekmore SP, Wells NF, Ruscetti FW, Keller JR, Hestdal K, Shimizu M, Rossio J, Alvord WG, Oppenheim JJ, Londo DL. The toxic and hematologic effects of interleukin-1 alpha administered in a phase I trial to patients with advanced malignancies. J Clin Oncol 1992; 10:1141–1152.

93. Tewari A, Buhles WC, Starnes HF. Preliminary report: effects of interleukin-1 on platelet counts. Lancet 1992; 1:712–714.

94. Crown J, Jakubowski A, Kemeny N, Gordon M, Gasparetto C, Wong G, Sheridan C, Toner G, Meisenberg B, Botet J, Applewhite J, Sinha S, Moore M, Kelsen W, Buhles W, Gabrilove J. A phase 1 trial of recombinant human interleukin-1β alone and in combination with myelosuppressive doses of 5-fluorouracil in patients with gastrointestinal cancer. Blood 1991; 78:1420–1427.

95. Smith JW II, Longo DL, Alvord WG, Janik JE, Sharfman WH, Gause B, Curti B, Creekmore S, Holmlund J, Fenton RG, Sznol M, Miller L, Shimizu M, Oppenheim JJ, Feim S, Hursey J, Powers

G, Urba WJ. The effects of interleukin 1α treatment on platelet recovery after high-dose carboplatin. N Engl J Med 1993; 328:756–761.

96. Emanuel PD, Wheller RH, Williams DE, Buchsbaum DL, LoBuglio AF. In vivo interleukin-1β (IL-1β) treatment markedly expands the human peripheral blood progenitor cell pool which retain responsiveness to other growth factors [abstr]. Blood 1992; 80:358.

97. Weisdorf DJ, Katsanis E, Verfaillie C, Ramsay NKC, Garrison L, Blazar BR. Interleukin-1α after autologous transplantation for lymphoma/Hodgkin's disease: clinical and hematologic effects. Blood 1992; 80:1321.

98. Neumunaitis J, Applebaum FR, Lilleby K, Buhles WC, Rosenfeld C, Zeigler ZR, Shadduck RK, Singer JW, Meyer W, Buckner CD. Phase 1 study of recombinant interleukin-1β in patients undergoing autologous bone marrow transplantation for acute myelogenous leukemia. Blood 1992; 83:3473–3479.

99. Wilson WH, Bryant G, Fox M, Miller L, Steinberg S, Goldspiel B, Urba W, O'Shaughnessy J, Smith J, Wittes R. Interleukin-1α administered before high-dose ifosfamide (I), CBDCA (C), and etoposide (E) (ICE) with autologous bone marrow rescue shortens neutrophil recovery: a phase I/II study. Proc Am Soc Clin Oncol 1993; 12:937 (abstr).

17

Kit-Ligand–Stem Cell Factor

Peter Besmer
Memorial Sloan-Kettering Cancer Center, New York, New York

I. INTRODUCTION

Mutations at both the *steel* (*Sl*) and the white spotting (*W*) loci in mice cause macrocytic anemia, infertility, and depigmentation (1,2). Because of their pleiotropic effects on several cell systems during embryonic development and in the postnatal animal, *W* and *Sl* mutations have been of great interest to both hematologists and developmental biologists for a long time. While mutations at the *W* locus are cell autonomous (i.e., the mutant gene is expressed and functions in the cellular targets of the mutation), mutations at the *steel* locus are not cell autonomous and they exert their effect in the microenvironment of the targets of *W* and *Sl* mutations. The complementary relation between the *Sl* and *W* mutations implied that their gene products function in one biochemical pathway. In agreement with this prediction, it was recently determined that the *Sl* and *W* loci encode a ligand–receptor pair: Kit-ligand and the Kit receptor tyrosine kinase, the gene product of the protooncogene c-*kit*. Pertinent reviews on Kit-ligand have been published elsewhere (3–7). This chapter will discuss insight into the function of the Kit-ligand–Kit receptor pair that arise from the molecular characterization of the c-*kit* receptor tyrosine kinase and its ligand and from the analysis of various alleles at the *Sl* and *W* loci, with particular emphasis on the important role of the Kit receptor system in several aspects of hematopoiesis.

II. THE c-*kit* RECEPTOR

The protooncogene c-*kit* is the cellular homologue of v-*kit*, the oncogene of the Hardy-Zuckerman 4–feline sarcoma virus (HZ4-FeSV), an acute transforming feline retrovirus isolated from an FeLV-associated feline fibrosarcoma (8). The *kit* sequences in the HZ4-FeSV specify a 370-amino acid polypeptide chain, with the characteristics of a tyrosine protein kinase. Cells transformed by the HZ4-FeSV express an 80-kDa protein with intrinsic tyrosine kinase activity containing FeLV *gag* as well as the *kit* determinants (9). c-*kit* cDNAs isolated from human and mouse libraries have been characterized (10,11). The c-*kit* protooncogene encodes a receptor tyrosine kinase of the platelet-derived growth factor (PDGF) receptor subfamily that includes the colony-stimulating M-CSF receptor. The c-*kit* protein has an extracellular domain, containing a signal peptide and five immunoglobulin repeats; a transmembrane domain; and a cytoplasmic domain, consisting of a juxtamembrane domain, a kinase domain that is divided into two subdomains by the so-called kinase insert segment, and a COOH-terminal domain (10,11). Whereas immunoglobulin folds I, II, III, and V are stabilized by intramolecular disulfide bonds, the immunoglobulin fold IV is not. The Kit protein contains N- and O-linked carbohydrate modifications, and the apparent molecular weight of the Kit protein varies in different cell types as a result of differential posttranslational modifications (12). A normal variant of the c-*kit* protein (Kit A$^+$), formed as the result of alternative usage of a 3'-splice site, contains a four-amino acid insert in the extracellular domain between amino acids 512 and 513 of the known murine c-*kit* sequence (13,14). Furthermore, soluble extracellular domain may be generated by posttranslational proteolytic processing (15–17).

The c-*kit* gene maps to human chromosome 4q12-13 and mouse chromosome 5 in the vicinity of the PDGF receptor α-chain gene and the *flk1* receptor kinase gene (10,18,19). Both the human and mouse c-*kit* genes have been characterized. They comprise 21 exons, covering 65 kb (20–24). The organization of the c-*kit* gene is reminiscent of the CSF-1 receptor gene, particularly for the exons and their boundaries, but not for the size of introns, thus demonstrating their close evolutionary relation. Although immunoglobulin folds I and II are contained in single exons, folds III, IV, and V are contained in two exons. The different exonic structures of the different immunoglobulin domains may reflect independent evolutionary origins of these domains. There are two alternative splice donor sites in the c-*kit* gene, in exons 9 and 14, giving rise to a four- and a single-amino acid insertion, respectively (13,14). The significance of these structural variants of Kit is not yet understood.

The realization that the c-*kit* protooncogene resides on mouse chromosome 5 (10) in the vicinity of the white spotting (*W*), the patch (*Ph*) and the rump white (*Rw*) loci initially raised the question of whether c-*kit* is encoded by either of these loci. Subsequently, the identity of the c-*kit* protooncogene with the white spotting locus was established by linkage analysis (25), the demonstration of rearrangements in the c-*kit* gene in some *W* alleles (26), and the finding of missense *W* mutations that inactivate the c-*kit* kinase (27,28).

III.   KIT-LIGAND OR STEM CELL FACTOR

The discovery of identity of c-*kit* and the *W* locus accelerated the quest for the ligand of the c-*kit* receptor. W mutant mice lack both connective tissue and gastrointestinal mucosa mast cells, indicating a role for c-*kit* in mast cell development. In vitro, bone marrow-derived mast cells (BMMC) require interleukin (IL)-3 and connective tissue mast cells isolated from the peritoneal cavity (CTMC) require both IL-3 and IL-4 for proliferation and survival in vitro. Although both BMMC and CTMC can be maintained by coculture with 3T3 fibroblasts in the absence of IL-3, BMMC from W mutant mice fail to do so, suggesting that fibroblasts produce the ligand for the c-*kit* receptor (29,30). The essential role of c-*kit* in BMMC provided an assay for the isolation of a soluble form of the ligand of the c-*kit* receptor, kit-ligand (KL), from Balb/c 3T3 fibroblasts and from a murine stromal cell line (31–33). However, the same protein was also isolated from buffalo rat liver-3A cells as a factor that promotes the formation of colonies from early hematopoietic progenitor (HPPC)-designated stem cell factor (SCF; 34,35).

The soluble KL protein has two intramolecular disulfide bonds and forms noncovalent dimers (31). On the basis of structural characteristics of the KL polypeptide chain as well as the conservation of exon boundaries in the M-CSF and the KL genes, KL belongs to the group of cytokines and growth factors that share a common four-helical bundle topology (36). More specifically, KL is predicted to be a member of the short-chain cytokine family, which includes IL-2, IL-3, IL-4, IL-5, IL-7, IL-9, IL-13, granulocyte–macrophage colony-stimulating factor (GM-CSF) and M-CSF (36). However, such topological characteristics remain to be demonstrated.

In the mouse and in humans, two alternatively spliced KL RNA transcripts encode two cell-associated KL protein products, KL-1 and KL-2, of 248 and 220 amino acids, respectively, which differ in their NH_2-terminal sequences of the transmembrane segment (37–39). Both the KL-1 and KL-2 proteins contain O-linked and N-linked carbohydrate modifications (40). The KL-1 and KL-2 RNA transcripts are expressed in a tissue-specific fashion (39). The KL-2 protein lacks sequences that include the major proteolytic cleavage site for the generation of the soluble KL protein from KL-1 (Fig. 1). Although KL-1 is efficiently processed by proteolytic cleavage to produce soluble KL; KL-2 is also processed to form soluble KL, but not as effectively. Therefore, KL-2 represents a differentially more stable cell-associated form of KL (39). The protease activities facilitating cleavage of KL-1 and KL-2 in COS-1 cells were shown to be distinct by using a panel of protease inhibitors (41). Interestingly, the protein kinase C-(PKC) inducer phorbol myristate acetate (PMA) accelerates proteolytic cleavage of both KL-1 and KL-2, suggesting that this process is subject to regulation (39). From the determination of NH_2-terminal and COOH-terminal amino acid sequences of soluble KL, the KL-1 polypeptide chain has a 25-amino acid signal peptide and a COOH-terminal cleavage site at amino acids (164,165). The KL-2 polypeptide chain lacks the exon 6 sequences, but has a secondary cleavage site presumably in exon 7. Therefore, differential expression of variant cell membrane-associated KL molecules and their proteolytic cleavage to generate soluble forms of KL provide different means to control and modulate c-*kit* function.

The KL gene maps on human chromosome 12 (12q22-24) and on mouse chromosome 10 at the steel (*Sl*) locus (32,35, 42–44). The structural organization of the KL genes has not been studied in detail, although exon boundaries in the coding sequences specifying the extracellular domain of rat and human KL have been reported, and they seem to conform with those of the M-CSF gene (45).

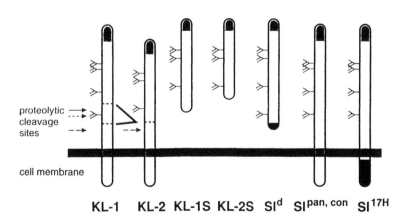

Figure 1 Schematic diagram of structural characteristics of normal and mutant (*Sl^d^*, *Sl^pan^*, *Sl^con^* and *Sl^17TH^*) KL protein products. Normal KL proteins: KL-1 and KL-2 (products of alternatively spliced transcripts) are membrane proteins and KL-1S and KL-2S are soluble proteins, produced by proteolytic cleavage of the KL-1 and KL-2 membrane forms. Dark-shaded areas in *Sl^d^* and *Sl^17H^* proteins indicate altered protein sequences.

Both the soluble forms of KL and M-CSF belong to the four-helix bundle cytokine family (36). Both KL and M-CSF are produced as transmembrane proteins that are proteolytically processed to release soluble forms of the factors and, therefore, exert their functions both in the membrane and the soluble forms. Also, both factors generate variant forms of the membrane-associated growth factor by alternative splicing, providing means for the regulation of the soluble and the membrane forms of the two factors in vivo during embryonic development and in the postnatal animal. These considerations imply a close evolutionary origin of the two factors. The Kit receptor, the M-CSF receptor, and the PDGF receptor a- and b-chains constitute a receptor tyrosine kinase subfamily, arising from two consecutive gene duplication events. The Kit–M-CSF and the PDGFa/b receptor systems presumably arose by a coevolutionary mechanism from an ancestral receptor–ligand pair. Whereas in the case of the PDGF receptors the two ligands in part still display cross-reactivity with the related receptors, with Kit and M-CSFR no such cross-reactivity has been noted.

IV. *W* AND *Sl* MUTANT PHENOTYPES

Mutations at both the *W* and the *Sl* locus of the mouse result in deficiencies in three cell systems: hematopoiesis, the pigmentary system, and germ cells during embryogenesis and in the adult animal (1,2). Many alleles are known at these loci with differing degrees of severity in the heterozygous and the homozygous state and in the different cellular targets (46). In hematopoiesis, during early development as well as in adult life, *W* and *Sl* mutations affect cells within the hematopoietic stem cell hierarchy, distinctive cell populations in the erythroid cell lineage and mast cells. W and Sl mutant mice, therefore, typically suffer from macrocytic anemia and they lack tissue mast cells.

A. Hematopoietic Phenotypes of *W* and *Sl* Mice

During embryonic development yolk-sac erythropoiesis, giving rise to nucleated erythrocytes, is affected in *W/W* (47), but not in *Sl/Sl*^{d} fetuses, whereas the number of nonnucleated erythrocytes in the fetal liver is significantly reduced in *W/W* and *Sl/Sl*^{d} embryos, and they are larger in size (48,49). However, the liver of *Sl/Sl* fetuses contain 30–40% of hematopoietic stem cells (HSC) found in +/+ controls implying that the generation of stem cells is not significantly affected by the mutation (50). It had been suggested that the *Sl* gene may not affect yolk sac erythropoiesis, but that the fetal liver is unable to promote normal differentiation from hematopoietic stem cells to progenitors and their differentiation into hemoglobinized erythroblasts. Postnatal animals carrying a double dose of the *W* or *Sl* alleles have an increased mean erythrocyte volume and reduced mean hematocrit percentages, with the most severe mutations causing postnatal lethality. The *W* and *Sl* mutations also affect erythroid homeostatic mechanisms (i.e., responses to hypoxia and blood loss are diminished; 51). A hallmark of mice carrying *W* and *Sl* mutations is a profound effect on mast cell development. Both *W/W*^{v} and *Sl/Sl*^{d} mice lack virtually all tissue mast cells (52,53).

Nonerythroid aspects of hematopoiesis in mice carrying *W* and *Sl* mutations are affected as well. The total cellularity of the marrow is approximately 75 and 50%, respectively, in *W/W*^{v} and *Sl/Sl*^{d} mice. The cell deficits include erythroid and myeloid progenitors and megakaryocytes. The number of megakaryocytes in both genotypes was approximately half, and the stage III megakaryocytes were larger. However, the numbers of circulating platelets and the mechanisms regulating thrombocytopoiesis are normal in mutant animals (54).

Both W and Sl mutant mice display an increased sensitivity to the lethal effects of whole-body irradiation. The $LD_{50/30}$ of W/W^v, $W/+$, $W^v/+$, and $+/+$ is 365, 690, 692, and 765 R, respectively (55). The radiosensitivity in Sl/Sl^d animals ($LD_{50/30} < 150$ R) is more pronounced than in animals carrying mutations at the W locus. Characteristically, the onset of hemopoietic regeneration was delayed in the mutant animals. The radiation sensitivity may stem either from increased sensitivity of progenitors in mutant mice or from an effect on the regeneration of the hematopoietic tissue. Evidence for increased sensitivity of progenitor cells or the suppression of radiation-induced apoptosis in these cells by KL will be discussed later in the chapter (56).

B. Cell Intrinsic and Microenvironmental Defects of *W* and *Sl* Mutations

The defects of W gene action are intrinsic to cellular targets of the mutation. The anemia of W/W^v is cured completely by implantation of normal bone marrow, spleen, and fetal liver cells into nonirradiated W/W^v recipients (57). After transplantation into mutant recipients, all hematopoietic tissues, including the thymus, marrow, spleen, and lymph nodes, are, at least partly, replaced by normal donor hematopoietic cells (58). In addition, the mast cell deficit in the skin, stomach wall, and cecum in mutant mice can also be "cured" by implantation of normal marrow (52).

Conversely, the defect of Sl gene action resides in the microenvironment of the targets of the mutation. The anemia of Sl/Sl^d mice cannot be cured by implantation of normal bone marrow; but marrow from Sl/Sl^d mice repopulated the hematopoietic system in lethally irradiated recipients and cured the anemia of W/W^v animals (59,60). Whereas, after transplantation of skin grafts from W/W^v donors into Sl/Sl^d recipients, the skin graft became repopulated with mast cells; in contrast, Sl/Sl^d skin failed to develop a mast cell population in both W/W^v and normal recipients (53).

C. Nonhematopoietic Phenotypes of *W* and *Sl* Mutations

During normal development, melanoblasts migrate from the neural crest, laterally and ventrally, to the periphery. They then enter the epidermal ectoderm, colonize hair follicles, and postnatally, amelanotic melanoblasts differentiate to become melanocytes. Several aspects of melanogenesis—the early migratory phase in embryonic development, the time when melanoblasts reach the epidermal ectoderm as well as differentiated melanocytes in hair follicles—appear to be affected by W and Sl mutations, causing differing degrees of white spotting (2,5). Primordial germ cells are derived from the posterior primitive streak and then migrate from the base of the allantois, through the hindgut endoderm, to the genital ridges; spermatogenesis and oogenesis then proceed, following distinct well-studied developmental programs. W and Sl mutations affect the survival, migration, and proliferation of primordial germ cells as well as steps in spermatogenesis and oogenesis, including the survival and proliferation of spermatogonia and oocyte growth, thereby, causing impaired fertility (61–64).

Taken together the defects in W and Sl mutant mice are consistent with a role of the c-*kit* receptor system in facilitating cell proliferation and survival of precursor cells, as well as promoting cell migration, cell adhesion, secretion, and other functions in differentiated cells (4,6).

V. THE c-*kit* AND KL EXPRESSION IN ORGANOGENESIS AND ADULT LIFE

The synchronous expression of the c-*kit* receptor and its ligand in close cellular environments is a good predictor for sites at which c-*kit* functions in vivo. Therefore, the examination of c-*kit* and KL expression during embryonic development and in the adult animal by RNA blot analysis, in situ hybridization, and immunohistochemistry has provided important insights into the understanding of c-*kit* function. In agreement with the cell autonomous nature of *W* mutations, c-*kit* is expressed in cellular targets of *W* and *Sl* mutations during embryogenesis and in the postnatal animal in melanogenesis, gametogenesis, and in cells of the hematopoietic system (27,65–70). Expression of KL is associated with migratory pathways of melanoblasts and germ cells, and homing sites of both germ cells and hematopoietic progenitors during embryonic development (71,77,78).

In hematopoietic tissues the Kit receptor is expressed in several immature and mature cell populations. Kit receptor expression is seen in lineage marker-negative (TER-119⁻, B200⁻, Mac-1⁻, Gr-1⁻) hematopoietic stem cells (HSC) in the fetal liver and in adult bone marrow (50,73). Furthermore, Kit receptor-positive, lin⁻ bone marrow cells reconstitute the hematopoietic system of lethally irradiated mice. Kit receptor expression persists in lymphoid, myeloid and erythroid progenitors (69,73); in progressively more mature cell populations, reduced c-*kit* expression is observed (138,139). In contrast tissue mast cells express high levels of c-*kit* (27,31,74). During embryonic development, KL expression is seen at sites of hematopoiesis (i.e., in the fetal liver and in blood islands). Similarly, in the postnatal animal, KL expression occurs at sites of active hematopoiesis in the microenvironment of hematopoietic cells in the bone marrow, the spleen, lymphnodes, and the thymus. KL is also expressed in numerous murine and human stromal cell lines, as well as in endothelial cells (i.e., human umbilical vein endothelial cells and bone marrow-derived microvascular endothelial cells; 75).

In the pigmentary system, c-*kit* expression is observed during embryonic development in melanoblasts leaving the neural crest, and its expression continues during the developmental stages and in the postnatal hair follicle in melanoblasts (60,70). KL expression associated with melanogenesis occurs during development in the dermatome overlaying the somites at the time of melanoblast migration, in the dermis, and in the hair-follicle (71; Manova K, unpublished data).

In gametogenesis in normal mice, c-*kit* is expressed at high levels in primordial germ cells (PGC) during their proliferative phase from E7½ to E13; subsequently c-*kit* expression subsides as the germ cells enter meiotic prophase (64,68). In ovarian development, beginning about embryonic day 18, the c-*kit* receptor is expressed at high levels in oocytes throughout postnatal development (65). In contrast, KL is expressed in granulosa cells of growing follicles, and the expression reaches its highest level in the three-layered follicles (76). In spermatogenesis in the postnatal testis in type A, intermediate and in type B spermatogonia, high c-*kit* expression levels are observed, and this expression is diminished in early spermatocytes. This expression profile is maintained throughout adult life (65,66). During fetal and postnatal spermatogenesis KL is expressed in Sertoli cells (76). Interestingly in the postnatal testis, the alternatively spliced variant KL-2 is the predominant KL protein product, suggesting an important role for membrane KL in spermatogonial development.

Both c-*kit* and KL are also expressed in tissues and cell types that are not known targets of *W* and *Sl* mutations. During embryonic development c-*kit* expression is seen in the neural tube, dorsal root ganglia, portions of the developing central nervous system, the olfactory epithelium, the digestive tract, the lung, and other tissues, whereas KL expression is seen in the floor plate of the neural tube, the thalamus, and in the olfactory epithelium (67,77,78; Manova K, unpublished data). In the developing nervous system, c-*kit* expression is typically seen in cells that have ceased to divide and have begun their differentiation. In the adult animal c-*kit* and KL expression are prominent in the lung and in the brain, including the hippocampus and the cerebellum, where c-*kit* expression is evident in basket, stellate, and Golgi interneurons, and KL expression in Purkinje neurons (78,79). Although, many mutations are known at the *W* and *Sl* loci, no neurological phenotypes and no histopathological changes have been noted in these mice. It may be that, as a result of redundancies, other signaling mechanisms compensate for the lack of c-*kit* function in neuronal cells. The elucidation of a role for the c-*kit* receptor system in the developing and the mature central nervous system is a challenging task of the future.

VI. MOLECULAR BASES OF *W* AND *Sl* MUTATIONS

An easily recognizable phenotype—coat color spotting—has made possible the isolation of many distinct mutations at the *W* and the *Sl* loci (1,2,46,80). Although most of the known *W* mutations occurred spontaneously, many of the Sl mutations were induced by γ-radiation protocols at the Harwell and Oak Ridge laboratories. These mutants provided an opportunity to characterize both the molecular basis of these mutations as well as their effects on different cell lineages and tissues, thereby furthering our understanding of c-*kit* function. The *W* and *Sl* mutations vary in their degree of severity on the affected cellular targets. In the homozygous state, several alleles, including the original *W* and *Sl* alleles, cause perinatal lethality, whereas others are viable and semifertile. Several different mouse, rat, and human *W* alleles and several mouse *Sl* alles have been characterized at the molecular level (Tables 1–3).

A. *W* Mutations

Murine W *Mutations*

The W^{19H} allele arose as a result of chrosomal deletion of approximately 2 cM, including the three receptor tyrosine kinase genes c-*kit*, the *Pdgfra*, as well as the *flk1* genes (25,81,82). Although mice homozygous for the W^{19H} mutation die early during development as a result of the deletion of at least three genes, heterozygous animals have a mild mutant phenotype, with a white ventral spot, a white blaze on the forehead, and depigmented extremities, but there are no effects on hematopoiesis and gametogenesis of the mutation. The W^{19H} mutation is a typical example of a c-*kit* null or loss of function mutation; therefore, c-*kit* null mutations have primarily recessive characteristics. Other examples of null mutations include the original *W* allele and the W^n allele (see Tables 1 and 2; Fig. 2). The *W* mutation was the first white-spotting mutation to be described (83). The *W* allele arose as a result of a splice donor site mutation that specifies a Kit protein lacking the transmembrane domain; hence, a protein that is not expressed on the cell surface and lacks tyrosine protein kinase activity (13,84). The W^n allele, a recently identified mutation, contains a missense mutation and specifies a Kit protein lacking kinase activity, which is not expressed on the cell surface owing to increased protein degradation (85).

Several *W* alleles ($W^{42,37,Jic,V,55,41}$) are known in which heterozygotes are affected more severely than *W*/+; they vary in severity in the homozygous state and affect the three principal cell lineages to a comparable degree (80). These alleles contain c-*kit* missense mutations (see Table 3 and Fig. 2), which impair c-*kit* kinase activity to differing degrees (29,84,86). In the W^{42} allele, an amino acid invariably conserved in protein kinases located in the catalytic fold of the enzyme, aspartic acid-790 is replaced by asparagine (28). The Kit^{W42} protein is expressed normally on the cell surface, but lacks kinase activity. Work on the mechanism of activation of several tyrosine kinase receptors, including the members of the PDGF receptor family and the c-*kit* receptor, implicates receptor dimers or oligomers as essential intermediates (87). The striking dominant phenotype of the W^{42} mutation may indicate that the mutant c-*kit* protein in receptor heterodimers interferes with KL-induced signal transmission, effectively reducing the number of active receptor dimer or oligomers on the cell surface. Consequently, this mutation gives rise to a more severe heterozygous mutant phenotype than null mutations and has the

Table 1 Molecular Bases of Murine *c-kit/W* Mutations

Mutation	Homozygous phenotypes	Heterozygous phenotypes	Molecular lesion	Type of mutation	Refs.
W^{19H}	Early lethal	Ventral spot, no effects on hematopoiesis or fertility	Deletion (approx. 2 cM)	Null, loss of function	25,26,81,82
W^x	Perinatal lethal; full effects on pigmentation, gametogenesis, hematopoiesis	Do[a]	Deletion	Do	26,100
W	Do	Do	Splice donor site mutation, deletion of TMS	Do	13,83,84
W^n	Do	Some irregular spotting, no effects on hematopoiesis and fertility	Ala-838 → Val; no kinase action; no cell-surface protein	Do	85
W^{42}	Do	Lacks pigment; anemia; reduced fertility	Asp-790 → Asn; no kinase action	Dominant negative; null	28,80
W^{37}	Do	Mottled fur, moderate anemia	Glu-582 → Lys; no kinase action	Do	80,84,86
W^{jic}		Do	Gly-595 → Arg; no kinase action	Do	85
$W^V = W^{55}$	Viable, black-eyed white, anemia, sterile	Coat color dilution, ventral spot, mild anemia	Thr-660 → Met; reduced kinase action	Partial, dominant negative	84,86,88
W^{41}	Viable, mottled fur and mild anemia	Mild anemia; spotting	Val-831 → Met; reduced kinase action	Weak, dominant negative	80,84,86
W^f	Viable, fertile, mottled-striped fur, mild anemia	Mild anemia; spotting	Arg-816 → Trp	Weak	85,89,90
W^{44}	Little pigment, reduced fertility, but not anemic	Some spotting	Intron insertion, reduced RNA level	Expression mutation	26,80
W^{57}	Mottled fur, no anemia, fertile	Some spotting	Reduced RNA levels	Expression mutation	86
W^{sh}	Black-eyed white, lack mast cells, but fertile and no anemia	White sash in lower trunk	DNA rearrangement 5' of *c-kit* exons, reduced/ increased RNA levels	Expression mutation	91,93,95,96
rW^s	Viable/perinatal–lethal, reduced fertility, black-eyed white, anemia, lack mast cells	Mild anemia, reduced mast cells, deletion of coat color and spotting	Deletion of amino acids 826-829	Partial, dominant negative	98,99

[a]Do = ditto.

Table 2 Molecular Bases of Human c-*kit*/Piebald Mutations

Mutation	Heterozygous phenotypes	Molecular lesion	Type of mutation	Refs.
Deletion mutations	Piebaldism, mental retardation, multiple congenital anomalies	Interstitial deletions lacking both the *Kit* and *Pdgfra* genes	Null, loss of function	102,103
Missense mutations	Depigmented patches on forehead, chest, and extremities; no other symptoms	Glu583Lys Phe584Leu Gly664Arg Arg 791Gly Gly812Val	Null and partial loss of function; dominant-negative	104–107
Splice site mutation	Variable phenotype	—	Null, loss of function	108
Frameshift mutations	Variable phenotype	Codon 561; codon 642	Null, loss of function	106
Frameshift mutations	Weak pigmentation phenotype	Codon 85; codons 250–251	Null, loss of function	108,109

Table 3 Molecular Bases of Murine KL/*Steel* Mutations

Mutation	Homozygous phenotypes	Heterozygous phenotypes	Molecular lesion	Type of mutation	Refs.
Sl, *Sl^J*, *Sl^gb*, *Sl^8H*, *Sl^10H*	Pre-/perinatal lethal; full effects on pigmentation; gametogenesis; hematopoiesis	Dilution of color on ventral side; occasional spotting	Deletion of KL-coding sequences	Null; loss of function	31,35,42,114–119
Sl^18H	Early lethal	Dilution of color on ventral side; occasional spotting	Deletion of KL-coding sequences	Null; loss of function	42
Sl^d	Viable; black-eyed white; sterile; anemic	Slight dilution of coat color, and spotting	Intragenic deletion, including TMS and COOH-terminal	Partial; lacks cell membrane forms	37,39,120,121
Sl^17H	Viable; black-eyed white; mild anemia; females fertile; males sterile	Lighter pigmentation	Splice site mutation; frameshift amino acids 238 and terminal at 265	Partial; impaired membrane forms	127,128
Sl^pan	Viable; black-eyed white; mild anemia; males fertile; females sterile	Lighter pigmentation	DNA rearrangement > 100-kb 5′ of KL exon sequences	Expression mutation; reduced RNA levels	125,129,130
sl^t	Same as *Sl^pan*, but more germ cells	Lighter pigmentation	Unknown		63
Sl^con	Same as *Sl^pan*, but fewer germ cells; follicle development normal	Lighter pigmentation	DNA rearrangement > 100-kb 5′ of KL exon sequences	Expression mutation; reduced and increased RNA levels	126,130

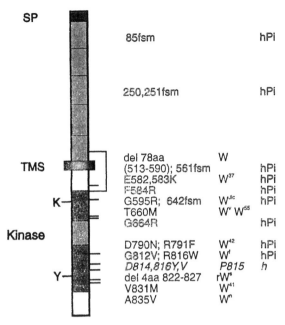

Figure 2 Schematic diagram of normal Kit receptor and of c-*kit*/W mutations, W, W^{42}, W^n, W^{37}, W^{Jic}, W^v/W^{57}, W^{41}, W^f, rW^s and human piebald (*Pi*) mutations. Amino acids at sites of mutations are indicated by single-letter code. The position of missense, frameshift, and gain of function mutations is indicated by codon position. Gain of function mutations are indicated by italic print. Human mutations are indicated using light print. Abbreviations: signal peptide (SP), transmembrane domain (TM), ATP binding site (ATP), kinase insert sequence (insert), and frameshift mutation (fsm).

hallmarks of a dominant-negative mutation. The W^{37} mutation inactivates the Kit kinase and causes somewhat reduced protein stability, diminishing the dominant-negative property of this mutation (80,84,86). Several mutations result in the partial inactivation of the Kit kinase (W^v, W^{55}, W^{41}, W^f) (84–86,90). Homozygotes carrying these alleles are viable. Although the W^v mutation causes complete depigmentation, anemia and infertility, the W^{41} and W^f mutations are mild, and homozygotes are semifertile.

Several c-*kit* expression mutations are known (W^{57}, W^{sh}, and *Ph*) that diminish or enhance c-*kit* gene expression. The W-sash (*sh*) mutation affects primarily mast cells and melanogenesis, but not other cellular targets of W and *Sl* mutations, including germ cells and erythroid cells (91,92). Characterization of the W/c-*kit* locus in the W^{sh} allele by megabase analysis revealed that the c-*kit* and *Pdgfra* genes are unlinked in W^{sh}, because of an inversion of a small segment of mouse chromosome 5 (93,96). As a result of this inversion in the W^{sh} allele, c-*kit* expression is shut off in some cell types, but not in others (93,95). Thus, positive up-stream regulatory elements controlling c-*kit* expression in mast cells are presumably displaced by the W^{sh} "inversion," and negative upstream c-*kit* control elements are also affected. Furthermore, misexpression of c-*kit* in diverse sites, such as dermatome and floor plate of the neural tube, in mutant embryos, may be caused by deletion or displacement of an upstream silencer (93,96). The inappropriate c-*kit* expression in the dermatome of W^{sh} mutant embryos may provide an explanation for the dominant pigmentation defect in these mice. The c-*kit* receptor misexpression in cells of the dermatome may bind or sequester KL and reduce its concentration. Consequently, KL may become limiting for the melanoblasts migrating over the dermatome, reducing their survival or proliferation and migration. Alternatively, coexpression of the Kit receptor and KL may close an autocrine loop in dermatomal cells, causing changes in the extracellular space.

Patch (*Ph*) is another c-*kit* expression mutation. Mice heterozygous for *Ph* display a pigmentation pattern very similar to $W^{sh/+}$ mice, whereas homozygous *Ph/Ph* mice die during embryonic development. The *Ph* mutation arose as a result of a deletion that includes *Pdgfra* (18,97). The 3′-deletion endpoint in the *Ph* allele, which is in juxtaposition with the c-*kit* gene is not known precisely, but it does not include c-*kit*-coding sequences. In *Ph/+* embryos, similar to embryos with the W^{sh} mutation, c-*kit* expression is enhanced in several cell types, including sites important for early melanogenesis (96). Therefore, both the W^{sh} and *Ph* mutations affect some of the same 5′ control elements in the c-*kit* gene. Also, c-*kit* misexpression during embryonic development may explain the pigmentation defect in mice with the *Ph* mutation (5,96).

Rat W Mutations

An interesting natural c-*kit*^{W} mutation in rats (W^s) involves a four-amino acid deletion, amino acids 826–829 in the rat c-*kit*-coding sequence (98,99). The four-amino acid deletion mutation is located in the vicinity of the major tyrosine phosphorylation site in Kit. The consequences of this mutation on receptor function are not very well understood. Although W^s/W^s rats are black-eyed whites, they are viable and fertile. Their initially strong anemia ameliorates with age, but the severe mast cell deficiency persists.

Human W Mutations

The piebald syndrome is an autosomal congenital disorder in humans that is characterized by depigmentation in the ventral aspect, forehead, and extremities. Piebald was first mapped between 4q11-4q12 (101). More recently, several piebald alleles were characterized at the molecular level as c-*kit* mutations. In a cytogenetically normal piebald patient, the c-*kit* and *Pdgfra* genes were deleted (102,103), and missense mutations have been identified in several piebald kindreds. The *E583K* mutation is identical with the murine W^{37} mutation (104). As in mice, this mutation causes a more severe pigmentation phenotype than null mutations. Missense mutations *G664R*, *F584L*, *R791G*, and *G812V* are also presumed dominant-negative mutations (105–107). In addition, four frameshift mutations mapping to codons 85, 250–251, 642, and 561 have been identified (106,108,109). No hematological problems are associated with any of the known Piebald mutations (110). Because piebald individuals are heterozygous for the mutation, this may be expected, but it is also possible that this may reflect real differences between humans and inbred mouse strains.

Gain of Function Mutations

The c-*kit* activating mutations have been identified in three murine, rat, and human mast cell lines. In the murine P815 mastocytoma cell line, the c-*kit* protein contains a missense mutation in which aspartic acid 814 is replaced with tyrosine, and the Kit kinase in these cells is constitutively activated (111,112). In the human mastocytoma cell line HMC-1, the Kit receptor is also constitutively activated (113). The HMC-1 Kit receptor contains two missense mutations: valine-560 is replaced by glycine and aspartic acid-816 by valine, and the latter is responsible for the constitutive activation of Kit. Therefore, mutation of corresponding amino acids in human and murine c-*kit* in the vicinity of the major autophosphorylation site result in constitutive activation of the Kit receptor.

B. *Sl* Mutations

Several severe *Sl* alleles (*Sl*, *Sl^{J,gb,8H,10H,12H,18H}*) (114–118,129) contain deletions that include the KL gene; therefore, they are KL loss of function mutations (32,35,42). These KL loss of function mutations result in severe anemia, causing perinatal lethality. Homozygotes for the Steel-Dickie allele (*Sl^d*) are viable, but they have moderate anemia, lack tissue mast cells, are black-eyed whites, and are sterile (120). The somewhat diminished effects on erythropoiesis (anemia) and viability in *Sl^d*/*Sl^d* animals imply some residual functional activity of KL. The *Sl^d* allele arose as a result of an intragenic deletion, including the transmembrane domain and COOH-terminus, generating a secreted KL protein product (see Fig. 1; 37,39,121). The biological activities of the soluble Sl^{d}KL protein in in vitro assays are indistinguishable from the normal protein. The mutant phenotypes of mice carrying the *Sl^d* allele suggest that, although the Sl^{d}KL protein sustains some activity, it is largely defective in facilitating proliferation and survival of target cells. In agreement with this notion, stromal cells obtained from long-term bone marrow cultures of *Sl*/*Sl^d* mice fail to support long-term hematopoietic cell maintenance and differentiation in vitro (122). In contrast, stromal cells, transfected to express the membrane forms of KL, support the long-term production of hematopoietic progenitor cells in vitro (123). This provides clear evidence that the membrane-bound form of KL plays a critical role in c-*kit* function.

Several *Sl* alleles, with mild phenotypic effects on hematopoiesis and melanogenesis, have significant numbers of germ cells in the gonad, but either female (*Sl^{pan}*, *Sl^t*, *Sl^{con}*) or male mice (*Sl^{17H}*) are sterile; revealing roles for the c-*kit* receptor in postnatal development of oocytes and spermatogonia (63,124–127). The *Sl^{17H}* allele contains a T > A transversion, 11 nucleotides preceeding the 3′-splice acceptor site in intron 7; consequently, exon 8 is skipped, generating a frameshift mutation in which all but one amino acid of the cytoplasmic domain of KL are replaced with 27 amino acids read in an alternative frame before a termination codon is reached (128). Although both males and females are affected equally during embryonic development by the mutation, postnatal germ cell development is affected differentially. The properties of the *Sl^{17H}* mutation provide some insight into the role of the cytoplasmic domain sequences of KL.

In contrast with *Sl^{17H}*, the *Sl^{pan}* allele is a KL expression mutation. Although the KL-coding sequences are normal in the *Sl^{pan}* allele, the levels of the KL transcripts are consistently reduced in most tissues analyzed in *Sl^{pan}*/*Sl^{pan}* mice (129). Similar to *Sl^{17H}*, *Sl^{pan}* affects both males and females during embryonic development (129,130). However, ovarian follicle development in homozygous mutant animals is arrested at the one-layered cuboidal stage. Therefore, a reduced level of KL in *Sl^{pan}*/*Sl^{pan}* ovarian follicle cells appears to arrest ovarian follicle development, implying an essential role for c-*kit* in oocyte growth and maturation. *Sl^t* and *Sl^{con}*, also affect female fertility (63,130). In *Sl^t*/*Sl^t* females follicles are arrested at a stage similar to that of *Sl^{pan}*/*Sl^{pan}*, but more germ cells are present.

C. Generation of *W* Phenotypes

Formal proof for the molecular basis of *W* mutations was obtained by generating *W* mutant phenotypes in normal mice. Two different approaches have been used to address this issue. In one instance, a germline modification was sought that would generate the *W* phenotypes by exploiting the dominant-negative characteristics of the *W^{42}* mutation. Transgenic mice that express the c-*kit^{W42}* gene products ectopically under the control of the human actin promotor were constructed to generate the *W* phenotypes (131). Transgenic mice expressing the c-*kit^{W42}* transgene show an effect on pigmentation and the number of tissue mast cells (5,131). Therefore, mice expressing the c-*kit^{W42}* transgene ectopically recapitulate some of the phenotypes of mice with *W* mutations, in agreement with the dominant-negative mutant characteristics of the c-*kit*^{W42} protein product.

In a second approach a monoclonal c-*kit* antibody, which is an antagonistic blocker of the c-*kit* receptor, was used to interfere with c-*kit* function in vivo (70). Injection of c-*kit* monoclonal antibody into pregnant mice during midgestation and postnatal animals inhibited development and survival of melanoblasts–melanocytes and affected spermatogenesis (66,70). In addition, injection of the antibody into adult mice resulted in a depletion of hematopoietic progenitor cells, and mature myeloid and erythroid cells in the bone marrow, providing evidence for an essential role of c-*kit* in intramarrow hematopoiesis (69).

VII.   CONTROL OF EXPRESSION OF KL–SCF AND c-*kit* GENE PRODUCTS

Analysis of the *W* and *Sl* mutant phenotypes and expression patterns of KL and c-*kit* imply complex regulatory mechanisms governing the expression of KL and c-*kit* in development, differentiation, homeostasis, and functional responses of various cell systems. Quantitative mechanisms of regulation, as well as qualitative mechanisms, facilitate the varied cellular responses of c-*kit* receptor function. They include control of transcription, RNA processing, posttranslational processing, and transmodulatory mechanisms.

A.   Regulation of KL–SCF RNA Expression

The temporal expression of KL transcripts in development and the postnatal animal indicate that mesenchymal cells in the microenvironment and in migratory pathways at sites of KL expression (71,76–79). Thus, KL-expressing cell types in vitro include stromal cells isolated from hemato-poietic tissues, such as the bone marrow, fetal liver, thymus, skin, as well as human umbilical endothelial vein cells, human bone marrow microvascular endothelial cells, and Sertoli cells (6,32,75,132–134). Cytokine production in many cell types may be transmodulated in vitro by treatment with other cytokines. Treatment of human bone marrow stromal cells with TNF-α and IL-1α—cytokines associated with chronic inflammation—down-regulates KL mRNA levels (135). In human umbilical vein endothelial cells, inflammatory stimuli, including various pathogens, LPS, IL-1α, and IL-1β, mediate the transient increase of KL transcript levels (133,136). In primary Sertoli cells follicle-stimulating hormone (FSH) and its intracellular mediator cAMP, increase KL mRNA levels (134).

Very little information is available on the mechanism of KL gene expression. The KL gene consists of at least nine exons, but the promotor region has not yet been identified and characterized (45,128). Insight about the control of KL transcription should come through studies of the promotor region of the KL gene as well as through the molecular characterization of the *Sl* expression mutations *Sl*^{pan}, *Sl*^{con}, *Sl*^{t}.

Two alternatively spliced KL transcripts specify two membrane-associated KL protein products (37–39). The expression of the alternatively spliced KL RNA transcript is controlled in a tissue-specific manner (39,76,79). The two KL transcripts are coexpressed in all tissues and cell types analyzed, but the ratio varies greatly. In Balb/c3T3 fibroblasts KL-1 is present in large excess. In the brain and thymus KL-1 is the preponderant transcript, but in spleen, bone marrow, heart, cerebellum, ovary, and testis, the KL-1 and KL-2 transcripts are seen in variable ratios (39). Furthermore, in postnatal development, in the cerebellum, ovary, and testis, KL-1/KL-2 ratios vary with age (76,79).

B. Posttranslational Processing of KL

Biosynthesis and proteolytic processing of the murine KL-1 and KL-2 protein products has been studied primarily in COS-1- cells (37,39). Pulse–chase experiments show that both the primary KL-1 and KL-2 translation products are modified by glycosylation and are expressed on the cell surface. The major soluble form of KL (KL-S1), is formed by proteolytic cleavage from the membrane form of KL-1 (see Fig. 1). Membrane-associated KL-2 also produces soluble biologically active KL (KL-S2), but at a diminished rate (43). Interestingly, the human KL-2 protein is differentially more resistant to cleavage than murine KL-2, presumably because the human KL-2 and mouse KL-2 polypeptide sequences differ in the proposed secondary proteolytic cleavage site, and modification of the murine KL-2 sequence to correct this difference renders this molecule more resistant to cleavage (137; Tajima Y, Vosseller K, unpublished data). The protein kinase C inducer phorbol 12-myristate 13-acetate (PMA) and calcium ionophore calcimycin (A-23187) accelerate the cleavage of both KL-1 and KL-2 at similar rates, indicating that cleavage can be regulated differentially (39). Proteolytic cleavage of KL-1 and KL-2 occurs on the cell surface, based on experiments in which (1) the protein was chased to the cell surface, (2) maturation was blocked in the ER–Golgi complex by employing low temperatures (16°–20°C), or treatment with brefeldin A (39; Huang E, Vosseller K, Besmer P, unpublished data).

The cytoplasmic domain sequences of KL are highly conserved across species, implying functional relevance. Evidence for a role of cytoplasmic domain sequences of KL comes from the Sl^{17H} allele, a splice site mutation that replaces the cytoplasmic domain with extraneous amino acids, and causes distinctive phenotypic defects (128). By using cytoplasmic domain deletion mutants and the Sl^{17H} allele, the role of the cytoplasmic domain sequences of KL in proteolytic processing and cell surface presentation has been investigated (Huang E, Vosseller K, Besmer P, unpublished data). The cytoplasmic domain KL is dispensable for proteolytic processing, but it does assist in proper transport of KL to the cell surface, where it is cleaved to release soluble factor.

C. Regulation of c-*kit* RNA Expression

The analysis of the temporal expression of c-*kit* RNA transcripts in development and the postnatal animal indicates expression of c-*kit* in targets of W and Sl mutations. In addition c-*kit* is expressed in other cell types, including cells of the CNS. In melanogenesis c-*kit* is expressed in melanoblasts from the time they leave the neural crest, and expression continues throughout development and in melanocytes in hair follicles. In the hematopoietic system c-*kit* expression progresses from low levels in noncycling, multipotent, primitive stem cells, to high levels in cycling progenitors of all hematopoietic cell lineages, and c-*kit* expression subsequently decreases to undetectable levels of mature blood cells in circulation, with the exception in tissue mast cells (72,138,139).

Transmodulation of c-*kit* expression by cytokines in hematopoietic cells has been occasionally reported, with conflicting results. IL-4 down-regulates c-*kit* RNA and protein expression in normal human CD34+ bone marrow cells, human leukemic myeloid cells, and in a mast cell leukemia cell line (140). IL-3 in murine mast cells and myeloid cells apparently down-regulates c-*kit* RNA levels (141), but in human hematopoietic cells, IL-3 does not alter or up-regulates c-*kit* RNA levels (143). In human AML, TNF-α up-regulates c-*kit* RNA transcript levels by a mechanism involving posttranscriptional stabilization of c-*kit* mRNA, rather than affecting the rate of c-*kit* transcription (142).

The structure of murine and of human c-*kit* genes and transcription start sites have been characterized by several groups (20–24,143,144). A single transcription start site for c-*kit* has been identified in hematopoietic cells, erythroid cell lines, and mast cells; and in melanocytes, small cell lung carcinoma, and glioblastoma (24,143,144). The 5′-upstream region does not contain a "TATA box," but has a high G plus C content, four potential Sp1-binding sites, four sites for basic helix–loop–helix proteins, one site for the transcription factor E47, three sites for AP-2, sites resembling binding sites for PU.1 or other *ets*-domain proteins, three *myb* sites, three GATA-1 sites, and one site for the core transcription factor *ETF* (24,143,144), and in addition, a putative melanocyte-specific element and a region strikingly similar to the promotor of the *TRP-1* gene, called M-box (22,145,146). The functional significance for these sequence signatures in the control of c-*kit* expression has not been investigated. The promotor region of the human gene has been analyzed by using CAT reporter constructs. In the human erythroleukemia cell line HEL, which produces high levels of endogenous c-*kit* RNA transcripts, but not in TALL-1 cells, a region from −180 to −22 of the human c-*kit* promotor promotes c-*kit* expression (144). The murine c-*kit* control region was investigated by using luciferase reporter constructs in the c-*kit* expressing HEL and C57.1 cell lines and in the c-*kit*-negative HL-60 and WEHI-3 cell lines. Positive elements controlling the expression of the murine c-*kit* gene were identified in the region between −105 and −44 (143). In addition negative regulatory elements appear to be present in the upstream c-*kit* control region (144). In agreement with the notion of negative upstream control elements in the c-*kit* promotor, transgenic mice failed to express a *lacZ* reporter construct containing 5 kb of the 5′-c-*kit* upstream region (24).

Important insight into the complex mechanisms controlling c-*kit* expression have come from the analysis of c-*kit* expression mutations, including W^{sh}, *Ph*, and W^{57} (discussed in Section VI.A). Analysis of the molecular bases of the W^{sh} mutation suggests that elements determining c-*kit* expression in mast cells and the embryonic digestive tract are situated far upstream from the c-*kit* transcription start site in between *Pdgfra* and c-*kit* (5,93,96). In addition, silencers located far upstream, block or suppress c-*kit* expression in mesenchymal cells during embryogenesis in the floor plate of the neural tube and the dermatomes over somites from embryonic days 10 to 13. In the W^{sh} and *Ph* mutations, these silencers are unlinked from the c-*kit* promotor by a chromosomal inversion or deletion, causing misexpression of c-*kit* in various sites during embryogenesis. Furthermore, the W^{sh} mutation blocks c-*kit* expression in mast cells and mesenchyme-derived cells in the lung.

Mutations at the murine *microphthalmia* locus affect neural crest-derived melanocytes, the retinal pigmented epithelium, and mast cells. Mice that are homozygous for severe mutations are depigmented and deaf as a result of a deficit of inner ear melanocytes; they develop small unpigmented eyes because of a defect in the retinal pigment epithelium; and they lack tissue mast cells. The Waardenburg syndrome type 2 is a human condition with pigmentary disturbances and hearing loss, resulting from mutations at the human *microphthalmia* locus (147). The *microphthalmia* gene encodes a transcription factor belonging to the basic helix–loop–helix–leucine zipper protein family. Mast cells derived from mice carrying microphthalmia mutations have severely reduced levels of c-*kit* transcripts, suggesting that the microphthalmia protein controls c-*kit* expression in mast cells and, presumably, also in melanocytes (148). In agreement with this hypothesis, the microphthalmia protein activates genes containing an M-Box (melanocyte-specific element CANNTG; 149,150), and CANNTG elements are present in the human (22) and in the mouse c-*kit* promotor region.

VIII.   CELLULAR RESPONSES MEDIATED BY KL–SCF

The Kit receptor mediates diverse cellular responses, including cell proliferation, survival, migration, adhesion, differentiation, secretion, and radioresistance. These activities provide the basis for the role KL displays in various cell lineages during embryogenesis and in the postnatal organism. Several cell systems, including bone marrow-derived mast cells, primordial germ cells, and human myeloid Mo7e cells, have been studied to characterize the functional activities of KL–SCF.

A. Cell Proliferation

KL/SCF, similar to CSF-1 and PDGF, mediates cell proliferation. Most often KL synergizes with other growth factors to generate a mitogenic response, although mast cells and Mo7 cells proliferate with KL as the sole mitogen. Bone marrow-derived mast cells and primary peritoneal mast cells (connective tissue mast cells) proliferate with natural KL (31) and with recombinant KL in serum-containing medium (151), as well as in serum-free medium (56). Human cord blood-derived mast cells proliferate in serum-free conditions, by contrast serum has an inhibitory influence (152). The Kit receptor number on murine mast cells is limiting for a full proliferative response (i.e., mast cells from $W/^+$ and $W^{42}/^+$ mice, containing reduced numbers of functional receptors, do not display the full proliferative response with saturating concentrations of KL (56). The presence of KL is required during the G_1 phase of the cell cycle, up to passage of the restriction point for BMMC to enter S phase. In synergy with IL-3, KL promotes a mitogenic response in BMMC at suboptimal levels of KL and IL-3. Furthermore, both IL-3 and IL-4 synergize in peritoneal mast cells to promote proliferation (153,154). In the hematopoietic cell hierarchy in multilineage progenitors—very early lineage progenitors in early and late erythropoiesis and in megakaryopoiesis—proliferation requires the synergistic action of several growth factors or cytokines, including KL–SCF.

Purified primordial germ cells in the presence of KL and LIF proliferate and survive for a limited time (155–157). In contrast, proliferation of primordial germ cells can be extended indefinitely in the presence of FGF in combination with KI and LIF, presumably because of the ability of FGF to block the differentiation program of the PGCs (158).

B. Cell Survival

Many cell types, after removal of growth stimuli, enter a period of a programmed cell death (apoptosis) mediated by growth factors and cytokines. In BMMC the induction of apoptosis by growth factor deprivation (KL, IL-3, and serum) follows zero order kinetics or gamma distribution that is characteristic of stochastic processes (56). BMMC undergoing apoptosis exhibit internucleosomal DNA fragmentation that is characteristic of apoptosing cells. Soluble KL promotes cell survival by suppressing the apoptosis in BMMC induced by growth factor deprivation and γ-radiation. The concentrations of soluble KL required for half-maximal cell survival are approximately 40% of the concentrations required to promote cell proliferation of BMMC (56,159). Furthermore, the Kit receptor number on murine mast cells is not limiting for a full survival response (i.e., mast cells from $W/^+$ and $W^{42}/^+$ mice, containing reduced numbers of functional receptors, display a full cell survival response with saturating concentrations of soluble KL; 56).

Evidence for a role of soluble KL in promoting cell survival has also come from numerous in vitro studies using different cell systems, including murine, but not human dormant hematopoietic stem cells (160,217), various murine and human hematopoietic cell types, natural killer cells (161), primordial germ cells (155,157,162), melanocytes (163), murine melanoblasts (164,165), and murine and avian dorsal root ganglia neurons (166,167). In addition, in vivo a survival function for KL in early melanoblasts and in tissue mast cells is evident (93,94,168). Also, in spermatogonia, a survival function of KL was revealed by treatment of mice with ACK2 *c-kit*-blocking antibody and identification of increased numbers of apoptotic cells by using the TUNEL method (66,169).

C. Cell Adhesion and Migration

Cell–extracellular matrix and cell–cell interactions, as well as cell migration, are important features of Kit function in development and in the mature animal. In agreement with these properties, two different mechanisms of adhesion have been described. First, similar to the CSF-1–CSF-1 receptor system adhesion may be mediated by the membrane-anchored versions of KL. Mast cells adhere to fibroblasts and COS cells expressing membrane forms of KL (37,170,171). Interestingly, *c-kit* receptor kinase activity is not required for this type of adhesion because mast cells obtained from W^{42}/W^{42} mice expressing *c-kit* receptors lacking kinase activity and normal mast cells do not differ in their ability to adhere. Second, KL induces adhesion of murine BMMC to a fibronectin matrix by an inside–out mechanism requiring *c-kit* kinase activity (172–174). The concentration of KL required to induce adhesion is approximately 20 times lower than that required for proliferation. This adhesion is mediated by $\alpha_5\beta_1$-integrins, and it is blocked by the RGDS peptide, which blocks the fibronectin–integrin interaction.

Cell migration may occur by means of a chemotactic response, facilitated by a gradient of soluble growth factor, or alternatively, by a haptotactic process facilitated by a gradient of cell-associated growth factor. In BMMC and connective tissue mast cells soluble KL induces a chemotactic response at concentrations that are 20- to 50-fold lower than that required for a proliferative response (175). Soluble KL also induces a chemotactic response in porcine aortic endothelial cells transfected to express the human *c-kit* receptor (87). In agreement with the mitogenic, survival, and chemotactic properties of KL, repeated subcutaneous injection of soluble KL into mast cell-deficient Sl/Sl^d mice results in the recruitment of large numbers of tissue mast cells at the site of injection (176).

D. Potentiation of Secretion

An important aspect of the function of mast cells is the release of the content of granules consisting of histamine, sertonin, proteases, and leukotrienes. Although activation of the FcεR1 receptor by IgE is the main mechanism for triggering the release of the contents of mast cell granules, KL alone or in synergy with IgE may induce secretion of granules. In vitro KL induces the direct release of granules from human lung and skin mast cells, as well as from mouse peritoneal mast cells, and in synergy with IgE stimulation of the FcεRI receptor, relatively low concentrations of KL potentiate or enhance mast cell secretion (177,178). In adoptive transfer experiments in vivo, using mast cell deficient W/W^v mice, KL induced mast cell activation and mast cell-dependent inflammation (179). Furthermore, in platelets, KL enhances epinephrin- and ADP-induced secretion of serotonin (180).

E. Maturation and Differentiation

A role for KL–c-*kit* in differentiation in several cell systems in embryonic development and in the adult organism has been implied. However in vitro experiments indicate a role for KL primarily as a survival and proliferation factor when present alone, and other cytokines are required additionally for differentiation (for reviews see Refs. 181–183). In vitro experiments with primordial germ cells (PGC) demonstrated a role for KL as a proliferation and survival factor alone and in combination with leukemia inhibitory factor (LIF) for up to 4 days in culture. Subsequently, PGCs would die as if they were intrinsically programmed and had progressed to the next differentiation stage that no longer required these factors. Interestingly, addition of FGF to the PGC cultures maintained their proliferation and blocked their differentiation (158). This is reminiscent of the differentiation of oligodendrocyte progenitors, for which PDGF controls an intrinsically clocked differentiation program that can be blocked by FGF (184,185).

Tissue mast cells display significant morphological heterogeneity, depending on their tissue environment. The different tissue mast cell populations differ primarily in the content of their granules (i.e., proteoglycans [heparin], proteases, histamine, and serotonin; for a review see Ref. 186). Although the granules of connective tissue and serosal and peritoneal mast cells contain heparin proteoglycans and histamine, mucosal- and bone marrow-derived mast cells contain no heparin, low levels of histamine, and a different protease expression profile. After long-term (3–4 weeks) coculture of BMMC with 3T3 fibroblasts, the mast cells assume a phenotype similar to that of connective tissue mast cells (i.e., they produce heparin proteoglycan, rather than chondroitin sulfate proteoglycans and the histamine levels are increased). In vitro BMMC acquired a more mature phenotype in the presence of high concentrations of KL as the sole growth factor over a 4–6 week time span (i.e., they produced heparin, contained increased histamin, and an altered protease expression profile; 151,187). Therefore, KL contributes to the phenotypic maturation of mast cells in vitro. However, this phenotypic conversion is incomplete, implying that an interplay of several factors is required for complete maturation.

IX. MECHANISM OF SIGNAL TRANSDUCTION

A. Mechanism of Receptor Signaling

Signaling that is mediated by tyrosine kinase receptors involves dimerization of ligand-bound receptor, activation of the intrinsic receptor kinase, autophosphorylation and phosphorylation of substrates, and association of the activated receptor with intracellular-signaling molecules. Association of the activated receptor with substrates may be mediated by interactions between short receptor peptide sequences containing phosphorylated tyrosine residues and docking domains, such as SH2 domains, of associating proteins. The associating proteins either exhibit catalytic activity by themselves, or they are subunits of a protein complex with catalytic activity. The activated proteins then either generate small second-messenger molecules, or modify other proteins, initiating signaling cascades that define functional responses. SH2-containing proteins known to interact with RTKs include phospholipase C-γ1 (PLC-γ1), the regulatory subunit (p85) of phosphatidylinositol 3′-kinase (PI$_3$-kinase), the Grb2 protein, GTPase-activating protein (GAP) of the *ras*, *src* family tyrosine kinases, and phosphatases. The molecular interactions between activated RTKs, p21*ras*, Raf and MAPK have been delineated recently for the EGF receptor kinase. Activation of PI$_3$-kinase, PLC-γ1, p21*ras*, and MAPK are thought to be essential elements in mediating growth and transformation.

KL–c-*kit* receptor signaling has been investigated in several cell systems that normally express c-*kit*, including BMMC, melanocytes, megakaryocytic MO7e cells, as well as in cells transfected to express c-*kit*, such as porcine aortic endothelial cells, CHO cells, and NIH3T3 cells. Soluble KL dimers induce c-*kit* receptor dimerization and kinase activation (87,189). The NH$_2$-terminal three immunoglobulin domains have been determined, by use of monoclonal antibodies and human–mouse receptor chimeras, to contain the ligand-binding pocket (190,191). Soluble dimeric KL first binds a single receptor molecule, followed by association of the receptor–ligand complex with a second receptor molecule. Recent experiments, in agreement with evidence from the related CSF-1 receptor, suggest that in the immunoglobulin domain 4, c-*kit* is responsible for stabilizing receptor dimers and activation of the kinase (192,193). Receptor autophosphorylation then provides docking sites for downstream-signaling molecules containing tyrosine phosphate recognition motifs (14,111). By analogy with other tyrosine receptor kinases, receptor autophosphorylation is thought to take place by an intermolecular mechanism.

Cellular proteins known to associate with the Kit receptor in vivo include the p85 subunit of phosphatidylinositide-3′ (PI 3′-kinase) (14,111), phospholipase Cγ-1 (PLCγ-1; weak association; 14,111), the GRB2 adaptor protein (weak association; 194), the c-*src* protein (194), the tyrosine protein kinase *tek* (195), and the tyrosine phosphatases PTP1C and Syp (196,197). The interactions of these proteins with the activated c-*kit* receptor have been further defined by demonstration of in vitro association of the activated receptor with various *gst*-fusion proteins containing different domains of these proteins. They include the GRB2 SH2 domain, the NH$_2$-terminal and COOH-terminal SH2 domains of the p85 subunit of PI 3′-kinase, the NH$_2$- and COOH-terminal SH2 domains of PLC-γ, the c-*src* SH2 domain (194), as well as the SH2 domain of the tyrosine phosphatase Syp (197). Mutational analysis of the c-*kit* receptor indicates that tyrosine-719 in the kinase insert region determines binding of the p85 subunit of PI 3′-kinase to the activated receptor (198).

In addition, activation of the Kit receptor causes phosphorylation or activation, without association with the receptor, of various cytoplasmic-signaling molecules, including SHC (199), p21*ras* (174,194,200), of the serine/theonine kinase Raf-1 (201), MAP kinase (163,174,194), of the serine kinases p90*rsk* and p70S6 kinase (202), phospholipase D (203), and of p95*vav* (204). Furthermore, early-response genes, induced by stimulation of bone marrow-derived mast cells with KL, include c-*fos*, c-*jun*, *jun*B, c-*myc*, and c-*myb* (174,205). Therefore, c-*kit* receptor activation triggers several distinct signaling cascades similar to other receptor tyrosine kinases.

Kit receptor activation in tissue mast cells and BMMC mediates diverse physiological responses, including proliferation, survival, differentiation, migration, adhesion to fibronectin, as well as enhancement of mediator release from granules. Therefore, mast cells are a very important model system, and they provide the opportunity to investigate the role of various signaling cascades in specifying different cellular responses that are induced by KL. W^{sh} is a c-*kit* expression mutation (91,92) and BMMC derived from W^{sh}/W^{sh} mice lack c-*kit* expression and do not respond to KL (93,95). In an attempt to dissect the signaling cascades mediated by KL–Kit, W^{sh}/W^{sh} BMMC have been reconstituted with mutant versions of the Kit receptor. Introduction of both murine Kit$_S$ and Kit$_L$ (isoform containing a four-amino–acid insert) into W^{sh}/W^{sh} BMMC restored KL-induced proliferation, survival, and adhesion of fibronectin, as well as activation of PI 3′-kinase, p21^{ras} and MAP kinase, and induced expression of c-*fos*, *jun*B, c-*myc*, and c-*myb* mRNA (174). Substitution of tyrosine-719 in the kinase insert with phenylalanine (Y719F), abolished PI 3′-kinase activation, diminished c-*fos* and *jun*B induction, impaired KL, induced adhesion of BMMC to fibronectin, and abolished KL-induced potentiation of secretion (serotonin release) (Vosseller K, et al., unpublished data). In addition, the Y719F mutation had partial effects on p21^{ras} activation, cell proliferation and survival, whereas MAP kinase activation was not appreciably affected (174,198). On the other hand, Y821F substitution impaired proliferation and survival without affecting PI 3′-kinase, p21^{ras}, and MAPK activation and induction of c-*myc*, c-*myb*, c-*fos*, and c-*jun* mRNA, whereas KL-induced cell adhesion to fibronectin and the secretory response remained intact. In agreement with a role for PI 3′-kinase in Kit-mediated cell adhesion, wortmannin, a specific inhibitor of PI 3′-kinase, blocked Kit-mediated cell adhesion in BMMC. Therefore, association of Kit with the p85 subunit of PI 3′-kinase and thus with PI 3′-kinase activity is necessary for a full mitogenic, as well as the cell adhesion and secretory responses in mast cells. In contrast, tyrosine-821 is essential for Kit-mediated mitogenesis and survival, but not cell adhesion and secretion.

Protein kinase C (PKC) is a negative regulator of KL–c-*kit* signaling (206). PKC phosphorylates the c-*kit* receptor on serine residues after stimulation by KL–SCF and inhibits the mitogenic response. However, PKC is required for the chemotactic response of KL in porcine aortic endothelial cells. Inhibition of PKC by calphostin in KL-stimulated cells results in an increased mitogenic response in porcine aortic endothelial cells. Both c-*kit* receptor phosphorylation and binding of the p85 PI$_{3'}$-kinase subunit is increased after PKC inhibition, but Raf-1 activation and MAP kinase activation is not increased (194). It has been proposed that the increased PI$_{3'}$-kinase activation may explain the increased mitogenic response.

B. Receptor Down-Regulation or Desensitization

Complex mechanisms are required to regulate the effects of growth factors that promote diverse cellular processes during embryonic development and in the adult animal. Thus, growth factor-induced signaling must be terminated effectively to precisely control the diverse cellular responses in target cells. Down-regulation of cell surface receptors is an important mechanism for the modulation of signals induced by growth factors. By reducing the receptor level, the cells become desensitized to further stimulation by the ligand. In mast cells, following ligand binding, c-*kit* receptors on the cell surface are down-regulated in a manner that depends on the concentration of the ligand and the time of incubation. KL-induced loss of the cell surface receptors is associated with internalization of the ligand. In addition, the ligand induces accelerated degradation of c-*kit* receptors, presumably by targeting receptor complexes to lysosomes, resulting in a decreased half-life of c-*kit* receptors (15). Also, the receptor is subjected to nonlysosomal degradation involving polyubiquitination of the protein (16,207). The mechanism of turnover of Kit in mast cells was studied by using mutant receptors generated in vitro, expressed in W mutant mast cells lacking endogenous c-*kit* expression, and the determination of the effects of the mutations on KL-induced internalization and ubiquitination and degradation of Kit (16). Inactivation of the Kit kinase reduced the rate of internalization of KL–Kit complexes, degradation of kinase-inactive receptor complexes was relatively slow, and receptor ubiquitination was absent. But abolishment of Kit receptor association and activation of PI-$_{3'}$ kinase and of tyrosine-821 autophosphorylation did not affect KL-induced internalization and ubiquitination or degradation of Kit.

In addition to the effects of PKC on c-*kit* receptor-signaling (discussed earlier), in mast cells, phorbol ester induced PKC as well as calcium ionophore down-regulation of the number of c-*kit* receptors on the cell surface (15). This down-regulation of the c-*kit* receptor is accompanied by release of the extracellular domain of the receptor into the culture medium, presumably by proteolytic cleavage near the transmembrane domain, but other mechanisms may contribute as well. The extracellular domain of the c-*kit* receptor released as a result of PKC-activated proteolytic cleavage may potentially have distinct physiological roles: (1) The soluble c-*kit* receptor may bind KL and control the access of ligand to the target cells. Ligand-binding activities of soluble receptors have been demonstrated in vitro, but their biological significance remains to be determined (17,199). (2) The c-*kit* receptor on target cells and the cell-surface KL on stromal cells is essential for cell adhesion, proliferation, differentiation, and migration. Proteolytic cleavage of the c-*kit* receptor or KL would allow target cells to disengage and migrate. (3) It is possible that the soluble form of the c-*kit* receptor may bind to cell-associated KL in stromal cells and trigger a cascade of signal transduction events in these cells. Significantly, the soluble form of Kit is produced by human umbilical endothelial vein cells (208) and is present in human serum (209).

X. IN VIVO AND IN VITRO ROLES OF KL–SCF IN HEMATOPOIESIS

A. Hematopoiesis

As revealed primarily through in vivo W and Sl mutant phenotypes, described in Section IV. A, KL/SCF displays distinctive activities in the hematopoietic cell hierarchy (i.e., in erythroid maturation, in mast cells, and in megakaryopoiesis; 1,6,7). Abnormal hematopoiesis is observed after acute injury or stress. Mice carrying W and Sl mutations exhibit increased sensitivity to irradiation, implying a role for KL–SCF in HSCs (55). Also, in vivo administration of KL increases the resistance to radiation in normal mice (239). Under conditions of hypoxia or erythrocyte depletion, rebound erythropoiesis is delayed in both W/W^v and Sl/Sl^d animals (51,210). Also, KL contributes to rebound thrombopoiesis following 5-fluorouracil treatment of mice (211).

In hematolymphoid tissues, the Kit receptor is expressed mostly in immature and in some mature cell populations. The c-kit expression occurs in lineage marker-negative HSCs in the fetal liver and in the adult bone marrow, and expression continues in lymphoid, myeloid, and erythroid progenitors. Significantly, Kit receptor-positive, lin$^-$ cells reconstitute the hematopoietic system of lethally irradiated mice (50,69,73). Although, nondividing c-kit$^+$ lin$^-$ rhodamine-dull cells repopulate the lymphohematopoietic system, the c-kit$^+$ lin$^-$ rhodamine-bright subset correlates with the number of day 12 colony-stimulating unit-spleen (CFU-S12) cells (212). In more mature hematopoietic and lymphoid cell populations c-kit expression is reduced or completely shut off (138,139). Although high Kit receptor expression is seen in tissue mast cells, somewhat lower levels occur in ADP-stimulated platelets (31,74,180,213). In agreement with a function of the Kit receptor in hematopoiesis, KL expression is observed in the microenvironment of hematopoietic cells in the bone marrow, spleen, lymph nodes, and thymus.

Numerous in vitro studies have defined many of the pleiotropic characteristics of KL–SCF in hematopoietic cells. Whereas KL–CSF alone does not promote proliferation in hematopoietic progenitors, in synergy with other factors, it is a powerful mitogen and survival factor. KL–SCF and two or more additional factors, including IL-1, IL-3, IL-6, G-CSF, and IL-12, are needed to stimulate proliferation of enriched primitive progenitors (214,215). Also, KL–SCF as well as IL-3 and G-CSF have been proposed to mediate the survival of murine primitive progenitors in culture (160). KL–SCF synergizes with IL-6, G-CSF, IL-11, and IL-12 to trigger entry into the cell cycle of dormant stem cells facilitating the formation of multipotential blast cell and multilineage colonies. In addition, KL–SCF and IL-3 support the survival of murine HSCs in G_0, but in human primitive progenitors such an activity could not be demonstrated (216,217). Several experimental approaches indicate that KL–SCF provides primarily a proliferative signal for hematopoietic progenitors. Addition of KL–SCF to cocktails of hematopoietic growth factors does not increase the number of blast colonies, but the size of the colonies is dramatically increased (218). In liquid culture KL–SCF synergizes with several hematopoietic growth and differentiation factors to produce various progenitors from more primitive HSCs (181,219). However, there is no good evidence to suggest that KL–SCF promotes self-renewal of HSCs. Therefore, KL–SCF importantly contributes to the optimal production of progenitors from HSCs in vitro. These synergistic activities of hematopoietic growth factors are also evident from in vivo experiments (220–222). Even though administration of G-CSF to mice increases circulating progenitors of all hematopoietic lineages, as well as blood granulocyte levels, simultaneous administration of KL–SCF and G-CSF enhances production of peripheral blood mononuclear cells and peripheralization of progenitors significantly over the levels seen with administration of G-CSF alone (221,222). In agreement with these findings in mice carrying W and Sl mutations the effects of administration of G-CSF are severely diminished. In contrast, administration of soluble KL–SCF alone results in a redistribution of hematopoietic progenitors from the bone marrow into the periphery and the spleen without affecting their number (222–224).

KL–SCF also has a role in more mature hematopoietic cell populations. It synergizes with EPO to facilitate erythroid development, and this is in agreement with the macrocytic anemia in mice lacking functional Kit receptors or KL–SCF (31,225,226). Also, in vivo administration of soluble KL–SCF reverses the macrocytic anemia in Sl/Sl^d mice and augments erythropoiesis in primates (35,221). In mast cells, KL–SCF plays an important role in survival, cell proliferation, maturation, cell adhesion, migration, and degranulation (see Sec. VIII). Whereas tissue mast cells express high levels of c-*kit*, levels in basophils are greatly reduced. Importantly, W and Sl mutant mice have a severe deficit in tissue mast cells (52,53). In megakaryopoiesis KL–SCF contributes at all stages of maturation of megakaryocytes (227). In both W and Sl mutant mice, megakaryocyte levels were approximately half, and the stage III megakaryocytes were larger; but the levels of circulating platelets and the regulation of thrombocytopoiesis are not altered in mutant animals (54).

B. Lymphopoiesis

Many c-*kit*-expressing cells are contained in the yolk-sac, starting at embryonic day 8–8.5, the fetal liver, and the fetal thymus (73). Stromal cells and epithelial cells from the fetal thymus and liver express KL. In the adult mouse c-*kit* expression is observed in early T- and B-cell development (i.e., pro-T and pro-B lymphocytes), whereas cells at the pre-T- and late pre-B-cell stage do not express detectable levels of c-*kit*. In vitro, KL enhances proliferative responses of lymphoid progenitors to other cytokines, including IL2, IL3, IL4, and IL7 (73,228). To elicit a proliferative response pro-T cells require KL and IL-2; pro-B cells KL and IL3 or IL4; and B220[+] pre-B cells, KL and IL7. However, KL is not required for differentiation of lymphoid progenitors to become mature T and B lymphocytes (73,229). By contrast KL is a potent comitogen for normal pro-B cells, B220[+] pre-B cells, and pro-T cells.

In lymphopoiesis in W and Sl mutant mice no effects by the mutations are observed on T and B lymphocytes in peripheral organs (1,6,230). However, the composition and number of intraepithelial T lymphocytes in the intestinal epithelium is affected in mice with W/W^v and Sl/Sl^d mutations (230). Beginning at 6 weeks of age the number of γδ-intraepithelial T lymphocytes decreased, whereas the number of αβ-intraepithelial T lymphocytes was increased. In agreement with a role for the Kit receptor in intestinal T lymphocytes, c-*kit* is expressed in intraepithelial T lymphocytes and KL in intestinal epithelial cells (230).

C. Potential Clinical Applications of KL–SCF

Because of the interesting pleiotropic properties of KL–SCF in hematopoiesis and in other cell systems it seems reasonable to expect beneficial effects from it in clinical settings. Sl/Sl^d mice have macrocytic anemia and reduced numbers of hematopoietic progenitors. Continuous long-term administration of KL–SCF to Sl/Sl^d mice increased the numbers of the progenitors and increased the hematocrit to normal levels (35,224). Human conditions with defects in hematopoietic progenitor populations including aplastic anemia; pure red cell anemia or Diamond-Blackfan anemia appear to be appropriate scenarios for KL–SCF intervention. Although in vitro studies with bone marrow from aplastic anemia patients failed to demonstrate a response to KL–SCF as a result of a possible defect in KL–Kit signaling, KL–SCF may be of use in some patients with Diamond-Blackfan anemia and constitutional red cell aplasia (231,232). Other possible applications may include patients with sickle cell disease. In vitro KL–SCF increases the amount fetal hemoglobin in progenitors of patients with sickle cell disease (233). A most promising clinical application of KL–SCF is its use to expand and mobilize HSCs into the periphery for harvesting for bone marrow transplantation from donors or before radical treatments that damage hematopoietic progenitors (234,235). Another important application may be the expansion in vitro of human hematopoietic progenitor cells in synergy with other hematopoietic growth factors (236). Complications from long-term application of KL–SCF may arise from increased mast cell numbers and mast cell activation (176,237,238). However, KL–SCF-induced mast cell hyperplasia is reversible after discontinuation of treatment, and it appears that these effects are manageable in the clinic. Although initial clinical applications have focused on the properties of KL–SCF in the hematolymphoid system, uses in other cell systems in which KL–SCF has physiological roles are possible.

XI.   CONCLUSIONS

The membrane growth factor KL–SCF is the cognate ligand of the Kit receptor tyrosine kinase, a close relative of the receptor tyrosine kinases M-CSFR and Flk-2/Flt3. KL–SCF and its receptor Kit play interesting pleiotropic roles in developmental processes, and in the mature organism, particularly in hematopoiesis, melanogenesis, and gametogenesis. Interestingly, in hematopoiesis, melanogenesis, and gametogenesis, c-*kit* plays a critical role in both the earliest precursors during embryonic development and in the postnatal organism and in more mature cells. In hematopoiesis, among other things, KL–SCF plays a critical role in hematopoietic stem cell homeostasis, erythropoiesis, and mast cells. Much of our knowledge of this receptor system stems from the analysis of a large number of germline mutations in the murine c-*kit* and KL–SCF genes, which have been obtained over a long period, because of their striking effect on coat pigmentation. From the analysis of mutant phenotypes and in vitro studies, KL–SCF mediates diverse cellular processes, including cell proliferation, survival, migration, adhesion to extracellular matrix, and secretion. Mechanistic studies of Kit receptor function in model cell systems, such as mast cells, have begun to unravel how the Kit receptor mediates these diverse cellular responses. Results from studies of the mechanism of KL–SCF signaling alone and in synergy with other factors in model cell systems will enable the study of KL–SCF-mediated processes in stem cell populations, including stem cells of the hematopoietic system, neural crest-derived melanoblasts, and germ cells, during development and in the postnatal organism. Finally, the wealth of knowledge on the function of KL–SCF in hematopoiesis and other cell systems facilitates the rapid application of this interesting hematopoietic growth factor in clinical settings.

ACKNOWLEDGMENTS

I would like to thank my collaborators Drs. Rosemary Bachvarova, Nelson Yee, Katia Manova, Georgina Berrozpe, Youichi Tajima, Gregory Stella, Keith Vosseller, and Inna Timokhina, and Drs. Werner Haas and Malcolm Moore for numerous discussions and comments on this manuscript. I would like to thank Mirna Rodriguez for her help preparing it. Support by grants from the National Cancer Institute, National Heart Lung and Blood Institute, and the American Cancer Society, for work carried out in the author's laboratory, is acknowledged.

REFERENCES

1.   Russell ES. Hereditary anemias of the mouse. Adv Genet (1979) 20:357–459.
2.   Silvers WK. The Coat Colors of Mice. New York: Springer-Verlag, 1979.

3. Bernstein A, Chabot B, Dubreuil P, Reith A, Nocka K, Majumder S, Ray P, Besmer P. The mouse *W*–c-*kit* locus. Ciba Found Symp 1990; 148:158–172.

4. Besmer P. The *kit* ligand encoded at the murine steel locus: a pleiotropic growth and differentiation factor. Curr Opin Cell Biol 1991; 3:939–946.

5. Besmer P, Manova K, Duttlinger R, Huang E, Packer AI, Gyssler C, Bachvarova R. The kit-ligand (steel factor) and its receptor c-*kit*/*W*: pleiotropic roles in gametogenesis and melanogenesis. Development 1993; (suppl):125–137.

6. Williams DE, deVries P, Namen AE, Widmer MB, Lyman SD. The steel factor. Dev Biol 1992; 151:368–376.

7. Galli SJ, Zsebo KM, Geissler EN. The kit ligand, stem cell factor. Adv Immunol 1994; 55:1–96.

8. Besmer P, Murphy JE, George PC, Qiu FH, Bergold PJ, Lederman L, Snyder HI, Brodeur D, Zuckerman EE, Hardy WD. v-*kit*: oncogene of a new acute transforming feline retrovirus (HZ4-FeSV)—relationship with the protein kinase gene family. Nature 1986; 320:415–421.

9. Majumder S, Ray P, Besmer P. Tyrosine protein kinase activity of the HZ4-feline sarcoma virus P80*gag-kit* transforming protein. Oncogene Res 1990; 5:329–335.

10. Yarden Y, Kuang W-J, Yang-Feng T, Coussens L, Munemitsu S, Dull TJ, Chen E, Schlessinger J, Francke U, Ullrich A. Human proto-oncogene c-*kit*: A new cell surface receptor tyrosine kinase for an unidentified ligand. EMBO J 1987; 6:3341–3351.

11. Qiu F, Ray P, Brown K, Parker PE, Jhanwar S, Ruddle FH, Besmer P. Primary structure of c-*kit*: Relationship with the CSF-1/PDGF receptor kinase family—oncogenic activation of v-*kit* involves deletion of extracellular domain and C terminus. EMBO J 1986; 7:1003–1011.

12. Majumder S, Brown K, Qiu FH, Besmer P. c-*kit*, a transmembrane kinase: identification in tissues and characterization. Mol Cell Biol 1988; 8:4896–4903.

13. Hayashi S, Kunisada T, Ogawa M, Yamaguchi K, Nishikawa S. Exon skipping by mutation of an authentic splice site of c-*kit* gene *W/W* mouse. Nucleic Acids Res 1991; 19:1267–71.

14. Reith AD, Ellis C, Lyman SD, Anderson DM, Williams DE, Bernstein A, Pawson T. Signal transduction by normal isoforms and W mutant variants of the kit receptor tyrosine kinase. EMBO J 1991; 10:2451–2459.

15. Yee NS, Langen H, Besmer P. Mechanism of kit ligand, phorbol ester, and calcium-induced down-regulation of c-*kit* receptors in mast cells. J Biol Chem 1993; 268:14189–14201.

16. Yee NS, Hsiau C-WM, Serve H, Vosseller K, Besmer P. Mechanism of down-regulation of c-*kit* receptor: roles of receptor tyrosine kinase, phosphatidylinositol 3'-kinase, and protein kinase C. J Biol Chem 1994; 269:31991–31998.

17. Brizzi MF, Blechman JM, Cavalloni G, Givol D, Yarden Y, Pegoraro L. Protein kinase C-dependent release of a functional whole extracellular domain of the mast cell growth factor (MGF) receptor by MGF-dependent human myeloid cells. Oncogene 1994; 9:1593–1586.

18. Stephenson DA, Mercola M, Anderson E, Wang C, Stiles CD, Bowen-Pope DF, Chapman VM. Platelet-derived growth factor receptor a-subunit gene (PDGFRA) is deleted in the mouse patch (*Ph*) mutation. Proc Natl Acad Sci USA 1991; 88:6–10.

19. Matthews W, Jordan CT, Gavin M, Jenkins NA, Copeland NG, and Lemischka IR. A receptor tyrosine kinase cDNA isolated from a population of enriched primitive hematopoietic cells and exhibiting genetic linkage to c-*kit*. Proc Natl Acad Sci USA 1991; 88:9026–9030.

20. André C, Martin E, Cornu F, Hu W-X, Wang X-P, Galibert F. Genomic organization of the human c-*kit* gene: evolution of the receptor tyrosine kinase subclass III. Oncogene 1992; 7:685–691.

21. Vandenbark GR, DeCastro CM, Taylor H, Dew-Knight S, Kaufman RE. Cloning and structural analysis of the human c-*kit* gene. Oncogene 1992; 7:1259–1266.

22. Giebel LB, Strunk KM, Holmes SA, Spritz RA. Organization and nucleotide sequence of the human KIT (mast/stem cell growth factor receptor) proto-oncogene. Oncogene 1992; 7:2207–2217.

23. Gokkel E, Grossman Z, Ramot B, Yarden Y, Rechavi G, Givol D. Structural organization of the murine c-*kit* proto-oncogene. Oncogene 1992; 7:1423–1429.

24. Chu TY. Characterization of the proto-oncogene c-*kit* encoded at the murine white spotting locus. PhD thesis, Cornell University Graduate School of Medical Sciences, 1992.

25. Chabot B, Stephenson DA, Chapman VM, Besmer P, Bernstein A. The proto-oncogene c-*kit* encoding a transmembrane tyrosine kinase receptor maps to the mouse *W* locus. Nature 1986; 335:88–89.

26. Geissler EN, Ryan MA, Housman DE. The dominant white spotting (*W*) locus of the mouse encodes the c-*kit* proto-oncogene. Cell 1988; 55:185–192.

27. Nocka K, Majumder S, Chabot B, Ray P, Cervonne M, Bernstein A, Besmer P. Expression of c-*kit* gene products in known cellular targets of *W* mutations in normal and W mutant mice—evidence for impaired c-*kit* kinase in mutant mice. Genes Dev 1989; 3:816–826.

28. Tan JC, Nocka K, Ray P, Traktman P, Besmer P. The dominant W^{42} phenotype results from a missense mutation in the c-*kit* receptor kinase. Science 1990; 247:209–212.

29. Levi-Schaffer F, Austen KF, Gravellese PM, Stevens RL. Coculture of interleukin 3-dependent mouse mast cells with fibroblasts results in a phenotypic change of the mast cells. Proc Natl Acad Sci USA 1986; 83:6485–6488.

30. Fujita J, Nakayama H, Onoue H, Kanakura Y, Nakano T, Asai H, Takeda S-I, Honjo T, Kitamura Y. Fibroblast dependent growth of mouse mast cells in vitro: duplication of mast cell depletion in mutant mice of W/W^v genotype. J Cell Physiol 1988; 134:78–84.

31. Nocka K, Buck J, Levi E, Besmer P. Candidate ligand for the c-*kit* transmembrane kinase receptor: KL, a fibroblast derived growth factor stimulates mast cells and erythroid progenitors. EMBO J 1990; 9:3287–3294.

32. Nocka K, Huang E, Beier DR, Chu TY, Buck J, Lahm HW, Wellner D, Leder P, Besmer P. The hematopoietic growth factor KL is encoded by the *SL* locus and is the ligand of the c-*kit* receptor, the gene product of the *W* locus. Cell 1990; 63:225–333.

33. William DE, Eisenmann J, Baird A, Rauch C, Van Ness K, March CJ, Park LS, Martin U, Mochizuki DJ, Boswell HS, Burgess GS, Cosman D, Lyman SD. Identification of a ligand for the c-*kit* proto-oncogene. Cell 1990; 63:167–174.

34. Zsebo KM, Wypych J, McNiece IK, Lu HS, Smith KA, Karkare SB, Sachdev RK, Yuschenkoff VN, Birkett NC, Williams RL, Satyagal VN, Tung W, Bosselman RA, Mendiaz EA, Langley KE. Identification, purification, and biological characterization of hematopoietic stem cell factor from buffalo rat liver conditioned medium. Cell 1990; 63:195–201.

35. Zsebo KM, Williams DA, Geissler EN, Broudy VC, Martin FH, Atkins HL, Hsu RY, Birkett NC, Okino KH, Murdock DC, Jacobsen FW, Takeishi T, Cattanach BM, Galli SJ, Suggs SV. Stem cell factor is encoded at the *Sl* locus of the mouse and is the ligand for the c-*kit* tyrosine kinase receptor. Cell 1990; 63:213–224.

36. Sprang SR, Bazan F. Cytokine structural taxonomy and mechanism of receptor engagement. Curr Opin Struct Biol 1993; 3:815–827.

37. Flanagan JG, Chan D, Leder P. Transmembrane form of the c-*kit* ligand growth factor is determined by alternative splicing and is missing in the Sl^d mutation. Cell 1991; 64:1025–1035.

38. Anderson DM, Williams DE, Tushinski R, Gimpel S, Eisenman J, Cannizzaro LA, Aronson M, Croce CM, Huebner K, Cosman D, Lyman SC. Alternate splicing of mRNAs encoding human mast cell growth factor and localization of the gene to chromosome 12q22-24. Cell Growth Differ 1991; 2:373–378.

39. Huang E, Nocka KC, Buck J, Besmer P. Differential expression and processing of two cell associated forms of the kit-ligand: KL-1 and KL-2—absence of cell membrane forms in Sl^d/Sl^d mice. Mol Biol Cell 1992; 3:349–362.

40. Lu HS, Clogston CL, Wypych J, Fausset PR, Lauren S, Mendiaz EA, Zsebo KM, Langley KE. Amino acid sequence and post-translational modification of stem cell factor isolated from buffalo rat liver cell-conditioned medium. J Biol Chem 1991; 266:8102–8107.

41. Pandiella A, Bosenberg MW, Huang EJ, Besmer P, Massague J. Cleavage of membrane anchored growth factors involves distinct protease activities regulated through common mechanisms. J Biol Chem 1992; 267:24028–24033.

42. Copeland NG, Gilbert DJ, Cho BC, Donovan PJ, Jenkins NA, Cosman D, Anderson D, Lyman SD,

Williams DE. Mast cell growth factor maps near the steel locus and is deleted in a number of steel alleles. Cell 1990; 63:175–183.

43. Geissler EN, Liao M, Brook JD, Martin FH, Zsebo KM, Housman DE, Galli SJ. Stem cell factor (SCF), a novel hematopoietic growth factor and ligand for c-*kit* tyrosine kinase receptor, maps on human chromosome 12 between 12q14.3 and 12qter. Somat Cell Mol Genet 1991; 17:207–214.

44. Mathew S, Murty VVVS, Hunziker W, Chaganti RSK. Subregional mapping of 13 single-copy genes on the long arm of chromosome 12 by fluorescence in situ hybridization. Genomics 1992; 14:775–779.

45. Martin FH, Suggs SV, Langley KE, et al. Primary structure and functional expression of rat and human stem cell factor DNAs. Cell 1990; 63:203–211.

46. Lyon MF, Searle AG. Genetic Variants and Strains of the Laboratory Mouse, 2nd ed. Oxford: Oxford University Press, 1989.

47. Russell ES, Thompson MW, McFarland EC. Analysis of effects of W and f genic substitutions on fetal mouse hematology. Genetics 1968; 58:259–270.

48. Chui DHK, Russell ES. Fetal erythropoiesis in steel mutant mice: I. A morphological study of erythroid cell development in fetal liver. Dev Biol 1974; 40:256–269.

49. Chui DHK, Loyer BV. Foetal erythropoiesis in steel mutant mice. II. Haemopoietic stem cells in foetal livers during development. Br J Haematol 1975; 29:553–565.

50. Ikuta K, Weissman IL. Evidence that hematopoietic stem cells express c-*kit* but do not depend on steel factor for their generation. Proc Natl Acad Sci USA 1992; 89:1502–06.

51. Harrison DE, Russell ES. The response of W/W^r and Sl/Sl^d anemic mice to haemopoietic stimuli. Br J Haematol 1972; 22:155–168.

52. Kitamura Y, Go S, Hatanaka K. Decrease of mast cells in W/W^v mice and their increase by bone marrow transplantation. Blood 1978; 52:447–452.

53. Kitamura Y, Go S. Decreased production of mast cells in Sl/Sl^d anemic mice. Blood 1978; 53:492–497.

54. Ebbe S, Phalen E. Regulation of megakaryocytes in W/W^r mice. J Cell Physiol 1978; 96:73–80.

55. Bernstein SE. Acute radiosensitivity in mice of differing W-genotype. Science 1962; 137:428–429.

56. Yee NS, Paek I, Besmer P. Role of *kit*-ligand in proliferation and suppression of apoptosis in mast cells: basis for radiosensitivity of white spotting and steel mutant mice. J Exp Med 1994; 179:1777–1787.

57. Russell ES. Abnormalities of erythropoiesis associated with mutant genes in mice. In Gordon AS, vol I. Regulation of Hematopoiesis, New York: Appleton, 1970:649–675.

58. Harrison DE, Astle CM. Population of lymphoid tissues in cured W-Anemic mice by donor cells. Transplantation 1976; 22:42–46.

59. McCulloch EA, Siminovich L, Till JL. Spleen-colony formation in anemic mice of genotype W/W^v. Science 1964; 144:844–846.

60. McCulloch EA, Siminovich L, Till JL, Russell ES, Bernstein SE. The cellular basis of the genetically determined hemopoietic defect in anemic mice of genotype Sl/Sl^d. Blood 1965; 26:399–410.

61. Bennett D. Developmental analysis of a mutation with pleiotropic effects in the mouse. J Morphol 1956; 98:199–233.

62. Nakayama H, Kuroda H, Onoue H, Fujita J, Nishimune Y, Matsumoto K, Nagano T, Suzuki F, Kitamura Y. Studies of Sl/Sl^d <> +/+ mouse aggregation chimeras II. Effect of the steel locus on spermatogenesis. Development 1988; 102:117–126.

63. Kuroda H, Terada N, Nakayama H, Matsumoto K, Kitamura Y. Infertility due to growth arrest of ovarian follicles in $SlSl^t$ mice. Dev Biol 1988; 126:7179.

64. Bachvarova RF, Manova K, Besmer P. Rose in gametogenesis of c-*kit* encoded at the W locus of mice. In: Bernfield M, ed. New York: Wiley-Liss, 1993: 1–18.

65. Manova K, Nocka K, Besmer P, Bachvarova RF. Gonadal expression of c-*kit* encoded at the W locus of the mouse. Development 1990; 110:1057–1069.

66. Yoshinaga K, Nishikawa S, Ogawa M, Hayashi S-I, Kunisada T, Fujimoto T, Nishikawa S-I. Role

of c-*kit* in spermatogenesis: identification of spermatogonia as a specific site of c-*kit* expression and function. Development 1991; 113:689.

67. Orr-Urtreger A, Avivi A, Zimmer Y, Givol D, Yarden Y, Lonai P. Developmental expression of c-*kit*, a proto-oncogene encoded by the *W* locus. Development 1990; 109:911–923.

68. Manova K, Bachvarova RF. Expression of c-*kit* encoded at the *W* locus of mice in developing embryonic germ cells and presumptive melanoblasts. Dev Biol 1991; 146:312–324.

69. Ogawa M, Matsuzaki Y, Nishikawa S, Hayashi S, Kunisada T, Sudo T, Kina T, Nakauchi H, Nishikawa S. Expression and function of c-*kit* in hemopoietic progenitor cells. J Exp Med 1991; 174:63–71.

70. Nishikawa S, Kusakabe M, Yoshinaga K, Ogawa M, Hayashi S-I, Kunisada T, Era T, Sakakura T, Nishikawa S-I. In utero manipulation of coat color formation by a monoclonal anti c-*kit* antibody: two distinct waves of c-*kit* dependency during melanogenesis. EMBO J 1991; 10:2111–2118.

71. Matsui Y, Zsebo KM, Hogan BLM. Embryonic expression of a haematopoietic growth factor encoded by the *Sl* locus and the ligand for c-*kit*. Nature 1990; 347:667–669.

72. Okada S, Nakauchi H, Nagayoshi K, Nishikawa S, Nishikawa S-I, Miura Y, Suda T. Enrichment and characterization of murine hematopoietic stem cells that express c-*kit* molecule. Blood 1991; 78:1706–1712.

73. Palacios R, Nishikawa S. Developmentally regulated cell surface expression and function of c-*kit* receptor during lymphocyte ontogeny in the embryo and adult mice. Development 1992; 115:1133–1147.

74. Ashman LK, Gadd SJ, Mayrhofer G, Spargo LDJ, Cole SR. A murine monoclonal antibody to an acute myeloid leukemia-associated cell surface antigen identifies tissue mast cells. In: McMichael A, ed. Leukocyte Typing III, White Cell Differentiation Antigens, Oxford: Oxford University Press, 1987:726.

75. Rafii S, Shapiro F, Rimarachin J, Nachman RL, Ferris B, Weksler B, Moore MA, Asch AS. Isolation and characterization of human bone marrow microvascular endothelial cells: hematopoietic progenitor cell adhesion. Blood 1994; 84:10–19.

76. Manova K, Huang EJ, Angeles M, DeLeon V, Sanchez S, Pronovost SM, Besmer P, Bachvarova RF. The expression of the c-*kit* ligand in gonads of mice supports a role for the c-*kit* receptor in oocyte growth and in proliferation of spermatogonia. Dev Biol 1993; 157:85–99.

77. Keshet E, Lyman SD, Williams DE, Anderson DM, Jenkins NA, Copeland NG, Parada LF Embryonic RNA expression patterns of the c-*kit* receptor and its cognate ligand suggest multiple functional roles in mouse development. EMBO J 1991; 10:2425–2435.

78. Motro B, Van der Kooy D, Rossant J, Reith A, Bernstein A. Contiguous patterns of c-*kit* and *steel* expression: analysis of mutations at the *W* and *Sl* loci. Development 1991; 113:1207–1221.

79. Manova K, Bachvarova RF, Huang E, Sanchez S, Velasquez E, McGuire B, Besmer P. c-*kit* receptor and ligand expression in postnatal development of the mouse cerebellum suggests a function for c-*kit* in inhibitory neurons. J Neurosci 1992; 12:4366–4376.

80. Geissler EN, McFarland EC, Russell ES. Analysis of pleiotropism at the dominant white-spotting (*W*) locus of the house mouse: a description of ten new *W* alleles. Genetics 1981; 97:337–361.

81. Lyon MF, Glenister PH, Loutit JF, Evans EP, Peters J. A presumed deletion covering the *W* and *Ph* loci of the mouse. Genet Res 1984; 44:161–168.

82. Nagle DL, Martin-DeLeon P, Hough RB, Bucan M. Proc Natl Acad Sci USA 1994; 91:7237–7241.

83. Durham FH. Further experiments on the inheritance of coat color in mice. J Genet 1911; 1:158–178.

84. Nocka K, Tan JC, Chiu E, Chu TY, Ray P, Traktman P, Besmer P. Molecular bases of dominant negative and loss of function mutations at the murine c-*kit*/white sptting locus: W^{37}, W^v, W^{41}, W. EMBO J 1990; 9:1805–1813.

85. Tsujimura T, Koshimizu U, Katoh H, Isozaki K, Kanakura Y, Tono T, Adachi S, Kasugai T, Tei H, Nishimune Y, Nomura S, Kitamura Y. Mast cell number in the skin of heterozygotes reflects the molecular nature of c-*kit* mutations. Blood 1993; 81:2530–2538.

86. Reith AD, Rottapel R, Giddens E, Brady C, Forrester L, Bernstein A. W mutant mice with mild or

severe developmental defects contain distinct point mutations in the kinase domain of the c-*kit* receptor. Genes Dev 1990; 4:390–400.

87. Blume-Jensen P, Claesson-Welsh L, Siegbahn A, Zsebo KM, Westermark B, Heldin C-H. Activation of the human c-*kit* product by ligand-induced dimerization mediates circular actin reorganization and chemotaxis. EMBO J 1991; 10:4121–4128.

88. Little CC, Cloudman AM. The occurance of a doinant spotting mutation in the house mouse. Proc Natl Acad Sci USA 1937; 23:535–537.

89. Guenet J-L, Marchal G, Milon G, Tambourin P, Wendling T. Fertile dominant spotting in the house mouse: a new allele at the *W* locus. J Hered 1979; 70:9–12.

90. Larue L, Dougherty N, Porter S, Mintz B. Spontaneous malignant transformation of melanocytes explanted from W^f/W^f mice with a *kit* kinase-domain mutation. Proc Natl Acad Sci USA 1992; 89:7816–7820.

91. Lyon MF, Glenister PH. A new allele sash (W^{sh}) at the *W*-locus and a spontaneous recessive lethal in mice. Genet Res 1982; 39:315–322.

92. Stevens J, Loutit JF. Mast cells in mutant mice (*W Ph mi*). Proc R Soc Lond B 1982; 215:405–409.

93. Duttlinger R, Manova K, Chu TY, Gyssler C, Zelenetz AD, Bachvarova RF, Besmer P. *W-SASH* affects positive and negative elements controlling c-*kit* expression—ectopic expression of c-*kit* at sites of kit-ligand expression affects melanogenesis. Development 1993; 118:705–717.

94. Steel KP, Davison DR, Jackson IJ. TRP-2/DT, a new early melanoblast marker, shows that steel growth factor (c-*kit* ligand) is a survival factor. Development 1992; 115:1111–1119.

95. Tono T, Tsujimura T, Koshimizu U, Kasugai T, Adachi S, Isozaki K, Nishikawa S, Morimoto M, Nishimune Y, Nomura S, Kitamura Y. c-*kit* gene was not transcribed in cultured mast cells of mast cell-deficient W^{sh}/W^{sh} mice that have a normal number erythrocytes and a normal c-*kit* coding region. Blood 1992; 80:1448–1453.

96. Duttlinger R, Manova K, Berrozpe G, Chu TY, DeLeon V, Timokina I, Chaganti RSK, Zelentz AD, Bachvarova RF, Besmer P. The W^{sh} and *Ph* mutations affect the c-*kit* expression profile: c-*kit* misexpression in embryogenesis impairs melanogenesis in W^{SH} and *Ph* mutant mice. Proc Natl Acad Sci USA 1995; (in press).

97. Smith EA, Seldin MF, Martinez L, Watson ML, Choudhury GG, Lalley PA, Pierce J, Aaronson S, Barker J, Naylor SL, Sakagouchi AY. Mouse platelet-derived growth factor receptor α gene is deleted in W^{19H} and patch mutations on chromosome 5. Proc Natl Acad Sci USA 1991; 88:4811–4815.

98. Niwa Y, Kasugai T, Ohno K, Morimoto M, Yamazaki M, Dohmae K, Nishimune Y, Kondo K, Kitamura Y. Anemia and mast cell depletion in mutant rats that are homozygous at "White Spotting (W^s)" locus. Blood 1991; 78:1936–1941.

99. Tsujimura T, Hirota S, Nomura S, Niwa Y, Yamazaki M, Tono T, Morii E, Kim H-M, Kondo K, Nishimune Y, Kitamura Y. Characterization of W^s mutant allele of rats: a 12-base deletion in tyrosine kinase domain of c-*kit* gene. Blood 1991; 78:1942–1946.

100. Russell ES, Lawson F, Schabtach G. Evidence for a new allele at the *W* locus of the mouse. J Hered 1957; 48:119–123.

101. Lacassie Y, Thurmon TF, Tracy MC, Pelias MZ. Piebald trait in a retarded child with interstitial deletion of chromosome 4. Am J Hum Genet 1977; 29:641–642.

102. Fleischman RA, Saltman DL, Stastny V, Zneimer S. Deletion of the c-*kit* proto-oncogene in the human developmental defect piebald trail. Proc Natl Acad Sci USA 1991; 88:10885–10889.

103. Spritz RA, Droetto S, Fukushima Y. Deletion of the Kit and PDGFRA genes in a patient with piebaldism. Am J Med Genet 1992; 44:492–495.

104. Fleischman RA. Human piebald trait resulting from a dominant negative mutant allele of the c-*kit* membrane receptor gene. J Clin Invest 1992; 89:1713–1717.

105. Giebel L, Spritz RA. Mutation of the KIT (mast/stem cell growth factor receptor) proto-oncogene in human piebaldism. Proc Natl Acad Sci USA 1991; 88:8696–8699.

106. Spritz RA, Giebel LB, Holmes SA. Dominant negative and loss of function mutations of the c-*kit* proto-oncogene in human piebaldism. Am J Hum Genet 1992; 50:261–269.

107. Spritz RA, Holmes SA, Itin P, Kuster W. Novel mutations of the *Kit* (mast/stem cell growth factor receptor) proto-oncogene in human piebaldism. J Invest Dermatol 1993; 101:22–25.

108. Spritz RA, Holmes SA, Ramesar R, Greenberg J, Curtis D, Beighton P. Mutations of the *Kit* (mast/stem cell growth factor receptor) proto-oncogene account for a continuous range of phenotypes in human piebaldism. Am J Genet 1992; 51:1058–1065.

109. Spritz RA, Holmes SA, Berg SZ, Nordlund JJ, Fukai K. A recurrent deletion in the KIT (mast/stem cell growth factor receptor) proto-oncogene is a frequent cause of human piebaldism. Hum Mol Genet 1993; 2:1499–1500.

110. Spritz RA. Lack of apparent hematologic abnormalities in human patients with c-*kit* (stem cell factor receptor) gen mutations. Blood 1992; 79:2497–2498.

111. Rottapel R, Reedijk M, Williams DE, Lyman SD, Anderson DM, Pawson T, Bernstein A. The steel/W signal transduction pathway: Kit autophosphorylation and its association with a unique subset of cytoplasmic signaling proteins is induced by the steel factor. Mol Cell Biol 1991; 11:3043–3051.

112. Tsujimura T, Furitsu T, Morimoto M, Isozai K, Nomura S, Matsuzawa Y, Kitamura Y, Kanakura Y. Ligand-independent activation of c-*kit* receptor tyrosine kinase in a murine mastocytoma cell line P-815 generated by a point mutation. Blood 1994; 83:2619–2626.

113. Furitsu T, Tsujimura T, Tono T, Ikeda H, Kitayama H, Koshimizu U, Sugahara H, Butterfield JH, Ashman LK, Kanayama Y, Kitamura Y, Matsuawa Y. Identification of mutations in the coding sequences of the proto-oncogene c-*kit* in a human mast cell leukemia cell line causing ligand-independent activation of c-*kit* product. J Clin Invest 1993; 92:1736–1744.

114. Sarvella PA, Russell LB. Steel, a new dominant gene in the house mouse. J Hered 1956; 47:123–128.

115. Beechey CV, Searle AG. Male fertile black-eyed white at *Sl* locus. Mouse News Lett 1985; 73:17.

116. Schaible RH. Mouse News Lett 1961; 24:38.

117. Schaible RH. Mouse News Lett 1963; 29:48–49.

118. Cattanach BM, Raspberry C, Beechey CV. A steel allele with preimplantation homozygous lethality. Mouse News Lett 1988; 80:156–157.

119. Evans EP, Burtenshaw M, Cattanach BM. Deletions at the *Sl* locus. Mouse News Lett 1990; 86:230.

120. Bernstein SE. Mouse News Lett 1960; 23:33.

121. Brannan CI, Lyman SD, Williams DE, Eisenman J, Anderson DM, Cosman D, Bedell MA, Jenkins NA, Copeland NG. Steel-Dickie mutation encodes a c-*kit* ligand lacking transmembrane and cytoplasmic domains. Proc Natl Acad Sci USA 1991; 88:4671–4674.

122. Dexter TM, Moore MAS. In vitro duplication and cure of hematopoietic defects in genetically anemic mice. Nature 1977; 269:412–414.

123. Toksoz D, Zsebo KM, Smith KA, Hu S, Brankow D, Suggs SV, Martin FH, Williams DA. Support of human hematopoiesis in long-term bone marrow cultures by murine stromal cells selectively expressing the membrane-bound and secreted forms of the human homolog of the steel gene product, stem cell factor. Proc Natl Acad Sci USA 1992; 89:7350–7354.

124. Cattanach BM, Rasberry C. A new steel allele with early postimplantation homozygous lethality. Mouse News Lett 1988; 80:157–158.

125. Beechey CV, Loutit JF, Searle AG. Panda a new steel allele. Mouse News Lett 1988; 74:92.

126. Beechey CV, Searle AG. Contrasted, a steel allele in the mouse with intermediate effects. Genet Res 1983; 42:183–191.

127. Peters J, Ball ST, Loutit JF. A new steel allele which does not lead to dilution of coat color. Mouse News Lett 1987; 77:125–126.

128. Brannan CI, Bedell MA, Resnick JL, Eppig JJ, Handel MA, Williams DE, Lyman SD, Donovan PJ, Jenkins NA, Copeland NG. Developmental abnormalities in *Steel^{17H}* mice result from a splicing defect in the steel factor cytoplasmic domain. Genes Dev 1992; 6:1832–1842.

129. Huang EJ, Manova K, Packer AI, Sanchez S, Bachvarova RF, Besmer P. The murine *steel–panda* mutation affects kit-ligand expression and growth of early ovarian follicles. Dev Biol 1993; 157:100–109.

130. Bedell MA, Brannan CI, Evans EP, Copeland NG, Jenkins NA, Donovan PJ. DNA rearrangements located over 100 kb 5' of the *Steel* (*Sl*)-coding region in *Steel–panda* and *Steel–constrasted* mice

deregulate *Sl* expression and cause female sterility by disrupting ovarian follicle development. Genes Dev 1995; 9:455–470.

131. Ray P, Higgins KM, Tan JC, Chu TY, Yee NS, Nguyen H, Lacy E, Besmer P. Ectopic expression of a c-kit^{W42} minigene in transgenic mice: recapitulation of W phenotypes and evidence for c-kit function in melanoblast progenitors. Genes Dev 1991; 5:2265–2273.

132. Deryugina EI, Müller-Sieburg CE. Stromal cells in long-term cultures: keys to the elucidation of hematopoietic development. Crit Rev Immunol 1993; 13:115–150.

133. Aye MT, Hasemi S, Leclair B, Zeibdawi A, Trudal E, Halpenny M, Fuller V, Cheng G. Expression of stem cell factor and c-kit mRNA in cultured endothelial cells, monocytes and cloned human bone marrow stromal cells (CFU-RF). Exp Hematol 1992; 20:523–527.

134. Rossi P, Marziali G, Albanesi C, Charlesworth A, Geremia R, Sorrentino V. A novel c-kit transcript, potentially encoding a truncated receptor, originates with a kit gene intron in mouse spermatids. Dev Biol 1992; 152:203–207.

135. Andrews DF, Bianco JA, Moran DA, Singer JW. Blood 1991; 78(suppl.1):303a.

136. Koenig A, Yaksian E, Reuter M, Huang M, Sykora KW, Corbacioglu S, Welte K. Differential regulation of stem cell factor mRNA expression in human endothelial cells by bacterial pathogens: an in vitro model for inflammation. Blood 1994; 83:2836–2843.

137. Majumdar MK, Feng L, Medlock E, Toksoz D, Williams DA. Identification and mutations of primary and secondary proteolytic cleavage sites in murine stem cell factor cDNA yields biologically active, cell-asociated protein. J Biol Chem 1994; 269:1237–1242.

138. Papayannopoulou T, Brice M, Broudy VC, Zsebo KM. Isolation of c-kit receptor expressing cells from bone marrow, peripheral blood, and fetal liver: functional properties and composite antigenic profile. Blood 1991; 78:1403–1412.

139. Katayama N, Shih J-P, Nishikawa S-I, Kina T, Clark SC, Ogawa M. Stage-specific expression of c-kit protein by murine hematopoietic progenitors. Blood 1993; 82:2353–2360.

140. Sillaber C, Strobl H, Bevec D, Ashman LK, Butterfield JH, Lechner K, Maurer D, Bettelheim P, Valent P. IL-4 regulates c-kit proto-oncogene expression in human mast and myeloid progenitor cells. J Immunol 1991; 147:4224–4228.

141. Welham MJ, Schrader JW. Modulation of c-kit mRNA and protein by hemopoietic growth factors. Mol Cell Biol 1991; 11:2901–2904.

142. Brach MA, Bühring H-J, Grub H-J, Ashman LK, Ludwg WD, Mertelsmann RH, Hermann. Functional expression of c-kit by acute myelogenous leukemia blasts is enhanced by tumor necrosis factor-α through posttranscriptional mRNA stabilization by a labile protein. Blood 1992; 80:1224–1230.

143. Yasuda H, Galli SJ, Geissler EN. Cloning and functional analysis of the mouse c-kit promoter. Biochem Biophys Res Commun 1993; 191:893–901.

144. Yamamoto K, Tojo A, Aoki N, Shibuya M. Characterization of the promoter region of the human c-kit proto-oncogene. Jpn J Cancer Res 1993; 84:1136–1144.

145. Jackson IJ, Chambers DM, Budd PS, Johnson R. The tyrosinase related protein-1 gene has a structure and promoter sequence very different from tyrosinase. Nucleic Acids Res 1991; 19:3799–3804.

146. Lowings P, Yavuzer U, Goding CR. Positive and negative elements regulate a melanocyte-specific promoter. Mol Cell Biol 1992; 12:3653–3662.

147. Tassabehji M, Newton VE, Read AP. Waardenburg syndrome type 2 caused by mutations in the human microphthalmia (*MITF*) gene. Nature Genet 1994; 8:251–255.

148. Ebi Y, Kanakura Y, Jippo-Kanemoto T, Tsujimura T, Furitsu T, Ikeda H, Adachi S, Kasugai T, Nomura S, Kanayanna Y, Yamatodani A, Nishikawa S-I, Matsuzawa Y, Kitamura T. Low c-kit expression of cultured mast cells of *mi/mi* genotype may be involved in their defective responses to fibroblasts that express the ligand for c-kit. Blood 1992; 80:1454–1462.

149. Hemesath TJ, Steingrimsson E, McGill G, Hansen MJ, Vaught J, Hodgkinson CA, Arnheiter H, Copland NG, Jenkins NA. *Microphthalmia*, a critical factor in melanocyte development, defines a discrete transcription factor family. Genes Dev 1994; 8:2770–2780.

150. Bentley NJ, Eisen T, Goding CR. Melanocyte specific expression of the human tyrosinase promoter:

activation by the microphthalmia gene product and role of the initiator. Mol Cell Biol 1994; 14:7996–8006.

151. Tsai M, Takeishi T, Thompson H, Langley KE, Zsebo KM, Metcalfe DD, Geissler EN, Galli SJ. Induction of mast cell proliferation, maturation, and heparin synthesis by the rat c-*kit* ligand, stem cell factor. Proc Natl Acad Sci USA 1991; 88:6382–6386.

152. Durand B, Migliaccio G, Yee NS, Eddleman K, Huima-Byron T, Migliaccio AR, Adamson JW. Long-term generation of human mast cells in serum-free cultures of $CD34^+$ cord blood cells stimulated with stem cell factor and interleukin-3. Blood 1994; 84:3667–3674.

153. Tsuji K, Zsebo KM, Ogawa M. Murine mast cell colony formation supported by IL-3, IL-4, and recombinant rat stem cell factor, ligand for c-*kit*. J Cell Physiol 1991; 148:362–369.

154. Takagi M, Nakahata T, Kubo T, Shiohara M, Kioke K, Miyajima A, Arai K-I, Nishikawa S-I, Zsebo KM, Komiyama A. Stimulation of mouse connective tissue-type mast cells by hemopoietic stem cell factor, a ligand for the c-*kit* receptor. J Immunol 1992; 148:3446–3453.

155. Dolci S, Williams DE, Ernst MK, Resnick JL, Brannan CL, Lock LF, Lyman SD, Boswell HS, Donovan PJ. Requirement for mast cell growth factor for promordial germ cell survival in culture. Nature 1991; 352:809–811.

156. Godin I, Deed T, Cooke J, Zsebo KM, Dexter M, Wylie CC. Effecs of the steel gene product on mouse primordial germ cells in culture. Nature 1991; 352:807–809.

157. Matsui Y, Toksoz D, Nishikawa S, Nishikawa S-I, William D, Zsebo K, Hogan BLM. Effect of steel factor and leukemia inhibitory factor on murine primordial germ cells in culture. Nature 1991; 353:750–752.

158. Matsui Y, Zseb K, Hogan BLM. Derivation of pluripotential embryonic stem cells from murine primoridal germ cells in culture. Cell 1992; 70:841–847.

159. Mekori YA, Oh CK, Metcalfe DD. IL-3-dependent murine mast cells undergo apoptosis on removal of IL-3. J Immunol 1993; 151:3775–3784.

160. Bodine DM, Crosier PS, Clark SC. Effects of hematopoietic growth factors on the survival of the primitive stem cells in liquid suspension culture. Blood 1991; 78:914.

161. Carson WE, Haldar S, Baiocchi RA, Croce CM, Caligiuri MA. The c-kit ligand suppresses apoptosis of human natural killer cells through the upregulation of bcl-2. Proc Natl Acad Sci USA 1994; 91:7553–7557.

162. Pesce M, Farrace MG, Piacentini M, Dolci S, De Filici M. Stem cell factor and leukemia inhibitory factor promote primordial germ cell survival by suppressing programmed cell death (apoptosis). Development 1993; 118:1089–1094.

163. Funasaka Y, Boulton T, Cobb M, Yarden Y, Fan B, Lyman SD, Williams DE, Anderson DM, Zakut R, Mishima Y, Halaban R. c-*kit* induces a cascade of protein tyrosine phosphorylation in normal human melanocytes in response to mast cell growth factor and stimulates MAP kinase but is down regulated in melanomas. Mol Biol Cell 1992; 3:197–209.

164. Murphy M, Reid K, Williams DE, Lyman SD, Barlett PF. Steel factor is required for maintenance, but not differentiation of melanocyte precursors in the neural crest. Dev Biol 1992; 153:396–401.

165. Morrison-Graham K, Weston JA. Transient steel factor dependence by neural crest-derived melanocyte precursors. Dev Biol 1993; 159:346–352.

166. Hirata T, Morii E, Morimoto M, Kasugai T, Tsujimura T, Hirota S, Kanakura Y, Nomura S, Kitamura Y. Stem cell factor induces outgrowth of c-*kit*-positive neurites and supports the survival of c-*kit*-positive neurons in dorsal root ganglia of mouse embryos. Development 1993; 119:49–56.

167. Carnahan JF, Patel DR, Miller JA. Stem cell factor in a neurotrophic factor for neural crest-derived chick sensory neurons. J Neurosc 1994; 14:1433–1440.

168. Iemura A, Tsai M, Ando A, Wershil BK, Galli SJ. The c-*kit* ligand, stel cell factor promotes mast cell survival by suppressing apoptosis. Am J Pathiol 1994; 144:321–328.

169. Packer AI, Besmer P, Bachvarova RF. Kit ligand mediates survival of type A spermatogonia and dividing spermatocytes in postnatal mouse testis. Mol Reprod Dev 1995; 42:303–310.

170. Fujita J, Onoue H, Ebi Y, Nakayanna H, Kanakura Y. In vitro duplication and in vivo cure of mast-cell

deficiency of *Sl/Sl^d* mutant mice by cloned 3T3 fibroblasts. Proc Natl Acad Sci USA 1989; 86:2888–2891.

171. Adachi S, Ebi Y, Nishikawa SI, Hayashi SI, Yamazaki M, Kasugai T, Yamamura T, Nomura S, Kitamura Y. Necessity of extracellular domain of *W* (c-*kit*) receptors for attachment of murine cultured mast cells in fibroblasts. Blood 1992; 79:650–656.

172. Dastych J, Melcalfe DD. Stem cell factor induces mast cell adhesion to fibronectin. J Immunol 1994; 152:213–219.

173. Kinashi T, Springer TA. Steel factor and c-*kit* regulate cell-matrix adhesion. Blood 1994; 83:1033–1038.

174. Serve H, Yee NS, Stella G, Sepp-Lorenzino L, Tan JC, Besmer P. Differential roles of P13-kinase and Kit tyrosine 821 in Kit receptor-mediated proliferation, survival and cell adhesion in mast cells. EMBO J 1995; 14:473–483.

175. Meininger CJ, Yano H, Rottapel R, Bernstein A, Zsebo KM, Zetter BR. The c-*kit* receptor ligand functions as a mast cell chemotractant. Blood 1991; 79:958–963.

176. Tsai M, Shih L, Newlands GFJ, Takeishi T, Langley KE, Zsebo KM, Miller HRP, Geissler EN, Galli SJ. The rat c-*kit* ligand, stem cell factor, induces the develoment of connective tissue-type mast cells in vivo. Analysis by anatomical distribution. J Exp Med 1991; 174:125–131.

177. Columbo M, Horowtz EM, Botana LM, MacGlashan DW Jr, Bochner BS, Gillis S, Zsebo KM, Galli SJ, Lichtenstein LM. The recombinant human c-kit receptor ligand, rhSCF, induces mediator release from human cutaneous mast cells and enhances IgE-dependent mediator release from both skin mast cells and peripheral blood basophils. J Immunol 1992; 149:599–608.

178. Coleman JW, Holliday MR, Kimber I, Zsebo KM, Galli SJ. Regulation of mouse peritoneal mast cell secretory function by stem cell factor, IL-3 or IL4. J Immunol 1993; 150:556–562.

179. Wershil BK, Tsai M, Geissler EN, Zsebo KM, Galli SJ. The rat c-*kit* ligand, stem cell factor, induces c-*kit* receptor-dependent mouse mast cell activation in vivo. Evidence that signaling through the c-kit receptor can induce expression of cellular function. J Exp Med 1992; 175:245–255.

180. Grabarek J, Groopman JE, Lyles LYR, Jiang S, Bennett L, Zsebo K, Avraham H. Human kit ligand (stem cell factor) modulates platelet activation in vitro. J Biol Chem 1994; 269:2171–21724.

181. Moore MAS. Clinical implications of positive and negative hematopoietic stem cell regulators. Blood 1991; 78:1–19.

182. Ogawa M. Differentiation and proliferation of hematopoietic stem cells. Blood 1993; 81:244–253.

183. Metcalf D. Hematopoietic regulators: redundancy or subtlety? Blood 1993; 82:355–3523.

184. Raff MC, Lillien LE, Richardson WD, Burne JF, Noble MD. Platelet-derived growth factor from astrocytes drives the clock that times oligodendrocyte development in culture. Nature 1988; 333:562–565.

185. Noble M, Murray K, Stroobant P, Waterfield MD, Riddle P. Platelet-derived growth factor promotes division and motility and inhibits premature differentiation of the oligodendrocyte/type-2 astrocyte progenitor cell. Nature 1988; 333:560–562.

186. Kitamura Y. Heterogeneity of mast cells and phenotypic changes between subpopulations. Annu Rev Immunol 1989; 7:59–76.

187. Gurish MF, Ghildyal N, McNel HP, Austen KF, Gillis S, Stevens RL. Differential expression of secretory granule proteases in mouse mast cells exposed to interleukin 3 and c-*kit* ligand. J Exp Med 1992; 175:1003–1012.

188. Kapeller R, Cantley LC. Phosphatidylinositol 3-kinase. Bioessays 1994; 16:565–576.

189. Lev S, Yarden Y, Givol D. Dimerization and activation of the kit receptor by monovalent and bivalent binding of the stem cell factor. J Biol Chem 1992; 267:15970–15977.

190. Blechman JM, Lev S, Brizzi MF, Leitner O, Pegoraro L, Givol D, Yarden Y. Soluble c-*kit* proteins and antireceptor monoclonal antibodies confine the binding site of the stem cell factor. J Biol Chem 1993; 268:4399–4406.

191. Lev S, Blechman J, Nishikawa S-I, Givol D, Yarden Y. Interspecies molecular chimeras of *kit* help define the binding site of the stem cell factor. Mol Cell Biol 1993; 13:2224–2234.

192. Carlberg K, Rohrschneider L. The effect of activating mutations on dimerization, tyrosine phosphor-

ylation and internalization of the macrophage colony stimulating factor receptor. Mol Biol Cell 1994; 5:81–95.

193. Blechman JM, Lev S, Barg J, Eisensten M, Vaks B, Vogel Z, Givol D, Yarden Y. The fourth immunoglobulin domain of the stem cell factor receptor couples ligand binding to signal transduction. Cell 1995; 80:103–113.

194. Blume-Jensen P, Ronnstrand L, Gout I, Waterfield D, Heldin C-H. Modulation of kit/SCF receptor-induced signaling by protein kinase C. J Biol Chem 1994; 269:21793–21802.

195. Tang B, Mano H, Yi T, Ihle J. Tec kinase associates with c-kit and is tyrosine phosphorylated and activated following stem cell factor binding. Mol Cell Biol 1994; 14:8432–37.

196. Yi T, Ihle JN. Association of hematopoietic cell phosphates with c-*Kit* after stimulation with c-*Kit* ligand. Mol Cell Biol 1993; 13:3350–3358.

197. Tauchi T, Feng G-S, Marshall MS, Shen R, Mantel C, Pawson T, Broxmeyer HE. The ubiquitously expressed Syp phosphatase interacts with c-*kit* and Grb2 in hematopoietic cells. J Biol Chem 1994; 269:25206–25211.

198. Serve H, Hsu Y, Besmer P. Tyrosine residue 719 of the c-*kit* receptor is essential for binding of Pl-3 kinase to c-*kit* and Pl-3 kinase activity. J Biol Chem 1994; 269:6026–6030.

199. Cutler RL, Liu L, Damen JE, Krystal G. Multiple cytokines induce tyrosine phosphorylation of Shc and its association with Grb2 in hemopoietic cells. J Biol Chem 1993; 268:21463–1465.

200. Duronio V, Welham M, Abraham S, Dryden P, Schrader J. p21^{ras} activation via hemopoietin receptors and c-*kit* requires tyrosine kinase activity but not tyrosine phosphorylation of p21^{ras} GTPase-activating protein. Proc Natl Acad Sci USA 1992; 89:1587–1591.

201. Lev S, Givol D, Yarden Y. A specific combination of substrates is involved in signal transduction by the *kit*-encoded receptor. EMBO J 1991; 10:647–654.

202. Tsai M, Chen RH, Tam SY, Blenis J, Galli SJ. Activation of MAP-kinases, pp90rsk and pp70-S6 kinase and proliferation in mouse mast cells. Eur J Immunol 1993; 23:3286–291.

203. Koike T, Hirai K, Morita Y, Nozawa Y. Stem cell factor-induced signal transduction in rat mast cells. J Immunol 1993; 151:359–366.

204. Alai M, Mui AL, Cutler RL, Bustelo XR, Barbacid M, Krystal G. Steel factor stimulates the tyrosine phosphorylation of the proto-oncogene product, p95^{vav}, in human hemopoietic cells. J Biol Chem 1992; 267:18021–18025.

205. Tsai M, Tam SY, Galli SJ. Distinct patterns of early response gene expression and proliferation in mouse mast cells stimulated by stem cell factor, interleukin-3, or IgE and antigen. Eur J Immunol 1993; 23:867–872.

206. Blume-Jensen P, Siegbahn A, Stabel S, Heldin C-H, Rönnstrand L. Increased Kit/SCF receptor induced mitogenicity but abolished cell motility after inhibition of protein kinase C. EMBO J 1993; 12:4199–4209.

207. Miyazawa K, Toyama K, Gotoh A, Hendrie PC, Mantel C, Broxmeyer HE. Ligand-dependent polyubiquitination of c-*kit* gene product: a possible mechanism of receptor down modulation in MO7e cells. Blood 1994; 83:137–145.

208. Broudy VC, Kovach NL, Bennett LG, Lin N, Jacobsen FW, Kidd PG. Human umbilical vein endothelial cells display high-affinity c-*kit* receptors and produce a soluble form of the c-*kit* receptor. Blood 1994; 83:145–152.

209. Wypych J, Bennet LG, Schwartz MG, Clogston CL, Lu HS, Broudy VC, Bartley TD, Parker VP, Langley KE. Soluble kit receptor in human serum. Blood 1995; 85:66–73.

210. Bernstein SE, Russell ES, Keighley GH. Two hereditary mouse anemias (*Sl/Sl^d* and *W/W^r*) deficient in response to erythropoietin. Ann NY Acad Sci 1968; 149:475–485.

211. Hunt P, Zsebo KM, Hokom MM, Hornkohl A, Birkett NC, delCastillo JC, Martin F. Evidence that stem cell factor is involved in the rebound thrombocytosis that follows 5-fluorouracil treatment. Blood 1992; 80:904–911.

212. Orlic D, Fischer R, Nishikawa S, Nienhuis AW, Bodine DM. Purification and characterization of heterogeneous pluripotent hematopoietic stem cell populations expressing high levels of c-*kit* receptor. Blood 1993; 82:762–770.

213. Lerner NB, Nocka KH, Cole SR, Qiu F, Strife A, Ashman LK, Besmer P. Monoclonal antibody YB5.B8 identifies the human c-*kit* protein product. Blood 1991; 77:1876–1883.

214. Migliaccio G, Migliaccio AR, Valinsky J, Langley K, Zsebo KM, Visser JW, Adamson JW. Stem cell factor induces proliferation and differentiation of highly enriched murine hematopoietic cells. Proc Natl Acad Sci USA 1991; 88:7420–7424.

215. Miura N, Okada S, Zsebo KM, Miura Y, Suda T. Rat stem cell factor and IL-6 preferentially support the proliferation of c-*kit* positive murine hematopoietic cells rather than their differentiation. Exp Hematol 1993; 21:143–149.

216. Leary AG, Zeng HQ, Clark SC, Ogawa M. Growth factor requirements for survival in G_0 and entry into the cell cycle of primitive human hemopoietic progenitors. Proc Natl Acad Sci USA 1992; 89:4013–4017.

217. Katayama N, Clark SC, Ogawa M. Growth factor requirement for survival in cell cycle dormancy of primitive murine lymphohematopoietic progenitors. Blood 1993; 81:610.

218. Metcalf D, Nicola NA. Direct proliferative actions of stem cell factor on murine bone marrow cells in vitro: effects of combination with colony-stimulating factors. Proc Natl Acad Sci USA 1991; 88:6239–6243.

219. Bernstein ID, Andrews RG, Zsebo KM. Recombinant human stem cell factor enhances the formation of colonies by $CD34^{+}lin^{-}$ cells and the generation of colony-forming cell progeny from $CD34^{+}lin^{-}$ cells cultured with interleukin-3, granulocyte colony-stimulating factor, or granulocyte–macrophage colony-stimulating factor. Blood 1991; 77:2316–2321.

220. Ulich TR, delCastillo J, McNiece IK, Yi ES, Alzona CP, Yin S, Zsebo KM. Stem cell factor in combination with granulocyte colony-stimulating factor (CSF) or granulocyte-macrophage CSF synergistically increases granulopoiesis in vivo. Blood 1991; 78:1954–1962.

221. Andrews RG, Knitter GH, Bartelmez SH, Langley KE, Farrar D, Hendren RW, Applebaum FR, Bernstein ID, Zsebo KM. Recombinant human stem cell factor, a c-*kit* ligand, stimulates hematopoiesis in primates. Blood 1991; 78:1975–1980.

222. Fleming WH, Alpern EJ, Uchida N, Ikuta K, Weissman IL. Steel factor influences the distribution and activity of murine hematopoietic stem cells in vivo. Proc Natl Acad Sci USA 1993; 90:3760–3764.

223. Molineux G, Migdalska A. Szmitkowski M, Zsebo K, Dexter TM. The effects of hematopoiesis of recombinant stem cell factor (ligand for c-*kit*) administered in vivo to mice either alone or in combination with granulocyte colony-stimulating factor. Blood 1991; 78:961–966.

224. Bodie DM, Orlic D, Birkett NC, Seidel NE, Zsebo KM. Stem cell factor increases colony-forming unit-spleen number in vitro in synergy with interleukin-6, and in vivo in Sl/Sl^d mice as a single factor. Blood 1992; 79:913–919.

225. McNiece IK, Langley KE, Zsebo KM. Recombinant human stem cell factor synergises with GM-CSF, G-CSF, IL-3 and epo to stimulate human progenitor cells of the myeloid and erythroid lineage. Exp Hematol 1991; 19:226–231.

226. Dai CH, Krantz SB, Zsebo KM. Human burst-forming units-erythroid need direct interaction with stem cell factor for further development. Blood 1991; 78:2493–2497.

227. Briddell RA, Brunno E, Cooper RJ, Brandt JE, Hoffman R. Effect of c-*kit* ligand on in vitro human megakaryopoiesis. Blood 1991; 78:2854–2859.

228. McNiece IK, Langley KE, Zsebo KM. The role of stem cell factor on early B cell development. J Immunol 1991; 146:3785–3790.

229. Rolink A, Streb M, Nishikawa S-I, Melchers F. The c-*kit* encoded tyrosine kinase regulates the proliferation of early pre-B cells. Eur J Immunol 1991; 21:2609–2612.

230. Puddington L, Olson S, Lefrancois L. Interactions between stem cell factor and c-kit are required for intestinal immune system homeostasis. Immunity 1994; 1:733–739.

231. Abkowitz JL, Sabo KM, Nakamoto B, Blau CA, Martin FH, Zsebo KM, Papayannopoulou T. Diamond-Blackfan anemia: in vitro response of erythroid progenitors to the ligand for c-*kit*. Blood 1991; 78:2198–2202.

232. Olivieri NF, Grunberger T, Ben-David Y, Ng J, Williams DE, Lyman S, Anderson DM, Axelrad AA,

Correa P, Bernstein A, Freedman MH. Diamond-Blackfan anemia: heterogeneous response of hematopoietic progenitor cells in vitro to the protein product of the *steel* locus. Blood 1991; 78:2211–2215.

233. Miller BA, Perrine SP, Bernstein A, Lyman SD, Williams DE, Bell LL, Oliveri NF. Influence of steel factor on hemoglobin synthesis in sickle cell disease. Blood 1992; 79:1861–1868.

234. Alter BP, Knobloch ME, He L, Gillio AP, O'Reilly RJ, Reilly LK, Weinberg RS. Effect of stem cell factor on in vitro erythropoiesis in patients with bone marrow failure syndrome. Blood 1992; 80:3000–3008.

235. Tong J, Gordon MS, Srour EF, Cooper RJ, Orazi A, McNiece I, Hoffman R. In vivo administration of recombinant methionyl human stem cell factor expands the number of human marrow hematopoietic stem cells. Blood 1993; 82:784–791.

236. Brugger W, Mocklin W, Heimfeld S, Berenson RJ, Mertelsmann R, Kanz L. Ex vivo expansion of enriched peripheral blood CD34+ progenitor cells by stem cell factor, interleukin-1β, IL-6, IL-3, interferon-γ and erythropoietin. Blood 1993; 81:2579–2584.

237. Costa JJ, Demetri GD, Hayes DF, Merica EA, Menchaca DM, Galli SJ. Proc Am Assoc Cancer Res 1993; 34:211 [abstr].

238. Demetri G, Costa J, Hayes D, Sledge G, Galli S, Hoffman R, Merica E, Rich W, Harkins B, McGuire B, Gordon M. Proc Am Soc Clin Oncol 1993; 12:A367 [abstr].

239. Zsebo KM, Smith KA, Hartley M, Greenblatt M, Cooke K, Rich W, McNiece K. Radioprotection of mice by recombinant rat stem cell factor. Proc Natl Acad Sci USA 1992; 89:9464.

18

Interleukin-5

Colin J. Sanderson
TVWT Institute for Child Health Research, Perth, Western Australia, Australia

I. INTRODUCTION

Interleukin-5 (IL-5) has two activities in vitro in the mouse: it stimulates the production of eosinophils (eosinophil differentiation factor; EDF) in bone marrow cultures, and it has activity on B cells (B-cell growth factor II; BCGFII; T-cell replacing factor, TRF; B-cell differentiation factor, BCDFn). In contrast, in humans the activity on eosinophils is similar to the activity in the mouse, but there is not yet any convincing evidence for an activity on B cells. Thus, in humans it is highly specific for the eosinophil–basophil lineage, and although other cytokines have similar activities on eosinophils and basophils in vitro, these other cytokines are also active on other cell lineages.

IL-5 is produced by T lymphocytes as a glycoprotein with an M_r of 40,000–45,000 and is unusual among the T-cell-produced cytokines in being a disulfide-linked homodimer. It is a highly conserved member of a group of cytokines that are closely linked on human chromosome 5 (5q31). This gene cluster includes the genes for IL-3, IL-4, IL-7, IL-9, IL-13, and granulocyte–macrophage colony-stimulating factor (GM-CSF). The critical role of IL-5 in the production of eosinophilia, coupled with a better understanding of the part played by eosinophils in the development of tissue damage in chronic allergy, suggests that IL-5 will be a major target for a new generation of antiallergy drugs.

II. GENE STRUCTURE AND EXPRESSION

The coding sequence of the IL-5 gene forms four exons (Fig. 1). The introns show areas of similarity between the mouse and human sequences, although the mouse has a considerable amount of sequences (including repeat sequences) that are not present in the human gene. The mouse gene includes a 738-base pair (bp) segment in the 3-untranslated region that is not present in the human gene; thus, the mouse mRNA is 1.6 kb whereas the human is 0.9 kb. Each of the exons contains the codons for an exact number of amino acids. The gene structure is also shared by other members of the cytokine gene cluster on chromosome 5 in humans (1–4) and

405

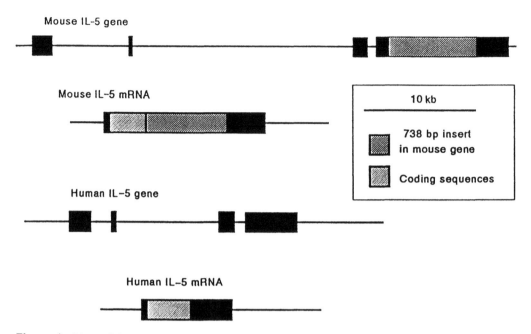

Figure 1 Maps of the human and mouse IL5 genomic genes and mRNA. Exons are shown as boxes. The insert in the 3' region of the mouse gene makes the mouse mRNA considerably longer than the human mRNA.

chromosome 11 in the mouse (5). Although there is no overall sequence homology at either the nucleotide or amino acid level among any of these cytokines, the localization and gene structural similarities suggest a common evolutionary origin (6); thus, they may be considered members of a gene family.

In T cells, transcription is induced by antigen, mitogens, and phorbol esters, and occurs for about 24 hr before the gene becomes silent again (7). Studies with T-cell clones in vitro suggest that IL-4 and IL-5 are often coexpressed in clones designated T_H2 (8), and this provides a possible explanation for the frequent association of eosinophilia with high IgE levels. However, anti-CD3 induced the expression of IL-4, IL-5, and GM-CSF mRNA in mouse T cells, whereas treatment with IL-2 induced IL-5 mRNA expression, but did not induce detectable IL-4 or GM-CSF (9), suggesting that independent control mechanisms exist for the regulation of IL-4 and IL-5. These differences are emphasized by the observation that cAMP and phorbol myristate acetate (PMA) have different effects on the induction of different cytokines in the mouse lymphoma EL4. PMA induces IL-2 and IL-4, but has only a low effect on IL-5 induction. Although cAMP markedly enhances the effect of PMA on the induction of IL-5, it has an inhibitory effect on IL-2, IL-3, IL-4, and IL-10 (10).

Corticosteroids inhibit IL-5 production both in vivo and in vitro (11–13). This may be an important mechanism of corticosteroid activity in asthma. Both progesterone and testosterone induce IL-5 transcription, whereas dexamethasone is an inhibitor (13). This latter study found that cyclosporine had no effect on the transcription of IL-5 in mouse lymphoma cells, whereas it does appear to inhibit in human peripheral blood mononuclear cells (14).

Little is known about the control of IL-5 transcription; computer searches reveal an array of potential protein-binding sites, and motifs analogous to those found in other genes. For example, the CLEO motif is highly conserved between IL-5 and GM-CSF (6), the presence of the CLEO element is necessary for the transcription of the GM-CSF promoter in vitro (15) and, therefore, may be important for IL-5 transcription.

III. PROTEIN STRUCTURE

Interleukin-5 is unusual among the T-cell–produced cytokines in being a disulfide-linked homodimeric glycoprotein that is highly homologous among species and in the cross-reactivity of the protein across a variety of mammalian species. Studies with mouse IL-5 indicate that the monomer has no biological activity and has no inhibitory activity, suggesting that no high-affinity interactions are formed with the IL-5 receptor (16). The dimer exists in an antiparallel (head-to-tail) configuration (17,18).

Mature human IL-5 monomer comprises 115 amino acids (M_r of 12,000 and 24,000 for the dimer). The secreted material has an M_r of 40,000–45,000, thus nearly half the native material consists of carbohydrate. Human IL-5 has one N-linked carbohydrate chain at position Asn-28, and one O-linked carbohydrate at position Thr-3 (19). Mouse IL-5 has an additional N-linked carbohydrate at Asn-55 (20), this site does not exist in human IL-5. The potential N-linked site at Asn-71 is apparently not glycosylated in either species.

The crystal structure of human IL-5 (21) shows it to be similar to the structures of other cytokines, and it most closely resembles IL-4 and GM-CSF, which consist of a bundle of four α-helices (A, B, C, D from the NH_2-terminus), with two overconnecting loops. The dimer structure of IL-5 forms an elongated ellipsoidal disk, made up of two domains about a twofold axis (Fig. 2). Each domain is made up of three helices from one monomer (A, B, C) and one helix (D′) from the other. The two monomers are held together by two sulfide bridges connecting Cys-44 of one molecule with Cys-86 of the other. In addition residues 32 to 35 form an antiparallel β-sheet with residues 89 to 92 of the other monomer. A large proportion of the monomer surface is at the interface of the two monomers. It is possible that the lack of biological activity in the monomer results from instability caused by exposure of hydrophilic residues normally concealed in the dimer.

IV. IL-5 RECEPTORS

The receptors for each of the three cytokines, IL-3, IL-5, and GM-CSF, consist of an α-chain, which is different for each ligand, and a βc-chain, which is common to each receptor complex (22). The α-chain forms the low-affinity interaction with its ligand, and the βc-chain serves to increase the affinity to give the high-affinity interaction (23,24). The βc-chain appears not to form a measurable interaction with any of the ligands in the absence of the α-chain. It seems likely that the cross-inhibition exhibited within the group is due to limiting numbers of βc-chains (25–27).

Murine eosinophils have been calculated to express approximately 50 high-affinity receptors for IL-5, and approximately 5000 low-affinity receptors (28). Several murine cell lines have been established that require the presence of IL-5 for growth, these are all of B-cell origin, and have relatively high numbers of IL-5 receptors. They represent an important tool in the study of the IL-5 receptor in the mouse, but no analogues of these B-cell lines have been described from human tissues. Initial work on the human receptor used a clone of the human promyelocytic leukemia cell line HL-60 (29). These cells express only a single population of high-affinity IL-5–binding sites. Similarly, only a single high-affinity receptor population has been identified on human eosinophils (30–32).

Both the α- and the βc-subunits share several features with other receptors, which has led

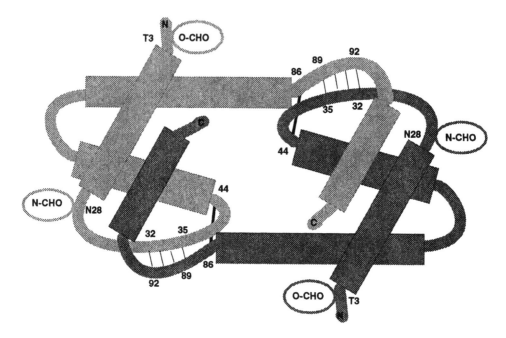

Figure 2 Diagram based on the crystal structure of the IL-5 showing the main structural features. One monomer is shown in light grey and the other in dark grey. The disulfide bridges connecting cysteines at positions 44 and 86 are shown as black lines. The interactions between antiparallel β-sheets are shown by fine black lines (residues 32–35 and 89–92). The attachment positions of O-linked carbohydrate (O-CHO) attached to residue 3 and N-linked carbohydrate (N-CHO) attached to residue 28 are indicated.

to their classification as members of the cytokine–hematopoietin superfamily. They have a modular structure build up of fibronectin-III–like domains, several conserved amino acids, and exist in various isoforms generated by alternative splicing. In each case the membrane-bound isoform results from splicing that retains a transmembrane domain (23). The other isoforms lack the transmembrane domain, resulting in soluble forms of the receptor. The possibility that the soluble forms of these receptors act as antagonists of the membrane-bound form to provide a negative control on hematopoiesis has been widely discussed. The soluble form of the human IL-5R α-chain is antagonistic in vitro (23), but there is not yet any evidence that it is active in vivo.

The cloning of cDNAs for both the human and mouse IL-5R α-chain revealed different isoforms. Remarkably, in humans the major transcript is a soluble isoform. The structure of the IL5-Rα gene provides an explanation for the isotypes observed (33,34). Figure 3 is a diagram of the gene, illustrating the transmembrane (TM) and soluble (S1) transcripts. The gene contains 14 exons. Three introns are located in the 5′-untranslated region; alternative splicing leads to skipping of exon 3 in some transcripts. Exon 4 corresponds to the signal peptide (S), and each of the three FN-III-like domains is encoded by exons 3 and 4; 5 and 6; and 7 and 8, respectively. Exon 11 encodes four amino acids, followed by a stop codon and a polyadenylation site. Exon 12 encodes a transmembrane domain, and the two remaining exons encode an intracytoplasmic tail. Three alternative events can occur at the 3′-end of exon 10:

1. No splicing: Translation continues adding two amino acids before a stop codon. This gives rise to one soluble isoform (S2).

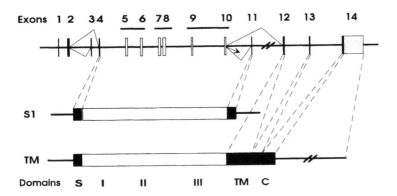

Figure 3 Diagram showing the exon structure (not to scale) of the IL-5R α-chain (top) and two of the alternatively spliced transcripts. S1 indicates the major transcript in human eosinophilic cells, a soluble form formed by splicing in exon 11 causing transcription to end before the exon encoding the transmembrane domain. TM indicates the transcript of the active form of the receptor formed by splicing out exon 11. Exons numbers are indicated at the top. Alternative splicing events are indicated by fine lines (the arrow indicates no splicing at this donor site, forming the soluble S2 transcript—see text). The structural domains of the protein are indicated at the bottom: S, signal peptide; I, II, and III, the three fibonectin type III-like domains derived from exons 5 and 6, 7 and 8, and 9 and 10, respectively; TM, transmembrane domain; C, intracytoplasmic tail.

2. Normal splicing leads to the inclusion of exon 11, and gives rise to a transcript truncated by the poly(A)site at the 3′-end of the exon. This soluble isoform (S1) is the most abundant transcript in HL60 cells and cultured eosinophils.

3. Alternative splicing skips exon 11, and gives rise to the functional membrane-bound form of the receptor (TM).

On the basis of cDNA sequences, the situation in the mouse is different. The soluble isoforms are formed by alternative splicing, which skips the exon encoding the transmembrane domain (exon 12 in humans), or both this exon and the next (exons 12 and 13 in humans). Although the structure of the murine gene has been reported, there is no sequence information across the zone analogous to the human exon 11; hence, it is not clear whether this exon exists in the mouse.

V. INTERACTION BETWEEN IL-5 AND THE RECEPTOR

There is an area of sequence similarity, determined using Dayhoff mutation indices, at the COOH-terminal region of several different cytokines, including IL-5 (6). This region lies in helix D of IL-5, IL-4, and GM-CSF. By making use of the 100-fold lower specific activity of hIL-5 when tested in a mouse assay system, a series of human–mouse hybrid IL-5 molecules were constructed and tested for biological activity and receptor displacement. The COOH-terminal third of the molecule determined the species specificity (17). This region contains eight residues that differ between human and mouse, two at the end of the C helix, three in the D helix, and the remaining three in the COOH-terminal tail. Determination of the residue(s) involved in species specificity will locate the region of interaction with the α-chain of the receptor more accurately.

Despite the twofold symmetry of the IL-5 dimer, which suggests two potential binding sites for the receptor, the α-chain of the receptor appears to bind to IL-5 in a 1:1 ratio (35).

Experiments with GM-CSF suggest that the βc-chain may bind in the region of residue 21 (36), and a series of mouse–human GM-CSF hybrids identified residue 20 as important in the interaction with the βc-chain. Furthermore, a hybrid in which the NH$_2$-terminal region of GM-CSF was substituted by the analogous region from IL-5, showed strong biological activity. This very elegant approach demonstrated that IL-5 uses the same βc-receptor subunit as GM-CSF, and implies an interaction between the βc-chain and the NH$_2$-terminal region of the cytokine molecule (37). A comparison of the structure of IL-5 and GM-CSF suggests that the interaction site between IL-5 and the βc-chain in the complex is Glu-13 (36).

VI. ACTIVITIES ON EOSINOPHILS

Eosinophilia is T-cell–dependent; therefore, it is not surprising that the controlling factor is a T-cell–derived cytokine (38). It is characteristic of a limited number of disease states, most notably parasitic infections and allergy. Clearly, as eosinophilia is not characteristic of all immune responses, it is obvious that the factors controlling eosinophilia must not be produced by all T cells. Similarly, because it is now clear that IL-5 is the main controlling cytokine for eosinophilia (see later discussion), then if IL-5 has other biological activities, it is likely that these will coincide with the production of eosinophils.

A. Biological Specificity of Eosinophilia

One of the features of eosinophilia that has attracted the curiosity of hematologists for several decades is the apparent independence of eosinophil numbers on the numbers of other leukocytes. Thus, eosinophils are present in low numbers in normal individuals, but can increase dramatically and independently of the number of neutrophils. Such changes are common during the summer months in individuals with allergic rhinitis (hay fever), or in certain parasitic infections. Clearly, such conditions will result in more broadly based leukocytosis when complicated by other infections. Although this specificity has been known for many years, somewhat surprisingly, it is not easy to find clear examples in the early literature. More recently, in experimental infection of volunteers with hookworms (*Necator americanus*), it was noted that an increase in eosinophils was the only significant change (39), and our own work with *Mesocestoides corti* in the mouse demonstrated massive increases in eosinophils, independent of changes in neutrophils (40). This biological specificity suggests a mechanism of control that is independent of the control of other leukocytes. This, coupled with the normally low numbers of eosinophils, provides a useful model for the study of the control of hematopoiesis by the immune system.

Liquid cultures of murine bone marrow produce neutrophils for extended periods without exogenous factors (41). It appears that the microenvironment of these cultures maintains the production of neutrophil precursors. No eosinophils are seen in the absence of exogenous factors, but IL-5 induces their production. This is lineage-specific, because only eosinophil numbers are increased in these cultures (38). In contrast, both IL-3 and GM-CSF induce eosinophils as well as other cell types, most notably neutrophils and macrophages in bone marrow cultures (42). The production of eosinophils is considerably higher when the bone marrow is taken from mice infected with *M. corti* than it is from normal marrows. This suggests that marrow from infected mice contains more eosinophil precursors than marrow from normal mice (38). Essentially similar results are obtained with human bone marrow cultures, in which IL-5 induces the production of eosinophils in liquid culture (43,44) and eosinophil colonies in semisolid medium (45).

B. Eosinophil Production In Vivo

Because IL-5 is normally a T-cell product and the gene is transcribed for only a relatively short time after antigen stimulation, transgenic mice have been produced in which IL-5 is constitutively expressed by all T cells (46). These mice have detectable levels of IL-5 in the serum. They show a profound and lifelong eosinophilia, with large numbers of eosinophils in the blood, spleen, and bone marrow. This indicates that the expression of IL-5 is sufficient to induce the full pathway of eosinophil differentiation. If other cytokines are required for the development of eosinophilia, then they must be either expressed constitutively, or their expression is secondary to the expression of the IL-5 gene. This clear demonstration that the expression of the IL-5 gene in transgenic animals is sufficient for the production of eosinophilia, provides an explanation for the biological specificity of eosinophilia. Therefore, it seems likely that because eosinophilia can occur without a concomitant neutrophilia or monocytosis, a mechanism must exist by which IL-5 is the dominant hematopoietic cytokine produced by the T-cell system in natural eosinophilia. An interesting observation with these transgenic animals was that, despite their massive, long-lasting eosinophilia, the mice remained normal. This illustrates that an increased number of eosinophils is not of itself harmful, and that the tissue damage seen in allergic reactions and other diseases must be due to agents that trigger the eosinophils to degranulate.

Another important approach to the understanding of the biological role of IL-5, comes from the administration of neutralizing antibody. Mice infected with *Trichinella spiralis* develop eosinophilia and increased levels of IgE; however, when treated with an anti–IL-5 antibody, no eosinophils are observed (8). Indeed, the number of eosinophils is lower than that seen in control animals. These experiments illustrate the unique role of IL-5 in the control of eosinophilia in this parasitic infection. They also show that the apparent redundancy seen in vitro, when both IL-3 and GM-CSF are also able to induce eosinophil production, does not operate in these infections. Furthermore, IL-5 plays no role in the development of IgE antibody (this activity is controlled by IL-4), or in the development of the granuloma seen surrounding schistosomes in the tissues (47).

C. Association Between IL-5 and Eosinophilia

Further evidence consistent with a key role for IL-5 in the control of eosinophilia comes from studies on eosinophilia in different diseases. If IL-5 were not the primary controlling factor in eosinophilia, it could be expected that cases of eosinophilia would exist without IL-5. As the following examples illustrate, eosinophilia in a wide variety of diseases is associated with IL-5 expression.

The development of eosinophilia in mice infected with *T. canis* is accompanied by the appearance of IL-5 mRNA in the spleen (48). A comparison of the production of cytokines by normal individuals and by eosinophilic patients infected with the filarial parasite *Loa loa* gave interesting results. Both groups produced similar levels of IL-3 and GM-CSF, but whereas the normals gave relatively little IL-5, the cells from infected patients produced high levels of IL-5 (49). Significant levels of IL-5 were also detected in the serum of patients with idiopathic hypereosinophilia (50), and in patients with the eosinophilic–myalgia syndrome, resulting from the ingestion of L-tryptophan (51). Patients with eosinophilia associated with Hodgkin's disease have IL-5 mRNA in the tumor cells (52), and a patient with angiolymphoid hyperplasia with eosinophilia (Kimura's disease) had constitutive expression of IL-5 mRNA in lymphnode tissue (53).

Asthma is associated with an inflammatory reaction involving infiltration of eosinophils, and in situ hybridization on mucosal bronchial biopsies indicates local expression of IL-5 that correlates with the number of infiltrating eosinophils (54).

D. Mechanism of Control of Eosinophilia

Although it has become clear that IL-5 is the controlling factor in eosinophilia (55), the mechanisms that allow a selective production of IL-5 and thus provide a selective increase in eosinophils remain unknown. Furthermore, in other diseases, eosinophils are not observed in significant numbers, suggesting that the expression of IL-5 is not induced in these diseases. The cellular abnormalities typical of particular diseases may reflect the induction of different cytokines by the immune system in response to different antigenic exposure.

It has been suggested that murine helper T cells fall into two groups (T helper-1 and T helper-2), the former producing predominantly interferon gamma (IFN-γ) and IL-2, and the latter producing IL-4 and IL-5 (33). This hypothesis provides an explanation for the frequently observed association between high levels of IgE antibody and eosinophilia, but does not provide an explanation for many of the complexities seen in the pathology of different diseases. For example, in the mouse IL-4 and IL-5 are thought to be the two major lymphokines in the control of antibody production; thus, this classification leaves open the question of how antibody responses are controlled in the absence of eosinophilia. Although the variety of cellular and antibody responses in different infections suggests some form of selection for T cells producing different patterns of cytokines, no clear mechanism has emerged to explain this selective response. The association between eosinophilia and helminth infections or allergic reactions could result from either features in common between the antigens involved in these reactions, or features in common between the tissues in which the immune system encounters these antigens. Clearly, understanding these processes will be an important step forward in our understanding of the immune system, and these preliminary attempts to understand functional subsets in T cells are a basis for future studies.

E. Activation of Eosinophils

The ability of eosinophils to perform in functional assays can be increased markedly by incubation with several different agents, including IL-5. The phenomenon of activation is apparently independent of differentiation. It appears to have a counterpart in vivo, as eosinophils from different individuals vary in functional activity. The ability of eosinophils to kill schisto-somula increases in proportion to the degree of eosinophilia (56,57). This is consistent with a common control mechanism for both the production and activation of eosinophils in these cases.

The first observations on selective activation of human eosinophils by IL-5 showed that the ability of purified peripheral blood eosinophils to lyse antibody-coated tumor cells, was increased when IL-5 was included in the assay medium (58). Similarly, the phagocytic ability of these eosinophils toward serum-opsonized yeast particles was increased in the presence of IL-5. There was a 90% increase in surface C3bi complement receptors, as well as an approximately 50% increase in the granulocyte functional antigens GFA1 and GFA2. Later studies demonstrated that IL-5 increases "polarization," including membrane ruffling and pseudopod formation, which appear to reflect changes in the cytoskeletal system. IL-5 also induces a rapid increase in superoxide anion production by eosinophils (59). In addition, IL-5 increases the survival of peripheral blood eosinophils (60).

A further interesting observation in this context was the demonstration that IL-5 is a potent inducer of immunoglobulin-induced eosinophil degranulation, as measured by the release of eosinophil-derived neurotoxin (EDN). IL-5 increased EDN release by 48% for secretory IgA and 136% for IgG. This enhancing effect appeared by 15 min and reached a maximum by 4 hr (61). The finding that secretory IgA can induce eosinophil degranulation is particularly important because eosinophils are frequently found at mucosal surfaces where IgA is the most abundant immunoglobulin.

F. Tissue Localization

Another aspect of the pathology of diseases characterized by eosinophilia, is the preferential accumulation of eosinophils in tissues. As the blood contains both eosinophils and neutrophils, there must exist a specific mechanism that allows the eosinophils to pass preferentially from the blood vessels to the tissues. It is likely that IL-5 plays a role in this, for the following reasons: The different tissue distribution of eosinophils in the two transgenic mice systems probably results from the different tissue expression of IL-5. By using the metallothionein promoter, transgene expression was demonstrated in the liver and skeletal muscle, and eosinophils were observed in these tissues (62). In contrast the CD2–IL-5 mice with IL-5 expression in T cells, did not have eosinophils in the liver or skeletal muscle (46). This suggests that eosinophils migrate into tissues where IL-5 is expressed. There are two mechanisms that might explain this: First IL-5 is chemotactic for eosinophils (63,64). Second, IL-5 upregulates adhesion molecules on eosinophils. Thus, it was demonstrated that IL-5 increased the expression of the integrin CD11b on human eosinophils (58), and this increased expression was accompanied by an increased adhesion to endothelial cells. Adhesion was inhibited by antibody to CD11b or CD18, suggesting that the integrins are involved in eosinophil adhesion to endothelial cells (65). More recently, it has been shown that eosinophils can use the integrin VLA-4 (CD49d/CD29) in adherence to endothelial cells. Here the ligand is VCAM-1. In contrast, neutrophils do not express VLA-4 and do not use this adherence mechanism (66).

VII. ACTIVITIES ON OTHER CELL TYPES

A. Human Tissues

The most pronounced effect of IL-5 on cells other than eosinophils in humans, is the effect on basophils. Although our studies suggested that IL-5 induced only eosinophils, a detailed study by electron microscopy of cells produced in human cord blood cultures revealed a small number of basophils (67). Other studies have shown that IL-5 primes basophils for increased histamine production and leukotriene generation (68,69), and basophils in the blood clearly express the IL-5 receptor (26). Thus, although the effect of IL-5 on the production of basophils may be minor, the priming effect on mature basophils appears to be of importance in the allergic response.

In view of the well-characterized activity of IL-5 on mouse B cells, it was surprising when no activity could be demonstrated in a wide range of human B-cell assay systems (70). This lack of activity of human IL-5 has been confirmed in many different systems (57,71). Unlike the demonstration of IL-5 receptors on human eosinophils and basophils, there are no reports of IL-5 receptors on human B cells. Until the true biological role of IL-5 in the mouse B-cell system is understood, it is unlikely that the human activity will be clarified.

B. Mouse Tissues

In the mouse, IL-5 has well-characterized activities on B cells in vitro. In the TRF assay IL-5 induces specific antibody production by mouse B cells primed with antigen in vivo (72). The BCGFII assay was based on the ability of IL-5 to induce DNA synthesis in normal splenic B cells in the presence of dextran sulfate, and later on the ability of IL-5 to increase DNA synthesis in the BCL1 cell (a mouse B-cell tumor) line (73). The BCDFn assay depends on the ability of IL-5 to induce BCL1 cells to secrete IgM (74).

Interleukin-5 is a late-acting factor in the differentiation of primary B cells, requiring a priming stimulus to make resting B cells responsive. This can be either polyclonal stimulants, such as dextran sulfate, bacterial lipopolysaccharide (LPS), anti-immunoglobulin, or specific antigen. Large splenic B cells, presumed to have been activated in vivo, when cultured with IL-5 for 7 days show markedly enhanced numbers of IgM- and IgG-producing cells (75). IL-5 in combination with antigen is sufficient to induce growth and differentiation of B cells at the single-cell level (76). Combinations of IL-2, IL-4, and IL-5 appear to regulate the amount of IgG1 isotype secretion by B cells (77,78). Neutralizing antibody to IL-5 inhibits the polyclonal antibody response induced by T-cell clones on B cells, suggesting a critical role for IL-5 in this system (79).

A possible role for IL-5 in the development of autoimmunity in mice was suggested by the observation that this cytokine stimulates B cells from NZB mice to produce high levels of IgM anti-DNA antibody (80). In another study, the B cells from autoimmune NZB/W mice were hyperresponsive to IL-5, whereas two other strains of mice that are prone to autoimmunity did not show this response. As NZB/W mice have elevated numbers of Ly-1-positive B cells, these were tested and showed a higher response to IL-5 than the negative cells, suggesting that the increased responsiveness to IL-5 in these mice may be due to the increased numbers of Ly-1 B cells (81). In support of this, freshly isolated peritoneal Ly-1 B cells express high levels of IL-5 receptor, and IL-5 increases the frequency of cells that produce autoantibodies (82). As these effects concern mainly the production of IgM, whereas autoimmune disease appears to be mainly due to IgG, the significance of these findings for autoimmunity is unclear. However, these experiments point to the possible restriction of IL-5 activity to the Ly-1 subpopulation of B cells.

A potentially interesting observation was the demonstration that IL-5 appears preferentially to enhance IgA production. When added to cultures in the presence of LPS, the highest increase over background occurs with the IgA-producing cells, with significant increases in IgM and IgG1 as well (83,84). The interpretation of these experiments is not straightforward, because the LPS itself induces a large effect, and the activity on IgA and IgG1 was small in comparison with the total levels of IgM produced. In a study of B cells from gut-associated lymphoid tissue (Peyer's patches), IL-5 increased the production of IgA, but maximum enhancement of IgA in these cultures requires IL-4 (85,86). This effect of IL-5 was due to the induction of a high rate of IgA synthesis in cells positive for surface IgA expression. No IgA secretion was induced in the surface IgA-negative cells (86–88). This suggests that IL-5 does not induce switching to IgA production, but acts after switching to enhance the production of IgA.

In contrast with these studies, which suggest a key role for IL-5 in the production of IgA, more recent studies have indicated that its effect is minor compared with the activity of other cytokines, and that it may only augment these activities. For example, IL-5 enhances IgA secretion from B cells isolated from Peyer's patches, but the effect was small compared with the effect of IL-6 (89). A combination of IL-5 and IL-6 had a greater effect than either cytokine alone (90). TGF-β has an important activity in the switching to IgA production in LPS-stimulated B cells, and although IL-5 enhances this effect, it is less active than IL-2. TGF-β is required early, whereas IL-5 appears to act only late in these cultures (91).

Important developments toward understanding the mechanism of action of IL-5 in the production of antibody have shown an increase in μ-chain mRNA in both B cells and BCL1 cells (92). IL-5 induces CL-3 (a B-cell line) cells into the IgM secretory state, and this is accompanied by an increase in the secretory form of μ mRNA. The action of IL-5 allows the cells to respond to IL-2 by amplification of J-chain mRNA. Thus, IL-5 and IL-2 are both necessary for IgM secretion (93). A possible mechanism for the effect of IL-2 on B cells is suggested by the observation that IL-5 increases the expression of the IL-2 receptor (94,95).

The significance of these activities in vitro remains unclear. Although transgenic mice expressing IL-5 on a metallothionein promoter develop autoimmunity (62), no effects on B cells or antibody levels were detected in transgenic mice expressing IL-5 under control of the CD2 locus control region (96). Similarly, treatment of mice with anti–IL-5 antibody completely blocked eosinophil production, but had no effect on antibody levels (97). This question of the biological role of IL-5 on the mouse antibody system will probably be resolved only by IL-5 gene knockout studies.

VIII. SOURCES OF IL-5

All of the original reports on the characterization, purification, and cloning of murine IL-5 used T-cell lines or lymphomas as the source of material, suggesting that T cells are an important source of the cytokine. The fact that eosinophilia is a T-cell–dependent phenomenon illustrates the central role of the T cell in producing IL-5. However, the demonstration that IL-5 as well as other cytokine mRNAs are produced by mast cell lines opens the possibility that these cells may serve to amplify the development of eosinophilia (98,99). Similarly, the observation that human Epstein-Barr virus-transformed B cells produce IL-5 raises the possibility that B cells may be an additional source of this cytokine (100). Furthermore, eosinophils themselves have been demonstrated to produce IL-5 (101). It is unclear whether the non–T-cell–derived IL-5 plays a significant biological role in the development of eosinophilia.

Eosinophilia has been observed in a significant proportion of a wide range of human tumors. In many cases the presence of eosinophils has been of positive prognostic significance (reviewed in Ref. 55). Clearly, it is important to understand the mechanism of production of these eosinophils. In a study of Hodgkin's disease with associated eosinophilia, all 16 patients gave a positive signal for IL-5 mRNA by in situ hybridization (52). This suggests that IL-5 may be responsible for the production of eosinophils in these cases, and raises the possibility that eosinophilia in other tumors may also be due to the production of IL-5 by the tumor cells.

REFERENCES

1. Campbell HD, Tucker WQ, Hort Y, Martinson ME, Mayo G, Clutterbuck EJ, Sanderson CJ, Young IG. Molecular cloning, nucleotide sequence, and expression of the gene encoding human eosinophil differentiation factor (interleukin 5). Proc Natl Acad Sci USA 1987; 84:6629–6633.
2. Sutherland GR, Baker E, Callen DF, Campbell HD, Young IG, Sanderson CJ, Garson OM, Lopez AF, Vadas MA. Interleukin-5 is at 5q31 and is deleted in the 5q⁻ syndrome. Blood 1988; 71:1150–1152.
3. van-Leeuwen BH, Martinson ME, Webb GC, Young IG. Molecular organization of the cytokine gene cluster, involving the human IL-3, IL-4, IL-5, and GM-CSF genes, on human chromosome. Blood 1989; 73:1142–1148.
4. Chandrasekharappa SC, Rebelsky MS, Firak TA, Le Beau MM, Westbrook CA. A long-range restriction map of the interleukin-4 and interleukin-5 linkage group of chromosome 5. Genomics 1990; 6:94–99.
5. Lee JS, Campbell HD, Kozak CA, Young IG. The IL-4 and IL-5 genes are closely linked and are part of a cytokine gene cluster on mouse chromosome 11. Somat Cell Genet 1989; 15:143–152.
6. Sanderson CJ, Campbell HD, Young IG. Molecular and cellular biology of eosinophil differentiation factor (interleukin-5) and its effects on human and mouse B cells. Immunol Rev 1988; 102:29–50.
7. Tominaga A, Matsumoto M, Harada N, Takahashi T, Kikuchi Y, Takatsu K. Molecular properties

and regulation of mRNA expression for murine T cell-replacing factor/IL-5. J Immunol 1988; 140:1175–1181.

8. Coffman RL, Seymour BW, Hudak S, Jackson J, Rennick D. Antibody to interleukin-5 inhibits helminth-induced eosinophilia in mice. Science 1989; 245:308–310.

9. Bohjanen PR, Okajima M, Hodes RJ. Differential regulation of interleukin 4 and interleukin 5 gene expression: a comparison of T-cell gene induction by anti-CD3 antibody or by exogenous lymphokines. Proc Natl Acad Sci USA 1990; 87:5283–5287.

10. Lee HJ, Koyona-Nagagawa N, Naito Y, Nishida J, Arai N, Arai K, Yokota T. cAMP activates the IL5 promoter synergistically with phorbol ester through the signally pathway involving protein kinase A in mouse thymoma EL-4. J Immunol 1993; 151:6135–6142.

11. Rolfe FG, Hughes JM, Armour CL, Sewell WA. Inhibition of interleukin-5 gene expression by dexamethasone. Immunology 1992; 77:494–499.

12. Corrigan CJ, Haczku A, Gemou-Engesaeth V, Doi S, Kikuchi Y, Takatsu K, Durham SR, Kay AB. CD4 T-lymphocyte activation in asthma is accompanied by increased serum concentrations of interleukin-5: effect of glucocorticoid therapy. Am Rev Respir Dis 1993; 147:540–547.

13. Wang Y, Campbell HD, Young IG. Sex hormones and dexamethasone modulate interleukin-5 gene expression in T lymphocytes. J Steroid Biochem Mol Biol 1993; 44:203–210.

14. Mori A, Suko M, Nishizaki Y, Kaminuma O, Matsuzaki G, Ito K, Etoh T, Nakagawa H, Tsuruoka N, Okudaira H. Regulation of interleukin-5 production by peripheral blood mononuclear cells from atopic patients with FK506, cyclosporin A and glucocorticoid. Int Arch Allergy Immunol 1994; 104:32–35.

15. Miyatake S, Shlomai J, Arai K, Arai N. Characterization of the mouse granulocyte–macrophage colony-stimulating factor (GM-CSF) gene promoter: nuclear factors that interact with an element shared by three lymphokine genes—those for GM-CSF, interleukin-4 (IL-4), and IL-5. Mol Cell Biol 1991; 11:5894–5901.

16. McKenzie ANJ, Ely B, Sanderson CJ. Mutated interleukin-5 monomers are biologically inactive. Mol Immunol 1991; 28:155–158.

17. McKenzie ANJ, Barry SC, Strath M, Sanderson CJ. Structure–function analysis of interleukin-5 utilizing mouse/human chimeric molecules. EMBO J 1991; 10:1193–1199.

18. Proudfoot AE, Davies JG, Turcatti G, Wingfield PT. Human interleukin-5 expressed in *Escherichia coli*: assignment of the disulfide bridges of the purified unglycosylated protein. FEBS Lett 1991; 283:61–64.

19. Minamitake Y, Kodama S, Katayama T, Adachi H, Tanaka S, Tsujimoto M. Structure of recombinant human interleukin 5 produced by Chinese hamster ovary cells. J Biochem (Tokyo) 1990; 107:292–297.

20. Kodama S, Tsujimoto M, Tsuruoka N, Sugo T, Endo T, Kobata A. Role of sugar chains in the in-vitro activity of recombinant human interleukin 5. Eur J Biochem 1993; 211:903–908.

21. Milburn M, Hassell AM, Lambert MH, Jordan SR, Proudfoot AEI, Grabar P, Wells TNC. A novel dimer configuration revealed by the crystal structure at 2.4 A resolution of human interleukin-5. Nature 1993; 363:172–176.

22. Lopez AF, Elliott MJ, Woodcock J, Vadas MA. GM-CSF, IL-3 and IL-5: cross-competition on human haemopoietic cells. Immunol Today 1992; 13:495–500.

23. Tavernier J, Devos R, Cornelis S, Tuypens T, Van der Heyden J, Fiers W, Plaetinck G. A human high affinity interleukin-5 receptor (IL5R) is composed of an IL5-specific x chain and a b chain shared with the receptor for GM-CSF. Cell 1991; 66:1175–1184.

24. Takaki S, Mita S, Kitamura T, Yonehara S, Yamaguchi N, Tominaga A, Miyajima A, Takatsu K. Identification of the second subunit of the murine interleukin-5 receptor: interleukin-3 receptor-like protein, AIC2B is a component of the high affinity interleukin-5 receptor. EMBO J 1991; 10:2883–2838.

25. Lopez AF, Eglinton JM, Gillis D, Park LS, Clark S, Vadas MA. Reciprocal inhibition of binding between interleukin 3 and granulocyte–macrophage colony-stimulating factor to human eosinophils. Proc Natl Acad Sci USA 1989; 86:7022–7026.

26. Lopez AF, Eglinton JM, Lyons AB, Tapley PM, To LB, Park LS, Clark SC, Vadas MA. Human interleukin-3 inhibits the binding of granulocyte–macrophage colony-stimulating factor and inter-leukin-5 to basophils and strongly enhances their functional activity. J Cell Physiol 1990; 145:69–77.

27. Nicola NA, Metcalf D. Subunit promiscuity among hemopoietic growth factor receptors. Cell 1991; 67:1–4.

28. Barry SC, McKenzie AN, Strath M, Sanderson CJ. Analysis of interleukin 5 receptors on murine eosinophils: a comparison with receptors on B13 cells. Cytokine 1991; 3:339–344.

29. Plaetinck G, der Heyden JV, Tavernier J, Fache I, Tuypens T, Fischkoff S, Fiers W, Devos R. Characterization of interleukin 5 receptors on eosinophilic sublines from human promyelocytic leukemia (HL-60) cells. J Exp Med 1990; 172:683–691.

30. Migita M, Yamaguchi N, Mita S, Higuchi S, Hitoshi Y, Yoshida Y, Tominaga M, Matsuda F, Tominaga A, Takatsu K. Characterization of the human IL-5 receptors on eosinophils. Cell Immunol 1991; 133:484–497.

31. Ingley E, Young IG. Characterization of a receptor for interleukin-5 on human eosinophils and the myeloid leukemia line HL-60. Blood 1991; 78:339–344.

32. Lopez AF, Vadas MA, Woodcock JM, Milton SE, Lewis A, Elliott MJ, Gillis D, Ireland R, Olwell E, Park LS. Interleukin-5, interleukin-3, and granulocyte–macrophage colony-stimulating factor cross-compete for binding to cell surface receptors on human eosinophils. J Biol Chem 1991; 267:24741–24747.

33. Tavernier J, Tuypens T, Plaetinck G, Verhee A, Fiers W, Devos R. Molecular basis of the membrane-anchored and two soluble isoforms of the human interleukin 5 receptor alpha subunit. Proc Natl Acad Sci USA 1992; 89:7041–7045.

34. Tuypens T, Plaetinck G, Baker E, Sutherland G, Brusselle G, Fiers W, Devos R, Tavernier J. Organization and chromosomal localization of the human interleukin 5 receptor alpha-chain gene. Eur Cytokine Netw 1992; 3:451–459.

35. Devos R, Guisez Y, Cornelis S, Vernee A, Van der Heyden J, Manneberg M, Lahm H-W, Fiers W, Tavernier J, Plaetinck G. Recombinant soluble human interleukin-5 (hIL-5) receptor molecules. Cross-linking and stoichiometry of binding to IL-5. J Biol Chem 1993; 268:6581–6587.

36. Lopez AF, Shannon MF, Hercus T, Nicola NA, Camereri B, Dottore M, Layton MJ, Eglinton L, Vadas MA. Residue 21 of human granulocyte–macrophage colony-stimulating factor is critical for biological activity and for high but not low affinity binding. EMBO J 1992; 11:909–916.

37. Scanafelt AB, Miyajima A, Kitamura T, Kastelelein RA. The amino-terminal helix of GM-CSF and IL-5 governs high affinity binding to their receptors. EMBO J 1991; 10:4105–4112.

38. Sanderson CJ, Warren DJ, Strath M. Identification of a lymphokine that stimulates eosinophil differentiation in vitro. Its relationship to IL3, and functional properties of eosinophils produced in cultures. J Exp Med 1985; 162:60–74.

39. Maxwell C, Hussian R, Nutman TB, Poindexter RW, Little MD, Schad GA, Ottesen EA. The clinical and immunologic responses of normal human volunteers to low dose hookworm (*Necator amer-icanus*) infection. Am J Trop Med Hyg 1987; 37:126–134.

40. Strath M, Sanderson CJ. Detection of eosinophil differentiation factor and its relationship to eosinophilia in *Mesocestoides corti*-infected mice. Exp Hematol 1986; 14:16–20.

41. Dexter TM, Allen TD, Lajtha LG. Conditions controlling the proliferation of haemopoietic stem cells in culture. J Cell Physiol 1977; 91:335–344.

42. Campbell HD, Sanderson CJ, Wang Y, Hort Y, Martinson ME, Tucker WQ, Stellwagen A, Strath M, Young IG. Isolation, structure and expression of cDNA and genomic clones for murine eosinophil differentiation factor. Comparison with other eosinophilopoietic lymphokines and identity with interleukin-5. Eur J Biochem 1988; 174:345–352.

43. Clutterbuck EJ, Sanderson CJ. Human eosinophil hematopoiesis studied in vitro by means of murine eosinophil differentiation factor (IL5): production of functionally active eosinophils from normal human bone marrow. Blood 1988; 71:646–651.

44. Clutterbuck EJ, Hirst EM, Sanderson CJ. Human interleukin-5 (IL-5) regulates the production of

eosinophils in human bone marrow cultures: comparison and interaction with IL-1, IL-3, IL-6, and GM-CSF. Blood 1989; 73:1504–1512.

45. Clutterbuck EJ, Sanderson CJ. The regulation of human eosinophil precursor production by cytokines: a comparison of rhIL1, rhIL3, rhIL4, rhIL6 and GM-CSF. Blood 1990; 75:1774–1779.

46. Dent LA, Strath M, Mellor AL, Sanderson CJ. Eosinophilia in transgenic mice expressing interleukin 5. J Exp Med 1990; 172:1425–1431.

47. Sher A, Coffman RL, Hieny S, Scott P, Cheever AW. Interleukin 5 is required for the blood and tissue eosinophilia but not granuloma formation induced by infection with *Schistosoma mansoni*. Proc Natl Acad Sci USA 1990; 87:61–65.

48. Yamaguchi Y, Matsui T, Kasahara T, Etoh S, Tominaga A, Takatsu K, Miura Y, Suda T. In vivo changes of hemopoietic progenitors and the expression of the interleukin 5 gene in eosinophilic mice infected with *Toxocara canis*. Exp Hematol 1990; 18:1152–1157.

49. Limaye AP, Abrams JS, Silver JE, Ottesen EA, Nutman TB. Regulation of parasite-induced eosinophilia: selectively increased interleukin 5 production in helminth-infected patients. J Exp Med 1990; 172:399–402.

50. Owen WF, Rothenberg ME, Petersen J, Weller PF, Silberstein D, Sheffer AL, Stevens RL, Soberman RJ, Austen KF. Interleukin 5 and phenotypically altered eosinophils in the blood of patients with the idiopathic hypereosinophilic syndrome. J Exp Med 1989; 170:343–348.

51. Owen WF Jr, Petersen J, Sheff DM, Folkerth RD, Anderson RJ, Corson JM, Sheffer AL, Austen KF. Hypodense eosinophils and interleukin 5 activity in the blood of patients with the eosinophilia–myalgia syndrome. Proc Natl Acad Sci USA 1990; 87:8647–8651.

52. Samoszuk M, Nansen L. Detection of interleukin-5 messenger RNA in Reed-Sternberg cells of Hodgkin's disease with eosinophilia. Blood 1990; 75:13–16.

53. Inoue C, Ichikawa A, Hotta T, Saito H. Constitutive gene expression of interleukin-5 in Kimura's disease. Br J Haematol 1990; 76:554–555.

54. Hamid Q, Azzawi M, Ying S, Moqbel R, Wardlaw AJ, Corrigan CJ, Bradley B, Durham SR, Collins JV, Jeffery PK, Quint DJ, Kay AB. Expression of mRNA for interleukin-5 in mucosal bronchial biopsies from asthma. J Clin Invest 1991; 87:1541–1546.

55. Sanderson CJ. IL5, eosinophils and disease. Blood 1992; 79:3101–3109.

56. David JR, Vadas MA, Butterworth AE, de Brito PA, Carvalho EM, David RA, Bina JC, Andrade ZA. Enhanced helminthotoxic capacity of eosinophils from patients with eosinophilia. N Engl J Med 1980; 303:1147–1152.

57. Hagan P, Wilkins HA, Blumenthal UJ, Hayes RJ, Greenwood BM. Eosinophilia and resistance to *Shistosoma haematobium* in man. Parasite Immunol 1985; 7:625–632.

58. Lopez AF, Begley CG, Williamson DJ, Warren DJ, Vadas MA, Sanderson CJ. Murine eosinophil differentiation factor. An eosinophil-specific colony stimulating factor with activity for human cells. J Exp Med 1986; 163:1085–1099.

59. Lopez AF, Sanderson CJ, Gamble JR, Campbell HD, Young IG, Vadas MA. Recombinant human interleukin 5 is a selective activator of human eosinophil function. J Exp Med 1988; 167:219–224.

60. Begley CG, Lopez AF, Nicola NA, Warren DJ, Vadas MA, Sanderson CJ, Metcalf D. Purified colony stimulating factors enhance the survival of human neutrophils and eosinophils in vitro: a rapid and sensitive microassay for colony stimulating factors. Blood 1986; 68:162–166.

61. Fijisawa T, Abu-Ghazaleh R, Kita H, Sanderson CJ, Gleich GJ. Regulatory effect of cytokines on eosinophil degranulation. J Immunol 1990; 144:642–646.

62. Tominaga A, Takaki S, Koyama N, Katoh S, Matsumoto R, Migita M, Hitoshi Y, Hosoya Y, Yamauchi S, Kanai Y, Miyazaki J-I, Usuku G, Yamamura K-I, Takatsu K. Transgenic mice expressing a B cell growth and differentiation factor gene (interleukin 5) develop eosinophilia and autoantibody production. J Exp Med 1991; 173:429–437.

63. Yamaguchi Y, Suda T, Suda J, Eguchi M, Miura Y, Harada N, Tominaga A, Takatsu K. Purified interleukin 5 supports the terminal differentiation and proliferation of murine eosinophilic precursors. J Exp Med 1988; 167:43–56.

64. Wang JM, Rambaldi A, Biondi A, Chen ZG, Sanderson CJ, Mantovani A. Recombinant human interleukin 5 is a selective eosinophil chemoattractant. Eur J Immunol 1989; 19:701–705.

65. Walsh GM, Hartnell A, Wardlaw AJ, Kurihara K, Sanderson CJ, Kay AB. IL-5 enhances the in vitro adhesion of human eosinophils, but not neutrophils, in a leucocyte integrin (CD11/18)-dependent manner. Immunology 1990; 71:258–265.

66. Walsh GM, Mermod JJ, Hartnell A, Kay AB, Wardlaw AJ. Human eosinophil, but not neutrophil, adherence to IL-1-stimulated human umbilical vascular endothelial cells is alpha 4 beta 1 (very late antigen-4) dependent. J Immunol 1991; 146:3419–3423.

67. Dvorak AM, Saito H, Estrella P, Kissell S, Arai N, Ishizaka T. Ultrastructure of eosinophils and basophils stimulated to develop in human cord blood mononuclear cell cultures containing recombinant human interleukin-5 or interleukin-3. Lab Invest 1989; 61:116–132.

68. Bischoff SC, Brunner T, De Weck AL, Dahinden CA. Interleukin 5 modifies histamine release and leukotriene generation by human basophils in response to diverse agonists. J Exp Med 1990; 172:1577–1582.

69. Hirai K, Yamaguchi M, Misaki Y, Takaishi T, Ohta K, Morita Y, Ito K, Miyamoto T. Enhancement of human basophil histamine release by interleukin 5. J Exp Med 1990; 172:1525–1528.

70. Clutterbuck E, Shields JG, Gordon J, Smith SH, Boyd A, Callard RE, Campbell HD, Young IG, Sanderson CJ. Recombinant human interleukin-5 is an eosinophil differentiation factor but has no activity in standard human B cell growth factor assays. Eur J Immunol 1987; 17:1743–1750.

71. Bende RJ, Jochems GJ, Frame TH, Klein MR, Van Eijk RVW, Van Lier RAW, Zeijlemaker WP. Effects of IL-4, IL-5, and IL-6 on growth and immunoglobulin production of Epstein-Barr virus-infected human B cells. Cell Immunol 1992; 143:310–323.

72. Takatsu K, Tominaga A, Harada N, Mita S, Matsumoto M, Takahashi T, Kikuchi Y, Yamaguchi N. T cell-replacing factor (TRF)/interleukin 5 (IL-5): molecular and functional properties. Immunol Rev 1988; 102:107–135.

73. Swain SL, McKenzie DT, Dutton RW, Tonkonogy SL, English M. The role of IL4 and IL5: characterization of a distinct helper T cell subset that makes IL4 and IL5 (Th2) and requires priming before induction of lymphokine secretion. Immunol Rev 1988; 102:77–105.

74. Vitetta ES, Brooks K, Chen Y, Isakson P, Jones S, Layton J, Mishra GC, Pure E, Weiss E, Word C, Yuan D, Tucker P, Uhr JW, Krammer PH. T cell drived lymphokines that induce IgM and IgG secretion in activated murine B cells. Immunol Rev 1984; 78:137–184.

75. O'Garra A, Warren DJ, Holman M, Popham AM, Sanderson CJ, Klaus GGB. Interleukin 4 (B-cell growth factor II/eosinophil differentiation factor) is a mitogen and differentiation factor for pre-activated murine B lymphocytes. Proc Natl Acad Sci USA 1986; 83:5228–5232.

76. Alderson MR, Pike BL, Harada N, Tominaga A, Takatsu K, Nossal GJ. Recombinant T cell replacing factor (interleukin 5) acts with antigen to promote the growth and differentiation of single hapten-specific B lymphocytes. J Immunol 1987; 139:2656–2660.

77. McHeyzer Williams MG. Combinations of interleukins 2, 4 and 5 regulate the secretion of murine immunoglobulin isotypes. Eur J Immunol 1989; 19:2025–2030.

78. Purkerson JM, Newberg M, Wise G, Lynch KR, Isakson PC. Interleukin 5 and interleukin 2 cooperate with interleukin 4 to induce IgG1 secretion from anti-Ig-treated B cells. J Exp Med 1988; 168:1175–1180.

79. Rasmussen R, Takatsu K, Harada N, Takahashi T, Bottomly K. T cell-dependent hapten-specific and polyclonal B cell responses require release of interleukin 5. J Immunol 1988; 140:705–712.

80. Howard M, Nakanishi K, Paul WE. B cell growth and differentiation factors. Immunol Rev 1984; 78:185–210.

81. Umland SP, Go NF, Cupp JE, Howard M. Responses of B cells from autoimmune mice to IL-5. J Immunol 1989; 142:1528–1535.

82. Wetzel GD. Interleukin 5 regulation of peritoneal Ly-1 B lymphocyte proliferation, differentiation and autoantibody secretion. Eur J Immunol 1989; 19:1701–1707.

83. Yokota T, Coffman RL, Hagiwara H, et al. Isolation and characterization of lymphokine cDNA clones

encoding mouse and human IgA-enhancing factor and eosinophil colony-stimulating factor activities: relationship to interleukin. Proc Natl Acad Sci USA 1987; 84:7388–7392.

84. Bond MW, Shrader B, Mosmann TR, Coffman RL. A mouse T cell product that preferentially enhances IgA production. II. Physicochemical characterization. J Immunol 1987; 139:3691–3696.

85. Lebman DA, Coffman RL. Interleukin 4 causes isotype switching to IgE in T cell-stimulated clonal B cell cultures. J Exp Med 1988; 168:853–862.

86. Murray PD, McKenzie DT, Swain SL, Kagnoff MF. Interleukin 5 and interleukin 4 produced by Peyer's patch T cells selectively enhance immunoglobulin A expression. J Immunol 1987; 139:2669–2674.

87. Kunimoto DY, Harriman GR, Strober W. Regulation of IgA differentiation in CH12LX B cells by lymphokines. IL-4 induces membrane IgM-positive CH12LX cells to express membrane IgA and IL-5 induces membrane IgA-positive CH12LX cells to secrete IgA. J Immunol 1988; 141:713–720.

88. Harriman GR, Kunimoto DY, Elliot JF, Paetkau V, Strober W. The role of IL-5 in IgA B cell differentiation. J Immunol 1988; 140:3033–3039.

89. Beagley KW, Eldridge JH, Lee F, Kiyono H, Everson MP, Koopman WJ, Hirano T, Kishimoto T, McGhee JR. Interleukins and IgA synthesis. Human and murine interleukin 6 induce high rate IgA secretion in IgA-committed B cells. J Exp Med 1989; 169:2133–2148.

90. Kunimoto DY, Nordan RP, Strober W. IL-6 is a potent cofactor of IL-1 in IgM synthesis and of IL-5 in IgA synthesis. J Immunol 1989; 143:2230–2235.

91. Sonoda E, Matsumoto R, Hitoshi Y, Ishii T, Sugimoto M, Araki S, Tominaga A, Yamaguchi N, Takatsu K. Transforming growth factor beta induces IgA production and acts additively with interleukin 5 for IgA production. J Exp Med 1989; 170:1415–1420.

92. Webb CF, Das C, Coffman RL, Tucker PW. Induction of immunoglobulin μ mRNA in a B cell transfectant stimulated with interleukin-5 and a T-dependent antigen. J Immunol 1989; 143:3934–3939.

93. Matsui K, Nakanishi K, Cohen DI, Hada T, Furuyama J, Hamaoka T, Higashino K. B cell response pathways regulated by IL-5 and IL-2. Secretory microH chain-mRNA and J chain mRNA expression are separately controlled events. J Immunol 1989; 142:2918–2923.

94. Loughnan MS, Sanderson CJ, Nossal GJ. Soluble interleukin 2 receptors are released from the cell surface of normal murine B lymphocytes stimulated with interleukin 5. Proc Natl Acad Sci USA 1988; 85:3115–3119.

95. Nakanishi K, Yoshimoto T, Katoh Y, et al. Both B151-T cell replacing factor 1 and IL-5 regulate Ig secretion and IL-2 receptor expression on a cloned B lymphoma line. J Immunol 1988; 140:1168–1174.

96. Sanderson CJ, Strath M, Mudway I, Dent LA. Transgenic experiments with interleukin-5. In: Gleich GJ, Kay AB, eds. Eosinophils: Immunological and Clinical Aspects. 17th ed. New York: Marcel Dekker, 1994:335–351.

97. Finkelman FD, Holmes J, Katona IM, Urban JF Jr, Beckmann MP, Park LS, Schooley KA, Coffman RL, Mosmann TR, Paul WE. Lymphokine control of in vivo immunoglobulin isotype selection. Annu Rev Immunol 1990; 8:303–333.

98. Plaut M, Pierce JH, Watson CJ, Hanley Hyde J, Nordan RP, Paul WE. Mast cell lines produce lymphokines in response to cross-linkage of Fc epsilon Rl or to calcium ionophores. Nature 1989; 339:64–67.

99. Burd PR, Rogers HW, Gordon JR, Martin CA, Jayaraman S, Wilson SD, Dvorak AM, Galli SJ, Dorf ME. Interleukin 3-dependent and -independent mast cells stimulated with IgE and antigen express multiple cytokines. J Exp Med 1989; 170:245–257.

100. Paul CC, Keller JR, Armpriester JM, Baumann MA. Epstein-Barr virus transformed B lymphocytes produce interleukin-5. Blood 1990; 75:1400–1403.

101. Broide DH, Paine MM, Firestein GS. Eosinophils express interleukin 5 and granulocyte macrophage-colony-stimulating factor mRNA at sites of allergic inflammation in asthmatics. J Clin Invest 1992; 90:1414–1424.

19

Macrophage Inflammatory Protein-1α

Mark A. Plumb and Eric Wright
Radiation and Genome Stability Unit, Medical Research Council, Harwell, Didcot, Oxfordshire, England

I. INTRODUCTION

Blood cells have finite life spans and are continuously produced by the proliferation and differentiation of precursor cells located primarily in the bone marrow. These precursor cells are derived from a common, self-maintaining population of pluripotential hematopoietic stem cells that is established during embryogenesis and functions for the lifetime of the individual. Within hematopoietic tissues, a complex network of inhibitory and stimulatory cytokines and a variety of cellular interactions act together to regulate the self-renewal and commitment to differentiation of the stem cells and the production of blood cells from lineage-restricted progenitor cells. We have a long-standing interest in the mechanisms controlling the proliferation of the stem cell compartment because aberrant proliferation is characteristic of several leukemias, and regulated stem cell proliferation is essential for bone marrow regeneration after radio- or chemotherapy and bone marrow transplantation.

Bone marrow transplantation experiments and in vitro clonogenic assays for colony-forming cells (CFC, also known as colony-forming units; CFU) have shown that the stem cell compartment is heterogeneous for self-renewal, differentiation potential, and long-term hematopoietic repopulation capacity. The most ancestral and, therefore, most primitive hematopoietic stem cell is identified as the long-term repopulating cell in transplantations and eventually gives rise to the more mature multipotent CFC. These more mature CFC are capable of multilineage differentiation in vitro (multi-CFC/CFC-mix, or CFU-GEMM) and in vivo (CFU-S), and are largely responsible for short-term marrow repopulation (for review see Ref. 1). The nature of proliferation control mechanisms acting on the stem cell compartment has been successfully investigated using various quantitative clonogenic assays. In normal steady-state bone marrow, murine stem cells assayed in vivo (as spleen colony-forming units; CFU-S) or in vitro (as multi-CFC or CFU-A) and human stem cells assayed in vitro (as CFU-A or CFU-GEMM) are proliferatively quiescent owing to specific inhibition of proliferation. When bone marrow is partially ablated by treatment with cytotoxic drugs or ionizing radiation, stem cells are stimulated to proliferate to regenerate the hematopoietic system as a consequence of specific stimulatory signals that are able to override the actions of inhibitory factors.

In this chapter we describe current advances in the cell and molecular biology of a pleiotropic cytokine that inhibits stem cell proliferation (SCI/MIP-1α).

II. MIP-1α AND THE REGULATION OF HEMATOPOIESIS

An inhibitor of CFU-S cell proliferation has been detected in normal bone marrow, and a stimulator of CFU-S cell proliferation detected in regenerating bone marrow (2–8). The inhibitor and stimulator activities are produced by distinct subpopulations of bone marrow macrophages (3,6,7). The identification of a transformed murine macrophage cell line (J774.2), which produces a reversible inhibitor of CFU-A cell proliferation (SCI; stem cell inhibitor), led to the copurification of an inhibitory activity composed of two small approximately 8-kDa polypeptides (9–11). The polypeptides turned out to be identical with two previously described related cytokines. Macrophage inflammatory protein-1α (MIP-1α) and MIP-1β, which are 59.8% identical at the amino acid level (11–13). MIP-1α was subsequently shown to be the active component and to be antigenically and functionally identical with the CFU-S inhibitory activity obtained from bone marrow cells (11,12).

MIP-1α inhibits the proliferation of the more mature cells in the stem cell compartment (CFU-S, CFU-A, CFU-GEMM), but is inactive on the less mature long-term reconstituting stem cells (11,14–16). MIP-1α, in the presence of an as yet unidentified accessory factor, will also inhibit the proliferation of clonogenic epidermal keratinocytes (17), suggesting that MIP-1α may be a more general inhibitor of stem cell proliferation. Although MIP-1β did not affect CFU-A cell proliferation at the concentrations (25 ng/ml) effective with MIP-1α (11), higher concentrations (0.2–1 μg/ml) of MIP-1β will inhibit CFU-A proliferation (18). However, MIP-1β will abrogate the ability of MIP-1α to suppress CFU-GEMM cell growth (19), consistent with the observation that they bind the same receptor, but a higher concentration (~25-fold; 20) of MIP-1β than MIP-1α is required to elicit an intracellular Ca^{2+} flux in transfected cells (see later). Thus, at low concentrations MIP-1β appears to be an antagonist of MIP-1α activity in the inhibition of CFU-S/CFU-A cell proliferation.

Similar to many other cytokines, MIP-1α (and MIP-1β) has pleiotropic effects, and there is evidence that it will stimulate myeloid progenitor (CFU-GM) colony formation in growth factor-limiting conditions (18,21).

The murine (Mu)MIP-1α gene has been knocked out by homologous recombination in embryonic stem (ES) cells, and homozygous transgenic mice were generated. Preliminary data suggests the mice are normal and fully viable (Cook D, Smithies O, personal communication). Presumably, the redundancy of cytokine function precludes the observation of gross abnormalities in the transgenic mouse, and because MIP-1α has many biological functions associated with a perturbation of the hematopoietic or immune systems, the gene knockout mice will be most informative when these systems are experimentally perturbed in vivo.

III. MIP-1α IS A MEMBER OF A LARGE FAMILY OF RELATED CYTOKINES

The MIP-1α belongs to a large family of cytokines that includes platelet factor 4 (PF4), RANTES (regulated on activation, normal T-expressed and secreted), and interleukin-8 (Il-8), that are defined on the basis of limited sequence homology, and in particular four conserved cysteine residues (22,23). This family can be further subdivided according to the positions of the first two cysteine residues; one branch, which includes MIP-1α, MIP-1β, MCAF (monocyte chemotactic and activating factor) also known as MCP-1 (human monocyte chemotactic protein 1), JE (murine homologue of MCP-1), TCA-3 (t-cell–activated-3) and hSISε/I-309 (human homologue of TCA-3)), contains adjacent cysteines (C-C) and has been termed SIS/RANTES (small, inducible, and secretable/RANTES) family (22,23); the second branch, which includes PF4, MIP-2 and IL-8, has the cysteine residues separated by one amino acid (C-X-C; 22,23). Collectively, the two branches have been termed both the PF4 superfamily, and more recently, termed "chemokines," which define the bipartite superfamily of soluble proteins that have been implicated in a wide range of acute and chronic inflammatory processes and other immuno-regulatory functions (20,24).

IV. MIP-1α AND THE IMMUNE RESPONSE

A role for MIP-1α in the immune response is implied by a number of observations:

A. T Cells

Gene expression of MIP-1α and MIP-1β is induced during T cell activation (25–27). The MIP-1α and the MIP-1β cytokines induce the adhesion of activated T cells to endothelium (28,29), a prerequisite for activated T-cell migration from the blood into tissues, which probably involves VCAM-1 (28,29). Human MIP-1α is chemotactic for CD3-activated CD8$^+$ T cells, whereas huMIP-1β is chemotactic for CD3-activated CD4$^+$ cells, and both were chemotactic for activated naive and memory T cells (29).

MIP-1α is also able to modulate the adhesion of other mature hematopoietic cells that are involved in the immune and inflammatory responses, because it will induce a rapid (30-min) and transient increase in circulating mature white blood cells (neutrophil, monocyte, and eosinophil), an effect that is enhanced by costimulation with G-CSF (16,30). The kinetics of this response suggest MIP-1α may be causing the release of preexisting mature white cells from the bone marrow into the blood. Similarly, MIP-1α increases the adhesion of committed human progenitor cell (BFU-E and CFU-GM) adhesion to marrow stroma, and it has been proposed this may result in the observed contact-mediated growth inhibitory effects of stroma on committed progenitors (31).

B. Macrophages and Mast Cells

Transformed human myelomonocytic cell lines (U937, THP1) and primary monocytes do not express basal levels of MIP-1α until they have been activated (25,26,32). MIP-1α modulates peritoneal macrophage function, for it stimulates the secretion of tumor necrosis factor (TNF), IL-1α, and IL-6, without inducing an oxidative burst (33). MIP-1α and MIP-1β are coordinately expressed by macrophages (34), and a complex of MIP-1α and MIP-1β (MIP-1) will enhance antibody-independent macrophage cytotoxicity and induce mature macrophage proliferation (33). However, the ability of MIP-1α to stimulate TNF, IL-1α, and IL-6 production in macrophages was blocked by an eightfold excess of MIP-1β (33), indicating again that MIP-1β is a MIP-1α antagonist in certain biological situations.

Basal levels of MIP-1α mRNA have been detected in cultured murine mast cells, and are induced during mast cell activation by the cross-linking of a multivalent antigen to IgE bound to the cell surface FcεR1 receptors (35). Mast cells have been implicated in a wide variety of biological responses, including protective and pathological immune responses, and mast cell activation induces degranulation, the release of histamine, heparin, proteoglycans, and proteases (for review see Ref. 36).

V. MIP-1α AND INFLAMMATION

The MIP-1α and MIP-1β mRNA levels and protein secretion in macrophages is superinduced by endotoxin stimulation (12,34,37–39). The complex of the two proteins (MIP-1) or purified MIP-1α, is chemotactic for neutrophils and monocytes, will act as a prostaglandin-independent pyrogen, and MIP-1α gene-expression is down-regulated by anti-inflammatory agents, such as prostaglandin E2 (12,37,40,41).

VI. THE MIP-1α RECEPTOR

The binding of MIP-1α to its receptor on human leukocytes elicits a transient elevation of intracellular Ca^{2+} levels, and the human MIP-1α receptor cDNA has been cloned in a functional expression vector that conferred a calcium response in *Xenopus* oocytes and human kidney 293 cells in response to the MIP-1α ligand (20,24). Sequence comparison with another known PF4 superfamily member receptor, the IL-8 receptor, suggests the MIP-1α receptor is a seven-trans-membrane-spanning receptor coupled to heterotrimeric G proteins. The receptor has a 33% sequence identity with the US28 binds open-reading frame of human cytomegalovirus, the protein encoded by US28 binds MIP-1α, and as cytomegalovirus infects myeloid and lymphoid cells, it has been suggested that the virus has hijacked the human gene which, when expressed in infected cells, would alter the cellular response to chemokines. The C-C chemokine receptor (C-C CKR1) is encoded by a single-copy gene located on chromosome 3p21 (20,24). Although the signal transduction pathway activated by MIP-1α binding to its receptor(s) has not been fully elucidated, MIP-1α will inhibit the activation of raf-1 kinase by GM-CSF and *steel* factor (SLF) in MO7e cells (42). Raf-1 is a component of the mitogen-activated protein kinase (MAP-kinase)-signaling cascade, so the inhibition of raf-1 kinase activity by MIP-1α may explain its growth-inhibitory properties.

Receptor-binding studies in human and murine cells indicate that several members of the C-C SIS/RANTES branch of cytokines (but not CXC) can bind the same receptor. For example, human and mouse MIP-1α and MIP-1β, RANTES, and MCAF will compete for MIP-1α binding to the receptor on transfected human kidney 393 cells and on untransfected human K562 cells and murine FDCPmix cells (20,24,43). The displacement of bound MIP-1α by MCAF and RANTES is much weaker if FDCPmix cells than in K562 cells, whereas MIP-1α is displaced equivalently by human and murine MIP-1α and MIP-1β in both FDCPmix and K562 cells (43). As MIP-1α and MIP-1β are implicated in the regulation of hematopoietic stem cell proliferation, but MCAF and RANTES are not, it has been suggested that two receptors exist, only one of which is implicated in stem cell regulation (24,43).

Although the affinity of binding of MIP-1α and MIP-1β to C-C CKR1-transfected 393 cells was essentially equivalent, MIP-1β binding did not elicit a Ca^{2+} flux at the concentrations (10 nM) that are effective for MIP-1α. A 25-fold higher concentration of MIP-1β (250 nM) did elicit a Ca^{2+} flux, but this was only 20% of the levels obtained with MIP-1α, and indicates that, at low concentrations, MIP-1β is an antagonist of MIP-1α-induced intracellular Ca^{2+} mobilization in 393 cells (20).

FDCPmix cells are an established primary murine multipotent hematopoietic line that responds to MIP-1α in vitro (43). Although no report of MIP-1α studies on murine CFUmix cells has come to our attention, the equivalent cells in humans (CFU-GEMM) do respond to MIP-1α (1). FDCPmix cells possess 18,000 MIP-1α receptors per cell, whereas Scatchard analysis of total murine bone marrow revealed a low-affinity site (K_d 16.5 nM, 40,000 receptors per cell) and a second site (560 receptors per cell) that had the same high-affinity binding (K_d

150 pM) as detected in FDCPmix cells (43). This again implies that there may be more than one MIP-1α receptor, which is expressed in a cell-specific manner and which is able to bind distinct subgroups of the SIS/RANTES superfamily. Furthermore, as MIP-1α and MIP-1β have antagonistic effects in certain biological situations (see foregoing), this may be achieved through competitive binding to the same receptor. However, in activated T-cell chemotaxis, as MIP-1α is specific for $CD8^+$ T cells, whereas MIP-1β is specific for $CD4^+$ T cells (28,29), there may also be distinct cell-specific receptors for MIP-1α and MIP-1β.

VII. THE CONTROL OF MIP-1α GENE EXPRESSION

Given that MIP-1α gene knockout mice appear to be normal and fully viable, and that several members of the C-C chemokines appear to bind the same receptor (resulting in redundant or antagonistic activities), it is difficult to separate the roles of the MIP-1α cytokine in hematopoiesis, the immune response, and inflammation. One approach is to attempt to define which cells express MIP-1α, and how that expression is regulated so that biological effects of MIP-1α in different circumstances can be placed in a physiological perspective. For example, bone marrow macrophages are presumably the source of that MIP-1α protein that inhibits the proliferation of the hematopoietic stem cell. It is tempting to speculate that this represents a negative-feedback loop in the control of hematopoiesis, with the number of mature end-cells (macrophages) modulating the rate of stem cell self-renewal–differentiation; thereby modulating the rate of production of progenitor and mature hematopoietic cells.

A. Cell-Specific and Inducible MIP-1α Gene Expression

The MIP-1α gene expression appears to be restricted to a subset of cells from the hematopoietic lineage. Basal levels of MIP-1α (and MIP-1β) mRNA have since been detected in macrophage cell lines, normal and cultured bone marrow, cultured peritoneal macrophages, mast cells, and Langerhans cells (Fig. 1A; 12,13,17,34,35,38,39,44). In macrophages, MIP-1α and MIP-1β mRNA levels are coordinately, rapidly, and transiently, induced by LPS, cycloheximide, serum, and phorbol esters (TPA), indicating that similar to other members of the SIS/RANTES family of genes, MIP-1α is an immediate early-response gene (34,38,45–47). Interestingly, transforming growth factor-beta (TGF-β), which is also an inhibitor of stem cell proliferation, down-regulates MIP-1α and MIP-1β mRNA levels in macrophage cells (48). Because bone marrow macrophages are the source of the stem cell inhibitor in vivo (2–8), this suggests that low basal levels of MIP-1α gene expression in bone marrow macrophages are sufficient for the control of stem cell proliferation in normal bone marrow, whereas the inducible expression observed in endotoxin- or serum-stimulated macrophages, and activated B and T cells (see later discussion) are associated with the immune response and inflammation in other tissues. Basal levels of MIP-1α mRNA have not been detected in murine erythroleukemia (MEL), fibroblast (NIH3T3, STO, mouse dermal), mouse epidermal keratinocyte, or B16 mouse melanoma cells (see Fig. 1; 17,34,38)

Phytohemagglutinin (PHA) and TPA treatment of primary (tonsillar and PBL) or transformed (Jurkat) T cells rapidly induces MIP-1α (and MIP-1β) mRNA levels (25–27). Similarly, human primary blood monocytes and myelomonocytic cells lines (HL60, THP1, and U937) do not express MIP-1α mRNA until they have been activated by TPA or IL-7, and can then be superinduced by LPS (25,26,32,49). However, it has been reported that human MIP-1α mRNA is detectable in PMA-treated primary fibroblast and human glioma (U105MG) cells (25,32). As shown in Figure 1B, low-stringency hybridization of MuMIP-1α cDNA antisense riboprobe will

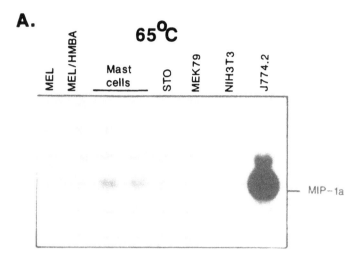

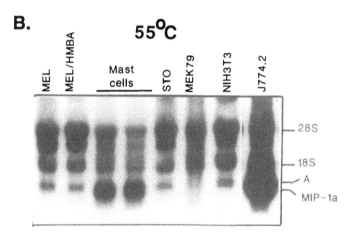

Figure 1 Northern blot analysis of total cellular RNA prepared from mouse primary and transformed cell lines. MEL, mouse erythroleukemia before (MEL) and after (MEL/HMBA) the induction of terminal differentiation; STO and NIH3T3, fibroblast cell lines; MEK79, mouse epidermal keratinocytes; J774, macrophage cell line. RNA was resolved by denaturing agarose gel electrophoresis, transferred to a nylon membrane and hybridized at either (A) 65°C or (B) 55°C to radiolabeled MIP-1α cDNA antisense riboprobe in 50% formamide as described (38).

detect a ubiquitously expressed RNA in Northern blots that almost comigrates with the MIP-1α mRNA detected at high-stringency of hybridization (see Fig. 1A). Thus, low-stringency hybridization of human MIP-1α cDNA to filter immobilized RNA could potentially result in cross-hybridization to other related RNA sequences. It is, therefore, unclear whether the detection of MIP-1α in human nonhematopoietic cells is a real difference between human and mouse cells.

B. MIP-1α/β Gene Structure and Organization

The human and murine MIP-1α and MIP-1β genes have been isolated and sequenced by several laboratories. The genes are comparatively small (~2000 bp), contain two introns and the first exon (~155 nt) encodes the 5′-untranslated region (5′UTR) and the signal peptide (amino acids 1–23) (32,50–53).

Murine MIP-1α and MIP-1β are encoded by single-copy genes (34,53–55) and are clustered on chromosome 11 within 5 cM of the *p53* (proximal) and *Hox-2* (distal) loci (56). Although the CIS/RANTES MIP-1α, MIP-1β, TCA3, and JE genes are clustered on chromosome 11, the precise molecular organization of the genes has yet to be established. However, as the MIP-1α and MIP-1β genes are linked, their intron and exon organization is similar (53–55), and they encode related peptides that exhibit 59.8% identity at the amino acid level (56.7% at the nucleotide level; 13), the two genes probably arose by gene duplication and divergence.

Three human MIP-1α genes, and two human MIP-1β genes have been identified (32,50,52,57). One human MIP-1α gene (*LD78α/AT464.1/GOS19-1*) is located within 14 kbp of one of the MIP-1β genes (*AT744.1*), and the two genes are transcribed from opposite strands of the DNA (50,52,57). A second MIP-1α gene (*LD78β,AT464.2*) appears to be linked to a second MIP-1β gene (*AT744.2*), suggesting that the *LD78β* and *AT744.2* genes arose by duplication of the *LD78α* and *AT744.1* genes as one unit, but although the *LD78β* gene is apparently functional, it contains an Alu-1 repeat sequence inserted in the promoter about 330-bp upstream from the CAP site (32,50,52,57). The third MIP-1α gene (*LD78γ,AT744.3*) is a 5′-truncated pseudogene, and as its sequence is more related to *LD78β* (99.7%) than LD78α (95.0%), it suggests it arose by a second duplication of *LD78β*, but without the stable duplication (or subsequent deletion) of both the 5′ sequence of the *LD78β* gene and the linked *AT744.2* MIP-1β gene (57). The human MIP-1α and MIP-1β genes, and other members of the SIS/RANTES family (MCP-1, I-309) are clustered on human chromosome 17 (q21.1-q21.3), and the *LD78β*, *LD78γ*, and *AT44.2* genes are variably amplified within the population, with the *LD78γ* pseudogene being most polymorphic (50,57,58).

VIII. TRANSCRIPTIONAL REGULATION OF THE MIP-1α GENE

The transcriptional regulation of the MIP-1α gene is being investigated in an attempt to define the molecular mechanisms that contribute to the observed cell-specific and inducible expression of the endogenous gene. The ability of promoter sequences to drive the transcription of a linked reporter gene has been assessed in transfected cells, and cis-acting sequences further defined by in vitro protein binding studies and site-directed mutagenesis.

A. Transfection Analyses of the MIP-1α Promoter

A minimal HuMIP-1α gene promoter (–220 bp) was sufficient to confer a maximal transcriptional response to PMA or PHA plus PMA in K562 and Jurkat cell lines (59). Similarly, a minimal MuMIP-1α gene promoter (~–200 bp) is essentially inactive in nonexpressing murine B16 (melanoma), MEL (erythroleukemia), and STO (embryonic fibroblast) cells, but is sufficient to confer basal promoter activity and a transcriptional response to LPS, serum, and PMA in macrophage (RAW264.7 and Wehi) cells (38,53). Interestingly, the cell-specificity and inducibility of the MIP-1β proximal promoter in the same transfected cells is very similar, with the exception that the MuMIP-β promoter is active and LPS-inducible in B16 cells (34,53).

The human (LD78α) and murine MIP-1α proximal promoters (–1 to –231 bp) exhibit 78% sequence identity (38,59). In vitro murine nuclear protein-binding sites in the MuMIP-1α promoter (38), and the sequences of the sites compared with consensus or known protein-binding sites in the promoters of other cytokine genes, are shown in Figure 2. Protein binding to the

A.

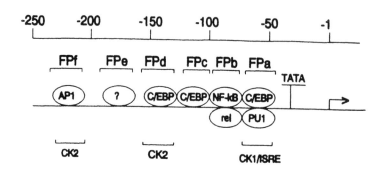

B. MuFPa

```
    C/EBP    : ATTGCGCAAT
    MuFPa    : gaTGaGgAAa
    HuGCSF   : ATTGCGCAAT
    MuGCSF   : tTTGtGaAAT

PU.1/SpiB   : AAAGAGGAAC
    MuFPa   : gAtGAGGAAa
```

C.   MuFPb

```
    KBpd     : GGGAATTCCC
    MuFPb    : aaattTTCCC
    IFNγ     : GaattTTCCa
    IL-6     : GGattTTCCC
```

D. MuFPc and MuFPd

```
    AGP/EBP  : ATCGTGTCATAA
    MuFPc    : ATgGTGTCATgA
    MuFPd    : ccCcTGTggTcA
```

E.   MuFPf

```
    AP1      : TGAG/CTCA
    MuFPf    : TGAG aCt
    TGFβ1    : TGAG aCg
    TGFβ2    : TGAG aCt
    MuMIPβ   : TtA  CTCA
```

Figure 2 (A) Schematic representation of the murine MIP-1α gene promoter transcription factor-binding sites, and relative positions of the CK1, CK2, and ISRE consensus sequences. (B–E) Sequence comparison of the murine transcription factor-binding sites (MuFPa–MuFPf) and known transcription factor-binding sites or consensus recognition sequences.

HuMIP-1α promoter has been analyzed with human cell nuclear extracts (59), but a methodical comparison of the protein binding to the two MIP-1α promoters has yet to be undertaken.

B.   EMSA Analyses of the MuMIP-1α Gene Promoter Protein-Binding Sites

Electrophoretic mobility shift assays (EMSAs) have further defined the nuclear proteins interacting with the MuMIP-1α proximal promoter.

Site muFPb (see Figs. 2A and C) binds two GTP-sensitive nuclear complexes, one of which (NFκB) is LPS-inducible in murine macrophages, and the other is a constitutively expressed member of the c-*rel* family of nuclear protooncogenes (38,60). The *rel*-related nucleoprotein complex is not detected in MEL or STO fibroblast cells, suggesting that its binding to site muFPb contributes to the cell-specific basal levels of transcription in macrophages, whereas LPS-induced NFκB binding to the same site is involved in the transcriptional response to LPS.

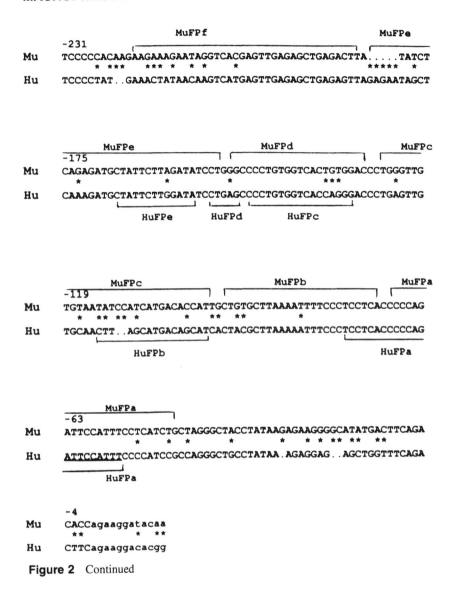

Figure 2 Continued

The anticoding strand of site *muFPa* contains an A plus G-rich sequence (GATGAGGAAA; see Fig. 2B) that resembles the recognition sequence for the c-*ets* family of nuclear protooncogenes. The cell-specificity of the nuclear-binding activity, and cross-linking experiments indicate that the protein bound to this site is PU.1 or SpiB (38). PU.1 gene expression is restricted to macrophages, B cells, and SFFV-transformed mouse erythroleukemia cells, whereas Spi-B expression is restricted to hematopoietic cells (61,62). Because the MuMIP-1α gene promoter is inactive in SFFV-transformed MEL cells (38), PU.1/Spi-B may be necessary, but is not sufficient, for cell-specific promoter function.

Sites *muFPa, muFPc,* and *muFPd* all bind at least three C/EBP-related nucleoprotein complexes, which are detectable in expressing and nonexpressing cell nuclear extracts (38). There are several C/EBP family members, each of which can homodimerize and heterodimerize with other family members, but C/EBPβ and C/EBPδ are expressed in macrophage cell lines, are LPS-inducible, and are a similar molecular weight (29–31 kDa) to the RAW264.7 cell nuclear proteins which are cross-linked to the MuMIP-1α promoter sites (38,63–66).

The C/EBPβ gene encodes three different C/EBPβ proteins, which can homo- and hetero-dimerize (C/EBPβ1, β2, and β3; 64,67). C/EBPβ1, C/EBPβ2, and C/EBPδ, all are transcriptional activators, whereas the smaller C/EBPβ3 lacks the COOH-terminal transcriptional activation domain and is a repressor of transcription (68). Furthermore, a C/EBP-related protein (C/EBP homologous protein, CHOP) is a dominant-negative inhibitor of C/EBP when dimerized with other members of the C/EBP family (69).

Thus, basal and inducible transcription of the MIP-1α promoter may be mediated in part by the competition for binding to the same sites by C/EBP complexes containing positive or negative transcriptional activities, and by unrelated (PU.1/SpiB) transcriptional regulators, and by dominant-negative binding to the transcription factors themselves.

Site muFPf is bound by API-related nuclear proteins, whereas the protein(s) binding to site muFPe have yet to be identified (see Fig. 2A and 2E). AP1 is a heterodimeric complex of two members of the c-*fos* and c-*jun* nuclear protooncogene families (70). AP1-related DNA-binding activity is induced by LPS stimulation of murine macrophage cells and is correlated with an increase in cellular c-*fos* mRNA and c-*jun* protein levels (34,71). AP-1 DNA-binding sites are also associated with the TPA-response element (TRE), and because MIP-1α and MIP-1β gene expression is induced in T cells, PBLs, and macrophages by TPA (26,39,51), it is likely that AP-1 activation contributes to this stimulation.

C. MIP-1α Promoter Consensus Sequences

Several sequence motifs in the MIP-1a promoters are conserved in the promoters of many cytokine genes, some of which are shown in Figure 3.

Two "cytokine-specific sequences" (cytokine 1 [CK-1] and cytokine 2 [CK-2] motifs; 72,73) are conserved in the promoters for the MIP-1α genes (see Fig. 2A and B), but as the motifs are present in sites (FPa, FPc, FPd, and FPf) that bind PU.1/SpiB, AP-1, or C/EBP-related proteins (38), specific nuclear protein binding to the CK-1 and CK-2 motifs has yet to be demonstrated.

The MuFPa sequence also resemble an interferon-stimulated response element (ISRE); (see Fig. 2C; 74–76). Preliminary results suggest that MIP-1α mRNA levels are induced by IFN-α/β, but not IFN-γ in murine macrophage cells (Grove M, unpublished results; 39), and IFN-γ weakly down-regulates HuMIP-1α gene expression in peripheral blood monocytes (25).

Several interferon regulatory factors (IRFs) have been identified, but interferon-stimulated gene factor 3 (ISGF3) is the primary transcription factor induced by IFN-α (74–76). ISGF3 is a complex of four proteins, but can be separated into two activities: ISGF3γ and ISGF3α. Both accumulate and complex in the nucleus following IFN-α treatment, and form the high-affinity ISRE-binding ISGF3 complex (75–77). The analysis of binding of IFN-α/β–inducible transcription factors to the MIP-1α promoter putative ISRE sequence has yet to be undertaken, for C/EBP- and PU.1/SpiB-related factors bind to the same site.

D. MIP-1β Gene Transcription

Although the MIP-1α and MIP-1β genes presumably arose by gene duplication and divergence, and both genes appear to be coordinately expressed in the same subset of cells, the promoter sequences of the two genes are not conserved.

The MuMIP-1β proximal promoter contains an ATF/CREB DNA-binding site

A.

```
CK1 consensus   :   GAGATTCCAC
MuMIP-1α        :   cAGATTCCAt   (MuFPa)
HuMIP-1α        :   cAGATTCCAt   (HuFPa)
MuMIP-1β        :   GAGATTCCgT
HuMIP-1β        :   GAaATTCCAC
HuGM-CSF        :   GAGATTCCAC
MuGM-CSF        :   GAGATTCCAC
HuIL-3          :   GAGgTTCCAt
MuIL-3          :   GAGgTTCCAt
HuIL-2          :   GgGATTTCAC
MuIL-2          :   GgGATTTCAC
HuG-CSF         :   GAGATTCCAC
MuG-CSF         :   GAGATTCCcC
```

B.

```
CK-2 consensus  :   TGTGGTCAC
MuMIP-1α        :   TGTGGTCAC    (MuFPd)
MuMIP-1α        :   ataGGTCAC    (MuFPf)
HuMIP-1α        :   TGTGGTCAC    (HuFPc)
MuMIP-1β        :   TGatGTCAC
HuMIP-1β        :   TGatGaCAt
HuGM-CSF        :   TGTGGTCAC
MuGM-CSF        :   TGTGGTCAC
IFNγ            :   TcT GTCAC
```

C.

```
ISRE consensus  :   CAGTTTCnnTTTCC
MuMIP-1α        :   CAGaTTCCATTTCC   (MuFPa)
HuMIP-1α        :   CAGaTTCCATTTCC   (HuFPa)
MuH-2Lᵈ         :   CAGTTTCACTTct
MuH-2Kᵇ         :   CAGTTTCACTTctg
HuHLA-A3        :   CAGTTTCT TTTCt
HuH-2Kᵇ         :   CAGTTTCACTTctg
Hu2',5'-oligo   :   tgGTTTCG TTTCC
Synthetase
```

Figure 3 Sequence comparison of (A) CK-1, (B) CK-2, and (C) ISRE consensus sequences that have been identified in human (Hu) or murine (Mu) MIP-1α and other cytokine gene regulatory elements.

(ATGAcaTCAT; –95 to –104 bp) that also weakly binds AP-1–related proteins, and is essential for basal and LPS-inducible promoter activity in macrophages (34). ATF/CREB-binding sites were initially described as cAMP response elements (CRE; 78,79), but endogenous MuMIP-1β (and MuMIP-1α) gene expression is not induced by cAMP; in fact, cAMP will partially inhibit the LPS-induced accumulation of MuMIP-1α/β mRNA in macrophage cells (39). The MuMIP 1β promoter is not cAMP-responsive, but mutagenesis of the ATF/CREB site from ATGAcaTCAT to CTGAcgTCAG confers a weak transcriptional response to cAMP in transfected macrophage cells (34). ATF/CREB is composed of a family of related transcription factors that are either cAMP-responsive (CREB and ATF1) or cAMP-independent (ATFa, ATF2, and ATF3; 78–83). A single (G–A) base pair change in the nucleotides separating the ATF/CREB-binding site inverted repeat (TGAnnTCA) has been implicated in determining whether the site is cAMP-dependent (tgaCGtca) or cAMP-independent (tgaCAtca) (81,83,84).

E. Leucine Zipper Interactions

Although the transcription of the MIP-1α and MIP-1β genes appear to be mediated by distinct transcription factors, several of them (Fos-, Jun-, ATF/CREB- and C/EBP-related factors) contain leucine zippers (bZIP proteins) which potentiate homo- or heterodimerization (85). There is considerable crosstalk between members of different bZip proteins, heterodimeric interactions that magnify the regulatory potential of a limited number of transcription factors (79,82,86–91).

Heterodimers between Fos, Jun, and C/EBPβ leucine zippers, and between the C/EBPβ leucine zipper and the Rel homology domain of the NFκB p50 subunit have been described (90,91). As the MIP-1α gene promoter contains C/EBP-, NFκB-, and AP-1-related binding sites, the interaction of the multiple heterodimeric complexes, which are presumably in competition with each other for high-affinity binding, provides an enormous potential for subtle positive or negative changes in transcription. Similarly, the AP-1 and ATF/CREB families are immunologically related, suggesting they may have originated from a common gene (82), and the affinity of binding to CRE- or TPA-response element (TRE)-related sequences is partly mediated by specific leucine-zipper-dependent heterodimeric complexes between members of the two families (72,87). Therefore, this may partly explain the observed similarities in the transcriptional regulation of the MIP-1α and MIP-1β genes.

IX. SIGNAL TRANSDUCTION PATHWAYS

Lipopolysaccharide binds a serum LPS-binding protein (LBP), and the LPS/LBP complex activates cells by either binding a cell surface receptor (CD14) or by being taken up by phagocytosis (92). It is generally agreed that LPS-induced phospholipase-γ (PLC-γ) activation results in a Ca^{2+} flux, which activates protein kinase C (PKC; 93). PKC can also be directly activated by TPA, so LPS and TPA induced MIP-1α gene expression in macrophages or T cells probably involves, at least partly, a common signal transduction pathway. Protein kinase A does not appear to be involved in the LPS response or in the induction of MIP-1α gene expression. In fact, prostaglandin E_2 suppresses the LPS-induced increase in MIP-1α mRNA levels in macrophages (39). LPS stimulation of macrophages causes the degradation of prostaglandin-induced cAMP levels, but also induces the production and secretion of prostaglandins. It is, therefore, possible that the prostaglandins eventually feedback and stimulate macrophage intracellular cAMP levels that activate protein kinase A, and thus down-regulate LPS-induced gene expression, resulting in the observed transient LPS response (39,93–95).

The signal transduction pathways that activate certain transcription factors are being elucidated, and as some of the transcription factors involved in the regulation of MIP-1α gene transcription are known, the signal transduction pathways involved can be inferred (also see Ref. 96).

A. NF-κB

At least five Rel-related proteins have been identified (p50, p65, p52, rel, and rel-B) that can form homo- and heterodimers, with NF-κB containing a p50/p65 heterodimer (97,98). In many cells, NF-κB is in very low or undetectable levels, but can be rapidly induced by a variety of agents, such as LPS, TPA, TNF, and protein synthesis inhibitors (97–99). Thus, MIP-1α gene expression and nuclear NF-kB are induced in PHA/TPA/LPS-stimulated human Jurkat, THP1, and U937 cells, and in cycloheximide-, LPS- or TPA-treated murine macrophages (26,27,39,100).

NF-κB is stored in the cytoplasm as a complex with inhibitor molecules (IκB), one of which, IκBα, is a product of the *MAD-3* gene (97,98). Protein kinase A and protein kinase C phosphorylation of IκBα in vitro dissociates it from the NF-κB molecule, and this presumably leads to the nuclear translocation of the active NF-κB complex in vivo (101). IκBα is constitutively synthesized, but is more labile than NF-κB, which may partly explain why most cells have low basal levels of nuclear NF-κB, and why protein synthesis inhibitors result in nuclear NF-κB translocation (97–99).

B. C/EBP

C/EBPβ is regulated largely at the posttranslational level, whereas C/EBPδ is regulated at the level of transcription. C/EBPβ activity is increased by phosphorylation of its transactivation domain by PKC, and by activated Ras-dependent mitogen-activated protein kinase (MAP kinase; 102,103). Phosphorylation of a serine residue located within the C/EBPβ leucine zipper occurs in response to increased cellular Ca^{2+} concentrations by calcium–calmodulin-dependent protein kinase (CaMKII; 104). C/EBPβ1/2 activity may, therefore, be increased by phosphorylation of its transactivation domain, when nonfunctional heterodimers between C/EBPβ1/2 and the dominant negative CHOP or CEBPβ3 proteins are dissociated, and when the phosphorylation of cytoplasmic C/EBPβ in response to LPS, TPA, and cAMP stimulates C/EBP translocation to the nucleus (105,106).

C. ATF/CREB

The cAMP-responsive members of the ATF/CREB family of bZip proteins (CREB and ATF1) are phosphorylated by cAMP-dependent protein kinase (PKA), whereas the cAMP-independent members mediate transcriptional activation by DNA tumor virus oncoproteins such as *E1A* and *Tax* (78,79,84,107,108). The cAMP-dependent and -independent ATF/CREB proteins contain potential PKA and casein kinase II (CKII) phosphorylation sites, but mutagenesis of these sites had little effect on the cAMP-independent ATF/CREB proteins (79). However, phosphorylation or DNA tumor virus gene products may activate ATF/CREB activity by stimulating heterodimer formation (78,108).

D. AP-1

There are at least five Fos-family (Fos, FosB, Fra1, Fra2, and delta-FosB) and three Jun-family (Jun, JunB, and JunD) members. AP-1 DNA-binding activity is modulated in several ways. For example, an inhibitor molecule (IP-1) has been identified that is inactivated in vitro by PKA, and in vivo by PKC, PKA, and Ca^{2+}-dependent kinase signal transduction pathways (109). Jun is phosphorylated at five sites in resting cells; phosphorylation at serines 63 and 73 stimulates the transactivation function of Jun, whereas phosphorylation of three clustered sites located upstream from its basic DNA-binding domain (Thr-231, Ser-249, and Ser-243) negatively regulate DNA-binding (110–112). The three clustered threonine and serine sites are rapidly phosphorylated by CKII and glycogen synthetase kinase 3 (GSK-3), whereas PKC activation or stimulation of oncoproteins such as Src, Ras, and Raf-1 results in the dephosphorylation of the three sites and the stimulation of AP-1 DNA-binding (110–112).

E. ISGF3

The ISGF3α proteins contain Src homology domains (SH2 and SH3 domains) which have been implicated in facilitating protein–protein interactions among tyrosine kinase-regulated proteins (75,113). IFN-α specifically induces a protein tyrosine kinase, and it is proposed that it phosphorylates a latent form of newly synthesized ISGF3α. Activated ISGF3α now forms a complex with ISGF3γ to allow the nuclear translocation of the active ISGF3 complex (75–77,114).

X. RNA STABILITY AND TRANSLATION

Many cytokine and early-response gene mRNAs are unstable before cell stimulation, and are transiently stabilized on cell stimulation. For example, as shown in Figures 4A and B, the stability of MIP-1α and MIP-1β mRNAs in actinomycin D-treated macrophage cells is increased by LPS stimulation of macrophage cells, as are several lymphokine mRNAs after α-CD3 or TPA stimulation of T cells (115). The 3′ untranslated regions (3′UTR) of early-response gene mRNAs have been implicated in the regulation of mRNA stability and, in particular, the presence of UUAUUUAU repeats has been shown to confer instability on mRNAs in vitro in a cell-specific manner (115–118). Cytoplasmic factors that bind AU-rich instability elements (AREs) in vitro have been identified, and distinct factors may mediate the instability of different mRNAs according to the precise sequence of the ARE (115,119). Therefore, it is likely that there is a family of ARE-binding proteins that has evolved to distinguish between subtypes of ARE-containing transcripts, and each ARE may have evolved its own distinct identity to suit the special function of the mRNA.

The murine and human MIP-1α mRNA 3′UTR AREs contain three to four AUUUA repeats and a single UUAUUUAU sequence (12,26). In contrast, the murine and human MIP-1β mRNA 3′UTR AREs contain only one copy each of the AUUUA and UUAUUUAU sequences (13,51). MuMIP-1α mRNA is significantly more stable in actinomycin D-treated control and LPS-stimulated RAW264.7 macrophage cells, although the stability of both mRNAs is increased after LPS-stimulation (see Fig. 4). Furthermore, when macrophage cells are cultured in serum-free media, intracellular levels of MuMIP-1α but not MIP-1β mRNA increase within 4 h (see Fig. 4C), suggesting there is a factor in serum that destabilizes MIP-1α but not MIP-1β mRNA in vivo. Detailed mutagenesis studies of the MIP-1α/β 3′UTRs have yet to be carried out, but there are clear differences in the stability of the mRNAs before and after cell stimulation and clear differences in the precise organization of the 3′UTR AREs. As the MIP-1α/β genes presumably arose by gene duplication and divergence, the ARE sequences have presumably diverged to suit the specific requirements of the two cytokine mRNAs in mouse and human.

The protein synthesis inhibitor cycloheximide up-regulates early-response gene mRNA levels, including that of MuMIP-1α and MuMIP-1β in RAW264.7 macrophage cells, and prolongs the normally transient period of early-response gene mRNA accumulation (34,120). Cycloheximide prevents the poly(A) tail shortening step of early response gene ARE-directed mRNA degradation pathway (120) and de novo synthesis of an RNA-binding protein is required for the destabilization of lymphokine mRNAs (115). However, cycloheximide may also stimulate transcription, because many transcription factors, such as NF-κB, are complexed with labile cytoplasmic inhibitors (see foregoing), so the induction of cellular mRNA levels by cycloheximide may be the result of transcriptional and posttranscriptional events.

In at least some cell types, cell stimulation leads to the transient accumulation of early-response gene mRNAs, but with no concomitant increase in the translation of the mRNAs

A.

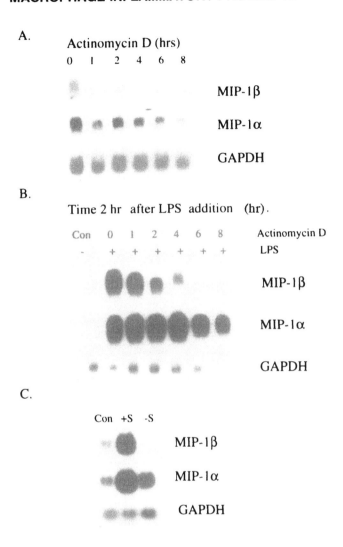

Figure 4 Northern blot analysis of total cellular RNA prepared from (A) control, or (B) 2-hr LPS-stimulated RAW264.7 cells treated with actinomycin D. Cells were harvested at the times indicated after the addition of actinomycin D; RNA was prepared and analyzed as described (38). (C) Northern blot analysis of RNA prepared from control (con.) RAW264.7 cells, from RAW254.7 cells stimulated with fresh serum for 4 hr (+S), or incubated in serum-free medium for 4 hr (−S). RNA was probed with either MIP-1α, MIP-1β, or GAPDH (loading control) probes, as shown.

(121–123). This suggests that the poly(A) shortening step can be uncoupled from the mRNA degradation step in the mRNA metabolic pathway. Although the MIP-1α and MIP-1β mRNA levels have been measured in several different cells following a number of cell-stimulatory signals and, with the exception of LPS stimulation of macrophages (37), it is unclear whether rates of mRNA translation or secretion reflect changes in mRNA levels in the biological systems studied. Similarly, it is not yet possible to exclude the possibility that other 3′UTR or 5′UTR sequences are involved in regulating MIP-1α/β metabolism or translation.

XI. CLINICAL PERSPECTIVES

As an inhibitor of stem cell proliferation, MIP-1α has potential clinical uses in the protection of multipotent hematopoietic stem cells from the cytotoxic effects of S-phase chemotherapeutic agents. Chemotherapeutic regimens frequently involve multiple doses, the first of which, by depleting actively proliferating hematopoietic progenitor cells, results in the semisynchronous recruitment of the normally quiescent stem cells into proliferation. Unless the bone marrow is allowed to fully recover, subsequent treatment(s) with the cytotoxic drug will necessarily kill a large proportion of the stem cells and result in bone marrow failure. The ability of MIP-1α to inhibit the proliferation of CFU-S cells in mice after a single treatment with S-phase cytotoxic drugs has, therefore, been evaluated. MIP-1α administered intravenously or by the intraperitoneal route did protect multipotent progenitor cells from the cytotoxic effects of a second dose of hydroxyurea or cytarabine (cytosine arabinoside) in vivo (15,16). Therefore, in the clinic, MIP-1α could theoretically allow intensification of the second dose of the cytotoxic drug or earlier treatment in the second round of chemotherapy.

One unexpected result of the studies on the myeloprotective effects of MIP-1α, was that high doses of MIP-1α appeared to enhance CFU-S numbers in the bone marrow (16), suggesting there was a secondary effect on the self-renewal properties of the CFU-S. More recent studies have confirmed this effect in both human and murine systems. MIP-1α, alone or with IL-3- and stromal-derived factors, can be used to expand the numbers of committed and primitive progenitors in vitro, and this is attributed to the direct induction of self-renewing cell divisions of primitive progenitor cells in addition to a block in differentiation (124–127). In the mouse, the affected cell is more primitive (pre-CFU-A) than that (CFU-A/CFU-S) which is inhibited by MIP-1α (126). There are also unpublished reports that MIP-1α will stimulate the recruitment of CFU-S cells into the peripheral blood in vivo (McCourt M, personal communication). The ability to amplify the numbers of primitive multipotent hematopoietic progenitor cells in vitro, or to recruit them into the peripheral blood, has obvious clinical potential both in gene therapy and in autologous or heterologous bone marrow and peripheral blood stem cell transplantation.

Acute and chronic leukemias represent a disorder in the regulation of proliferation or differentiation of hematopoietic stem or progenitor cells, and the response of primitive chronic myeloid (CML) and acute myeloid (AMLs) cells to MIP-1α has been examined to further define the mechanism of disregulated proliferation and to explore potential clinical applications (128–131). Whereas MIP-1α blocked the activation of high-proliferation potential progenitors in long-term normal bone marrow cultures, it had no effect on long-term cultures of neoplastic CML cells, providing a clear correlation between this defective response to an inhibitor of proliferation and the deregulated proliferative behavior of the CML cells (128,129). In contrast, one report suggests that MIP-1α did inhibit the proliferation of early and mature AML progenitors (130), although another report suggests the response of AML cells to MIP-1α is highly heterogeneous (131). Interestingly, whereas very low levels of MIP-1α mRNA are detectable in normal human bone marrow nucleated cells (NBMC), high levels of MIP-1α mRNA have been detected in the NBMC of patients with preleukemic disorders, such as aplastic anemia and myelodysplastic syndrome (25), and in fresh unstimulated leukemic (ANLL and ALL) cells (26), supporting the argument that leukemic cells are largely unresponsive to MIP1-α. This raises several interesting possibilities: (1) bone marrow failure observed in some leukemic and preleukemic disorders is a consequence of the overexpression of MIP-1α, although MIP-1α translation and secretion by (pre)-leukemic cells has yet to formally proved; (2) a number of quite distinct lesions can lead to a similar deregulated proliferation in different leukemias; and

(3) MIP-1α has the potential to selectively protect normal, but not neoplastic multipotent progenitor cells in chemotherapeutic regimens in certain leukemias.

NOTE ADDED IN PROOF

Since going to press, several publications have appeared which indicate that MIP-1α and its receptor play an important role in HIV-1 infection, HIV-1 replication, and the progression of AIDS. In addition, a detailed description of the MIP-1α gene knockout mouse has been published (132).

A. Receptor Cloning and Function

Several chemokine receptors have recently been cloned (133–135), and the biochemistry of the ligand-receptor interaction linked to the activation of Raf-1 protein kinase and inositol 1,4,5, triphosphate generation (136,137).

B. HIV-1 Infection and the Progression of AIDS

MIP-1α, MIP-1β, and Rantes were first shown to inhibit HIV-1 replication (138,139). Most recently, it has been demonstrated that CC-CKR-5 is a co-receptor for entry of HIV-1 into $CD4^+$ cells (140–142), and that patients who show a resistance to HIV-1 infection and AIDS progression have CKR5 receptor mutations (143,144). These observations clearly have enormous implications in the treatment of and/or prevention of AIDS.

REFERENCES

1. Wright EG, Lord BI. Haemopoietic tissue. Bailliere's Clin Haematol 1992; 5:499–507.
2. Wright E, Sheridan P, Moore MA. An inhibitor of murine stem cell proliferation produced by normal human bone marrow. Leuk Res 1980; 4:309–319.
3. Wright EG, Garland JM, Lord BI. Specific inhibition of haemopoietic stem cell proliferation: characteristics of the inhibitor producing cells. Leuk Res 1980; 4:537–545.
4. Lord BI, Mori KJ, Wright EG, Lajtha LG. An inhibitor of stem cell proliferation in normal bone marrow. Br J Haematol 1976; 34:441–445.
5. Wright EG, Lord BI. Regulation of CFU-S proliferation by locally produced endogenous factors. Biomedicine 1977; 27:215–218.
6. Wright EG, Ali AM, Riches AC, Lord BI. Stimulation of haemopoietic stem cell proliferation: characteristics of the stimulator-producing cells. Leuk Res 1982; 6:531–539.
7. Wright EG, Lorimore SA. The production of factors regulating the proliferation of haemopoietic spleen colony-forming cells by bone marrow macrophages. Cell Tissue Kinet 1987; 20:191–203.
8. Lord BI, Mori KJ, Wright EG. A stimulator of stem cell proliferation in regenerating bone marrow. Biomedicine 1977; 27:223–226.
9. Eckmann L, Freshney M, Wright EG, Sproul A, Wilkie N, Pragnell IB. A novel in vitro assay for murine haemopoietic stem cells. Br J Cancer 1988; 58:36–40.
10. Pragnell IB, Wright EG, Lorimore SA, Adam J, Rosendaal M, DeLamarter JF, Freshney M, Eckmann L, Sproul A, Wilkie N. The effect of stem cell proliferation regulators demonstrated with an in vitro assay. Blood 1988; 72:196–201.
11. Graham GJ, Wright EG, Hewick R, Wolpe SD, Wilkie NM, Donaldson D, Lorimore S, Pragnell IB. Identification and characterisation of an inhibitor of haemopoietic stem cell proliferation. Nature 1990; 344:442–444.
12. Davatelis G, Tekamp-Olson P, Wolpe SD, Hermsen K, Luedke C, Gallegos C, Coit D, Merryweather J, Cerami A. Cloning and characterisation of a cDNA for murine macrophage inflammatory protein

(MIP), a novel monokine with inflammatory and chemokinetic properties. J Exp Med 1988; 167:1939–1944.

13. Sherry B, Tekamp-Olson P, Gallegos C, Bauer D, Davatelis G, Wolpe SD, Masiarz A, Coit D, Cerami A. Resolution of the components of macrophage inflammatory protein 1, and cloning and characterisation of one of those components, macrophage inflammatory protein 1β. J Exp Med 1988; 168:2251–2259.

14. Quesniaux VFJ, Graham GJ, Pragnell I, Donaldson D, Wolpe SD, Iscove NN, Fagg B. Use of 5-fluoruracil to analyse the effects of macrophage inflammatory protein 1α on long term reconstituting stem cells in vivo. Blood 1993; 81:1497–1504.

15. Dunlop DJ, Wright EG, Lorimore S, Graham GJ, Holyoake T, Kerr DJ, Wolpe SD, Pragnell IB. Demonstration of stem cell inhibition and myeloprotective effects of SCI/rhMPI1α in vivo. Blood 1993; 79:2221–2225.

16. Lord BI, Dexter TM, Clements JM, Hunter MA, Gearing AJH. Macrophage inflammatory protein protects multipotent hematopoietic cells from the cytotoxic effects of hydroxyurea in vivo. Blood 1992; 79:2605–2609.

17. Parkinson EK, Graham GJ, Daubersies P, Burns JE, Heufler C, Plumb M, Schuler G, Pragnell IB. Hemopoietic stem cell inhibitor (SCI/MIP-1α) also inhibits clonogenic epidermal keratinocyte proliferation. J Invest Dermatol 1993; 101:113–117.

18. Broxmeyer HE, Sherry B, Lu L, Cooper S, Oh K-O, Tekamp-Olson P, Kwon BS, Cerami A. Enhancing and suppressing effects of recombinant murine macrophage inflammatory proteins on colony formation in vitro by bone marrow myeloid progenitor cells. Blood 1990; 76:1110–1116.

19. Broxmeyer HE, Sherry B, Cooper S, Ruscetti FW, Williams DE, Arosio P, Kwon BS, Cerami A. Macrophage inflammatory protein (MIP)-1β abrogates the capacity of MIP-1α to suppress myeloid progenitor cell growth. J Immunol 1991; 147:2586–2594.

20. Neote KD, DiGregorio D, Mak JY, Horuk R, Schall TJ. Molecular cloning, functional expression, and signalling characteristics of a C-C chemokine receptor. Cell 1993; 72:415–425.

21. Broxmeyer HE, Sherry B, Lu L, Cooper S, Carow C, Wolpe SD, Cerami A. Myelopoietic enhancing effects of murine macrophage inflammatory proteins 1 and 2 on colony formation in vitro by murine and human bone marrow granulocyte/macrophage progenitor cells. J Exp Med 1989; 170:1583–1594.

22. Schall TJ. Biology of the RANTES/SIS cytokine family. Cytokine 1991; 3:165–183.

23. Plumb M, Graham GJ, Grove M, Reid A, Pragnell IB. Molecular aspects of a negative regulator of haemopoiesis. Br J Cancer 1991; 64:990–992.

24. Gao J-L, Kuhns DB, Tiffany HL, McDermott D, Li X, Francke U, Murphy PM. Structure and functional expression of the human inflammatory protein 1α/RANTES receptor. J Exp Med 1993; 177:1421–1427.

25. Maciejewski JP, Liu JM, Green SW, Walsh CE, Plumb M, Pragnell IB, Young NS. Expression of the stem cell inhibitor (SCI) gene in bone marrow and peripheral blood from normal individuals and patients with bone marrow failure. Exp Haematol 1992; 20:1112–1117.

26. Yamamura Y, Hattori T, Obaru K, Sakai K, Asou N, Takatsuki K, Ohmoto Y, Nomiyama H. Synthesis of a novel cytokine and its gene (LD78) expressions in haemopoietic fresh tumour cells and cell lines. J Clin Invest 1989; 84:1707–1712.

27. Obaru K, Fukuda M, Maeda S, Shimada K. A cDNA clone used to study mRNA inducible in human tonsillar lymphocytes by a tumor promoter. J Biochem 1986; 99:885–894.

28. Tanaka Y, Adams DH, Hubscher S, Hirano H, Siebenlist U, Shaw S. T-cell adhesion induced by proteoglycan immobilised cytokine MIP-1β. Nature 1993; 361:79–82.

29. Taub DD, Conlon K, Lloyd AR, Oppenheim JJ, Kelvin DJ. Preferential migration of activated CD4+ and CD8+ T cells in response to MIP-1α and MIP-1β. Science 1993; 260:355–358.

30. McCourt M, Comer MB, Edwards RM, Hunter MG, Wood LM. BB-10010 a demultimerised variant of LD78 produces a rapid increase in mature white blood cells. Blood 1994; 84, suppl 1:136a.

31. Migas J, Hurley R, Verfaillie CM. MIP-1α and TGF-β increase adhesion of normal committed progenitors to marrow stroma. Exp Hematol 1994; 22:736.

32. Nakao M, Nomiyama H, Shimada K. Structures of human genes coding for cytokine LD78 and their expression. Mol Cell Biol 1990; 10:3646–3658.

33. Fahey TJ III, Tracey KJ, Tekamp-Olson P, Cousens LS, Jones WG, Shires GT, Cerami A, Sherry B. Macrophage inflammatory protein 1 mediates macrophage function. J Immunol 1992; 148:2764–2769.

34. Proffitt J, Crabtree G, Grove M, Daubersies P, Bailleul B, Wright E, Plumb M. ATF/CREB- and AP1-binding sites and a CK1 motif are required for cell specific and LPS-inducible transcription of the MIP-1β immediate early gene. Gene 1995; 152:173–179.

35. Burd PR, Rogers HW, Gordon JR, Martin CA, Jayaraman S, Wilson SD, Dvorak AM, Galli SJ, Dorf ME. Interleukin 3-dependent and -independent mast cells stimulated with IgE and antigen express multiple cytokines. J Exp Med 1989; 170:245–257.

36. Gordon JR, Burd PR, Galli SJ. Mast cells as a source of multifunctional cytokines. Immunol Today 1990; 11:458–464.

37. Wolpe SD, Davatelis G, Sherry B, Beutler B, Hesse DG, Nguyen HT, Moldawer L, Nathan CF, Lowry SF, Cerami A. Macrophages secrete a novel heparin binding protein with inflammatory and neutrophil chemokinetic properties. J Exp Med 1988; 167:570–581.

38. Grove M, Plumb M. C/EBP, NF-kB, and C-*Ets* family members and the transcriptional regulation of the cell-specific and inducible macrophage inflammatory protein 1α immediate early gene. Mol Cell Biol 1993; 13:5276–5289.

39. Martin CA, Dorf ME. Differential regulation of interleukin-6, macrophage inflammatory protein-1, and JE/MCP-1 cytokine expression in macrophage cell lines. Cell Immunol 1991; 135:245–258.

40. Davatelis G, Wolpe SD, Sherry B, Dayer J-M, Chicheportiche R, Cerami A. Macrophage inflammatory protein 1: a prostaglandin-independent endogenous pyrogen. Science 1989; 243:1066–1068.

41. Minano FJ, Sansibrian M, Vizcaino M, Paez X, Davatelis G, Fahey T, Sherry B, Cerami A, Myers RD. Macrophage inflammatory protein-1; unique action on the hypothalamus to evoke fever. Brain Res Bull 1990; 24:849–852.

42. Aronica S, Mantel C, Marshall M, Sarris A, Zhang X, Broxmeyer HE. IP-10 and MIP-1α inhibit synergistically enhanced growth factor-stimulated activation of RAF-1 kinase in factor-dependent MO7e cells. Blood 1994; 84(suppl 1):122a.

43. Graham GJ, Zhou L, Weatherbee JA, Tsang M,   L-S, Napolitano M, Leonard WJ, Pragnell IB. Characterization of a receptor for macrophage inflammatory 1α and related proteins on human and murine cells. Cell Growth Differ 1993; 4:137–146.

44. Heufler C, Topar G, Koch F, Trockenbacher B, Kampgen E, Romani N, Schuler G. Cytokine gene expression in murine epidermal cell suspensions: interleukin 1β and macrophage inflammatory protein 1α are selectively expressed in Langerhans cells but are differentially regulated in culture. J Exp Med 1992; 176:1221–1226.

45. Koerner TJ, Hamilton TA, Introna M, Tannenbaum CS, Bast RC Jr, Adams DO. The early competence genes JE and KC are differentially regulated in murine peritoneal macrophages in response to lipopolysaccharide. Biochem Biophys Res Commun 1987; 149:969–974.

46. Rollins BJ. JE/MCP-1: an early-response gene encodes a monocyte-specific cytokine. Cancer Cells 1991; 3:517–524.

47. Rameh LR, Armelin MCS. Downregulation of JE and KC genes by glucocorticoids does not prevent the G_0-G_1 transition in BALB/3T3 cells. Mol Cell Biol 1992; 12:4612–4621.

48. Maltman J, Pragnell IB, Graham GJ. Transforming growth factor β: is it a downregulator of stem cell inhibition by macrophage inflammatory protein 1α? J Exp Med 1993; 178:(in press).

49. Ziegler SF, Tough TW, Franklin TL, Armitage RJ, Alderson MR. Induction of macrophage inflammatory protein-1β gene expression in human monocytes by lipopolysaccharide and IL-7. J Immunol 1991; 147:2234–2239.

50. Irving SG, Zipfel PF, Balke J, McBride OW, Morton CC, Burd PR, Siebenlist U, Kelly K. Two inflammatory mediator cytokine genes are closely linked and variably amplified on chromosome 17q. Nucleic Acids Res 1990; 18:3261–3270.

51. Napolitano M, Modi WS, Cevario SJ, Gnarra JR, Seuanez HN, Leonard WJ. The gene encoding the Act-2 cytokine. J Biol Chem 1991; 266:17531–17536.

52. Blum S, Forsdyke RE, Forsdyke DR. Three human homologues of a murine gene encoding an inhibitor of stem cell proliferation. DNA Cell Biol 1990; 9:589–602.

53. Widmer U, Monogue KR, Cerami A, Sherry B. Genomic cloning and promoter analysis of macrophage inflammatory protein (MIP)-2, MIP-1α, and MIP-1β, members of the chemokine superfamily of proinflammatory cytokines. J Immunol 1993; 150:4996–5012.

54. Grove M, Lowe S, Graham G, Pragnell I, Plumb M. Sequence of the murine haemopoietic stem cell inhibitor/macrophage inflammatory protein 1α gene. Nucleic Acids Res 1990; 18:5561.

55. Widmer U, Yang Z, van Deventer S, Manogue KR, Sherry B, Cerami A. Genomic structure of murine macrophage inflammatory protein-1α and conservation of potential regulatory sequences with a human homolog, LD78. J Immunol 1991; 146:4031–4040.

56. Wilson SD, Billings PR, D'Eustachio P, Fournier REK, Geissler E, Lalley PA, Burd PR, Housman DE, Taylor BA, Dorf ME. Clustering of cytokine genes on mouse chromosome 11. J Exp Med 1990; 171:1301–1314.

57. Hirashima M, Ono T, Nakao M, Nishi H, Kimura A, Nomiyama H, Hamada F, Yoshida MC, Shimada K. Nucleotide sequence of the third cytokine LD78 gene and mapping of all three LD78 gene loci to human chromosome 17. DNA Seq 1992; 3:203–212.

58. Donlon TA, Krensky AM, Wallace MR, Collins FS, Lovett M, Clayberger C. Localisation of a human T-cell-specific gene RANTES (D17S136E) to chromosome 17 q11.2-q12. Genomics 1989; 6:548-556.

59. Nomiyama H, Hieshima K, Hirokawa K, Hattori T, Takatsuki K, Miura R. Characterization of cytokine LD78 gene promoters: positive and negative transcriptional factors bind to a negative regulatory element common to LD78, interleukin-3, and granulocyte–macrophage colony-stimulating factor gene promoters. Mol Cell Biol 1993; 13:2787–2801.

60. Vincenti MP, Burrell TA, Taffet SM. Regulation of NF-kappa B activity in murine macrophages: effect of bacterial lipopolysaccharide and phorbol ester. J Cell Physiol 1992; 150:204–213.

61. Ray D, Bosselut R, Ghysdael J, Mattei M-G, Tavitian A, Moreau-Gachelin F. Characterisation of Spi-B, a transcription factor related to the putative oncoprotein Spi-1/PU.1. Mol Cell Biol 1992; 12:4297–4304.

62. Klemsz MJ, McKercher SR, Celada A, Van Beveren C, Maki RA. The macrophage and B cell-specific transcription factor PU.1 is related to the *ets* oncogene. Cell 1990; 61:113–124.

63. Akira S, Isshiki H, Sugita T, Tanabe O, Kinoshita S, Nishio Y, Nakajima T, Hirano T, Kishimoto T. A nuclear factor for IL-6 expression (NF-IL6) is a member of a C/EBP family. EMBO J 1990; 9:1897–1906.

64. Williams SC, Cantwell CA, Johnson PF. A family of C\EBP-related proteins capable of forming covalently linked leucine zipper dimers in vitro. Genes Dev 1991; 5:1553–1567.

65. Thomassin H, Hamel D, Bernier D, Guertin M, Belanger L. Molecular cloning of two C/EBP-related proteins that bind to the promoter and the enhancer of the α1-fetoprotein gene. Further analysis of the C/EBPβ and C/EBPγ. Nucleic Acids Res 1992; 20:3091–3098.

66. Poll V, Mancini FP, Cortese R. IL-6BP, a nuclear protein involved in interleukin-6 signal transduction, defines a new family of leucine zipper proteins related to C/EBP. Cell 1990; 63:643–653.

67. Cao Z, Umek RM, McKnight SL. Regulated expression of three C\EBP isoforms during adipose conversion of 3T3-L1 cells. Genes Dev 1991; 5:1538–1552.

68. Kinoshita S, Akira S, Kishimoto T. A member of the C/EBP family, NF-IL6b, forms a heterodimer and transcriptionally synergises with NF-IL6. Proc Natl Acad Sci USA 1992; 89:1473–1476.

69. Ron D, Habener JF. CHOP, a novel developmentally regulated nuclear protein that dimerises with transcription factors C/EBP and LAP and functions as a dominant-negative inhibitor of gene transcription. Genes Dev 1992; 6:439–453.

70. Curran T, Franza BR. Fos and Jun: the AP-1 connection. Cell 1988; 55:395–397.

71. Fujihara M, Muroi M, Muroi Y, Ito N, Suzuki T. Mechanism of lipopolysaccharide-triggered junB activation in a mouse macrophage-like cell line (J774). J Biol Chem 1993; 268:14898–14905.

72. Shannon MF, Gamble JR, Vadas MA. Nuclear proteins interacting with the promoter region of the human granulocyte/macrophage colony-stimulating factor gene. Proc Natl Acad Sci USA 1988; 85:674–678.

73. Nimer SD, Morita EA, Martis MJ, Wachsman W, Gasson JC. Characterisation of the human granulocyte–macrophage colony-stimulating factor promoter by genetic analysis: correlation with DNase1 footprinting. Mol Cell Biol 1988; 8:1979–1984.

74. Nelson N, Marks MS, Driggers PH, Ozato K. Interferon consensus sequence-binding protein, a member of the interferon regulatory factor family, suppresses interferon-induced gene transcription. Mol Cell Biol 1993; 13:588–599.

75. Fu X-Y. A transcription factor with SH2 and SH3 domains is directly activated by an interferon α-induced cytoplasmic protein tyrosine kinase(s). Cell 1992; 70:323–335.

76. Kessler DS, Veals SA, Fu X-Y, Levy DE. Interferon-α regulates nuclear translocation and DNA-binding affinity of ISGF3, a multimeric transcriptional activator. Genes Dev 1990; 4:1753–1765.

77. David M, Larner AC. Activation of transcription factors by interferon-α in a cell-free system. Science 1992; 257:813–815.

78. Yamamoto KK, Gonzalez GA, Biggs WH, Montminy MR. Phosphorylation-induced binding and transcriptional efficacy of nuclear factor CREB. Nature 1988; 334:494–498.

79. Flint KJ, Jones NC. Differential regulation of three members of the ATF/CREB family of DNA-binding proteins. Oncogene 1991; 6:2019–2026.

80. Gaire M, Chatton B, Kedinger C. Isolation and characterisation of two novel, closely related ATF cDNA clones from HeLa cells. Nucleic Acids Res 1990; 18:3467–3473.

81. Georgopoulos K, Morgan BA, Moore DD. Functionally distinct isoforms of the CRE-BP DNA-binding protein mediate activity of a T-cell enhancer. Mol Cell Biol 1992; 12:747–757.

82. Hai T, Liu F, Allegretto EA, Karin M, Green MR. A family of immunologically related transcription factors that includes multiple forms of ATF and AP-1. Genes Dev 1988; 2:1216–1226.

83. Kaszubska W, van Huijsduijnen RH, Ghersa P, DeRaemy-Schenk A-M, Chen BPC, Hai T, DeLamarter JF, Whelan J. Cyclic AMP-independent ATF family members interact with NF-κB and function in the activation of the E-selectin promoter in response to cytokines. Mol Cell Biol 1993; 13:7180–7190.

84. Du W, Maniatis T. An ATF/CREB binding site protein is required for virus induction of the human interferon β gene. Proc Natl Acad Sci USA 1992; 89:2150–2154.

85. Alber T. Structure of the leucine zipper. Curr Opin Genet Dev 1992; 2:205–210.

86. Hai T, Curran T. Cross-family dimerization of transcription factors Fos/Jun and ATF/CREB alters binding specificity. Proc Natl Acad Sci USA 1991; 88:3720–3724.

87. Sassone-Corsi P., Ransone LJ, Verma IM. Cross-talk in signal transduction: TPA-inducible factor jun/AP-1 activates cAMP-responsive enhancer elements. Oncogene 1990; 5:427–431.

88. Benbrook DM, Jones NC. Heterodimer formation between CREB and JUN proteins. Oncogene 1990; 5:295–302.

89. Chatton B, Bocco JL, Goetz J, Gaire M, Lutz Y, Kedinger C. Jun and Fos heterodimerise with ATFa, a member of the ATF/CREB family and modulate its transcriptional activity. Oncogene 1994; 9:375–385.

90. Hsu W, Kerppola TK, Chen P-L, Curran T, Chen-Kiang S. Fos and Jun repress transcription activation by NF-IL6 through association at the basic zipper region. Mol Cell Biol 1994; 14:268–276.

91. LeClair KP, Blanar MA, Sharp PA. The p50 subunit of NF-κB associates with the NF-IL6 transcription factor. Proc Natl Acad Sci USA 1992; 89:8145–8149.

92. Wright SD, Ramos RA, Tobias PS, Ulevitch PJ, Mathison JC. CD14, a receptor for complexes of lipopolysaccharide (LPS) and LPS binding protein. Science 1990; 249:1431–1433.

93. Chen TY, Lei MG, Suzuki LT, Morrisson DC. Lipopolysaccharide receptors and signal transduction pathways in mononuclear phagocytes. Curr Top Microbiol Immunol 1992; 181:169–188.

94. Kunkel SL, Spengler M, May MA, Spengler R, Larrick J, Remick D. Prostaglandin E_2 regulates macrophage-derived tumour necrosis factor gene expression. J Biol Chem 1988; 263:5380–5384.

95. Okonogi K, Gettys TW, Uhing RJ, Tarry WC, Adams DO, Prpic V. Inhibition of prostaglandin

E$_2$-stimulated cAMP accumulation by lipopolysaccharide in murine peritoneal macrophages. J Biol Chem 1991; 266:10305–10312.

96. Kishimoto T, Taga T, Akira S. Cytokine signal transduction. Cell 1994; 76:253–262.

97. Blank V, Kourilsky P, Israel A. NF-kB and related proteins: Rel/dorsal homologies meet ankyrin repeats. Trends Biol Sci 1992; 17:135–140.

98. Nolan GP, Baltimore D. The inhibitory ankyrin and activator Rel proteins. Curr Opin Genet Dev 1992; 2:211–220.

99. Rice NR, Ernst MK. In vivo control of NF-κB activation by IκBα. EMBO J 1993; 12:4685–4695.

100. Zipfel PF, Balke J, Irving SG, Kelly K, Siebenlist U. Mitogenic activation of human T cells induces two closely related genes which share structural similarities with a new family of secreted factors. J Immunol 1989; 142:1582–1590.

101. Shirakawa F, Mizel SB. In vitro activation and nuclear translocation of NF-κB catalysed by cyclic AMP-dependent protein kinase and protein kinase C. Mol Cell Biol 1989; 9:2424–2430.

102. Nakajima T, Kinoshita S, Sasagawa T, Sasaki K, Naruto M, Kishimoto T, Akira S. Phosphorylation at threonine-235 by a *ras*-dependent mitogen-activated protein kinase cascade is essential for transcription factor NF-IL6. Proc Natl Acad Sci USA 1993; 90:2207–2211.

103. Trautwein C, Caelles C, van der Geer P, Hunter T, Karim M, Chojkier M. Transactivation by NF-IL6/LAP is enhanced by phosphorylation of its activation domain. Nature 1993; 364:544–547.

104. Wegner M, Cao Z, Rosenfeld MG. Calcium-regulated phosphorylation within the leucine zipper of C/EBPβ. Science 1992; 256:370–373.

105. Metz R, Ziff E. cAMP stimulates the C/EBP-related transcription factor rNFIL-6 to trans-locate to the nucleus and induce c-*fos* transcription. Genes Dev 1991; 5:1754–1766.

106. Katz S, Kownz-Leutz E, Muller C, Meese K, Ness SA, Leutz A. The NF-M transcription factor is related to C/EBPβ and plays a role in signal transduction, differentiation and leukaemogenesis of avian myelomonocytic cells. EMBO J 1993; 12:1321–1332.

107. Liu F, Green MR. A specific member of the ATF transcription factor family can mediate transcription activation by the adenovirus E1a protein. Cell 1990; 61:1217–1224.

108. Wagner S, Green MR. HTLV-1 *Tax* protein stimulation of DNA binding of bZip proteins by enhancing dimerisation. Science 1993; 262:395–399.

109. Auwerz J, Sassone-Corsi P. AP1 (Fos-Jun) regulation by IP-1: effect of signal transduction pathways and cell growth. Oncogene 1992; 7:2271–2280.

110. Smeal T, Binetruy B, Mercola D, Grover-Bardwick A, Heidecker G, Rapp UR, Karin M. Oncoprotein-mediated signalling cascade stimulates c-*jun* activity by phosphorylation of serines 63 and 73. Mol Cell Biol 1992; 12:3507–3513.

111. Lin A, Frost J, Deng T, Smeal T, Al-Alawi N, Kikkawa U, Hunter T, Brenner D, Karin M. Casein kinase II is a negative regulator of c-*jun* DNA binding and AP-1 activity. Cell 1992; 70:777–789.

112. Boyle WJ, Smeal T, Defize LHK, Angel P, Woodgett JR, Karin M, Hunter T. Activation of protein kinase C decreases phosphorylation of c-*jun* at sites that negatively regulate its DNA-binding activity. Cell 1991; 64:573–584.

113. Pawson T, Gish GD. SH2 and SH3 domains: from structure to function. Cell 1992; 71:359–362.

114. Velazquez L, Fellous M, Stark GR, Pellegrini S. A protein tyrosine kinase in the interferon α/β signaling pathway. Cell 1992; 70:313–322.

115. Bohjanen PR, Petryniak B, June CH, Thompson CB, Lindsten T. An inducible cytoplasmic factor (AU-B) binds selectively to AUUUA multimers in the 3′ untranslated region of lymphokine mRNA. Mol Cell Biol 1991; 11:3288–3295.

116. Caput D, Beutler B, Hartog K, Thayer R, Brown-Shimer S, Cerami A. Identification of a common nucleotide sequence in the 3′ untranslated region of mRNA molecules specifying inflammatory mediators. Proc Natl Acad Sci USA 1986; 83:1670–1674.

117. Shaw G, Kamen R. A conserved AU sequence from the 3′-untranslated region of GM-CSF mRNA mediates selective mRNA degradation. Cell 1986; 46:659–667.

118. Chen C-YA, Chen T-M, Shyu A-B. Interplay of two functionally and structurally distinct domains of

the c-*fos* AU-rich element specifies its mRNA-destabilising function. Mol Cell Biol 1994; 14:416–426.

119. Brewer G. An A + U-rich element RNA-binding factor regulates c-*myc* mRNA stability in vitro. Mol Cell Biol 1991; 11:2460–2466.

120. Laird-Offringa IA, de Wit C, Elfferich P, Van der Eb AJ. Poly(A) tail shortening is the translation dependent step in c-*myc* mRNA degradation. Mol Cell Biol 1990; 10:6132–6140.

121. Schindler R, Clark BD, Dinarello CA. Dissociation between IL-1β mRNA and protein synthesis in human peripheral blood mononuclear cells. J Biol Chem 1990; 265:10232–10237.

122. Han JH, Beutler B. Endotoxin responsive sequences control cachectin/tumour necrosis factor biosynthesis at the translational level. J Exp Med 1990; 171:465–475.

123. Kruys V, Marinx O, Shaw G, Deschamps J, Huez G. Translational blockade imposed by cytokine-derived UA-rich sequences. Science 1989; 245:852–855.

124. Han CS, Dugan MJ, Verfaillie CM, Wagner JE, McGlave PB. In vitro expansion of umbilical cord blood committed and primitive progenitors. Exp Hematol 1994; 22:685.

125. Verfaillie CM, Catanzaro P, Vanoverbeek K, Miller JS. Adult marrow LTC-IC proliferate ex vivo when cultured in stroma-conditioned media containing MIP-1a + IL3. Exp Hematol 1994; 22:785.

126. Holyoake TL, Freshney MG, McNair L, McKay PJ, Parker A, Steward WP, Fitzsimons E, Pragnell IB. MIP-1α appears to promote self-renewal of primitive murine haemopoietic progenitors. Exp Hematol 1994; 22:817.

127. Miller JS, Verfaillie CM. Ex vivo culture of CD34+Lin−DR− cells in stroma derived soluble factors, MIP-1α and IL3 maintains not only myeloid but also lymphoid progenitors in a novel switch culture assay. Blood 1994; 84(suppl 1):419a.

128. Eaves CJ, Cashman JD, Wolpe SD, Eaves AC. Unresponsiveness of primitive chronic myeloid leukemia cells to macrophage inflammatory protein 1α, an inhibitor of primitive normal hematopoietic cells. Proc Natl Acad Sci USA 1993; 90:12015–12019.

129. Cashman JD, Eaves AC, Eaves CJ. The tetrapeptide AcSDKP specifically blocks the cycling of primitive normal but not leukemic progenitors in long-term culture: evidence for an indirect mechanism. Blood 1994; 84:1534–1542.

130. Ferrajoli A, Talpaz M, Zipf TF, Hirsch-Ginsberg C, Estey E, Wolpe SD, Estrov Z. Inhibition of acute myelogenous leukemia progenitor proliferation by macrophage inflammatory protein 1α. Leukemia 1994; 8:798–805.

131. Defard M, Lemoine FM, Khoury E, Bonnet M-L, Pontvert-Delucq S, Naiman M, Guigon M. Heterogeneity in the response of acute leukemia cells to the negative regulators serapenide, macrophage inflammatory protein 1α (MIP1α) and transforming factor β (TGFβ). Blood 1994; 84(suppl 1):617a.

132. Cook DN, Beck MA, Coffman TM, Kirby SL, Sheridan JF, Pragnell IB, Smithies O. Requirement of MIP-1α for an inflammatory response to viral infection. Nature 1995; 269:1583–1585.

133. Gao J-L, Murphy PM. Cloning and differential tissue-specific expression of three mouse β chemokine receptor-like genes, including the gene for functional macrophage inflammatory protein-1α receptor. J Biol Chem 1995; 270:17494–17501.

134. Hoogewerf A, Black D, Proudfoot AE, Wells TN, Power CA. Molecular cloning of murine CC CKR-4 and high affinity binding of chemokines to murine and human CC CKR-4. Biochem Biophys Res Comm 1996; 218:337–343.

135. Raport CJ, Gosling J, Schweickart VL, Gray PW, Charo IF. Molecular cloning and functional characterization of a novel human CC chemokine receptor (CCR5) for RANTES, MIP-1β and MIP-1α. J Biol Chem 1996; 271:1761–1766.

136. Aronica SM, Mantel C, Gonin R, Marshall MS, Sarris A, Cooper S, Hague N, Zhang XF, Broxmeyer HE. Interferon-inducible protein 10 and macrophage inflammatory protein-1 alpha inhibit growth factor stimulation of Raf-1 kinase activity and protein synthesis in a human growth factor-dependent haemopoietic cell line. Blood 1995; 270:21998–22007.

137. Heyworth CM, Pearson MA, Dexter TM, Wark G, Owen-lynch PJ, Whetton AD. Macrophage

inflammatory protein-1 alpha mediated growth inhibition in a haemopoietic stem cell line is associated with inositol 1,4,5 triphosphate generation. Growth Factors 1995; 12:165–172.

138. Cocchi F, DeVico AL, Garzino-Demo A, Arya SK, Gallo RC, Lusso P. Identification of RANTES, MIP-1 alpha and MIP-1 beta as the major HIV-suppressive factors produced by CD8[+] T cells. Science 1995; 270:1811–1815.

139. Balter M. Elusive HIV-suppressor factors found. Science 1995; 270:1560–1561.

140. Deng H, Liu R, Ellmeier W, Choe S, Unutmaz D, Burkhart M, Di Marzio P, Marmon S, Sutton RE, Hill CM, Davis CB, Peiper SC, Schall TJ, Littman DR, Landau NR. Identification of a major co-receptor for primary isolates of HIV-1. Nature 1996; 381:661–666.

141. Dragic T, Litwin V, Allaway GP, Martin SR, Huang Y, Nagashima KA, Cayanan C, Maddon PJ, Koup RA, Moore JP, Paxton WA. HIV-1 entry into CD4[+] cells is mediated by the chemokine receptor CC-CKR-5. Nature 1996; 381:667–673.

142. Weiss RA, Clapham PR. Hot fusion of HIV. Nature 1996; 381:647–648.

143. Dean M, Carrington M, Winkler C, Huttley GA, Smith MW, Allikmets R, Goedert JJ, Buchbinder SP, Vittinghoff E, Gomperts E, Donfield S, Vlahov D, Kaslow R, Saah A, Rinaldo C, Detels R, Hemophilia Growth and Development Study, Multicenter AIDS Cohort Study, Multicenter Hemophilia Cohort Study, San Francisco City Cohort, ALIVE Study, O'Brien SJ. Genetic restriction of HIV-1 infection and progression of AIDS by a deletion of the CKR5 structural gene. Science 1996; 273:1856–1861.

144. Cohen J. Receptor mutations help slow disease progression. Science 1996; 273:1797–1798.

145. Schmidtmayerova H, Sherry B, Bukrinsky M. Chemokines and HIV replication. Nature 1996; 382:767.

20

Transforming Growth Factor-β, Tumor Necrosis Factor-α, and Macrophage Inflammatory Protein-1α Are Bidirectional Regulators of Hematopoietic Cell Growth

Jonathan R. Keller, Francis W. Ruscetti, Krzysztof J. Grzegorzewski and Robert H. Wiltrout
SAIC—Frederick and National Cancer Institute, Frederick Cancer Research and Development Center, Frederick, Maryland

Stephen H. Bartelmez and Ewa Sitnicka
University of Washington, Seattle, Washington

Sten Eirik W. Jacobsen*
The Norwegian Radium Hospital, Oslo, Norway

I. INTRODUCTION

The proliferation and differentiation of hematopoietic cells is positively regulated both in vitro and in vivo by a growing family of cytokines (1–3). Similarly, hematopoietic cell growth can also be negatively regulated by the interferons (INF; 4,5), tumor necrosis factor (TNF; 6,7), macrophage inflammatory protein-1 alpha (MIP-1α; 8), transforming growth factor-beta (TGF-β; 9–11), and others (12). Thus, hematopoietic progenitor cell proliferation can be partly regulated by the balance between opposing positive and negative growth signals. Understanding the effects of positive and negative growth factors on hematopoietic development may provide important insights into the mechanisms that maintain the critical mass of developing hematopoi-

The content of this publication does not necessarily reflect the views or policies of the Department of Health and Human Services, nor does mention of trade names, commercial products, or organizations imply endorsement by the U. S. Government.
Current affiliation: University Hospital of Lund, Lund, Sweden.

etic cells, including the regulation of the survival, proliferation, and self-renewal of the most primitive quiescent pluripotential stem cells, the less primitive progenitor–stem cells, and the most committed–mature progenitors. This chapter will focus on the most recent advances from our laboratories, together with those of others on the effects of TGF-β, TNF-α, and MIP-1α on hematopoietic cell growth.

II. THE BIOLOGICAL EFFECTS OF TGF-β ON HEMATOPOIETIC PROGENITOR CELL GROWTH IN VITRO

A. Summary

Transforming growth factor-β is a member of a growing family of related polypeptides that have wide-ranging effects on various cell types and tissues (13,14). TGF-β was originally characterized by its ability to stimulate the proliferation of rodent fibroblasts (15,16), but has since been shown to be a multifunctional regulator of cell growth, the effects of which depend on the cell type targeted and the other cytokines present (17–19). All TGF-β family members show amino acid sequence similarity in the biologically active form, which is a disulfide-linked dimer of subunits containing 110–140 amino acids and the conserved spacing of seven cysteine residues (13,20). The subunits are derived from the COOH-terminal portion of a larger precursor that is cleaved at a cluster of basic residues, which releases the biologically active domain (20). Association of the biologically active form with the propeptide yields a latent complex that is unable to bind to receptor and can be activated later in the extracellular milieu (20).

The results of many studies have shown that, although TGF-β alone has no effect on hematopoietic cell growth, TGF-β is a bifunctional regulator of hematopoietic cell growth that can either inhibit (21–27) or stimulate (28–30) the growth of murine and human hematopoietic progenitor cells (summarized in Table 1). The effects of TGF-β are direct and selective and are dependent on the other growth factors present (22,23,28,31), the differentiation state of target cells (22,23,27,28,31), and the concentration of TGF-β in culture (31). To illustrate this, TGF-β

Table 1 Effects of TGF-β on Human and Murine Hematopoietic Colony Formation

Cytokine	Effect of TGF-β	
	Murine	Human
IL-3	I	I
GM-CSF	S	S (day 7 CFU)
		I (day 14 CFU)
G-CSF	N.E.	N.E.
CSF-1	N.E. (unfractionated BMCs)	N.A.
	I (Lin⁻ progenitors)	
EPO (CFU-E)	N.E.	N.E.
IL-3 + EPO (BFU-E, CFU-GEMM)	I	I
SLF + EPO		
or		
SLF + IL-3 + EPO		

I, inhibition; S, stimulation; N.E., no effect; N.A., not applicable; CFU, colony-forming unit; E, erythroid; meg, megakaryocyte; CFU-GEMM, granulocyte–erythroid–macrophage–megakaryocyte; BFU, burst-forming unit.

is a potent inhibitor of interleukin (IL)-3-induced murine bone marrow colony formation, but has little or no effect on granulocyte colony-stimulating factor (G-CSF)- or erythropoietin (EPO)-induced growth and differentiation of more committed hematopoietic progenitors (22). Furthermore, TGF-β completely inhibits primitive multiwedge progenitor growth (CFU-GEMM BFU-E) which is stimulated by the combination of EPO and IL-3 or steel factor (SLF), but has no effect on EPO-induced CFU-E colony formation. Interestingly, although TGF-β inhibits IL-3-induced colony formation, it enhances the growth of granulocyte–macrophage (GM)-CSF-induced colony formation (30). Combining the results of many studies, TGF-β directly (in single-cell assays) and selectively inhibits the growth of primitive hematopoietic progenitors and either has no effect or enhances the growth of committed progenitors (see Table 1).

Interestingly, MIP-1α, a member of the chemokine family, has remarkably similar biological effects on bone marrow cell growth in vitro (9,32); therefore, a direct comparison with MIP-1α was made on several purified primitive hematopoietic progenitors.

B. Direct Bidirectional Effects of TGF-β on Purified Primitive Hematopoietic Progenitor Cells: Comparison with MIP-1α

In contrast with effects of TGF-β on the growth of normal bone marrow cells in vitro, TGF-β inhibits the growth of growth factor-dependent progenitor cell lines, regardless of the cytokine used to promote cell proliferation, including IL-3, G-CSF, CSF-1, SLF, IL-4, IL-6, as well as GM-CSF, which synergizes with TGF-β on normal bone marrow cells (22,33). This observation is consistent with the hypothesis that TGF-β inhibits the growth of primitive hematopoietic progenitors, regardless of the cytokine used to promote proliferation, for the factor-dependent cell lines are representative of primitive progenitors blocked in their ability to differentiate. Because many of the progenitor cell lines express low levels of Thy-1 antigen and lack lineage specific antigens (Lin⁻), which are characteristics of primitive hematopoietic progenitor cell subsets (34), Lin⁻Thy-1^{lo} cells were purified from normal murine bone marrow to examine whether TGF-β might inhibit, rather than stimulate, the proliferation of these cells in response to GM-CSF in single-cell assays. TGF-β directly enhances the proliferation of isolated Lin⁻ cells in response to GM-CSF, and directly inhibits the growth of isolated Lin⁻Thy-1^{lo} cells in response to GM-CSF (summarized in Table 2). Furthermore, TGF-β inhibited the growth of Lin⁻Thy-1^{lo} cells, regardless of the cytokine used to stimulate single-cell proliferation, even if multiple cytokines were used, such as SLF plus IL-3 (35). Thus, similar to the results obtained with cell lines, TGF-β can inhibit GM-CSF–induced proliferation of normal purified primitive hematopoietic progenitors that is predicted by the expression of Thy-1 antigen.

In comparison, MIP-1α had similar direct effects on cytokine-induced proliferation of Lin⁻Thy-1^{lo} progenitor cells, but was a consistently less potent inhibitor than TGF-β. For example, although TGF-β inhibits between 75 and 90% or more of the growth factor-induced proliferation of single Lin⁻Thy-1^{lo} cells, MIP-1α inhibits between 25 and 40%. A significant difference between TGF-β and MIP-1α (not previously reported) was observed on the growth of isolated Lin⁻ cells in response to IL-3. Specifically, MIP-1α directly enhanced, whereas TGF-β directly inhibited, IL-3–induced growth of isolated Lin⁻ cells. MIP-1α forms inactive multimers in low-salt and dilute protein solutions, which can produce variable results among studies (36). The overall observed direct effects of TGF-β and MIP-1α on GM-CSF-induced growth of Lin⁻ cells was enhancement; however, both cytokines inhibited GM-CSF–induced proliferation of the Thy-1^{lo}- purified subpopulation of Lin⁻ cells (see Table 2).

The effect of TGF-β and MIP-1α were further compared on two other purified murine bone marrow progenitor populations separated from Lin⁻ cells and known to be enriched for cells with

Table 2 Direct Effects of TGF-β and MIP-1α on CSF-Dependent Growth of Purified Murine Bone Marrow Progenitors

Cells	Growth Factors					
	IL-3		GM-CSF		CSF-1	
	TGF-β	MIP-1α	TGF-β	MIP-1α	TGF-β	MIP-1α
Lin⁻	I	S	S	S	I	N.E.
Lin⁻Thy-1^{lo}	I	I	I	I	I	I
Lin⁻Thy-1^{Neg}	I	S	S	S	N.D.	N.D.

Murine Lin⁻ cells were further purified by FACS separation and seeded as single cells in Terasaki plates. A minimum of 1200 wells were scored per group. Cultures were supplemented with IL-3 (20 ng/ml), GM-CSF (20 ng/ml), CSF-1 (50–100 ng/ml) plus or minus TGF-β (20 ng/ml) or MIP-1α (50 ng/ml).
I, inhibition; S, stimulation; N.E., no effect; N.D., not determined

long-term reconstituting activity (LTRC) including stem cells purified on the basis of low Hoechst 33342 and low rhodamine 123 staining (Ho/Rh123; 37), and those that express stem cell antigen (Sca-1⁺ Ly-6 A/E; 34). TGF-β directly inhibits (80–90% in single cell assays) the proliferation of isolated HO/Rh123 Lin⁻ cells and Lin⁻Sca-1⁺ cells stimulated by the combination of SLF plus IL-3, whereas MIP-1α has no effect. Similar results were obtained when 50 Ho/Rh123 cells were expanded in liquid culture with IL-3, SLF, and IL-6 in the presence or absence of TGF-β or MIP-1α. Specifically, after an 8-day incubation in the combination of cytokines, the number of high-proliferative potential colony-forming cells (HPP-CFC) in control cultures expanded from 35 to 450, the number of GM-CFC from none to roughly 70,000, and the total cellularity to approximately 300,000. The addition of TGF-β to these cultures completely inhibited the proliferation and expansion of the initial cell input. In comparison, MIP-1α had little effect on HPP-CFC expansion, but showed a decrease in GM-CFC output to 25,000 and increase in total cellularity to 600,000 cells, relative to untreated cultures. The increase in total cellularity observed in the presence of MIP-1α was due to an increase in committed myeloid cells. Thus, although MIP-1α has little effect on the primitive HPP-CFC, it enhances the proliferation of the more-committed progeny, possibly through the synergistic action of MIP-1α with IL-3, resulting in increased total cellularity and a decrease in the total GM-CFC.

Thus, TGF-β and MIP-1α are direct bidirectional modulators of murine hematopoietic cell growth, the effects of which are dependent on the differentiation state of the target progenitor cells and the other cytokines present. Furthermore, TGF-β inhibits the growth of a broad spectrum of progenitor cells, including, Lin⁻Thy-1^{lo}, Lin⁻Sca-1⁺, low Ho/Rh123 cells, CFU-S-d12, CFU-A, HPP-CFC-1, HPP-CFC-2, CFU-GEMM, CFU-Mix, whereas MIP-1α has no effect on Lin⁻Sca-1⁺, low Ho/Rh123, HPP-CFC, CFU-GEMM, CFU-Mix, but inhibits CFU-A, CFU-Sd12, and Lin⁻Thy-1^{lo} cells. In addition, TGF-β enhances hematopoietic colony formation only in response to GM-CSF, whereas MIP-1α enhances colony formation in response to GM-CSF, IL-3, and, in some reports, to CSF-1 (32). Thus, TGF-β and MIP-1α partially overlap in their activities on murine hematopoietic progenitor growth, but they also show distinct effects, with TGF-β inhibiting the proliferation of several stem cell populations, whereas MIP-1α shows no effect.

Direct inhibitory effects of TGF-β on purified CD-34⁺ human progenitors has also been reported. When using purified peripheral blood CD-34+CD-33− stem cells, plated at 100 cells per dish, Sergiamo et al. demonstrated that TGF-β could inhibit 80–90% of the colony formation induced by GM-CSF plus IL-3 plus EPO (38). Subsequently, Lu et al. demonstrated that TGF-β directly inhibits (> 80%) the growth of purified CD-34^{Hi} bone marrow progenitors in single-cell assays in response to IL-3 plus SLF plus EPO (39).

Therefore, the overall biological effects of cytokines on the growth of hematopoietic cells in vitro is the summation or the combined effects of opposing stimulatory and inhibitory effects on cell growth.

C. Modulation of Hematopoietic Growth Factor Receptors by TGF-β

Recent attention has been focused on the mechanism(s) whereby TGF-β functions as an antiproliferative signal. These studies have shown that TGF-β can act at various levels, including inhibiting the expression of c-*myc* (40–42), preventing the phosphorylation of the retinoblastoma gene *RB* (40) and affecting the activity and expression of the cyclins and cyclin-dependent kinases (43,44). The effect of TGF-β at the level of hematopoietic growth factor (HGF) receptor expression has also been examined to determine whether modulation of HGF receptors could play a role in TGF-β–mediated effects on hematopoietic progenitor cell growth (summarized in Table 3). The effects of TGF-β on IL-1, IL-3, GM-CSF, G-CSF, and SLF receptor expression (receptor numbers and affinity) have been examined on hematopoietic cell lines that are growth-inhibited by TGF-β and resemble normal progenitor cells in that they maintain their dependence on growth factors for survival and proliferation, but do not terminally differentiate. In contrast with the rapid (5- to 10-min) down-modulation of receptors by their cognate ligands, TGF-β trans–down-modulates receptors for other ligands such as IL-1, GM-CSF, IL-3, G-CSF, and SLF receptors, which begins 6–12 hr after TGF-β addition and is maximum between 72 and 96 hr, depending on the ligand examined; IL-1 and SLF receptor trans–down-modulation is maximum by 24 hr (33,45,46). This results in a 65–80% reduction in receptor number, without an effect on receptor affinity by scatchard analysis.

The TGF-β–mediated trans–down-modulation of HGF receptors on factor-dependent cell lines could be sustained for up to 8 days in vitro and is completely reversible, regardless of the

Table 3 The effects of TGF-β on Hematopoietic Progenitor Cell Growth and Receptor Expression

Growth factor	Target cells	Growth proliferation + TGF-β	Receptor expression + TGF-β
IL-3	Cell lines	I	↓
IL-3	Lin⁻ progenitors	I	↓
GM-CSF	Cell lines	I	↓
GM-CSF	Lin⁻ progenitors	S	↑
G-CSF	Cell lines	I	↓
G-CSF	Lin⁻	N.E.	N.E.
CSF-1	Cell lines	I	N.D.
CSF-1	Lin⁻ progenitors	I	↓
SLF	Cell lines	I	↓
SLF	Lin⁻ progenitors	I	↓
IL-1	Cell lines	↓	↓
IL-1	Lin⁻ progenitors	N.D.	N.R.

duration of the TGF-β treatment. Furthermore, cell viability was not affected by the prolonged treatment with TGF-β. Comparing cell proliferation in [^{3}H]thymidine assays, total cell number in culture, and receptor expression, receptor trans–down-modulation precedes effects on DNA synthesis and cell growth. Taken together, TGF-β–mediated trans–down-modulation of growth factor receptors may act to keep progenitor cells in a quiescent state (unresponsive to positive acting growth factors) that is below the threshold for cytokine-induced growth as long as TGF-β is present.

Multiple mechanisms could account for the observed trans–down-modulation of HGF receptors by TGF-β, including effects on gene transcription, receptor trafficking, mRNA stability, and modifications of preexisting receptors. With FDC-P1 cells, we observed that the reduction in c-*kit* cell surface expression was preceded by a marked reduction in c-*kit* mRNA levels that could be seen by 2 hr and was maximum 6 hr after treatment with TGF-β (33). This inhibition was partly explained by the steady-state levels of c-*kit* mRNA that showed a half-life of 2–4 hr for control cells, compared with 0.5–1.5 hr for TGF-β–treated cells (33). In this regard, c-*kit* mRNA has several AU-rich motifs (47), which might be involved in the control mRNA stability (48,49). Also, the addition of TGF-β to SLF-responsive AML blast cells decreased the expression of cell surface c-*kit* and c-*kit* mRNA (50). Furthermore, transcriptional runon assays showed AML cells constitutively transcribed c-*kit* and that treatment with TGF-β did not alter this activity. Thus, one mechanism by which TGF-β regulates the responsiveness of hematopoietic progenitors to SLF is through decreased c-*kit* mRNA stability, leading to decreased cell surface expression of c-*kit*.

TGF-β has a bidirectional effect on freshly aspirated bone marrow cells that is dependent on the cytokine used and the differentiation state of the target cells. To determine whether the effects of TGF-β on the growth of normal bone marrow cells in vitro also correlated with receptor expression, we examined the effects of TGF-β on HGF receptor expression on unfractionated (light-density bone marrow cells; LDBM) and progenitor-enriched (lineage-negative; Lin⁻)bone marrow cells (51). TGF-β inhibited IL-3 (51), CSF-1 (51), and SLF (33) receptor expression, but enhanced GM-CSF, and had no effect on G-CSF-specific binding (51) to Lin⁻ progenitor cells, which directly correlates with the effect of TGF-β on HGF-induced colony formation for each cytokine, respectively. Interestingly, although TGF-β had the same effect on IL-3-, SLF-, GM-CSF-, and G-CSF-specific binding on LDBM cells, there was no significant effect of TGF-β on the binding of CSF-1 to LDBM cells. The difference between TGF-β's effects on CSF-1 receptor expression on LDBM cells, versus Lin⁻ progenitors, directly correlates with effects on CSF-1–induced colony formation in vitro, in that TGF-β had little or no effect on CSF-1–induced colony formation of LDBM cells, whereas it inhibited 40–50% of CSF-1–induced Lin⁻ progenitor cell colony formation (TGF-β inhibits 55% of CSF-1 binding to Lin⁻ cells after 24 hr). The effects of TGF-β on HGF receptor expression consistently preceded effects on [^{3}H]thymidine incorporation (51). For example, TGF-β–mediated inhibition of IL-3–stimulated bone marrow proliferation was preceded by reduced IL-3 receptor expression, and up-regulation of GM-CSF receptor expression preceded the synergistic effect of TGF-β on GM-CSF–stimulated proliferation.

The biological significance of growth factor receptor modulation has been questioned, because the level of receptor occupancy required to elicit a biological response has been reported to range from 10 to 50% (52,53). We demonstrated that treatment of IL-3–dependent DA-1 cells with TGF-β reduced the number of IL-3 receptors from 498 to 178 sites per cell and from 80 to 37 receptors per cell on LDBM. Thus, TGF-β reduces the mean number of receptors, which could put many progenitor cells below the threshold receptor level needed to elicit a response. However, most bone marrow cell populations are heterogeneous for HGF receptor expression, ranging from cells without detectable receptor expression, to cells with very high receptor expression (54,55). Thus, observed decreases in IL-3 receptor expression could result from a complete loss of receptor expression on progenitor cells or from a decrease in the number of receptors per cell, or from both mechanisms. Thus, TGF-β–mediated inhibition of hematopoietic cell growth could result from a turning on or off of receptor expression or from modulation of the number of receptors per cell. Mechanisms other than receptor modulation have been implicated in TGF-β–mediated regulation of progenitor cell growth; however, receptor modulation could be involved in maintaining the quiescence of primitive progenitor cells over an extended period.

D. Potential Physiological Role of TGF-β on Stem Cells

The foregoing results demonstrate that TGF-β can inhibit purified stem cell populations with LTRC activity and suggest that TGF-β might play a physiological role in regulating the maintenance of stem cell quiescence and the cycling status of normal hematopoietic cells. To address these questions, several studies have examined the effects of TGF-β on hematopoietic cell growth in long-term bone marrow cultures (LTBMCs), an in vitro culture system that contains a complex mixture of hematopoietic cells and stromal–accessory cells that can sustain the growth and proliferation of primitive progenitors over many weeks in culture (56). Both mRNA for TGF-β and active and latent TGF-β protein have been detected in the stroma of LTBMCs (57). Several important results have been obtained from these studies: first, the addition of exogenous TGF-β at the initiation of the culture completely arrests hematopoiesis, such that no clonogenic progenitors are found after 2 weeks (58); second, the addition of TGF-β inhibits the increased cycling of primitive progenitors initiated in the culture by the addition of IL-1 or medium change containing fresh horse serum (59); third, the addition of neutralizing TGF-β antibodies 3 days after the addition of IL-1, when progenitors were maximally cycling, allowed the cells to remain in S-phase, whereas progenitor cells in control cultures underwent cell cycle arrest (57); and fourth, the addition of neutralizing TGF-β antibodies to LTBMCs resulted in an increased output of mature myeloid cells, mature progenitors and CFUs d14 (60). Thus, the production of TGF-β by stromal cells results in negative-feedback to maintain stem cell quiescence and to inhibit progenitor cell cycling, and cytokines such as IL-1 can transiently override this inhibitory effect. The recent purification of low Ho/Rho 123 murine stem cells which can give long-term donor repopulation with as few as 5 cells and upon stimulation with SCF + IL-3 + IL-6 100% of the cells develop large HPP (> 100,000 cells) in vitro has allowed definitive demonstration that TGF-β directly and reversibly inhibits the cell division of stem cells containing long-term donor repopulation ability (61).

Interestingly, recent evidence suggests another mechanism that involves the autocrine production of TGF-β by hematopoietic progenitor cells and can negatively regulate the cycling status of progenitors. Specifically, the addition of TGF-β antisense oligonucleotides or TGF-β–neutralizing antibodies enhanced the number of CFU-GEMM, CFU-GM, and BFU-E colonies from purified CD34$^+$ cells in response to the combination of IL-3 plus IL-6 plus G-CSF and EPO (62,63). In some experiments, more than half of the detectable CD34$^+$ colony-forming progenitors were maintained in a growth factor-unresponsive state by autocrine TGF-β production. Similar effects of TGF-β-neutralizing antibodies have been observed on the growth of light-density bone marrow cells obtained from mice previously treated with 5-FU (progenitor-enriched population; 64). In particular, with antibodies that neutralize TGF-β, the autocrine production of TGF-β blocks the growth factor response of primitive stem cells to SLF or SLF plus IL-11, and this effect was direct in single-cell assays. Furthermore, neutralizing TGF-β antibodies shortened the time of appearance of colonies in soft agar, increased the total number and size of developing colonies, and increased the total number of progenitors expanded in liquid cultures in response to growth factors. The mechanism by which TGF-β–neutralizing antibodies enhance growth factor responsiveness of progenitor cells is unknown; however, as will be discussed in the following section, effects on increased growth factor receptor expression may play a role. Thus, the regulation of TGF-β production can occur either at the level of stromal cells (paracrine), or at the level of hematopoietic progenitor cells (autocrine), or both.

E.   Effect of TGF-β In Vivo and Potential Clinical Relevance

It is known that slowly proliferating or quiescent hematopoietic progenitor–stem cells are less sensitive to cell cycle-active cancer therapeutic agents and radiation, whereas actively cycling progenitor cells are killed (65). Furthermore, high-dose or accelerated chemotherapy has been positively correlated with the increased tumor response (67). Thus, given the results obtained in vitro, TGF-β could improve the dose-limiting toxicity of these agents by protecting cycling bone marrow progenitors, either in vivo or ex vivo, in bone marrow-purging protocols. Bone marrow cells obtained from mice previously treated with TGF-β locoregionally showed a dose- and time-dependent inhibition of IL-3–induced bone marrow colony formation in vitro, with primitive progenitors (CFU-GEMM) more potently inhibited than the committed progenitors (CFU-c) (67). Migdalska et al. (68) examined more primitive progenitors in vivo and showed that IP administration of 5 μg/mouse TGF-β reversibly inhibited the cycling of day 8 and day 12 colony-forming unit spleen (CFU-S) progenitor cells. Also, Carlino et al. demonstrated that a 14-day SC administration of TGF-β resulted in a decrease in red blood cell and platelet counts, elevated white cell counts, and slightly elevated circulating CFU-S in the peripheral blood, with no significant changes in granulocytes (69,70). These effects were reversible on suspension of the treatment. Morphological analysis of the spleen and bone marrow in treated mice revealed a reduction in the red pulp in the spleen; however, there was a concomitant increase in granulopoiesis in the spleen and bone marrow. Interestingly, the observed increase in granulopoiesis in vivo directly correlates with studies demonstrating that TGF-β can enhance granulopoiesis in vitro in the presence of GM-CSF, possibly through the up-regulation of GM-CSF receptors (30). In two other studies examining the single administration of TGF-β, TGF-β enhanced the total number of GM-CSF–responsive bone marrow progenitor cells in vivo, without effecting erythroid lineage progenitors (71,72). These studies differed in that one required the sequential administration of TGF-β, followed by GM-CSF (71), and the other study required only TGF-β (72). Interestingly, similar to the results obtained in vitro, TGF-β up-regulates the expression of GM-CSF receptors on bone marrow cells in vivo that was maximum by 40 hr, and led to a rational design for the sequential administration of the two cytokines (71). Taken together, TGF-β acts as a multifunctional regulator of hematopoietic cell growth, both in vitro and in vivo, by selectively inhibiting primitive stem cell growth, resulting in the decreased production of red cells and platelets, while promoting granulopoiesis.

The studies presented in the foregoing suggested that TGF-β could protect critical stem and progenitor cells from the myelotoxic effects of cell cycle-active drugs; therefore, mice were pretreated with TGF-β before the administration of a high dose (60% lethal) of 5-FU (73). Unexpectedly, the prior administration of TGF-β potentiated the toxicity of high doses of 5-FU (90–100% lethal); however, there was a marked increase in bone marrow progenitors before death in mice that received TGF-β and 5-FU, versus those that received only 5-FU. Interestingly, the 5-FU toxicity that was potentiated by the pretreatment with TGF-β was reversed (90% survival versus 90–100% lethal) by the administration of a suboptimal number of bone marrow cells that had little effect on mice that received only 5-FU. Because chemotherapeutic agents are administered to patients in repeated cycles, the effect of TGF-β was then compared in a two-hit model of 5-FU in which a sublethal dose of 5-FU was administered, followed either on day 8 or day 12 with a second higher dose of 5-FU. When TGF-β was administered 3 days before the second dose of 5-FU on day 12, 90% of the mice survived versus 40% for the mice that received only 5-FU. Thus, these data provide the first preclinical evidence that negative regulators may be practically useful for chemoprotection in vivo. However, the administration of TGF-β before the first administration of 5-FU required a minimal amount of normal transplanted bone marrow cells (short-term marrow-repopulating activity), whereas no bone marrow support was required when 5-FU was used during bone marrow rebound in two-hit regimens. The mechanisms for these differences is currently unknown; however, because the rebound marrow is highly enriched for cycling progenitors, when compared with normal bone marrow, it is possible there is a critical number of STMRA that are protected in the rebound marrow, versus the control, that can promote survival. These data provide the rational basis to examine other chemotherapeutic agents, such as cyclophosphamide and doxorubicin, and provides a rational basis to examine the combination of negative regulators with positive regulators allowing dose escalation and acceleration of hematopoietic recovery.

III. TUMOR NECROSIS FACTOR-α IS A MULTIFUNCTIONAL REGULATOR OF HEMATOPOIESIS: DIRECT AND INDIRECT EFFECTS

Tumor necrosis factor-α (TNF-α) is a polypeptide growth factor with pleiotropic biological effects on a wide range of cell types (reviewed in Refs. 74–77). The active native form of TNF-α is a homotrimer, with a molecular mass of 52 kDa (78). TNF-α was originally identified, in the serum of endotoxin-treated mice, as an activity capable of inducing necrosis of transplantable tumors (79). Numerous studies have examined the effects of TNF-α on hematopoietic cell growth, and seemingly conflicting data have been reported relative to the effects of TNF-α on in vitro hematopoiesis, ranging from potent inhibition (6,80–92), to stimulation (88–92). The initial studies of TNF-α on hematopoiesis suggested that TNF-α could potently inhibit the growth of multipotent progenitor cells, granulocyte–macrophage (GM) colonies, as well as erythroid colony-forming cells in normal bone marrow (6,80–87). In addition, it was demonstrated that TNF-α and IFN-γ could synergistically suppress hematopoietic progenitor cell growth (6,80,82), similar to the synergy observed between the cytotoxicities of these two cytokines (reviewed in Ref. 77). These early studies also demonstrated that progenitors from patients with acute leukemia were more sensitive to the inhibitory effects of TNF-α than normal bone marrow progenitors (6,82). These studies, for the most part, were performed on unfractionated or partially enriched progenitors and, thus, did not address whether the observed effects of TNF-α were mediated directly or indirectly by stimulating the production of other cytokines by accessory cells. In particular, TNF-α potently stimulates the production of multiple HGFs, including IL-1, IL-6, CSF-1, GM-CSF, and G-CSF (93–97).

The use of single-cell assays have enabled researchers to more accurately determine the ability of TNF-α to directly and bifunctionally (inhibit or stimulate) influence the growth of hematopoietic progenitor cells (88–92,98,99). Caux et al. demonstrated that TNF-α could synergistically enhance GM-CSF– or IL-3–stimulated colony formation, whereas TNF-α inhibited the response to G-CSF using purified CD34+ cells (88). Limiting dilution studies suggested that the stimulatory effects were directly mediated on the targeted progenitors. A study by Backx et al. found that TNF-α inhibited EPO-stimulated erythroid colony formation (90). TNF-α stimulates GM-CSF–stimulated proliferation of acute myeloblastic leukemia progenitor cells; however, endogenously produced IL-1 mediates this effect (98). A study on enriched murine bone marrow progenitors also demonstrated the ability of TNF-α to synergistically enhance the proliferation observed in response to GM-CSF and IL-3 (91). However, at variance with the human studies on normal progenitors, these studies suggested that the stimulatory effects were indirect through stimulation of cytokine production, whereas the direct effect of TNF-α was inhibitory, regardless of the other cytokines present (91). In later studies on human CD34$^+$ bone marrow progenitors, it was confirmed by single-cell assays that TNF-α can directly stimulate GM colony formation in combination with GM-CSF or IL-3 and can inhibit the response to G-CSF and CSF-1 (99). In agreement with the studies of Backx et al. (90), we observed that the stimulatory effects on GM-CSF– or IL-3–induced colony growth of human CD34$^+$ bone marrow progenitors could be observed only at low TNF-α concentrations, whereas high TNF-α concentrations were inhibitory (99). Thus, differences between the direct and indirect effects of TNF-α on murine and human bone marrow progenitors could be due to differences in culture conditions or in dose responses. However, all these studies used heterogeneous populations of progenitor cells; thus, the net effect of a bifunctional regulator, such as TNF-α, might differ. Furthermore, because the stimulatory effects of TNF-α are restricted to low concentrations of TNF-α, it cannot be excluded that they are mediated through stimulation of an autocrine (progenitor cell) production of cytokines. Nonetheless, it appears that the ability of TNF-α to synergize with GM-CSF and IL-3 is more pronounced on human than on murine hematopoietic progenitors. Interestingly, TNF-α and GM-CSF also synergize to stimulate the formation of dendritic cells from CD34$^+$ human cord blood cells (100,101).

Synergism is a phenomena observed predominantly on primitive hematopoietic progenitors (reviewed in Refs. 1,102,103), and has led the authors to conclude that TNF-α should be considered an enhancer, rather than an inhibitor, of early myelopoiesis (92). However, in our opinion, it remained possible that the specific growth factor that TNF-α interacts with is the critical determinant for the qualitative response (stimulatory or inhibitory) elicited, and that the maturity of the targeted progenitor could be a determinant for the magnitude of synergy observed between TNF-α and GM-CSF or IL-3. Several recent studies support this hypothesis. In one study, the multifactor-dependent growth of primitive human HPP-CFCs from CD34$^+$-purified bone marrow was always inhibited by TNF-α when HPP-CFC were stimulated by combinations, including G-CSF, whereas HPP-CFC colonies were enhanced by TNF-α when stimulated in combination with GM-CSF plus IL-3 (99). These studies also confirmed that the stimulatory effect of TNF-α could be observed only at low concentrations of TNF-α, whereas higher concentrations of TNF-α inhibit progenitor cell growth, regardless of the cytokines used to stimulate progenitor proliferation.

Thus far, only TGF-β has been identified as a negative regulator of SLF-stimulated hematopoiesis (33,35). Studies on murine, as well as human bone marrow progenitor cells, addressing specifically the interaction of TNF-α with SLF, revealed that TNF-α is also a potent inhibitor of SLF-stimulated normal myelopoiesis in vitro. For example, TNF-α directly inhibits the weak growth-promoting activity of SLF on murine Lin$^-$Sca-1$^+$ bone marrow cells in a concentration-dependent manner and inhibits the synergistic stimulation of Lin$^-$Sca-1$^+$ cells in response to SLF in combination with IL-3, G-CSF, and IL-6, suggesting that TNF-α can antagonize the potent stimulatory effects of SLF on primitive murine hematopoietic progenitor cells. Similar studies on CD34$^+$ bone marrow cells confirmed that TNF-α is a potent direct inhibitor of SLF-stimulated proliferation of human hematopoietic progenitor cells (104). In these studies, TNF-α (20 ng/ml) completely blocked GM colony formation observed in response to

SLF plus G-CSF or SLF plus CSF-1, whereas TNF-α reduced SLF plus GM-CSF- or SLF plus IL-3-stimulated colony formation to that observed in the absence of SLF (104). Thus, the ability of TNF-α to inhibit SLF-stimulated progenitor cell growth, as well as recent experiments showing that TNF-α also inhibits Flt-3–induced hematopoietic cell growth (105), another stimulator of early hematopoiesis (106,107), further supports the conclusion that the growth factor with which TNF-α interacts is an important determinant of the response elicited by TNF-α on normal hematopoietic progenitor cells.

Most previous reports have demonstrated an inhibitory effect of TNF-α on in vitro erythropoiesis (6,82,85,89,90), and TNF-α has been implicated in the pathogenesis of anemia in chronic diseases (reviewed in Ref. 108). Some studies have suggested that the inhibitory effects of TNF-α on CFU-E colonies are indirect through stimulation of IFN-β by accessory cells (109,110). However, the inhibitory effects of TNF-α on BFU-E colony formation of CD34[+] human bone marrow cells were directly mediated in single-cell experiments (111). In addition, human BFU-E colony formation was suppressed by TNF-α, regardless of the cytokines stimulating erythroid colony formation, for inhibition was observed in response to EPO alone or in combination with IL-3, IL-9, or SLF. Because erythroid colony formation has an absolute requirement for EPO, this might simply reflect the suppression of EPO-induced erythropoiesis. Interestingly, the ED_{50} for TNF-α–induced inhibition was 1000-fold higher for BFU-E colony formation in response to SLF plus EPO versus EPO alone (20 ng/ml and 0.02 ng/ml, respectively). Thus, the more primitive progenitors recruited by SLF plus EPO may be less sensitive to TNF-α than those responding to EPO alone, or alternatively, the combined action of multiple stimulatory cytokines might override the inhibitory actions of TNF-α, described as follows. Table 4 summarizes the interaction (synergism or inhibition) of TNF-α with different HGFs on the proliferation of murine and human hematopoietic progenitor cells.

A. TNF-α Regulation of Hematopoiesis Signaling Through Two Distinct Receptors

The discovery and cloning of two distinct TNF receptors (TNFR, p55 and p75), which belong to an emerging larger family of cytokine receptors (112–114) capable of signaling separately, led to the development of agonistic antibodies and TNF mutant proteins with selective activity on either of the two receptors (115–118). Although most cell types express both TNF receptors, TNF-α–induced effects are mediated predominantly through the TNFR55, such as cytotoxicity, proliferation of fibroblasts, and prostaglandin synthesis (reviewed in Refs. 109,110). In contrast, responses signaled through the TNFR75 are much more restricted and limited to stimulation of T cells and induction of GM-CSF production (109,110,115). The relative role of these two receptors in signaling the bifunctional effects of TNF-α on the growth of hematopoietic progenitor cells has been investigated using agonistic antibodies or TNF mutant proteins signaling specifically through either TNFR55 or TNFR75 (summarized in Table 5).

TNF-α enhances GM-CSF–induced proliferation but inhibits G-CSF–stimulated growth through TNFR55, but not through TNFR75, on human myeloid leukemia blast cells (119). The lack of TNFR75 involvement was not due to differential expression of the two receptors, because both receptors were expressed in most AML patients studied (119). In agreement with these studies, experiments with normal murine bone marrow progenitor cells found that the inhibitory effects of TNF-α on G-CSF-- or IL-3–stimulated colony formation of committed progenitor cells was mediated through TNFR55 (120). In contrast, the inhibitory effects of TNF-α on purified Lin⁻Sca-1⁺ progenitor cells was signaled through TNFR75, as shown by use of TNFR agonist antibodies, thereby implicating a role of TNFR75 in early hematopoiesis (120).

Table 4   Direct In Vitro Interactions of TNF-α
with Stimulatory HGFs on the Proliferation of
Hematopoietic Progenitor Cells: Proliferative
Effect of TNF-α on Hematopoietic Progenitors

Cytokine	Human	Mice
G-CSF	I	I
CSF-1	I	I
EPO	I	I
SCF	I	I
Flt-3 ligand	I	I
GM-CSF	S/I[a]	I
IL-3	S/I[a]	I

The table summarizes the inhibitory (I) and stimulatory (S) effects of TNF-α on the proliferation
of human and murine hematopoietic progenitor
cells stimulated by various stimulatory HGF.
[a]For GM-CSF and IL-3-stimulated human bone
marrow progenitor cells, enhancement by TNF-α is
observed predominantly on more primitive progenitors and at low TNF-α concentrations, whereas
higher concentrations of TNF-α either have no
effect or are inhibitory.

This was confirmed by comparing the effects of murine and human TNF-α on proliferation of
murine progenitors, for human TNF-α binds and activates only the murine TNFR55 (113,115).
The involvement of TNFR75 in signaling growth inhibition was observed on Lin⁻Sca-1⁺
progenitors cultured individually, regardless of the cytokines stimulating their growth. Cytokine
stimulation of HPP-CFC colony growth was also specifically inhibited by the TNFR75 agonist,
but not the TNFR55 agonist (120). A similar study on human CD34⁺ bone marrow progenitor
cells demonstrated that the suppressive effects of TNF-α on G-CSF–induced colony formation,
as well as the synergistic effects on GM-CSF–– and IL-3–stimulated clonal proliferation, were
exclusively signaled through TNFR55 (99). However, in agreement with the murine studies,
TNF-α–induced inhibition of human HPP-CFC colony formation involved signaling through the
TNFR75. Interestingly, the synergistic enhancement of TNF-α on GM-CSF plus IL-3–stimulated
HPP-CFC growth was mediated only through TNFR55 (99). Also, TNF-α can enhance the
colony formation of Lin⁻Sca-1⁺ progenitor cells, in combination with SLF plus IL-7, and this
effect was mediated through TNFR55. Thus, both TNFR55 and TNFR75 are involved in
mediating the pleiotropic effects of TNF-α on in vitro hematopoiesis. Most hematopoietic effects
of TNF-α appear to be signaled solely through TNFR55, whereas TNFR75 seems to be involved
in signaling inhibition of primitive progenitor cells (see Table 5).

Although the mechanisms by which TNF-α can bifunctionally affect the growth of hematopoietic progenitor cells remains unclear, TNF-α can modulate the cell surface expression
of numerous cytokine receptors on hematopoietic progenitor cells. Specifically, TNF-α can potently signal down-regulation of the expression of G-CSF receptors on normal as well as leukemic myeloid progenitor cells through the TNFR55 receptor (91,119,120,122), which correlates
with growth suppression induced by G-CSF. In contrast, TNF-α up-regulates the expression of

Table 5 Involvement of TNFR55 and TNFR75 in TNF-α-Induced Regulation of Hematopoietic Progenitor Cell Proliferation

Progenitor population	Involvement of TNF receptors		
	TNF-α response	TNFR55	TNFR75
Committed myeloid (murine)	I	+	−
Committed myeloid (human)	S	+	−
Primitive myeloid (murine)	I	(+)[a]	+
Primitive myeloid (human)	S	+	−
AML blasts	I	+	−
AML blasts	S	+	−

The involvement of TNFR55 and TNFR75 in signaling stimulatory (S) and inhibitory (I) effects of TNF-α on the clonal growth of normal committed and primitive hematopoietic progenitor cells (derived from human and murine studies) as well as on the proliferation of AML blasts (human only).

[a]Whereas no involvement of TNFR55 was observed on primitive murine Lin⁻Sca-1⁺ progenitors (120), TNFR55 as well as TNFR75 was involved in mediating inhibition of human HPP-CFCs (99).

IL-3 and GM-CSF receptors on normal and leukemic progenitor cells (91,119,123,124), and this is correlated with an enhanced proliferative response. Interestingly, the up-regulation of GM-CSF and IL-3 receptor expression is due to an up-regulation of their common β-subunit (124).

In a recent study, TNF-α up-regulated the c-*kit* mRNA and cell surface expression on acute myeloid leukemic blasts, and this corresponded to a synergistic interaction between SLF and TNF-α (125). In contrast, TNF-α inhibited SLF-induced proliferation of normal human and murine bone marrow progenitors (104), and this was accompanied by down-regulation of c-*kit* expression (104). Whether the direct interactions between TNF-α and SCF are in fact different on normal (inhibition) and leukemic (synergy) progenitors remains uncertain, in particular in light of the previous finding that AML blasts can produce cytokines potentially capable of synergizing with SCF (98). Taken together, all the effects of TNF-α on the growth of hematopoietic progenitor cells cannot be explained by its ability to bidirectionally modulate the expression of receptors for other cytokines. However, the described studies collectively suggest that receptor modulation might be one mechanism involved in mediating the pleiotropic effects of TNF-α on hematopoiesis.

In vivo studies have demonstrated enhancing as well as inhibitory effects of TNF-α on hematopoiesis (reviewed in Ref. 108). Furthermore, TNF-α confers myeloprotection of lethally irradiated mice (126–128). However, it remains uncertain to what degree these results are due to a direct inhibitory effect of TNF-α on primitive hematopoietic progenitors, or are due to its ability to stimulate the production of cytokines. Thus, the complex physiological role of TNF-α awaits more studies before any potential clinical use can be determined. Of particular interest are ongoing studies of mice with selective knockout of either TNFR55 or TNFR75 (129,130).

B. Recent Advances in Understanding the Role of TNF-α in Hematopoiesis

Since this review was completed during the fall of 1994, a number of new studies have shed new light on the potential role of TNF and its two receptors in hematopoiesis. We have demonstrated that TNF-α also can bidirectionally affect the viability/apoptosis of primitive murine hematopoietic progenitor cells in vitro (131). Of particular interest is a recent report demonstrating an increased number of candidate stem cells in TNFR55-deficient mice, supporting that TNF-α might play a role in inhibiting the proliferation and expansion of early progenitor/stem cells (132). In addition, these studies demonstrated that TNFR55 is critically involved in mediating TNF-induced growth inhibition of primitive progenitor cells (132).

IV. BALANCE BETWEEN OPPOSING POSITIVE AND NEGATIVE GROWTH SIGNALS

It has been proposed that hematopoietic cell growth can be partly regulated by a complex interaction between stimulatory and inhibitory growth factors. Several studies have suggested that cytokines that stimulate proliferation can override the effects of TGF-β (133–135). However, in these studies, combining stimulatory cytokines resulted in synergy and increased the cloning efficiency of progenitor cells; thus, it remained possible that TGF-β had little, or no effect, on the additionally recruited progenitors, resulting in the conclusion that stimulatory cytokines can override the effects of inhibitory cytokines. Whereas various growth factors can directly synergize to promote the growth of isolated Lin⁻Thy-1^{lo} cells, the combination of SLF plus IL-3 promotes maximum cloning efficiency in single-cell assays (roughly 38 colonies in 300 single cells), and TGF-β inhibits 90% of the proliferation to 4 in 300 single cells. Furthermore, the addition of a five factors (IL-1, IL-6, GM-CSF, M-CSF, and G-CSF) to the combination of SLF plus IL-3 does not increase the overall cloning efficiency of single Lin⁻Thy-1^{lo} cells (Table 6). However, this combination of cytokines increases the number of TGF-β–unresponsive progenitors to 27–29 in 300 single cells, resulting in only a 25% inhibition of single-cell growth. Furthermore, increasing the concentration of TGF-β from 20 to 50 or 100 ng/ml does not change

Table 6   Direct Cooperation Between Negative Growth Regulators on Multigrowth Factor-Stimulated Lin⁻Thy-1^{lo} Progenitor Cells

	Inhibitor addition				
Growth factors	TGF-β	TNF-α	INF-γ	MIP-1α	Growth inhibition (%)
IL-3 + SLF	+	−	−	−	90
Multifactor	+	−	−	−	25
Multifactor	−	+	−	−	22
Multifactor	−	−	+	−	17
Multifactor	−	−	−	+	0
Multifactor	+	+	−	−	25
Multifactor	+	−	−	+	25
Multifactor	+	−	+	+	59
Multifactor	+	−	+	+	59
Multifactor	+	+	+	−	76

Lin⁻Thy-1^{lo} cells were seeded as single cells in Terasaki plates. A minimum of 1200 wells were scored for each group. The cultures were supplemented with IL-3 plus SLF or multiple growth factors that included IL-3/SLF plus IL-1, IL-6, GM-CSF, G-CSF, and CSF-1. The cultures were also supplemented with the indicated inhibitors [plus inhibitor (+), or no inhibitor (−)].

the observed result (135). Additional experiments were designed to determine which of the growth factors, in combination with IL-3 plus SLF, could overcome the effects of TGF-β. G-CSF compared with the other cytokines as a third factor to the combination of SLF plus IL-3 increased the number of TGF-β–resistant clones to 8 in 300 single cells (80% inhibition). IL-1 compared with the other cytokines, as a fourth factor to the combination of SLF plus IL-3 plus G-CSF, further increased the number of TGF-β–resistant clones to 11–13 in 300 cells (65–70% inhibition). Thus, multiple stimulatory cytokines can directly cooperate to partially override the inhibitory effects of TGF-β. These studies also demonstrate that multiple signals are required to release hematopoietic stem cells from the inhibitory actions of TGF-β, which could partly explain why multiple stimulatory cytokines have overlapping biological activities.

Because multiple simulators could partially overcome the effects of single inhibitors, we examined whether multiple inhibitors, as TNF-α, MIP-1α, and INF-γ, could directly cooperate with TGF-β to inhibit the proliferation of hematopoietic progenitor cells. The addition of TNF-α, MIP-1α, or INF-γ as single inhibitors to isolated Lin⁻Thy-1^{lo} cells, stimulated by the seven-factor combination, resulted in 22% inhibition, 17% inhibition, and no effect, respectively, compared with 25% inhibition for TGF-β alone. Although MIP-1α or TNF-α in combination with TGF-β showed no more inhibition than TGF-β alone, the addition of INF-γ plus TGF-β resulted in significantly more inhibition (from 25% inhibition to 59% inhibition) of single-cell growth than TGF-β alone. Furthermore, TNF-α, MIP-1α, potentiated the inhibition observed with TGF-β plus INF-γ, resulting in 76% inhibition of single-cell growth, stimulated by seven cytokines. The inability of MIP-1α to cooperate with TGF-β can be partially explained by the observation that TGF-β reversibly down-regulates MIP-1α receptors 50–70% after 24 hr (137).

These data support the concept that hematopoiesis is regulated by a balance between opposing positive and negative growth regulators and demonstrate that TNF-α and INF-γ can directly cooperate with TGF-β to inhibit multifactor-stimulated progenitors that are unresponsive to either inhibitor alone. Thus, similar to stimulatory cytokines, inhibitory cytokines can synergize to affect the growth of hematopoietic progenitors.

REFERENCES

1. Ogawa M. Differentiation and proliferation of hematopoietic stem cells. Blood 1993; 81:2844–2853.
2. Metcalf D. The molecular control of cell division, differentiation commitment and maturation in haemopoietic cells. Nature 1989; 339:27–30.
3. Nicola NA. Hemopoietic cell growth factors and their receptors. Annu Rev Biochem 1989; 58:45–77.
4. Broxmeyer HE, Lu L, Platzer E, Feit C, Juliano L, Rubin BY. Comparative analysis of human gamma, alpha and beta interferon on human multipotential (CFU-GEMM) progenitor cells. J Immunol 1983; 131:1300.
5. Raefsky EL, Platanias LC, Zoumbos NC, Young NS. Studies of interferon as a regulator of hematopoietic cell proliferation. J Immunol 1985; 135:2507.
6. Broxmeyer HE, Williams DE, Lu L, Cooper S, Anderson SL, Beyer GS, Hoffman R, Rubin BY. The suppressive influences of human tumor necrosis factors on bone marrow hematopoietic progenitor cells from normal donors and patients with leukemia: synergism of tumor necrosis factor and interferon-γ. J Immunol 1986; 136:4487–4493.
7. Akahane K, Hosoi T, Urabe A, Kawakami M, Takaku F. Effects of recombinant human tumor necrosis factor (rhTNF) on normal human and mouse hemopoietic progenitor cells. Int J Cell Cloning 1987; 5:16.
8. Broxmeyer HE, Sherry B, Lu L, Cooper S, Oh D, Tekamp-Olson P, Kwon BS, Cerami A. Enhancing and suppressing effects of recombinant murine macrophage inflammatory proteins on colony formation in vitro by bone marrow myeloid progenitor cells. Blood 1990; 76:1110.

9. Keller J, Jacobsen SE, Dubois C, Hestdal K, Ruscetti F. Transforming growth factor β and its role in hematopoiesis. Int J Cell Cloning 1992; 10:2–11.

10. Ruscetti F, Dubois C, Jacobsen SE, Keller JR. Transforming growth factor β and interleukin 1: a paradigm for opposing regulation of hematopoiesis. In: Lord B, Dexter TM, eds. Growth Factors in Haemopoiesis. London: WB Saunders, 1992:703–721.

11. Broxmeyer HE. Suppressor cytokines and regulation of myelopoiesis. Am J Pediatr Hematol Oncol 1992; 14:22–35.

12. Axelrad A. Some hemopoietic negative regulators. Exp Hematol 1990; 18:143–150.

13. Roberts AB, Sporn MB, eds. In: The Handbook of Experimental Pharmacology. Peptide Growth Factors and Their Receptors. Heidelberg: Springer-Verlag, 1989:419–472.

14. Sporn MB, Roberts AB. TGF-β: problems and prospects. Cell Regul 1990; 1:875–882.

15. DeLarco JE, Todaro GJ. Growth factors from murine sarcoma virus transformed cells. Proc Natl Acad Sci USA 1978; 75:4001.

16. Roberts AB, Anzano MA, Meyers CA, Lamb LC, Smit LM, Sporn MB. New class of transforming growth factor potentiated by epidermal growth factor: isolation from non-neoplastic tissue. Proc Natl Acad Sci USA 1981; 78:5339.

17. Battegay EJ, Raines EW, Seifert RA, Bowen-Pope DF, Ross R. TGF-β induces bimodal proliferation of connective tissue cells via complex control of an autocrine PDGF loop. Cell 1990; 63:515.

18. Centrella M, McCarthy TL, Canalis E. Transforming growth factor β is a bifunctional regulator of replication and collagen synthesis in osteoblast-enriched cell cultures from fetal rat bone. J Biol Chem 1987; 262:2869.

19. Moses HL, Tucker RF, Leof EB, Coffey RJ, Halper J, Shipley GD. Type beta transforming growth factor is a growth stimulator and a growth inhibitor. Cancer Cells 1987; 3:65.

20. Massague J. The transforming growth factor-β family. Annu Rev Cell Biol 1990; 6:597–641.

21. Ohta M, Greenberger JS, Anklesaria P, Bassols A, Massagué J. Two forms of transforming growth factor-β distinguished by multipotential haematopoietic progenitor cells. Nature 1987; 329:539.

22. Keller JR, Mantel C, Sing GK, Ellingsworth L, Ruscetti SK, Ruscetti FW. Transforming growth factor $β_1$ selectively regulates early murine hematopoietic progenitors and inhibits the growth of IL-3-dependent myeloid leukemia cell lines. J Exp Med 1988; 168:737.

23. Sing GK, Keller JR, Ellingsworth LR, Ruscetti FW. Transforming growth factor β selectively inhibits normal and leukemic human bone marrow cell growth in vitro. Blood 1988; 72:1504.

24. Hino M, Tojo A, Miyazono K, Urabe A, Takaku F. Effects of type β transforming growth factors on hematopoietic progenitor cells. Br J Haematol 1988; 70:143.

25. Keller JR, Sing GK, Ellingsworth LR, Ruscetti FW. Transforming growth factor β: possible roles in the regulation of normal and leukemic hematopoietic cell growth. J Cell Biochem 1989; 39:175.

26. Ishibashi T, Miller SL, Burnstein SA. Type beta transforming growth factor is a potent inhibitor of murine mega-karyocytopoiesis in vitro. Blood 1989; 69:1737.

27. Keller JR, McNiece IK, Sill KT, Ellingsworth LR, Quesenberry PJ, Sing GK, Ruscetti FW. Transforming growth factor β directly regulates primitive murine hematopoietic cell proliferation. Blood 1990; 75:596.

28. Ottmann OG, Pelus LM. Differential proliferative effects of transforming growth factor-β on human hematopoietic progenitor cells. J Immunol 1988; 140:2661.

29. Aglietta M, Stacchini A, Severino A, Sanavio F, Ferrando ML, Piacibello W. Interaction of transforming growth factor-beta 1 with hematopoietic growth factors in the regulation of human normal and leukemic myelopoiesis. Exp Hematol 1989; 17:296.

30. Keller JR, Jacobsen SEW, Sill KT, Ellingsworth LR, Ruscetti FW. Stimulation of granulopoiesis by transforming growth factor β: synergy with granulocyte/macrophage-colony-stimulating factor. Proc Natl Acad Sci USA 1991; 88:7190.

31. Jacobsen SEW, Keller JR, Ruscetti FW, Kondaiah P, Roberts AB, Falk LA. Bidirectional effects of TGF-β on colony-stimulating factor induced human myelopoiesis in vitro: differential effects of distinct TGF-β isoforms. Blood 1992; 78:2239.

32. Broxmeyer HE, Sherry B, Lu L, Cooper S, Carob C, Wolpe SD, Cerami A. Myelopoietic enhancing

effects of murine macrophage inflammatory proteins 1 and 2 on colony formation in vitro by murine and human bone marrow granulocyte/macrophage progenitor cells. J Exp Med 1989; 170:1583.

33. Dubois CM, Ruscetti FW, Stankova J, Keller JR. Transforming growth factor-β regulates c-*kit* message stability and cell-surface protein expression in hematopoietic progenitors. Blood 1994; 83:3138–3145.

34. Spangrude GJ, Heimfield S, Weissman IL. Purification and characterization of mouse hematopoietic stem cells. Science 1988; 241:58.

35. McNiece IK, Bertoncello I, Keller JR, Ruscetti FW, Hartlely CA, Zsebo KM. Transforming growth factor β inhibits the action of stem cell factor on mouse and human hematopoietic progenitors. Int J Cell Cloning 1992; 10:80.

36. Mantel C, Kin KY, Cooper S, Kwon B, Broxmeyer HE. Polymerization of murine macrophage inflammatory protein 1α inactivates its myelosuppressive effects in vitro: the active form is a monomer. Proc Natl Acad Sci USA 1993; 90:2232.

37. Wolf NS, Kone A, Priestley GV, Bartelmez SH. In vivo and in vitro characterization of long-term repopulating primitive hematopoietic cells isolated by sequential Höechst 33342–rhodamine 123 FACS selection. Exp Hematol 1993; 21:614.

38. Sargiacomo M, Valtieri M, Gabrianelli M. Pure human hematopoietic progenitors: direct inhibitory effect of transforming growth factors-β2. In: Anagostou A, Dainiak N, Najman A, eds. Negative Regulators of Hematopoiesis. New York: Ann NY Acad Sci 1991; 628:84–94.

39. Lu L, Xiao M, Grigsby S, Wang WX, Wu B, Shen R-N, Broxmeyer HE. Comparative effects of suppressive cytokines on isolated single CD34^{+++} stem/progenitor cells from human bone marrow and umbilical cord blood plated with and without serum. Exp Hematol 1993; 21:1442–1446.

40. Pietenpol JA, Stein RW, Moran E, Yaciuk P, Schlegel R, Lyons RM, Pittelkow MR, Münger K, Howley PM, Moses HL. TGF-β1 inhibition of c-*myc* transcription and growth in keratinocytes is abrogated by viral transforming proteins with pRB binding domains. Cell 1990; 61:777–785.

41. Laiho M, De Caprio JA, Ludlow JW, Livingston DM, Massagué J. Growth inhibition by TGF-β linked to suppression of retinoblastoma protein phosphorylation. Cell 1990; 62:175.

42. Pietenpol JA, Holt JT, Stein RW, Moses HL. Transforming growth factor-β1 suppression of c-*myc* gene transcription: role in inhibition of keratinocyte proliferation. Proc Natl Acad Sci USA 1990; 87:3578.

43. Ewen ME, Sluss HK, Whitehouse LL, Livingston DM. TGFβ inhibition of Cdk4 synthesis is linked to cell cycle arrest. Cell 1993; 74:1009–1020.

44. Koff A, Ohtsuki M, Polyak K, Roberts JM, Massagué J. Negative regulation of G$_1$ in mammalian cells: inhibition of cyclin E-dependent kinase by TGF-β. Science 1993; 260:536.

45. Dubois CM, Ruscetti FW, Palaszynski EW, Falk LA, Oppenheim JJ, Keller JR. Transforming growth factor β is a potent inhibitor of interleukin 1 (IL-1) receptor expression: proposed mechanism of inhibition of IL-1 action. J Exp Med 1990; 172:737.

46. Jacobsen SEW, Ruscetti FW, Dubois CM, Lee J, Boone TC, Keller JR. Transforming growth factor-β trans-modulates the expression of colony stimulating factor receptors on murine hematopoietic progenitor cell lines. Blood 1991; 77:1706.

47. Qiu F, Ray P, Brown K, Barker P, Jhanwar S, Ruddle FH, Besmer P. Primary structure of c-*kit*: relationship with the CSF-1/PDGF receptor kinase family—oncogenic activation of v-*kit* involves deletion of extracellular domain and C terminus. EMBO J 1988; 7:1003.

48. Brawerman G. mRNA decay: finding the right targets. Cell 1989; 57:9.

49. Shaw G, Ramen R. A conserved AU sequence from the 3′ untranslated region of GM-CSF mRNA mediates selective mRNA degradation. Cell 1986; 46:659.

50. de Vos S, Brach MA, Asano Y, Ludwig W-D, Bettelheim P, Gruss H-J, Herrmann F. Transforming growth factor-β1 interferes with the proliferation-inducing activity of stem cell factor in myelogenous leukemia blasts through functional down-regulation of the c-*kit* proto-oncogene product. Cancer Res 1993; 53:3638–3642.

51. Jacobsen SEW, Ruscetti FW, Roberts AB, Keller JR. TGF-β is a bidirectional modulator of cytokine receptor expression on murine bone marrow cells. J Immunol 1993; 151:4534–4544.

52. Park LS, Friend D, Gillis S, Urdal DL. Characterization of the cell surface receptor for granulocyte–macrophage colony-stimulating factor. J Biol Chem 1986; 261:4177.

53. Nicola NA, Peterson L, Hilton DJ, Metcalf D. Cellular processing of murine colony stimulating factor (multi-CSF, GM-CSF, G-CSF) receptors by normal hemopoietic cells and cell lines. Growth Factors 1988; 1:41.

54. Nicola NA. Hemopoietic cell growth factors and their receptors. Annu Rev Biochem 1989; 58:45.

55. Jacobsen SE, Ruscetti FW, Dubois CM, Wine J, Keller JR. Induction of colony-stimulating factor receptor expression on hematopoietic progenitor cells. Proposed mechanism for growth factor synergism. Blood 1992; 80:678.

56. Dexter TM, Allen TD, Lajitha LG. Conditions controlling the proliferation of hematopoietic stem cells in vitro. J Cell Physiol 1971; 91:335–344.

57. Eaves CJ, Cashman JD, Kay RJ, et al. Mechanisms that regulate the cell cycle status of very primitive hematopoietic cells in long-term human marrow cultures. II. Analysis of positive and negative regulators produced by stromal cells within the adherent layer. Blood 1991; 78:110–117.

58. Kincade PW. Differential effects of TGF-β_1 on lymphohemopoiesis in long-term bone marrow cultures. Blood 1989; 74:1711–1717.

59. Cashman JD, Eaves AC, Raines EW, Ross R, Eaves CJ. Mechanisms that regulate the cell cycle status of very primitive hematopoietic cells in long-term human marrow cultures. I. Stimulatory role of a variety of mesenchymal cell activators and inhibitor role of TGF-β. Blood 1990; 75:96–101.

60. Waegell WO, Hegley HR, Kincade PW, Dasch JR. Growth acceleration and stem cell expansion in Dexter-type cultures by neutralization of TGF-β. Exp Hematol (in press).

61. Sitnicka E, Ruscetti FW, Priestly GV, Wolf NS, Bartlemez SH. Transforming growth factor β directly and reversibly inhibits the initial cell divisions of long-term repopulation hematopoietic stem cells. Blood 1996; 88:82–88.

62. Hatzfeld J, Li M-L, Brown EL, Sookdeo H, Levesque J-P, O'Toole T, Gurney C, Clark SC, Hatzfeld A. Release of early human hematopoietic progenitors from quiescence by antisense transforming growth factor β1 or Rb oligonucleotides. J Exp Med 1991; 174:925–929.

63. Cardoso AA, Li M-L, Batard P, Hatzfeld A, Brown EL, Levesque J-P, Sookdeo H, Panterne B, Sansilvestri P, Clark SC, Hatzfeld J. Release from quiescence of CD34+CD38⁻ human umbilical cord blood cells reveals their potentiality to engraft adults. Proc Natl Acad Sci USA 1993; 90:8707–8711.

64. Ploemacher RE, van Soest PL, Boudewijn A. Autocrine transforming growth factor β1 blocks colony formation and progenitor cell generation by hematopoietic stem cell stimulated with steel factor. Stem Cells 1993; 11:336–347.

65. Hodgson GS, Bradley TR. Properties of haematopoietic stem cells surviving 5-fluorouracil treatment. Evidence for a pre-CFU-S cell? Nature 1979; 281:381.

66. Ayash LJ, Antman K, Cheson BD. A perspective on dose-intensive therapy with autologous bone marrow transplantation for solid tumors. Oncology 1991; 5:25–33.

67. Goey H, Keller J, Back T. Inhibition of early murine hematopoietic progenitor cell proliferation after in vivo locoregional administration of transforming growth factor-β1. J Immunol 1989; 143:877–883.

68. Migdalska A, Molineux G, Demuynck H, Evans GS, Ruscetti F, Dexter TM. Growth inhibitory effects of transforming growth factor-β1 in vivo. Growth Factors 1991; 4:239–245.

69. Carlino JA, Higley HR, Creson JR, Avis PD, Ogawa Y, Ellingsworth LR. Transforming growth factor β1 systemically modulates granuloid, erythroid, lymphoid, and thrombocytic cells in mice. Exp Hematol 1992; 20:943.

70. Miller KM, Carlino JA, Ogawa Y, Avis PD, Carroll KG. Alterations in erythropoiesis in TGF-β1-treated mice. Exp Hematol 1992; 20:951–956.

71. Hestdal K, Jacobsen SE, Ruscetti F, Longo DL, Oppenheim J, Keller J. Increased granulopoiesis after sequential administration of transforming growth factor β1 and granulocyte–macrophage CSF. Exp Hematol 1993; 21:799–805.

72. Bursuker I, Neddermann KM, Petty BA, Schacter B, Spitalny GL, Tepper MA, Pasternak RD. In

vivo regulation of hemopoiesis by transforming growth factor beta 1: stimulation of GM-CSF- and M-CSF-dependent murine bone marrow precursors. Exp Hematol 1992; 20:431–435.

73. Grzegorzewski K, Ruscetti FW, Usui N, Damia G, Longo DL, Carlino JA, Keller JR, Wiltrout RH. Recombinant transforming growth factor β1 and β2 protect mice from acutely lethal doses of 5-flurouracil and doxorubicin. J Exp Med 1994, 180:104–107.

74. Le J, Vilcek J. Biology of disease: tumor necrosis factor and interleukin 1: cytokines with multiple overlapping biological activities. Lab Invest 1987; 56:234–248.

75. Tracey KJ, Vlassara H, Cerami A. Peptide regulative factors. Cachectin/tumor necrosis factor. Lancet 1989; 1:1122–1126.

76. Fiers W. Tumor necrosis factor. Characterization at the molecular, cellular and in vivo level. FEBS Lett 1991; 285:199–212.

77. Balkwill FR. Tumor necrosis factor and cancer. Prog Growth Factor Res 1992; 4:121–137.

78. Wingfield P, Pain RH, Craig S. Tumor necrosis factor is a compact trimer. FEBS Lett 1987; 211:179–184.

79. Carswell EA, Old LJ, Kassel RL, Green S, Fiore N, Williamson B. An endotoxin-induced serum factor that causes necrosis of tumors. Proc Natl Acad Sci USA 1975; 72:3666–70.

80. Degliantoni G, Murphy M, Kobayashi M, Francis MK, Perussia B, Trinchieri G. Natural killer (NK) cell-derived hematopoietic colony-inhibiting activity and NK cytotoxic factor. Relationship with tumor necrosis factor and synergy with immune interferon. J Exp Med 1985; 162:1512–1530.

81. Peetre C, Gullberg U, Nilsson E, Olsson I. Effects of recombinant tumor necrosis factor on proliferation and differentiation of leukemic and normal hematopoietic cells in vitro. J Clin Invest 1986; 78:1694–1700.

82. Roodman GD, Bird A, Hutzler D, Montgomery W. Tumor necrosis factor-alpha and hematopoietic progenitors: effects of tumor necrosis factor on the growth of erythroid progenitors CFU-E and the hematopoietic cell lines K562, HL60, and HEL cells. Exp Hematol 1987; 15:928–935.

83. Munker R, Koeffler P. In vitro action of tumor necrosis factor on myeloid leukemia cells. Blood 1987; 69:1102–1108.

84. Wisniewski D, Strife A, Atzpodien J, Clarkson BD. Effects of recombinant human tumor necrosis factor on highly enriched hematopoietic progenitor cell populations from normal bone marrow and peripheral blood and bone marrow from patients with chronic myeloid leukemia. Cancer Res 1987; 47:4788–4794.

85. Murase T, Hotta T, Saito H, Ohno R. Effect of recombinant human tumor necrosis factor on the colony growth of human leukemia progenitor cells and normal hematopoietic progenitor cells. Blood 1987; 69:467–472.

86. Murphy M, Perussia B, Trinchieri G. Effects of recombinant tumor necrosis factor, lymphotoxin, and immune interferon on proliferation and differentiation of enriched hematopoietic precursor cells. Exp Hematol 1988; 16:131–138.

87. Williams DE, Cooper S, Broxmeyer HE. Effects of hematopoietic suppressor molecules on the in vitro proliferation of purified murine granulocyte–macrophage progenitor cells. Cancer Res 1988; 48:1548–1550.

88. Caux C, Saeland S, Favre C, Duvert V, Mannoni P, Bancheran J. Tumor necrosis factor-α strongly potentiates interleukin-3 and granulocyte–macrophage colony-stimulating factor-induced proliferation of human CD34[+] hematopoietic progenitor cells. Blood 1990; 75:2292–2298.

89. Caux C, Favre C, Saeland S, Duvert V, Durand I, Mannoni P, Bancheran J. Potentiation of early hematopoiesis by tumor necrosis factor-α is followed by inhibition of granulopoietic differentiation and proliferation. Blood 1991; 78:635–644.

90. Backx B, Broeders L, Bot FJ, Löwenberg B. Positive and negative effects of tumor necrosis factor on colony growth from highly purified normal marrow progenitors. Leukemia 1990; 5:66–70.

91. Jacobsen SEW, Ruscetti F, Dubois C, Keller JR. Tumor necrosis factor α directly and indirectly regulates hematopoietic progenitor cell proliferation: role of colony-stimulating factor receptor modulation. J Exp Med 1992; 175:1759–1772.

92. Caux C, Durand I, Moreau I, Duvert V, Saeland S, Bancheran J. Tumor necrosis factor α cooperates

with interleukin 3 in the recruitment of a primitive subset of human CD34$^+$ progenitor. J Exp Med 1993; 177:1815–1820.

93. Munker R, Gasson J, Ogawa M, Koeffler HP. Recombinant human TNF induces production of granulocyte–monocyte colony-stimulating factor. Nature 1986; 323:79–82.

94. Vogel SN, Douches SD, Kaufman EN, Neta R. Induction of colony stimulating factor in vivo by recombinant interleukin 1α and recombinant tumor necrosis factor α. J Immunol 1987; 138:2143–2148.

95. Koeffler HP, Gasson J, Ranyard J, Souza L, Shepard M, Munker R. Recombinant human TNF stimulates production of granulocyte colony-stimulating factor. Blood 1989; 70:55–59.

96. Oster W, Lindeman A, Horn S, Mertelsmann R, Herrman F. Tumor necrosis factor (TNF)-α but not TNF-β induces secretion of colony stimulating factor for macrophages (CSF-1) by human monocytes. Blood 1987; 70:1700–1703.

97. Zucali JR, Broxmeyer HE, Gross MA, Dinarello CA. Recombinant human tumor necrosis factors α and β stimulate fibroblasts to produce hematopoietic growth factors in vitro. J Immunol 1988; 140:840–844.

98. Hoang T, Levy B, Onetto N, Haman A, Rodrigues-Cimadevilla JC. Tumor necrosis factor α stimulates the growth of the clonogenic cells of acute myeloblastic leukemia in synergy with granulocyte/macrophage colony-stimulating factor. J Exp Med 1989; 170:15–26.

99. Rusten L, Jacobsen FW, Lesslauer W, Loetscher H, Smeland EB, Jacobsen SEW. Bifunctional effects of TNF-α on the growth of mature and primitive human hematopoietic progenitor cells: Involvement of p55 and p75 TNF receptors. Blood 1994; 83:3152–3159.

100. Caux C, Dezutter-Dambuyant C, Schmitt D, Banchereau J. GM-CSF and TNF-α cooperate in the generation of dendritic Langerhans cells. Nature 1992; 360:258–261.

101. Santiago-Schwarz F, Belilos E, Diamond B, Carsons SE. TNF in combination with GM-CSF enhances the differentiation of neonatal cord blood stem cells into dendritic cells and macrophages. J Leukoc Biol 1992; 52:274–281.

102. Moore MAS. Clinical implications of positive and negative hematopoietic stem cell regulators. Blood 1991; 78:1–19.

103. Metcalf D. Hematopoietic regulators: redundancy or subtlety? Blood 1993; 82:3515–2523.

104. Rusten LS, Smeland EB, Jacobsen FW, Lien E, Lesslauer W, Loetscher H, Dubois CM, Jacobsen SEW. TNF-α inhibits stem cell factor-induced proliferation of human bone marrow progenitor cells in vitro: role of p55 and p75 TNF receptors. J Clin Invest 1994; 94:165–172.

105. Jacobsen SEW, Veiby OP, Myklebust J, Okkenhaug C, Lyman SD. Ability of flt3 ligand to stimulate the in vitro growth of primitive murine hematopoietic progenitors is potently and directly inhibited by transforming growth factor-β and tumor necrosis factor-α. Blood 1996; 87:5016–5026.

106. Lyman SD, James L, Bos TV, de Vries P, Brasel K, Gliniak B, Hollingsworth LT, Picha KS, McKenna HJ, Splett RR, Fletcher FA, Maraskovsky E, Farrah T, Foxworthe D, Williams DE, Beckmann MP. Molecular cloning of a ligand for the Flt3/Flk-2 tyrosine kinase receptor: a proliferative factor for primitive hematopoiesis. Cell 1993; 75:1157–1167.

107. Jacobsen SEW, Okkenhaug C, Myklebust J, Veiby OP, Lyman SD. The flt3 ligand potently and directly stimulates the growth and expansion of primitive murine bone marrow progenitor cells in vitro: Synergistic interactions with interleukin (IL) 11, IL-12, and other hematopoietic growth factors. J Exp Med 1995; 181:1357–1363.

108. Means RT, Krantz SB. Progress in understanding the pathogenesis of the anemia of chronic disease. Blood 1992; 80:1639–1647.

109. Means RT, Dessypris EN, Krantz SB. Inhibition of human colony-forming-unit erythroid by tumor necrosis factor requires accessory cells. J Clin Invest 1990; 86:538–541.

110. Means RT, Krantz SB. Inhibition of human erythroid colony-forming units by tumor necrosis factor requires beta interferon. J Clin Invest 1993; 91:416–419.

111. Rusten LS, Jacobsen SEW. TNF-α directly inhibits human erythropoiesis in vitro: Role of p55 and p75 TNF receptors. Blood 1995; 85:989–996.

112. Loetscher H, Brockhaus M, Dembic Z, Gentz R, Gubler U, Hohmann HP, Lahm HW, van Loon

APGM, Pan Y-CE, Schlaeger EJ, Steinmetz M, Tabuchi H, Lesslauer W. Two distinct tumor necrosis factor receptors: members of a new cytokine receptor gene family. Oxford Surv Eukaryot Genes 1991; 7:119–142.

113. Tartaglia LA, Goeddel DV. Two TNF receptors. Immunol Today 1992; 13:151–153.

114. Smith CA, Farrah T, Goodwin RG. The TNF receptor superfamily of cellular and viral proteins: activation, costimulation, and death. Cell 1994; 76:959–962.

115. Tartaglia LA, Weber RF, Figari IS, Reynolds C, Palladino MA, Goeddel DV. The two different receptors for tumor necrosis factor mediate distinct cellular responses. Proc Natl Acad Sci USA 1991; 88:9292–9296.

116. Brockhaus M, Schoenfeld H-J, Schlaeger E-J, Hunziker W, Lesslauer W, Loetscher H. Identification of two types of tumor necrosis factor receptors on human cell lines by monoclonal antibodies. Proc Natl Acad Sci USA 1990; 87:3127–3131.

117. van Ostade X, Vandenabeele P, Everaerdt B, Loetscher H, Gentz R, Brockhaus M, Lesslauer W, Tavernier J, Brouckaert P, Fiers W. Human TNF mutants with selective activity on the p55 receptor. Nature 1993; 361:266–269.

118. Vandenabeele P, Declercq W, Vercammen D, van de Craen M, Grooten J, Loetscher H, Brockhaus M, Lesslauer W, Fiers W. Functional characterization of the human tumor necrosis factor receptor p75 in a transfected rat/mouse T cell hybridoma. J Exp Med 1992; 176:1015–1024.

119. Delwel R, van Buitenen C, Lövenberg B, Touw I. Involvement of tumor necrosis factor (TNF) receptors p55 and p75 in TNF responses of acute myeloid leukemia blasts in vitro. Blood 1992; 80:1798–1803.

120. Jacobsen FW, Rothe M, Rusten L, Goeddel DV, Smeland EB, Veiby OP, Slørdal L, Jacobsen SEW. Novel role of the 75-kDa tumor necrosis factor (TNF) receptor: Inhibition of early hematopoiesis. Proc Natl Acad Sci USA 1994; 91:10695–10699.

121. Fahlman C, Jacobsen FW, Veiby OP, McNiece IK, Blomhoff HK, Jacobsen SEW. TNF-α potently enhances in vitro macrophage production from primitive murine hematopoietic progenitor cells in combination with stem cell factor and interleukin-7: novel stimulatory role of p55 TNF receptors. Blood 1994; 84:1528–1533.

122. Elbaz O, Budel LM, Hoogerbrugge H, Touw IP, Delwel R, Mahmoud LA, Lövenberg B. Tumor necrosis factor downregulates granulocyte-colony-stimulating factor receptor expression of human acute myeloid leukemia cells and granulocytes. J Clin Invest 1991; 87:838–841.

123. Elbaz O, Budel LM, Hoogerbrugge H, Touw IP, Delwel R, Mahmoud LA, Lövenberg B. Tumor necrosis factor regulates the expression of granulocyte–macrophage colony-stimulating factor and interleukin-3 receptors on human acute myeloid leukemia cells. Blood 1991; 77:989–995.

124. Watanabe Y, Kitamura T, Hayashida K, Miyaijima A. Monoclonal antibody against the common β subunit (βC) of the human interleukin-3 (IL-3), IL-5, and granulocyte–macrophage colony-stimulating factor receptors shows upregulation of βC by IL-1 and tumor necrosis factor-α. Blood 1992; 80:2215–2220.

125. Brach MA, Buhring H-J, Gruss H-J, Ashman LK, Ludwig W-D, Mertelsmann RH, Herrmann F. Functional expression of c-*kit* by acute myelogenous leukemia blasts is enhanced by tumor necrosis factor-α through posttranscriptional mRNA stabilization by a labile protein. Blood 1992; 80:1224–1230.

126. Neta R, Oppenheim JJ, Douches SD. Interdependence of the radioprotective effects of human recombinant interleukin 1α, tumor necrosis factor α, granulocyte colony-stimulating factor, and murine recombinant granulocyte–macrophage colony-stimulating factor. J Immunol 1988; 140:108–111.

127. Slørdal L, Muench MO, Warren DJ, Moore MAS. Radioprotection by murine and human tumor-necrosis factor: dose-dependent effects on hematopoiesis in the mouse. Eur J Haematol 1989; 43:428–434.

128. Slørdal L, Warren DJ, Moore MAS. Protective effects of tumor necrosis factor on murine hematopoiesis during cycle-specific cytotoxic chemotherapy. Cancer Res 1990; 50:4216–4220.

129. Pfeffer K, Matsuyama T, Kunkig TM, Wakeham A, Kishihara K, Shahinian A, Wiegmann K, Ohashi

PS, Kronke M, Mak TW. Mice deficient for the 55 KD tumor necrosis factor receptor are resistant to endotoxic shock, but succumb to *L. monocytogenes* infection. Cell 1993; 73:457–467.

130. Rothe J, Lesslauer W, Lötscher H, Lang Y, Koebel P, Köntgen F, Althage A, Zinkernagel R, Steinmetz M, Bluethmann H. Mice lacking the tumor necrosis factor receptor 1 are resistant to TNF-mediated toxicity but highly susceptible to infection by *Listeria monocytogenes*. Nature 1993; 364:798–801.

131. Jacobsen FW, Veiby OP, Stokke T, Jacobsen SEW. TNF-α bidirectionally modulates the viability of primitive murine hematopoietic progenitor cells in vitro. J Immunol 1996; 157:1193–1199.

132. Zhang Y, Harada A, Bluethmann H, Wang JB, Nakao S, Mukaida N, Matsushima K. Tumor necrosis factor (TNF) is a physiological regulator of hematopoietic progenitor cells: Increase of early hematopoietic progenitor cells in TNF receptor p55–deficient mice in vivo and potent inhibition of progenitor cell proliferation by TNFα in vitro. Blood 1995; 86:2930–2937.

133. Del Rizzo DF, Eskinazi D, Axelrad AA. Interleukin-3 opposes the action of negative regulatory protein (NRP) and of transforming growth factor beta (TGF-β) in their inhibition of DNA synthesis of the erythroid stem cell BFU-E. Exp Hematol 1990; 18:138.

134. Gabrilove JL, Wong G, Bollenbacher E, White K, Kojima S, Wilson EL. Basic fibroblast growth factor counteracts the suppressive effect of transforming growth factor-β1 on human myeloid progenitor cells. Blood 1993; 81:909.

135. Kishi K, Ellingsworth LR, Ogawa M. The suppressive effects of type β transforming growth factor (TGF-β) on primitive murine hematopoietic progenitors are abrogated by interleukin-6 and granulocyte colony stimulating factor. Leukemia 1989; 3:687.

136. Jacobsen SEW, Ruscetti FW, Ortiz M, Gooya JM, Keller JR. The growth response of Lin-Thy-1+ hematopoietic progenitors to cytokines is determined by the balance between synergy of multiple stimulators and negative cooperation of multiple inhibitors. Exp Hematol 1994; 22:985–989.

137. Maltman J, Pragnell IB, Graham GJ. Transforming growth factor β: is it a downregulator of stem cell inhibition by macrophage inflammatory protein 1α? J Exp Med 1993; 178:925–932.

21

Leukemia Inhibitory Factor

Paul M. Waring*

The Walter and Eliza Hall Institute of Medical Research, Melbourne, Victoria, Australia

I. MOLECULAR AND CELLULAR BIOLOGY OF LIF

A. Discovery and Nomenclature

Leukemia inhibitory factor (LIF) has a curious history of discovery and rediscovery by different groups working on apparently unrelated factors, who in the late 1980s came to the unsettling conclusion that they all had purified the same protein or cloned the same gene. As a result of this confusing start, the molecule that we now call leukemia inhibitory factor is also referred to by several other names (Table 1), each reflecting a different property of the protein.

The original discovery of an activity that can, at least partially, be attributed to LIF was based on the observation that cells of the murine M1 myeloblastic line could be induced to differentiate in vitro into mature granulocytes and macrophages when treated with conditioned media from embryo and spleen cell cultures (14). By using M1 cells as a bioassay, an activity, then known as differentiation-inducing factor (DIF) or D-factor, was partially purified from mouse fibroblast L929 and mouse Ehrlich ascites tumor cell-conditioned media (2,3). The same group also showed that granulocyte colony-stimulating factor (G-CSF) induced granulocytic differentiation of M1 cells (15). Coincidentally, another group had identified a different activity, termed MG1-2A, that also induced M1 cell differentiation, and this later proved to be interleukin-6 (IL-6;16). By early 1988, it had become clear that there were at least three distinct molecules (D-factor, G-CSF, and IL-6) that were capable of inducing M1 cell differentiation (17).

By employing a dual assay system using M1 and WEHI-3BD$^+$ cells, Metcalf et al. were able to identify an M1 cell differentiation–inducing activity, which unlike G-CSF and IL-6, was incapable of inducing granulocytic differentiation in WEHI-3BD$^+$ cells (1,17). The activity was purified from lipopolysaccharide (LPS)-stimulated Krebs II ascites tumor conditioned medium,

---

Current affiliation: Melbourne University, Parkville, Victoria, Australia.

Table 1 Defining Properties and Synonyms of LIF

Synonym	Species	Defining biological activity	Refs.
LIF	Murine	Induction of M1 differentiation	1
D-factor	Murine	Induction of M1 differentiation	2,3
DIF	Human	Induction of M1 differentiation	4
DIA	Rat	Inhibition of ES cell differentiation	5
DRF	Murine	Inhibition of EC cell differentiation	6
HSF III	Human	Stimulation of acute-phase protein synthesis	7
CNDF/CDF	Rat	Induction of cholinergic phenotype	8
MLPLI	Human	Inhibition of lipoprotein lipase	9
HILDA	Human	Stimulation of DA-1a cell proliferation	10
GATS	Human	Growth stimulatory activity for TS1 cells	11
	Bovine	Inhibition of aortic endothelial cell growth	12,13

and the amino acid sequence of the purified activity proved to be unique (18). Because of its concomitant action of inhibiting M1 cell proliferation and clonogenicity (Fig. 1), this new molecule was termed leukemia inhibitory factor. With use of nucleotide probes based on the amino acid sequence data from purified murine LIF, a cDNA for LIF was isolated from a T-lymphocyte library (20), and murine and human genomic LIF clones were isolated (21). Subsequent sequencing of D-factor proved it to be almost identical with LIF (22).

B. The LIF Molecule

Leukemia inhibitory factor is a disulfide-linked monomeric glycoprotein of 179-amino acid residues, with predicted molecular weights for both murine and human forms of 20,000 D (18). The reported molecular weights for native LIF range from 38,000 to 67,000 D (2,5,23), and because deglycosylation reduced the molecular weight to 20,000–25,000 D (2), this heterogeneity appeared to be due to extensive and variable glycosylation. In as much as recombinant nonglycosylated and variously glycosylated forms of LIF all exhibited similar specific activities (approx 10^8 units/mg protein;20, 24–26), the carbohydrate moiety does not appear to be necessary for the biological actions of LIF, although the glycosylation pattern appears to influence the stability of the molecule both in vitro and in vivo (27,28).

The nucleotide sequence of the LIF gene and the inferred amino acid sequences from mouse, human, sheep, pig and rat LIF have been determined (29–31). Comparisons of these sequences showed that LIF is highly conserved, with all six cysteine residues present at identical positions, suggesting that intramolecular disulfide bonds are vital to the integrity and the activity of the LIF molecule. Six of the seven potential N-linked glycosylation sites are also conserved, suggesting that glycosylation is important. The X-ray crystallographic structure of LIF has been determined (32,33), and compromises a four-α-helix bundle structure common to several members of the hematopoietic cytokine family. Oncostatin M (OSM) shares a 30% amino acid sequence homology with LIF (34), and because the positions and pairing of cysteinyl residues and disulfide bonds of LIF are closely homologous with those of OSM (33–35), it appears that the two molecules are structurally similar. The use of a pattern-search algorithm suggested that there are primary sequence and secondary structural similarities between LIF, OSM, ciliary neurotropic factor (CNTF), IL-6, and G-CSF (33,34,36), which, given their functional similarities, led to the suggestion that they are all members of the same cytokine family (34).

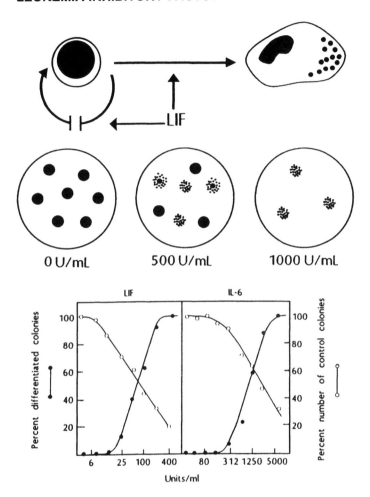

Figure 1 Schematic representation of the effect of LIF on M1 cells: The top panel represents the action of LIF on undifferentiated M1 myeloblastic cells induced to differentiate into macrophages, with consequent reduced clonogenicity. The middle and lower panels represents M1 cells grown in semisolid agar cultures. In the absence of LIF, all M1 colonies are compact, being composed of myeloblastic cells. With increasing LIF concentrations, stochastic differentiation of M1 cells results in the formation of some dispersed colonies. At high LIF concentration there is a reduction in the number of colonies owing to suppression of clonogenicity. (Partially reproduced from Ref. 19.)

C. The LIF Gene

Leukemia inhibitory factor is encoded by a single-copy gene that has a complete nucleotide sequence of 8.7 kb for murine LIF and 7.6 kb for human LIF (29). Both genes comprise three exons, and two introns that encode transcripts 4.2 kb in length, which include long (3.2-kb) 3′-untranslated sequences (29,37,38). The first exon encodes a short 5′-untranslated region of the message and the first six residues of the preprotein leader sequence, with the remainder of the leader sequence and the first third of the mature LIF protein being encoded by the second exon and the COOH-terminal two-thirds of LIF and the 3′-untranslated region encoded by exon 3 (29).

There are alternative copies of the first exon of the murine LIF gene, which are transcribed from separate upstream promoter regions (39). Ribonuclease protection analysis showed two alternative transcripts that differed by 15 nucleotides (39). The alternative exon 1 (exon 1b) within intron 1, at a position 500-bp downstream from exon 1, alters the NH_2-terminal portion of the leader sequence and apparently changes the extracellular fate of the encoded LIF protein. The transcript termed D-LIF encodes a diffusible form of LIF that is the biologically active secreted form of the molecule, and the transcript termed M-LIF encodes a matrix-associated form. The matrix-associated form of LIF, however, might be unique to the mouse, for neither human, sheep, nor pig LIF genes encode the alternative exon (30).

The LIF gene has been localized to chromosome 22q12.1-12.2 in humans (40,41) and to proximal chromosome 11 (11A1-2) in the mouse (42,43). The human LIF gene was localized to the same cytogenetic band as the Ewing's sarcoma breakpoint, the neurofibromatosis type II (NF-2) gene (40,41,44,45), and the gene for OSM (46–49) which is situated within 19 kb of the LIF locus. Equivalent functional, structural, and genetic linkage of related cytokines has also been observed for interleukin-1α and IL-β and tumor necrosis factors (TNF)α and TNF-β.

The 5'-flanking region of the mouse LIF gene contains four TATA-like elements, putative promoter, and negative regulatory and enhancer-like elements (29,50,51). The LIF promoter region contains elements resembling AP-2, SP-1, CpG islands, and CAAT transcription-binding domains, and a NF-IL-6-like binding sequence (29,52,53); the "enhancer" region contains several minimal steroid-response elements (51). The 3'-untranslated region contains long CT-rich repeat sequences and AU-rich regions, sequences known to be involved in mRNA stability. In addition, the 200-bp region immediately proximal to the poly(A) addition signal was almost completely conserved between mouse and human LIF (29), suggesting a role for this sequence in LIF mRNA end formation or mRNA stability.

D. Regulation of LIF Expression and Protein Secretion

The transcriptional machinery for the LIF gene is likely to involve many of the same *cis*-acting elements and trans-acting factors that are essential for activation of other cytokine genes; however, little is now known about the mechanisms that underlie the functional regulation of LIF gene expression.

The levels of constitutive LIF mRNA transcription and protein secretion appear to be extremely low in vivo. LIF transcripts were undetectable by Northern blotting in a survey of normal mouse tissues and organs (54), but were detected by semiquantitative RT-PCR, RNase protection, or in situ hybridization in embryonic brain (55,56), cholinergic sympathetic neurons (55), lung (57,58), skin, intestine (58), gestational and nongestational endometrium, decidua (59–61), and placenta (61,62). Tissues explants cultured in vitro, however, produced large amounts of LIF mRNA (54) and protein (63,64), suggesting that in vitro culture is a powerful inductive signal for LIF production.

A large number of rodent and human normal embryonic and adult cells, and virally transformed or tumor cell lines constitutively produce LIF or express LIF mRNA in vitro (Table 2). In many of these cell types the abundance of LIF or its transcripts could be induced or enhanced by a variety of stimuli, including LPS (1,17,63,65,66,69,84,92,94,95,106,111); T-cell receptor stimuli, such as concanavalin A, phytohemagglutinin (PHA), or lectin (20,23,38,96,116); phorbol esters, such as phorbol myristate acetate (PMA) or tetradecanoylphorbol acetate (TPA) (4,10, 11,38,69,72,82,83,85,86,116); cytokines, such as IL-1 and TNF (11,63,72,82,83,85,86,89,92–96,98,101,106,107,115), transforming growth factor (TGF)-β (63,72,85,86,93,94,96,106,107); growth factors (e.g., basic fibroblast growth factor [FGF] and epidermal growth factor [EGF]; (82); retinoic acid (89); vitamin D_3 (65,66); calcium ionophore A23187 (calcimycin;86);

Table 2 Sources of LIF

Hematopoietic
    Monocyte/macrophages
        Activated primary peripheral blood monocytes (65–68), monocytic and myelomonocytic
           leukemias or macrophage-derived cell lines (4,29,65,69,70), primary human CML cells in blast
           crisis (71)
    T lymphocytes
        Activated normal T lymphocytes (72), virally transformed T lymphocytes (69), alloreactive T
           lymphocytes (73–76), lectin, phytohemagglutinin, and concanavalin A-treated (23,38), IL-2–
           dependent T-cell clones (20,29), leukemic T-lymphocytoid cells (70), T-cell lymphoma cell
           lines (10,77),
    Erythroleukemic cells (70)
    Megakaryoblastic cells (78)
    Multiple myeloma cells (79)
    Mast cell lines (80)
    Hodgkin's disease-derived cell lines (81)
Mesenchymal cells
    Fibroblasts (2,6,11,12,23,29,72,82–86)
    Osteoblasts and osteosarcoma cell lines (87–94)
    Endothelial cells (72)
    Preadipocytes (95)
    Bone marrow stromal cells (83,96–101)
    Thymic stromal and epithelial cell lines (100,102,103)
    Hepatosarcoma cells (11)
    Cardiac myocytes (8,31,104)
    Chondrocytes (63,105,106)
    Synoviocytes (106,107)
    Endometrial stromal cells (61)
Epithelial cells
    Squamous cells: cultured and virally transformed keratinocytes (108), squamous cell carcinomas
        (7,108,109), bladder carcinoma cell line 5637 (24,76,108)
    Glandular cells: lung, colonic, pancreatic, and breast adenocarinomas (38,108); endometrium (59–61)
    Hepatocytes (5,38,69,110,111)
    Neuroepithelial (112)
    Melanoma (9,11,108,113,114)
    Astrocytes (85,115)
    Pituitary follicular cells (13)

parathyroid hormone (90); doxorubicin (Adriamycin) (117); OKT3 (116); and by several viruses, such as measles virus (11), cytomegalovirus (115), and human T-cell lymphotropic virus type 1 (HTLV-1; 118,119).

The production of LIF is mediated, at least partly, by protein kinase C (PKC), as evidenced by the facts that LIF production can be induced by PMA, and IL-1 and TNF-induced LIF production can be inhibited by prolonged phorbol ester preincubation (86). Inhibitory mechanisms of LIF production also exist, and a putative negative regulatory element in the LIF promoter has been indentified (50). Glucocorticoids inhibited or reduced LPS or IL-1α-induced LIF expression (69,82,107); cyclosporine blocked transcriptional activation of LIF by mitogen and antigenically stimulated T cells (120); and the anti-inflammatory cytokines IL-4 and interferon (IFN)-γ reduced IL-1-induced LIF production by synoviocytes (107); and IL-4 reduced constitutive LIF production by bone marrow stromal cells (98).

E. The LIF Receptor

To understand how LIF mediates its biological activities, it has been necessary to characterize its cell surface receptor. Autoradiographic studies, using radiolabeled LIF, showed that, in normal fetal and adult murine tissues, LIF receptors (LIF-Rs) were displayed on osteoblasts, hepatocytes, adipocytes, endothelial, and trophoblast cells; and in the choroid plexus, small intestinal villi, splenic follicular marginal cells, adrenal cortical cells, Leydig cells, and in the metanephros (27; Hilton D, personal communication). In vitro binding studies revealed the presence of both low- and high-affinity LIF-Rs. High-affinity LIF-Rs were expressed by cells of all biologically responsive types (Table 3), had a K_d of 20–80 pM, were characterized by rapid association kinetics, and at 4°C by very slow dissociation kinetics (131). Low-affinity LIF-Rs (K_d 1–3 nM) were identified on activated macrophages and on membranes prepared from cells expressing high-affinity receptors, and differed from the high-affinity receptors only in the rate at which the dissociation of LIF occurred (124). LIF exhibits partial species-specificity, and although human LIF is able to bind to the murine LIF-R and exert biological effects on murine cells, murine LIF exhibits relatively low affinity for the human LIF-R (132) and reduced bioactivity (33). Curiously, the binding affinity for human LIF to murine soluble LIF-R is much greater than the binding affinity of murine LIF (92,133); this difference in binding affinity being due to markedly different dissociation kinetics (133–135).

The cDNA clones for the human and murine LIF-R genes were isolated (136), and the cloned human LIF-R gene encoded a 200-kDa glycoprotein that specifically bound human, but not murine, LIF with low affinity. The predicted amino acid sequence indicated that the LIF-Rα chain belonged to the hematopoietin (or cytokine) receptor family, which also includes the receptors for growth hormone, prolactin, IL-2 (β and γ chains), IL-3, IL-4, IL-5, IL-6, IL-7, IL-9, IL-12, G-CSF, GM-CSF, erythropoietin, CNTF, gp130, KH97/AIC2B (137), and c-*Mpl*, the ligand for which was recently identified as a Meg-CSF/thrombopoietin-like protein (138). All members of this family contain one or two extracellular hematopoietin domains, which possess

Table 3 Cell Types Demonstrated to Possess High-Affinity LIF Receptors

Monocyte/macrophages (24,121,122)
Megakaryocytes (27,123)
Hepatocytes (124,125)
Preadipocytes: 3T3-L1 (124)
Placental choriocarcinoma cell lines: JAR and JEG-3 (122)
Erythroleukemia cells: TF-1 (126)
Embryonal carcinoma and embryonal stem cell lines (24,122)
Neural crest, neuroepithelial, and neuroblastoma cells (122,124,127)
Osteoblasts and osteosarcoma cells (89,92,122,128)
Kidney epithelium (129)
Melanoma (122)
Colonic adenocarcinoma (122)
Breast epithelium (122)
Myeloma cells (130)
Ewing's sarcoma cells (127)

conserved cysteinyl residues and a five-residue motif, WSXWS; a transmembrane region and intracytoplasmic portions that lack intrinsic tyrosine kinase domains or any other recognizable catalytic function. The LIF-Rα chain possesses additional immunoglobulin-like domains (as do CNTF-Rα, IL-6-Rα, G-CSF-R, and gp130) and fibronectin type III domains (as do G-CSF-R and gp 130), the functions of which are currently unknown. Ligands for all members of the hematopoietin receptor family are predicted to comprise a four–α-helix bundle structure, and their receptors are predicted to form a sandwich of two antiparallel β-sheets, similar to that recently determined for growth hormone and its receptor (139). Similar to ligands in this family, LIF is predicted to occupy two binding sites on the extracellular hematopoietin domains of each receptor chain. However, the classic two-site model has been challenged by data derived from the use of mouse–human LIF chimeric molecules that have identified an additional binding site on the human LIF molecule that conferred high-affinity binding to the murine LIF-R (33,132–135).

The LIF-Rα chain gene has been localized to human chromosome 5p12-13 and mouse chromosome 15, clustered with the genes for the growth hormone, prolactin, and IL-7 receptors (140). Various LIF-Rα chain gene transcripts are known to be produced (10,5,3, and 2 kb), and their pattern of expression varied depending on the tissue (141–143). Binding studies in B9 and COS-7 cells transfected with LIF-Rα cDNA, showed that COS-7 cells expressing LIF-Rα bound LIF with low affinity (136), whereas high-affinity LIF binding was conferred in B9 cells (144,145). This indicated that expression of the LIF-Rα constituted the low-affinity receptor and suggested the presence of a high-affinity–converting β-subunit. Molecular cloning of the β-subunit demonstrated identity with gp130 (145).

The LIF-R belongs to a promiscuous receptor system, comprising the specificity-conferring LIF-R (136), IL-6-R (146), CNTF-R (147), and the IL-11-R (148) α-chains and a common

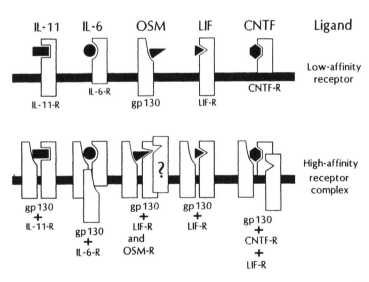

Figure 2 Diagramatic representation of the structure of the low- and high-affinity receptors for IL-11, IL-6, oncostatin M, LIF, and CNTF: Low-affinity receptors are formed by a ligand-binding receptor chain, designated the α-chain, except OSM, which binds the gp130 β-chain with low-affinity. Dimerization with a gp130 β-chain results in formation of high-affinity receptors for IL-11 and LIF. Specific receptors for CNTF and OSM comprise trimers formed by the high-affinity LIF-R and an additional ligand-binding chain (CNTF-Rα and OSM-Rα), whereas the high-affinity IL-6 receptor comprises a trimer of the IL-6-Rα chain and two gp130 β-chains. (Partially reproduced from Ref. 153.)

affinity-converting β-subunit, gp130, which is common to the LIF (145,149); IL-6 (150), CNTF (149,151), IL-11 (152), and OSM (145), high-affinity receptors (Fig. 2). High-affinity receptors for LIF (136), CNTF (154), and OSM (145), all share the LIF-Rα chain. Following ligand binding to the low-affinity subunit, receptor assembly is thought to occur sequentially, with receptor activation resulting from ligand-induced homo- or heterodimerization of the signal-transducing components (155). The G-CSF-R, gp130, and LIF-Rα subunits, all are capable of signal transduction (156) and show a high degree of similarity to each other, particularly in their proximal cytoplasmic regions that contain domains (termed boxes 1 and 2) essential for signaling (156–159). The G-CSF-R is thought to form a homodimer when G-CSF is bound (160), whereas activation of the IL-6 receptor involves homodimerization of gp130 in the presence of IL-6 and either the cellular or soluble form of the IL-6-Rα chain (161). The LIF-Rα and gp130 form a heterodimer when LIF or OSM are bound (145) or in the presence of CNTF and the soluble CNTF-Rα chain (126,149). These observations led to the notion that the LIF-Rα chain ought to be considered a second β-subunit (155; and is referred to as the LIF-Rβ chain by authors from that laboratory).

The sharing of the LIF-Rα chain by the high-affinity LIF, OSM, and CNTF receptors raises the question of specificity of action of LIF. Indeed LIF, OSM, and CNTF exhibit several shared biological effects, yet each appears to retain some unique actions (162). The functional CNTF-R complex comprises three different chains, CNTF-Rα, gp130, and LIF-Rα (149,151,154). Owing to the limited distribution of the CNTF-Rα chain, the actions of CNTF are restricted to neurons, hepatocytes, and embryonal stem cells. As with the high-affinity LIF-R, the OSM-R is formed by the heterodimerization of the LIF-Rα chain and gp130 (144,145), with gp130 being both signal transducer (163) and the low-affinity binding subunit for OSM (145,164), indicating that LIF and OSM should exactly duplicate each another's functions. However, because some OSM-responsive cells, such as endothelial, A375 melanoma, A549 epithelial, and Kaposi's sarcoma cells (12,144,162,165) do not respond to LIF and because LIF and OSM induce different patterns of tyrosine-phosphorylated proteins in certain cell types (165–167), the receptor config-uration and signaling pathways between OSM and LIF must be different, at least in these cell types. These observations led to the characterization and recent cloning of a specific OSM-Rβ chain, which is 30% identical to the LIF-Rα chain and interacts with the low affinity OSM/gp130 complex to increase affinity and transduce a signal (Mosley B, personal communication).

Specificity for LIF, therefore, appears to be conferred by the absence of CNTF-Rα and OSM-Rα chains on target cells bearing both the LIF-Rα and gp130 subunits. Both membrane-anchored and soluble forms exist for the IL-6 (150), LIF (136), and CNTF receptors (126,147), which adds another dimension of complexity to this receptor system. Soluble receptors represent truncated forms of membrane-bound low-affinity receptors, lacking their transmembrane and cytoplasmic domains, but still exhibiting cytokine-binding affinities similar to their membrane-bound counterparts. Each of these three soluble receptors appears to be generated by different mechanisms, with soluble LIF-R being a result of alternative promoter usage by alternative splicing of mRNA transcripts (136), the soluble IL-6-R being formed through proteolytic cleavage of full-length receptors (168), and the soluble CNTF-R being generated by the action of phosphatidylinositol-specific phospholipase C (126,147). The cytoplasmic region of the IL-6-Rα is very short and is dispensable for signaling (146,150), and CNTF lacks such a region altogether (147). Both the soluble IL-6-Rα (150) and soluble CNTF-Rα (126), when occupied by their respective ligands, can function as soluble heterodimeric proteins, and each confers ligand-induced signal transduction on cells that express only gp130, or gp130 and LIF-Rα, respectively. Because these signal-transducing subunits are expressed on a wide variety of cells, soluble receptors may act as intercellular communicators and donors of a responsive phenotype on otherwise unresponsive cells.

Soluble IL-6-Rα (169,170), CNTF-Rα (126), and LIF-Rα (133,171), and a soluble form of gp130 (172), all occur naturally and have been demonstrated in biological fluids. The 3-kb LIF-Rα transcript appears to encode the soluble LIF-R, and is expressed predominantly by the liver (141–143), particularly during pregnancy (142,143). The soluble LIF-R is an acid-unstable, heat-labile protein that is present at high levels (1μg/ml) in the serum of mice, and its concentration is elevated during pregnancy (133,143,171). The physiological significance of this circulating protein is uncertain, however, because soluble LIF-R blocked binding of LIF to its cellular receptor (144) and inhibited the biological action of LIF in vitro (133), it may serve to modulate concentrations of "active" LIF in the extracellular milieu, perhaps to attentuate any unwanted systemic action of LIF. However, the soluble LIF-R, which is an isoform of the low-affinity receptor subunit, does not effectively compete for ligand against the high-affinity cellular receptor, suggesting that it would not serve as an effective physiological competitor (Gearing D, personal communication). Furthermore, soluble LIF-R has not been identified in human serum or other body fluids, although human serum does contain other molecules that are capable of binding LIF, for example, α_2-macroglobulin (173) and soluble gp130 (172), which potentially may interfere with LIF assays.

In addition to binding to cellular and soluble receptors, LIF also binds to extracellular matrix (ECM). Preliminary studies indicate that a major matrix-associated LIF-binding protein is the soluble LIF-R (174). Interestingly, LIF associated with this ECM-localized protein is biologically active. The significance of this observation is uncertain, but may be linked to specific functions of the M-LIF isoform. Studies in transgenic mice (175) indicate that there are important functional differences between the matrix-associated and diffusable forms of LIF. What these putative matrix-associated functions are remains to be determined, but may be related to LIF's ability to alter the activity of metalloproteases, collagenase, stromelysin, and gelatinase (176–178), enzymes that help regulate ECM protein metabolism.

F. Intracellular Signaling

The pleiotropic actions of LIF raise the question of how LIF initiates mitogenic and differentiation-inhibiting signals in one cell type, yet can induce the opposing effect in another cell type? Presumably, the response of the cell following ligand–receptor binding is dependent on the programming of the intrinsic signaling pathways of the particular cell type, with each distinct cell type interpreting and responding to the signal in its own way. Until recently, little was known about the signaling pathways downstream from gp130 following stimulation with LIF, IL-6, CNTF, or OSM. It is now known that the intracellular-signaling pathways, activated by a diverse array of structurally unrelated and related cytokines, converge on nodal points just inside the cell membrane and within the cytoplasm and nucleus. Signaling appears to be initiated by ligand-induced dimerization of the signal transducer subunits, allowing their cytoplasmic portions to interact, resulting in the activation of tyrosine kinase, recruitment of a number of latent intracytoplasmic proteins by tyrosine phosphorylation, rapid activation of DNA binding of various transcription factors, and transcriptional activation of early-response genes and genes for a variety of secreted and structural proteins (179).

Most members of the hematopoietic cytokine receptor family induce tyrosine phosphorylation of cellular proteins, yet none of their receptor components possess any intrinsic tyrosine kinase domains in their cytoplasmic portions. Both Jak-1 and Jak-2, members of the Jak–Tyk family of nonreceptor kinases, are constitutively bound to the same proximal cytoplasmic domains of gp130 and LIF-Rα (180,181), that were previously shown to be essential for signaling (156,158,159). Stimulation with LIF, CNTF, OSM, IL-11, or IL-6 activated various members of this kinase family, particularly Jak-2 (180–183), which are also activated by several other immune–hematopoietic cytokines (including IFNs, EPO, GH, prolactin, IL-3, GM-CSF, G-CSF; 179). It appears that members of the Jak–Tyk family associate with signal-transducing subunits of the hematopoietic cytokine receptor superfamily, which thereby, acquire potential tyrosine kinase activity. In this manner, they may function similarly to the tyrosine kinase type of growth factor receptors, which include c-*kit*, and receptors for EGF, insulin, platelet-derived growth factor (PDGF), and macrophage colony-stimulating factor (M-CSF) (184).

Although multiple-signaling pathways no doubt exist, so far three distinct signal transduction pathways have been demonstrated to connect LIF receptors to the nucleus, each possibly triggered by kinases associated with gp130–LIF-Rα heterodimers. The best-studied system is the pathways leading to activation of the acute-phase proteins, although it is expected that variations on this theme also exist in other cell types. One pathway leads directly to tyrosine phosphorylation of the cytoplasmic acute-phase response factor (APRF; 185), which rapidly migrates to the nucleus and binds to the hexanucleotide motif CTGGGA, present in the promoters of various acute-phase proteins. APRF is also activated by IL-6, IL-11, CNTF, and OSM, but notably not by IFN-γ, TNF, or IL-1 (180,185–187) and shows strong homology to p91 (Stat91), a component of the IFN-stimulated gene factor-3 (ISGF3), which is also activated by LIF, IL-6, CNTF, OSM, and IL-11 (127,180, 188–190). The second pathway involves the sequential phosphorylation of the Ras/Raf/MEK/mitogen-activated protein (MAP) kinase cascade (127,182,191), which ultimately leads to activation of the transcriptional factor NF-IL-6, which binds to response elements within the promoters of various target genes. This pathway is also used by the lymphocyte receptors (192) and is also activated by CNTF, IL-6, IL-11, and OSM (127, 165,166,182). Third, LIF signaling is in part mediated by the NF-κB pathway (193), which is activated also by LPS, IL-1, and IL-6. Activation of this pathway results in dissociation of the NF-κB–IkB complex through phosphorylation of the specific inhibitor IκB, with subsequent translocation of NF-κB into the nucleus where it binds to κB-binding sites also within the promoters of acute-phase protein and proinflammatory cytokine (IL-6, GM-CSF, TNF) genes.

Table 4 Functional Comparison of LIF with Related Cytokines

	LIF	IL-1	TNF-α	IL-6	IL-11	OSM	CNTF
Endogenous pyrogen	+	+	+	+			+
Inhibition of lipoprotein lipase	+	+	+	+	+	+	
Stimulation of acute-phase proteins	+	+	+	+	+	+	+
M1 leukemic differentiation	+			+		+	
Megakaryocyte progenitor proliferation	+	+		+	+		
Neuronal differentiation	+			+	+	+	+
Stimulation of bone resorption	+	+	+	+	+		
Inhibition of ES cell differentiation	+			+(+ sIL-6R)		+	+

The demonstration that the patterns of tyrosine phosphorylation in common target cell types are similar for LIF, IL-6, IL-11, CNTF, and OSM (127), irrespective of whether signaling was initiated by homodimerization (as with IL-6) or heterodimerization (as with CNTF, LIF, and OSM) of the "β" signal transducer subunits, indicates that these pathways are shared and, presumably, accounts for their functional relatedness (Table 4). In cell types in which unique actions exist (e.g., CNTF alone promotes survival of ciliary neurons, and OSM alone stimulates Kaposi's sarcoma cell proliferation) or where more generic actions are absent (e.g., lack of IL-6 responsiveness in most neuronal cells), these restricted actions presumably result from modulation of these or other pathways by the specific receptor α-chains (i.e., CNTF-Rα, OSM-Rα, and IL-6-Rα chains). Of further interest is the observation that most of these pathways are also part of the signaling mechanisms used by other cytokines, both within hematopoietic cytokine receptor family and in others; for example, tyrosine kinase (e.g., SCF), immunoglobulin (e.g., IL-1α and β) families, which may, partly, explain some of their functional similarity and cooperative interactions with LIF. Each receptor family is likely to be further modified by other associated tyrosine kinase, for example, Fes kinase for the IL-3, GM-CSF receptor family; Lck, Lyn, and Fyn for IL-2-R (179); and Hck for gp130 (194). Several other known proteins undergo tyrosine phosphorylation in response to LIF, including phospholipase Cγ, phosphoinositol 3-kinase, PTPID, pp120, SHC, GRB2 (127), a 160-kDa protein (195), a 27-kDa heat-shock protein (hsp27;196), a ribosomal S6 protein kinase (pp90[rsk]; 182,191), although it is unclear how these fit into the present scheme. Interestingly, tyrosine phosphorylation of hsp27 has also been reported for TNF and IL-1 (196).

Similar to several other cytokines and peptide hormones, LIF activates early-response genes (e.g., *jun*B, c-*jun*, c-*fos*; 195,197–201), and induces interferon regulatory factor 1 (197) in responsive cells. LIF can also induce the production of most proinflammatory cytokines and modulate expression of their receptors. LIF induces mRNA expression or protein production for IL-1β (202), TNF (203), IL-6 (19,193,202,204,205), IL-8 (202,206), and GM-CSF (203) in a variety of cell types. The promoters of most of these cytokine genes contain binding sites for the inducible transcriptional factors NF-κB, NF-IL-6, IL-6-RF, and APRF, most of which are known to be activated by LIF. In addition, LIF down-regulated expression of IL-6-R mRNA and the number of IL-6 (205) and TNF receptors (207) on M1 cells, indicating that LIF can modulate the responsiveness of cells to other cytokines.

Many of the pathways discussed in the foregoing appear to converge on a common set of transcriptional factors, several of which can be present in the promoter regions of the same genes. The combinatorial action of these various transcriptional factors is the presumed basis for the synergistic or inhibitory interactions of various cytokine combinations. As will be discussed in the following section, these phenomena provide a framework for a mechanistic understanding of many of the biological actions of LIF and its interactions with other cytokines in complex in vitro systems and in vivo.

II. BIOLOGICAL ACTIONS OF LIF

As indicated in Tables 1 and 4, LIF is a pleiotropic growth factor, with effects in a diverse range of in vitro and in vivo experimental systems. This rapidly growing body of data indicate that LIF has several important physiological actions and may play a role in the pathogenesis of several disease processes. For the purpose of this chapter, these have been divided into the following systems:

A. Hematopoiesis

There are several lines of evidence that suggest that LIF may play a role in hematopoiesis, although in many instances, a direct effect of LIF has been difficult to demonstrate, being revealed only by its ability to augment the action of other regulators or by the study of LIF-deficient (LIF^-/LIF^-) mice generated by gene-targeting techniques.

First, LIF appears to have effects on the hematopoietic stem cell compartment, an action that may link it to the regulation of other stem cell systems, including primordial germ cells and embryonic stem cells. For instance, LIF potentiates IL-3 or stem cell factor (SCF)-stimulated normal human CD34$^+$ bone marrow (208,209) and cord blood (210) blast cell colony formation and enhances the growth of purified mouse Thy-1$^+$ or c-kit^+, lineage-negative, bone marrow cells (211), although it had no demonstrable effect alone. LIF has also been reported to support the in vitro survival of hematopoietic stem cells and to improve retroviral transfer from cocultivated vector-producing cells (212–214), an effect of potential benefit to gene therapy technology. This effect of LIF is likely due to the stimulation of dormant hematopoietic stem cells (in G$_0$) to undergo cell division (209).

Second, with the exception of the megakaryocytic lineage, LIF appears to have limited effects on hematopoietic progenitors and their mature progeny. For instance, LIF had no observable in vitro colony-stimulating activity when added to cultures of normal murine marrow, either alone (215) or in combination with GM-CSF or multi-CSF (211). Similar results were also demonstrated with normal human marrow (216), although others reported that LIF alone promoted the growth of granulocyte, erythrocyte, monocyte, megakaryocyte, and eosinophil progenitor cell colonies (217). Some of the latter results, however, are difficult to ascribe to a direct effect of LIF, because cells of the granulocyte and eosinophil lineages do not possess LIF-Rs. For similar reasons, it is also doubtful whether human interleukin for DA-1a cells (HILDA) truely had the reported ability to chemoattract and activate human as well as mouse eosinophils (72,74). Although LIF-Rs have not been demonstrated on erythroid cells, LIF promotes erythroid burst activity (72,74) and stimulates the proliferation of cells of murine c-myc-transformed (218) and human TF-1 (219) erythroleukemic lines and therefore, may play some role in erythropoiesis. On the contrary, despite the possession of LIF receptors, albeit of low affinity, by normal monocytes or macrophages, no proliferative, differentiative, of functional activation properties of LIF have been described for these cells, and the role of LIF in this lineage still remains elusive.

Third, there is evidence of local and regulated production of LIF by the stromal cells of the hematopoietic microenvironment (83,96–101), which argues for the idea that there may be a concomitant physiological role for the stroma-derived LIF in hematopoiesis. In addition, LIF has been implicated in bone marrow stromal cell commitment to adipocyte and osteoblastic differentiation pathways (220), as well as in modulating the remodeling and metabolism of marrow bone and adipocyte elements. LIF is also expressed by thymic stroma and epithelial cells (100,102,103), and because these cells appear to produce factors that promote differentiation of CD3$^-$4$^-$8$^-$ thymocytes into more mature stages, it is possible that LIF may play a role in the selection of self-restricted, antigen-specific T cells. Interestingly, LIF was recently reported to stimulate the proliferation of an IL-9-dependent mouse helper T-cell clone (11), giving further support to the view that LIF may be involved in T-cell growth.

Perhaps the most conclusive evidence that LIF plays a role in hematopoiesis comes from recent studies of LIF^-/LIF^- mice. LIF-deficient mice had dramatically reduced numbers of stem cells in the spleen and bone marrow, and this stem cell deficit could be repaired by administration of LIF (221). These observations indicated that LIF was indeed required for the survival of the normal pool of stem cells. Furthermore, the number of committed progenitors was also decreased in the spleen, but not in the bone marrow. B-lymphocyte numbers and subsets in the bone marrow, spleen, and peritoneal cavity were normal, as were T cells in the thymus. There was, however, a dramatic reduction in thymic, but not peripheral, T-cell responses to concanavalin A or allogenic stimulation, which suggested impairment of thymic T-cell stimulation. The B-cell compartment remained intact in LIF^-/LIF^- mice, which contrasts with the action of IL-6, but is consistent with the lack of reports on the action of LIF on B lymphocytes, and the lack of demonstrable LIF receptors on these cells (122,125). Curiously, however, LIF has recently been shown similar to IL-6, to stimulate the proliferation and support the long-term growth of human myeloma cells (130,122).

The stem cell and splenic progenitor cell deficiency seen in the LIF^-/LIF^- mice suggested that LIF does play a role in the regulation of hematopoiesis in vivo. Reconstitution of wild-type mice with engrafted marrow or spleen cells from LIF-deficient mice restored normal splenic and medullary stem cell populations and thymic T-cell responses, indicating that the defect lay in the hematopoietic and thymic microenvironment. Therefore, LIF appears to play an important role in the microenvironmental regulation of hematopoietic stem cells (similar to SCF) and maturing thymocytes. However, the circulating blood cell levels in the LIF-deficient mice were normal (as they are in Sl/Sl^d mice), indicating that although LIF may maintain a subset of the pool of hematopoietic stem cells, compensatory mechansims exist to maintain steady-state hematopoiesis.

Given that LIF has effects on hematopoiesis, its role in leukemogenesis and the potential for treatment of leukemia have been intensively studied, but with varied results. LIF induced macrophage differentiation and suppressed clonogenicity of M1 cells in vitro, an action that involves the reprogramming of the transcriptional activity of proliferating M1 myeloblastic cells toward their conversion into differentiated macrophages (223–225). Because differentiated M1 cells loose their leukemogenicity when injected into syngeneic SL mice (226), this observation is of potential therapeutic interest. Indeed the injection of LIF (28) into mice inoculated with undifferentiated M1 cells prolonged their survival, supporting the notion that LIF may be effective in the therapy of myeloid leukemia (227). LIF, however, had the opposite effect on another myeloid leukemia cell line DA-1a, cells of which proliferated and failed to differentiate in the presence of HILDA/LIF (10,38,73–76). In the human myeloid cell lines HL-60 and U937, LIF inhibited clonal growth when used in combination with GM-CSF or G-CSF, although it did not exert such an inhibitory by itself (228,229). In combination with cytarabine (Ara-C), LIF induced monocytic differentiation of U937 cells (230). In other myelogenous human and murine leukemic cell lines—ML-1 (231), FDCP-1, 32D, WEHI-3B D$^+$ (215)—LIF had no significant effect on growth or differentiation. Likewise, LIF has also been reported to have no effect on the growth or differentiation of human acute myeloblastic leukemia (AML) cells (232,233), although others have reported that LIF suppressed the self-renewal of AML blasts (234) and prolonged their doubling time (235), these contradictory results presumably reflect the heterogeneity of leukemic cells and patients. An indirect leukemogenic role of LIF may still exist since LIF mRNA expression by bone marrow cultures from patients with chronic myelogenous leukemia (CML) is increased (71,97,98), and the level of expression correlated with advanced disease being more common in patients with CML in blast crisis, as opposed to the chronic phase of the disease (71).

B. Megakaryocytopoeisis and Thrombopoiesis

Extensive evidence indicates that megakaryocytopoiesis is regulated by humoral and microenvironmental influences. Several growth factors including IL-1α, IL-1β, IL-3, IL-6, IL-7, IL-11,

CM-CSF, EPO, SCF, and Mpl-ligand, in addition to LIF, are capable of stimulating megakaryo-cytopoiesis and platelet production and may be useful as therapeutic agents for the treatment of thrombocytopenia (138,236). Mice injected with LIF showed a sequential increase in numbers of megakaryocyte (Meg) progenitors, mature megakaryocytes, and platelets (237). Comparable or greater elevations in platelet levels (approximately 50%) have also been shown in rhesus (238) and cynomolgus monkeys (239) injected with LIF, which significantly shortened the period of thrombocytopenia in a non-human primate model of radiation-induced marrow aplasia (240). Curiously, although LIF had no effect on murine Meg-CFCs in vitro, it did enhance IL-3-stim-ulated Meg-CFC proliferation (123). Similarly, LIF enhanced IL-3-stimulated growth of mega-karyocytes derived from human marrow (241) and cord blood (242) $CD34^+$ cells, but had no effect by itself. LIF appears to act as an "accessory" growth factor, although in serum-free liquid cultures of murine marrow, LIF alone increased acetylcholinesterase activity, total megakaryo-cyte mass, megakarocyte diameter and ploidy, but not megakaryocyte number (243), indicating that similar to IL-6 and IL-11, LIF alone promoted megakaryocyte maturation, at least in vitro. However, the absence of an early release of platelets with in vivo LIF treatment, as occurs with IL-6, suggests that LIF does not have a prominent action on megakaryocyte maturation in vivo.

Despite these reproducible in vitro and in vivo effects, the role of LIF in steady-state megakaryocytopoiesis and platelet homeostasis ought to be questioned, because platelet counts in LIF^-/LIF^- mice were normal (221). These observations were similar to findings in Sl/Sl^d and W/W^v mice, which, despite their inability to produce SCF or its receptor c-*kit*, continued to maintain normal platelet levels through a compensatory process of disturbed steady-state megakaryocytopoiesis (244,245).

C. Acute-Phase Response

The *acute-phase response* is the term employed to describe the systemic host response to microbial invasion, tissue injury, and immunological reactions. Collectively, the acute-phase response is characterized by fever, neutrophil leukocytosis, increase in erythrocyte sedimentation rate (ESR), increases in secretion of several hormones (particularly, ACTH and glucocorticoste-roids), activation of complement and clotting cascades, decreases in serum levels of iron and zinc, a negative nitrogen balance, and a dramatic increase in the concentration of certain plasma proteins named acute-phase proteins. The acute-phase proteins are secreted primarily by the liver, and the pattern varies from species to species.

The hepatic acute-phase response is thought to be mediated by cytokines produced by monocytes and macrophages activated by the inflammatory response. IL-1, TNF, IL-6, and IFN-γ are the principal mediators of the phase response. Cultured keratinocytes and squamous carci-noma cell lines were subsequently found to release additional factors, termed hepatocyte-stim-ulating factors (HSF)-I, -II, and -III, which acted on liver cells in a manner similar to IL-6 (246,247). HSF-III was purified from COLO-16-conditioned media and proved to have phys-iochemical properties similar to LIF (7,109). Recombinant human LIF substituted for HSF-III and HepG2 cells and its action was neutralized by HSF-III antibodies, suggesting that HSF-III was an epidermal cell-derived form of LIF (7). In vivo studies largely confirmed the in vitro results. Injection of LIF into mice produced an elevated ESR, hypoabluminemia (237), and hyperfibrinogenemia (248), and injection of LIF into nonhuman primates resulted in the eleva-tion of C-reactive protein (CRP), α_1-antitrypsin, haptoglobin, and ceruloplasmin (238). Recently, IL-11 (249), OSM (250), and CNTF (251–253) have also been shown to be acute-phase stimulants.

Type 1 acute-phase proteins (haptoglobin, hemopexin, complement C3, α_1-acid glycopro-tein, serum amyloid A, and CRP) are regulated mainly by the IL-1 and IL-6 (254), whereas type

2 proteins (α_1-antitrypsin, α_1-antichymotrypsin, thiostatin, fibrinogen, and α_2-macroglobulin) respond mainly to IL-6, LIF (254), IL-11 (249), and OSM (250). It appears that the cytokines regulating type 2 acute-phase proteins function, in part, through a common-signaling pathway involving the IL-6 response elements present in the fibrinogen and α_2-macroglobulin gene promoters (249,250,254). In contrast, type 1 acute-phase protein genes possess binding sites for NF-IL-6 (254). The regulation of acute-phase proteins is characterized by the interplay of the various cytokines, resulting in additive, synergistic, or inhibitory actions, depending on the combinations, suggesting that the integrated control of acute-phase protein synthesis is regulated by the combinational effect of nuclear transcription factors. The relative contribution that each cytokine plays in the in vivo acute-phase response is, therefore, difficult to assess, but recent studies in IL-6-deficient mice indicate that IL-6 is clearly important (255,256).

The action of LIF on hepatocytes is likely to be direct because hepatocytes possess specific high-affinity receptors for LIF (125). Interestingly, the systemic effects of LIF may be modulated by the very acute-phase proteins that they induce, since α_2-macroglobulin has been identified as a cytokine-binding plasma protein and is a carrier protein for LIF (173). LIF also appears to be an endogenous pyrogen. Rabbits given intravenous bolus doses of purified endotoxin-free LIF developed a dose-dependent, monophasic febrile response beginning within 15 min of injection (Fig. 3). Both LIF and CNTF (257) share this action with other known pyrogens, notably IL-1, TNF, IL-6, IFN-γ, IFN-α, and macrophage inhibitory protein (MIP)-1.

D. Development

Leukemia inhibitory factor is responsible for the maintenance of the pluripotential state of embryonic stem cells (ES) and embryonic carcinoma cells (EC) cells in vitro (5,24). ES cells are permanent cell lines established directly from the inner cell mass of the preimplantation embryo and, when injected into blastocyts, contribute to all cell lineages and generate chimeric animals that are mosaic in all their tissues. As mosaicism extends to the germline, they are increasingly being exploited as vectors for the introduction of both dominant and recessive genetic mutations into the murine germline. Embryonal carcinoma (EC) cells are the malignant counterpart of ES cells and are pluripotent stem cells derived from teratomas and teratocarcinomas. Previously, ES and EC cells could be maintained as pluripotent cells in vitro only in the presence of feeder cells or by conditioned media of certain cell lines that contained a soluble factor known as differentiation-retarding factor (DRF;6) or differentiation inhibitory activity (DIA; 5,110). In the absence of DRF or DIA, cultured EC and ES cells spontaneously differentiated into various cell types, which eventually died. LIF is identical with DIA and substitutes for feeder cells (5,24). Both ES and EC cells possess high-affinity LIF-Rs suggesting a direct action of LIF (5,24,124,258). Similar effects have recently been reported for IL-6 and sIL-6R (259), OSM (260), and CNTF (261,262). ES cells isolated and cultured in the presence of LIF for over 100 cells generations have retained their undifferentiated state, their normal karyotype, and the potential to form chimeric mice when injected into blastocysts (5,24). Such cells were capable of colonizing the germline (263–265) and have been used in the generation of "knockout" mice.

In the absence of LIF, ES cells in suspension culture form aggregates, called embryoid bodies, which spontaneously differentiate into a wide range of primitive embryonic cell types including extraembryonic visceral (e.g., yolk-sac) and parietal endodermal, mesodermal, and ectodermal derivatives. Addition of LIF blocked the differentiation of EC cells (258,266,267) and the formation of embryoid bodies from ES cells (60). Development was inhibited in each germline, including the parietal endoderm (258), primitive ectoderm (60), and mesoderm (60,267,268), while permitting the differentiation of the primitive endoderm and neuroectoderm

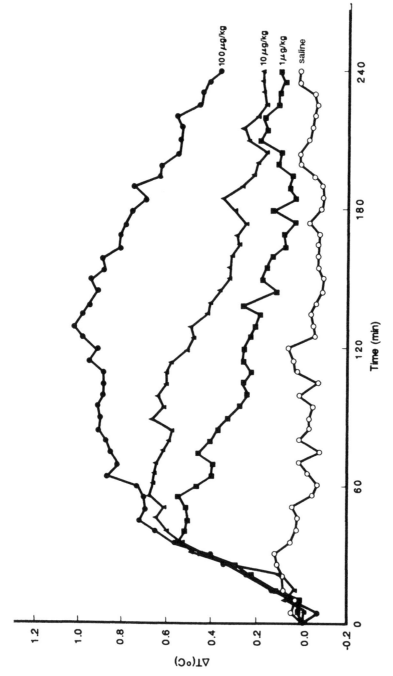

Figure 3 Pyrogenic effect of LIF: Rabbits were restrained in a constant ambient temperature (23–26.5°C) environment and basal temperatures were recorded every 5 min by indwelling rectal thermoprobes, calibrated to 0.1°C precision. After 90 min of stable temperature (<0.2°C) recording, rabbits (three per group) were injected IV with 100 µg/kg (●), 10 µg/kg (▲), 1 µg/kg (■) of purified, endotoxin-free (<0.5 pg/kg body weight) E. coli-derived recombinant human LIF in 5-ml sterile nonpyrogenic 0.9% saline, or saline alone (○), prewarmed to body temperature.

(267). LIF mRNA expression was undetectable in developing blastocysts (269) or present at only very low levels (82,258) in ES and EC cells. However, LIF expression was greatly increased as the ES cell underwent spontaneous differentiation and could be induced by several growth factors, particularly the fibroblast growth factor (FGF) family (82). These findings suggest that the expression of LIF is both developmentally programmed and controlled by the action of other growth factors implicated in the control of early embryogenesis.

The role of LIF in embryonic development is unknown, but some clues have come from studies examining the distribution of LIF expression. LIF mRNA expression was demonstrated in the egg cylinder-stage mouse embryo (82,270) and later in the yolk-sac (59). Given the postulated role of LIF as a regulator of adult hematopoietic stem cells, and its expression in yolk-sac, the earliest site of hematopoiesis, an effect of LIF on early hematopoiesis is likely. LIF may also play a role in later development, as LIF blocked nephrogenesis in cultured mouse kidney rudiments by blocking induction of collecting duct formation by the metanephric mesenchyme and by arresting epithelialization of the mesenchymal elements to form nephrons (271). Interestingly, differentiated kidney epithelial cells possess binding sites for LIF (129), and in vivo radiolabeling studies using ^{125}I-LIF showed intense staining in the metanephros (Hilton DJ, personal communication). In other fetal tissues, LIF mRNA expression has also been documented in the rat footpad (55)—the target of cholinergic neurons that innervate sweat glands.

Recent studies have confirmed that LIF appears to play a crucial role in early development. Targeted overexpression of M-LIF or D-LIF cDNAs in early embryogenesis was performed in transgenic mice (175). Chimeric mouse embryos overexpressing D-LIF appeared normal whereas chimerae overexpressing M-LIF resulted in an embryonic lethal mutation in all three clones injected. The embryos appeared normal up to day 6.5 pc, but then underwent developmental arrest just before gastrulation. The embryos appeared to lack cells derived from the primitive ectoderm and were devoid of detectable markers for early mesoderm and because the abnormal phenotype occurred at the time when ectoderm normally delaminated to give mesoderm, the likely defect was failure of mesodermal differentiation. Recent studies by RNase protection analysis showed that D-LIF and M-LIF have overlapping, but nonidentical, patterns of expression (58). These observations show that there is a profound functional difference between the soluble and matrix-bound forms of LIF, the latter presumably vital for inductive signals in mesodermal development.

E. Reproduction

Despite the importance of germ cells to the survival of species, surprisingly little is known of the control of their proliferation, migration, and entry into mitotic arrest or meiosis. Survival of primordial germ cell (PGC) both in vivo (272) and in vitro (273–276) requires the presence of SCF. PGC proliferation in vitro requires the presence of SCF plus feeder layers, such as STO cells (277) (which produce LIF). Both SCF and LIF inhibit PGC apoptosis in vitro (276), and LIF sustains the survival of proliferating PGCs on feeder layers of Sertoli cell line TM$_4$ (278) and synergizes with SCF to stimulate proliferation of PGCs (274,277). In the presence of SCF, LIF, and FGF-2, PCGs continued to proliferate in vitro, giving rise to large colonies of ES cells (279,280), which were capable of differentiating into embryoid bodies and a wide variety of mature cell types both in vitro and in vivo; generated teratocarcinomas in nude mice; and also contributed to chimerae when injected into host blastocyst (277). LIF alone, however, does not appear to be essential for germ cell migration and proliferation, because *LIF⁻/LIF⁻* mice are fertile (221,281). The contribution of LIF to the development of long-term PGC cultures may provide an important tool for the study of the development of the mammalian germline, and analysis of such key issues as recombination, pluripotency, sexual differentiation, and imprinting.

The ovarian steroids estrogen and progesterone regulate the cellular and molecular changes that occur during reproductive events. Cycles of proliferation, differentiation, and cell death are a direct result of changes in hormone concentrations. How these hormones exert their effects is not completely understood and often results from the action of intermediaries such prostaglandins, leukotrienes, or growth factors. Several avenues of investigation have recently shown that cytokines, including LIF, play important and sometimes crucial roles in the control of several reproductive events.

Little is known about the role of cytokines in the endometrial cycle, but they are clearly important, as expression of TNF, IL-1, IL-6, and GM-CSF mRNA are cyclic and parallel changes in ovarian hormone concentrations (282). Furthermore, ovariectomized mice showed little or no cytokine activity in the uterus, which could be induced by systemic administration of estrogen or progesterone (282). Uterine expression of LIF mRNA has been studied during the estrous cycle in virgin femal mice (59, 60), and was low during diestrous, proestrous, and metestrous 2, and expression of the D-LIF form was elevated during estrous metestrous 1, peaking after ovulation (60). LIF mRNA expression was also detected at high levels in human secretory endometrial tissue and decidua (61), suggesting a role for LIF in the reproductive cycle as well as during pregnancy.

The preimplantation stage of development is a period of exponential yet automonous growth, and there is evidence that the embryo produces its own growth factors in an autocrine manner. Recent studies indicate that LIF plays a crucial role in the control of mammaliam preimplantation embryogenesis and implantation. First, mouse preimplantation (3.5 day) blastocysts express LIF (269, 283). Given that ES cells derived from the inner cell mass of the blastocyst possess high-affinity LIF-Rs, it is possible that autocrine LIF production may be important in maintaining pluripotency until implantation and maternal support have been secured. With these actions in mind, the ability of LIF to reduce embryo wastage in in vitro fertilization procedures has been investigated, and although LIF improved the viability of cultured ovine embryos (284), it did not improve implantation rates when added to the embryo transfer medium (285).

Second, LIF plays a pivotal role in mediating estrogen-initiated blastocyst implantation in the progesterone-primed uterus. The mechanisms that direct implantation are poorly understood, but it is clear that changes in the preimplanation embryo are synchronized with sex hormone-induced changes in the uterus. Transcription of the LIF gene has been detected at high levels in the murine uterus just before blastocyst implantation (59,60), a finding that is in stark contrast to the very low levels in other normal adult murine tissues. LIF mRNA expression occurred as a distinct burst at day 4 that always preceded implantation (59,60) and was localized by in situ hybridization to the glandular endometrium. Curiously, LIF expression also occurred at day 4 in pseudopregnant mice (59,60), but not in mice undergoing experimentally induced delayed implantation. In the latter situation, LIF expression could be restored following a single injection of β-estriol, indicating that LIF expression was under maternal control (59). Of profound significance was the observation that female *LIF⁻/LIF⁻* mice were infertile owing to failure of blastocyst implantation (281). By day 7 of gestation in *LIF⁻/LIF⁻* mice, the blastocysts were viable, but had not implanted (281). When harvested from the uterine cavity and transferred to pseudopregnant wild-type females, the blastocysts implanted and developed to term. Furthermore, wild-type embryos failed to implant into LIF-deficient mothers, indicating that the defect was maternal. An embryonic defect was excluded, as homozygous mutant embryos implanted and developed normally in a normal uterine environment, also indicating that autocrine LIF production by the preimplantation blastocyst was not essential for their survival.

The mechanisms by which maternal expression of LIF governs implantation is unknown, but the action of LIF may be to prepare the uterus for implantation by inducing a decidual reaction. By day 7 of gestation the uteri of the LIF-deficient mice were poorly vascularized and the endometrial stromal cells had failed to decidualize, a response that should have started on the fifth day of gestation (281). This block could be overcome by administration of exogenous LIF. Because LIF expression has been documented in murine decidual natural killer cells (286) and granulated metrial gland cells (GMC) (287), there are potential sources of LIF production within the decidua. However, it is not known whether decidual cells possess LIF-Rs and whether LIF has a functional effect on these cells.

Several growth factors appear to act on trophoblastic cells and are important for placental development and differentiation (288). LIF, too, may play a role in placental development because LIF mRNA expression is present in the extraembryonic trophoectoderm and placenta (61,62,69) and placental trophoblastic cells possess LIF-Rs (27,122). Furthermore, elevated serum levels of soluble LIF-R were present during pregnancy in mice (133,171). Exactly what these potential roles may be is uncertain, but may relate to trophoblastic differentiation, placental invasion, or maternal–fetal immunoregulation. Recently, it was shown that LIF reduced production of chorionic gonadotropin (hCG) by cultured human placental cytotrophoblasts, suggesting that LIF may shift trophoblastic differentiation away from hCG-producing syncytial phenotype toward an extravillous anchoring or invading phenotype (Arici A, personal communication). Notably, LIF-R –/–, which die in the perinatal period, exhibit abnormal placental architecture (Ware C, personal communication).

F. Bone Metabolism

Bone is a dynamic tissue, which to maintain the structural integrity of the skeleton, undergoes a lifelong renewal process. This process of bone remodeling involves the continuous removal of bone, followed by synthesis of new bone matrix and subsequent mineralization. Removal of old bone by osteoclastic resorption and new bone formation by osteoblasts is finely balanced by both humoral and locally produced regulators. Osteoblasts seem to regulate the resorptive response of osteoclasts, presumably to maintain finely balanced bone homeostasis.

The term "osteoclast-activating factor" (OAF) was used to explain bone resorption that accompanied certain malignant tumors that were associated with hypercalcemia. Subsequently, it was realized that several factors, including parathyroid hormone-related protein (PTH-rp), TNF-α and -β, IL-1, IL-6, IL-11, and DIF, all shared properties originally attributed to OAF, and were responsible for this bone-resorbing activity. The bone-resorbing action of DIF was first recognized as an activity from conditioned media of murine mitogen-treated spleen cells (23) and osteoblastic MC3T3-E1 cells (87) and proved to be biologically identical with the murine differentiation-stimulating factor on M1 cells, D-factor (2,3). Subsequent purification and cloning of human D-factor (22) and murine DIF (4) revealed that both factors were identical with LIF.

The availability of recombinant LIF allowed examination of the action of LIF on bone metabolism and several conflicting effects have been reported. Recombinant LIF increased, in a prostaglandin-dependent manner, the numbers of osteoclasts and $^{45}Ca^{2+}$-release from neonatal mouse calvaria (92,289,290). In contrast, LIF inhibited basal bone resorption in fetal rat long-bone cultures (291) and fetal mouse metacarpal bones (292). The reasons for these conflicting effects are unknown, but similar differences in these in vitro systems have also been reported for PTH and TGF-β (293), and may relate to the number, maturity, or location of osteoclasts at different stages of development. In adult mice, LIF promoted bone resorption in vivo (294,295) and caused hypercalcemia (237), an action that probably involved osteoblast-dependent osteoclastic bone resorption, for osteoblasts, but not osteoclasts, displayed high-affinity LIF-Rs (89,128). LIF induces the formation of osteoclast-like cells in bone marrow cultures (87,296), supporting the view that the bone-resorbing effects of LIF are more due to the enhanced generation of new osteoclasts, rather than to activation of mature osteoclasts present in bone (292), although others have reported that LIF had no effect on osteoclast generation (297). It seems, therefore, that LIF acts by stimulating osteoblasts to promote osteoclastic bone resorption, in much the same manner that has also been proposed for TNF-α and IL-1.

LIF also exerts anabolic effects on bone. In vitro, LIF promoted DNA synthesis in neonatal mouse calvaria (289,290) and human bone (298) and stimulated the proliferation of isolated fetal rat (293) and human (299) osteoblasts. Growth inhibitory effects, however, were observed in MC3T3-E1 (92,300) and UMR-106 osteoblastic cells (293) and fetal mouse metacarpal bones (292). LIF has also been implicated in the regulation of osteoblast differentiation and, again, conflicting results have been reported. Alkaline phosphatase (AP) activity and mRNA expression are markers of osteoblastic differentiation, and LIF has been variously reported to have no effect on AP activity in normal human bone or isolated osteoblasts (298,299) and rat RCT-1 cells (128), but enhanced retinoic acid-induced osteoblastic diffentiation in RCT-1 cells (128). Similar results were also obtained with MC3T3-E1 cells (92); however, others have reported an inhibitory effect on AP activity using the same cell line (91,300), and on fetal mouse metacarpal bones (292).

The seemingly contradictory in vitro effects of LIF on osteoblastic and osteoclastic activity appears to reflect and depend on the developmental stage of the respective cells. A clearer picture of the potential role of LIF in bone metabolism has come from in vivo studies. Mice chronically exposed to high levels of LIF developed marked osteoblastic proliferation in their long bones with excess new bone formation, together with evidence of bone resorption and metastic calcification (294,295). The local effect of LIF on bone was further assessed in vivo by subcutaneous injection of LIF over the calvariae of normal adult mice and resulted in thickening of the bone and periostium, fibrosis, increased osteoclast recruitment, increased resorption area and increased osteoblast numbers and new bone formation (301). These changes were similar to those seen in osteodystrophia fibrosa, a condition caused by prolonged and excessive exposure to PTH, which also bears some resemblance to Paget's disease of bone. Because both conditions are characterized by an imbalance of both osteoclastic and osteoblastic activity and as LIF production by osteoblasts can be induced by PTH (90), it seems likely that LIF may play an important role in the bone changes seen in hyperparathyroidism.

The role of LIF in humoral hypercalemia of malignancy (HHM) may also be important. Tumors associated with HHM produce IL-1 and TNF-α (302), and animals injected with tumor cells transfected with cDNA for TNF-α (303), IL-6 (304), or LIF (294,295) developed hypercalcemia and increased bone resorption, as well as leukocytosis and cachexia, which commonly accompany this syndrome. LIF may also contribute to the hypercalcemia and osteolytic bone lesions that accompany adult T-cell leukemia (ATL) associated with infection by the human T-lymphotropic leukemia virus (HTLV). In contrast with the transient induction of cytokines during inflammation, some viral infections result in constitutive expression. The HTLV type 1 constitutively activates production of PTH-rp, IL-2, IL-3, IL-6, TGF-β, GM-CSF, OSM, TNF-α and TNF-β, and LIF (119) through its transactivator protein *Tax*. Leukemic T cells from HTLV-1-infected individuals and HTLV-1 and HTLV-2-infected T cells from normal individuals abnormally express LIF (118,119) as well as PTH-rp, IL-1, TNF-β, and IL-6, all of which have

OAF-like properties. Similarly, LIF may contribute to the osteolytic changes that accompany deposits of multiple myeloma.

Also, LIF has been implicated in chondrocyte biology, findings of potential importance to the pathogenesis of arthritis. LIF stimulates cartilage resorption (305), and chondrocyte proliferation (105), and is capable of being produced by articular cartilage and by isolated chondocytes in vitro (63,106). LIF has recently been reported to stimulate the production of the chemokines, monocyte chemoattractant protein-1 and IL-8, and to increase collagenase and stromelysin production by human chondrocytes (176,306), thus potentially contributing to joint tissue destruction that is a feature of the inflammatory arthritides.

G. Lipoprotein Lipase Inhibitor

Lipoprotein lipase (LPL), an enzyme produced by many parenchymal tissues, resides on the luminal surface of capillary walls, where it promotes lipolysis by mediating the exchange of lipids between circulating lipoproteins (307). By generating fatty acids in white adipose tissue, brown fat, cardiac muscle, and lactating breast, LPL provides the major source of acyl groups used for fat storage, thermogenesis, fuel, and milk synthesis by these tissues, respectively. The activity of LPL responds rapidly to changes in the nutritional state, to physical exercise, cold adaptation, severe infections, and tumors, and it is influenced by hormones (insulin and prolactin) and growth factors (TNF, IL-1, IFN-γ, TGF-β, IL-6, IL-11, OSM, and LIF). The action of LIF on LPL was discovered by investigation of a human melanoma cell line, SEKI, that induced severe weight loss in tumor-bearing nude mice. SEKI cells secreted a factor, then known as melanoma lipoprotein lipase inhibitor (MLPLI), which inhibited the activity of lipoprotein lipase. The amino acid sequence of purified MLPLI revealed complete homology with human LIF (9). SEKI cells were shown to produce LIF mRNA (113), and LIF was subsequently shown to decrease LPL activity in cultured adipocytes (308).

Suppression of LPL activity is thought to curtail the ability of adipocytes to extract fatty acids from plasma lipoproteins for storage, leading to hypertriglyceridemia and an inability to replace used fat stores. In certain chronic infections and malignancies, these changes are thought to contribute to the development of a chronic catabolic state manifest by weight loss, asthenia, anorexia, anemia, hypoglycemia, and severe wasting of both protein and fat stores. The weight loss that accompanies this condition, termed cachexia, often persists inexplicably in spite of adequate caloric intake. There are several lines of evidence that some of the factors that are capable of suppressing LPL activity, including LIF, can induce cachexia. First, administration of TNF or IL-1α (309), IFN-γ (310), CNTF (311), or LIF (237) to mice, each rapidly induced body weight loss. Second, mice injected with tumorigenic cell lines transfected with vectors expressing human TNF (312) or IFN-γ (313), or murine IL-6 (304) or LIF genes (294,295) developed cachexia, whereas mice injected with the parental cell lines did not. Lastly, administration of anti-TNF (302), anti-IFN-γ (310), or anti–IL-6 antibodies (314) significantly prevented the development of cachexia in tumor-bearing mice; however, such studies have not been performed with LIF antibodies. All these observations promote these proteins as strong candidates for causing the cancer cachexia syndrome.

H. Neuropoietic Effects

In the nervous system, neuronal growth, survival, and phenotype are determined, in part, through the action of instructive signals generated by neurotrophic and neuropoietic factors (315). Neurotrophic factors, such as nerve growth factor (NGF) and brain-derived neurotrophic factor (BDNF), operate as retrograde, target-derived signals that determine survival and promote the growth of innervating neurons. Neuropoietic factors, such as cholinergic differentiation factor (CDF) and CNTF, in addition to providing trophic support, also influence cholinergic neurotransmitter phenotype. During development, noradrenergic sympathetic neurons that innervate eccrine sweat glands acquire cholinergic function, indicating that target tissues also influence the phenotype of the neurons that innervate them.

CDF was characterized (316) and purified from media conditioned by neonatal rat heart cells (8) and skeletal muscle (317), by its ability to switch the neurotransmitter phenotype of postmitotic noradrenergic sympathetic neurons to a cholinergic phenotype, without affecting their growth. Its amino acid sequence was established as identical with LIF, and LIF was shown to be able to duplicate the actions of CDF in culture (31). CDF–LIF induces expression of the cholinergic-associated neuropeptides, substance P, somatomedin, somatostatin, vasoactive intestinal polypeptide; cholecystokinin, and enkephalin in neuronal cells (318–325). In addition, CDF–LIF possesses neurotrophic activities promoting survival or growth of embryonic and postnatal sensory and sympathetic neurons (326–329), embryonic motoneurons (330), cholinergic neurons (331), and the differentiation of spinal cord neurons (332), including cholinergic neurons (333). The demonstration that LIF can be taken up by the endings of sensory neurons and retrogradely transported back to the neuronal cell body (334,335), suggests that LIF is also a target-derived neurotrophic factor.

CDF–LIF has several activities in common with CNTF, a point of particular interest because the LIF-Rα and gp130 subunits are common to both of their receptors. Both CDF–LIF and CNTF are required for survival of embryonic mononeurons (330,336), and in sympathetic neurons, they both act as cholinergic differentiation factors (322,332, 337), induce the same set of neurotransmitter synthetic enzymes and neuropeptides (321–325), and reduce the expression of muscarinic receptors in sympathetic neurons (338). OSM regulates the same neuropeptides as LIF and CNTF in neuroblastoma cells (339); however, IL-6 appears to have no effect on neurotransmitter phenotype or neurotransmitter gene expression. With the exception of IL-6, these familial resemblances have led to the proposal that these cytokines should be considered part of a neuropoietic cytokine family (340). However, there are important functional differences, because neither CDF–LIF nor OSM promote the survival of ciliary neurons (104,339), a property unique to CNTF.

LIF-Rs have been demonstrated by autoradiography on sensory and autonomic ganglia (334,341); specific regions of the cerebral motor, sensory, and limbic systems (341); and also on neuroblastoma cell lines (122) and Ewing's sarcoma cells (Kanourakis G, personal communication). mRNAs for the LIF-Rα, gp130, and CNTFRα chains have also been demonstrated in sensory ganglia (328,329) and Ewing's sarcoma cells (127). Expression of CDF–LIF has been documented in rat footpads containing sweat glands, the superior colliculus, visual cortex, and hippocampus (55,56), and in embryonic skeletal muscles (342). In vivo, LIF has been demonstrated to induce neurotransmitter switching in sympathetic neurons innervating the pancreatic islets in transgenic mice expressing under control of the insulin promoter (343). However, because the addition of neutralizing anti-LIF antibodies to footpad extracts did not inhibit their cholinergic activity (344) and because sweat gland innervation was phenotypically normal in LIF-deficient mice (345), it is likely that LIF is not an obligatory cholinergic-switching factor for sweat gland neurons during development.

There are several observations that indicate potential LIF-mediated interactions between the glial and neuronal elements of the nervous system. Firstly, cultured astrocytes (115,116) and sympathetic ganglia (346) express LIF mRNA or protein. Secondly, glial secretion of LIF induces substance P production by neurons (347). Third, immunoprecipitation using anti-LIF antibodies of ganglionic nonneuronal cell-conditioned medium removes the ability of these cells to inhibit adrenergic phenotype in sensory neurons (348). Furthermore, LIF as well as CNTF, OSM, and IL-6, all induce astrocytic differentiation in vitro (349,350), and LIF also supports the generation, maturation, and survival of oligodendrocytes in vitro (351,352), thus indirectly maintaining neuronal viability.

LIF also appears to be involved in the specialized response to injury in the central and peripheral nervous systems, as has been reported for CNTF (353). Brain CDF–LIF mRNA expression is increased in mice with acute encephalomyelitis (354), in rats injected with the neurotoxin kainic acid (56,355), and nerve expression is rapidly up-regulated following nerve injury (356,357). LIF is retrogradely transported in sensory and motor neurons in response to crush injury (356) and, interestingly, neuropeptide induction in ganglia of axotomized nerves is suppressed in *LIF-/LIF*-mice (345), indicating that LIF is important in determining the neuro-transmitter changes induced by nerve injury. Furthermore, local administration of LIF prevented degeneration of sensory and motor neurons following axotomy in mice (358,359).

There is also growing evidence that LIF influences skeletal and cardiac myogenesis, innervation, and repair. First, LIF stimulated myoblast proliferation in vitro (360,361), and was chemotactic for muscle precursor cells (362). Second, LIF is a candidate muscle-derived neurotrophic factor, as LIF produced by cardiac and skeletal muscle (8,317), promoted survival of motoneurons (330) and induced expression of peripherin intermediate filaments in neuronal cells (363), properties ascribed to soluble factors produced by muscle cells. Lastly, LIF mRNA expression was induced by muscle injury, and administration of LIF to the injury site stimulated muscle regeneration (364). LIF also induced the differentiation of cultured cardiac neural crest cells into neurons, rather than cardiac myocytes (365), and improved their ability to migrate and to support normal heart development (366).

I. Endothelial Effects

Two groups independently purified factors produced by human diploid fibroblast cell lines (12) and bovine pituitary follicular cells (13), that were potent inhibitors of cultured bovine endothe-lial cell (BAEC) growth. Each purified a protein that, on amino acid sequencing, proved to be LIF (12). LIF was mitogenic for endothelial cells derived from capillaries in the adrenal cortex (13) and for SV40T–immortalized human endothelial cell line (367), properties also similar to those of OSM. These activities may be relevant to wound healing, tumor growth, and diabetes. Recently, LIF expression has been demonstrated in activated macrophages of atheromatous plaques (368), and because administration of LIF rapidly induced hypotriglyceridemia and progressive delipidation of plasma lipoproteins in mice (369) and reduced the extent of aortic atherosclerosis in rabbits fed a cholesterol-rich diet (370), LIF may be of importance in the pathogenesis of atherosclerosis.

LIF may also play a role in the regulation of the fibrinolytic system, the activity of which is a consequence of a balanced production of tissue plasminogen activator (t-PA) and its inhibitor, plasminogen activator inhibitor-1 (PAI-1) by endothelial cells. This balance can be altered by proinflammatory cytokines, including LIF, which increased endothelial cell expression of the t-PA and PAI-1 genes (367), again implicating LIF in the inflammatory cascade pathways.

III. THE PATHOLOGICAL EFFECTS OF LIF

As discussed in the foregoing and summarized in Table 4, LIF bears a remarkable functional similarity to various cytokines, particularly TNF, IL-1, and IL-6. These three cytokines have been extensively implicated in the pathogenesis of human disease, and recent experimental and clinical studies performed in our laboratory, and elsewhere, suggest that LIF is involved in the same spectrum of human disease.

A. The Pathological Effects of LIF in Experimental Animals

The in vivo pathological effects of LIF were investigated in mice engrafted with cells of the immortalized myeloid cell line FDC-P1, that had been transfected with LIF cDNA (294,295). These mice (termed FD/LIF mice) subsequently developed constitutively elevated serum LIF levels and a fatal syndrome with a complex pathology (Fig. 4), comprising cachexia, neutrophil leukocytosis, thymic atrophy, splenomegaly, megakaryocytic hyperplasia, thrombocytosis, hepatic necrosis, excess bone formation and dystrophic calcification, reflecting many of the in vitro activities described earlier.

The in vivo pathological effect of the administration of LIF has been studied in several species. In mice (237) and rhesus monkeys (238), low doses of LIF (24–50 µg/kg per day) administered for up to 2 weeks induced an acute-phase response and thrombocytosis, but little evidence of systemic toxicity. Mice administered 300 µg/kg per day of rLIF for up to 2 weeks (28,237) developed body weight loss largely due to loss of body fat, hypercalcemia, thrombocytosis, an elevated ESR, hypoalbuminemia, thymic atrophy, splenomegaly, and behavioral changes manifest as hyperactivity. Some of these changes recapitulated those observed in the FD/LIF mice. High doses of rLIF (1200 µg/kg per day) were tolerated for 24 hr without any

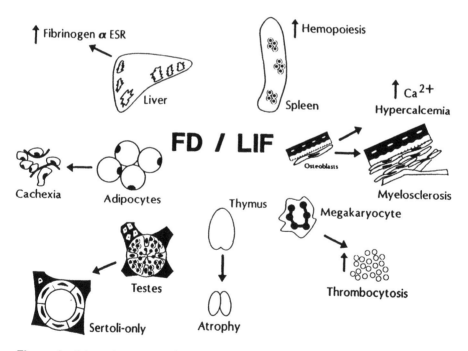

Figure 4 Schematic representation of the pathology present in FD/LIF mice: DBA/2 mice were injected IV with 10^6 FD/LIF cells, which engrafted in the bone marrow, spleen, and lymph nodes. Within 1–2 months, the mice developed elevated circulating LIF levels and a complex pathology including an acute-phase response, hypercalcemia, thrombocytosis, splenomegaly, fibro-osseous endostosis of long bones, extramedullary hematopoiesis, megakaryocytic hyperplasia, thymic cortical atrophy, weight loss with reduced subcutaneous and intra-abdominal adipose tissue, and marked suppression of spermatogenesis.

obvius adverse effects (371); however, with continued administration mice became anorexic, hypothermic, dehydrated, and wasted, and were moribund within 4 days (369).

Transgenic mice bearing transgenes for M-LIF and D-LIF directed by a CMV promoter (175), and D-LIF (372) directed by the human CD2 enhancer, have been generated. The M-LIF form resulted in an embryonic lethal phenotype, the defect thought to be a consequence of arrested mesodermal differentiation (175). Mice overexpressing the soluble form of LIF, specifically in T lymphocytes, developed a fatal disease characterized by lymphadenopathy, splenomegaly, reduction in adipose tissue, male sterility, an acute-phase response, and splenic megakaryocytic hyperplasia, changes strongly reminiscent of the pathology seen in FD/LIF mice and mice administered high doses of LIF. These changes were suggestive of the effects of constitutive systemic LIF production. The disease culminated in fatal acute pulmonary edema, with pleural effusions and ascites. In addition, thymic and pulmonary B-cell hyperplasia, polyclonal hypergammaglobulinemia, and mesangial proliferative glomerulonephritis occurred resembling the changes observed in IL-6 transgenic mice (373), although it is uncertain whether these changes represent a direct effect of LIF. Curiously, the mice developed a bizarre interconversion of thymic and lymph node architectures and cellular components. Thymic hyperplasia with distortion of thymic epithelial cell architecture, loss of cortical $CD4^+CD8^+$ lymphocytes and the development of numerous B-cell follicles were observed, changes closely resembling normal lymph nodes. Reverse changes were observed in peripheral lymph nodes, which showed paracortical expansion, reduction in B-cell follicles, and marked plasmacytosis of the medullary cords, the former resulting from a vastly expanded population of $CD4^+CD8^+$ lymphocytes. In conjunction with the stromal-induced defect in thymic T-cell activation observed in LIF^-/LIF^- mice (221), these changes further support the belief that LIF plays a crucial role in mediating thymic stromal–lymphocyte interactions.

B. The Association of LIF in Human Disease

As a result of its detection in disease-associated body fluids, LIF, similar to TNF, IL-1, IL-6, IL-8, the CSFs, and INF-γ, has emerged as a potentially important mediator of host defense and has been widely implicated in the pathogenesis of various human diseases. These studies had to await the development of highly sensitive and specific assays capable of detecting native human LIF protein. Polyclonal (31,374) and monoclonal (116,375,376) antibodies to human LIF have been developed, and recently enzyme-linked immunosobentassays (ELISAs; capable of detecting 10 pg/ml) and a radioimmunoassays (RIA; capable of detecting 25 pg/ml) have become available. Less sensitive, but highly specific radioreceptor competition assays (sensitivity 0.1–1 ng/ml) have also been developed (92,377).

With these assays, it has been possible to build up a disease profile for LIF (Fig. 5). LIF was readily detectable in exudates associated with acute inflammatory conditions: namely, bacterial meningitis (377), intra-amniotic infection (64), and in malignant effusions (377). LIF was commonly found in synovial fluid collected from patients with rheumatoid arthritis and other inflammatory arthritides (92,106,378), a finding of potential significance to the development of joint erosions because LIF is capable of inducing cartilage destruction and bone resorption. However, LIF was undetectable in serum or urine of healthy individuals, pregnant women (373,377,379), or volunteers administered an intravenous injection of *Escherichia coli* endotoxin (4 ng/kg) (Waring PM, Suffredini A, unpublished data, 1992), but could be detected in some individuals with septic shock (377). As has been shown for TNFα, IL-1, IL-6, and IL-8, LIF was transiently present, sometimes at very high concentrations, in the serum and urine of patients with meningococcal septic shock, and the serum levels correlated with disease severity

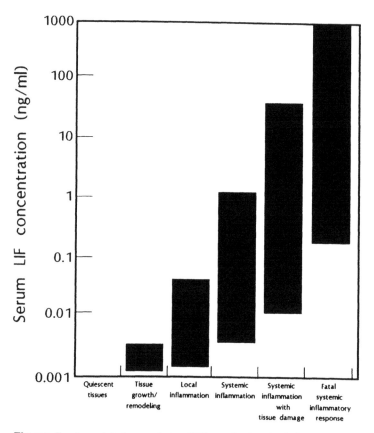

Figure 5 A model for escalating LIF production in inflammatory disease: In quiescent tissue, LIF production is negligible. LIF production progressively increases in processes involving tissue repair and remodeling or local inflammation. In acute systemic inflammation (sepsis syndrome), plasma LIF concentrations are moderately elevated, and in septic shock, approximately half of the subjects had plasma LIF levels higher than 1 ng/ml. Extremely high plasma LIF concentrations (up to 1 µg/ml) are commonly present in the initial phases of fulminant meningiococcemia. Within this disease spectrum, plasma LIF concentrations may increase 10^5–10^6 times basal levels.

(380). Lower serum LIF levels were also detected in patients with chronic myeloid leukemia (< 10–50 pg/ml) or temporal arteritis (< 10–5000 pg/ml) (374,381). Urinary LIF concentrations were elevated in patients with severe meningococcemia, renal and metabolic disorders, urinary infections (377,380), and acute renal transplant rejection (379,382). In some of the renal transplant rejection patients, the elevated urinary LIF levels were associated with T-cell–depletion therapy with OKT3 monoclonal antibodies (382), suggesting that LIF may contribute to the transient OKT3-induced flu-like syndrome that is associated with T-cell activation and the systemic release of lymphokines in this reaction.

It is likely that LIF, in conjunction with TNF-α, IL-1, and IL-6, may induce some of the systemic effects, such as acute-phase response, fever, and thrombocytosis, common to many of these diseases. It also seems reasonable to speculate that LIF may also be involved in cachexia associated with various chronic inflammatory diseases, cancer, and AIDS. In the latter, it is noteworthy that LIF, as well as TNF, IL-1, and IL-6, stimulated HIV replication in monocytes (383). The most compelling data supporting a causal role for LIF in disease come from studies of endotoxic shock. In mice, LIF mRNA expression by many tissues (54,57), and elevated serum concentrations of LIF can be transiently induced by administration of LPS or live *E. coli* (227,369). Curiously, mice injected with high doses of LIF, did not develop acute toxicity (369,371), and failed to develop the typical histological changes of shock (369). Similarly, intravenous infusion of LIF caused no hemodynamic alteration in anethetized piglets (Waring PM, Henning R, unpublished data, 1992). However, passive immunization of mice with anti-LIF polyclonal antibodies protected mice ag··inst lethal doses of endotoxin (57), an action that was associated with suppression of the host cytokine response, particularly IL-1 and IL-6, but not TNF. Thus, although LIF alone does not induce a septic shock-like state, it appears likely to participate in the process leading to septic shock, probably by acting in concert with other cytokines and inflammatory mediators. Paradoxically, LIF also protected mice against a lethal dose of LPS or live *E. coli* (369,371), when administered before challenge, probably by inducing endotoxin tolerance. LIF also enhanced host resistance against radiation injury (384,385) and oxygen toxicity (386), effects that may be related to induction of manganous superoxide dismutase (MnSOD) a mitochondrial enzyme induced by LPS, TNF, IL-1, IL-6, or LIF (386) that is involved in protective mechanisms against cellular injury from reactive oxygen species.

IV. CONCLUSION

One of the most striking properties of LIF is its extraordinary functional diversity, with effects on development, reproduction, metabolism, hematopoiesis, immunity, inflammation, and the local response to injury within the nervous system. The induction of LIF production appears to be tightly linked to these events and operates in a manner such that LIF is released in close proximity to its target cell population. LIF's pleiotropic actions reflect the pattern of LIF-R expression on these target cells and the design of their intracellular-signaling pathways. Curiously, many of LIF's actions are shared by structually related and unrelated cytokines. The explanation for this functional redundancy appears to reside in shared components of their receptors, intracellular signaling, and transcriptional pathways. The latter appear to operate in a combinatorial fashion, permitting synergistic or inhibitory interactions, thereby coordinating their combined actions. Furthermore, many of these cytokines modulate the level of each other's production and receptor expression; consequently, their production and actions are inextricably linked.

Although important functions and pathological effects of LIF have been revealed by its overexpression in mice, an understanding of its true physiological role has proved elusive. Despite its multitude of well-documented developmental effects, LIF-deficient mice appear to develop normally. To date, the only obligatory function of LIF, as tested by loss of function in LIF-knockout mice, has been failure of implantation. As seen in the hematopoietic and neural systems, powerful compensatory mechanisms appear to redress the loss of LIF, presumably by the action of one or more cytokines with an analagous function. Interestingly, IL-6-deficient mice, similar to their LIF-deficient cousins, were also apparently normal. However they failed to develop protective immune and inflammatory responses to viral and bacterial infection and injury (255,256). Given all the parallel biological actions of IL-6 and LIF and their presence in the same wide range of inflammatory disease states, it is highly likely that LIF plays an analogous role as a host distress signal to injury or infection.

ACKNOWLEDGMENTS

I would like to thank David Gearing and Linda Richards for helpful comments and Douglas Hilton for assistance. The author was a recipient of a Medical Postgraduate Research Fellowship from the National Health and Medical Research Council, Canberra, Australia.

REFERENCES

1. Hilton DJ, Nicola NA, Metcalf D. Purification of a murine leukemia inhibitory factor from Krebs ascites cells. Anal Biochem 1988; 173:359–367.

2. Tomida M, Yamamoto-Yamaguchi Y, Hozumi M. Purification of a factor inducing differentiation of mouse myeloid leukemic M1 cells from conditioned medium of mouse fibroblast L929 cells. J Biol Chem 1984; 259:10978–10982.

3. Tomida M, Yamamoto-Yamaguchi Y, Hozumi M. Characterization of a factor inducing differentiation of mouse myeloid leukemic cells purified from conditioned medium of mouse Ehrlich ascites tumor cells. FEBS Lett 1984; 178:291–296.

4. Abe T, Murakami M, Sato T, Kajiki M, Ohno M, Kodaira R. Macrophage differentiation inducing factor from human monocytic cells is equivalent to murine leukemia inhibitory factor. J Biol Chem 1989; 264:8941–8945.

5. Smith AG, Heath JK, Donaldson DD, Wong GG, Moreau J, Stahl M, Rogers D. Inhibition of pluripotential embryonic stem cell differentiation by purified polypeptides. Nature 1988; 336:688–690.

6. Koopman P, Cotton RGH. A factor produced by feeder cells which inhibits embryonal carcinoma cell differentiation. Characterization and partial purification. Exp Cell Res 1984; 154:233–242.

7. Baumann H, Wong GG. Hepatocyte-stimulating factor III shares structural and functional identity with leukemia inhibitory factor. J Immunol 1989; 143:1163–1167.

8. Fukada K. Purification and partial characterization of a cholinergic neuronal differentiation factor. Proc Natl Acad Sci USA 1985; 82:8795–8799.

9. Mori M, Yamaguchi K, Abe K. Purification of a lipoprotein lipase-inhibiting protein produced by a melanoma cell line associated with cancer cachexia. Biochem Biophys Res Commun 1989; 160:1085–1092.

10. Gascan H, Godard A, Ferenz C, Naulet J, Praloran V, Peyrat M-A, Hewick R, Jacques Y, Moreau J-F, Soulillou J-P. Characterization and NH$_2$-terminal sequence of natural human interleukin for DA cells: leukemia inhibitory factor. J Biol Chem 1989; 264:21509–21515.

11. Van Damme J, Uyttenhove C, Houssiau F, Put W, Proost P, Van Snick J. Human growth factor for murine interleukin (IL)-9 responsive T cell lines: co-induction with IL-6 in fibroblasts and identification as LIF/HILDA. Eur J Immunol 1992; 22:2801–2808.

12. Brown TJ, Shoyab M. Differential effects of leukemia inhibitory factor (LIF) and oncostatin M (ONCO M) on endothelial cell (EC) physiology [abstr]. J Cell Biochem 1991; 15F(suppl):247.

13. Ferrara N, Winer J, Henzel WJ. Pituitary follicular cells secrete an inhibitor of aortic endothelial cell growth: identification as leukemia inhibitory factor. Proc Natl Acad Sci USA 1992; 89:698–702.

14. Ichikawa Y. Differentiation of a cell line of myeloid leukemia. J Cell Physiol 1969; 74:223–234.

15. Tomida M, Yamamoto-Yamaguchi Y, Hozumi M, Okabe T, Takaku F. Induction by recombinant human granulocyte colony-stimulating factor of differentiation of mouse myeloid leukemic M1 cells. FEBS Lett 1986; 207:271–275.

16. Shabo Y, Lotem J, Rubinstein M, Revel M, Clark SC, Wolf SF, Kamen R, Sachs L. The myeloid blood cell differentiation-inducing protein MGI-2A is interleukin-6. Blood 1988; 72:2070–2073.

17. Hilton DJ, Nicola NA, Gough NM, Metcalf D. Resolution and purification of three distinct factors produced by Krebs ascites cells which have differentiation-inducing activity on murine myeloid leukemic cell lines. J Biol Chem 1988; 263:9238–9243.

18. Simpson RJ, Hilton DJ, Nice EC, Rubira MR, Metcalf D, Gearing DP, Gough NM, Nicola NA.

Structural characterization of a murine myeloid leukaemia inhibitory factor. Eur J Biochem 1988; 175:541–547.

19. Metcalf D. Actions and interactions of G-CSF, LIF, and IL-6 on normal and leukemic murine cells. Leukemia 1989; 3:349–355.

20. Gearing DP, Gough NM, King JA, Hilton DJ, Nicola NA, Simpson RJ, Nice EC, Kelso A, Metcalf D. Molecular cloning and expression of cDNA encoding a murine myeloid leukaemia inhibitory factor (LIF). EMBO J 1987; 6:3995–4002.

21. Gough NM, Gearing DP, King JA, Willson TA, Hilton DJ, Nicola NA, Metcalf D. Molecular cloning and expression of the human homologue of the murine gene encoding myeloid leukemia-inhibitory factor. Proc Natl Acad Sci USA 1988; 85:2623–2627.

22. Lowe DG, Nunes W, Bombara M, McCabe S, Ranges GE, Henzel W, Tomida M, Yamamoto-Yamaguchi Y, Hozumi M, Goeddel DV. Genomic cloning and heterologous expression of human differentiation-stimulating factor. DNA 1989; 8:351–359.

23. Abe E, Tanaka H, Ishimi Y, Miyaura C, Hayashi T, Nagasawa H, Tomida M, Yamaguchi, Hozumi M, Suda T. Differentiation-inducing factor purified from conditioned medium of mitogen-treated spleen cell cultures stimulates bone resorption. Proc Natl Acad Sci USA 1986; 83:5958–5962.

24. Williams RL, Hilton DJ, Pease S, Willson TA, Stewart CL, Gearing DP, Wagner EF, Metcalf D, Nicola N, Gough NM. Myeloid leukaemia inhibitory factor maintains the development of embryonic stem cells. Nature 1988; 336:684–687.

25. Gearing DP, Nicola NA, Metcalf D, Foote S, Willson TA, Gough NM, Williams RL. Production of leukemia inhibitory factor in *Escherichia coli* by a novel procedure and its use in maintaining embryonic stem cells in culture. Biotechnology 1989; 7:1157–1161.

26. Gough NM, Hilton DJ, Gearing DP, Willson TA, King JA, Nicola NA, Metcalf D. Biochemical characterization of murine leukaemia inhibitory factor produced by Krebs ascites and by yeast cells. Blood Cells 1988; 14:431–442.

27. Hilton DJ, Nicola NA, Waring PM, Metcalf D. Clearance and fate of leukemia inhibitory factor (LIF) after injection into mice. J Cell Physiol 1991; 148:430–439.

28. Yamamoto-Yamaguchi Y, Tomida M, Hozumi M. Prolongation by differentiation-stimulating factor, leukemia inhibitory factor on the survival time of mice implanted with mouse myeloid leukemia cells. Leuk Res 1992; 16:1025–1029.

29. Stahl J, Gearing DP, Willson TA, Brown MA, King JA, Gough NM. Structural organization of the genes for murine and human leukemia inhibitory factor. Evolutionary conservation of coding and non-coding regions. J Biol Chem 1990; 265:8833–8841.

30. Willson TA, Metcalf D, Gough NM. Cross-species comparison of the sequence of the leukaemia inhibitory factor gene and its protein. Eur J Biochem 1992; 204:21–30.

31. Yamamori T, Kukada K, Aebersold R, Korsching S, Fann M-J, Patterson PH. The cholinergic neuronal differentiation factor from heart cells is identical to leukemia inhibitory factor. Science 1989; 246:1412–1416.

32. Betzel C, Visanji M, Dauter Z, Fourme R, Weber W, Marnitz U, Boone T, Pope J, Miller J, Hawkins N, Samal B. Crystallization and preliminary X-ray analysis of leukemia inhibitory factor. FEBS Lett 1993; 336:236–238.

33. Robinson RC, Grey LM, Staunton D, Vankelecom H, Vernallis AB, Moreau J-F, Stuart DI, Heath JK, Jones EY. The crystal structure and biological function of leukemia inhibitory factor: implications for receptor binding. Cell 1994; 77:1101–1116.

34. Rose TM, Bruce AG. Oncostatin M is a member of a cytokine family that includes leukemia-inhibitory factor, granulocyte colony-stimulating factor, and interleukin 6. Proc Natl Acad Sci USA 1991; 88:8641–8645.

35. Nicola NA, Cross B, Simpson RJ. The disulfide bond arrangement of leukemia inhibitory factor: homology to oncostatin M and structural implications. Biochem Biophys Res Commun 1993; 190:20–26.

36. Bazan JF. Neuropoeitic cytokines in the hematopoietic fold. Neuron 1991; 7:197–208.

37. Gearing DP, King JA, Gough NM. Complete sequence of the murine myeloid leukemia inhibitory factor (LIF). Nucleic Acids Res 1988; 16:9857.

38. Moreau J-F, Donaldson DD, Bennett F, Witek-Giannotti J, Clark SC, Wong GG. Leukaemia inhibitory factor is identical to the myeloid growth factor human interleukin for DA cells. Nature 1988; 336:690–692.

39. Rathjen PD, Toth S, Willis A, Heath JK, Smith AG. Differentiation inhibiting activity is produced in matrix-associated and diffusible forms that are generated by alternate promoter usage. Cell 1990; 62:1105–1114.

40. Budarf M, Emanuel BS, Mohandas T, Goeddel DV, Lowe DG. Human differentiation-stimulating factor (leukemia inhibitory factor, human interleukin DA) gene maps distal to the Ewing sarcoma breakpoint on 22q. Cytogenet Cell Genet 1989; 52:19–22.

41. Sutherland GR, Baker E, Hyland VJ, Callen DF, Stahl J, Gough NM. The gene for human leukemia inhibitory factor (LIF) maps to 22q12. Leukemia 1989; 3:9–13.

42. Kola I, Davey A, Gough NM. Localization of the murine leukemia inhibitory factor gene near the centromere on chromosome 11. Growth Factors 1990; 2:235–240.

43. Bottorff D, Stone JC. The murine leukemia inhibition factor gene (LIF) is located on proximal chromosome 11, not chromosome 13. Mamm Genome 1992; 3:681–684.

44. Selleri L, Hermanson GG, Eubanks JH, Lewis KA, Evans GA. Molecular localization of the t(11;22)(q24;q12) translocation of Ewing sarcoma by chromosomal in situ suppression hybridization. Proc Natl Acad Sci USA 1991; 88:887–891.

45. Watson CJ, Gaunt L, Evans G, Patel K, Harris R, Strachan T. A disease-associated germline deletion maps the type 2 neurofibromatosis (NF2) gene between the Ewing sarcoma region and the leukaemia inhibitory factor locus. Hum Mol Genet 1993; 2:701–704.

46. Giovannini M, Selleri L, Hermanson GG, Evans GA. Localization of the human oncostatin M gene (OSM) to chromosome 22q12, distal to the Ewing's sarcoma breakpoint. Cytogenet Cell Genet 1993; 62:32–34.

47. Giovannini M, Djabali M, Mc Elligott D, Selleri L, Evans GA. Tandem linkage of genes coding for leukemia inhibitory factor (LIF) and oncostatin M gene (OSM) on human chromosome 22. Cytogenet Cell Genet 1993; 64:240–244.

48. Jeffery E, Price V, Gearing DP. Close proximity of the genes for leukemia inhibitory factor and oncostatin M. Cytokine 1993; 5:107–111.

49. Rose TM, Lagrou MJ, Fransson I, Werelius B, Delattre O, Thomas G, De Jong PJ, Todaro GJ, Dumanski JP. The genes for oncostatin M (OSM) and leukemia inhibitory factor (LIF) are tightly linked on human chromosome 22. Genomics 1993; 17:136–140.

50. Stahl J, Gough NM. Delineation of positive and negative control elements within the promoter region of the murine leukaemia inhibitory factor (LIF) gene. Cytokine 1993; 5:386–393.

51. Hsu LW, Heath JK. Identification of two elements involved in regulating expression of the murine leukaemia inhibitory factor gene. Biochem J. 1994; 302:103–110.

52. Gearing DP. Leukemia inhibitory factor: does the cap fit? Ann NY Acad Sci 1991; 628:9–18.

53. Kaspar P, Dvorák M, Bartunek P. Identification of CpG island at the 5′ end of murine leukemia inhibitory factor gene. FEBS Lett 1993; 319:159–162.

54. Brown MA, Metcalf D, Gough NM. Leukaemia inhibitory factor and interleukin-6 are expressed at very low levels in the normal adult mouse and are induced by inflammation. Cytokine 1993; 6:300–309.

55. Yamamori T. Localization of cholinergic differentiation factor/leukemia inhibitory factor mRNA in the rat brain and peripheral tissues. Proc Natl Acad Sci USA 1991; 88:7298–7302.

56. Minami M, Kuraishi Y, Satoh M. Effects of kainic acid on messenger RNA levels of IL-1beta, IL-6, TNF alpha and LIF in the rat brain. Biochem Biophys Res Commun 1991; 176:593–598.

57. Block MI, Berg M, McNamara MJ, Norton JA, Fraker DL, Alexander HR. Passive immunization of mice against D factor blocks lethality and cytokine release during endotoxemia. J Exp Med 1993; 178:1085–1090.

58. Robertson M, Chambers I, Rathjen P, Nichols J, Smith A. Expression of alternate forms of

differentiation inhibiting activity (DIA/LIF) during murine embryogenesis and in neonatal and adult tissues. Dev Genet 1993; 14:165–173.

59. Bhatt H, Brunet LJ, Stewart CL. Uterine expression of leukemia inhibitory factor coincides with the onset of blastocyst implantation. Proc Natl Acad Sci USA 1991; 88:11408–11412.

60. Shen MM, Leder P. Leukemia inhibitory factor is expressed by the pre-implantation uterus and selectively blocks primitive endoderm formation in vitro. Proc Natl Acad Sci USA 1992; 89:8240–8244.

61. Kojima K, Kanzaki H, Iwai M, Hatayama H, Fujimoto M, Inoue T, Horie K, Nakayama H, Fujita J, Mori T. Expression of leukemia inhibitory factor in human endometrium and placenta. Biol Reprod 1994; 50:882–887.

62. Kohchi C, Noguchi K, Tanabe Y, Mizuno D-I, Soma G-I. Constitutive expression of TNF-α and -β genes in mouse embryo: roles of cytokines as regulator and effector on development. Int J Biochem 1994; 26:111–119.

63. Campbell IK, Waring PM, Novak U, Hamilton JA. Production of leukemia inhibitory factor by human articular chondrocytes and cartilage in response to interleukin-1 and tumor necrosis factor-α. Arthritis Rheum 1993; 36:790–794.

64. Waring PM, Romero R, Laham N, Gomez R, Rice G. Leukemia inhibitory factor: association with intra-amniotic infection. Am J Obstet Gynecol 1994; 171:1335–1341.

65. Anegon I, Moreau J-F, Godard A, Jacques Y, Peyrat M-A, Hallet M-M, Wong G, Soulillou J-P. Production of human interleukin for DA cells (HILDA)/leukemia inhibitory factor (LIF) by activated monocytes. Cell Immunol 1990; 130:50–65.

66. Anegon I, Grolleau D, Soulillou J-P. Regulation of HILDA/LIF gene expression in activated human monocytic cells. J Immunol 1991; 147:3973–3980.

67. Grolleau D, Soulillou JP, Anegon I. Control of HILDA/LIF gene expression in activated human monocytes. Ann NY Acad Sci 1991; 628:19–30.

68. Geng Y, Zhang B, Lotz M. Protein tyrosine kinase activation is required for lipopolysaccharide induction of cytokines in human blood monocytes. J Immunol 1994; 151:6692–6700.

69. Baffet G, Cui M-Z, Fletcher RG, Fey GH. Production of leukemia inhibitory factor (LIF) by cultured macrophages and hepatoma cell lines [abstr]. J Cell Biochem 1991; 15G(suppl):S216.

70. Wu KF, Zhu YM, Rao Q, Zhao JM. Expression of transforming growth factor-β, tumor necrosis factor-α, and leukemia inhibitory factor mRNAs in rodent and human hematopoietic cells. Ann NY Acad Sci 1991; 628:151–152.

71. Falcinelli F, Ciurnelli R, Zannelli B, Onorato M, Falzetti F, Mannoni P, Martelli MF, Tabilio A. mRNA expression of the human interleukin for DA-1a cells (HILDA) in chronic myeloid leukemia cells [abstr]. Exp Hematol 1991; 19:542.

72. Lübbert M, Mantovani L, Lindemann A, Mertelsmann R, Herrmann F. Expression of leukemia inhibitory factor is regulated in human mesenchymal cells. Leukemia 1991; 5:361–365.

73. Moreau JF, Bonneville M, Peyrat MA, Jacques Y, Soulillou JP. Capacity of alloreactive human T clones to produce factor(s) inducing proliferation of the IL3-dependent DA-1 murine cell line. I. Evidence that this production is under IL 2 control. Ann Inst Pasteur Immunol 1986; 137:25–37.

74. Moreau JF, Bonneville M, Godard A, Gascan H, Gruart V, Moore MAS, Soulillou JP. Characterization of a factor produced by human T cell clones exhibiting eosinophil-activating and burst-promoting activities. J Immunol 1987; 138:3844–3849.

75. Moreau JF, Bonneville M, Godard A, Peyrat MA, Capron M, Moore MA, Soulillou JP. Allogeneic T lymphocyte clones derived from rejected human kidney produce high levels of a new lymphokine active on murine interleukin 3-sensitive cell lines. Transplant Proc 1987; 19:300–302.

76. Godard A, Gascan H, Naulet J, Peyrat M-A, Jaques Y, Soulillou J-P, Moreau JF. Biochemical characterization and purification of HILDA, a human lymphokine active on eosinophils and bone marrow cells. Blood 1988; 71:1618–1623.

77. Gascan H, Lemetayer J. Induction of leukemia inhibitory factor secretion by interleukin-1 in a human T lymphoma cell line. Lymphokine Cytokine Res 1991; 10:115–118.

78. Sasaki H, Kajigaya Y, Hirabayashi Y, Funabiki T, Inoue T, Yokota T, Ikuta K, Arai K, Matsuyama

S. Establishment and characterization of a murine megakaryoblastic cell line growing in protein-free culture (L8057Y5). Leukemia 1991; 5:408–415.

79. Portier M, Zhang X-G, Ursule E, Lees D, Jourdan M, Bataille R, Klein B. Cytokine gene expression in human multiple myeloma. Br J Haematol 1993; 85:514–520.

80. Marshall JS, Gauldie J, Nielsen L, Bienenstock J. Leukemia inhibitory factor production by rat mast cells. Eur J Immunol 1993; 23:2116–2120.

81. Gruss H-J, Brach MA, Drexler H-G, Bonifer R, Mertelsmann RH, Herrmann F. Expression of cytokine genes, cytokine receptor genes, and transcription factors in cultured Hodgkin and Reed-Sternberg cells. Cancer Res 1992; 52:3353–3360.

82. Rathjen PD, Nichols J, Toth S, Edwards DR, Heath JK, Smith AG. Developmentally programmed induction of differentiation inhibiting activity and the control of stem cell populations. Genes Dev 1990; 4:2308–2318.

83. Miyagi T, Akashi M, Yamato K, Miyoshi I, Koeffler HP. D-factor: modulation of expression in fibroblasts. Leuk Res 1991; 15:441–451.

84. Lorenzo JA, Jastrezebski SL, Kalinowski JF, Downie E, Korn JH. Tumor necrosis factor α stimulates production of leukemia inhibitory factor in human dermal fibroblast cultures. Clin Immunol Immunopathol 1994; 70:260–265.

85. Aloisi F, Rosa S, Testa U, Bonsi P, Russo G, Peschle C, Levi G. Regulation of leukemia inhibitory factor synthesis in cultured human astrocytes. J Immunol 1994; 152:5022–5031.

86. Elias JA, Zheng T, Whiting NL, Marcovici A, Trow TK. Cytokine–cytokine synergy and protein kinase C in the regulation of lung fibroblast leukemia inhibitory factor. Am J Physiol 1994; 266:L426–L435.

87. Abe E, Ishimi Y, Takahashi N, Akatsu T, Ozawa H, Yamaha H, Yoshiki S, Suda T. A differentiation-inducing factor produced by the osteoblastic cell line MC3T3-E1 stimulates bone resorption by promoting osteoclast formation. J Bone Miner Res 1988; 3:635–645.

88. Shiina-ishimi Y, Abe E, Tanaka H, Suda T. Synthesis of colony-stimulating factor (CSF) and differentiation-inducing factor (D-factor) by osteoblastic cells, clone MC3T3-E1. Biochem Biophys Res Commun 1986; 134:400–406.

89. Allan EH, Hilton DJ, Brown MA, Evely RS, Yumita S, Metcalf D, Gough NM, Ng KW, Nicola NA, Martin TJ. Osteoblasts display receptors for and responses to leukemia inhibitory factor. J Cell Physiol 1990; 145:110–119.

90. Greenfield EM, Gornik SA, Horowitz MC, Donahue HJ, Shaw SM. Regulation of cytokine expression in osteoblasts by parathyroid hormone: rapid stimulation of interleukin-6 and leukemia inhibitory factor mRNA. J Bone Miner Res 1993; 8:1163–1171.

91. Hakeda Y, Sudo T, Ishizuka S, Tanaka K, Higashino K, Kusuda M, Kodama H, Kumegawa M. Murine recombinant leukemia inhibitory factor modulates inhibitory activity of 1,25 dihydroxy-vitamin D3 on alkaline phosphatase activity in MC3T3-E1 cells. Biochem Biophys Res Commun 1991; 175:577–582.

92. Ishimi Y, Abe E, Jin CH, Miyaura C, Hong HM, Oshida M, Kurosawa H, Yamaguchi Y, Tomida M, Hozumi M, Suda T. Leukemia inhibitory factor/differentiation-stimulating factor (LIF/D-factor): regulation of its production and possible roles in bone metabolism. J Cell Physiol 1992; 152:71–78.

93. Phillips DL, Glasebrook AL. Human osteosarcoma cells secrete leukemia inhibitory factor, but not interleukin-6 [abstr]. Mol Biol Cell 1992; 3:331.

94. Marusic A, Kalinowski JF, Jastrzebski S, Lorenzo JA. Production of leukemia inhibitory factor mRNA and protein by malignant and immortalized bone cells. J Bone Miner Res 1993; 8:617–624.

95. Nishikawa M, Ozawa K, Tojo A, Yoshikubo T, Okano A, Tani K, Ikebuchi K, Nakauchi H, Asano S. Changes in hematopoiesis-supporting ability of C3H10T1/2 mouse embryo fibroblasts during differentiation. Blood 1993; 81:1184–1192.

96. Wetzler M, Talpaz M, Lowe DG, Baiocchi G, Gutterman JU, Kurzrock R. Constitutive expression of leukemia inhibitory factor RNA by human bone marrow stromal cells and modulation by IL-1, TNF-α, and TGF-β. Exp Hematol 1991; 19:347–351.

97. Wetzler M, Kurzrock R, Lowe DG, Kantarjian H, Gutterman JU, Talpaz M. Alteration in bone

marrow adherent layer growth factor expression: a novel mechanism of chronic myelogenous leukemia progression. Blood 1991; 78:2400–2406.

98. Wetzler M, Estrov Z, Talpaz M, Kim J, Alphonso M, Srinivasan R, Kurzrock R. Leukemia inhibitory factor in long-term adherent layer cultures: increased levels of bioactive protein in leukemia and modulation by IL-4, IL-1β and TNF-α. Cancer Res 1994; 54:1837–1842.

99. Paul SR, Yang YC, Donahue RE, Goldring S, Williams DA. Stromal cell-associated hematopoiesis: immortalization and characterization of a primate bone marrow-derived stromal cell line. Blood 1991; 77:1723–1733.

100. Watson JD, McKenna HJ. Novel factors from stromal cells: bone marrow and thymus microenvironments. Int J Cell Cloning 1992; 10:144–152.

101. Derigs HG, Boswell HS. LIF mRNA expression is transcriptionally regulated in murine bone marrow stromal cells. Leukemia 1993; 7:630–634.

102. Sakata T, Iwagami S, Tsuruta Y, Yamaguchi M, Michishita M, Nagata K, Takai Y, Ogata M, Teraoka H, Fujiwara H. Constitutive expression of leukemia inhibitory factor (LIF) mRNA and production of LIF by a cloned murine thymic stromal cell line. Thymus 1992; 19:89–95.

103. Le PT, Lazorick S, Whichard LP, Yang Y-C, Clark SC, Haynes BF, Singer KH. Human thymic epithelial cells produce IL-6, granulocyte–monocyte-CSF, and leukemia inhibitory factor. J Immunol 1990; 145:3310–3315.

104. Rao MS, Landis SC, Patterson PH. The cholinergic neuronal differentiation factor from heart cell conditioned medium is different from the cholinergic factors in sciatic nerve and spinal cord. Dev Biol 1990; 139:65–74.

105. Moats T, Villiger P, Lotz M. Leukemia inhibitory factor is expressed by human articular chondrocytes and can stimulate their proliferation [abstr]. Arthritis Rheum 1991; 34:C52a.

106. Lotz M, Moats T, Villiger PM. Leukemia inhibitory factor is expressed in cartilage and synovium and can contribute to the pathogenesis of arthritis. J Clin Invest 1992; 90:888–896.

107. Hamilton JA, Waring PM, Filonzi EL. Induction of leukemia inhibitory factor in human synovial fibroblasts by IL-1 and tumor necrosis factor-α. J Immunol 1993; 150:1496–1502.

108. Gascan H, Anegón I, Praloran V, Naulet J, Godard A, Soulillou J-P, Jacques Y. Constitutive production of human interleukin for DA cells/leukemia inhibitory factor by human tumor cell lines derived from various tissues. J Immunol 1990; 144:2592–2598.

109. Baumann H, Won K-A, Jahreis GP. Hepatocyte-stimulating factor III and interleukin-6 are structurally and immunologically distinct but regulate the production of the same acute phase plasma proteins. J Biol Chem 1989; 264:8046–8051.

110. Smith AG, Hooper ML. Buffalo rat liver cells produce a diffusible activity which inhibits the differentiation of murine embryonal carcinoma and embryonic stem cells. Dev Biol 1987; 121:1–9.

111. Mezzasoma L, Biondi R, Benedetti C, Floridi C, Ciurnelli R, Falcinelli F, Onorato M, Scaringi L, Marconi P, Rossi R. In vitro production of leukemia inhibitory factor (LIF) by HepG2 hepatoblastoma cells. J Biol Regul Homeost Agents 1993; 7:126–132.

112. Gough NM, Willson TA, Stahl J, Brown MA. Molecular biology of the leukaemia inhibitory factor gene. In: Bock G, Marsh J, Widdows K, eds. Polyfunctional Cytokines: IL-6 and LIF. Ciba Found Symp 1992; 167:24–46.

113. Mori M, Yamaguchi K, Honda S, Nagasaki K, Ueda M, Abe O, Abe K. Cancer cachexia syndrome developed in nude mice bearing melanoma cells producing leukemia-inhibitory factor. Cancer Res 1991; 51:6656–6659.

114. Mattei S, Colombo MP, Melani C, Silvani A, Parmiani G, Herlyn M. Expression of cytokine/growth factors and their receptors in human melanoma and melanocytes. Int J Cancer 1994; 56:853–857.

115. Wesselingh SL, Gough NM, Finlay-Jones JJ, McDonald PJ. Detection of cytokine mRNA in astrocyte cultures using the polymerase chain reaction. Lymphokine Res 1990; 9:177–185.

116. De Groote D, Fauchet F, Jadoul M, Dehart I, Raher S, Gevaert Y, Lopez M, Gathy R, Franssen JD, Radoux D, Franchimont P, Soulillou JP, Jacques Y, Godard A. An ELISA for the measurement of human leukemia inhibitory factor in biological fluids and culture supernatants. J Immunol Methods 1994; 167:253–261.

117. Mirshahi SS, Vasse M, Soria C, Moreau JF, Taupin JL, Mirshahi M, Pujade-Lauraine E, Bernadou A, Soria J. Incubation of monocytes with Adriamycin increases secretion of hepatocyte stimulating factor for fibrinogen biosynthesis. Blood Coagul Fibrinolysis 1993; 4:149–152.

118. Umemiya Okada T, Natazuka T, Matsui T, Ito M, Taniguchi T, Nakao Y. Expression and regulation of the leukemia inhibitory factor/D factor gene in human T-cell leukemia virus type 1 infected T-cell lines. Cancer Res 1992; 52:6961–6965.

119. Lal RB, Rudolph D, Buckner C, Pardi D, Hooper WC. Infection with human T-lymphotropic viruses leads to constitutive expression of leukemia inhibitory factor and interleukin-6. Blood 1993; 81:1827–1832.

120. Bentouimou N, Moreau J-F, Peyrat M-A, Soulillou J-P, Hallet M-M. The effects of cyclosporine on HILDA/LIF gene expression in human T cells. Transplantation 1993; 5:163–167.

121. Hilton DJ, Nicola NA, Metcalf D. Specific binding of murine leukemia inhibitory factor to normal and leukemic monocytic cells. Proc Natl Acad Sci USA 1988; 85:5971–5975.

122. Godard A, Heymann D, Raher S, Anegon I, Peyrat M-A, Le Mauff B, Mouray E, Gregoire M, Virdee K, Soulillou J-P, Moreau J-F, Jacques Y. High and low affinity receptors for human interleukin for DA cell/leukemia inhibitory factor on human cells. Molecular characterization and cellular distribution. J Biol Chem 1992; 267:3214–3222.

123. Metcalf D, Hilton D, Nicola NA. Leukemia inhibitory factor can potentiate murine megakaryocyte production in vitro. Blood 1991; 77:2150–2153.

124. Hilton DJ, Nicola NA. Kinetic analyses of the binding of leukemia inhibitory factor (LIF) to receptors and membranes and in detergent solution. J Biol Chem 1992; 267:10238–10247.

125. Hilton DJ, Nicola NA, Metcalf D. Distribution and characterization of receptors for LIF on haemopoietic and hepatic cells. J Cell Physiol 1991; 146:207–215.

126. Davis S, Aldrich TH, Ip NY, Stahl N, Scherer S, Farruggella T, Di Stefano PS, Curtis R, Panayotatos N, Gascan H, Chevalier S, Yancopoulos GD. Released form of CNTF receptor α component as a soluble mediator of CNTF responses. Science 1993; 259:1736–1739.

127. Boulton TG, Stahl N, Yancopoulos GD. Ciliary neurotrophic factor/leukemia inhibitory factor/interleukin 6/oncostatin M family of cytokines induces tyrosine phosphorylation of a common set of proteins overlapping those induced by other cytokines and growth factors. J Biol Chem 1994; 269:11648–11655.

128. Rodan SB, Wesolowski G, Hilton DJ, Nicola NA, Rodan GA. Leukemia inhibitory factor binds with high affinity to preosteoblastic RCT-1 cells and potentiates the retinoic acid induction of alkaline phosphatase. Endocrinology 1990; 127:1602–1608.

129. Tomida M, Yamamoto-Yamaguchi Y, Hozumi M, Holmes W, Lowe DG, Goeddel DV. Inhibition of development of Na^+-dependent hexose transport in renal epithelial LLC-PK1 cells by differentiation-stimulating factor for myeloid leukemic cells/leukemia inhibitory factor. FEBS Lett 1990; 268:261–264.

130. Zhang X-G, Gu J-J, Lu Z-Y, Yasukawa K, Yancopoulos GD, Turner K, Shoyab M, Taga T, Kishimoto T, Bataille R, Klein B. Ciliary neurotropic factor, interleukin 11, leukemia inhibitory factor, and oncostatin M are growth factors for human myeloma cell lines using the interleukin 6 signal transducer GP130. J Exp Med 1994; 177:1337–1342.

131. Hilton DJ. LIF: Lots of interesting functions. Trends Biochem Sci 1992; 17:72–76.

132. Owczarek CM, Layton MJ, Metcalf D, Lock P, Willson TA, Gough NM, Nicola NA. Interspecies chimaeras of leukaemia inhibitory factor define a major human receptor binding determinant. EMBO J 1993; 12:3487–3495.

133. Layton MJ, Cross BA, Metcalf D, Ward LD, Simpson RJ, Nicola NA. A major binding protein for leukemia inhibitory factor in normal mouse serum: identification as a soluble form of the cellular receptor. Proc Natl Acad Sci USA 1992; 89:8616–8620.

134. Layton MJ, Owczarek CM, Metcalf D, Lock PA, Willson TA, Gough NM, Hilton DJ, Nicola NA. Complex binding of leukemia inhibitory factor to its membrane-expressed and soluble receptors. Proc Soc Exp Biol Med 1994; 206:295–298.

135. Layton MJ, Lock P, Metcalf D, Nicola NA. Cross-species receptor binding characteristics of human

and mouse leukemia inhibitory factor suggest a complex binding interaction. J Biol Chem 1994; 269:17048–17055.

136. Gearing DP, Thut CJ, VandenBos T, Gimpel SD, Delaney PB, King J, Price V, Cosman D, Beckmann MP. Leukemia inhibitory factor receptor is structurally related to the IL-6 signal transducer, gp130. EMBO J 1991; 10:2839–2848.

137. Cosman D. The hematopoietin receptor superfamily. Cytokine 1993; 5:95–106.

138. De Sauvage FJ, Hass PE, Spencer SD, Malloy BE, Gurney AL, Spencer SA, Darbonne WC, Henzel WJ, Wong SC, Kuang W-J, Oles KJ, Hultgren B, Solberg LA Jr, Goeddel DV, Eaton DL. Stimulation of megakaryocytopoiesis and thrombopoiesis by the c-*Mpl* ligand. Nature 1994; 369:533–538.

139. DeVos AM, Ultsch M, Kossiakoff AA. Human growth hormone and extracellular domain of its receptor: crystal structure of the complex. Science 1992; 255:306–312.

140. Gearing DP, Druck T, Huebner K, Overhauser J, Gilbert DJ, Copeland NG, Jenkins NA. The leukemia inhibitory factor receptor (LIFR) gene is located within a cluster of cytokine receptor loci on mouse chromosome 15 and human chromosome 5p12-p13. Genomics 1993; 18:148–150.

141. Chambers I, Robertson M, Li M, Smith A. Analysis of the expression of the DIA/LIF receptor [abstr]. J Cell Biochem 1993; 17B(suppl):73.

142. Tomida M, Yamamoto-Yamaguchi Y, Hozumi M. Pregnancy associated increase in mRNA for soluble D-factor: LIF receptor in mouse liver. FEBS Lett 1993; 334:193–197.

143. Tomida M, Yamamoto-Yamaguchi Y, Hozumi M. Three different cDNAs encoding mouse D-factor/LIF receptor. J Biochem 1994; 115:557–562.

144. Gearing DP, Bruce AG. Oncostatin M binds to the high-affinity leukemia inhibitory factor receptor. New Biol 1992; 4:61–65.

145. Gearing D, Comeau MR, Friend DJ, Gimpel SD, Thut CJ, McCourty J, Brasher KK, King JA, Gillis S, Mosley B, Ziegler SF, Cosman D. The IL-6 signal transducer, gp130: an oncostatin M receptor and affinity converter for the LIF receptor. Science 1992; 255:1434–1437.

146. Yamasaki K, Taga T, Hirata Y, Yawata H, Kawanishi Y, Seed B, Taniguchi T, Hirano T, Kishimoto T. Cloning and expression of the human interleukin-6 receptor (BSF-2/IFNβ2) receptor. Science 1988; 241:825–828.

147. Davis S, Aldrich TH, Valenzula DM, Wong V, Furth ME, Squinto SP, Yancopoulos GD. The receptor for ciliary neurotrophic factor. Science 1991; 253:59–63.

148. Hilton DJ, Hilton AA, Raicevic A, Rakar S, Harrison-Smith M, Gough NM, Begley CG, Metcalf D, Nicola NA, Willson TA. Cloning of a murine IL-11 receptor α-chain; requirement for gp130 for high affinity binding and signal transduction. EMBO J 1994; 13:4765–4775.

149. Ip NY, Nye SH, Boulton TG, Davis S, Taga T, Li Y, Birren SJ, Yasukawa K, Kishimoto T, Anderson DJ, Stahl N, Yanocopoulos GD. CNTF and LIF act on neuronal cells via shared signaling pathways that involve the IL-6 signal transducing receptor component gp130. Cell 1992; 69:1121–1132.

150. Taga T, Hibi M, Hirata Y, Yamasaki K, Yasukawa M, Matsuda T, Hirano T, Kishimoto T. Interleukin-6 triggers the association of its receptor with a possible signal transducer, gp130. Cell 1989; 58:573–581.

151. Davis S, Aldrich TH, Stahl N, Pan L, Taga T, Kishimoto T, Ip NY, Yancoupoulos GD. LIFRβ and gp130 as heterodimerizing signal transducers of the tripartite CNTR receptor. Science 1993; 260:1805–1808.

152. Fourcin M, Chevalier S, Lebrun JJ, Kelly P, Pouplard A, Wijdenes J, Gascan H. Involvement of gp130/interleukin-6 receptor transducing component in interleukin-11 receptor. Eur J Immunol 1994; 24:277–280.

153. Kishimoto T, Akira S, Taga T. Interleukin-6 and its receptor: a paradigm for cytokines. Science 1992; 258:593–597.

154. Stahl N, Davis S, Wong V, Taga T, Kishimoto T, Ip NY, Yancopoulos GD. Cross-linking identifies leukemia inhibitory factor-binding protein as a ciliary neurotropic receptor component. J. Biol Chem 1993; 268:7628–7631.

155. Stahl N, Yancopoulos GD. The alpha, betas, and kinases of cytokine receptor complexes. Cell 1993; 74:587–590.

156. Baumann H, Symes AJ, Comeau MR, Morella KK, Wang Y, Friend D, Ziegler SF, Fink JS, Gearing DP. Multiple regions within the cytoplasmic domains of the leukemia inhibitory factor receptor and gp130 cooperate in signal transduction in hepatic and neuronal cells. Mol Cell Biol 1994; 14:138–146.

157. Fukunaga R, Ishizaka-Ikeda A, Pan C-X, Seto Y, Nagata S. Functional domains of the granulocyte colony-stimulating factor receptor. EMBO J 1991; 10:2855–2865.

158. Murakami M, Narazaki M, Hibi M, Yawata H, Yasukawa K, Hamaguchi M, Taga T, Kishimoto T. Critical cytoplasmic region of the IL-6 signal transducer, gp130, is conserved in the cytokine receptor family. Proc Natl Acad Sci USA 1991; 88:11349–11353.

159. Baumann H, Gearing D, Ziegler SF. Signaling by the cytoplasmic domain of hematopoietin receptors involves two distinguishable mechanisms in hepatic cells. J Biol Chem 1994; 269:16297–16304.

160. Fukunaga R, Ishizaka-Ikeda E, Nagata S. Purification and characterization of the receptor for murine granulocyte colony-stimulating factor. J Biol Chem 1990; 265:14008–14015.

161. Murakami M, Hibi M, Nakagawa T, Yasukawa K, Yamanishi K, Taga T, Kishimoto T. IL-6-induced homodimerization of gp130 and associated activation of a tyrosine kinase. Science 1993; 260:1808–1810.

162. Piquet-Pellorce C, Grey L, Mereau A, Heath JK. Are LIF and related cytokines functionally equivalent? Exp Cell Res 1994; 213:340–347.

163. Liu J, Modrell B, Aruffo A, Marken JS, Taga T, Yasukawa K, Murakami M, Kishimoto T, Shoyab M. Interleukin-6 signal transducer gp130 mediates oncostatin M signaling. J Biol Chem 1992; 267:16763–16766.

164. Liu J, Modrell B, Aruffo A, Scharnowske S, Shoyab M. Interactions between oncostatin M and the IL-6 signal transducer, gp130. Cytokine 1994; 6:272–278.

165. Amaral MC, Miles S, Kumar G, Nel AE. Oncostatin-M stimulates tyrosine protein phosphorylation in parallel with the activation of p42^{MARK}/ERK-2 in Kaposi's cells. Evidence that this pathway is important in Kaposi cell growth. J Clin Invest 1993; 92:848–857.

166. Thoma B, Bird TA, Friend DJ, Gearing DP, Dower SK. Oncostatin M and leukemia inhibitory factor trigger overlapping and different signals through partially shared receptor complexes. J Biol Chem 1994; 269:6215–6222.

167. Yin T, Yang Y-C. Mitogen-activated protein kinases and ribosomal S6 protein kinases are involved in signaling pathways shared by interleukin-11, interleukin-6, leukemia inhibitory factor, and oncostatin M in mouse 3T3-L1 cells. J Biol Chem 1994; 269:3731–3738.

168. Müllberg J, Schooltink H, Stoyan T, Gunther M, Graeve L, Buse G, Maekiewicz A, Heinrich PC, Rose-John S. The soluble interleukin-6 receptor is generated by shedding. Eur J Immunol 1993; 23:473–480.

169. Novick D, Engelmann H, Wallach D, Rubenstein M. Soluble cytokine receptors are present in normal human urine. J Exp Med 1989; 170:1409–1414.

170. Honda M, Yamamoto S, Cheng M, Yasukawa K, Susuki H, Saito T, Osugi Y, Tokunaga T, Kishimoto T. Human soluble IL-6 receptor: its detection and enhanced release by HIV infection. J Immunol 1992; 148:2175–2180.

171. Yamaguchi-Yamamoto T, Tomida M, Hozumi M. Pregnancy associated increase in differentiation-stimulating factor (D-factor)/leukemia inhibitory factor (LIF)-binding substance(s) in mouse serum. Leuk Res 1993; 17:515–522.

172. Narazaki M, Yasukawa K, Saito T, Ohsugi Y, Fukui H, Koshihara Y, Yancopoulos GD, Taga T, Kishimoto T. Soluble form of the interleukin-6 signal transducing receptor component gp130 in human serum possessing a potential to inhibit signals through membrane anchored gp130. Blood 1993; 82:1120–1126.

173. Mereau A, Thio M, Heath JK. Interaction of leukemia inhibitory factor with at least three distinct binding proteins [abstr]. J Cell Biochem 1992; 16F:80.

174. Mereau A, Grey L, Piquet-Pellorce C, Heath JK. Characterization of a binding protein for leukemia inhibitory factor localized in extracellular matrix. J Cell Biol 1993; 122:713–719.

175. Conquet F, Peyrieras N, Tiret L, Brûlet P. Inhibited gastrulation in mouse embryos overexpressing the leukemia inhibitory factor. Proc Natl Acad Sci USA 1992; 89:8195–8199.

176. Lotz M, Ganu V, Melton R, Udwaadia N, Geng Y, Villiger PM. Leukemia inhibitory factor can promote connective tissue degradation [abstr]. Osteoarthritis Cartilage 1993; 1:7–8.

177. Roeb E, Graeve L, Hoffman R, Decker K, Edwards DR, Heinrich PC. Regulation of tissue inhibitor of metalloproteinases-1 gene expression by cytokines and dexamethasone in rat hepatocyte primary cultures. Hepatology 1993; 18:1437–1442.

178. Richards CD, Shoyab M, Brown TJ, Gaudie J. Selective regulation of metalloproteinase inhibitor (TIMP-1) by oncostatin M in fibroblasts in culture. J Immunol 1993; 150:5596–5603.

179. Kishimoto T, Taga T, Akira S. Cytokine signal transduction. Cell 1994; 76:253–262.

180. Lütticken C, Wegenka UM, Yuan J, Buschmann J, Schindler C, Ziemiecki A, Harpur AG, Wilks AF, Yasukawa K, Taga T, Kishimoto T, Barbieri G, Pellegrini S, Sendtner M, Heinrich PC, Horn F. Association of transcription factor APRF and protein kinase Jak1 with the interleukin-6 signal transducer gp130. Science 1994; 263:89–92.

181. Stahl N, Boulton TG, Farruggella T, Ip NY, Davis S, Witthuhn BA, Quelle FW, Silvennoinen O, Barbieri G, Pellegrini S, Ihle JN, Yancopoulos GD. Association and activation of Jak–Tyk kinases by CNTF–LIF–OSM–IL-6β receptor components. Science 1994; 263:92–95.

182. Yin T, Yasukawa K, Taga T, Kishimoto T, Yang Y-C. Identification of a 130-kilodalton tyrosine-phosphorylated protein induced by interleukin-11 as JAK2 tyrosine kinase, which associates with gp130 signal transducer. Exp Hematol 1994; 22:467–472.

183. Narazaki M, Witthuhn BA, Yoshida K, Silvennoinen O, Yasukawa K, Ihle JN, Kishimoto T, Taga T. Activation of JAK2 kinase mediated by the interleukin 6 signal transducer gp130. Proc Natl Acad Sci USA 1994; 91:2285–2289.

184. Ullrich A, Schlessinger J. Signal transduction by receptors with tyrosine kinase activity. Cell 1990; 61:203–212.

185. Akira S, Nishio Y, Inoue M, Wang X-J, Wei S, Matsusaka T, Yoshida K, Sudo T, Naruto M, Kishimoto T. Molecular cloning of APRF, a novel IFN-stimulated gene factor 3 p91-related transcription factor involved in gp130-mediated signaling pathway. Cell 1994; 77:63–71.

186. Wegenka UM, Buschmann J, Lütticken C, Heinrich PC, Horn F. Acute-phase response factor, a nuclear factor binding to acute phase response elements, is rapidly activated by interleukin-6 at the posttranslational level. Mol Cell Biol 1993; 13:276–288.

187. Wegenka UM, Lütticken C, Buschmann J, Yuan J, Lottspeich F, Müller-Esterl W, Schindler C, Roeb E, Heinrich PC, Horn F. The interleukin-6-activated acute-phase response factor is antigenically and functionally related to members of the signal transducer and activator of transcription (STAT) family. Mol Cell Biol 1994; 14:3186–3196.

188. Bonni A, Frank DA, Schindler C, Greenberg ME. Characterization of a pathway for ciliary neurotrophic factor signaling to the nucleus. Science 1993; 262:1575–1579.

189. Lamb P, Kessler LV, Suto C, Levy DE, Seidel HM, Stein RB, Rosen J. Rapid activation of proteins that interact with the interferon γ activation site in response to multiple cytokines. Blood 1994; 83:2063–2071.

190. Feldman GM, Petricoin EF III, David M, Larner AC, Finbloom DS. Cytokines that associate with the signal transducer gp130 activate the interferon-induced transcription factor p91 by tyrosine phosphorylation. J Biol Chem 1994; 269:10747–10752.

191. Schiemann WP, Nathanson NM. Involvement of protein kinase C during activation of the mitogen-activated protein kinase cascade by leukemia inhibitory factor. J Biol Chem 1994; 269:6376–6382.

192. Weiss A, Littman DR. Signal transduction by lymphocyte antigen receptors. Cell 1994; 76:263–274.

193. Gruss H-J, Brach MA, Herrmann F. Involvement of nuclear factor-κB in induction of the interleukin-6 gene by leukemia inhibitory factor. Blood 1992; 80:2563–2570.

194. Ernst M, Gearing DP, Dunn AR. Functional and biochemical association of Hck with the LIF/IL-6 receptor signal transducing subunit gp130 in embryonic stem cells. EMBO J 1994; 13:1574–1584.

195. Lord KA, Abdollahi A, Thomas SM, DeMarco M, Brugge JS, Hoffman-Liebermann B, Liebermann DA. Leukemia inhibitory factor and interleukin-6 trigger the same immediate early response,

including tyrosine phosphorylation, upon induction of myeloid leukemia differentiation. Mol Cell Biol 1991; 11:4371–4379.

196. Michishita M, Satoh M, Yamaguchi M, Hirayoshi K, Okuma M, Nagata K. Phosphorylation of the stress protein hsp27 is an early event in murine myelomonocytic leukemic cell differentiation induced by leukemia inhibitory factor/D-factor. Biochem Biophys Res Commun 1991; 176:979–984.

197. Abdollahi A, Lord KA, Hoffman-Liebermann B, Liebermann DA. Interferon regulatory factor 1 is a myeloid differentiation primary response gene induced by interleukin 6 and leukemia inhibitory factor: role in growth inhibition. Cell Growth Differ 1991; 2:401–407.

198. Abdollahi A, Lord KA, Hoffman-Liebermann B, Liebermann DA. Sequence and expression of a cDNA encoding MyD118: a novel myeloid differentiation primary response gene induced by multiple cytokines. Oncogene 1991; 6:165–167.

199. Yamamori T. CDF/LIF selectively increases c-fos and jun-B transcripts in sympathetic neurons. Neuroreport 1991; 2:173–176.

200. Mouthon M-A, Navarro S, Katz A, Breton-Gorius J, Vainchenker W. c-jun and c-fos are expressed by human megakaryocytes. Exp Hematol 1992; 20:909–915.

201. Baumann H, Marinkovic-Pajovic S, Won KA, Jones VE, Campos SP, Jahreis GP, Morella KK. The actions of interleukin-6 and leukemia inhibitory factor in liver cells. In: Bock G, Marsh J, Widdows K, eds. Polyfunctional Cytokines: IL-6 and LIF. Ciba Found Symp 1992; 167:80–99.

202. Villiger PM, Geng Y, Lotz M. Induction of cytokine expression by leukemia inhibitory factor. J Clin Invest 1993; 91:1575–1581.

203. Lotem J, Sachs L. Regulation of leukaemic cells by interleukin 6 and leukaemia inhibitory factor. In: Bock G, Marsh J, Widdows K, eds. Polyfunctional Cytokines: IL-6 and LIF. Ciba Found Symp 1992; 167:80–99.

204. Miyaura C, Jin CH, Yamaguchi Y, Tomida M, Hozumi M, Matsuda T, Hirano T, Kishimoto T, Suda T. Production of interleukin 6 and its relation to the macrophage differentiation of mouse myeloid leukemia cells (M1) treated with differentiation-inducing factor and 1-alpha,25-dihydroxyvitamin D3. Biochem Biophys Res Commun 1989; 158:660–666.

205. Yamaguchi M, Masahiro M, Hirayoshi K, Yasukawa K, Okuma M, Nagata K. Down-regulation of interleukin 6 receptors of mouse myelomonocytic leukemic cells by leukemia inhibitory factor. J Biol Chem 1992; 267:22035–22042.

206. Musso T, Gusella GL, Bosco MC, Longo D, Varesio L. Leukemia inhibitory factor (LIF) induces IL 8 expression in human monocytes [abstr]. J Cell Biochem 1993; 17B(suppl):95.

207. Michishita M, Yoshida Y, Uchino H, Nagata K. Induction of tumor necrosis factor-α and its receptor during differentiation in myeloid leukemic cells along monocytic pathway. A possible regulatory mechanism for TNF-α production. J Biol Chem 1990; 265:8751–8759.

208. Leary AG, Wong GG, Clark SC, Smith AG, Ogawa M. Leukemia inhibitory factor differentiation-inhibiting activity/human interleukin for DA cells augments proliferation of human hematopoietic stem cells. Blood 1990; 7:1960–1964.

209. Leary AG, Zeng HQ, Clark SC, Ogawa M. Growth factor requirements for survival in G0 and entry into the cell cycle of primitive human hemopoietic progenitors. Proc Natl Acad Sci USA 1992; 89:4013–4017.

210. Mancini GC, Durand B, Samal B, Migliaccio AR, Migliaccio G, Adamson JW. Effects of macro-phage inhibitory protein 1 (MIP-1α) and leukemia inhibitory factor (LIF) in long-term cultures of human CD34+ cord blood cells [abstr]. Exp Hematol 1992; 20:753.

211. Gooya JG, Ruscetti FW, Keller JR. The direct synergistic effects of leukemia inhibitory factor on hematopoietic progenitor cell growth [abstr]. J Cell Biochem 1994; 18A(suppl):19.

212. Fletcher FA, Williams DE, Maliszeski C, Anderson D, Rives M, Belmont JW. Murine leukemia inhibitory factor enhances retroviral-vector infection efficiency of hematopoietic progenitors. Blood 1990; 76:1098–1103.

213. Fletcher FA, Moore KA, Ashkenazi M, De Vries P, Overbeek PA, Williams DE, Belmont JW. Leukemia inhibitory factor improves survival of retroviral vector-infected hematopoietic stem cells in vitro, allowing efficient long-term expression of vector-encoded human adenosine deaminase in

vivo—retrovirus mediated gene transmission to hematopoietic stem cell; potential use in gene therapy of diseases affecting blood-forming tissues. J Exp Med 1991; 174:837–845.

214. Dick JE, Kamel-Reid S, Murdoch B, Doedens M. Gene transfer into normal human hematopoietic cells using in vitro and in vivo assays. Blood 1991; 78:624–634.

215. Metcalf D, Hilton DJ, Nicola NA. Clonal analysis of the actions of the murine leukemia inhibitory factor on leukemic and normal murine hemopoietic cells. Leukemia 1988; 2:216–221.

216. Schaafsma MR, Falkenburg JHF, Duinkerken N, Moreau J-F, Soulillou JP, Willemze R, Fibbe WE. Human interleukin for DA cells (HILDA) does not affect the proliferation and differentiation of hematopoietic progenitor cells in human long-term bone marrow cultures. Exp Hematol 1992; 20:6–10.

217. Verfaillie C, McGlave P. Leukemia inhibitory factor/human interleukin for DA cells; a growth factor that stimulates the in vitro development of multipotential human hematopoietic progenitors. Blood 1991; 77:263–270.

218. Cory S, Maekawa T, McNeall J, Metcalf D. Murine erythroid cell lines derived with c-*myc* retroviruses respond to leukemia-inhibitory factor, erythropoietin, and interleukin 3. Cell Growth Differ 1991; 2:165–172.

219. Kitamura T, Tnage T, Terasawa T, Chiba S, Kuwaki T, Miyagawa K, Piao YF, Miyazono K, Urabe A, Takaku F. Establishment of a human cell line that proliferates dependently on GM-CSF, IL-3 or erythropoietin. J Cell Physiol 1989; 140:323–334.

220. Gimble JM, Wanker F, Wang C-S, Bass H, Wu X, Kelly K, Yancopoulos GD, Hill MR. Regulation of bone marrow stromal cell differentiation by cytokines whose receptors share the gp130 protein. J Cell Biochem 1994; 54:122–133.

221. Escary J-L, Perreau J, Duménil D, Ezine S, Brûlet P. Leukaemia inhibitory factor is necessary for maintenance of haematopoietic stem cells and thymocyte stimulation. Nature 1993; 363:361–364.

222. Nishimoto N, Ogata A, Shima Y, Tani Y, Ogawa H, Nakagawa M, Sugiyama H, Yoshizaki K, Kishimoto T. Oncostatin M, leukemia inhibitory factor, and interleukin 6 induce the proliferation of human plasmacytoma cells via the common signal transducer gp130. J Exp Med 1994; 179:1343–1347.

223. Hoffman-Leibermann B, Leibermann DA. Interleukin-6- and leukemia inhibitory factor-induced terminal differentiation of myeloid leukemia cells is blocked at an intermediate stage by constitutive c-*myc*. Mol Cell Biol 1991; 11:2375–2381.

224. Hoffman-Lieberman B, Liebermann DA. Suppression of c-*myc* and c-*myb* is tightly linked to terminal differentiation induced by IL6 or LIF and not growth inhibition in myeloid leukemia cells. Oncogene 1991; 6:903–909.

225. Selvakumaran M, Leiebermann DA, Hoffman-Liebermann B. Deregulated c-*myb* disrupts interleukin-6- or leukemia inhibitory factor-induced myeloid differentiation prior to c-*myc*: role in leukemogenesis. Mol Cell Biol 1992; 12:2493–2500.

226. Ichikawa Y. Further studies on the differentiation of a cell line of myeloid leukemia. J Cell Physiol 1970; 76:175–184.

227. Metcalf D. Suppression of myeloid leukemic cells by normal hemopoietic regulators. In: Yamamura Y, ed. Proceedings of the First International Mochida Memorial Symposium: Recent progress in cytokine research. Tokyo: Mochida Memorial Foundation for Medical Research and Pharmaceutical Research, 1989:119–134.

228. Maekawa T, Metcalf D. Clonal suppression of HL60 and U937 cells by recombinant human leukemia inhibitory factor in combination with GM-CSF and G-CSF. Leukemia 1989; 3:270–276.

229. Maekawa T, Metcalf D, Gearing DP. Enhanced suppression of human myeloid leukemic cell lines by combinations of IL-6, LIF, GM-CSF and G-CSF. Int J Cancer 1990; 45:353–358.

230. Brach MA, Riedel D, Mertelsmann RH, Hermann F. Synergistic effect of recombinant human leukemia inhibitory factor (LIF) and 1-β-D arabinofuranosylcytosine (Ara-C) on proto-oncogene expression and induction of differentiation in human U937 cells. Leukemia 1990; 4:646–649.

231. Samal BB, Stearns GW, Boone TC, Arakawa T. Comparative analysis of the effects of recombinant

cytokines on the growth and differentiation of ML-1, a human myelogenous leukemic cell line. Leuk Res 1993; 17:299–304.

232. Kerangueven F, Sempere C, Tabilio A, Mannoni P. Effects of transforming growth factor beta, tumor necrosis factor alpha, interferon gamma and LIF-HILDA on the proliferation of acute myeloid leukemia cells. Eur Cytokine Netw 1990; 1:99–107.

233. Takanashi M, Motoji T, Masuda M, Oshimi K, Mizoguchi H. The effects of leukemia inhibitory factor and interleukin 6 on the growth of acute myeloid leukemia cells. Leuk Res 1993; 17:217–222.

234. Chen G-J, Tomida M, Hozumi M, Nara N. Effect of recombinant human D-factor on the growth of leukemic blast progenitors from acute myeloblastic leukemic patients. Jpn J Cancer Res 1992; 83:1341–1346.

235. Wang C, Lischner M, Minden MD, McCulloch EA. The effects of leukemia inhibitory factor (LIF) on blast stem cells of acute myeloblastic leukemia. Leukemia 1990; 4:548–552.

236. Hoffman R. Regulation of megakaryocytopoiesis. Blood 1989; 74:1196–1212.

237. Metcalf D, Nicola NA, Gearing DP. Effects of injected leukaemia inhibitory factor on hematopoietic and other tissues in mice. Blood 1990; 76:50–56.

238. Mayer P, Giessler K, Ward M, Metcalf D. Recombinant human leukemia inhibitory factor induces acute phase proteins and raises blood platelet counts in nonhuman primates. Blood 1993; 81:3226–3233.

239. Zeidler C, Kanz L, Hurkuck F, Rittmann KL, Wildfang I, Kadoya T, Mikayama T, Souza L, Welte K. In vivo effects of interleukin-6 on thrombocytopoiesis in healthy and irradiated primates. Blood 1992; 80:2740–2745.

240. Farese AM, Myers LA, MacVittie TJ. Therapeutic efficacy of recombinant human (rh) leukemia inhibitory factor (LIF) relative to the interleukins (IL) 6 and 3 in a nonhuman-primate model of radiation-induced marrow aplasia [abstr]. J Cell Biochem 1994; 18A(suppl):34.

241. Debili N, Masse J-M, Katz A, Guichard J, Breton-Gorius J, Vainchecker W. Effects of the recombinant haematopoietic growth factors interleukin-3, interleukin-6, stem cell factor, and leukemia inhibitory factor on the megakaryocytic differentiation of $CD34^+$ cells. Blood 1993; 82:84–95.

242. Warren MK, Guertin M, Rudzinski I, Seidman MM. A new culture and quantitation system for megakaryocyte growth using cord blood $CD34^+$ cells and the GPIIb/IIIa marker. Exp Hematol 1993; 21:1473–1479.

243. Burstein SA, Mei R-L, Henthorn J, Friese P, Turner K. Leukemia inhibitory factor and interleukin-11 promote maturation of murine and human megakaryocytes in vitro. J Cell Physiol 1992; 153:305–312.

244. Ebbe S, Phalen E, Stohlman F Jr. Abnormalities of megakaryocytes in Sl/Sl^d mice. Blood 1973; 42:865–871.

245. Ebbe S, Phalen E, Stohlman F Jr. Abnormalities of megakaryocytes in W/W^V mice. Blood 1973; 42:857–864.

246. Baumann H, Hill RE, Sauder DN, Jaheris GP. Regulation of major acute phase plasma proteins by hepatocyte-stimulating factors of human squamous carcinoma cells. J Cell Biol 1986; 102:370–383.

247. Baumann H, Onorato U, Gauldie J, Jahreis GP. Distinct sets of acute phase plasma proteins are stimulated by separate human hepatocyte-stimulating factors and monokines in rat hepatoma cells. J Biol Chem 1987; 262:9756–9768.

248. Waring PM, Wall D, Dauer R, Parkin D, Metcalf D. The effects of leukaemia inhibitory factor on platelet function. Br J Haematol 1993; 83:80–87.

249. Baumann H, Schendel P. Interleukin-11 regulates the hepatic expression of the same plasma protein genes as interleukin-6. J Biol Chem 1991; 266:20424–20427.

250. Richards CD, Brown TJ, Shoyab M, Baumann H, Gauldie J. Recombinant oncostatin M stimulates the production of acute phase proteins in HepG2 cells and rat primary hepatocytes in vitro. J Immunol 1992; 148:1731–1736.

251. Schooltink H, Stoyan T, Roeb E, Heinrich PC, Rose-John S. Ciliary neurotrophic factor induces acute-phase protein expression in hepatocytes. FEBS Lett 1992; 314:280–284.

252. Nesbitt JE, Fuentes NL, Fuller GM. Ciliary neurotrophic factor regulates fibrinogen gene expression

in hepatocytes by binding to the interleukin-6 receptor. Biochem Biophys Res Commun 1993; 190:544–550.

253. Baumann H, Ziegler SF, Mosley B, Morella KK, Pajovic S, Gearing DP. Reconstitution of the response to leukemia inhibitory factor, oncostatin M, and ciliary neurotrophic factor in hepatoma cells. J Biol Chem 1993; 268:8414–8417.

254. Hocke GM, Barry D, Fey GH. Synergistic action of interleukin-6 and glucocorticoids is mediated by the interleukin-6 response element of the rat α_2 macroglobulin gene. Mol Cell Biol 1992; 12:2282–2294.

255. Kopf M, Baumann H, Freer G, Freudenberg M, Lamers M, Kishimoto T, Zinkernagel R, Bluethmann H, Kohler G. Impaired immune and acute-phase responses in interleukin-6-deficient mice. Nature 1994; 368:339–342.

256. Libert C, Takahashi N, Cauwels A, Brouckaert P, Bluethmann H, Fiers W. Response of interleukin-6-deficient mice to tumor necrosis factor-induced metabolic changes and lethality. Eur J Immunol 1994; 24:2237–2242.

257. Shapiro L, Zhang XX, Rupp RG, Wolff SM, Dinarello CA. Ciliary neurotrophic factor is an endogenous pyrogen. Proc Natl Acad Sci USA 1993; 90:8614–8618.

258. Brown GS, Brown MA, Hilton D, Gough NM, Sleigh MJ. Inhibition of differentiation in a murine F9 embryonal carcinoma cell subline by leukemia inhibitory factor. Growth Factors 1992; 7:41–52.

259. Yoshida K, Chambers I, Nichols J, Smith A, Saito M, Yasukawa K, Shoyab M, Taga T, Kishimoto T. Maintenance of the pluripotential phenotype of embryonic stem cells through direct activation of gp130 signalling pathways. Mech Dev 1994; 45:163–171.

260. Rose TM, Weinford DM, Gunderson NL, Bruce AG. Oncostatin M (OSM) inhibits the differentiation of pluripotent embryonic stem cells in vitro. Cytokine 1994; 6:48–54.

261. Conover JC, Ip NY, Poueymirou WT, Bates B, Goldfarb MP, DeChiara TM, Yancopoulos GD. Ciliary neurotrophic factor maintains the pluripotentiality of embryonic stem cells. Development 1993; 119:559–565.

262. Wolf E, Kramer R, Polejaeva I, Thoenen H, Brem G. Efficient generation of chimaeric mice using embryonic stem cells after long-term culture in the presence of ciliary neurotrophic factor. Transgenic Res 1994; 3:152–158.

263. Pease S, Braghetta P, Gearing D, Grail D, Williams RL. Isolation of embryonic stem (ES) cells in media supplemented with recombinant leukemia inhibitory factor (LIF). Dev Biol 1990: 141:344–352.

264. Pease S, Williams RL. Formation of germ-line chimeras from embryonic stem cells maintained with recombinant leukemia inhibitory factor. Exp Cell Res 1990; 190:209–211.

265. Nichols J, Evans EP, Smith AG. Establishment of germ-line-competent embryonic stem (ES) cells using differentiation inhibiting activity. Development 1990; 110:1341–1348.

266. Hirayoshi K, Tsuru A, Yamashita M, Tomida M, Tamamoto-Yamaguchi Y, Yasukawa K, Hozumi M, Goeddel DV, Nagata K. Both D factor/LIF and IL-6 inhibit the differentiation of mouse teratocarcinoma F9 cells. FEBS Lett 1991; 282:401–404.

267. Pruitt SC, Natoli TA. Inhibition of differentiation by leukemia inhibitory factor distinguishes two induction pathways in P19 embryonal carcinoma cells. Differentiation 1992; 50:57–65.

268. Pruitt SC. Primitive streak mesoderm-like cell lines expressing PAX-3 and Hox genes autoinducing activities. Development 1994; 120:37–47.

269. Conquet F, Brûlet P. Developmental expression of myeloid leukemia inhibitory factor gene in preimplantation blastocysts and in the extraembryonic tissue of mouse embryos. Mol Cell Biol 1990; 10:3801–3805.

270. Smith AG, Nichols J, Robertson M, Rathjen PD. Differentiation inhibiting activity (DIA/LIF) and mouse development. Dev Biol 1992; 151:339–351.

271. Bard JB, Ross AS. LIF, the ES-cell inhibition factor, reversibly blocks nephrogenesis in cultured mouse kidney rudiments. Development 1991; 113:193–198.

272. Williams DE, De Vries P, Namen AE, Widmar MB, Lyman SD. The steel factor. Dev Biol 1992; 151:368–376.

273. Dolci S, Williams DE, Ernst MK, Resnick JL, Branan CI, Lock LF, Lyman SD, Boswell HS, Donovan PJ. Requirement for mast cell growth factor for primordial germ cell survival in culture. Nature 1991; 352:809–811.

274. Dolci S, Pesce M, De Felici M. Combined actions of stem cell factor, leukemia inhibitory factor, and cAMP on in vitro proliferation of mouse primordial germ cells. Mol Reprod Dev 1993; 35:134–139.

275. Godin I, Deed R, Cooke J, Zsebo K, Dexter M, Wylie CC. Effects of the *steel* gene product on mouse primordial germ cells in culture. Nature 1991; 352:807–809.

276. Pesce M, Farrace MG, Piacenti M, Dolci S, De Felici M. Stem cell factor and leukemia inhibitory factor promote primordial germ cell survival by suppressing programmed cell death (apoptosis). Development 1993; 118:1089–1094.

277. Matsui YD, Toksoz D, Nishikawa S, Nishikawa S-I, Williams D, Zsebo K, Hogan BLM. Effect of steel factor and leukaemia inhibitory factor on murine primordial germ cells in culture. Nature 1991; 353:750–752.

278. De Felici M, Dolci S. Leukemia inhibitory factor sustains the survival of mouse primordial germ cells cultured on TM4 feeder layers. Dev Biol 1991; 147:281–284.

279. Matsui Y, Zsebo K, Hogan BLM. Derivation of pluripotential embryonic stem cells from murine primordial germ cells in culture. Cell 1992; 70:841–847.

280. Resnick JL, Bixler LS, Cheng L, Donovan PJ. Long-term proliferation of mouse primordial germ cells in culture. Nature 1992; 359:550–551.

281. Stewart CL, Kaspar P, Brunet LJ, Bhatt H, Gadi I, Köntgen F, Abbondanza SJ. Blastocyst implantation depends upon maternal expression of leukemia inhibitory factor. Nature 1992; 359:76–79.

282. De M, Sanford TR, Wood GW. Interleukin-1, interleukin-6, and tumor necrosis factor α are produced in the mouse uterus during the estrous cycle and are induced by estrogen and progesterone. Dev Biol 1992; 151:297–305.

283. Murray R, Lee F, Chiu C-P. The genes for leukemia inhibitory factor and interleukin-6 are expressed in mouse blastocysts prior to the onset of hemopoiesis. Mol Cell Biol 1990; 10:4953–4956.

284. Fry RC, Batt PA, Fairclough RJ, Parr RA. Human leukemia inhibitory factor improves the viability of cultured ovine embryos. Biol Reprod 1992; 46:470–474.

285. Fry RC. The effect of leukaemia inhibitory factor (LIF) on embryogenesis. Reprod Fertil Dev 1992; 4:449–458.

286. Saito S, Nishikawa K, Morii T, Enomoto M, Narita N, Motoyoshi K, Ichijo M. Cytokine production by CD16–CD56 (bright) natural killer cells in the human early pregnancy decidua. Int Immunol 1993; 5:559–563.

287. Croy BA, Guilbert LJ, Brown MA, Gough NM, Stinchcomb DT, Reed N, Wegmann TG. Characterization of cytokine production by the metrial gland and granulated metrial gland cells. J Reprod Immunol 1991; 19:149–166.

288. Guilbert L, Robertson SA, Wegmann TG. The trophoblast as an integral component of a macrophage–cytokine network. Immunol Cell Biol 1993; 71:49–57.

289. Reid IR, Lowe C, Cornish J, Skinner SJM, Hilton DJ, Willson TA, Gearing DP, Martin TJ. Leukemia inhibitory factor: a novel bone-active cytokine. Endocrinology 1990; 126:1416–1420.

290. Lowe C, Cornish J, Martin TJ, Reid IR. Effects of leukemia inhibitory factor on bone resorption and DNA synthesis in neonatal mouse calvaria. Calcif Tissue Int 1991; 49:394–397.

291. Lorenzo JA, Sousa SL, Leahy CL. Leukemia inhibitory factor (LIF) inhibits basal bone resorption in fetal rat long bone cultures. Cytokine 1990; 2:266–271.

292. Van Beek E, Van der Wee-Pals L, Van de Ruit M, Nijweide P, Papapoulos W, Lowik C. Leukemia inhibitory factor inhibits osteoclastic resorption, growth, mineralization, and alkaline phosphatase activity in fetal mouse metacarpal bones in culture. J Bone Miner Res 1993; 8:191–198.

293. Lowe C, Cornish J, Callon K, Martin TJ, Reid IR. Regulation of osteoblast proliferation by leukemia inhibitory factor. J Bone Miner Res 1991; 6:1277–1278.

294. Metcalf D, Gearing DP. A myelosclerotic syndrome in mice engrafted with cells producing high levels of leukemia inhibitory factor (LIF). Leukemia 1989; 3:847–852.

295. Metcalf D, Gearing DP. Fatal syndrome in mice engrafted with cells producing high levels of the leukemia inhibitory factor. Proc Natl Acad Sci USA 1989; 86:5948–5952.

296. Tamura T, Udagawa N, Takahashi N, Miyaura C, Tanaka S, Yamada Y, Koishihara Y, Ohsugi Y, Kumaki K, Taga T, Kishimoto T, Suda T. Soluble interleukin-6 receptor triggers osteoclast formation by interleukin 6. Proc Natl Acad Sci USA 1993; 15:11924–11928.

297. Shinar DM, Sato M, Rodan GA. The effect of hemopoietic growth factors on the generation of osteoclast-like cells in mouse bone marrow cultures. Endocrinology 1990; 126:1728–1735.

298. Evans DB, Gerber B, Feyen JHM. Effects of recombinant human leukemia inhibitory factor on human bone cells [abstr]. Calcif Tissue Int 1992; 50(suppl 1):26.

299. Evans DB, Gerber B, Feyen JH. Recombinant human leukemia inhibitory factor is mitogenic for human bone-derived osteoblast-like cells. Biochem Biophys Res Commun 1994; 199:220–226.

300. Noda M, Vogel RL, Hasson DM, Rodan GA. Leukemia inhibitory factor suppresses proliferation, alkaline phosphatase activity, and type I collagen messenger ribonucleic acid level and enhances osteopontin mRNA level in murine osteoblast-like (MC3T3E1) cells. Endocrinology 1900; 127:185–190.

301. Cornish J, Callon K, King A, Reid IR. The effect of leukemia inhibitory factor on bone in vivo. Endocrinology 1993; 132:1359–1366.

302. Yoneda T, Alsina MA, Chavez JB, Bonewald L, Nishimura R, Mundy GR. Evidence that tumour necrosis factor plays a pathogenetic role in the paraneoplastic syndromes of cachexia, hypercalcemia, and leucocytosis in a human tumour in nuce mice. J Clin Invest 1991; 87:977–985.

303. Johnson RA, Boyce BF, Mundy GR, Roodman GD. Tumors producing human tumor necrosis factor induce hypercalcemia and osteoclastic bone resorption in nude mice. Endocrinology 1989; 124:1424–1247.

304. Black K, Garrett IR, Mundy GR. Chinese hampster ovarian cells transfected with the murine interleukin-6 gene cause hypercalcemia as well as cachexia, leukocytosis and thrombocytosis in tumor-bearing nude mice. Endocrinology 1991; 128:2657–2659.

305. Carroll GJ, Bell MC. Leukaemia inhibitory factor stimulates proteoglycan resorption in porcine articular cartilage. Rheumatol Int 1993; 13:5–8.

306. Villiger PM, Terkeltaub R, Lotz M. Monocyte chemoattractant protein-1 (MCP-1) expression in human articular cartilage. J Clin Invest 1992; 90:488–496.

307. Eckel RH. Lipoprotein lipase. A multifunctional enzyme relevant to common metabolic diseases. N Engl J Med 1989; 320:1060–1068.

308. Marshall MK, Doerrler W, Feingold KR, Grunfeld C. Leukemia inhibitory factor induces changes in lipid metabolism in cultured adipocytes. Endocrinology 1994; 135:141–147.

309. Fong Y, Moldawer LL, Marano M, Wei H, Barber A, Manogue K, Tracey KJ, Kuo G, Fischman DA, Cerami A, Lowry SG. Cachectin/TNF or IL-1-α induces cachexia with redistribution of body proteins. Am J Physiol 1989; 256:R659–R665.

310. Langstein HN, Doherty GM, Fraker DL, Buresh CM, Norton JA. The roles of γ-interferon and tumour necrosis factor α in an experimental model of cancer cachexia. Cancer Res 1991; 15:2302–2306.

311. Henderson JT, Seniuk NA, Richardson PM, Gauldie J, Roder JC. Systemic administration of ciliary neurotrophic factor induces cachexia in rodents. J Clin Invest 1994; 93:2632–2638.

312. Oliff A, Defeo-Jones D, Boyer M. Tumors secreting human TNF/cachectin induce cachexia in mice. Cell 1987; 50:555–563.

313. Matthys P, Dukmans R, Proost P, Van Damme J, Heremans H, Sobins H, Billiau A. Severe cachexia in mice inoculated with interferon-γ-producing tumor cells. Int J Cancer 1991; 49:77–82.

314. Strassmann G, Fong M, Kenny JS, Jacob CO. Evidence for involvement of interleukin 6 in experimental cancer cachexia. J Clin Invest 1992; 89:1681–1684.

315. Patterson PH, Nawa H. Neuronal differentiation factors/cytokines and synaptic plasticity. Cell 1993; 72:123–137.

316. Patterson PH, Chun LLY. The induction of acetylcholine synthesis in primary cultures of dissociated sympathetic neurons. I. Effect of conditioned medium. Dev Biol 1977; 56:263–280.

317. Weber MJ, Raynaud B, Delteil C. Molecular properties of a cholinergic differentiation factor from muscle conditioned medium. J Neurochem 1985; 45:1541–1547.

318. Nawa H, Patterson PH. Separation and partial characterization of neuropeptide-inducing factors in heart cell conditioned medium. Neuron 1990; 4:269–277.

319. Nawa H, Nakanishi S, Patterson PH. Recombinant cholinergic differentiation factor (leukemia inhibitory factor) regulates sympathetic neuron phenotype by alterations in the size and amounts of neuropeptide mRNAs. J Neurochem 1991; 56:2147–2150.

320. Freidin M, Kessler JA. Cytokine regulation of substance P expression in sympathetic neurons. Proc Natl Acad Sci USA 1991; 88:3200–3203.

321. Fann MJ, Patterson PH. A novel approach to screen for cytokine effect on neuronal gene expression. J Neurochem 1993; 61:1349–1355.

322. Rao MS, Tyrrell S, Landis SC, Patterson PH. effects of ciliary neurotrophic factor (CNTF) and depolarization on neuropeptide expression in cultured sympathetic neurons. Dev Biol 1992; 150:281–293.

323. Lewis SE, Rao MS, Symes AJ, Dauer WT, Fink JS, Landis SC, Hyman SC. Coordinate regulation of choline acetyltransferase, tyrosine hydroxylase, and neuropeptide mRNAs by ciliary neurotropic factor and leukemia inhibitory factor in cultures sympathetic neurons. J Neurochem 1993; 63:429–438.

324. Mulderry PK. Neuropeptide expression by newborn and adult rat sensory neurons in culture: effects of nerve growth factor and other neurotrophic factors. Neuroscience 1994; 59:673–688.

325. Zurn AD, Werren F. Development of CNS cholinergic neurons in vitro; selective effects of CNTF and LIF on neurons from mesencephalic cranial motor nuclei. Dev Biol 1994; 163:309–315.

326. Murphy M, Reid K, Hilton DJ, Bartlett PF. Generation of sensory neurons is stimulated by leukemia inhibitory factor. Proc Natl Acad Sci USA 1991; 88:3498–3501.

327. Murphy M, Reid K, Brown MA, Bartlett PF. Involvement of leukemia inhibitory factor and nerve growth factor in the development of dorsal root ganglion neurons. Development 1993; 117:1173–1182.

328. Thaler CD, Suhr L, Ip N, Katz DM. Leukemia inhibitory factor and neurotrophins support overlapping populations of rat nodose sensory neurons in culture. Dev Biol 1994; 161:338–344.

329. Kotzbauer PT, Lampe PA, Estus S, Milbrandt J, Johnson EM Jr. Postnatal development of survival responsivness in rat sympathetic neurons to leukemia inhibitory factor and ciliary neurotrophic factor. Neuron 1994; 12:763–773.

330. Martinou J-C, Martinou I, Kato AC. Cholinergic differentiation factor (CDF/LIF) promotes survival of isolated rat embryonic motoneurons in vitro. Neuron 1992; 8:737–744.

331. Kushima Y, Hatanaka H. Interleukin-6 and leukemia inhibitory factor promote the survival of acetylcholinesterase-positive neurons in culture from embryonic rat spinal cord. Neurosci Lett 1992; 143:110–114.

332. Richards LJ, Kilpatrick TJ, Bartlett PF, Murphy M. Leukemia inhibitory factor promotes the neuronal development of spinal cord precursors from the neural tube. J Neurosci Res 1992; 33:476–484.

333. Michikawa M, Kikuchi S, Kim SU. Leukemia inhibitory factor (LIF) mediated increase of choline acetyltransferase activity in mouse spinal cord neurons in culture. Neurosci Lett 1992; 140:75–77.

334. Hendry I, Murphy M, Hilton DJ, Nicola NA, Bartlett PF. Binding and retrograde transport of leukemia inhibitory factor by the sensory nervous system. J Neurosci 1992; 12:3427–3434.

335. Ure DR, Campenot RB. Leukemia inhibitory factor and nerve growth factor are retrogradely transported and processed by cultured rat sympathetic neurons. Dev Biol 1994; 162:339–347.

336. Oppenheim RW, Prevette D, Qin-Wei Y, Collins F, MacDonald J. Control of embryonic motoneuron survival in vivo by ciliary neurotrophic factor. Science 1991; 251:1616–1618.

337. Saadat S, Sendtner M, Rohrer H. Ciliary neurotrophic factor induces cholinergic differentiation of rat sympathetic neurons in culture. J Cell Biol 1989; 108:1807–1816.

338. Ludlam WH, Kessler JA. Leukemia inhibitory factor and ciliary neurotrophic factor regulate expression of muscarinic receptors in cultured sympathetic neurons. Dev Biol 1993; 155:497–506.

339. Rao MS, Symes A, Malik N, Shoyab M, Fink JS, Landis SC. Oncostatin M regulates VIP expression in a human neuroblastoma cell line. Neuroreport 1992; 3:865–868.

340. Patterson PH. The emerging neuropoietic cytokine family: first CDF/LIF, CNTF and IL-6; next ONC, MGF, GCSF? Curr Opin Neurobiol 1992; 2:94–97.

341. Qiu L, Bernd P, Kukada K. Cholinergic neuronal differentiation factor (CDF)/leukemia inhibitory factor (LIF) binds to specific regions of the developing nervous system in vivo. Dev Biol 1994; 163:516–520.

342. Patterson PH, Fann M-J. further studies of the distribution of CDF/LIF mRNA. In: Bock G, Marsh J, Widdows K, eds. Polyfunctional Cytokines: IL-6 and LIF. Ciba Found Symp 1992; 167:125–140.

343. Bamber BA, Masters BA, Hoyle GW, Brinster RL, Palmiter RD. Leukemia inhibitory factor induces neurotransmitter switching in transgenic mice. Proc Natl Acad Sci USA 1994; 91:7839–7843.

344. Rao MS, Patterson PH, Landis SC. Multiple cholinergic differentiation factors are present in footpad extracts: comparison with known cholinergic factors. Development 1992; 116:731–744.

345. Rao MS, Sun Y, Escary JL, Perreau J, Tresser S, Patterson PH, Zigmond RE, Brulet P, Landis SC. Leukemia inhibitory factor mediates an injury response but not a target-directed developmental transmitter switch in sympathetic neurons. Neuron 1993; 11:1175–1185.

346. Zigmond RE, Hyatt-Sachs H, Baldwin C, Qu XM, Sun Y, McKeon TW, Schreiber RC, Vaidyanathan U. Phenotypic plasticity in adult sympathetic neurons: changes in neuropeptide expression in organ culture. Proc Natl Acad Sci USA 1992; 89:1507–1511.

347. Shadiack AM, Hart RP, Carlson CD, Jonakait GM. Interleukin-1 induces substance P in sympathetic ganglia through the induction of leukemia inhibitory factor. J Neurosci 1993; 13:2601–2609.

348. Fan G, Katz DM. Non-neuronal cells inhibit catecholamine differentiation of primary sensory neurons: role of leukemia inhibitory factor. Development 1993; 118:83–93.

349. Yoshida T, Satoh M, Nakagaito Y, Kuno H, Takeuchi M. Cytokines affecting survival and differentiation of an astrocyte progenitor cell line. Dev Brain Res 1993; 76:147–150.

350. Nishiyama K, Collodi P, Barnes D. Regulation of glial fibrillary acidic protein in serum-free mouse embryo (SFME) cells by leukemia inhibitory factor and related peptides. Neurosci Lett 1993; 163:114–116.

351. Barres BA, Schmid R, Sendtner M, Raff MC. Multiple extracellular signals are required for long-term oligodendrocyte survival. Development 1993; 118:283–295.

352. Mayer M, Bhakoo K, Noble M. Ciliary neurotrophic factor and leukemia inhibitory factor promote the generation, maturation and survival of oligodendrocytes in vitro. Development 1994; 120:143–153.

353. Sendtner M, Kreutzberg GW, Thoenen H. Ciliary neurotrophic factor prevents the degeneration of motor neurons after axotomy. Nature 1990; 345:440–441.

354. Wesselingh SL, Levine B, Fox RJ, Choi S, Griffin DE. Intracerebral cytokine mRNA expression during fatal and nonfatal alphavirus encephalitis suggests a predominant type 2 T cell response. J Immunol 1994; 152:1289–1297.

355. Higuchi M, Ito T, Imai Y, Iwaki T, Hattori M, Kohsaka S, Niho Y, Sakaki Y. Expression of the alpha 2-macroglobulin-encoding gene in rat brain and cultured astrocytes. Gene 1994; 141:155–162.

356. Curtis R, Scherer SS, Somogyi R, Adryan KM, Ip NY, Zhu Y, Lindsay RM, DiStefano PS. Retrograde axonal transport of LIF is increased by peripheral nerve injury: correlation with increased LIF expression in distal nerve. Neuron 1994; 12:191–204.

357. Banner LR, Patterson PH. Major changes in the expression of the mRNAs for cholinergic differentiation factor/leukemia inhibitory factor and its receptor after injury to adult peripheral nerves and ganglia. Proc Natl Acad Sci USA 1994; 91:7109–7113.

358. Cheema SS, Richards L, Murphy M, Bartlett PF. Leukemia inhibitory factor prevents the death of axotomised sensory neurons in the dorsal root ganglia of the neonatal rat. J Neurosci Res 1994; 37:213–218.

359. Cheema SS, Richards LJ, Murphy M, Bartlett PF. Leukemia inhibitory factor rescues motoneurones from axotomy-induced death. Neuroreport 1994; 5:989–992.

360. Austin L, Burgess AW. Stimulation of myoblast proliferation in culture by leukemia inhibitory factor and other cytokines. J Neurol Sci 1991; 101:193–197.

361. Austin L, Bower J, Kurek J, Vakakis N. Effects of leukaemia inhibitory factor and other cytokines on murine and human myoblast proliferation. J Neurol Sci 1992; 112:185–191.

362. Robertson TA, Maley MA, Grounds MD, Papadimitriou JM. The role of macrophages in skeletal muscle regeneration with particular reference to chemotaxis. Exp Cell Res 1993; 207:321–331.

363. Djabali K, Zissopoulou A, de Hoop MJ, Georgatos SD, Dotti CG. Peripherin expression in hippocampal neurons induced by muscle soluble factors(s). J Cell Biol 1993; 123:1197–1206.

364. Barnard W, Bower J, Brown MA, Murphy M, Austin L. Leukemia inhibitory factor (LIF) infusion stimulates skeletal muscle regeneration after injury: injured muscle expresses LIF mRNA. J Neurol Sci 1994; 123:108–113.

365. Kirby ML, Kumiski DH, Mishima N. Altered expression of smooth muscle and neuronal phenotypes by cardiac neural crest cells cultured in leukemia inhibitory factor (LIF) [abstr]. Mol Biol Cell 1992; 3:285a.

366. Kirby ML, Kumiski DH, Myers T, Cerjan C, Mishima N. Backtransplanation of chick cardiac neural crest cells cultured in LIF rescues heart development. Dev dynam 1993; 198:296–311.

367. Ribeiro M, Ades EW, Evatt BL, Hooper WC. Modulation of endothelial cell gene expression by leukemia inhibitory factor [abstr]. Blood 1992; 80(suppl 1):324a.

368. Gillett NA, Lowe D, Lu L, Chan C, Ferrara N. Leukemia inhibitory factor expression in human carotid plaques: possible mechanism for inhibition of large vessel endothelial regrowth. Growth Factors 1993; 9:301–305.

369. Waring PM, Waring LJ, Billington T, Metcalf D. Leukemia inhibitory factor protects against lethality in experimental E. coli septic shock in mice. Proc Natl Acad Sci USA 1995; 92: 1337–1341.

370. Moran CS, Campbell JH, Simmons DL, Campbell GR. Human leukemia inhibitory factor inhibits development of experimental atherosclerosis. Arterioscler Thromb 1994; 14:1356–1363.

371. Alexander HR, Wong GGH, Doherty GM, Venzon DJ, Fraker DL, Norton JA. Differentiation factor/leukemia inhibitory factor protection against lethal endotoxemia in mice: synegistic effect with interleukin 1 and tumor necrosis factor. J Exp Med 1992; 175:1139–1142.

372. Shen MM, Skoda R, Cardiff RD, Campos-Torres J, Leder P, Ornitz DM. Expression of LIF in transgenic mice results in altered thymic epithelium and apparent interconversion of thymic and lymph node morphologies. EMBO J 1994; 13:1375–1385.

373. Suematsu S, Matsuda T, Aozasa K, Akira S, Nakano N, Ohno S, Miyazaki J-I, Yamanura K-I, Hirato T, Kishimoto T. IgG1 plasmacytosis in interleukin 6 transgenic mice. Proc Natl Acad Sci USA 1989; 86:7547–7551.

374. Lecron JC, Roblot P, Chevalier S, Morel F, Alderman E, Gombert J, Gascan H. High circulating leukaemia inhibitory factor (LIF) in patients with giant cell arteritis: independent regulation of LIF and IL-6 under corticosteroid therapy. Clin Exp Immunol 1993; 92:23–26.

375. Kim KJ, Alphonso M, Schmelzer CH, Lowe D. Detection of human leukemia inhibitory factor by monoclonal antibody based ELISA. J Immunol Methods 1992; 156:9–17.

376. Godard A, Fauchet F, Raher S, Jadoul M, Thuillier B, Dehart J, Soulillou JP, Baudrihaye M, Jacques Y, De Groote D. Generation of monoclonal antibodies against HILDA/LIF and their use in the quantitative assay of the cytokine. Cytokine 1993; 5:16–23.

377. Waring P, Wycherley K, Cary D, Nicola N, Metcalf D. Leukemia inhibitory factor levels are elevated in septic shock and in various inflammatory body fluids. J Clin Invest 1992; 90:2031–2037.

378. Waring PM, Carroll GJ, Kandiah DA, Buirski G, Metcalf D. Increased levels of leukemia inhibitory factor in synovial fluid from patients with rheumatoid arthritis and other inflammatory arthritides. Arthritis Rheum 1993; 36:911–915.

379. Blancho G, Moreau J-F, Anegon I, Soulillou J-P. HILDA/LIF is present in the urine of rejecting kidney graft recipients. Transplant Int 1992; 5:57–58.

380. Waring PM, Waring LJ, Metcalf D. Circulating leukemia inhibitory factor levels correlate with disease severity in meningococcemia. J Infect Dis 1994; 170:1224–1228.

381. Roblot P, Lecron JCl, Chevalier S, Alderman E, Gombert J, Becq-Giraudon B. Taux élevés de LIF au cours de la maladie de Horton. Régulation indépendante du LIF at de l'IL6 sous corticothérapie [abstr]. Rev Méd Interne 1992; 7:S412.

382. Taupin JL, Morel D, Moreau JP, Gualde N, Potaux L, Bezian J-H. HILDA/LIF urinary excretion during acute kidney rejection. Transplantation 1992; 53:655–658.

383. Broor S, Kusari AB, Zhang B, Seth P, Richman DD, Carson DA, Wachsman W, Lotz M. Stimulation of HIV replication in mononuclear phagocytes by leukemia inhibitory factor. J Acquir Immune Defic Syndr 1994; 7:647–654.

384. Neta R, Wong GHW, Pilcher M. LIF and IL-4 used after lethal irradiation protect mice from death [abstr]. Lymphokine Res 1990; 9:568.

385. Wong GHW, Neta R, Goeddel DV. Protective roles of MnSOD, TNF-α, TNF-β and D-factor (LIF) in radiation injury. In: Eicosanoids and other bioactive lipids in cancer, inflammation and radiation injury. 2nd International Conference. Klinikum Steglitz, Berlin 1992: 353–358.

386. Tsan M-F, White JE, Wong GHW. D-factor and growth hormone enhance tumor necrosis factor-induced increase of Mn superoxide dismutase mRNA and oxygen tolerance. Cytokine 1992; 4:101–105.

22

Interleukin-12 in Hematopoiesis

Sten Eirik W. Jacobsen* and Erlend B. Smeland
The Norwegian Radium Hospital, Oslo, Norway

I. ROLE OF HEMATOPOIETIC GROWTH FACTORS IN HEMATOPOIESIS

Hematopoiesis is a lifelong process in which mature hematopoietic cells with limited life spans are continuously replaced by a pool of pluripotent hematopoietic stem cells capable of forming all myeloid and lymphoid lineages in the blood (reviewed in Refs. 1–4). These stem cells give rise to large populations of intermediate progenitors, with different degrees of multipotentiality and committment toward specific lineages (1–4). The enormous production of hematopoietic cells is illustrated by the fact that approximately 1 trillion cells are replaced every day in a 70-kg man (reviewed in Ref. 3).

It has long been speculated that strict and complex control mechanisms must be involved in the regulation of hematopoiesis under steady-state conditions, as well as following hemato-poietic injuries requiring enhanced turnover in the hematopoietic hierarchy. From studies on the isolation, characterization, and cloning of numerous hematopoietic growth factors (HGFs), it has become increasingly clear that the complex regulation of hematopoietic progenitor cell growth and differentiation is regulated, at least partly, through the opposing actions of stimulatory and inhibitory cytokines (reviewed in Refs. 2–4).

The classic family of colony-stimulating factors (CSFs), which include granulocyte-CSF (G-CSF), granulocyte–macrophage-CSF (GM-CSF), macrophage-CSF (M-CSF or CSF-1), and interleukin-3 (IL-3) remain unique in their ability to stimulate the in vitro colony formation of hematopoietic progenitor cells in the absence of other HGFs (2–4). Although procedures for purification and biological evaluation of murine primitive multi- or pluripotent bone marrow progenitors (reviewed in Ref. 1) have been established, it has become clear that only more mature or committed progenitors can grow in response to a single cytokine (2–4). In contrast,

---

**Current affiliation*: University Hospital of Lund, Lund, Sweden.

primitive progenitors can proliferate optimally only in response to combinations of cytokines (2–4). To stimulate the proliferation of these early progenitors, such cytokine combinations must include at least one CSF or stem cell factor (SCF), as well as one or more synergistic HGFs. Thus, stimulatory HGFs can be grouped into four major categories (summarized in Table 1). The first group consists of the four classic CSFs (5) as well as thrombopoietin (TPO) (6), erythropoietin (EPO; 7) and IL-5 (8), which can stimulate the proliferation of committed progenitors as single HGFs. The synergistic cytokines may be divided into two classes. The first could be named "SCF-like cytokines" or "stem cell growth factors," because their effects are predominantly on primitive progenitors. This class now includes SCF (also called mast cell growth factor, steel factor, and kit ligand; 9–12), as well as the recently cloned *flt3/flk-2* ligand (FL; 13–15). These two cytokines are characterized by their weak ability to stimulate the growth of hematopoietic progenitor cells in the absence of other HGFs, and their potent ability to synergize with the CSFs and the group of "pure synergistic HGFs" (9–22). The purely synergistic class of HGFs includes multiple interleukins and leukemia inhibitory factor (LIF; see Table 1). In the absence of HGFs from either class I or class II, these cytokines have no effect on the proliferation of hematopoietic progenitor cells, but, as single cytokines, can affect the viability of progenitors (3,4,23–33). The last group of HGFs capable of synergistically enhancing the proliferation of hematopoietic progenitor cells, includes transforming growth factor-β (TGF-β), tumor necrosis factor-α (TNF-α), interferon gamma (IFN-γ), macrophage inflammatory protein-1α (MIP-1α), and IL-4. These cytokines are all bifunctional regulators of hematopoiesis, in that they can directly inhibit or stimulate the growth of hematopoietic progenitor cells, depending on the specific growth factors with which they interact, the maturity of the targeted progenitor, and the active concentration of the cytokine in culture (34–42). In addition, several other cytokines affect hematopoiesis, including hepatocyte growth factor (HGF) and basic fibroblast growth factor (bFGF) (43–44).

The list of cytokines isolated and potentially capable of stimulating (or inhibiting) in vitro hematopoiesis is rapidly expanding, and it remains a challenge to determine their role (or lack thereof) in the complex network of cytokines regulating hematopoiesis by direct and indirect mechanisms.

One of the recently cloned cytokines is IL-12, which originally was identified by its ability to stimulate NK/lymphokine-activated killer (LAK) cells. The striking homology between IL-12, IL-6, and G-CSF, as well as the IL-6 receptor, suggested that IL-12 might also have activities on hematopoietic progenitor cells similar to the well-characterized hematopoietic effects of IL-6 and G-CSF (4,5,26,27). In fact, several recent studies on primitive murine bone marrow progenitor cells confirm the ability of IL-12 to stimulate the growth of primitive hematopoietic

Table 1 Hematopoietic Growth Factors (HGFs)

Class	Members
I. Colony-stimulating HGFs	G-CSF, CSF-1, GM-CSF, IL-3, IL-5, EPO, TPO
II. Stem cell HGFs	SCF, FL
III. Purely synergistic HGFs	IL-1, IL-6, IL-7, IL-9, IL-10, IL-11, IL-12, LIF
IV. Bifunctional HGFs	TNF-α, MIP-1α, TGF-β, IFN-γ, IL-4

This table is based on the pattern by which different cytokines affect the proliferation of primitive hematopoietic progenitor cells. Stimulation of proliferation of primitive hematopoietic progenitors can occur by the combined stimulation of (at least) two class I cytokines, two class II cytokines, or one class I and one class II cytokine. In contrast, growth stimulation of primitive progenitors by class III and class IV HGFs can be observed only when combined with one (or more) of the class I or class II cytokines.

progenitor cells (45–49). These studies, which we will summarize and discuss in this chapter clearly indicate that IL-12 belongs to the class of purely synergistic HGFs (see Table 1).

II. STRUCTURE OF IL-12 AND ITS RECEPTOR

Interleukin-12 is a heterodimer consisting of two glycoproteins, with relative molecular masses of 35 (p35) and 40 kDa (p40), which are linked by a disulfide bond (50–53). Interestingly, the two subunits are encoded by unrelated genes that are located on different chromosomes in humans (50,51,54). The IL-12 heterodimer has features of a cytokine (p35) coupled to a soluble cytokine receptor (p40) (50,51,55). Thus, the p35 subunit shows homology to the IL-6/G-CSF family of α-helical cytokines (56). In contrast, the p40 subunit is a member of the hemato-poietin receptor superfamily, with a WSXWS motif as well as four conserved cystein residues (50,51,55). It is most closely related to the IL-6 receptor and the receptor for ciliary neurotrophic factor and, similar to the IL-6 receptor, has an NH_2-terminal immunoglobulin-like domain.

Many cytokine receptors exist in both membrane-bound and soluble forms. The soluble cytokine receptors can be generated by alternative splicing of mRNA or by proteolytic cleavage (56). However, a cell membrane form of p40 has not yet been described (51,57). Complexes of soluble cytokine receptors and their respective cytokine are usually not able to activate cells, and soluble cytokine receptors, therefore, will often function as cytokine inhibitors, at least in vitro. However, IL-6 and truncated, soluble IL-6 receptor can signal on cells that express gp130, the signal transduces of the IL-6 receptor complex (56) (Fig. 1). This parallels the situation with IL-12, if one assumes that p40 evolutionally represents a soluble cytokine receptor.

Transfection studies with cDNAs encoding the p35 and the p40 subunits, have demonstrated that both subunits are required to obtain biological activity (52,58). It has also recently been proposed that p40 alone can inhibit the effects of IL-12 (59). Comparison of the murine and human genes encoding the two IL-12 subunits has shown that the genes encoding p40 and p35

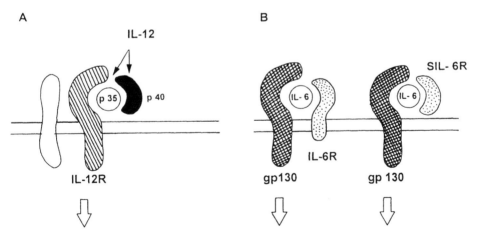

Figure 1 Structure of IL-12 and IL-6 receptors. IL-12 is a heterodimer with two covalently linked subunits, p35 and p40. p35 is distantly related to IL-6 and G-CSF, whereas p40 is a member of the hematopoietin receptor superfamily and is most closely related to the IL-6 receptor. gp130 has been identified as the signal transducer of several cytokine receptors, including IL-6R. IL-6 bound to the naturally occurring transmembrane form (B) or to an in vitro truncated form of the IL-6R (B), can signal on cells expressing gp130 (reviewed in Ref. 56). Although gp130 does not appear to be part of the IL-12 receptor complex, the recently cloned IL-12R shows homology to gp130 and is also a member of the hematopoietin receptor superfamily (63). It is possible that still another, yet unidentified, molecule (A) is involved in the generation of high-affinity IL-12 receptors.

show 70 and 67% overall homology at the amino acid level, respectively (51,60). Accordingly, murine IL-12 acts on both murine and human cells, whereas human IL-12 has no effect on murine cells (50,51,60). This has been ascribed to properties of the p35 subunit (51,60).

IL-12 binds to cell surface receptors that are expressed on resting NK cells and on activated NK cells and activated T cells (50,51,61,62), although no detectable binding was observed on B cells and monocytes (50). The biological effects of IL-12 on hematopoiesis (see later discussion) suggest that IL-12 receptors are also expressed on hematopoietic progenitor cells, although no receptor studies have yet confirmed this. Similar to many cytokine receptors, high-, intermediate-, and low-affinity IL-12 receptors have been described (reviewed in Ref. 50). Cross-linking studies have suggested a molecular mass of 110 kDa for the IL-12 receptor (61), and an associated protein of 85 kDa (61). Recently, the gene encoding one component of the IL-12 receptor has been cloned and encodes a transmembrane protein belonging to the hematopoietin receptor superfamily, with closest homology to the gp130 signal transducer (63). It appears that this receptor binds IL-12 with low affinity (50,63), and it is possible that yet another molecule is necessary for the creation of high-affinity receptors, analogous to several other cytokine receptor complexes (56; see Fig. 1).

III. IL-12 EXPRESSION AND PRODUCTION

Interleukin-12 is primarily produced by B cells and macrophages (51,53,54,64). Most Epstein-Barr virus (EBV) transformed B cell lines, as well as normal B cells, produce IL-12 (50,51). Activated monocytes represent a major source of IL-12, and they can be induced to secrete IL-12 on stimulation with IFN-γ or with bacteria/bacterial products (51,65,66). In contrast, IL-4 and IL-10 inhibit IL-12 secretion in human and murine monocytes (67). In bone marrow, granulocytes and possibly immature myeloid cells can produce IL-12, in addition to monocytes–macrophages and B cells (64).

Interestingly, the genes encoding the two IL-12 subunits are differently regulated. Thus, p35 is constitutively expressed in many tissues (51). In contrast, p40 mRNA is primarily found in lymphoid tissues, and can be subjected to regulation (50,51,60). Therefore, it appears that several tissues express p35 but not p40. In addition, cell types coexpressing both IL-12 subunits seem to produce higher levels of p40 than of p35 (51), and may secrete free p40 in addition to IL-12 heterodimers (51). This may represent a level of control of IL-12 action, as free p40 has been suggested to inhibit IL-12 binding to its receptor (59).

IV. EFFECTS OF IL-12 ON LYMPHOCYTES

Interleukin-12 plays an important role in the regulation of immune and inflammatory responses. It was originally described as a natural killer (NK) cell-stimulating factor (NKSF) and as a cytotoxic lymphocyte maturation factor (CLMF), and has pleiotropic effects on NK cells and T cells (50–53,68). Thus, it has been demonstrated to stimulate the proliferation and to induce cytotoxicity of NK cells and T cells (50–53,68–72). IL-12, therefore, serves as an important link between the nonspecific (innate) and the specific immune system. It can modulate the effects of IL-12 on proliferation and cytotoxic activity of T cells and NK cells, the net effects being dependent on the concentrations used and the cell type studied (50,51). Recently, IL-12 has been suggested to modulate T-cell maturation of thymocytes (73). Importantly, IL-12 also enhances the expansion of mature human $T_{H}1$ cells in vitro and drives T cells toward $T_{H}1$ development, thereby opposing the effects of IL-4, which stimulates the production of $T_{H}2$ cells (reviewed in Refs. 74,75). In vivo, the ability of various pathogens, such as gram-negative bacteria, *Staphy*-

lococcus aureus, Toxoplasma gondii, Mycobacterium tuberculosis, Listeria monocytogens, and *Leishmania major* to induce a T_H1 response correlates with their capacity to induce the production of IL-12 (75).

A central effect of IL-12 is its ability to induce secretion of cytokines (50–51). Thus, IL-12 potently induces cytokine production in T cells and NK cells at very low (picomolar) concentrations (50,51). In particular, IL-12 is a potent inducer of IFN-γ from both T cells and NK cells (50–52,76–79). It can also stimulate NK cells to produce TNF-α and GM-CSF, but less efficiently (50). IL-12 synergizes with IL-2 in stimulating IFN-γ production (52,77). In addition, IL-12 can promote increased expression of cytokine receptors and certain adhesion molecules, such as CD2, CD16, CD54, and CD56, on NK cells (50,51,76,80,81). IL-12 inhibits IL-4–induced IgE synthesis by IFN-γ–dependent and –independent mechanisms (82). Whether the effects on isotype selection can be partly mediated by direct effects of IL-12 on B cells, however, is unclear.

IL-12 seems to be an important mediator of the early immune response to intracellular pathogens, and it has been demonstrated to possess antimicrobial and antifungal activity in murine models after in vivo administration (50,51,83–87). Moreover, IL-12 has potent antitumor effects against different tumors in mice (88). Neutralizing antibodies against IFN-γ abrogate the antitumor effect in a murine tumor model system, suggesting an important role of IFN-γ in the antitumor effects of IL-12 (88). However, it has been observed that in vivo effects of IL-12 may be strongly dependent on the concentrations used and that high concentrations of IL-12 may have net inhibitory effects on the function and number of immune cells (50,51).

V. EFFECTS OF IL-12 ON HEMATOPOIESIS IN VITRO

In the following, we will review the effects of IL-12 on hematopoiesis in vitro. The studies are on murine cells, unless otherwise specified. Three groups have recently investigated the potential role of IL-12 in in vitro murine hematopoiesis (45–49). In a study from our laboratory, purified recombinant murine (rMu) IL-12 had no ability alone to stimulate colony formation of murine bone marrow progenitors present in unfractionated bone marrow, lineage-depleted (Lin⁻) bone marrow, or a population of purified primitive Lin⁻Sca-1⁺ progenitors (45). This was confirmed by other studies that used enriched populations of primitive progenitors (46–49).

Because IL-12 shows homology to IL-6 and G-CSF which predominantly act as purely synergistic HGFs on primitive murine progenitor cells (reviewed in Refs. 2–4), it was natural to next investigate the effects of rMuIL-12 in combination with other cytokines on the proliferation of the primitive murine Lin⁻Sca-1⁺ bone marrow progenitor cells. This progenitor cell population represents only 0.05–0.1% of the unfractionated bone marrow, but is highly enriched in pluripotent hematopoietic progenitor cells and is able to proliferate in vitro in response to defined growth factor combinations (19,89,90). Studies on this stem cell population demonstrated the potent ability of IL-12 to synergize with CSFs (all but GM-CSF) as well as SCF to enhance colony formation (45). IL-12 synergized with G-CSF, but this combination was not a strong proliferative stimulus for Lin⁻Sca-1⁺ progenitor cells (45). In contrast, the combined stimulation of CSF-1 plus IL-12, and in particular IL-3 plus IL-12 resulted in a potent enhancement of colony formation of Lin⁻Sca-1⁺ progenitor cells (45). In addition to increasing the number of CSF-responsive Lin⁻Sca-1⁺ cells, IL-12 dramatically increased the size of the colonies formed (45), suggesting that IL-12 acts to recruit additional progenitors to proliferate, as well as to synergistically enhance the proliferative response of individual progenitors.

Stem cell factor, the ligand for the c-*kit* tyrosine kinase receptor encoded at the steel (*Sl*) locus in mice, is a key regulator of hematopoiesis (9–12). Mutations in the *Sl* locus results in defects in hematopoietic stem cell development, as well as deficiencies in pigmentation and fertility (91–93). Since the cloning of SCF in 1990 (9–11), numerous studies have demonstrated that SCF, in combination with other cytokines, can potently enhance the in vitro and in vivo proliferation of murine and human hematopoietic progenitor cells (reviewed in Refs. 9–11), 16–22). Multiple studies demonstrate that IL-12 also falls into the large group of cytokines capable of synergizing with SCF to enhance the proliferation of primitive murine hematopoietic progenitor cells (45–49). In the studies from our laboratory (45), IL-12 enhanced SCF-induced proliferation of Lin$^-$Sca-1$^+$ progenitors in a concentration-dependent fashion (maximum stimulation at 20–200 ng/ml), resulting in an eightfold increase in the number of proliferating progenitors. This effect was specifically mediated by IL-12, as a neutralizing anti–IL-12 antibody completely abrogated the effect of purified IL-12 (45). As in the synergy with the CSFs, IL-12 also potently enhanced the size of the clones formed by SCF-stimulated Lin$^-$Sca-1$^+$ progenitors. Accordingly, the total cell production in response to SCF plus IL-12 was 25-fold higher than in the presence of SCF alone (45).

High-proliferative potential colony-forming cells (HPP-CFCs), which represent some of the most primitive progenitors capable of growing in vitro and which are highly enriched in the Lin$^-$Sca-1$^+$ bone marrow cells (94,95), were also stimulated by IL-12 when combined with CSFs, IL-1, IL-6, or SCF (45).

In agreement with the studies from our laboratory, Hirayama et al. found that IL-12 enhanced the number of SCF-responsive Lin$^-$Sca-1$^+$ bone marrow progenitors from normal as well as of 5-fluorouracil (5-FU)-treated mice (48). However, IL-12 enhanced the number of SCF-responsive Lin$^-$Sca-1$^+$ progenitors only twofold, as compared with the eightfold enhancement observed by us (45). Although, the reason for this difference in magnitude of synergy between SCF and IL-12 remains uncertain, a likely explanation is that different concentrations of IL-12 were used in the two studies. Specifically, our studies demonstrated that optimal synergy with SCF was observed at 20–200 ng/ml rMuIL-12 (45). The synergistic effect of IL-12 dropped off rapidly below 20 ng/ml, and at 2 ng/ml only a twofold increase in colony formation was observed. Later experiments with different lots of rMuIL-12 have confirmed that optimal enhancement of SCF-stimulated proliferation requires approximately 50 ng/ml of IL-12 (Jacobsen SEW, unpublished observations). Thus, Hirayama et al., who did not report performing a concentration–response and who consistently used rMuIL-12 at 10 ng/ml might, in fact, have used suboptimal concentrations of IL-12 (48). Alternatively, or additionally, the difference between the two studies could be based on the much higher cloning frequency of the Lin$^-$Sca-1$^+$ progenitors in response to SCF alone in the studies of Hirayama et al. than in the studies from our laboratory (45,48).

The requirement for high concentrations of IL-12 in vitro for optimal synergy with SCF (or IL-3) alone is not unique, because other synergistic cytokines, such as IL-6 and IL-11, are also required at very high concentrations for optimal synergism with SCF. This, however, does not necessarily imply that lower concentrations of IL-12 or other HGFs are not active on hematopoietic progenitor cells. In fact, recent studies suggest that combining multiple cytokines with SCF can lower the threshold by which these cytokines elicit their synergistic effects (96). That this also holds true for IL-12 was recently shown by Ploemacher et al. (46), who, in agreement with us, demonstrated that only high concentrations of rMuIL-12 (> 30 ng/ml) could enhance the proliferation of primitive murine progenitors when acting in combination with only one cytokine (IL-3). In contrast, a 100-fold lower concentration (0.3 ng/ml) of IL-12 synergistically enhanced the colony formation observed in response to IL-3 plus SCF (46), suggesting that IL-12 might stimulate hematopoiesis at probably more "physiological" concentrations.

The complexity of assessing the interactions of cytokines on the proliferation of primitive hematopoietic progenitor cells is also underscored by studies demonstrating that hematopoietic progenitor cells can produce potent inhibitory cytokines, such as TGF-β, even in the absence of cytokine stimulation (97). Although this might represent an important mechanism for down-regulating hematopoiesis to keep primitive progenitors out of cycle, it could also potentially mask the actions of stimulatory cytokines. Thus, the lack of synergy between SCF and IL-12, reported in initial studies by Ploemacher et al. (46), was later explained by the target population of progenitor cells being able to produce TGF-β by an autocrine mechanism (49). Accordingly, in the presence of a neutralizing TGF-β antibody, potent synergy was observed between SCF and IL-12 (49).

IL-12 stimulates NK cells to production of other cytokines (76–79), which might indirectly mediate the effects of IL-12 on hematopoietic progenitor cells. However, single-cell studies from our laboratory as well as in the studies of Hirayama et al. clearly demonstrated that the stimulatory effect of IL-12 was directly mediated on the proliferating progenitors (45,48). Although it still remains possible that IL-12 can act through an autocrine loop by stimulating the progenitors to HGF production, which then stimulates proliferation, these data provide evidence that primitive progenitors express functional IL-12 receptors.

Collectively, the studies described in the foregoing demonstrate that IL-12 can potently stimulate the proliferation of primitive murine hematopoietic progenitor cells in combination with CSFs or SCF. Interestingly, we have also recently observed that IL-12 potently and directly synergizes with FL to enhance the proliferation of Lin⁻Sca-1⁺ progenitor cells (98).

Several studies have also examined the ability of IL-12 to affect the differentiation of the progeny of primitive hematopoietic progenitors (45–49). The studies of Hirayama et al, and Ploemacher et al. demonstrated that IL-12 stimulated the formation of multilineage myeloid colonies as well as granulocyte–macrophage (GM) colonies (46–48). Furthermore, although IL-12 potently enhanced SCF-stimulated proliferation of Lin⁻Sca-1⁺ progenitors, it did not affect the relative production of mature myeloid cells (mainly granulocytes and some macrophages; 45). This was observed in bulk liquid culture, but also confirmed by picking colonies from Lin⁻Sca-1⁺ cells cultured individually (45). Whereas these studies also demonstrated that IL-12 stimulated the production of megakaryocytes and erythroid cells from primitive progenitors, it is unclear whether IL-12 can stimulate the growth of pure megakaryocyte and erythroid progenitor cells. It was not surprising that IL-12 exclusively stimulated the formation of myeloid progeny, as all synergistic effects of IL-12 were observed with myeloid growth factors, whereas IL-12 in combination with "lymphoid" cytokines, such as IL-2 and IL-7, had no proliferative effect on these primitive progenitors (45). In fact, until recently it has been a major limitation of available in vitro assays for primitive hematopoietic progenitor cells that they support only the outgrowth of mature myeloid cells and not lymphopoiesis, even though the progenitors in vivo have been demonstrated to also have lymphoid potential (89,90). Although the reason for this remains uncertain, it seems likely that the outgrowth of mature lymphoid cells from primitive progenitors require additional as yet unidentified cytokines or cofactors, or cell-to-cell interactions. In fact, recent studies have demonstrated that combined myeloid and lymphoid development is possible when primitive progenitors are cocultured on murine stromal cell layers. In addition, the two-step clonal culture system recently developed by Hirayama et al., allows the in vitro assessment of the lymphoid potential of these primitive progenitors in the absence of stromal cells (99). This is possible by plating the primitive progenitors in the presence of SCF plus IL-7, as well as either IL-6, G-CSF, IL-11, or LIF (99). If the progenitors are stimulated continuously by these cytokine combinations, they will ultimately form only myeloid cells (99). However, if replaced by day 12, and stimulated by only SCF plus IL-7, a high proportion of the progeny demonstrate an ability to form pre-B-cell colonies (99). Such studies demonstrated that

IL-12 also falls into the group of early-acting cytokines capable of synergizing with SCF to stimulate the growth of primitive progenitors with combined lymphoid and multilineage myeloid potential (48).

In our laboratory we have also tried to determine whether IL-12 can affect the proliferation of more committed GM progenitors. The initial experiments demonstrated that IL-12 could enhance the colony formation of unfractionated Lin⁻ bone marrow cells observed in response to SCF, CSF-1, GM-CSF, and G-CSF, but not IL-3 (45). The interpretation of these experiments did however have limitations, because although studying the effects of cytokines on the whole population of Lin⁻ progenitors is usually reflective of the dominating Lin⁻Sca-1⁻ population, it might, as for SCF-responsive progenitors, be predominantly an effect on the low-frequency Lin⁻Sca-1⁺ progenitors (19). Thus, we have recently compared the effects of IL-12 alone and in combination with SCF or IL-3 on the growth of Lin⁻Sca-1⁻ and Lin⁻Sca-1⁺ progenitors cultured individually (Table 2). These experiments demonstrated, in agreement with others, that almost all SCF-responsive Lin⁻ progenitors are Lin⁻Sca-1⁺ (19). Specifically, the frequency of SCF plus IL-12-responsive progenitors was 20-fold higher in the Lin⁻Sca-1⁺ than in the Lin⁻Sca-1⁻ population (see Table 2). In addition, the size of the colonies formed in response to SCF plus IL-12 was much larger in the Lin⁻Sca-1⁺ than the Lin⁻Sca-1⁻ progenitor cell compartment (data not shown). In comparison, IL-3-responsive progenitors were present at similar frequencies in the two population, but the synergy between IL-3 and IL-12 was observed exclusively on Lin⁻Sca-1⁺ progenitors (see Table 2). However, in other experiments we have observed that IL-12 can synergize with G-CSF or CSF-1 to enhance the proliferation of Lin⁻Sca-1⁻ as well as Lin⁻Sca-1⁺ progenitors (Jacobsen SEW, unpublished observations). Thus, although the predom-

Table 2 Effects of IL-12 on Colony Formation of Lin⁻Sca-1⁻ and Lin⁻Sca-1⁺ Bone Marrow Progenitor Cells

	Colonies/300 cells			
	Lin⁻Sca-1⁻		Lin⁻Sca-1⁺	
Cytokine	−IL-12	+IL-12	−IL-12	+IL-12
SCF	1(1)	3(2)	9(3)	64(12)
IL-3	17(3)	17(4)	23(6)	42(7)

Lin⁻Sca-1⁻ and Lin⁻Sca-1⁺ bone marrow cells were separated as previously described (45), and plated in microtiter plates at a concentration of one cell per well in 20 μl IMDM (Gibco, Paisly UK) supplemented with 20% FCS (Sera Lab, Sussex, UK), L-glutamine, and antibiotics. Purified recombinant (r) cytokines were added at predetermined optimal concentrations; rrSCF 100 ng/ml (Amgen Corp., Thousand Oaks CA), rMuIL-3 20 ng/ml (Peprotech Inc., Rocky Hill NJ), and rMuIL-12 50 ng/ml (Genetics Institute, Cambridge MA). Wells were scored for cell growth (> 10 cells) after 14 days incubation at 37°C, 5% CO_2 in air. Three hundred wells were plated per group in each experiment. Results represent the mean (SEM) of four separate experiments.

inant synergistic effects of IL-12 appear to be on the most primitive lymphohematopoietic progenitors, it can also directly synergize with other HGFs to enhance the proliferation of more committed myeloid progenitors cells. Figure 2 summarizes the levels at which IL-12 has been demonstrated to affect proliferation in the hematopoietic hierarchy.

Both TNF-α and TGF-β represent bifunctional regulators of hematopoiesis, with primarily inhibitory effects on the growth of hematopoietic progenitor cells in vitro (34–38). We have recently investigated the ability of these two cytokines to modulate the stimulatory effects of IL-12 on hematopoiesis, and found that both TGF-β and TNF-α were potent inhibitors of the proliferation of Lin⁻Sca-1⁺ progenitor cells in response to IL-12 plus SCF and IL-12 plus IL-3 (Jacobsen SEW, unpublished observation).

Work in our laboratory suggest that IL-12, in combination with SCF and IL-3, can synergistically enhance the in vitro proliferation of human CD34⁺ bone marrow progenitor cells (Jacobsen SEW, et al., unpublished observation). The magnitude of synergy, however, is much less pronounced than reported for primitive murine progenitor cells. This could be because CD34⁺ progenitors consist predominantly of more committed progenitors; thus, they are much less prone to synergy. Also recent studies by Bellone and Trichieri, suggest that IL-12 can stimulate cytokine-stimulated growth of human bone marrow progenitor cells in combination with SCF and IL-3 (100).

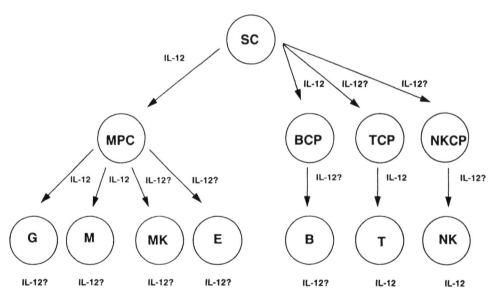

Figure 2 Effects of IL-12 on lymphohematopoietic cells in vitro. The figure is based on characterized effects of IL-12 on mature T and NK cells as well as synergistic effects on development of myeloid and B-lymphoid cells from stem cells. Effects on committed granulocyte–macrophage progenitors are less pronounced than on primitive stem cells. Although increases in colonies containing erythroid and mega-karyocytic cells have been shown, so far no direct effect of IL-12 on committed erythroid and mega-karyocytic progenitor cells has been shown. Effects on isotype selection of mature B cells may be indirectly mediated; Question mark represents possible target cells for IL-12, for which a direct or definite effect of IL-12 has not been reported. SC, stem cell; MPC, myeloid progenitor cell; BCP, B-cell progenitor; TPC, T-cell progenitor; NKPC, natural killer cell progenitor; G, granulocyte; M, macrophage; MK, megakaryo-cyte; E, erythrocytes; B, B cells; T, T cells; NK, natural killer cells.

VI. HEMATOPOIETIC EFFECTS OF IL-12 IN VIVO: POTENTIAL PHYSIOLOGICAL AND CLINICAL IMPLICATIONS

Recent studies by Gately et al. (50,101) have demonstrated that IL-12 treatment of mice also affects myelopoiesis in vivo, in that it results in enhanced extramedullary hematopoiesis in the spleen and liver, and a concomitant suppression of hematopoiesis in the bone marrow (50,101). Quantitation of the number of CFU-GM and BFU-E in the spleen and bone marrow following daily injections (IP) of 1 μg IL-12 resulted in a time-dependent dramatic increase in these progenitors in the spleen, whereas in the bone marrow there was a potent reduction of the both CFU-GM and BFU-E (50). However, whether these observations were due to a true expansion of already existing hematopoietic progenitor cells in the spleen or to a mobilization of progenitors from the bone marrow to extramedullary sites, such as the spleen, was not investigated. In addition, these studies did not explore the effects of IL-12 on more primitive progenitors (such as CFU-GEMM and HPP-CFC) which IL-12 has been demonstrated to affect primarily in vitro, nor the effects of IL-12 on the number of progenitor cells in peripheral blood. Thus, further studies need to be undertaken to determine the hematopoietic effects of IL-12 in vivo. However, it is clear from these studies that mice given 1 μg/day of IL-12 develop lymphopenia and neutropenia by 2 days (101). The reason for this might be the potent ability of IL-12 in vitro as well as in vivo to stimulate the production of other cytokines, such as IFN-γ or TNF-α (50–51,76–79), that are primarily inhibitors of in vivo hematopoiesis. Thus, these studies suggest that the overriding effects of IL-12 in vivo on hematopoiesis might be inhibitory (indirect), rather than stimulatory (direct). Figure 3 summarizes how IL-12 might affect hematopoiesis directly and indirectly.

That the ability of IL-12 to directly enhance in vitro hematopoiesis is purely synergistic (45–49), suggests that stimulatory effects of IL-12 on hematopoiesis in vivo might be more pronounced if IL-12 is administered in conjunction with one or more other HGFs, such as SCF or a CSF. In addition, such a combined administration might result in the requirement for much lower doses of IL-12 and, thereby, potentially reduce the indirect inhibitory effects of IL-12. Although this might appear speculative and remains to be demonstrated, IL-12 (46) as well as other HGFs are required at much lower concentrations in vitro to exert their synergistic effects on hematopoiesis when combined with other HGFs (96). Also, the stimulatory effects of SCF on in vivo hematopoiesis can be obtained at much lower doses when combined with G-CSF (102). Consequently, the serious toxicities of SCF are reduced potentially, allowing the clinical use of an agent that otherwise might have been excluded from clinical application.

In the last few years, it has become clear that transplantation of mobilized progenitors in peripheral blood might reduce the costs and side effects of transplantations (103,104). The finding that IL-12 might mobilize bone marrow progenitors from the bone marrow to peripheral sites could, therefore, be clinically important. In addition, IL-12 might also become important as a factor capable of in vitro expansion of hematopoietic progenitor cells.

The physiological role of IL-12 as well as numerous other HGFs in hematopoiesis remains to be determined. This can be done only by the deletion or inactivation of the IL-12 gene or, alternatively, by the use of neutralizing antibodies that will efficiently neutralize IL-12. Given the high degree of overlapping activities between IL-12 and other regulators of early hematopoiesis, such as IL-6, IL-11, and G-CSF (2–4), our prediction would be that a knockout of the IL-12 gene would have little or no effect on myelopoiesis. However, as suggested by others (4), such knockout studies are not always conclusive or simple to interpret. First, the lack of effect with deletion of a gene in normal healthy animals, does not exclude the possibility that a deletion

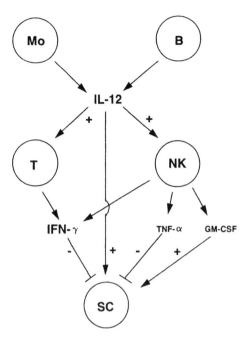

Figure 3 Effects of IL-12 on hematopoiesis: IL-12 generally stimulates T cells and NK cells in vivo, and induces cytokine production, especially of IFN-γ. In addition other cytokines with potential effects on hematopoiesis, such as TNF-α and GM-CSF, are also likely to be produced. IL-12 has been demonstrated to have a direct synergistic stimulatory effects on hematopoietic progenitor cells. In contrast, IFN-γ and TNF-α are bifunctional regulators of hematopoiesis, but with primarily inhibitory effects. Thus, effects of IL-12 on hematopoiesis in vivo might represent a combination of direct stimulatory and indirect inhibitory effects. Mo, monocytes; B, B cells; T, T cells; NK, natural killer cells; SC, stem cells.

can become critical under different circumstances, such as myelosuppression. Second, the possibility of activating alternative pathways to compensate for a deficiency, suggests that the timing for studying the effects of knockouts can be critical.

In conclusion, IL-12 is a potent stimulator of the in vitro proliferation of primitive lymphohematopoietic progenitor cells. Studies of the in vivo effects of IL-12 on hematopoiesis suggest that IL-12 might mobilize progenitor cells from the bone marrow to extramedullary sites. In addition, the development of neutropenia and anemia in IL-12-treated mice implies that the net hematopoietic effects of IL-12 in vivo might be inhibitory, potentially through its ability to stimulate the production of inhibitory cytokines. However, further studies are required to elucidate the physiological role and potential clinical applications of IL-12 in hematopoiesis.

NOTE ADDED IN PROOF

Since this review was completed during the fall of 1994, many new studies have been published, further investigating the potential role of IL-12 in hematopoiesis. Of particular interest are studies performed by Eng et al. in IFN-γ receptor-deficient mice, which demonstrated that the in vivo suppressive effects of IL-12 in wild type mice is a consequence of enhanced IFN-γ production, and that in the absence of IFN-γ production the effect of IL-12 on hematopoiesis is rather stimulatory (105). Other studies have demonstrated that IL-12 administration protects wild type mice from lethal bone marrow toxicity induced by irradiation, and that this is likely to be a result of enhanced hematopoiesis (106). Recent studies in IL-12–deficient mice revealed deficiencies in IFN-γ production and certain Th1 responses, but the mice displayed no defects in any organs or abnormal hematological parameters, suggesting that IL-12 does not play any critical role in steady-state hematopoiesis (107).

Progress has also been made in the understanding of signal transduction through the IL-12 receptor. Binding of IL-12 to its receptor induces tyrosine phosphorylation of a range of intracellular substrates (108,109). Moreover, like many other cytokines, IL-12 seems to activate the JAK-STAT pathway. Specifically, JAK2 and TYK2 are phosphorylated in response to IL-12 in human T and NK cells (108). Bacon and coworkers recently demonstrated that IL-12 activates STAT4, which is a member of the STAT family of transcription factors involved in the signal transduction cascades induced by many cytokines known to activate JAK kinases (110).

ACKNOWLEDGMENTS

We thank Drs. Makio Ogawa, Giorgio Trinchieri, Maurice Gately, and Leiv Rusten for important input to this review.

REFERENCES

1. Sprangrude GJ, Smith L, Uchida N, Ikuta K, Heimfeld S, Friedman J, Weissman IL. Mouse hematopoietic stem cells. Blood 1991; 78:1395–1402.
2. Moore M. Clinical implications of positive and negative hematopoietic stem cell regulators. Blood 1991; 78:1–19.
3. Ogawa M. Differentiation and proliferation of hematopoietic stem cells. Blood 1993; 81:2844–2853.
4. Metcalf D. Hematopoietic regulators: redundancy or subtlety? Blood 1993; 82:3515–3523.
5. Metcalf D. The Hematopoietic Colony Stimulating Factors. Amsterdam: Elsevier, 1984.
6. Kaushansky K. Thrombopoietin: The primary regulator of platelet production. Blood 85:419–431.
7. Spivak JC. The mechanism of action of erythropoietin. Int J Cell Cloning 1986; 4:139–146.
8. Sanderson CJ. Interleukin-5, eosinphils, and disease. Blood 1992; 79:3101–3109.
9. Williams DE, Eisenman J, Baird A, Rauch C, Van Ness K, March CJ, Park LS, Martin U, Mochizuki DY, Boswell HS, Burgess GS, Cosman D, Lyman SD. Identification of a ligand for the c-*kit* proto-oncogene. Cell 1990; 63:167–174.
10. Zsebo KM, Wypych J, McNiece IK, Lu HS, Smith KA, Karkare SB, Sachdev RK, Yuschenkoff VN, Birkett NC, Williams LR, Satyagal VN, Tung W, Bosselman RA, Mendiaz EA, Langley KE. Identification, purification, and biological characterization of hematopoietic stem cell factor from buffalo rat liver-conditioned medium. Cell 1990; 63:195–201.
11. Huang E, Nocka K, Beier D, Chu T-U, Buch J, Lahm H-W, Wellner D, Leder P, Besmer P. The hematopoietic growth factor KL is encoded by the *Sl* locus and is the ligand of the c-*kit* receptor, the gene product of the *W* locus. Cell 1990; 63:225–233.
12. Galli SJ, Zsebo KM, Geissler EN. The kit ligand, stem cell factor. Adv Immunol 1994; 55:1–96.
13. Lyman SD, James L, Bos TV, de Vries P, Brasel K, Gliniak B, Hollingsworth LT, Picha KS, McKenna HJ, Splett RR, Fletcher FA, Maraskovsky E, Farrah T, Foxworthe D, Williams DE, Beckmann MP. Molecular cloning of a ligand for the FLT3/FLK-2 tyrosine kinase receptor: a proliferative factor for primitive hematopoietic cells. Cell 1993; 75:1157–1167.
14. Hannum C, Culpepper J, Campbell D, McClanahan T, Zurawski S, Bazan JF, Kastelein R, Hudak S, Wagner J, Mattson J, Luh J, Duda G, Martina N, Peterson D, Menin S, Shanafeld A, Muench M, Kelner G, Namikawa R, Rennik D, Roncarplo M-G, Zlotnic A, Rosnet O, Dubreuil P, Birnbaum D,

Lee F. Ligand for FLT3/FLK2 receptor tyrosine kinase regulates growth of hematopoietic stem cells and is encoded by variant RNAs. Nature 1994; 368:643–648.

15. Small D, Levenstein M, Kim E, Carow C, Amin S, Rockwell P, Witte L, Burrow C, Ratajczak MZ, Gewirtz AM, Civin CI. STK-1 the human homolog of FLK2/FLT3, is selectively expressed in CD34[+] human bone marrow cells and is involved in the proliferation of early progenitor/stem cells. Proc Natl Acad Sci USA 1994; 91:459–463.

16. de Vries P, Brasel KA, Eisenman JR, Alpert AR, Williams DE. The effect of recombinant mast cell growth factor on purified murine hematopoietic stem cells. J Exp Med 1991; 173:1205–1211.

17. Metcalf D, Nicola NA. Direct proliferative actions of stem cell factor on murine bone marrow cells in vitro: effects of combination with colony-stimulating factors. Proc Natl Acad Sci USA 1991; 88:6239–6243.

18. Broxmeyer HE, Hangoc G, Cooper S, Anderson D, Cosman D, Lyman SD, Williams DE. Influence of murine mast cell growth factor (c-kit ligand) on colony formation by marrow hematopoietic progenitor cells. Exp Hematol 1991; 19:143–146.

19. Williams N, Bertoncello I, Kavnoudias H, Zsebo K, McNiece I. Recombinant rat stem cell factor stimulates the amplification and differentiation of fractionated mouse stem cell populations. Blood 1992; 79:58–64.

20. Tsuji K, Zsebo KM, Ogawa M. Enhancement of murine blast cell colony formation in culture by recombinant rat stem cell factor, ligand for c-*kit*. Blood 1991; 78:1223–1229.

21. Carow C, Hanoc G, Cooper SH, Williams DE, Broxmeyer HE. Mast cell growth factor (c-*kit* ligand) supports the growth of human multipotential progenitor cells with a high replating potential. Blood 1991; 78:2216–2221.

22. McNiece IK, Langley KE, Zsebo KM. Recombinant human stem cell factor synergises with GM-CSF, G-CSF, IL-3 and Epo to stimulate human progenitor cells of the myeloid and erythroid lineages. Exp Hematol 1991; 19:226–231.

23. Leary AG, Wong GG, Clark SC, Smith AG, Ogawa M. Leukemia inhibitory factor differentiation-inhibiting activity/human interleukin for DA cells augments proliferation of human hematopoietic stem cells. Blood 1990; 75:1960–1964.

24. Mochizuki DY, Eisenman JR, Conlon PJ, Larsen AD, Tushinski RJ. Interleukin-1 regulates hematopoietic activity, a role previously ascribed in hemopoietin 1. Proc Natl Acad Sci USA 1987; 84:5267–5271.

25. Zsebo KM, Wypych J, Yuschenkoff VN, Lu H, Hunt P, Dukes PP, Langley KE. Effects of hematopoietin-1 and interleukin 1 activities on early hematopoietic cells of the bone marrow. Blood 1988; 71:962–968.

26. Ikebuchi K, Wong GG, Clark S, Ihle JN, Hirai Y, Ogawa M. Interleukin 6 enhancement of interleukin 3-dependent proliferation of multipotential hematopoietic progenitors. Proc Natl Acad Sci USA 1987; 84:9035–9039.

27. Ikebuchi K, Ihle JN, Hirai Y, Wong GG, Clark SC, Ogawa M. Synergistic factors for stem cell proliferation: further studies of the target stem cells and the mechanism of stimulation by interleukin-1, interleukin-6, and granulocyte colony-stimulating factor. Blood 1988; 72:2007–2014.

28. Jacobsen FW, Veiby OP, Skjønsberg C, Jacobsen SEW. Novel role of interleukin 7 in myelopoiesis: stimulation of primitive murine hematopoietic progenitor cells. J Exp Med 1993; 178:1777–1782.

29. Donahue RE, Yang Y-C, Clark SC. Human p40 T-cell growth factor (interleukin-9) supports erythroid colony formation. Blood 1990; 75:2271–2275.

30. Williams DE, Morrissey PJ, Mochizuki DY, de Vries P, Anderson D, Cosman D, Boswell HS, Cooper S, Grabstein KH, Broxmeyer HE. T-cell growth factor p40 promotes the proliferation of myeloid cell lines and enhances erythroid burst formation by normal murine bone marrow cells in vitro. Blood 1990; 76:906–911.

31. Rennic D, Hunte B, Dang W, Thompson-Snipes L, Hudak S. Interleukin-10 promotes the growth of megakaryocyte, mast cell, and multilineage colonies: analysis with committed progenitors and Thy1[lo]Sca1[+] stem cells. Exp Hematol 1994; 22:136–141.

32. Musashi M, Yang Y-C, Paul SR, Clark SC, Sudo T, Ogawa M. Direct and synergistic effects of interleukin 11 on murine hematopoiesis in culture. Proc Natl Acad Sci USA 1991; 88:765–769.

33. Musashi M, Clark SC, Sudo T, Urdal DL, Ogawa M. Synergistic interactions between interleukin-11 and interleukin-4 in support of proliferation of primitive hematopoietic progenitors of mice. Blood 1991; 78:1448–1451.

34. Keller JR, Mantel C, Sing GK, Ellingsworth LR, Ruscetti SK, Ruscetti FW. Transforming growth factor β1 selectively regulates early murine hematopoietic progenitors and inhibits the growth of IL-3-dependent myeloid leukemia cell lines. J Exp Med 1988; 168:737–750.

35. Keller JR, Jacobsen SE, Sill KT, Ellingsworth LR, Ruscetti FW. Stimulation of granulopoiesis by transforming growth factor B: synergy with granulocyte/macrophage-colony-stimulating factor. Proc Natl Acad Sci USA 1991; 88:7190–7194.

36. Broxmeyer HE, Williams DE, Lu L, Cooper S, Anderson SL, Beyer GS, Hoffman R, Rubin BY. The suppressive influences of human necrosis factors on bone marrow hematopoietic progenitor cells from normal donors and patients with leukemia: synergism of tumor necrosis factor and interferon-γ. J Immunol 1986; 136:4487–4493.

37. Caux C, Sealand S, Favre C, Duvert V, Mannoni P, Banchereau J. Tumor necrosis factor-α strongly potentiates interleukin-3 and granulocyte–macrophage colony-stimulating factor-induced proliferation of human CD34+ hematopoietic progenitor cells. Blood 1990; 75:2292–2298.

38. Jacobsen SEW, Ruscetti F, Dubois CM, Keller JR. Tumor necrosis factor-α directly and indirectly regulates hematopoietic progenitor cell proliferation: role of colony-stimulating factor receptor modulation. J Exp Med 1992; 175:1759–1772.

39. Broxmeyer HE, Sherry B, Lu L, Cooper S, Carow C, Wolpe SD, Cerami A. Myelopoietic enhancing effects of murine macrophage inflammatory proteins 1 and 2 on colony formation in vitro by murine and human bone marrow granulocyte/macrophage progenitor cells. J Exp Med 1989; 170:1583–1594.

40. Graham GJ, Wright EG, Hewick R, Wolpe SD, Wilkie NM, Donaldson D, Lorimore S, Pragnell IB. Identification and characterization of an inhibitor of hematopoietic stem cell proliferation. Nature 1990; 344:442–444.

41. Peschel C, Paul WE, Ohara J, Green I. Effects of cell stimulatory factor-1/interleukin-4 on hematopoietic progenitor cells. Blood 1987; 70:254–263.

42. Rennic D, Yang G, Muller-Sieburg C, Smith C, Arai N, Takabe Y, Gemmel L. Interleukin 4 (B-cell stimulatory factor 1) can enhance or antagonize the factor-dependent growth of hematopoietic progenitor cells. Proc Natl Acad Sci USA 1987; 84:6889–6893.

43. Kmiecik TE, Keller JR, Rosen E, Vande Woude GF. Hepatocyte growth factor is a synergistic factor for the growth of hematopoietic progenitor cells. Blood 1992; 80:2454–2457.

44. Gabrilove JL, Wong G, Bollenbacher E, White K, Kojima S, Wilson L. Basic fibroblast growth factor counteracts the suppressive effect of transforming growth factor-β1 on human myeloid progenitor cells. Blood 1993; 81:909–915.

45. Jacobsen SEW, Veiby OP, Smeland EB. Cytotoxic lymphocyte maturation factor (interleukin 12) is a synergistic growth factor for hematopoietic stem cells. J Exp Med 1993; 178:413–418.

46. Proemacher RE, van Soest PL, Boudewijn A, Neben S. Interleukin-12 enhanches interleukin-3 dependent multilineage hematopoietic colony formation stimulated by interleukin-11 or steel factor. Leukemia 1993; 7:1374–1380.

47. Ploemacher RE, Soest PL, Voorwinden H, Boudewjin A. Interleukin-12 synergizes with interleukin-3 and steel factor to enhance recovery of murine hematopoietic stem cells in liquid culture. Leukemia 1993; 7:1381–1388.

48. Hirayama F, Katayama N, Neben S, Donaldson D, Nickbarg EB, Clark SC, Ogawa M. Synergistic interaction between interleukin-12 and steel factor in support of proliferation of murine lympho-hematopoietic progenitors in culture. Blood 1994; 83:92–98.

49. Ploemacher RE, van Soest PL, Boudewijn A. Autocrine transforming growth factor β1 blocks colony

formation and progenitor cell generation by hematopoietic stem cells stimulated with steel factor. Stem Cells 1993; 11:336–347.

50. Gately MK, Gubler Euli, Brunda MJ, Nadeau RR, Anderson TD, Lipman JM, Sarmiento U. Interleukin-12: a cytokine with therapeutic potential in oncology and infectious diseases. Ther Immunol 1994; 1:187–196.

51. Wolf SF, Sieburth D, Sypek J. Interleukin 12: a key modulator of immune function. Stem Cells 1994; 12:154–168.

52. Kobayashi M, Fitz L, Ryan M, Hewick RM, Clark SC, Chan S, Loudon R, Sherman F, Perussia B, Trinchieri G. Identification and purification of natural killer cell stimulatory factor (NKSF), a cytokine with multiple biological effects on human lymphocytes. J Exp Med 1989; 170:827–845.

53. Stern AS, Podlaski FJ, Hulmes JD, Pan JE, Quinn PM, Wolitzky AG, Familetti PC, Stremlo DL, Truitt T, Chizzonite R, Gately MK. Purification to homogeneity and partial characterization of cytotoxic lymphocyte maturation factor from human B-lymphocyte cells. Proc Natl Acad Sci USA 1990; 87:6808–6812.

54. Sieburth D, Jabs EW, Warrington JA, Li Y, Lasota Y, LaForgia S, Kelleher K, Wasmuth JJ, Wolf SF. Assignment of NKSF/IL-12, a unique cytokine composed of two unrelated subunits, to chromosomes 3 and 5. Genomics 1992; 14:59–62.

55. Gearing DP, Cosman D. Homology of the p40 subunit of natural killer cell stimulatory factor (NKSF) with the extracellular domain of the interleukin-6 receptor. Cell 1991; 66:9–10.

56. Kishimoto T, Taga T, Akira S. Cytokine signal transduction. Cell 1994; 76:253–262.

57. Merberg DM, Wolf SF, Clark SC. Sequence similarity between NKSF and the IL-6/G-CSF family. Immunol Today 1992; 13:77–78.

58. Gubler U, Chua AO, Schoenhaut DS, Dwyer CM, McComas W, Motyka R, Nabavi N, Wolitzky, Quinn PM, Familletti PC, Gately MK. Coexpression of two distinct genes is required to generate secreted bioactive cytotoxic lymophocyte maturation factor. Proc Natl Acad Sci USA 1991; 88:4143–4147.

59. Mattner F, Fisher S, Guckes S, Jin S, Kaulen H, Schmitt E, Rude E, Germann T. The interleukin-12 subunit p40 specifically inhibits effects of the interleukin-12 heterodimer. Eur J Immunol 1993; 23:2202–2208.

60. Schoenhaut DS, Chua AO, Wolitzky AG, Quinn PM, Dweyer CM, McComas W, Familletti FC, Gately MK, Gubler U. Cloning and expression of murine IL-12. J Immunol 1992; 148:4433–3440.

61. Chizzonite R, Truitt T, Desai BB, Nunes P, Podlaski FJ, Stern AS, Gately MK. IL-12 receptor: characterization of the receptor on phytohemagglutinin-activated human lymphoblasts. J Immunol 1992; 148:3117–3124.

62. Desai BB, Quinn PM, Wolitzky AG, Mongini PKA, Chizzonite R, Gately MK. IL-12 receptor: distribution and regulation of the receptor expression. J Immunol 1992; 148:3125–3132.

63. Chua AO, Chissonite R, Desai BB, Triutt TP, Nunes P, Minetti LI, Warrier RR, Presky DH, Levine JF, Gately MK, Gubler U. Expression cloning of a human IL-12 receptor component: a new member of the cytokine receptor superfamily with strong homology to gp130. J Immunol 1994; 153:128–136.

64. D'Andrea A, Rengarasju M, Valiante NM, Chehimi J, Kubin M, Aste M, Chan SH, Kobayashi M, Young D, Nickbarg E, Chizzonite R, Wolf SR, Trinchieri G. Production of natural killer cell stimulatory factor (interleukin 12) by peripheral blood mononuclear cells. J Exp Med 1992; 176:1387–1398.

65. Gazzinelli S, Hieny TA, Wynn SF, Wolf SF, Sher A. Interleukin-12 is required for the T-lymphocyte-independent induction of interferon by an intracellular parasite and induces resistance in T-cell deficient hosts. Proc Natl Acad Sci USA 1993; 90:6115–6119.

66. Hsieh C, Macatonia SE, Tripp CS, Wolf SF, O'Garra A, Murphy KM. Development of T_H1 CD4$^+$ T cells through IL-12 produced by *Listeria*-induced macrophages. Science 1993; 260:547–549.

67. Tripp CS, Wolf SF, Unanue ER. Interleukin 12 and tumor necrosis factor α are costimulators of interferon γ production by natural killer cells in severe combined immunodeficiency mice

with listeriosis, and interleukin 10 is a physiologic antagonist. Proc Natl Acad Sci USA 1993; 90:3725–3729.

68. Gately ME, Desay BB, Wolitzky AG, Quinn PM, Dwayer CM, Podlaski FJ, Familetti PC, Sinigaglia F, Chizonnite R, Gubler U, Stern AS. Regulation of human lymphocyte proliferation by a heterodimeric cytokine, IL-12 (cytokine lymphocyte maturation factor). J Immunol 1991; 147:874–882.

69. Gately MK, Wloitzky AG, Quinn PM, Chizzonite R. Regulation of human cytolytic lymphocyte responses by interleukin-12. Cell Immunol 1992; 143:127–142.

70. Mehrotra PT, Wu D, Crim JA, Mostowski HS, Siegel JP. Effects of IL-12 on the generation of cytotoxic activity in human CD8$^+$ T lymphocytes. J Immunol 1993; 151:2444–2452.

71. Gately MK, Desai B, Wolitzky HS, Quinn PM, Dweyer CM, Podlaski FJ, Familletti PC, Sinigaglia F, Chizzonite R, Gubler U, Stern AS. Regulation of human lymphocyte proliferation by a heterodimeric cytokine, IL-12 (cytokine lymphocyte maturation factor). J Immunol 1991; 147:874–882.

72. Perussia B, Chan SH, D'Andrea A, Tsuji K, Santoli D, Pospisil M, Young D, Wolf SF, Trinchieri G. Natural killer (NK) cell stimulatory factor or IL-12 has differential effects on the proliferation of TCR-$\alpha\beta^+$, TCR-$\gamma\delta^+$ T lymphocytes, and NK cells. J Immunol 1992; 149:3495–3502.

73. Godfrey DI, Gately M, Zlotnik A. Influence of IL-12 on a intrathymic T cell development. J Immunol 1993; 150:11A.

74. Trinchieri G. Interleukin-12 and its role in the generation of T$_H$1 cells. Immunol Today 1993; 14:335–338.

75. Scott P. IL-12: initiation cytokine for cell-mediated immunity. Science 1993; 260:496–97.

76. Naume B, Gately M, Espevik T. A comparative study of IL-12 (cytokine lymphocyte maturation factor)$^-$, IL-2, and IL-7-induced effects of immunomagnetically purified CD56$^+$ NK cells. J Immunol 1992; 148:2429–2436.

77. Chan SH, Perussia B, Gupta JW, Kobayashi M, Pospisil M, Young HA, Wolf SF, Young D, Clark SC, Trinchieri G. Induction of interferon γ production by natural killer cell stimulatory factor: characterization of the responder cells and synergy with other inducers. J Exp Med 1991; 173:869–879.

78. Naume B, Johnsen A-C, Espevik T, Sundan A. Gene expression and secretion of cytokine receptors from highly purified CD56$^+$ natural killer cells stimulated with interleukin-2, interleukin-7 and interleukin-12. Eur J Immunol 1993; 23:1831–1838.

79. Chan SH, Kobayashi M, Santoli D, Perussia B, Trinchieri G. Mechanism of IFN-γ induction by natural killer cell stimulatory factor (NKSF/IL-12). J Immunol 1992; 148:92–98.

80. Naume B, Gately MK, Desai BB, Sundan A, Espevik T. Synergistic effects of interleukin-4 and interleukin-12 on NK cell proliferation. Cytokine 1993; 5:38–46.

81. Robertson MJ, Soiffer RJ, Wolf SF, Manley TJ, Donahue C, Young D, Herrmann SH, Ritz J. Response of human natural killer cells to natural killer cell stimulatory factor: cytolytic activity and proliferation of NK cells are differentially regulated by NKSF. J Exp Med 1992; 175:789–796.

82. Kiniwa M, Gately M, Gubler U, Chizzonite R, Fargeas C, Delespesse G. Recombinant interleukin-12 suppresses the synthesis of immunoglobulin E by interleukin-4 stimulated human lymphocytes. J Clin Invest 1992; 90:262–266.

83. Locksley RM. Interleukin 12 in host defense against microbial pathogens. Proc Natl Acad Sci USA 1993; 90:5879–5880.

84. Heinzel FP, Schoenhaut DS, Rerko RM, Rosser LE, Gatelt MK. Recombinant interleukin-12 cures mice infected with *Leishmania major*. J Exp Med 1993; 177:1505–1509.

85. Sypek JP, Chung CL, Mayor SEH, Subramanyam JM, Goldman SJ, Sieburth DS, Wolf SF, Schaub RG. Resolution of cutaneous leishmaniasis: interleukin-12 initiates a protective T helper type 1 immune response. J Exp Med 1993; 177:1797–1802.

86. Gazzinelli RT, Hieny S, Wynn TA, Wolf S, Sher A. Interleukin-12 is required for the T-lymphocyte-independent induction of interferon γ by an intracellular parasite and induces resistance in T-cell-deficient hosts. Proc Natl Acad Sci USA 1993; 90:6115–6119.

87. Tripp CS, Stanley FW, Unanue ER. Interleukin 12 and tumor necrosis factor α are costimulators of interferon γ production by natural killer cells in severe combined immunodeficiency mice

with listeriosis, and interleukin 10 is a physiologic antagonist. Proc Natl Acad Sci USA 1993; 90:3725–3729.

88. Brunda MJ, Luistro L, Warrier RR, Wright RB, Hubbard BR, Murphy M, Wolf SF, Gately MK. Antitumor and antimetastatic activity of interleukin 12 against murine tumors. J Exp Med 1993; 178:1223–1230.

89. Sprandrude GJ, Heimfeld S, Weissman I. Purification and characterization of mouse hematopoietic stem cells. Science 1988; 241:58–62.

90. Sprangrude GJ, Scollay R. A simplified method for enrichment of mouse hematopoietic stem cells. Exp Hematol 1990; 18:920–926.

91. Bennett D. Developmental analysis of a mutant with pleiotropic effects in the mouse. J Morphol 1956; 98:199–234.

92. Russel ES. Hereditary anemias of the mouse: a review for geneticists. Adv Genet 1979; 20:357–459.

93. Silvers WK. White-spotting, patch and rump-white. In: The Coat Colors of Mice: A Model for Gene Action and Interaction. New York: Springer-Verlag, 1979:206–241.

94. Lowry PA, Zsebo KM, Deacon DH, Eichman CE, Quesenberry PJ. Effects of rrSCF on multiple cytokine responsive HPP-CFC generated from SCA$^+$Lin$^-$ murine hematopoietic progenitors. Exp Hematol 1991; 19:994–996.

95. Li CL, Johnson GR. Rhodamine-123 reveals heterogeneity within murine Lin$^-$,Sca$^+$ hematopoietic stem cells. J Exp Med 1992; 175:1434–1447.

96. Lowry PA, Deacon D, Whitefield P, McGrath HE, Quesenberry PJ. Stem cell factor induction of in vitro murine hematopoietic colony formation by "subliminal" cytokine combinations: the role of "anchor factors." Blood 1992; 80:663–669.

97. Hatzfeld J-P, Li M-L, Brown E, Sookdeo H, Levesque J-P, O'Toole T, Gurney C, Clark SC, Hatzfeld A. Release of early human hematopoietic progenitors from quiescence by antisense transforming growth factor β1 or < rb oligonucleotides. J Exp Med 1991; 174:925–929.

98. Jacobsen SEW, Okkenhaug C, Myklebust J, Veiby OP, Lyman SD. The flt3 ligand potently and directly stimulates the growth and expansion of primitive murine bone marrow progenitor cells in vitro: Synergistic interactions with interleukin (IL) 11, IL-12, and other hematopoietic growth factors. J Exp Med 1995; 181:1357–1363.

99. Hirayama F, Shih J-P, Awgulewitsch A, Warr GY, Clark SC, Ogawa M. Clonal proliferation of murine lymphohemopoietic progenitors in culture. Proc Natl Acad Sci USA 1992; 89:5907–5911.

100. Bellone G, Trincheri G. Dual stimulatory and inhibitory effect of NK cell stimulatory factor/IL-12 on human hematopoiesis. J Immunol 1994; 153:930–937.

101. Gately MK, Warrier RR, Honasoge S, Carvajal DM, Faherty DA, Connaughton SE, Anderson TD, Sarmiento U, Hubbard BR, Murphy A. Administration of recombinant IL-12 to normal mice enhances cytolytic lymphocyte activity and induces production of IFN-γ in vivo. Int Immunol 1994; 6:157–167.

102. Briddel RA, Hartley CA, Smith KA, McNiece IK. Recombinant rat stem cell factor synergizes with recombinant human granulocyte colony-stimulating factor in vivo in mice to mobilize peripheral blood progenitor cells that have enhanced repopulating potential. Blood 1993; 82:1720–1723.

103. Zander AR, Lyding J, Bielack S. Transplantation with blood stem cells. Blood 1991; 17:301–309.

104. Lowry PA. Tabbara IA. Peripheral hematopoietic stem cell transplantation: current concepts. Exp Hematol 1992; 20:937–942.

105. Eng VM, Car BD, Schnyder B, Lorenz M, Lugli S, Aguet M, Anderson TD, Ryffel B, Quesniaux VFJ. The stimulatory effects of interleukin 12 (IL-12) on hematopoiesis are antagonized by IL-12–induced interferon γ in vivo. J Exp Med 1995; 181:1893–1898.

106. Neta R, Stiefel SM, Finkelmann F, Herrmann S, Ali N. IL-12 protects bone marrow from and sensitizes intestinal tract to ionizing radiation. J Immunol 1994; 153:4230–4238.

107. Magram J, Connaughton SE, Warrier RR, Carvajal DM, Wu CY, Ferrante J, Stewart C, Sarmiento U, Faherty DA, Gately MK. IL-12–deficient mice are defective in IFN-γ production and type 1 cytokine responses. Immunity 1996; 4:471–481.

108. Bacon CM, McVicar DW, Ortaldo JR, Rees RC, O'Shea JJ, Johnston JA. Interleukin 12 (IL-12) induces tyrosine phosphorylation of JAK2 and TYK2: differential use of Janus family tyrosine kinases by IL-2 and IL-12. J Exp Med 1995; 181:399–404.

109. Pignata C, Sanghera JS, Cossette L, Pelech DL, Ritz J. Interleukin-12 induces tyrosine phosphorylation and activation of 44-kD mitogen-activated protein kinase in human T cells. Blood 1994; 83:184–190.

110. Bacon CM, Petricoin EF, Ortaldo JR, Rees RC, Larner AC, Johnston JA, O'Shea JJ. Interleukin 12 induces tyrosine phosphorylation and activation of STAT4 in human lymphocytes. Proc Natl Acad Sci USA 1995; 92:7307–7311.

23

Ex Vivo Expansion of Hematopoietic Stem and Progenitor Cells

Ahmad-Samer Al-Homsi and Peter J. Quesenberry
University of Massachusetts Medical Center, Worcester, Massachusetts

Knowledge is only fascinating because incomplete.

I. INTRODUCTION

Since first accomplished, bone marrow (BM) transplantation has been established as a valuable therapeutic measure for selected patients with an ever-increasing variety of inherited and malignant illnesses. Introduction of peripheral blood stem cells (PBSC) and, more recently, umbilical cord blood (UCB) as alternative sources for hematopoietic progenitors (HP), has already illuminated further transplantation horizons. However, in parallel to these advances, a variety of obstacles have surfaced. First, collection of PBSC by BM harvest or by peripheral blood pheresis is still associated with morbidity. Second, the number of HPs obtained may be insufficient, even with repeated BM aspirations or peripheral blood phereses. Recently, several studies have shown a correlation between the speed of hematological recovery and treatment-related mortality and the CD-34$^+$ cell dose in both autologous and allogeneic PBSC and BM transplantations (1–4). Furthermore, although theoretical calculations predict that there are enough HPs in a single collection of UCB for an adult recipient, this is still to be demonstrated in patients who weigh more than 40 kg (5). Finally, whatever the source of HPs, generation of functional blood cells in vivo after transplantation is delayed 1–3 weeks, resulting in significant morbidity and mortality related to infectious and hemorrhagic complications.

Stimulated by the foregoing obstacles, scientists, over the last decade, have focused on expanding HP in vitro. The work of Dexter and co-workers, in the late 1970s, that introduced in vitro cell cultures capable of supporting prolonged survival of HP and of producing large numbers of differentiated progeny, was an important step in this path (6). The earlier colony assays were limited because hematopoiesis could not be maintained for longer than 4 weeks (7). Since these pioneer experiments, much work on expanding ex vivo HP production has been accomplished.

II. DEFINING HEMATOPOIETIC STEM CELLS

To select, track, and assess the functionality of hematopoietic stem cells (HSC) in ex vivo expansion experiments, various systems have been used:

A. Cell Cycle Characteristics

Most HSCs are in G_0 phase. This can be demonstrated by their resistance to S phase-specific agents such as 5-fluorouracil (5-FU) and 4-hydroperoxycyclophosphamide [4-HC; Table 1 (8)].

B. Surface Phenotype

The murine and human HSCs can be portrayed by various surface markers (see Table 1). Cell populations highly enriched in primitive HPs can be isolated using these surface markers in multiparameter cell-sorting systems. Rhodamine is, in fact, a supravital dye that primitive hematopoietic precursors can pump out of the cell using the P-glycoprotein efflux mechanism, which is also the mechanism responsible for multidrug resistance (12). Rhodamine[dull], as opposed to rhodamine[bright], cells constitute most of the primitive CD-34$^+$ cells (12). In addition, a membrane marker (PKH2) has been used to track primitive hematopoietic cells in culture. Nondividing cells remain PKH2[bright], whereas cells undergoing proliferation become PKH2[dim] (13).

C. Genomic Traits

More recently, it was noticed that during cell division, small parts of the telomeric DNA, at the far 3' end of the lagging strand, are not replicated and, therefore, are lost (14). In vitro studies showed that each doubling in HSCs is associated by a 35–45-base pair (bp) loss in telomeric DNA (15). From these observations, telomere length of HPs could be considered a parameter for their proliferative potential.

Table 1 Pharmacological and Phenotypic Characteristics of Stem Cells

Murine	Human
5-FU-resistant	5-FU-resistant
Lin$^-$	4-HC-resistant
CD-34$^+$ (9)	CD-34$^+$
Sca-1$^+$	Lin$^-$
Thy-1$^+$	HLA-DR^{-a}
c-*kit*$^+$	c-*kit*$^+$
Rhodamine[dull]	Thy-1$^+$
Hoechst[low]	CD-45RA[low]
	CD-71[low]
	CD-59$^+$ (10)
	Rhodamine[dull]

[a]UCB stem cells are HLA-DR^{+11} (11).

D. Hematopoietic Stem Cell Assays

The spleen colony-forming unit (CFU-S) (Table 2) assay is an in vivo assay based on the ability of HSC to form macroscopic splenic colonies when injected into lethally irradiated mice (16). However, only day 12- to day 13-appearing colonies may be capable of self-renewal in serial transplant studies (17). In comparison, high-proliferative–potential colony-forming cell (HPP–CFC) assay is an in vitro assay where quiescent and 5-FU- or 4-HC-resistant cells form colonies larger than 0.5 mm in diameter containing more than 50,000 cells each (18). On the other hand, blast colony-forming cells (CFU-Bl) are CD-34$^+$/Lin$^-$/HLA-DR$^-$ cells that, also in vitro, remain dormant for more than 2 weeks before forming small colonies of more than 25 blast-like cells (8,18,19). Both HPP–CFC and CFU-Bl assays are based on detection of proliferation in the presence of cytokines. In contrast, long-term culture-initiating cell (LTC-IC) and cobblestone area-forming cell (CAFC) assays are based on culture on an irradiated stromal layer (8–18). The LTC-IC can sustain cultures for weeks and CAFC can form foci containing more than five dark-phase cells with characteristic cobblestone arrangement beneath the stromal layer (20). However, recent studies indicate that the LTC-IC may be a rapidly dividing line that is not responsible for engraftment in immunodeficient mice. Furthermore, this test is reproduced only with great difficulty among laboratories. Lastly, delta assays consist of serial recultures of the output cells under the same initial conditions. The cumulative expansion of cells can provide an

Table 2 Assays for Stem Cells

Assay	Comments
Spleen colony-forming units (CFU-S)	Hierarchically heterogenous.
High proliferative potential colony-forming cells (HPH-CFC)	Hierarchically heterogeneous; some self-renewal capacity; require up to seven cytokines for in vitro colony formation.
Blast colony-forming cells (CFC-Bl)	Similar to HPP-CFC; each can give rise to multiple HPP-CFC of some sort; probably equivalent to larger multifactor-responsive HPP-CFC.
Long-term culture-initiating cells (LTC-IC)	Proliferating cell that does not appear to be the repopulating cell in immunodeficient mice; difficult to standardize and probably no advantage over HPP-CFC.
Cobblestone area-forming cells (CAFC)	Also very primitive assay; can be done in few laboratories, but is not easily reproducible; similar to LTC-IC.

Progenitors
  Colony-forming units–granulocytes–erythroid–monocyte–megakaryocyte (CFU-GEMM)
  Colony-forming units–granulocyte–monocyte (CFU-GM)
  Burst-forming units–erythroid (BFU-E)
  Colony-forming units–erythroid (CFU-E)
  Colony-forming units–megakaryocyte (CFU-MK)
	Committed: Give rise to colonies exhibiting multiple different lineages; in general, cycling cells.

index of the proliferative potential of the starting cell population. When carried out by limiting dilution analysis techniques, these assays may quantify LTC-IC (21).

E. In Vivo Experiments

Despite all of the foregoing advances, the relevance of the different ex vivo assays of stem cells is still unclear. The only convincing evidence of "stemness" of HP cells resides solely in their ability to repopulate ablated BM in vivo. Toward this end, the transplantation of human HC into immunodeficient mice has provided an unprecedented opportunity to assess the functionality of these cells. The severe combined immunodeficiency (*scid*) and the *bg/nu/xid* mice were the earliest models used for human HP transplantation. The first is unable to produce mature T and B lymphocytes. The second is constructed by crossing mice bearing three recessive mutations (nude, beige, and xid), resulting, respectively, in poor development of the thymus and, therefore, an inability to produce differentiated T lymphocytes, in deficient cytotoxic T lymphocytes and natural killer (NK) cells; and in deficient B-cell response to thymus-dependent antigens and in lymphokine-activated killer (LAK) cells. In these two models, human HPs were injected intravenously, intraperitoneally, or more recently, into human fetal bone fragments of disparate HLA-types that have been preimplanted into the mouse peritoneal cavity (scid–hu model; 22). Low levels of radiation were generally required for engraftment (23). Other limitations of these recipients included cell–dose dependency, requirement of cytokines for better engraftment (23), and failure to engraft with sorted CD-34$^+$ cells (23). A more recent alternative to these models has been provided by crossing the *scid* mutation with the NOD/LtSz strain. The resulting mice lack T and B lymphocytes and have reduced NK levels, a defective complement system, and impaired macrophage activity (24). Higher levels of engraftment were obtained using this model (24). Other improvements in recipients, such as creation of transgenic mice for hematopoietic growth factors, are still being introduced. For now, a competitive repopulation assay, in which two distinct populations of cells are competing against each other for engraftment, provides the most sensitive model (25). Because many cell cycles are required for repopulation, if one population of cells has a slightly longer cell cycle, the intermixed competitor group will acquire a large advantage.

III. TECHNIQUES OF EXPANSION

Two different approaches have been used to expand HP in vitro (26,27).

A. Stroma-Free Cultures Supplemented with Growth Factors

In this most-studied method for human cells, HP enriched in CD-34$^+$ cells, are cultured in liquid media in the presence of multiple cytokines (26,27). Selection of CD-34$^+$ cells is achieved using either fluorescence-activated cell-sorting (FACS) systems or solid-phase immunoselection devices, such as immunoaffinity adsorption columns or immunomagnetic cell selection equipment (28–30). The FACS instruments are more precise, achieving a purity of selected CD-34$^+$ cells of 96–100%, compared with 64–98% for solid-phase immunoselection methods. However the FACS is a time-consuming and less practical method for large clinical-scale cultures.

The CD-34$^+$-sorting has three justifications: First, reduction in tumor cell contamination of HP after such sorting (31). Indeed, studies on patients with breast cancer or multiple myeloma showed a 2-log reduction in the number of tumor cells contaminating PBSC after enrichment with CD-34$^+$ cells. The clinical relevance of this reduction, however, is yet to be demonstrated. Also, the contaminating tumor cells do not expand in in vitro cultures, but rather, fade (32).

Second, purification of CD-34$^+$ is essential for optimal expansion of UCB cells (33). The expansion for unselected UCB, in suspension culture under influence of stem cell factor (SCF), granulocyte colony-stimulating factor (G-CSF), and megakaryocyte growth and development factor (MGDF), was severalfold less than the expansion of selected UCB cells. Third, CD-34$^+$ selection has the advantage of reducing the number of cells to be targeted if ex vivo expansion of HP is to be employed in gene therapy (34).

Bone marrow, PBSC, UCB, and fetal liver, all have been successfully used to initiate HP expansion in stroma-free cultures (12). When mobilized PBSC were used, a negative correlation was seen between the amount of cytotoxic and radiation therapy required by patients before the peripheralization procedure and the yield of CD-34$^+$ cells (35). Furthermore, although cumulative myelotoxicity was not reflected by a decrease in the mean terminal restriction fragment (TRF) of CD-34$^+$ cells collected after mobilization (36), suggesting a quantitative, but not qualitative problem, Shapiro et al. (37) still noticed a negative effect of previous treatments on the expansion potential of HP. Likewise, the regimen used for mobilization of PBSC was of relevance. Intensive cytotoxic drug regimens resulted in increased entry of HPs into the circulation (12). Addition of SCF to a mobilization regimen of cyclophosphamide and G-CSF also enhanced the peripheralization of CD-34$^+$ cells and LTC-IC (39). With increasing doses of SCF, the CD-34$^+$ cells increased by threefold and the LTC-IC increased from 1:10,916 nucleated cells to 1:3,150 nucleated cells.

Umbilical cord blood has drawn special attention over the last years because of the extensive proliferative capacity of its HP. When CD-34$^+$ cells with a high density of CD-34 antigen (CD-34^{3+}) were sorted from normal UCB and cultured by 1, 5, or 10 cell(s) per well or by 5000 cells per dish, 48% of the initiated cultures were positive for growth by day 14. The expansion of the total nucleated cells reached 5000-fold for one-cell per well cultures. On the other hand, the expansion for progenitor cells attained 160-fold in the 5000-cell per dish cultures (39). In addition, single cells that gave rise to large colonies, could be serially replated at least five to six times, demonstrating the self-renewal capacity of these cells (40). Moreover, although LTC-IC decline rapidly in adult HP in stroma-free cultures, these cells, when derived from UCB, can be expanded up to 20-fold (41). Moore et al. directly compared BM- or PBSC-derived HP with UCB-derived HP and found that the latter had a much greater expansion capacity than either adult source (42). Finally, Mayani et al. were able to achieve an unparalleled 4700-fold increase in HP by culturing a subpopulation of UCB-derived CD-34$^+$ cells with low or undetectable expression of CD-45RA and CD-71 antigens and with intermediate or high expression of Thy-1 antigen (CD-34$^+$/CD-45RA^{lo}/CD-71^{lo}/Thy-1$^+$ cells) (43).

Various assortments of cytokines have been used in stroma-free cultures. Data from different studies are difficult to compare because of different experimental conditions, including different input of cells, different culture conditions, and different endpoints (Table 3). Haylock et al. compared several growth factor combinations in large-volume stroma-free cultures of human peripheral blood mononuclear cells (PBMNC), collected after mobilization with G-CSF or cyclophosphamide and enriched in CD-34$^+$ cells. They identified a combination of interleukin-1 (IL-1), IL-3, IL-6, SCF, granulocyte–monocyte colony-stimulating factor (GM-CSF), and G-CSF, as the most potent combination measured by generation of nucleated cells and of nascent CFU-GM (28). A mixture of IL-1, IL-3, IL-6, SCF, and erythropoietin (Epo) resulted in optimal expansion of CD-34$^+$ cell-enriched human PBMNC for both nucleated cells and clonogenic progenitors (45). Also addition of interferon gamma (IFN-γ) to the five-cytokine combination was synergistic (45). Conversely, when leukemia-inhibiting factor (LIF) or GM-CSF were added, the number of nucleated cells increased, whereas the number of clonogenic progenitors decreased. At the same time, G-CSF had no synergistic effect. In contrast, the addition of LIF factor

Table 3 Summary of Human Stroma-Free In Vitro HP Expansion Studies

Ref.	Input	Cytokines	Test	Duration (d)	Fold-expansion
44	PB CD-34$^+$	IL-1, 3, 6, SCF, and Epo	TNC	12	30–100
			CFU-GM, BFU-E, and CFU-mixed	12	20–100
			LTC-IC	12	>1
45	PB CD-34$^+$	IL-1, 3, 6, SCF, and Epo	TNC	16	76–995
			CFU-GM, BFU-E, and CFU-mixed	12–14	46–930
				14	
			CD-34$^+$Lin$^-$	24	3–21
			4-HC-resistant		>1
28	PB CD-34$^+$	IL-1, 3, 6, SCF, GM-, and G-CSF	TNC	21	1324
			CFU-GM	14	66
46	PB CD-34$^+$ or Lin$^-$	IL-3, 6, SCF, and Epo	LTC-IC	4	<1
30	PB CD-34$^+$	PIXY321	TNC	12	7.4–207
			CFU	12	1–11.6
			CD-34$^+$	12	6–64
42	PB CD-34$^+$	IL-1, 3, and SCF	TNC	14	223
			CFU-GM	14	109
47	BM CD-34$^+$/ HLA-DR$^-$/ CD-15$^-$/Rh^{dull}	SCF and PIXY321	HPP-CFC	21	2
				28	5.5
42	BM CD-34$^+$	IL-3, 6 and SCF	TNC	14	59
			CFU-GM	14	
41	UCB CD-34$^+$	IL-1, 3, and SCF	LTC-IC	7–14	15–20
43	UCB CD-34$^+$/ CD45RA^{lo}/ CD-71^{lo}/Thy-1$^+$	IL-3, SCF, PIXY321, GM-, M-, G-CSF, and Epo	CFU-C	29	31,000
			HPP-CFC	29	652
			CD-34$^+$/Thy-1$^+$	21	20

PB, peripheral blood; IL-1, interleukine-1; SCF, stem cell factor; Epo, erythropoietin; TNC, total nucleated cells; CFU-C, colony-forming units–myeloid; GM-CSF, granulocyte–monocyte colony-stimulating factor; G-CSF, granulocyte colony-stimulating factor; PIXY321, fusion protein of IL-3 and GM-CSF; Rh, rhodamine.

to a combination of IL-3, IL-6, IL-11, SCF, GM-CSF, and G-CSF resulted in a suppressive effect on the differentiation and lineage restriction of human CD-34$^+$/Lin$^-$/Thy-1$^+$ BM cells with less total nucleated cell expansion and greater CD-34$^+$/Lin$^-$ cell increase (48). More recently, the effect of *flt3* ligand was studied. Using c-*kit*$^+$ murine bone marrow cells, this novel cytokine alone resulted in a sevenfold increase in the number of nucleated cells. A synergistic effect with IL-1, IL-3, SCF, and LIF was also noted. On the other hand, the same ligand failed to maintain HPP-CFC alone and showed no synergistic effect when added to IL-3 and SCF (49). When human BM CD-34$^+$ cells were studied, the addition of *flt3* ligand to a combination of IL-3, IL-6, and SCF resulted in an enhanced amplification of CFU-GM by factor of 10 and of LTC-IC by a factor of 2 (50).

In addition to the cytokine mixture, attention has been given to other culture conditions, including cell concentration, culture media, and oxygen tension. In one study (51), mobilized peripheral blood progenitor concentrations of 10^6/mL resulted in better cell recovery than concentrations of 10^7/mL. Simple media, such as medium 199, gave less optimal results than did the more complex media, such as IMDM (51). In another study, the efficacy of a serum-free medium, when compared with xenogeneic serum-containing media, was shown on expansion of early progenitors in stroma-free cultures of human bone marrow cells (52). In addition, when UCB-derived HPs were used, cord blood plasma was superior to peripheral blood plasma, at least as measured by the total cell-fold expansion (53). On the other hand, the effect of oxygen tension is arguable. Whereas a beneficial effect of reduced tensions (5%) on human UCB clonogenic progenitor expansion was shown (54), no such effect was noted when murine HPs were studied (55).

Although the capability to achieve a substantial augmentation in the number of nucleated cells has been amply demonstrated by stroma-free cultures, whether true stem cell expansion happens in these cultures is still undetermined. In fact, the amplification of the cultured cells seems to be accompanied by terminal differentiation, conceivably driven by the added cytokines, which may be inducing active cell cycling. The number of the more primitive progenitors rarely increases and usually declines shortly after initiation of culture. Obviously, if used after marrow ablative therapies, this loss in cell "stemness" may promote late graft failure. As a result, various studies have looked more specifically into the expansion of the most primitive HPs. In one study (44), the absolute number of CD-34$^+$/HLA-DR$^-$ or CD-38$^-$ cells increased after expansion. The same group used limiting dilution assays for LTC-IC to demonstrate that early progenitor cells could be preserved in culture, contradicting several other studies, with the exception of those using UCB progenitors as starting cells (44). Srour et al. (47) initiated a stromal cell-free long-term culture of CD-34$^+$/HLA-DR$^-$/CD-15$^-$/rhodamine ^{dull} bone marrow cells with repeated addition of SCF and a fusion protein of IL-3 and GM-CSF (PIXY321). They showed, by phenotypic analysis and by secondary cloning of HPP-CFC, that it is possible to expand the number of primitive progenitor cells (47). Unfortunately, these studies, by looking into phenotypic and in vitro characteristics of the expanded cells, cannot guarantee that long-term repopulation of lethally damaged BM by cultured HPs will occur. Murine studies attempted to reply more convincingly (Table 4). HP transplantation in mice, using suspension cultures of BM with different combinations of cytokines, was compared with transplantation using fresh BM. Transplantation with cultured BM resulted in accelerated recovery of blood counts of lethally irradiated or 5-FU-treated recipients; the combination of IL-1 with SCF was the most effective. Although all mice survived at least 30 days, the recipients receiving the IL-1- and SCF-expanded BM required 200-fold fewer cells, while leading to predominantly donor-derived hematopoietic reconstitution for 280 days after primary transplantation and another 71 days after secondary transplantation (56). When the same authors looked into HPP-CFC assays at 40 weeks after transplantation, there was no difference between the recipients of cultured BM in comparison with those receiving an equivalent number of fresh BM cells (56,57). However, BM cultures were of short duration (7 days) and donor animals were pretreated with 5-FU. Similarly, the radioprotective abilities of BM, collected after 5-FU treatment and stimulated with IL-3 and IL-6 or with IL-3 and SCF, were higher than the same number of fresh BM harvested after 5-FU treatment (58). In competitive assays, significant capability of long-term reconstitution of the cultured BM progenitors was also observed as long as culture duration remained brief (58). However, the results of these two series of experiments were challenged (60,61). Two competi-

Table 4 Summary of Murine Stroma-Free In Vitro Expansion Studies

Ref.	Study	5-FU pretreatment	Duration of culture (d)	Cytokines	Outcome
56	Comparison of C to NC BM	Yes	7	IL-1 and IL-3	Accelerated hematological recovery; maintained HPP-CFC at 5 and 10 wk
57	Comparison of C to NC BM	Yes	7	IL-1 and SCF	Accelerated hematological recovery; maintained of HPP-CFC at 40 wk; maintained reconstitution ability in secondary hosts
53	Comparison of C and NC BM	Yes	3	IL-3 and IL-6, or IL-3 and SCF	Improved radioprotective ability; maintained long-term reconstitution ability in competitive experiments
59	Comparison of C to NC Lin⁻/Sca-1⁺/WGA⁺ BM cells	No	14	IL-6, SCF, and Epo, with or without IL-3	Maintained (but not increased) in vivo reconstituting ability
60,61	Comparison of C to NC BM	No	2	IL-3, IL-6, IL-11, and SCF	HPP-CFC expansion; impaired engraftment in competitive experiments

C, cultured; NC, noncultured; WGA, wheat germ agglutinin.

tive transplantation models were used in which HP cells collected from nonpretreated donors and cultured in the presence of cytokines competed with fresh marrow cells in lethally irradiated or nonmyeloablated recipients. In one competitive transplant model Ly5.1 cells competed with cultured Ly5.2 cells in irradiated Ly5.1/Ly5.2 congenic hosts: Engraftments of 70% at 12 weeks and 93% at 22 weeks for Ly5.1 cells were observed. When cultured Ly5.2 cells competed with fresh Ly5.1 cells, 5 and 4% engraftments at 12 and 22 weeks, respectively, were noted. In the second model, BALB/c BM of the opposite gender was used. Cultured male cells led to 13 and 2% engraftment at 10 and 14 weeks. Noncultured male cells led to 70 and 90% engraftment at 10 and 14 weeks. It was concluded that, although HPP-CFC were expanded in their marrow culture, ex vivo expansion of murine BM progenitors resulted in impaired engraftment in irradiated hosts (60,61).

B. Continuous Perfusion-Based Cultures

The continuous perfusion method of HP expansion is based on creation of a physiological environment similar to the one that occurs during transplant engraftment or in vivo expansion (26,27). Such cultures can be conducted using unselected hematopoietic cell populations. The stromal layer is derived from the initial unselected cell population itself or is preformed from

BM or UCB in CD-34$^+$ cell-enriched cultures (62,63). Continuous perfusion or frequent medium exchange is used in parallel (bioreactor systems) for optimal results (62,64). In contrast with stroma-free cultures, addition of exogenous cytokines in this method does not seem as necessary for LTC-IC maintenance (65,66). In fact, cultured stromal elements produce multiple cytokines, including SCF, IL-3, and IL-6 (67). Furthermore, in a stroma noncontact culture system, in which HPs are cultured separated from stroma by a microporous membrane that prevents HP–stroma contact, but allows free passage of diffusible factors (27), addition of different combinations of cytokines resulted in exhaustion of LTC-IC, and it was assumed that a persistent stimulation of HP by large doses of growth factors can overcome the antidifferentiation capacity of unknown stromal factors (27). Addition of IL-3 and macrophage inhibitory protein 1α (MIP-1α) resulted in maintenance of 100% of LTC-IC for at least 8 weeks, suggesting that maintenance of HP ex vivo requires not only growth-promoting cytokines, but also differentiation-prohibiting factors (27).

In sum, in comparison with stroma-free cultures, perfusion-based systems allow more limited amplification of HP, but better maintenance of LTC-IC (Table 5).

IV. CLINICAL TRIALS AND POTENTIAL APPLICATIONS

The first trials that returned cultured HPs to humans were reported in 1984 (71). Although limited, this study demonstrated the feasibility and safety of this approach. Multiple reports on transplantation of expanded HPs have thereafter appeared in the literature. In a phase I–II trial (72), ten patients with advanced tumors underwent high-dose cytotoxic therapy rescued by a starting population of 11×10^6 PBSC, mobilized by antitumor chemotherapy and G-GSF and expanded ex vivo in stroma-free cultures under the influence of IL-1, IL-3, IL-6, SCF, and Epo.

Table 5 Summary of Continuous Perfusion-Based Ex Vivo HP Expansion Studies

Ref.	Input	Cytokines	Readout	Duration (d)	Fold-expansion	Comments
62	Human BM	IL-3, SCF, GM-CSF, and Epo	TCN	14	10	
			CFU-GM	14	21	
			LTC-IC	14	7.5	
68	Human CD-34$^+$- enriched BM with pre- formed stroma	IL-3, SCF, GM-CSF, and Epo	TNC	14	3	
			CFU-GM	14	5	
			LTC-IC	14	3	
69	Human CD-34$^+$- enriched.	None	LTC-IC	7	3.57	
				14	4	
	UCB with stroma; non- contat system	IL-3 and MIP-1α	LTC-IC	7	5.92	
				14	4.87	
70	Human CD-34$^+$/ Lin$^-$/Thy-1$^+$- enriched BM with PMVEC	IL-3, IL-6, SCF, and GM-CSF	CAFC	21	>1	
			scid–hu assays	21		No loss in ca- pability of engraftment

PMVEC, porcine endothelial microvascular endothelial cell.

No side effects related to the infusion of the expanded cells were noted. Sustained hematological recovery was seen in all patients with a delay comparable with that observed in historical controls rescued with uncultured cells. Four patients, however, received simultaneous administration of cultured and uncultured cells (72). More recently, nine patients with breast cancer were treated in a similar fashion using PIXY321 as a cytokine supplement to their culture medium (30). The number of the infused cells varied widely from 0.8 to 156.6×10^6/kg. Again, no adverse effects caused by infusion of the cultured cells were noted, and blood recovery was prompt, but not shortened. Interestingly, the patients who had the quickest hematological recovery were those who received the highest doses of cultured cells. Also, the infusion of the cultured cells was associated with an initial in vivo augmentation in neutrophil count.

Although no acute toxicity related to the administration of ex vivo expanded HPs was seen in any of the published studies, no conclusions can be drawn about the capacity of cultured cells to guarantee long-term BM repopulation. Indeed, the preparative regimens given in all of these trials were not ablative. Moreover, by single-dose infusion, and at the cell dose used in these trials, the hematological recovery does not seem accelerated. It remains true that the potentials for clinical applications of ex vivo expansion of HP are immense. First, the problem of inadequacy of HP encountered in some cases, despite repeated BM aspirations or blood pheresis, would be resolved. Fewer BM aspirations or a smaller volume of pheresis products might suffice. This would result in reduced morbidity and cost of these procedures. In addition, expanded UCB HP would guarantee adequate hematopoietic reconstitution when used for any adult recipients. One UCB product could even be used for more than one adult. This is especially relevant at a time when tandem high-dose therapy is being introduced in clinical practice. Second, the collected HP could conceivably be divided in two parts: The first could be infused un-manipulated to reestablish the stem cell compartment and to guarantee long-term engraftment. The second, cultured ex vivo to produce a large number of mature granulocytes, could be used for prophylactic transfusions of neutrophils at defined intervals, or as adjuvant therapy to antibiotics in life-threatening infections. In fact, the role of white blood cell transfusion should be totally reassessed in the light of the current ability to produce many neutrophils by ex vivo expansion. Several studies have demonstrated that daily transfusions of $2-3 \times 10^{10}$ cells is effective and could be life-saving in selected patients with opportunistic infections (73). For self-transfusion of white blood cells, it is likely that the required number of cells for effective transfusion is even fewer. In any event, Haylock et al. (28) were able to obtain about 2×10^{11} functionally normal neutrophils from a single PBSC collection by stroma-free culture using a combination of SCF, IL-1, IL-3, IL-6, GM-CSF, and G-CSF. This amount of white blood cells would be enough for effective transfusions over at least 10 days. Indeed, such a transfusional approach could also be generalized to patients receiving antitumor chemotherapy who are considered at risk for severe neutropenia. Third, expansion of nonmyeloid HP could also have interesting applications. Megakaryocytic progenitors could be selected from HP and grown in culture (74). Although proplatelet production does not seem to occur in these in vitro cultures (75), possibly because of low cell densities, or because of lack of other signals (76), expansion of megakaryocytic progenitors could possibly resolve the problem of prolonged thrombo-cytopenia after HP transplantation that may result, at least occasionally, from insufficient numbers of transfused megakaryocyte progenitors. The role of thrombopoietin in these expansion experiments is still to be explored. On the other hand, expansion of lymphocytic precursors could open new directions to adaptive immunotherapy by promoting T lymphocytes to antitumor activity. Stem cell factor recruits and expands early progenitors responsive to GM-CSF and tumor necrosis factor-α (TNF-α) (77), effecting up to a 1000-fold expansion of colony-forming unit–dentritic cells (CFU-CD). Fourth, ex vivo expansion of stem cells could facilitate transduction of these cells by adeno- or retroviruses for gene therapy purposes.

Several key issues still need to be addressed before we know whether ex vivo expansion of HP constitutes a clinical advance. Perhaps the main unanswered question is, if cultured HP are to be used after myeloablative therapy, are these cells truly capable of assuring long-term hematopoiesis? Only clinical trials, possibly in which the expanded cells are tagged by viral or membranous markers, would give a definitive answer to this question. In addition, the optimal expansion conditions for the starting population of cells, methods of culture, and culture conditions, are still to be identified. Finally, the application of expansion techniques in an allogeneic setting remains ambiguous owing to the theoretical fear of increased incidence of graft-versus-host disease secondary to an increase in the number of administered of T lymphocytes.

REFERENCES

1. Bensinger WI, Longin K, Appelbaum F, et al. Peripheral blood stem cells (PBSCs) collected after recombinant granulocyte colony stimulating factor (rhG-CSF): analysis of factors correlating with the tempo of engraftment after transplantation. Br J Haematol 1994; 87:825.

2. Weaver C, Hazelton B, Birch R, Palmer P, Allen C, Schwartzberg L, West W. An analysis of engraftment kinetics as a function of the CD34 content of peripheral blood progenitor cell collections in 692 patients after the administration of myeloablative chemotherapy. Blood 1995; 86:4961.

3. Korbling M, Huh YO, Durett A, et al. Allogeneic blood stem cell transplantation: peripheralization and yield of donor-derived primitive hematopoietic progenitor cells CD34$^+$, Th-1^{dim}) and lymphoid subsets, and possible predictors of engraftment and graft-versus-host disease. Blood 1995; 86:2842.

4. Mavroudis D, Read E, Cottler-Fox M, et al. CD34$^+$ cell dose predicts survival, posttransplant morbidity, and rate of hematologic recovery after allogeneic marrow transplants for hematologic malignancies. Blood 1996; 88:3223.

5. Broxmeyer HE. Questions to be answered regarding umbilical cord blood hematopoietic stem and progenitor cells and their use in transplantation. Transfusion 1995; 35:694.

6. Dexter TM, Allen TD, Lajtha LG. Conditions controlling the proliferation of haematopoietic stem cells in vitro. J Cell Physiol 1977; 91:335.

7. Bradley TR, Metcalf D. The growth of mouse bone marrow cells in vitro. Aust J Exp Biol Med Sci 1966; 44:287.

8. Gordon MY. Human haematopoietic stem cell assays. Blood Rev 1993; 7:190.

9. Krause DS, Ito T, Fackler MJ, Smith OM, Collector MI, Sharhis SJ, May WS. Characterization of murine CD34, a marker for hematopoietic stem and progenitor cells [abstr]. Exp Hematol 1994; 22:784.

10. Hill B, Rozler E, Travis M, et al. High-level expression of a novel epitope of CD59 identifies a subset of CD34$^+$ bone marrow cells highly enriched for pluripotent stem cells. Exp Hematol 1996; 24:936.

11. Apperely JF. [Meeting report] Umbilical cord blood progenitor transplantation. Bone Marrow Transplant 1994; 14:187.

12. Moore MAS. Expansion of myeloid stem cells in culture. Semin Hamatol 1995; 32:183.

13. Srour EF, Siena S, Bregni M, Kosak S, Traycoff CM, Grisby S, Gianni AM. Long-term hematopoietic culture-initiating cells mobilized into peripheral blood with high-dose cyclophosophamide and recombinant cytokines do not undergo additional ex vivo expansion [abstr]. Exp Hematol 1994; 22:784.

14. Sandell LL, Zakian VA. Loss of a yeast telomere: arrest, recovery, and chromosome loss. Cell 1993; 75:729.

15. Vaziri H, Dragowska W, Allosopp RC, et al. Evidence for a mitotic clock in human hematopoietic stem cells: loss of telomeric DNA with age. Proc Natl Acad Sci USA 1994; 91:9857.

16. Till JE, McCulloch EA. A direct measurement of the radiation sensitivity of normal mouse bone marrow cells. Radiat Res 1961; 14:213.

17. Magli MC, Iscove NN, Odartchenko N. Transient nature of early haematopoietic spleen colonies. Nature 1982; 295:527.

18. Moore MAS. Clinical implications of positive and negative hematopoietic stem cell regulators. Blood 1991; 78:1.

19. Gordon MY, Dowding CR, Riley GP, Greaves MF. Characterization of stroma-dependent blast colony-forming cells in human marrow. J Cell Physiol 1987; 130:150.

20. Breems DA, Blokland EAW, Neben S, Ploemacher RE. Frequency analysis of human primitive haematopoietic stem cell subsets using a cobblestone area forming cell assay. Leukemia 1994; 8:1095.

21. Sutherland HJ, Landsorp PM, Henkelman DH, Eaves AC, Eaves CJ. Functional characterization of individual human hematopoietic stem cells cultured at limiting dilution on supportive marrow stromal layers. Proc Natl Acad Sci USA 1990; 87:3584.

22. Kyoizumi S, Baum CM, Kaneshima H, McCune JM, Yee EJ, Namikawa R. Implantation and maintenance of functional human bone marrow in scid-hu mice. Blood 1992; 79:1704.

23. Dick JE. Immune-deficient mice as models of normal and leukemic human hematopoiesis. Cancer Cells 1991; 3:39.

24. Lowry PA, Shultz LD, Greiner DL, et al. Improved engraftment of human cord blood stem cells in NOD/LtSz–scid.scid mice following irradiated or multiple day injections into unirradiated recipients. Biol Blood Marrow Transplant 1995; 2:15.

25. Harrison DE. Evaluating functional abilities of primitive hematopoietic stem cell populations. Curr Top Microbiol Immunol 1992; 177:13.

26. Emerson SG. Ex vivo expansion of hematopoietic precursors, progenitors, and stem cells: the next generation of cellular therapeutics. Blood 1996; 87:3082.

27. Verfaillie CM. Can human hematopoietic stem cells be cultured ex vivo? Stem Cell 1994; 12:466.

28. Haylock DN, To LB, Dowse TL, Juttner CA, Simmons PJ. Ex vivo expansion and maturation of peripheral blood CD-34[+] cells into myeloid lineage. Blood 1990; 80:1405.

29. Brugger W, Heimfeld S, Berenson RJ, Mertelsmann R, Kanz L. Reconstitution of hematopoiesis after high-dose chemotherapy by autologous progenitor cells generated ex vivo. N Engl J Med 1995; 7:333.

30. Williams SF, Lee WJ, Bender JG, et al. Selection and expansion of peripheral blood CD-34[+] cells in autologous stem cell transplantation for breast cancer. Blood 1996; 87:1687.

31. Shpall EJ, Jones RB, Bearman SJ, et al. Transplantation of enriched CD-34[+] autologous marrow into breast cancer patients following high dose chemotherapy: influence of CD-34[+] cell peripheral blood progenitors and growth factors on engraftment. J Clin Oncol 1994; 12:28.

32. Vogel W, Behringer D, Scheding S, Kanz I, Brugger W. Ex vivo expansion of CD-34[+] peripheral progenitor cells: implications for the expansion of contaminating endothelial tumor cells. Blood 1996; 88:2707.

33. Briddell RA, Kerm BP, Zilm KL, Stoney B, McNiece IK. Purification of CD34[+] cells is essential for optimal expansion of umbilical cord blood cells [abstr]. Exp Hematol 1996; 24:1041.

34. Moore MAS. Ex vivo expansion and gene therapy using cord blood CD-34[+] cells. J Hematol 1993; 2:221.

35. Hass R, Mohle R, Fruhauf S, et al. Patient characteristics associated with successful mobilization and autografting of peripheral blood progenitor cells in malignant lymphoma. Blood 1994; 83:3787.

36. Kronenwett R, Maurea S, Hass R. Telomere length of blood-derived mononuclear cells from cancer patients during G-CSF-enhanced marrow recovery. Bone Marrow Transplant 1996; 18(Suppl 1):510.

37. Shapiro F, Yao TJ, Raptis G, et al. Optimization of conditions for ex vivo expansion of CD34[+] cells from patients with stage IV breast cancer. Blood 1994; 84:3567.

38. Weaver A, Ryder D, Dexter TM, Crowther D, Testa NG. Increasing numbers of long term culture-initiating cells (LTC-IC) are mobilized into peripheral blood with increasing doses of stem cell factor (SCF) in combination with filgrastim (G-CSF) [abstr]. Exp Hematol 1996; 24:1041.

39. Xiao M, Boxmeyer HE, Horie M, Grigsby S, Lu L. Extensive proliferative capacity of single isolated CD-34[3+] human cord blood cells in suspension culture. Blood Cells 1994; 20:455.

40. Lu L, Xiao M, Shen RN, Grigsby S, Boxmeyer HE. Enrichment, characterization, and responsiveness

of single primitive CD-34^{3+} human umbilical cord blood hematopoietic progenitors with high proliferative and replating potential. Blood 1993; 81:41.

41. Moore MAS, Hoskins I. Ex vivo expansion of cord blood-derived stem cells and progenitors. Blood Cells 1994; 20:468.

42. Moore MAS, Schneider JG, Shapiro F, Bengla C. Ex vivo expansion of CD34$^+$ hematopoietic progenitors. In: Gross S, ed. Advances in Bone Marrow Purging and Processing, 4th International Symposium. New York: Wiley-Liss, 1994:217.

43. Mayani H, Lansdorp PM. Thy-1 expression is linked to functional properties of primitive hematopoietic progenitor cells from human umbilical cord blood. Blood 1994; 83:2410.

44. Henschler R, Brugger W, Luft T, Frey T, Mertelsmann R, Kanz L. Maintenance of transplantation potential in ex vivo expanded CD34$^+$-selected human peripheral blood progenitor cells. Blood 1994; 84:2898.

45. Brugger W, Mocklin V, Heimfeld S, Berneson RJ, Mertelsmann R, Kanz L. Ex vivo expansion of enriched peripheral blood CD34$^+$ progenitor cells by stem cell factor, interleukin-1 beta (IL-1 beta), IL-6, IL-3, interferon-gamma, and erythropoietin. Blood 1993; 8:2579.

46. Dooley DC, Plunkett JM, Oppenlander BK, Novak FP. Expansion of myeloid progenitors is accompanied by rapid loss of long-term culture initiating cells during ex vivo cultivation of steady state peripheral blood [abstr]. Exp Hematol 1994; 22:758.

47. Srour EF, Brandt JE, Briddell RA, Grisby S, Leemhuis T, Hoffman R. Long-term generation and expansion of human primitive hematopoietic progenitor cells in vitro. Blood 1992; 81:661.

48. Brandt JE, Sundy S, Hoffman R, Tsukamoto A, Tushinski R. Leukemia inhibitory factor suppresses lineage restriction and differentiation during in vitro expansion of progenitor cells from adult human marrow [abstr]. Exp Hematol 1994; 22:725.

49. DeVries P, Brasel KA, McKenna HJ, Beckmann MP, Gliniak BC, Williams DE, Lyman SD. The effect of soluble FLT3 ligand on murine pluripotent hematopoietic stem cells [abstr]. Exp Hematol 1994; 22:724.

50. Douay L, Poloni A, Korabi L, Giarranta MC, Firat H, Gorin NC. Ex vivo expansion of hematopoietic stem/progenitor cells in serum-free stroma-free conditions: the interest of FLT3 ligand, MGDF, and G-CSF [abstr]. Exp Hematol 1996; 24:1039.

51. Chang Q, Hanks S, Akard L, Thompson J, Harvey K, English D, Jansen J. Maturation of mobilized peripheral progenitor cells: preclinical and phase I clinical studies. J Hematol 1995; 4:289.

52. Biddle W, Lebkowski J, Wysocki M, et al. Cultivation and ex vivo expansion of human CD-34$^+$ progenitor cells under serum-free culture conditions. Prog in Clin and Biol Res 1994; 389:351.

53. Ruggieri L, Heimfeld S, Boxmeyer HE. Cytokine-dependent ex vivo expansion of early subsets of CD-34$^+$ cord blood myeloid progenitors is enhanced by cord blood plasma, but expansion of the more mature subsets of progenitors is favored. Blood Cells 1994; 20:436.

54. Koller MR, Bender JG, Papoutsakis ET, Miller WM. Beneficial effects of reduced oxygen tension and perfusion in long-term hematopoietic cultures. Ann NY Acad Sci 1992; 665:105.

55. Muench MO, Gasparetto C, Moore MA. The in vitro growth of murine high proliferative potential-colony forming cells is not enhanced by growth in a low oxygen atmosphere. Cytokine 1992; 4:488.

56. Muench MO, Moore MAS. Accelerated recovery of peripheral blood cell counts in mice transplanted with in vitro cytokine-expanded hematopoietic progenitors. Exp Hematol 1992; 20:611.

57. Muench MO, Firpo MT, Moore MAS. Bone marrow transplantation with interleukine-1 plus *kit* ligand ex vivo expanded bone marrow accelerates hematopoietic reconstitution in mice without the loss of stem cell lineage and proliferative potential. Blood 1993; 81:3463.

58. Varas F, Bernard A, Bueren JA. Analysis of the differential rate of 5-FU treated murine marrow under ex vivo expansion with ILs 3 + 6 and IL-3 + SCF. Exp Hematol 1992; 22:786.

59. Rebel VI, Dragowska W, Eaves CJ, Humphries RK, Landorp PM. Amplification of Sca-1$^+$ Lin$^-$ WGA$^+$ cells in serum-free cultures containing steel factor, interleukin-6, and erythropoietin with maintenance of cells with long-term in vivo reconstitution potential. Blood 1994; 83:128.

60. Peters SO, Kittler ELW, Ramshaw HS, Quesenberry PJ. Ex vivo expansion of murine marrow cells

with interleukine-3 (IL-3), IL-6, IL-11, and stem cell factor leads to impaired engraftment in irradiated hosts. Blood 1996; 87:30.

61. Peters SO, Kittler ELW, Ramshaw HS, Quesenberry PJ. Murine marrow cells expanded with IL-3, IL-6, IL-11, and SCF acquire an engraftment defect in normal hosts. Exp Hematol 1995; 23:461.

62. Koller MR, Emerson SG, Palsson BO. Large scale expansion of human stem and progenitor cells from bone marrow mononuclear cells in continuous perfusion cultures. Blood 1993; 82:378.

63. Ye ZQ, Burkholder JK, Qiu P, Schultz JC, Shahidi NT, Yang NS. Establishment of an adherent cell feeder layer from human umbilical cord blood for support of long-term hematopoietic progenitor cell growth. Proc Natl Acad Sci USA 1994; 91:12144.

64. Schwartz RM, Palsson BO, Emerson SG. Rapid medium perfusion rate significantly increases the productivity and longevity of human bone marrow cultures. Proc Natl Acad Sci USA 1991; 88:6760.

65. Sutherland HJ, Hogge DE, Cook D, Evans CJ. Alternative mechanisms with and without steel factor support primitive human hematopoiesis Blood 1993; 81:1465.

66. Wineman JP, Nishikawa SI, Muller-Sieburg CE. Maintenance of high levels of pluripotent hematopoietic stem cells in vitro: effect of stromal cells and c-*kit*. Blood 1993; 81:365.

67. Kittler ELW, McGrath H, Temeles D, Crittenden RB, Kister VK, Quesenberry P. Biological significance of constitutive and subliminal growth factor production by bone marrow stroma. Blood 1992; 79:3168.

68. Koller MR, Palsson MA, Manchel I, Palsson B. Long-term culture-initiating cell is dependent on frequent medium exchange combined with stromal and other accessory cell effects. Blood 1995; 86:1784.

69. Han CS, Dugan MJ, Verfaillie CM, Wagner JE, McGlave PB. In vitro expansion of umbilical cord blood committed and primitive progenitors [abstr]. Exp Hematol 1994; 22:723.

70. Brandt J, Galy A, Luens K, et al. Maintenance of human hematopoietic stem cells during ex vivo progenitor cell expansion on a porcine endothelial cell line [abstr]. Exp Hematol 1996; 24:1029.

71. Spooncer E, Dexter TM. Transplantation of long term culture bone marrow cells. Transplantation 1984; 35:624.

72. Brugger W, Heimfeld S, Berneson RJ, Mertelsmann R, Kanzm L. Reconstitution of hematopoiesis after high-dose chemotherapy by autologous progenitor cells generated ex vivo. N Engl J Med 1995; 333:283.

73. Strauss RG. Clinical perspectives of granulocyte transfusion: efficacy to date. J Clin Apheresis 1995; 10:114.

74. Guerriero R, Testa U, Gabbianelli M, et al. Unilineage megakaryocytic proliferation and differentiation of purified hematopoietic progenitors in serum-free liquid culture. Blood 1995; 86:3725.

75. Nagler A, Eldor A, Naparstek E, Mumcuoglu M, Salvin S, Deutch VR. Ex vivo expansion of megakaryocyte precursors by preincubation of marrow allografts with interleukin-3 and granulocyte–macrophage colony-stimulating factor in vitro. Exp Hematol 1995; 23:1268.

76. Debili N, Wendling F, Katz A, Guichard J, Breton-Gorius J, Hunt P, Vainchenker W. The *Mpl* ligand or thrombopoietin or megakaryocyte growth and differentiative factor has both direct proliferative and differentiative activities on human megakaryocyte progenitors. Blood 1995; 86:2516.

77. Szabolcs P, Feller ED, Moore MAS, Young JW. Progenitor recruitment and in vitro expansion of immunostimulatory dendritic cells from human CD34[+] bone marrow cells by c-*kit* ligand, GM-CSF, and TNF alpha. In: Bancherau J, Schmitt D, eds. Dendritic Cells in Fundamental and Clinical Immunology. New York: Plenum Press; 1995:17.

24
Clinical Use of Cytokines

Darryl W. Maher
Ludwig Institute for Cancer Research, Melbourne Tumour Biology Branch, Parkville, Victoria, Australia

I. INTRODUCTION

The advent of recombinant DNA technology has made it possible to produce clinical-grade cytokines in sufficient quantity for therapeutic use. A number of cytokines are now licensed for clinical use, and many more are in developmental stages. Cytokines (Table 1) used therapeutically include the interferons, some of the interleukins, the colony-stimulating factors, erythropoietin, tumor necrosis factor, and some neurotrophic factors, among others. The biology of the individual factors is described in detail elsewhere in this book. It must be remembered, however, that in the whole animal, many of these factors have overlapping activities and can stimulate or inhibit others in a complex cytokine network. The administration of one cytokine could have effects directly on target cells, or indirectly by perturbation of the network. Cytokines have been used for their trophic actions, antiproliferative effects, antiviral properties, hematopoietic supportive properties, and immunomodulatory actions. Administration has generally been systemic by subcutaneous injection, although local and regional uses have also been applied. In most cases, cytokines have been given by daily injection, but three times weekly or even weekly regimens have been used successfully, despite plasma half lives usually measured in hours. Cytokines vary widely in their tolerability and toxicity, reflecting their degree of pleiotropy. Those with restricted lineage activity, such as granulocyte colony-stimulating factor (G-CSF), tend to be better tolerated than cytokines with multiple actions, such as tumor necrosis factor-α (TNF-α) and interleukin-1 (IL-1).

II. CYTOKINES

A. Interferons

Interferon alfa [the spelling preferred for the recombinant drug form] is currently licensed in the United States for hairy cell leukemia, AIDS-related Kaposi's sarcoma, genital warts, hepatitis B

Table 1 Cytokines Licensed in One or More Western Countries for Clinical Use

Cytokine	Clinical indications
G-CSF[a]	Hematopoietic support post chemotherapy; post autologous bone marrow transplantation; chronic neutropenia
GM-CSF[b]	Hematopoietic support post chemotherapy; post autologous bone marrow transplantation
Erythropoietin	Anemia associated with renal failure
Interferon alfa	Hairy cell leukemia; chronic myeloid leukemia; essential thrombocytosis; follicular non-Hodgkin's lymphoma; AIDS-related Kaposi's sarcoma; genital warts; hepatitis B and hepatitis C; renal cell cancer, and cutaneous T-cell lymphoma
Interferon beta	Multiple sclerosis
Interleukin-2	Renal cell cancer

[a]G-CSF, granulocyte colony-stimulating factor.
[b]GM-CSF, granulocyte–macrophage colony-stimulating factor.

and hepatitis C, and additionally in Europe for the treatment of chronic myeloid leukemia, renal cell cancer, and cutaneous T-cell lymphoma.

The interferons were first identified for their antiviral properties, but subsequently, have been shown to have many other actions. Nevertheless, interferon alfa has been extensively used in the treatment of serious chronic viral infections, such as hepatitis B (HBV) and hepatitis C (HCV). Both diseases are characterized by a tendency to chronic infection, progression to cirrhosis (a fibrotic response to chronic injury in the liver), hepatic cancer, and liver failure. Treatment with interferon alfa reduces the disease activity in hepatitis B in approximately one-third to one-half of the patients (1), with clearing of viral DNA (HBVDNA), and approximately 50% of patients with chronic HCV infection also respond (2). Up to 17% of patients are unable to tolerate treatment owing to side effects, such as malaise and lethargy, and response rates are better in patients with early disease before the onset of cirrhosis (3). Unfortunately, over 50% of patients relapse within 6–12 months of completing therapy (2).

In addition to their role as antiviral agents in hepatitis B, hepatitis C, and also genital warts, the interferons, because of their antiproliferative properties, have been successfully used as anticancer agents. The ability of interferon alfa to induce remissions in hairy cell leukemia was first described in 1984 (4). Subsequently, interferon alfa was shown to have antiproliferative activity in the chronic phase of chronic myeloid leukemia, in low-grade lymphoma, myeloma, and some solid tumors, including renal cell cancer and melanoma.

Chronic myeloid leukemia (CML) is a disease characterized by myeloid proliferation owing to the presence of an abnormal clone of hematopoietic stem cells with a characteristic 9:22 chromosomal translocation, detectable by cytogenetics. Studies conducted initially at the M. D. Anderson Hospital in Texas showed that interferon alfa not only achieved myeloid cytoreduction in chronic-phase chronic myeloid leukemia, but also induced cytogenetic remissions in 19% of cases (5). These cytogenetic remissions were often durable and maintained with ongoing therapy. This was the first nontransplant treatment approach resulting in cytogenetic remissions in a substantial proportion of patients with this disease. A recently reported randomized trial of interferon alfa in chronic myeloid leukemia confirmed the benefits of this treatment by showing a significantly prolonged time to progression to an accelerated or blastic phase and improved survival, compared with conventional chemotherapy (6). The survival curves from this study are shown in Figure 1.

Interferon alfa has also been evaluated in the treatment of multiple myeloma, a malignant clonal disorder characterized by an excess of plasma cells, bone lysis, and monoclonal im-

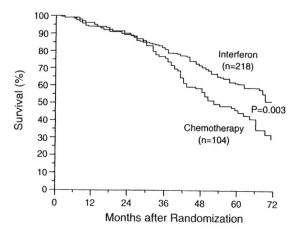

Figure 1   Overall survival of patients with chronic myeloid leukemia randomly assigned to treatment with either interferon alfa or standard chemotherapy. (From Ref. 6.)

munoglobulin in serum. In a randomized study of interferon versus placebo, in addition to standard chemotherapy, the response rate was significantly increased, but there was no difference in overall survival, except for the subset of patients with IgA or Bence Jones only myeloma (7). There is also some evidence that maintenance interferon alfa prolongs the quiescent or plateau phase of multiple myeloma after a course of chemotherapy (8); hence, its use has been incorporated into several protocols for maintenance treatment of multiple myeloma after high-dose chemotherapy supported with bone marrow autografts. Further studies are required to determine whether interferon alfa should be used routinely in the management of patients with multiple myeloma.

Low-grade non-Hodgkin's lymphoma (NHL) was also reported in 1984 to respond to treatment with interferon alfa (9). As these diseases are generally responsive to chemotherapy and because of preclinical evidence for synergistic antitumor activity of interferon plus chemo-therapy, the French Adult Lymphoma Study Group conducted a randomized trial of interferon alfa plus chemotherapy, versus chemotherapy alone, in patients with advanced follicular (low-grade) NHL. This demonstrated a higher rate of response (85 vs. 69%) and improved survival at 3 years (86 vs. 69%) in the interferon-treated group (10). It is now clear in both chronic-phase chronic myeloid leukemia and follicular non-Hodgkin's lymphoma that improved patient sur-vival can be achieved by treatment with interferon alfa.

The other interferons, including interferon gamma and interferon beta, are also undergoing clinical trials. Interferon gamma has been tried in renal cell carcinoma (11), melanoma (12), and other cancers alone and in combination with other cytokines or chemotherapy. It appears to be particularly useful in the prevention of infections in patients with chronic granulomatous disease (12) and has shown some efficacy in rheumatoid arthritis (13). Interferon beta appears to benefit patients with relapsing–remitting multiple sclerosis, by significantly lowering exacerbation rates in a dose-dependent manner, with a twofold reduction in the frequency of moderate and severe attacks, and improvements in the appearance on magnetic resonance imaging (MRI) scans, at higher doses (14,15).

B.   Colony-Stimulating Factors

In vitro and in vivo animal studies of the colony-stimulating factors indicated the obvious clinical potential of these cytokines in the therapeutic support of patients undergoing chemotherapy and bone marrow transplantation. To this end, the four colony-stimulating factors and other hemato-poietic growth factors including, erythropoietin (EPO), interleukin-1, interleukin-6, and interleu-kin-11, have all been used clinically. Two of these factors, granulocyte-colony stimulating factor (G-CSF) and granulocyte–macrophage colony-stimulating factor (GM-CSF) have been licensed in the United States and Europe.

G-CSF

Granulocyte colony-stimulating factor (generic name, filgrastim or lenograstim) has been licensed in most Western countries for the prevention of neutropenia and associated infection after myelosuppressive chemotherapy and also to accelerate recovery from autologous bone marrow transplantation. Phase I studies, commenced in 1987, demonstrated this factor potently increased neutrophil numbers and had minimal toxicity (16–18). In phase I/II trials, G-CSF reduced the duration of neutropenia after myelosuppressive chemotherapy and accelerated recovery, allowing the next cycle of chemotherapy to be delivered on time (19,20), or even earlier than planned (21). Indeed, many studies have shown that the dose intensity of chemother-apy can be safely escalated beyond standard doses with the use of G-CSF (21,22). In 1991, Crawford et al. (23) published the results of a pivotal randomized phase III trial of G-CSF versus placebo after myelosuppressive chemotherapy for small-cell lung cancer. The results of this trial showed that G-CSF, given preventively, reduced the period of severe neutropenia (Fig. 2) and reduced the incidence of infection associated with neutropenia after moderately intensive chemotherapy by approximately 50%. It was the results of this study that led to the granting of a license for the use of G-CSF after chemotherapy. Other studies by Trillet-Lenoir (19), also in small-cell lung cancer, and Pettengell et al. (24) in lymphoma, supported the findings that G-CSF could reduce by 50% the incidence of fever and neutropenia in patients receiving myelosuppress-ive chemotherapy. None of these randomized studies impinged on survival, although despite the cost of G-CSF, they did appear to save money by avoiding hospital admission (25).

Another approach to the problem of chemotherapy-induced neutropenia has been to use G-CSF at presentation with fever (infection), rather than giving it preventively. A randomized trial of G-CSF in patients with established neutropenia and infection following cancer chemo-therapy has recently been completed (26). This study showed that the administration of G-CSF during the infected neutropenic period resulted in accelerated neutrophil recovery, but did not significantly reduce the days of fever or time in the hospital, when used in addition to standard therapy with intravenous broad-spectrum antibiotics. The risk of prolonged hospitalization,

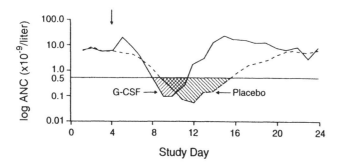

Figure 2 Median absolute neutrophil count (ANC) during cycle 1 of chemotherapy in patients randomly assigned to preventive G-CSF treatment ($n = 92$) or placebo ($n = 102$). (From Ref. 23.)

however, appeared to be reduced by G-CSF treatment, and patients with severe neutropenia benefited the most (26).

Both G-CSF and GM-CSF accelerate the neutrophil recovery after autologous bone marrow transplantation, when compared with historical controls (27,28), and in a controlled randomized trial (29,30). During the phase I studies of both G-CSF and GM-CSF, however, the novel observation was made that these factors could stimulate the release of progenitor cells from the bone marrow into the peripheral blood (31–33). This finding has led to use of G-CSF and GM-CSF to enable the harvesting of stem cells by leukapheresis for transplantation. Autologous bone marrow transplantation may be largely supplanted by the use of peripheral blood stem cells (PBSCs) after high-dose chemotherapy, in preference to bone marrow. This is because infusion of PBSCs provides more rapid platelet recovery than bone marrow transplantation after high-dose chemotherapy (34). Furthermore, the use of G-CSF alone or during recovery from myelosuppressive chemotherapy stimulates the release of large numbers of progenitor cells into the blood, which can be collected by leukapheresis and used to support patients after high-dose chemotherapy (35,36). This approach is revolutionizing the management of patients undergoing high-dose chemotherapy and is enabling patients with less responsive tumors, such as breast cancer, to have multiple cycles of high-dose chemotherapy supported with G-CSF-stimulated PBSCs (37).

G-CSF is very effective in the treatment of cyclic and other forms of severe chronic neutropenia (38,39). Young patients with severe chronic neutropenia, who suffer recurrent severe life-threatening infections, have had dramatic reductions in the frequency of infections and, consequently, improvements in their quality of life by G-CSF treatment (40). G-CSF appears to increase neutrophil counts and, in cyclic neutropenia, results in a much shorter nadir as well as more frequent cycling at an overall higher neutrophil level (41). It appears to have only limited usefulness in acquired neutropenic states caused by myelodysplasia and aplastic anemia (42–44).

GM-CSF

Granulocyte–macrophage colony-stimulating factor has been used for many of the same indications as G-CSF to support myelopoiesis after chemotherapy and to accelerate engraftment after marrow transplantation. It is not effective in cyclic or congenital neutropenia (45). Although many of its clinical effects are similar to those of G-CSF, including the side effect of bone pain, it is associated with more toxicity. In particular, fever, rash, and transient dyspea resulted in a large randomized trial of GM-CSF given preventively after chemotherapy (46). In this study of 182 patients with non-Hodgkin's lymphoma receiving myelosuppressive chemotherapy, GM-CSF at 400 µg/day or placebo, was given for 7 days after chemotherapy. Although this trial clearly showed GM-CSF accelerated neutrophil recovery, its ability to significantly reduce the incidence of infection was diluted because 24% of patients assigned to GM-CSF discontinued the cytokine treatment because of adverse events, compared with 6% of the placebo patients. Therefore, although those patients able to tolerate at least 70% of the planned treatment appeared to benefit, the overall group did not significantly reduce their risk of infection by planning to undergo preventive GM-CSF treatment (46). GM-CSF was approved for use in the United States after autologous bone marrow transplantation following the results of a randomized double-blind, placebo-controlled trial (29). One hundred and twenty-eight patients received GM-CSF, 250 µg/m^2, or placebo daily for 21 days after autologous bone marrow transplantation for hematological malignancy. GM-CSF significantly reduced time to neutrophil recovery, but not platelet recovery. This was associated with a reduction in duration of intravenous antibiotic therapy and hospitalization (29). Long-term follow-up showed no difference in tumor recurrence rates or survival between the treatment groups, and no evidence of late hematopoietic failure (47).

GM-CSF appears to play a role in host antitumor responses in animal models (48); therefore, it has the potential not only to provide supportive care for patients undergoing chemotherapy, but also possibly anticancer effects in its own right. However, there has as yet been no evidence for antitumor properties of GM-CSF in patients. Studies using GM-CSF in gene therapy protocols offer potential benefits, based on animal studies (48) and are being evaluated in patients with melanoma. The immune-activating properties of GM-CSF have also been evaluated in trials of GM-CSF in vaccinations for the prevention of influenza and other infectious diseases.

Interleukin-3

Interleukin-3 is a multilineage hematopoietic growth factor, otherwise known as multi-CSF, which synergizes with lineage-specific factors, such as erythropoietin (49), G-CSF (50), and thrombopoietin (51), to promote colony formation in vitro. Therefore, it has the potential to stimulate myeloid, erythroid, and platelet lineages in vivo, which would be of considerable theoretical value after high-dose chemotherapy and in the treatment of marrow failure states. Preclinical studies in primates suggested IL-3 could increase production of neutrophils, especially in combination with GM-CSF given sequentially after IL-3 (52). Phase I trials began in Germany in the late 1980s in patients with cancer (53), after chemotherapy (54,55), myelodysplasia (56), and aplastic anemia occurred (57). Toxicity was generally mild to moderate, with the main side effects being fever, headache, facial flushing, and injection site reactions (53). IL-3 showed a modest neutrophil-stimulating effect in comparison with G-CSF and GM-CSF, but also increased platelet counts and reticulocyte counts in some patients (53). A proportion of patients with myelodysplastic syndrome and patients with aplastic anemia appeared to benefit from IL-3 therapy (56,57). Large randomized trials of IL-3 given postchemotherapy are underway to determine its clinical value as a hematopoietic supportive agent. IL-3 is also being tested in combination with other cytokines to take advantage of its synergistic properties that are apparent in vitro (49–51). Preclinical studies in primates suggest sequential treatment with IL-3, followed by GM-CSF, may be more effective than either cytokine alone in protecting the marrow from the myelotoxic effects of chemotherapy, and also in promoting engraftment of autologous marrow after high-dose myeloablative chemotherapy (52). Combinations of IL-3 with GM-CSF and other later-acting factors, given together or sequentially to mobilize peripheral blood progenitor cells, are also being evaluated.

M-CSF

Macrophage colony-stimulating factor (M-CSF) has undergone phase I and II trials in patients with cancer and in patients with fungal infections. Efficacy in these clinical situations has not been established. In a randomized trial conducted in Japan, M-CSF, administered after allogeneic or syngeneic bone marrow transplant, accelerated engraftment, without increasing the incidence of graft-versus-host disease (GVHD) and without influencing the rate of leukemic relapse (58). Although quite well tolerated, a clinical role for M-CSF has yet to be found (59).

C. Platelet-Stimulating Cytokines

The use of the colony-stimulating factors has helped overcome the problems of neutropenia after chemotherapy. With the use of these agents, more emphasis has been placed on the problem of thrombocytopenia associated with dose-intensive chemotherapy. The most effective means of accelerating platelet recovery after high-dose chemotherapy has been the infusion of peripheral blood stem cells (PBSCs). Platelet-stimulating cytokines currently being evaluated include IL-3, IL-6, IL-11, and leukemia inhibitory factor (LIF). IL-3 does appear to accelerate platelet recovery after chemotherapy (60), but it is not known if it significantly reduces the need for platelet transfusion after high-dose chemotherapy. IL-6 has been evaluated in phase I trials (61) and clearly stimulates an increase in platelets in patients who have not received chemotherapy, and it appears to lessen the nadir after chemotherapy, although this needs to be confirmed in a randomized trial. It would appear that all the foregoing platelet-stimulating cytokines tend to induce a late thrombocytosis, with a lag period of some 7–10 days, which is too long to ensure rapid platelet recovery after chemotherapy. It remains to be seen whether strategies can be devised that will result in platelet stimulation that is as effective as the infusion of PBSCs. Recently, mpl ligand has been cloned (51,62,63) and has the biological activities of the long sought "thrombopoietin" and will obviously attract considerable clinical attention.

D. Erythropoietin

Erythropoietin (EPO) has been in clinical use for over 7 years. It specifically stimulates red blood cell production and has been particularly useful in maintaining hemoglobin levels in patients with chronic renal failure. In a dose-dependent manner, EPO substantially increases the hematocrit of both dialysis (64) and predialysis (65) patients with renal failure and associated anemia. Over 95% of patients respond, and the dose can be adjusted to maintain the hematocrit at a level associated with improved well-being and quality of life (66). Its efficacy has been confirmed in a double-blind, placebo-controlled trial (67). Administration of EPO three times per week, either intravenously or subcutaneously, is associated with few adverse effects, the main one being a mild to moderate increase in blood pressure (64,66).

EPO is also being used to successfully stimulate red cell production in patients who are to have autologous blood transfusions for surgical procedures (68). It has also been used in the treatment of anemia of chronic disease and, in particular, has been effective in patients with anemia associated with cancer, including patients with solid tumors and other malignancies, such as myeloma and lymphoma (69). Approximately 25% of patients with myelodysplastic syndrome and associated transfusion-dependent anemias also respond (70,71). EPO has also been tested after allogeneic bone marrow transplantation, for which it appeared to enhance erythroid engraftment, but did not reduce transfusion requirements nor duration of hospital stay (72).

E. Stem Cell Factor

Stem cell factor (SCF), otherwise known as steel factor or c-*kit* ligand, has considerable potential clinically, as it acts on primitive hematopoietic precursors and stimulates the release of PBSCs in mice and baboons (73). It is currently being trialed for its ability to mobilize stem cells alone, and in combination with G-CSF, in patients undergoing chemotherapy (37) and has potential uses in patients with hypoplastic marrow states, including acquired chronic neutropenia and aplastic anemia, in which depletion of hematopoietic stem cells is central to the disease process. Results of phase I trials suggest that allergic-type reactions are dose-limiting and may be caused by SCF's effects on mast cells (74,75).

F. Interleukin-2

Interleukin-2 (IL-2) is a potent stimulus for the generation of cytotoxic T cells, LAK cells and other natural cells in vitro. Therefore, IL-2 administration may augment host antitumor responses that involve these cells.

Interleukin-2 has been approved for use in renal cell cancer in the United States, where it has approximately a 22% response rate (76). However, the administration of high doses is associated with significant toxicity, including a capillary leak syndrome, associated with hypotension, and respiratory failure, requiring inotropic support in an intensive care ward. IL-2 administration is also associated with fever, chills, and fatigue, cytopenias, diarrhea, fluid retention, and renal insufficiency (76). Its toxicity and marginal response rates have deterred many clinicians from its use. It has also been used in lower, better-tolerated doses that result in measurable immunological changes (77,78), although clinical efficacy at these doses is unproved. More recently, combination treatment of low-dose IL-2 and interferon alfa has been tried and appears to provide response rates similar to high-dose IL-2, without the associated toxicity, in renal cell cancer (79).

IL-2 has been tried in the treatment of metastatic melanoma, with response rates of approximately 20% (76). Although this may be only marginally above spontaneous remission rates, some patients achieved response durations of longer than 12 months. Further studies combining IL-2 with interferon alfa and combination chemotherapy appear to improve the response rates (80), but randomized studies are required to confirm these promising data. It is still unclear which is the optimal dose and regimen for IL-2, either alone or in combination with other cytokines or chemotherapy for the treatment of melanoma, renal cell cancer, and other tumors.

IL-12 has recently entered phase I trials. Originally described as natural killer cell stimulatory factor (NKSF), it appears more potent than IL-2 in augmenting host antitumor responses and is very effective in promoting tumor rejection in animal models (81).

G. Other Interleukins and Tumor Necrosis Factor

Interleukin-1 (IL-1) accelerates platelet recovery after chemotherapy, but is associated with significant toxicity (82). Interleukin-4 (IL-4) has undergone phase I trials in patients with cancer (83,84) and it is currently being evaluated in phase II studies in various malignancies. Other interleukins, such as IL-7 and IL-10, are in very early stages of development.

Tumor necrosis factor-alpha (TNF-α) has potent antiproliferative activity on tumors in preclinical studies. Unfortunately, its clinical use is associated with severe toxicity (85,86). To overcome this problem of systemic toxicity, TNF-α has been used in a number of local and regional settings, such as in isolated limb perfusion for malignant melanoma (87), and it is also being evaluated in gene therapy protocols (88).

III. FUTURE DIRECTIONS

Other promising cytokines with potential clinical value are the neurotrophic factors and epithelial and fibroblast growth factors. Cytokines, such as nerve growth factor (NGF), brain-derived neurotrophic factor (BDNF), ciliary neurotrophic factor (CNTF), leukemia inhibitory factor (LIF), and insulin-like growth factor-I (IGF-I) have trophic actions for motor neurons, dopaminergic neurons, and cholinergic neurons, and trials of some of these agents are ongoing in disorders, such as motor neuron disease, Parkinson's disease, Alzheimer's dementia, and nerve injury. Acceleration of wound and fracture healing may be possible with factors such as transforming growth factor-beta (TGF-β; 89), platelet-derived growth factor (PDGF; 90,91), and epidermal growth factor (EGF; 92). In addition to evaluating the many cytokines available as potential clinical agents, the use of cytokine antagonists is also being evaluated. Examples of these include naturally occurring IL-1 receptor antagonist, which can be used in neutralizing IL-1. Soluble cytokine receptors are also being developed as potential antagonists of cytokines, and the use of certain cytokines themselves that are immunomodulatory or down-regulatory.

REFERENCES

1. Di Bisceglie AM, Fong TL, Fried MW, Swain MG, Baker B, Korenman J, Bergasa NV, Waggoner JG, Park Y, Hoofnagle JH. A randomized, controlled trial of recombinant alpha-interferon therapy for chronic hepatitis B. Am J Gastroenterol 1993; 88:1887–1892.

2. Tine F, Magrin S, Craxi A, Pagliaro L. Interferon for non-A, non-B chronic hepatitis. A meta-analysis of randomised clinical trials. J Hepatol 1991; 13:192–199.

3. Pagliaro L, Craxi A, Cammaa C, Tine F, Di Marco V, Lo Iacono O, Almasio P. Interferon-alpha for chronic hepatitis C: an analysis of pretreatment clinical predictors of response. Hepatology 1994; 19:820–828.

4. Quesada JR, Reuben J, Manning JT, Hersh EM, Gutterman JU. alpha-Interferon for induction of remission in hairy cell leukemia. N Engl J Med 1984; 310:15.

5. Talpaz M, Kantarjian H, Kurzrock R, Trujillo JM, Gutterman JU. Interferon-alpha produces sustained cytogenetic responses in chronic myelogenous leukemia. Philadelphia chromosome-positive patients. Ann Intern Med 1991; 114:532–538.

6. The Italian Cooperative Study Group on Chronic Myeloid Leukemia. Interferon alfa-2a as compared with conventional chemotherapy for the treatment of chronic myeloid leukemia. N Engl J Med 1994; 330:820–825.

7. Osterborg A, Bjoerkholm M, Bjoereman M, et al. Natural interferon-alpha in combination with melphalan/prednisone versus melphalan/prednisone in the treatment of multiple myeloma stages II and III: a randomized study from the myeloma group of central Sweden. Blood 1993; 81:1428–1434.

8. Mandelli F, Avvisati G, Amadori S, Boccadoro M, Gernone A, Lauta VM, Marmont F, Petrucci MT, Tribalto M, Vegna ML, Dammacco F, Pileri A. Maintenance treatment with recombinant interferon alfa-2b in patients with multiple myeloma responding to conventional induction chemotherapy. N Engl J Med 1990; 322:1430–1434.

9. Foon KA, Sherwin SA, Abrams PG, Longo DL, Fer MF, Stevenson HC, Ochs JJ, Bottino GC, Schoenberger CS, Zeffren J, Jaffe ES, Oldham RK. Treatment of advanced non-Hodgkin's lymphoma with recombinant leukocyte A interferon. N Engl J Med 1984; 311:1148–1152.

10. Solal Celigny P, Lepage E, Brousse N, Reyes F, Haioun C, Leporrier M, Peuchmaur M, Bosly A, Parlier Y, Brice P, Coiffier B, Gisselbrecht C, for the Groupe D'Etude des Lymphomes de l'Adulte. Recombinant interferon alfa-2b combined with a regimen containing doxorubicin in patients with advanced follicular lymphoma. Groupe d'Etude des Lymphomes de l'Adulte. N Engl J Med 1993; 329:1608–1614.

11. Bruntsch U, de Mulder PH, ten Bokkel Huinink WW, Clavel M, Drozd A, Kaye SB, Renard J, van Glabbeke M. Phase II study of recombinant human interferon-gamma in metastatic renal cell carcinoma. J Biol Response Mod 1990; 9:335–338.

12. Kirkwood JM, Ernstoff MS, Trautmann T, Hebert G, Nishida Y, Davis CA, Balzer J, Reich S, Schindler J, Rudnick SA. In vivo biological response to recombinant interferon-gamma during a phase I dose–response trial in patients with metastatic melanoma. J Clin Oncol 1990; 8:1070–1082.

13. Lemmel EM, Gaus W, Hofschneider PH. Multicenter double blind trial of interferon-gamma versus placebo in the treatment of rheumatoid arthritis. Arthritis Rheum 1991; 34:1621–1622.

14. Interferon beta-1b is effective in relapsing–remitting multiple sclerosis. I. Clinical results of a multicenter, randomized, double-blind, placebo-controlled trial. The IFNB Multiple Sclerosis Study Group [see comments]. Neurology 1993; 43:655–661.

15. Paty DW, Li DK. Interferon beta-1b is effective in relapsing-remitting multiple sclerosis. II. MRI analysis results of a multicenter, randomized, double-blind, placebo-controlled trial. UBC MS/MRI Study Group and the IFNB Multiple Sclerosis Study Group [see comments]. Neurology 1993; 43:662–667.

16. Morstyn G, Campbell L, Souza LM, Alton NK, Keech J, Green M, Sheridan W, Metcalf D, Fox R. Effect of granulocyte colony stimulating factor on neutropenia induced by cytotoxic chemotherapy. Lancet 1988; 1:667–672.

17. Bronchud MH, Scarffe JH, Thatcher N, Crowther D, Souza LM, Alton NK. Phase I/II study of recombinant human granulocyte colony-stimulating factor in patients receiving intensive chemotherapy for small cell lung cancer. Br J Cancer 1987; 56:809–813.

18. Gabrilove JL, Jakubowski A, Fain K, Grous J, Scher H, Sternberg C, Yagoda A, Clarkson B, Bonilla MA, Oettgen HF, Alton K, Boone T, Altrock B, Welte K, Souza L. Phase I study of granulocyte colony-stimulating factor in patients with transitional cell carcinoma of the urothelium. J Clin Invest 1988; 82:1454–1461.

19. Trillet-Lenoir V, Green J, Manegold C, Von Pawel J, Gatzemeier U, Lebeau B, Depierre A, Johnson P, Decoster G, Tomita D, Ewen C. Recombinant granulocyte colony stimulating factor reduces the infectious complications of cytotoxic chemotherapy. Eur J Cancer 1992; 29A:319–324.

20. Gabrilove JL, Jakubowski A, Scher H, Sternberg C, Wong G, Grous J, Yagoda A, Fain K, Moore MAS, Clarkson B, Oettgen HF, Alton K, Welte K, Souza L. Effect of granulocyte colony-stimulating factor on neutropenia and associated morbidity due to chemotherapy for transitional-cell carcinoma of the urothelium. N Engl J Med 1988; 318:1414–1422.

21. Bronchud MH, Howell A, Crowther D, Hopwood P, Souza L, Dexter TM. The use of granulocyte colony-stimulating factor to increase the intensity of treatment with doxorubicin in patients with advanced breast and ovarian cancer. Br J Cancer 1989; 60:121–125.

22. Neidhart J, Mangalik A, Kohler W, Stidley C, Saiki J, Duncan P, Souza L, Downing M. Granulocyte colony-stimulating factor stimulates recovery of granulocytes in patients receiving dose-intensive chemotherapy without bone marrow transplantation. J Clin Oncol 1989; 7:1685–1692.

23. Crawford J, Ozer H, Stoller R, Johnson D, Lyman G, Tabbara I, Kris M, Grous J, Picozzi V, Rausch G, Smith R, Gradishar W, Yahanda A, Vincent M, Stewart M, Glaspy J. Reduction by granulocyte colony-stimulating factor of fever and neutropenia induced by chemotherapy in patients with small-cell lung cancer. N Engl J Med 1991; 325:164–170.

24. Pettengell R, Gurney H, Radford JA, Deakin DP, James R, Wilkinson PM, Kane K, Bentley J, Crowther D. Granulocyte colony-stimulating factor to prevent dose-limiting neutropenia in non-Hodgkin's lymphoma: a randomized controlled trial. Blood 1992; 80:1430–1436.

25. Glaspy JA, Bleecker G, Crawford J, Stoller R, Strauss M. The impact of therapy with filgrastim (recombinant granulocyte colony-stimulating factor) on the health care costs associated with cancer chemotherapy. Eur J Cancer 1993; 29A Suppl 7:S23–S30.

26. Maher DW, Lieschke GJ, Green M, Bishop J, Stuart-Harris R, Wolf M, Sheridan WP, Kefford R, Cebon J, Olver I, McKendrick J, Toner G, Bradstock K, Lieschke M, Cruickshank S, Tomita DK, Hoffman EW, Fox RM, Morstyn G. Filgrastim (rmetHuG-CSF) in patients with chemotherapy induced febrile neutropenia: a double-blind, placebo-controlled trial. Ann Intern Med 1994; 121:492–501.

27. Sheridan WP, Morstyn G, Wolf M, Dodds A, Lusk J, Maher D, Layton JE, Green MD, Souza L, Fox RM. Granulocyte colony stimulating factor (G-CSF) and neutrophil recovery following high dose chemotherapy and autologous bone marrow transplantation. Lancet 1989; 2:891–895.

28. Brandt SJ, Peters WP, Atwater SK, Kurtzberg J, Borowitz MJ, Jones RB, Shpall EJ, Bast RC Jr, Gilbert CJ, Oette DH. Effect of recombinant human granulocyte–macrophage colony-stimulating factor on hematopoietic reconstitution after high-dose chemotherapy and autologous bone marrow transplantation. N Engl J Med 1988; 318:869–875.

29. Nemunaitis J, Rabinowe SN, Singer JW, Bierman PJ, Vose JM, Freedman AS, Onetto N, Gillis S, Oette D, Gold M, Buckner CD, Hansen JA, Ritz J, Appelbaum FR, Armitage JO, Nadler LM. Recombinant granulocyte–macrophage colony-stimulating factor after autologous bone marrow transplantation for lymphoid cancer. N Engl J Med 1991; 324:1773–1778.

30. Advani R, Chao NJ, Horning SJ, Blume KG, Ahn DK, Lamborn KR, Fleming NC, Bonnem EM, Greenberg PL. Granulocyte–macrophage colony-stimulating factor (GM-CSF) as an adjunct to autologous hemopoietic stem cell transplantation for lymphoma. Ann Intern Med 1992; 116:183–189.

31. Duhrsen U, Villeval JL, Boyd J, Kannourakis G, Morstyn G, Metcalf D. Effects of recombinant granulocyte colony-stimulating factor on hematopoietic progenitor cells in cancer patients. Blood 1988; 72:2074–2081.

32. Villeval J-L, Dührsen U, Morstyn G, Metcalf D. Effect of recombinant human granulocyte–macrophage colony stimulating factor on progenitor cells in patients with advanced malignancies. Br J Haematol 1990; 74:36–44.

33. Socinski MA, Cannistra SA, Elias A, Antman KH, Schnipper L, Griffin JD. Granulocyte–macrophage colony-stimulating factor expands the circulating haemopoietic progenitor cell compartment in man. Lancet 1988; 1:1194–1198.

34. To LB, Dyson PG, Branford A, Russell JA, Haylock DN, Ho JQK, Kimber RJ, Juttner CA. Peripheral blood stem cells collected in very early remission produce rapid and sustained autologous hematopoietic reconstitution in acute non-lymphoblastic leukaemia. Bone Marrow Transplant 1987; 2:103–108.

35. Sheridan WP, Begley CG, Juttner CA, Szer J, To LB, Maher D, McGrath KM, Morstyn G, Fox RM. Effect of peripheral-blood progenitor cells mobilised by filgrastim (G-CSF) on platelet recovery after high-dose chemotherapy. Lancet 1992; 339:640–644.

36. Pettengell R, Test NG, Swindell R, Crowther D, Dexter TM. Transplantation potential of hematopoietic cells released into the circulation during routine chemotherapy for non-Hodgkin's lymphoma. Blood 1993; 82:2239–2248.

37. Basser R, Begley CG, Maher D, To B, Juttner C, Fox R, Cebon J, Grigg A, Szer J, Marty J, Sheridan W, Collins J, Russell I, Green M. The use of peripheral blood progenitor cells (PBPC) mobilized by stem cell factor (SCF) and filgrastim (G-CSF) to support multiple cycles of high-dose chemotherapy in untreated women with poor prognosis breast cancer [abstract]. Br J Haematol 1994; 87 (suppl 1):91.

38. Hammond WP, Price TH, Souze LM, Dale DC. Treatment of cyclic neutropenia with granulocyte colony-stimulating factor. N Engl J Med 1989; 320:1306–1311.

39. Dale DC, Bonilla MA, Davis MW, Nakanishi AM, Hammond WP, Kurtzberg J, Wang W, Jakubowski A, Winton E, Lalezari P, Robinson W, Glaspy JA, Emerson S, Gabrilove J, Vincent M, Boxer LA. A randomized controlled phase III trial of recombinant human granulocyte colony-stimulating factor (filgrastim) for treatment of severe chronic neutropenia. Blood 1993; 81:2496–2502.

40. Jones EA, Bolyard AA, Dale DC. Quality of life of patients with severe chronic neutropenia receiving long-term treatment with granulocyte colony-stimulating factor. JAMA 1993; 270:1132–1133.

41. Dale DC, Bolyard AA, Hammond WP. Cyclic neutropenia: natural history and effects of long-term treatment with recombinant human granulocyte colony-stimulating factor. Cancer Investi 1993; 11:219:223.

42. Kobayashi Y, Okabe T, Ozawa K, Chiba S, Hino M, Miyazono K, Urabe A, Takaku F. Treatment of myelodysplastic syndromes with recombinant human granulocyte colony-stimulating factor: a preliminary report. Am J Med 1989; 86:178–182.

43. Negrin RS, Haeuber DH, Nagler A, Olds LC, Donlon T, Souza LM, Greenberg PL. Treatment of myelodysplastic syndromes with recombinant human granulocyte colony-stimulating factor. A phase I–II trial. Ann Intern Med 1989; 110:976–984.

44. Greenberg P, Taylor K, Larson R, Koeffler P, Negrin R, Saba H, Ganser A, Jakubowski A, Gabrilove J, Mufti G, Cruz J, Hammond W, Broudy V, Langley GR, Keating A, Vardiman J, Lamborn K, Brown S. Phase III randomized multicenter trial of G-CSF vs. observation for myelodysplastic syndromes (MDS) [abstract]. Blood 1993; 82(suppl):196a.

45. Freund MRF, Luft S, Schöber C, Heussner P, Schrezenmaier H, Porzsolt F, Welte K. Differential effect of GM-CSF and G-CSF in cyclic neutropenia. Lancet 1990; 336:313.

46. Gerhartz HH, Engelhard M, Meusers P, et al. Randomized, double-blind, placebo-controlled, phase III study of recombinant human granulocyte–macrophage colony-stimulating factor as adjunct to induction treatment of high-grade malignant non-Hodgkin's lymphomas. Blood 1993; 82:2329–2339.

47. Rabinowe SN, Neuberg D, Bierman PJ, Vose JM, Nemunaitis J, Singer JW, Freedman AS, Mauch P, Demetri G, Onetto N, Gillis S, Oette D, Buckner D, Hansen JA, Ritz J, Armitage JO, Nadler LM, Appelbaum FR. Long-term follow-up of a phase III study of recombinant human granulocyte–macrophage colony-stimulating factor after autologous bone marrow transplantation for lymphoid malignancies. Blood 1993; 81:1903–1908.

48. Dranoff G, Jaffee E, Lazenby A, Golumbek P, Levitsky H, Brose K, Jackson V, Hamada H, Pardoll D, Mulligan RC. Vaccination with irradiated tumor cells engineered to secrete murine granulocyte–

macrophage colony-stimulating factor stimulates potent, specific, and long-lasting anti-tumor immunity. Proc Natl Acad Sci USA 1993; 90:3539–3543.

49. Migliaccio G, Migliaccio AR, Visser JWM. Synergism between erythropoietin and interleukin-3 in the induction of hematopoietic stem cell proliferation and erythroid burst colony formation. Blood 1988; 72:944–951.

50. Metcalf D, Nicola NA. The clonal proliferation of normal mouse hematopoietic cells: enhancement and suppression by colony-stimulating factor combinations. Blood 1992; 79:2861–2866.

51. Lok S, Kaushansky K, Holly RD, et al. Cloning and expression of murine thrombopoietin cDNA and stimulation of platelet production in vivo [see comments]. Nature 1994; 369:565–568.

52. Mayer P, Valent P, Schmidt G, Liehl E, Bettelheim P. The in vivo effects of recombinant human interleukin-3: demonstration of basophil differentiation factor, histamine-producing activity, and priming of GM-CSF-responsive progenitors in nonhuman primates. Blood 1989; 74:613–621.

53. Ganser A, Lindemann A, Seipelt G, Ottmann OG, Herrmann F, Eder M, Frisch J, Schulz G, Mertelsmann R, Hoelzer D. Effects of recombinant human interleukin-3 in patients with normal hematopoiesis and in patients with bone marrow failure. Blood 1990; 76:666–676.

54. Biesma B, Willemse PH, Mulder NH, Sleijfer DT, Gietema JA, Mull R, Limburg PC, Bouma J, Vellenga E, de Vries EG. Effects of interleukin-3 after chemotherapy for advanced ovarian cancer. Blood 1992; 80:1141–1148.

55. Postmus PE, Gietema JA, Damsma O, Biesma B, Limburg PC, Vellenga E, de Vries EG. Effects of recombinant human interleukin-3 in patients with relapsed small-cell lung cancer treated with chemotherapy: a dose-finding study. J Clin Oncol 1992; 10:1131–1140.

56. Ganser A, Seipelt G, Lindemann A, Ottmann OG, Falk S, Eder M, Herrmann F, Becher R, Hoffken K, Buchner T, Klausmann M, Frisch J, Schulz G, Mertelsmann R, Hoelzer D. Effects of recombinant human interleukin-3 in patients with myelodysplastic syndromes. Blood 1990; 76:455–462.

57. Ganser A, Lindemann A, Seipelt G, Ottmann OG, Eder M, Falk S, Herrmann F, Kaltwasser JP, Meusers P, Klausmann M, Frisch J, Schulz G, Mertelsmann R, Hoelzer D. Effects of recombinant human interleukin-3 in aplastic anemia. Blood 1990; 76:1287–1292.

58. Masaoka T, Shibata H, Ohno R, Katoh S, Harada M, Motoyoshi K, Takaku F, Sakuma A. Double-blind test of human urinary macrophage colony-stimulating factor for allogeneic and syngeneic bone marrow transplantation: effectiveness of treatment and 2-year follow-up for relapse of leukaemia. Br J Haematol 1990; 76:501–505.

59. Nemunaitis J, Meyers JD, Buckner CD, Shannon Dorcy K, Mori M, Shulman H, Bianco JA, Higano CS, Groves E, Storb R, Hansen J, Appelbaum FR, Singer JW. Phase I trial of recombinant human macrophage colony-stimulating factor in patients with invasive fungal infections. Blood 1991; 78:907–913.

60. D'Hondt V, Weynants P, Humblet Y, Guillaume T, Canon JL, Beauduin M, Duprez P, Longueville J, Mull R, Chatelain C, Symann M. Dose-dependent interleukin-3 stimulation of thrombopoiesis and neutropoiesis in patients with small-cell lung carcinoma before and following chemotherapy: a placebo-controlled randomized phase Ib study [see comments]. J Clin Oncol 1993; 11:2063–2071.

61. Weber J, Yang JC, Topalian SL, Parkinson DR, Schwartzentruber DS, Ettinghausen SE, Gunn H, Mixon A, Kim H, Cole D, Levin R, Rosenberg SA. Phase I trial of subcutaneous interleukin-6 in patients with advanced malignancies. Clini Oncol 1993; 11:499–506.

62. Wendling F, Maraskovsky E, Debili N, Florindo C, Teepe M, Titeux M, Methia N, Breton Gorius J, Cosman D, Vainchenker W. cMpl ligand is a humoral regulator of megakaryocytopoiesis [see comments]. Nature 1994; 369:571–574.

63. Kaushansky K, Lok S, Holly RD, Broudy VC, Lin N, Bailey MC, Forstrom JW, Buddle MM, Oort PJ, Hagen FS, Roth GJ, Papayannopoulou T, Foster DC. Promotion of megakaryocyte progenitor expansion and differentiation by the c-*Mpl* ligand thrombopoietin [see comments]. Nature 1994; 369:568–571.

64. Eschbach JW, Egrie JC, Downing MR, Browne JK, Adamson JW. Correction of the anemia of end-stage renal disease with recombinant human erythropoietin: results of a combined phase I and II clinical trial. N Engl J Med 1987; 316:73–78.

65. Lim VS, DeGowin RL, Zavala D, Kirchner PT, Abels R, Perry P, Fangman J. Recombinant human erythropoietin treatment in pre-dialysis patients. Ann Intern Med 1989; 110:108–114.

66. Eschbach JW, Abdulhadi MH, Browne JK, et al. Recombinant human erythropoietin in anemic patients with end-stage renal disease: results of a phase III multicenter clinical trial. Ann Intern Med 1989; 111:992–1000.

67. Double-blind, placebo-controlled study of the therapeutic use of recombinant human erythropoietin for anemia associated with chronic renal failure in predialysis patients. The US Recombinant Human Erythropoietin Predialysis Study Group [published erratum appears in Am J Kidney Dis 1991 Sept; 18(3):420]. Am J Kidney Dis 1991; 18:50–59.

68. Biesma DH, Marx JJM, Kraaijenhagen RJ, Franke W, Messinger D, Van de Wiel A. Lower homologous blood requirement in autologous blood donors after treatment with recombinant human erythropoietin. Lancet 1994; 344:367–370.

69. Case DCJ, Bukowski RM, Carey RW, Fishkin EH, Henry DH, Jacobson RJ, Jones SE, Keller AM, Kugler JW, Nichols CR, Salmon SE, Silver RT, Storniolo AM, Wampler GL, Dooley CM, Larholt KM, Nelson RA, Abels RI. Recombinant human erythropoietin therapy for anemic cancer patients on combination chemotherapy. J Natl Cancer Inst 1993; 85:801–806.

70. Goy A, Belanger C, Casadevall N, Picard F, Guesnu M, Jaulmes D, Poisson D, Varet B. High doses of intravenous recombinant erythropoietin for the treatment of anaemia in myelodysplastic syndrome. Br J Haematol 1993; 84:232–237.

71. Razzano M, Caslini C, Cortelazzo S, Battistel V, Rambaldi A, Barbui T. Clinical and biological effects of erythropoietin treatment of myelodysplastic syndrome. Leuk Lymphoma 1993; 10:127–134.

72. Biggs JC, Atkinson KA, Booker V, Concannon A, Dart GW, Dodds A, Downs K, Szer J, Turner J, Worthington R, for the Australian Bone Marrow Transplant Study Group. A prospective randomised double blind trial of the in vivo use of recombinant human erythropoietin in bone marrow transplantation from HLA identical sibling donors. Bone Marrow Transplant 1994 (in press).

73. Andrews RG, Bertelmez SH, Knitter G, Langley K, Farrar D, Hendren RW, Bernstein ID, Apelbaum FR, Zsebo K. Recombinant human stem cell factor (SCF) stimulates increased numbers of peripheral blood and marrow CFU-GM, BFU-E, CFU-MIX, HPP-CFC, and CD34+ cells in baboons. Blood 1991; 78(suppl):261a–1036.

74. Demetri G, Costa J, Hayes D, Sledge G, Galli S, Hoffman R, Merica E, Rich W, Harkins B, McGuire B, Gordon M. A phase I trial of recombinant methionyl human stem cell factor (SCF) in patients with advanced breast carcinoma pre- and postchemotherapy (CHEMO) with cyclophosphamide (C) and doxorubicin (A) [abstract]. Proc Am Soc Clin Oncol 1993; 12:142.

75. Crawford J, Lau D, Erwin R, Rich W, McGuire B, Meyers F. A phase I trial of recombinant methionyl stem cell factor (SCF) in patients with advanced non-small cell lung carcinoma [abstr]. Proc Am Soc Clin Oncol 1993; 12:135.

76. Rosenberg SA, Lotze MT, Yang JC, Aebersold PM, Linehan WM, Seipp CA, White DE. Experience with the use of high-dose interleukin-2 in the treatment of 652 cancer patients. Ann Surg 1989; 210:474–484.

77. Vlasveld LT, Hekman A, Vyth-Dreese FA, Rankin EM, Scharenberg JGM, Voordouw AC, Sein JJ, Dellemijn TAM, Rodenhuis S, Melief CJM. A phase I study of prolonged continuous infusion of low dose recombinant interleukin-2 in melanoma and renal cell cancer. Part II: Immunological aspects. Br J Cancer 1993; 68:559–567.

78. Hladik F, Tratkiewicz JA, Tilg H, Vogel W, Schwulera U, Kronke M, Aulitzky WE, Huber C. Biologic activity of low dosage IL-2 treatment in vivo: molecular assessment of cytokine network interaction. J Immunol 1994; 153:1449–1454.

79. Vogelzang NJ, Lipton A, Figlin RA. Subcutaneous interleukin-2 plus interferon alfa-2a in metastatic renal cancer: an outpatient multicenter trial. J Clin Oncol 1993; 11:1809–1816.

80. Richards JM, Mehta N, Ramming K, Skosey P. Sequential chemoimmunotherapy in the treatment of metastatic melanoma. J Clin Oncol 1992; 10:1338–1343.

81. Brunda MJ, Luistro L, Warrier RR, Wright RB, Hubbard BR, Murphy M, Wolf SF, Gately MK.

Antitumor and antimetastatic activity of interleukin 12 against murine tumors. J Exp Med 1993; 178:1223–1230.

82. Smith JW, Longo DL, Alvord WG, Janik JE, Sharfman WH, Gause BL, Curti BD, Creekmore SP, Holmlund JT, Fenton RG, Sznol M, Miller LL, Shimizu M, Oppenheim JJ, Fiem SJ, Hursey JC, Powers GC, Urba WJ. The effects of treatment with interleukin-1 alpha on platelet recovery after high-dose carboplatin. N Engl J Med 1993; 328:756–761.

83. Ghosh AK, Smith NK, Prendiville J, Thatcher N, Crowther D, Stern PL. A phase I study of recombinant human interleukin-4 administered by the intravenous and subcutaneous route in patients with advanced cancer: immunological studies. Eur Cytokine Netw 1993; 4:205–211.

84. Prendiville J, Thatcher N, Lind M, McIntosh R, Ghosh A, Stern P, Crowther D. Recombinant human interleukin-4 (rhuIL-4) administered by the intravenous and subcutaneous routes in patients with advanced cancer—a phase I toxicity study and pharmacokinetic analysis. Eur J Cancer A 1993; 29A:1700–1707.

85. Mittelman A, Puccio C, Gafney E, Coombe N, Singh B, Wood D, Nadler P, Ahmed T, Arlin Z. A phase I pharmacokinetic study of recombinant human tumor necrosis factor administered by a 5-day continuous infusion. Invest New Drugs 1992; 10:183–190.

86. Logan TF, Kaplan SS, Bryant JL, Ernstoff MS, Krause JR, Kirkwood JM. Granulocytopenia in cancer patients treated in a phase I trial with recombinant human tumor necrosis factor. J Immunother 1991; 10:84–95.

87. Lejeune FJ, Lienard D, Leyvraz S, Mirimanoff RO. Regional therapy of melanoma. Eur J Cancer 1993; 29A:606–612.

88. Rosenberg SA, Aebersold P, Cornetta K, Kasid A, Morgan RA, Moen R, Karson EM, Lotze MT, Yang JC, Topalian SL, Merino MJ, Culver K, Miller AD, Blaese RM, Anderson WF. Gene transfer into humans—immunotherapy of patients with advanced melanoma, using tumor-infiltrating lymphocytes modified by retroviral gene transduction [see comments]. N Engl J Med 1990; 323:570–578.

89. Nielsen HM, Andreassen TT, Ledet T, Oxlund H. Local injection of TGF-beta increases the strength of tibial fractures in the rat. Acta Orthop Scand 1994; 65:37–41.

90. Mustoe TA, Cutler NR, Allman RM, Goode PS, Deuel TF, Prause JA, Bear M, Serdar CM, Pierce GF. A phase II study to evaluate recombinant platelet-derived growth factor-BB in the treatment of stage 3 and 4 pressure ulcers. Arch Surg 1994; 129:213–219.

91. Robson MC, Phillips LG, Thomason A, et al. Recombinant human platelet-derived growth factor-BB for the treatment of chronic pressure ulcers [see comments]. Ann Plast Surg 1992; 29:193–201.

92. Pastor JC, Calonge M. Epidermal growth factor and corneal wound healing. A multicenter study. Cornea 1992; 11:311–314.

Index

About the Editors

JOHN M. GARLAND is Senior Lecturer in the Division of Cancer Cell and Molecular Biology, FORCE Cancer Research Laboratory, Institute of Clinical Science, University of Exeter, Wonford, England. The author or coauthor of numerous publications in his field of expertise, he is a member of the International Society for Experimental Hematology and the Genetic Society. Dr. Garland received the MB.Ch.B. (1966), B.Sc. (1968), and Ph.D. (1971) degrees from Edinburgh University, Scotland.

PETER J. QUESENBERRY is Director of the Cancer Center and Professor of Medicine as well as Cell Biology at the University of Massachusetts Medical Center, Worcester. The author or coauthor of over 200 book chapters and journal articles and the editor of the journal *Experimental Hematology* and coeditor of the *Year Book of Hematology*, he is a member of numerous clinical research societies. Dr. Quesenberry received the B.A. (1960) and M.D. (1964) degrees from the University of Virginia, Charlottesville.

DOUGLAS J. HILTON is a Queen Elizabeth II Postdoctoral Fellow at the Walter and Eliza Hall Institute of Medical Research, Cytokine Program Manager at AMRAD Operations, as well as Director of the Cooperative Research Center for Cellular Growth Factors, Melbourne, Australia. The author or coauthor of over 50 journal publications and the holder of numerous patents for novel cytokine discoveries, Dr. Hilton received the B.Sc. Honors degree (1986) from the University of Melbourne and the Ph.D. degree (1990) from the University of Melbourne for research conducted at the Walter and Eliza Hall Institute of Medical Research, Melbourne.

Milton Keynes UK
Ingram Content Group UK Ltd.
UKHW052027071024
449327UK00027B/2465